Leon A. Schein, EdD, LCSW
Henry I. Spitz, MD
Gary M. Burlingame, PhD
Philip R. Muskin, MD
Editors
with Shannon Vargo

Psychological Effects of Catastrophic Disasters
Group Approaches to Treatment

More pre-publication
REVIEWS, COMMENTARIES, EVALUATIONS . . .

"It would be difficult to imagine a more timely, more relevant, or more necessary collective body of work than that put forth in *Psychological Effects of Catastrophic Disasters: Group Approaches to Treatment.* Although many mental health professionals were experienced and comfortable using various forms of group therapy with trauma patients prior to the events of 9/11, what we learned from that disaster challenged the effectiveness of our interventions. There were far more unanswered questions about how to best treat survivors of terrorist disasters than there were evidence-based treatments. This book pulls together and synthesizes the thinking of the top-thought leaders in the field and offers solid theoretical rationale, concrete clinical examples, and directions for the future. This is a must-read for mental health students and for practicing clinicians who want to provide their trauma patients with the most current and scientifically sound treatments."

Jerilyn Ross, LICSW
Director, The Ross Center
for Anxiety & Related Disorders,
Author of *Triumph Over Fear:
A Book of Help and Hope
for People with Anxiety,
Panic Attacks and Phobias*

"I could not more highly recommend this book. The most striking thing about it is its comprehensiveness. After an overview chapter, the book consists of ten chapters that address the bases for working with the psychological sequelae of terrorist disasters, ten chapters that describe the various models for working with victims in group formats, and a chapter that predicts future directions in the field. The editors have done a commendable job of highlighting the approaches that have demonstrated some empirical validity, even as they acknowledge that research in this area is in a nascent stage and describe the limitations on how rigorous this research can ever be. The contributor list is a veritable Who's Who of those who are at the forefront of this work. Although some may wish to read the book in its entirety, its format lends itself to selective reading that is targeted to the reader's areas of interest. The editors have done a very impressive job of pulling together the 'state of the art' in this new area of group work, which sadly has become so pertinent in recent years."

Harold S. Bernard, PhD, ABPP, CGP, FAGPA
Immediate Past President, American Group Psychotherapy Association;
Clinical Associate Professor of Psychiatry, New York University
School of Medicine

More pre-publication
REVIEWS, COMMENTARIES, EVALUATIONS . . .

"This timely book fills an important gap in the professional literature on disaster psychiatry. The distinguished multidisciplinary editors deserve congratulations for bringing together leading clinicians and researchers on traumatic stress. Drawing on the lessons learned from the World Trade Center terrorist attack (among others), the authors build a solid foundation for understanding the complex biopsychosocial dimensions of postdisaster syndromes. The further emphasis on the impact of cultural, spiritual, and community factors is especially valuable, as these issues are frequently overlooked in other works.

As the title suggests, several chapters are devoted to a variety of group treatments for survivors of disasters. Essential information is clearly presented to enable the reader to select the most appropriate and specific therapeutic modality for particular patient populations. Another strength of the book is its reliance on evidence-based data to support treatment recommendations. To their credit, the editors have not forgotten that psychotherapists working with trauma survivors need assistance to lessen the burdens of compassion fatigue.

Every mental health professional who seeks to help the victims of past, present, and future disasters will want to study this book."

Spencer Eth, MD
Professor and Vice Chairman of Psychiatry, New York Medical College;
Medical Director and Senior Vice President, Behavioral Health Centers,
Saint Vincent Catholic Medical Centers, New York City

The Haworth Press
New York • London • Oxford

Psychological Effects
of Catastrophic Disasters
Group Approaches
to Treatment

THE HAWORTH PRESS
Haworth Series in *Family and Consumer Issues in Health*
F. Bruce Carruth, PhD, Senior Editor

Addicted and Mentally Ill: Stories of Courage, Hope, and Empowerment by Carol Bucciarelli

Compassion and Courage in the Aftermath of Traumatic Loss: Stones in My Heart Forever by Kathryn Bedard

Other Titles of Related Interest

Healing 9/11: Creative Programming by Occupational Therapists edited by Pat Precin

Surviving 9/11: Impact and Experiences of Occupational Therapy Practitioners edited by Pat Precin

The Trauma of Terrorism: Sharing Knowledge and Shared Care, An International Handbook edited by Yael Danieli, Danny Brom, and Joe Sills

Trauma Practice in the Wake of September 11, 2001 edited by Steven N. Gold and Jan Faust

On the Ground After September 11: Mental Health Responses and Practical Knowledge Gained edited by Yael Danieli and Robert L. Dingman

Psychological Effects of Catastrophic Disasters
Group Approaches to Treatment

Leon A. Schein, EdD, LCSW
Henry I. Spitz, MD
Gary M. Burlingame, PhD
Philip R. Muskin, MD
Editors

with Shannon Vargo

The Haworth Press
New York • London • Oxford

For more information on this book or to order, visit
http://www.haworthpress.com/store/product.asp?sku=5571

or call 1-800-HAWORTH (800-429-6784) in the United States and Canada
or (607) 722-5857 outside the United States and Canada

or contact orders@HaworthPress.com

PUBLISHER'S NOTE
The development, preparation, and publication of this work has been undertaken with great care. However, the Publisher, employees, editors, and agents of The Haworth Press are not responsible for any errors contained herein or for consequences that may ensue from use of materials or information contained in this work. The Haworth Press is committed to the dissemination of ideas and information according to the highest standards of intellectual freedom and the free exchange of ideas. Statements made and opinions expressed in this publication do not necessarily reflect the views of the Publisher, Directors, management, or staff of The Haworth Press, Inc., or an endorsement by them.

PUBLISHER'S NOTE
Identities and circumstances of individuals discussed in this book have been changed to protect confidentiality.

The Haworth Press, Inc., 10 Alice Street, Binghamton, NY 13904-1580.

Cover design by Marylouise E. Doyle.

Library of Congress Cataloging-in-Publication Data

Psychological effects of catastrophic disasters : group approaches to treatment / Leon A. Schein . . . [et al.] editors, with Shannon Vargo.
    p. ; cm.
    Includes bibliographical references and index.
    ISBN-13: 978-0-7890-1840-3 (hard : alk. paper)
    ISBN-10: 0-7890-1840-3 (hard : alk. paper)
    ISBN-13: 978-0-7890-1841-0 (soft : alk. paper)
    ISBN-10: 0-7890-1841-1 (soft : alk. paper)
    1. Terrorism—Psychological aspects. 2. Victims of terrorism—Psychology. 3. Post-traumatic stress disorder—Treatment. 4. Psychic trauma—Treatment. 5. Group psychotherapy. I. Schein, Leon A.
    [DNLM: 1. Stress Disorders, Traumatic—therapy. 2. Psychotherapy, Group. 3. Terrorism—psychology. WM 172 P974545 2005]

RC569.5.T47P79 2005
616.85'21—dc22

2005012129

The Talmud says that the "world awaits the physical and emotional return of the mourner." We dedicate this book to members of our communities who perished, survived, witnessed, or suffered personal losses, with the events of September 11, 2001, and of Hurricane Katrina (August 29-30, 2005) and for whom a return to the fullness of life has not as yet been possible. This book is also dedicated to our colleagues for whom "repairing the world" is an essential value in tirelessly caring for the survivors of these and future horrific and heartbreaking disasters.

CONTENTS

PART III: GROUP MODELS

PART IV: FUTURE DIRECTIONS

David W. Foy
Daryl A. Schrock

ABOUT THE EDITORS

Leon A. Schein, EdD, is Clinical Assistant Professor of Psychiatry, Department of Psychiatry and Behavioral Science, New York Medical College, New York, New York; St. Vincent's Hospital, New York, New York; Co-Director and Dean of Curriculum, Eastern Group Psychotherapy Training Program, New York, New York.

Henry I. Spitz, MD, is Director, Group Therapy Training Program, New York State Psychiatric Institute, New York, New York; Clinical Professor of Psychiatry, Columbia University College of Physicians & Surgeons, New York, New York.

Gary M. Burlingame, PhD, is Professor, Department of Clinical Psychology, Brigham Young University, Provo, Utah.

Philip R. Muskin, MD, is Professor of Clinical Psychiatry, Department of Psychiatry, Columbia University College of Physicians and Surgeons; Chief of Service, Consultation Liaison Psychiatry, Columbia University Medical Center, New York, New York.

CONTRIBUTORS

Melissa J. Brymer, PsyD, is Manager, Terrorist and Disaster Branch, University of California, Los Angeles/Duke University National Center for Child Traumatic Stress, Department of Psychiatric and Biobehavioral Sciences, David Geffen School of Medicine, University of California, Los Angeles.

Marylene Cloitre, PhD, is Cathy and Stephen Graham Professor of Child and Adolescent Psychiatry, New York University School of Medicine, New York, New York; Director, Institute for Trauma and Stress, New York University Child Study Center, New York, New York.

Judith A. Cohen, MD, is Professor, Department of Psychiatry, Drexel University College of Medicine, Pittsburgh, Pennsylvania; Medical Director, Center for Traumatic Stress in Children and Adolescents, Allegheny General Hospital, Pittsburgh, Pennsylvania.

Yael Danieli, PhD, is Director, Group Project for Holocaust Survivors and Their Children, New York, New York; Past-President, Senior Representative to the United Nations of the International Society for Traumatic Stress Studies.

D. Rob Davies, PhD, is Intern, Department of Psychology, Brigham Young University, Provo, Utah.

Kent D. Drescher, PhD, is Assessment and Data Management Coordinator, National Center for Post-Traumatic Stress Disorders, Veterans Affairs Palo Alto Health Care System, Palo Alto, California.

George S. Everly Jr., PhD, CTS, is Professor of Psychology, Core Faculty, Loyola College, Baltimore, Maryland; Associate Faculty, Bloomberg School of Public Health, Johns Hopkins University, Baltimore, Maryland; Co-Founder, International Critical Incident Stress Foundation, (ICISF) Ellicott City, Maryland.

Sherry A. Falsetti, PhD, is Associate Professor and Director of Behavioral Sciences, Family Health Center, Department of Family and Community Medicine, College of Medicine, University of Illinois, Rockford, Illinois.

David W. Foy, PhD, is Professor of Psychology, Graduate School of Education and Psychology, Pepperdine University, Culver City, California; Adjunct Professor of Psychology, Headington Program in International Trauma, Graduate School of Psychology, Fuller Theological Seminary, Pasadena, California.

Sara A. Freedman, MSc, is Research Coordinator, Center for the Treatment of Traumatic Stress and Anxiety Disorders, Department of Psychiatry, Hadassah University Hospital, Jerusalem, Israel.

John B. Fulton, BS, is Doctoral Student, Department of Psychology, Brigham Young University, Provo, Utah.

Shirley M. Glynn, PhD, is Clinical Research Psychologist, Veteran's Administration Greater Los Angeles Health Care System, Los Angeles, California; Associate Research Psychologist, Department of Psychiatry and Biobehavioral Sciences, David Geffen School of Medicine, University of California, Los Angeles.

Bonnie J. Gorscak, PhD, is Research Therapist and Clinical Supervisor, Bereavement and Grief Program, Western Psychiatric Institute and Clinic, University of Pittsburgh School of Medicine, Pittsburgh, Pennsylvania.

Craig L. Katz, MD, is Director, Acute Care Psychiatric Services, Mount Sinai Medical Center, New York, New York; Clinical Assistant Professor, Mount Sinai School of Medicine, New York, New York; President, Disaster Psychiatry Outreach, New York, New York.

Terence M. Keane, PhD, is Associate Chief of Staff for Research and Development, VA Boston Healthcare System, Boston, Massachusetts.

Steven R. Lawyer, PhD, is Post-Doctoral Fellow, National Crime Victims Research and Treatment Center, Medical University of South Carolina, Charleston, South Carolina.

Christopher M. Layne, PhD, is Assistant Professor, Department of Psychology, Brigham Young University, Provo, Utah; Member, University of California, Los Angeles/Duke University National Center for Child Traumatic Stress, Los Angeles, California.

Randall D. Marshall, MD, is Associate Professor of Clinical Psychiatry, Columbia University College of Physicians and Surgeons, New York, New York; Director, Trauma Studies and Services, New York State Psychiatric Institute, New York, New York; Associate Director, Anxiety Disorders Clinic, New York State Psychiatric Institute, New York, New York.

Nadine Melhem, MPH, PhD, is Senior Statistical Coordinator, Advanced Center for Intervention and Services Research for Early Onset Mood and Anxiety Disorders, Department of Child and Adolescent Psychiatry, Western Psychiatric Institute and Clinic, University of Pittsburgh School of Medicine, Pittsburgh, Pennsylvania.

Jeffrey T. Mitchell, PhD, CTS, is Clinical Associate Professor of Emergency Health Services, University of Maryland, Baltimore County, Maryland; Adjunct Faculty, Emergency Management Institute, Federal Emergency Management Agency, Emmitsburg, Maryland; President Emeritus, International Critical Incident Stress Foundation, (ICISF), Ellicott City, Maryland.

Kathleen Nader, DSW, is Director, Two Suns, Cedar Park, Texas.

Linda A. Piwowarczyk, MD, MPH, is Co-Director, Boston Center for Refugee Health and Human Rights, Boston University Medical Center, Boston, Massachusetts; Assistant Professor, Boston University School of Medicine, Boston, Massachusetts.

Robert S. Pynoos, MD, MPH, is Co-Director, UCLA/Duke University National Center for Child Traumatic Stress, Los Angeles, California; Director, Trauma Psychiatry Program, Department of Psychiatry and Biobehavioral Sciences, David Geffen School of Medicine, University of California, Los Angeles; Professor in Residence, Department of Psychiatry and Biobehavioral Sciences, David Geffen School of Medicine, University of California, Los Angeles.

Beverley Raphael, AM, MBBS, MD, FRANZCP, FRCPsych, FASSA, Hon. MD (N'Cle, NSW), is Professor of Population Mental Health and Disasters, University of Western Sydney Medical School; Professor of Psychological Medicine, Australian National University; Emerita Professor of Psychiatry, University of Queensland.

Heidi S. Resnick, PhD, is Professor, Clinical Psychology, Department of Psychiatry and Behavioral Sciences, Medical University of South Carolina, Charleston, South Carolina.

Steven B. Rudin, MD, is Clinical Instructor of Psychiatry, Columbia University College of Physicians and Surgeons, New York, New York; Psychiatrist, Trauma Studies and Services, Anxiety Disorders Clinic, New York State Psychiatric Institute, New York, New York; Medical Director of Partial Hospitalization Program, New York State Psychiatric Institute, New York, New York.

William R. Saltzman, PhD, is Associate Professor, College of Education, California State University, Long Beach, California; Co-Director, School Crisis and Intervention Unit, UCLA/Duke University National Center for Child Traumatic Stress, University of California, Los Angeles.

Merritt D. Schreiber, PhD, is Program Manager, Terrorism and Disaster Branch, University of California, Los Angeles/Duke University National Center for Child Traumatic Stress, Department of Psychiatric and Biobehavioral Sciences, David Geffen School of Medicine, University of California, Los Angeles.

Daryl A. Schrock, MA, is Doctoral Candidate, Headington Program in International Trauma, Graduate School of Psychology, Fuller Theological Seminary, Pasadena, California.

M. Katherine Shear, MD, is Professor of Psychiatry, University of Pittsburgh School of Medicine, Pittsburgh, Pennsylvania; Director, Bereavement and Grief Program, Western Psychiatric Institute and Clinic, University of Pittsburgh School of Medicine, Pittsburgh, Pennsylvania.

Rebecca E. Spitz, BA, is Manhattan Reporter, NY1 News, New York, New York.

Alan M. Steinberg, PhD, is Associate Director, UCLA/Duke University National Center for Child Traumatic Stress, Los Angeles, California; Director of Research, Trauma Psychiatry Program, Department of Psychiatry and Biobehavioral Sciences, University of California, Los Angeles Neuropsychiatric Institute and Hospital, David Geffen School of Medicine, University of California, Los Angeles.

K. Chase Stovall-McClough, PhD, is Assistant Professor, Department of Psychiatry, New York University School of Medicine, New York, New York; Child Study Center, Institute for Trauma and Stress, New York, New York.

Karen Stubenbort, PhD, is Clinical Assistant Professor, Department of Psychiatry, Drexel University College of Medicine, Pittsburgh, Pennsylvania; Senior Psychiatric Clinician, Center for Traumatic Stress in Children and Adolescents, Allegheny General Hospital, Pittsburgh, Pennsylvania.

Rivka Tuval-Mashiach, PhD, is Lecturer, Department of Psychology, Bar-Ilan University, Ramat-Gan, Israel; Research Fellow, Center for Traumatic Stress, Hadassah University Hospital, Jerusalem, Israel.

William S. Unger, PhD, is Chief, PTSD Clinic, Veterans Affairs Medical Center, Providence, Rhode Island; Assistant Clinical Professor, Brown University School of Medicine, Providence, Rhode Island.

Serena Y. Volpp, MD, MPH, is Clinical Instructor of Psychiatry, New York University School of Medicine, New York, New York; Associate Unit Chief, Residency Training Unit, Bellevue Hospital, New York, New York.

Jared S. Warren, PhD, is Assistant Professor, Department of Psychology, Brigham Young University, Provo, Utah.

Melissa S. Wattenberg, PhD, is Director, Psychiatric Relapse Prevention Program, Outpatient Clinic, VA Boston Healthcare System, Boston, Massachusetts; Staff Psychologist, Outpatient Clinic, PTSD Clinic, VA Boston Healthcare System, Boston, Massachusetts; Instructor, Tufts University School of Medicine, Boston, Massachusetts; Adjunct Clinical Professor, Physicians' Assistant Department, Massachusetts College of Pharmacy and Health Sciences, Boston, Massachusetts.

Daniel S. Weiss, PhD, is Professor of Medical Psychology, Department of Psychiatry, School of Medicine, University of California, San Francisco.

Sally Wooding, BA (Hons), PhD (Clin Psych), MAPS, MCCP, is Clinical Psychologist and Senior Project Officer, Centre for Mental Health, NSW Health Department, Sydney, Australia.

Rachel Yehuda, PhD, is Professor of Psychiatry, Mount Sinai School of Medicine, New York, New York; Director, Division of Traumatic Studies, Mount Sinai School of Medicine, New York, New York.

Allan Zuckoff, PhD, is Assistant Professor of Psychiatry, University of Pittsburgh School of Medicine, Pittsburgh, Pennsylvania; Bereavement and Grief Program, Addiction Medicine Services, Western Psychiatric Institute and Clinic, Pittsburgh, Pennsylvania.

Preface

Immediately following the terrorist events of September 11, 2001, social workers, psychologists, psychiatrists, and nurses mobilized at family centers, hospitals, schools, and social service agencies to provide assistance and support to those who felt bereft, alone, and confused, and whose personal and emotional anchors had been suddenly wrenched from their moorings. Though many in our communities have demonstrated significant capacities to recover, many continue to exhibit diminished resilience and impairment in reestablishing lives as they existed prior to September 11. Children, adolescents, parents, spouses, partners, adults, emergency service personnel, and mental health professionals have been profoundly affected.

This book emerged as an outgrowth of the conference Group Approaches for the Psychological Effects of Terrorist Disasters, conducted May 2 through 4, 2002, sponsored by the American Group Psychotherapy Association and the Eastern Group Psychotherapy Society. Generous financial support for the conference was provided by the New York Times Company Foundation. For registrants of the conference, this was their first exposure to a comprehensive exploration of group models that might be used to ameliorate the biopsychosocial effects of a trauma precipitated by a terrorist event.

The trauma experienced after the events of September 11 raised serious questions among professionals concerning the most efficacious group interventions to be used considering the large numbers of individuals to be treated and the variety of mental health issues to be addressed. The unpredictability of the events of September 11, as well as their unique intensity, shattered the American illusion of being invulnerable to terrorist attacks on our soil. Although the biopsychosocial reactions to a terrorist attack may resemble the signs and symptoms of traumas caused by rape, vehicle accidents, or natural disasters, the psychological effects precipitated by a terrorist attack may necessitate additional considerations for those mental health professionals involved in treating individuals who have been affected. Understanding the impact of this form of trauma may require a paradigm shift in our individual and collective perspective.

Although disagreement may arise regarding the diagnosis and the psychological effects of exposure to a large-scale traumatic circumstance, little doubt exists as to the necessity of introducing services and resources to diminish their impact. The certainty of the clinical need for services and resources following a terrorist event does not alleviate the responsibility to approach the problem with a sound clinical methodology. This book provides a comprehensive practice focus for psychiatrists, psychologists, social workers, nurses, and clergy who conduct group interventions for those affected by the psychological effects of terrorist disasters.

Prior to the events of September 11, interest in the study of terrorist events, their effects, and effective treatment modalities was very limited. Current research regarding the psychological effects of terrorist disasters and appropriate group interventions is still in a nascent state. Although the use of groups for this purpose has been widespread, evidence-based data regarding the efficacy of group intervention are limited. Understandable challenges exist in implementing empirical studies that test the efficacy or effectiveness of group-based trauma models. From our perspective, these limitations demand a more encompassing, generous definition of "evidence-based treatment" that embraces efficacy, effectiveness, and descriptive research designs.

The majority of group models presented in this book are those for which substantive empirical data have been recorded in response to extraordinary conditions. For the purpose of comprehensiveness, we have included models that are widely practiced but for which questions have been raised about their substantive empirical data. Because the models explicated in this volume were not developed specifically for the terrorist disaster of September 11, the challenge will be to evaluate their ongoing effectiveness as they are applied to this and future events.

In developing this book the editors were concerned with the indiscriminate use of group interventions without regard to the population, treatment focus, conceptual foundations, and methodological efficacy. The group models represented in this volume, for the most part, consider these parameters. The majority of the group models are presented at a level of description approximating that of a clinical manual. The specific and diverse models represent the best of those for which empirical validation is available. Although a single book cannot include all of the available extant literature, we believe that the authors have provided a comprehensive foundation in thinking about this subject within a substantive framework.

By incorporating relevant foundational information (Chapters 2 through 11) with specific group interventions, our hope is that readers will be equipped to integrate and apply the subject matter to their unique circumstances. A common structure was followed in presenting the specific group models (Chapters 12 through 21) to enable the reader to comparatively understand how each protocol conceptually and clinically addresses the psychological and psychosocial effects of a significant traumatic event.

This book is divided into four sections. Part I provides an overview of the book. Part II discusses foundations and broad issues that potentially affect the outcome of group treatment. Part III includes group models that address the particular needs of children, adolescents, parents, emergency service personnel, and mental health practitioners. Part IV considers future directions.

How does one read this book? The reader can peruse a particular chapter that applies to a group modality he or she is thinking about implementing or is already using but hoping to improve its effectiveness. It can also be read front to back, because the chapters build on one another. This volume can also be a quick reference and not a voluminous tome that the reader must commit to studying. This is a definitive reference that touches on most of the current group treatment methods for traumatic events and does so with specific detail and examples to illustrate the interventions. The authors are leading experts, internationally renowned, highly respected, and at the forefront of this field. We cannot emphasize too strongly that this book would not have been possible without the extraordinary collaboration among the editors and authors.

ADDENDUM

As this book was in its final stages of preparation prior to publication, the devastating human impact of Hurricane Katrina on the Gulf Coast became apparent. Although this book was initially undertaken to focus on the psychological trauma of September 11, 2001, we cannot overlook the magnitude of the catastrophic losses and their emotional effects incurred by those affected by Katrina. We hope that this book can provide some understanding of the trauma and the application of group methodologies that address and short- and long-term needs of those in the Gulf Coast region.

Acknowledgments

The royalties from this book shall be donated to organizations which provide direct services to those who continue to be affected by the events of September 11, 2001, and Hurricane Katrina. Maintaining a focus on the difficult nature of the content of this book, as an outgrowth of September 11, 2001, was made easier by the pursuit of excellence by all those who collaborated on this project.

The editors would like to express profound appreciation to the authors for their thoughtfulness, extensive cooperation, generosity of spirit, original thinking, quality of effort, and expertise in the production of this book;

appreciation to Michelle Hammer for her work during the early stages of this book;

and sincere thanks to Shannon Vargo, whose unique talents, persistence, and constructive suggestions were invaluable in the organization and clarity of the text.

Finally, we must express special appreciation to Suzanne Schein for her love, heartfelt support, and willingness to accompany me from the earliest hours after September 11 to the completion of this book (LAS); Susan Spitz, Becky, and Jake, the loving foundation of my life (HIS); Chad, Shaun and Brooke Burlingame, my companions through life's journey (GMB); Matthew and Marlene Muskin who provide consistent love and inspiration in a sometimes uncertain world (PRM). To these sustaining people in our lives, we gratefully acknowledge their support for enduring the extensive time needed for the preparation of this book.

PART I:
OVERVIEW

Chapter 1

Trauma, Terror, and Fear: Mental Health Professionals Respond to the Impact of 9/11—An Overview

Terence M. Keane
Linda A. Piwowarczyk

Extraordinary events require extraordinary responses. The events of September 11, 2001, were extraordinary by any definition or standard; the mental health communities in New York City, Washington, DC, and across the United States responded to these tragedies with alacrity and zeal. Collaboration and cooperation transcended traditional boundaries of profession, location, organization, and funding sources in an attempt to respond effectively to these attacks on our country. The goal was to help those suffering in the aftermath of one of the most destructive peacetime attacks on America in its history. The purpose of this book is to provide an overview of the lessons learned from these signal events for all mental health professionals. At the same time, it looks toward the future; in various chapters the lessons learned are integrated into theoretical models with a vision to future developments in the emerging field of early mental health interventions. Thus, this volume employs the lessons learned to direct future initiatives, an objective that can guide collaborations and planning well into the new century.

Terrorism is defined as

> any systematic threat or use of unpredicted violence by organized groups to achieve a political objective. Terrorism's impact has been magnified by the deadliness and technological sophistication of modern day weapons and the ability of mass communications to inform the world of such acts. (Merriam-Webster, 2000)

Written more than a year before the fatal attacks in the United States, the definition is strikingly eerie as it portends well the 9/11 terrorist events in our country. The attacks on New York City and the Pentagon and the crash of a passenger plane in the fields of Pennsylvania all readily fit this definition, and the impact of these events affected America in ways that no prior singular attack on the United States had. On television stations worldwide for many days after the attacks, the almost constant availability of footage of the aircraft hitting the Twin Towers exposed the world, and especially all Americans, to the brutality of terrorism in unprecedented ways.

The response of the mental health community was immediate and decisive: key mental health professions prepared their constituents for assisting in efforts of early intervention and promotion of recovery. Expertise stemming from past experiences with other disasters, traumatic events, terrorism, war, and stress informed professionals through the availability of conceptual models, evaluation strategies, and interventions. The present volume addresses the broad spectrum of knowledge derived from the 9/11 attacks and other types of traumatic events, integrating the knowledge in ways that at once assemble the treatments available while denoting specific needs that must still be addressed empirically and procedurally. Most important, the volume brings contemporary conceptual models and intervention methods to the care of those exposed to terrorism. The ultimate goal of the text, therefore, is to ensure that future generations of professionals learn from the experiences of those involved at multiple levels in the mental health profession's responses to the terrorism of 9/11.

CONCEPTUAL MODELS

When traumatic events occur, we must understand why some people develop lasting psychological problems whereas others, seemingly exposed to the same fundamental experiences, do not. To explain these distinct response patterns, researchers generally focus on individual difference models. For example, Keane and Barlow (2002) extended to post-traumatic stress disorder (PTSD) Barlow's (2002) model for the development of anxiety disorders. This triple vulnerability model specified the importance of (1) psychological vulnerabilities, (2) biological vulnerabilities, and (3) the characteristics of the traumatic event. The advantage of this type of individual difference model is that it points to the interactive importance of past experiences, biological constitution, and psychological factors (e.g., mood, arousal, and cognitive processes) that exist prior to the occurrence of a traumatic life event. A second advantage of this model is that it aligns itself with the same precipitating factors of other anxiety disorders, such as panic disorder, generalized anxiety disorder, and social anxiety disorder. From a heuristic perspective, this permits one to understand the common features of these various anxiety disorders as well as the features that differentiate them. In addition, it establishes a theoretical model that can be put to scientific test.

Other components of the triple vulnerability model include conditioned learning at the time of the traumatic event; this conditioned learning establishes the mechanism by which false alarms occur. Cues in one's environment come to elicit the emotional responses associated with the traumatic event. These flashback experiences are emotionally painful for survivors and precipitate anxious apprehension about the event, or even a triggered flashback, recurring. This apprehension is part of the preoccupation with the traumatic event that is apparent in most survivors who develop PTSD.

Mitigating factors determine who does and does not go on to develop PTSD once exposed to a traumatic event. Empirical studies suggest that one's coping style and level of social support are key variables in the aftermath of a traumatic event. Our model proposes that the presence of an active, instrumental coping style and the assurance that one can ask for and receive social support from friends and family may alter the trajectory of one's response to traumatic events.

Yet this model focuses largely on the individual level of analysis. This is a strength, but it is also a potential limitation in our efforts to understand the ultimate outcomes of terrorist attacks. Viewing the outcomes largely from the individual level may not provide the best prediction of impact. For example, the level of analysis following 9/11 might be the family, the neighborhood, the city, the state, the country, or even beyond. In the case of terrorism, the target is often the larger population. Is the impact at the societal level simply the mean (i.e., average) of the individual impact? This is not likely, as each level interacts with those that precede and follow it. How children react is known to be a function of how their parents react; how families react is in part a function of how the neighborhood responds. This interactive notion of how outcomes emerge following large-scale disasters is accepted as a tenet in the community psychology literature (Norris, 2001). Although conceptual models that work at the individual level, such as that proposed by Keane and Barlow (2002), serve multiple purposes when it is vital for the level of analysis to go beyond the individual and move to higher ordinates, it is imperative that interactive models be developed and considered. Models that incorporate individual, group, and societal levels of analysis will undoubtedly be more predictive of the cumulative response of communities to the adversity experienced via disaster.

The current text attempts to accomplish complex model building in several places. Borrowing on public health strategies for understanding and intervening in community-wide disasters, these models ultimately will assist mental health professionals in interfacing most effectively with extant non-mental health resources following traumatic events. Appreciating who is most vulnerable following a traumatic event, based on levels of exposure and known risk factors, is key to the delivery of scarce mental health resources to those in greatest need and at greatest risk for developing prolonged psychological distress. In addition, it highlights one of the central findings in all of disaster research: many people exposed to great adversity readjust over time, demonstrating the remarkable resiliency of human beings.

Future model development that includes societal, community, family, and individual levels of analysis would greatly assist public policy experts in understanding and developing strategies for the optimal allocation of resources. Studies that would empirically examine variables at each of these levels of analysis would be especially welcome at this time (see Green et al., 2003).

LEVELS OF INTERVENTION

Much of our knowledge about effective treatments for psychological distress stems from randomized-control clinical trials with individuals who meet criteria for a particular disorder. Typically, these trials include participants who meet certain inclusion criteria and do not meet other criteria. It is widely known that people can develop a wide range of psychological disorders following traumatic events such as a terrorist attack. These disorders include PTSD, depression, panic, and substance abuse. Frequently, these disorders appear concurrently and represent high levels of comorbidity (Keane & Kaloupek, 1997; Keane & Wolfe, 1990), yet the evidence for treatment efficacy rarely addresses individual cases with multiple comorbidities.

In addition, virtually all the treatment outcome data that exist apply to treatments delivered at the individual level. Few randomized studies examine the role of group therapy in the treatment of PTSD (Keane, 1997). Among the lessons learned from 9/11, one is unremitting: addressing a major public health disaster such as a terrorist attack is impossible with only individual-level treatments. Alternatives are necessary. Group treatments, the subject of this book, are commonly employed in clinical settings, and virtually all mental health professionals receive training in some form of group therapy. An expressed goal of this text is to bring together the long clinical history surrounding the use of group-based interventions with the needs of the public when disaster strikes. How should groups be assembled? For how long should groups continue? Who should be included in groups? What type of intervention yields maximal results? When following the disaster should the interventions be provided? What types of interventions are effective in the short-term as preventive actions? What types of interventions can be provided in the mid- to long-term? Should couples and families be seen together with others? What professional qualifications are needed to run group therapy? These are only a few of the fundamental questions that many of the chapters in this text address.

However, group therapy alone may also be inadequate when extraordinary events occur and needs are great. When questions about the meaning of life and the loss of life predominate, what are the most appropriate and effective interventions? Clearly, public health models and messages must serve as the backbone of any comprehensive approach when thousands and even millions are affected by a traumatic event or series of events. Much as the impact of the 9/11 assaults was heightened by the presence of telecommunications, our approaches to managing the mental health impact of such events must make intelligent use of television, radio, and other forms of telecommunication. Broadcasts that include valuable information for managing one's own reactions and those of one's family and friends can form a preventive strategy that will reach large numbers of the affected population readily and quickly. In addition, telecommunications can minimize and address the nefarious effects of rumors and misinformation rapidly and definitively.

Similarly, the Internet and World Wide Web constitute an important and relatively new resource for responding to the needs of those affected by disaster and terrorism. Web sites such as those of the National Center for PTSD (www.ncptsd.org), the International Society for Traumatic Stress Studies (www.istss.org), the American Psychological Association (www.apa.org), and the American Psychiatric Association (www.psych.org) all contain important information for trauma in general; all responded to the events of 9/11 with enhanced sites to assist those suffering from the aftermath of the assaults.

Each of these methods can mitigate the effects of terrorism, disaster, and other types of traumatic events. Individual treatments, group treatments, public health strategies, and the use of technology to deliver relevant and needed information on coping and other resources comprise a comprehensive strategy for intervening in mass disasters. Conceptualizing the approach to assistance as one requiring multiple levels of intervention can result in a coordinated and systematic effort that will maximally utilize available resources and reach the largest number of people.

In a related text, Green and colleagues (2003) attempted to develop intervention strategies for entire countries adversely affected by mass trauma and disasters. Baron, Jensen, & de Jong (2003) devised a set of principles based on an inverted triangle that placed the highest priority at the level of interventions reaching the largest numbers of people with the lowest levels of intensity, resources, and cost. Such efforts as public health interventions coupled with national policies and procedures to minimize the impact of mass disasters are portrayed in this model as optimal, especially when re-sources are scarce. In places such as Rwanda, Burundi, Congo, Cambodia, Gaza, and Sudan, interventions based on public health models using West-ern and indigenous strategies are seen as the highest priorities. Specific cul-turally based interventions that address the psychological impact of trau-matic events are important features to include in any blending of Western and indigenous interventions as they may improve acceptance of the inter-vention while enhancing the effectiveness of the treatment. This is particu-larly the case when few professionally trained mental health professional resources are available in a region or a country.

For example, when Kuwait was recovering from the invasion by Iraq in 1990-1991, considerable financial resources were available, but due to cul-tural variables, resources were modest in terms of trained clinical psychia-trists, psychologists, and social workers in that country. Alternative inter-ventions were required. Efforts to work extensively through family and religious networks became the priority. Telecommunications also served as an important medium for assisting families and individuals in the recovery process following exposure to that war.

CONTROVERSIES

For many decades mental health professionals have been actively in-volved clinically in the immediate aftermath of disasters. From a research perspective, involvement of mental health professionals possibly dates back to Lindeman (1944) who studied the psychological impact of the Coconut Grove nightclub fire in downtown Boston. The field of crisis intervention, beginning in the 1960s with the community mental health movement in the United States, was the clear forebear of our current clinical involvement. Employing psychological first aid and screening strategies, these initial for-ays into the field of disaster mental health set the stage for more comprehen-sive involvement. Yet this involvement has in recent years been fraught with controversy.

The stage for controversy is in the specific nature of mental health pro-fessionals' involvement following disasters and similar critical incidents. Most people recognize and acknowledge the need for psychosocial support for those exposed to a critical incident, but there is less agreement about the components of any particular intervention. It appears from the literature that what is needed is a clearly defined process rather than any technique. All involved in the debate agree that a single intervention of whatever type is likely to be inadequate to prevent the occurrence of untoward psychological outcomes. What the process should be and how the process unfolds over time constitute the nature of the discussion, debate, and controversy in this field (Litz, 2003).

Mitchell (1983), in a landmark publication, presented the process of critical incident stress debriefing (CISD). Borrowing from military and paramilitary organizational efforts, this approach to psychosocial intervention quickly became an international phenomenon with cities, towns, and communities recognizing the need for mental health interventions in the wake of disasters and some organizations even mandating CISD for employees following exposure to a critical incident.

Eventually, Everly and Mitchell (1999) expanded their notion of mental health intervention to include a broader process that they termed critical incident stress management. This approach represents a more comprehensive set of interventions that emerges over a significant time period following a traumatic event.

Raphael and colleagues (Raphael, Meldun, & McFarlane, 1995; Raphael & Wilson, 2000) challenge the value of CISD, in particular due to a lack of empirical evidence to support the widespread application of CISD. She is joined with reviews by Rose, Bisson, & Wesseley (2002) supported by the Cochrane Collaboration indicating the absence of studies recommending widespread adoption of the CISD model of intervention.

Litz and colleagues (Litz, 2003; Litz, Gray, Bryant, & Adler, 2002) comprehensively reviewed the literature on the topic of early mental health interventions and concluded that a comprehensive, multistage model for intervention is needed to optimize outcomes and to prevent the development of psychopathology. Reviewing the literature on cognitive-behavioral approaches (CBT), he concludes that interventions that address cognitive distortions, promote emotional processing, and equip individuals with knowledge and skills to managetheir reactions are likely to be the most effective in preventing the development of PTSD (see Bryant, Harvey, Dang, Sackville, & Basten, 1998).

Chapters in this text address the important discussion points surrounding this issue. From definitional problems to the quality of key studies in the area, the debate is multifaceted and important. Ultimately, the importance of this debate is that already additional research has followed it. More is needed. In a blue ribbon panel assembled in the aftermath of 9/11 and sponsored by many national agencies, the participants concluded that empirical study is necessary for the field of early intervention to grow and thrive. With the high frequency of traumatic events in all societies internationally, work in this area should be a priority. Whatever the outcome of the debate, it is clear that mental health's involvement in the aftermath of disasters and traumatic events is here to stay. The contribution of mental health professionals to survivors and to emergency workers is now well accepted. Its development and evolution will undoubtedly depend on multiple studies of interventions and their effects. Changes and modifications in techniques and approaches should be premised on empirical study. These changes should optimally reflect an iterative technique that evaluates components of the process as well as the process as a whole. A concerned and caring society will want the best possible services to be provided to those individuals and communities adversely affected by mass trauma. How to provide those services, who should provide them, when they should be provided, and what particular services to be provided are the questions for the next decade of research.

ON-SITE ROLES FOR MENTAL HEALTH PROVIDERS

In addition to providing mental health services directly during crises and disasters, clinicians are beginning to develop important adjunct roles during fieldwork at disaster sites as well. These roles include assessment (Wilson & Keane, 1997) and intervention. Norris et al. (2002) cite numerous roles that mental health professionals assume already. Not surprisingly, the first duty, as it is for all professionals, is to *protect* survivors from further injury. As well, clinicians assume the role of protecting from injury those who are working at the site. The second role is to *direct* survivors to the appropriate locations to receive services available. Third is to *connect* survivors to family, friends, co-workers, and others who have survived. Fourth, clinicians are increasingly responsible for *detecting* the need for additional services through a variety of on-site rapid screening strategies. Clinicians must be able to detect the presence of extreme reactions that might also require immediate psychiatric assistance. The fifth role of clinicians is to *select* those in need of services and be familiar with the resources available in order to refer individuals to these services as quickly as possible. Finally, the sixth role identified is to *validate* the extreme nature of the incident and to normalize the person's reactions to the event. One key component of this cognitive assessment is to try to evaluate the extent to which the survivor is distorting responsibility for the event or its aftermath. For many survivors of traumatic events, self-blame contributes to more persistent psychological distress and can in time become very difficult to challenge therapeutically.

CLINICIAN SELF-CARE

For mental health clinicians working with trauma-exposed populations, it is vital to address one's own psychological needs in order to minimize the possibility of developing mood- and anxiety-related problems. Recommendations for those working in crisis situations include the following:

- maintaining one's usual schedule of sleep and work to the extent possible;
- ensuring adequate nutritional consumption at regular intervals;
- paying attention to physical exercise especially if the work involved no physical exertion;
- taking adequate breaks with others involved in similar work;
- avoiding or minimizing the use of mood-altering substances such as alcohol, cigarettes, and other drugs;
- taking advantage of opportunities to discuss and debrief about the impact of the work on one's well-being; and
- using humor when appropriate for communicating and sharing experiences with others.

Adhering to these guidelines will reduce the likelihood that a clinician at the site of a traumatic incident for prolonged time periods will develop mood or anxiety syndromes. Following these suggestions will permit one to continue functioning for a sustained period of time in a difficult work atmosphere. Taking care of oneself in the short and immediate term increases the likelihood that one will be available to assist others for a longer period of time. Burnout among emergency workers is high, and despite the best intentions of impassioned clinicians, the work does take its emotional toll, potentially compromising one's ability to contribute to the success of both current and future interventions.

CHAPTER CONTENT

Each of the chapters contained in this book addresses an important set of information pertaining to the successful treatment of survivors of terrorism, mass trauma, or individual trauma. The field of trauma studies is nearing twenty-five years old, and many questions have been asked, with some answers now available. In some instances the answers are qualified and tentative, while in others more definitive principles are emerging from empirical study. It is now clear that immediate and short-term interventions for trauma exposure must be studied by employing the most rigorous methodological standards so that future generations of survivors and clinicians will receive the benefit of our experiences.

Represented in this book are chapters describing the ever-expanding knowledge of the biological correlates of exposure to trauma and the development of PTSD. In addition, one chapter comprehensively reviews the existing literature on the effects of psychopharmacological interventions in treating trauma responses. Each of these two areas has burgeoned in the past dozen years with greater knowledge deriving from infrahuman studies as well as studies with human participants. Studies focusing on multiple systems concurrently will assist in more fully understanding the structure and function of various neurobiological systems of the body under stress.

Psychotherapy researchers have made great progress in the treatment of chronic PTSD in the past twenty years. Beginning with the first randomized clinical trial of PTSD (Keane, Fairbank, Caddell, & Zimering, 1989) using cognitive-behavioral approaches, the field has grown dramatically, as is reflected in the need for the development of best-practice guidelines by the ISTSS (Foa, Keane, & Friedman, 2000). The present book represents much of the progress in the field of trauma treatment, especially as it is translated into group intervention methods.

Interventions for families and children must inevitably form a part of the response to traumatic events such as terrorism and disasters. The present text highlights the importance of this work with chapters written by some of the top people in the field of trauma treatment in children. Each of the teams, affiliated with such institutions as the University of California-Los Angeles, Drexel University Medical School, University of Medicine and Dentistry of New Jersey, and Brigham Young University, is actively involved in the National Center for Child Traumatic Stress established by the U.S. Congress in 2000 to address the lack of knowledge of effective treatments for PTSD and related disorders in children exposed to traumatic events. From the establishment of the national center and the respective treatment development centers and treatment practice centers will come rapid advancement in the care and treatment of PTSD in children, adolescents, and their families. Recent institution of a disaster treatment center at the University of Oklahoma Health Sciences Center will ensure that some of this progress will be reflected in the needs of children and families exposed to mass disasters.

Many group therapeutic approaches may be helpful for people exposed to traumatic events. The current text attempts to address the needs of therapists by focusing on many of the key approaches used in contemporary American society: cognitive-behavioral therapy (CBT), psychodynamic, present day (problem-focused) therapy, and grief/bereavement treatment. These approaches, naturally, can be applied to adults, adolescents, and children. The format for the interventions can be didactic, emotional process-ing, skill building, and supportive in nature. Although much remains to be learned about the efficacy of all of these approaches, they are all rooted in sound theoretical frameworks and typically some evidence supports the work, even if only at an observational level of analysis.

One area often ignored in the treatment of PTSD is the role of spiritual-ity. This omission is especially egregious in models that focus exclusively on biological factors. Important work is now ongoing examining the impact of trauma exposure on one's religious and spiritual beliefs. Some clinicians are also examining the role of spirituality in the recovery process. The fact that clinicians often differ markedly from their clients in their views on the importance of religion in one's life may also contribute to the paucity of in-formation available to assist clinicians in treating trauma-exposed clients. It is clear that more information on this topic is needed. How clients view their spirituality and their religion is often a function of fundamental develop-mental factors. These factors may constitute risk or protective variables. Further exploration of how one's spirituality is affected by trauma exposure and how spirituality can assist in the recovery process is a topic of growing interest and importance in the field. In light of the appearance of large num-bers of individuals that are victims of the clergy sexual abuse crisis, more attention to this matter is paramount (McMackin & Keane, 2004).

Finally, the attacks of 9/11 on the United States may well represent an end to our sense of security from external attacks. Chapter 3 written by col-leagues from Israel, examines the impact of living in a society that is under continuing siege from its enemies. Israelis have lived under ominous condi-tions since the nation was formed more than fifty years ago, but the past five years represent a continually elevated alarm level for all the population of that country. Although we can hope that the level of chronic alarm in the United States will never reach that of Israel, the recent events in America and our involvement in military actions in the Middle East suggest that ter-rorism might become increasingly common across national borders. Les-sons from Israel for managing the impact of terrorism on our soil are indeed relevant and valuable for us to consider implementing as a country in the event that the number of terrorist attacks increases.

SUMMARY

This overview suggests that the role of mental health clinicians in the af-termath of disaster and terrorism is growing in importance. The events of 9/11 underscore the value of having a trained and ready mental health workforce. Training in psychological first aid, crisis intervention, and early interventions may well become a part of graduate and postgraduate training in clinical programs in psychology, psychiatry, and social work. The scope and the depth of the psychological impact coming from these events would warrant this type of investment of time and resources.

Among the lessons learned from 9/11 is the inability of any single organization to immediately and effectively establish a comprehensive registry of all those affected directly and indirectly by the attacks. In employment units such as the Fire Department of New York, New York City Police Department, schools, unions, and corporations, it was relatively straightforward to establish such registries. Attempts to combine these registries, however, failed, largely due to political issues and concerns for confidentiality. Although these interests are real, a greater good might emerge from collaboration and cooperation among the various stakeholders. This greater good might inform public policy and create multiple venues for assisting those most seriously affected by exposure to mass disasters. Coordinated efforts that cross geographic, political, and social lines would be welcome additions to societal efforts to understand the impact of mass violence and to provide the best possible interventions in service to recovery. It is clear that most survivors of traumatic events are resilient and do recover, but a significant minority of survivors may endure prolonged periods of distress. It is for these people that interventions should be developed with the hope of minimizing the impact of trauma exposure, limiting disability and dysfunction, and maximizing optimal emotional recovery. It is to these goals that this book is dedicated.

REFERENCES

Barlow, D.H. (2002). *Anxiety and its disorders: The nature and treatment of anxiety and panic.* New York: Guilford Press.

Baron, N., Jensen, S.B., & de Jong, J.T. (2003). Refugees and internally displaced people. In B. Green, M. Friedman, J. de Jong, S. Solomon, T.M. Keane, J. Fairbank, B. Donelan, & E. Frey-Wouters (Eds.), *Trauma interventions in war and peace: Prevention, practice, and policy* (pp. 243-270). New York: Kluwer Academic/Plenum Publishers.

Bryant, R.A., Harvey, A.G., Dang, S.T., Sackville, T., & Basten, C. (1998). Treatment of acute stress disorder: A comparison of cognitive-behavioral therapy and supportive counseling. *Journal of Consulting and Clinical Psychology, 66,* 862-866.

Everly, G.S., & Mitchell, J.T. (1999). *Critical incident stress management: A new era and standard of care in crisis intervention.* Ellicott City, MD: International Critical Incident Stress Foundation.

Foa, E.B., Keane, T.M., & Friedman, M.J. (Eds.). (2000). *Effective treatments for PTSD: Practice guidelines from the International Society for Traumatic Stress Studies.* New York: The Guilford Press.

Green, B.L., Friedman, M.J., de Jong, J., Solomon, S., Keane, T.M., Fairbank, J.A., Donelan, B., & Frey-Wouters, E. (2003). *Trauma interventions in war and peace: Prevention, practice, and policy.* New York: Kluwer Academic/Plenum Publishers.

Keane, T.M. (1997). Psychological and behavioral treatments for post-traumatic stress disorder. In P. Nathan & J. Gorman (Eds.), *A guide to treatments that work* (pp. 398-407). New York: Oxford University Press.

Keane, T.M., & Barlow, D.H. (2002). Posttraumatic stress disorder. In D.H. Barlow (Ed.), *Anxiety & its disorders: The nature and treatment of anxiety and panic* (pp. 418-453). New York: Guilford Press.

Keane, T.M., Fairbank, J.A., Caddell, J.M., & Zimering, R.T. (1989). Implosive (flooding) therapy reduces symptoms of PTSD in Vietnam combat veterans. *Behavior Therapy, 20,* 245-260.

Keane, T.M., & Kaloupek, D.G. (1997). Comorbid psychiatric disorders in PTSD: Implications for research. In R. Yehuda & A. McFarlane (Eds.), *Psychobiology of posttraumatic stress disorder* (pp. 24-34). New York: Annals of New York Academy of Science.

Keane, T.M., & Wolfe, J. (1990). Comorbidity in post-traumatic stress disorder: An analysis of community and clinical studies. *Journal of Applied Social Psychology, 20,* 1776-1788.

Lindeman, E. (1944). Symptomatology and management of acute grief. *American Journal of Psychiatry, 101,* 141-148.

Litz, B.T. (2003). *Early interventions for psychological trauma.* New York: Guilford Press.

Litz, B.T., Gray, M., Bryant, R.T., & Adler, A. (2002). Early intervention for trauma: Current status and future directions. *Clinical Psychology: Science and Practice, 9,* 112-134.

McMackin, R., & Keane, T.M. (2004). Understanding postraumatic stress disorder and clergy sexual abuse. *Human Development, 25,* 21-25.

Merriam-Webster Dictionary of the English Language. (2000). Springfield, MA: Merriam-Webster.

Mitchell, J.T. (1983). When disaster strikes. . . The critical incident stress debriefing process. *Journal of Emergency Medical Services, 8,* 36-39.

Norris, F. (2001). *Fifty thousand voices speak.*

Norris, F.H., Friedman, M.J., Watson, P.J., Byrne, C.M., Dizz, E., & Kaniasty, K. (2002). 60,000 disaster victims speak: Part I. An empirical review of the empirical literature. *Psychiatry, 65*(3), 207-239.

Raphael, B., Meldun, L., & McFarlane, A.C. (1995). Does debriefing after psychological trauma work? *British Medical Journal, 310,* 1479-1480.

Raphael, B., & Wilson, J.P. (2000). *Psychological debriefing: Theory, practice, & evidence.* Cambridge, UK: Cambridge University Press.

Rose, S., Bisson, J., & Wesseley, S. (2002). Psychological debriefing for preventing posttraumatic stress disorder (PTSD). *The Cochrane Library, Issue 1.* Oxford: Update Software.

Wilson, J. P., & Keane, T.M. (1997). *Assessing psychological trauma and PTSD.* New York: Guilford Press.

PART II:
FOUNDATIONS

Chapter 2

Images of Trauma:
The Aftermath of Terrorism and Disasters

Henry I. Spitz
Rebecca E. Spitz

INTRODUCTION

No one is immune to the psychological impact of terrorism and catastrophic disasters. The timing, form, and severity of these reactions may
vary, but everyone exposed to a large-scale traumatic event feels its impact
on some level. Quite naturally, those individuals and groups who have received the most attention during such crises are those that have been at the
epicenter of the events. Direct trauma victims and their families, and those
who have experienced profound injury and loss as a by-product of the traumatic event, are often the main focus of early interventions. These people
require immediate assistance. Others on the periphery are either helped
later or run the risk of being relegated to the status of "forgotten people."

The "ripple effect" of psychological trauma is often seriously underap-
preciated. The focus of this chapter will be to call attention to the human toll
taken on those individuals and groups who, by virtue of their proximity to
or distance from traumatic events, constitute a psychologically vulnerable
population in the aftermath of terrorist acts or natural disasters of catastrophic proportion.

It is the authors' belief that an underserved trauma population exists. By
increasing awareness in the mental health community of the multiplicity of
subtle and glaring forms trauma responses can take, more appropriate and
timely therapeutic interventions can be employed to assist *all* trauma
survivors.

Throughout this text, the categories of acute and delayed stress reactions,
ranging from normal immediate responses to traumatic circumstances
through severe and persistent trauma reactions such as post-traumatic stress
disorder (PTSD) and a range of dissociative states, will be described in detail. For the purposes of this chapter, it may be assumed that the symptoma-

tology of normal and abnormal stress and trauma adaptations and the forms of therapeutic intervention are similar for those on the periphery as they are for direct trauma victims.

A critical difference lies in the ability to recognize and treat those in need of psychological services who, for a variety of personal reasons including feeling a lack of entitlement, guilt, shame, cultural, vocational, and religious beliefs, among others, may not be comfortable coming forward to identify themselves as needing professional attention. We have designated this group as the "forgotten people," those who are affected by the ripple effect of major catastrophes, and they form one of the focal points of the material presented in this chapter.

CASE ILLUSTRATIONS

The poignant and dramatic ways that direct and indirect trauma has shown itself is vividly illustrated in the words and stories of people whose voices express the profound impact terrorism and disaster has had on their lives. For some, the initial response is horrific but self-limiting, the so-called "normal" or acute stress responses to overwhelming circumstances. For others, clinically recognizable states of stress and trauma-induced emotional problems emerge or are rekindled.

The vignettes in this chapter are taken from, but are not limited to, the emotional fallout of the September 11 attacks in New York City, Washington, DC, and the plane crash site in Pennsylvania. The cause of trauma in these cases was terrorism, but parallels can be drawn to other traumatized populations under different circumstances. Those who have experienced natural disasters, including fires, floods, and earthquakes; observed loss of life from large- and small-scale acts of violence, including combat; and survived prior traumas of rape, physical assault, and sexual abuse in childhood all share similar emotional responses.

What follows is a representative sampling of the "voices" heard after the September 11 attacks based on a model of closeness or distance, both geographical and emotional, from the epicenter of the attacks at Ground Zero.

"The Eye of the Storm": Narrow Escapes

Ed

Ed was in a conference room on the twenty-fifth floor of the North Tower when the first plane struck. He and his colleagues ran to the window to see smoke and paper floating in the air, reminiscent of a ticker-tape parade. The group tried to evacuate but in the stairwell was told by the Port Authority there was no danger. Ed elected to leave anyway. By the time he ran to his apartment near Canal Street the towers were collapsing behind him. To date, he refuses to talk about this with anyone, including his fiancée.

Manny

Manny worked on the eighty-first floor but was waiting for the elevator on the seventy-ninth when the plane tore into the building. Doused in jet fuel, he was immediately set ablaze. Already burned over more than 80 percent of his body, Manny knew he could not wait for help to reach him. He staggered to the stairwell, where he walked down one step at a time into a waiting ambulance. Manny lost a finger, has had several skin grafts, and wears compression garments to minimize the significant swelling which still exists. His face is badly disfigured, but when asked if it bothers him responds, "Why would I be upset? I'm lucky. I'm alive."

The Business Trainees

Joe, Sam, Annie, and Phil, young business trainees, got to their jobs at a large financial corporation in Tower One, as usual, at 6 a.m. They took a breakfast break at around 8 a.m. and were seated in a nearby coffee shop when the first plane hit. Stunned, together they fled to Joe's home, which was closest to where they were. It was only when they turned on the television that they first understood the magnitude of what had happened.

Carole

Carole had taken another high-powered job outside of the World Trade Center and was coming in late on 9/11 to clear her desk. Moments after the initial strike, she emerged from the subway surrounded by falling debris. For a moment she thought a parade was taking place. She was swept up in a stampeding mob running toward safety.

Those who had close calls continue to ruminate about why they were spared when their close friends and colleagues, fine and decent people, were not. Some have become quite spiritual or turned to religion in earnest, others sought treatment, and the rest, not feeling entitled to ask for more than what they have already been fortunate enough to get (their survival), have gone underground with their feelings.

"In The Pit": Rescue and Recovery

John

"My name is John, and I'm a carpenter and an emergency medical technician. I was at the World Trade Center before the second plane hit. I remained on-site for thirty-one hours after the collapse. I felt like I had smoke inhalation, but after time it wasn't going away; it was getting worse. My voice changed. I had a raspy cough, and I was out of breath by just going up and down stairs. As I lay down at night I felt like I was drowning.

"After that I came to the clinic. This was about a year ago. Over the past year my lungs and airways are recovering. My lungs can hold more breath than a year ago. I learned a lot about asthma and airways. I've learned to live with a medical condition. My life and health will never be as it was September tenth, but I'm getting closer every day. I'm glad to say that I'm better than I was September twelfth. This is my goal to get better, to get to be September tenth again."

"A year ago I was actually carrying oxygen with me to help me breathe. Sometimes the coughing would make it necessary to take some oxygen to relieve the problem. Constantly out of breath, now I'm more able to run up and down stairs more easily. Before I would get exhausted just cutting the lawn, that kind of stuff. Normal, everyday occurrences were taking their toll. Now I can do all the things I want to do. I'm not at September tenth by any respect, but I'm not at September twelfth either."

Vincent

"Life after September eleventh has never been normal for me again. I was a CEO of an aerospace company for twenty years and answered the call in my role as a volunteer firefighter on September eleventh and made it there before sunrise on the twelfth. I lost one of my best friends down there, Jeff of Ladder Three, and I was digging for a long time down there, just trying to help in any way I could. We came down with a lot of problems during our stay down there; we just didn't realize what was happening to us over the period of time.

"I believe there are something like twenty-seven symptoms; I have twenty-four of them, and some of them are pretty severe. I'll be going for sinus surgery on November sixth to try to relieve these headaches I've been having since January. So many people don't realize this is happening, don't care this is happening, or are part of the 'MTV group' which is like 'let's do a little sound bite and go on to the next thing.'

"This disaster is still ongoing. We're living this every day of our lives and it's not getting any easier. I haven't gone back to work. I have three children and I'm living my life by borrowing money from my eighty-three-year-old mother. I've never borrowed a dime in my life from anybody and it's extremely embarrassing to live like this.

"Many of us can't go back to life as normal. It really won't ever be the same. That people are taking this seriously [is the one thing that's giving us hope]. And maybe these symptoms, in time, will be relieved."

Jimmy

"I'm proud to be a union rep for the transit workers who were down at Ground Zero. We had three thousand Transit Workers Union members working on the pile. Not too many folks know that. We also had about one thousand other individuals whose work brought them into the area that day or subsequently.

"On the morning of nine-eleven one of our bus operators was ordered to keep his bus one block from the towers as they were burning. And after those towers came down rather than do what most everyone else did which was leave the area, this gentleman stayed there. What he did was he went around and he picked the injured up out of the street and he filled his bus up with injured and bleeding people. He got all of those people to the hospital; he evacuated them. Unfortunately, from his time spent there he suffered depression and posttraumatic stress, but he's a true hero."

"The Ripple Effect"

Many people not directly involved as victims or rescue workers are nonetheless affected by traumatic events. Some by virtue of their proximity to the trauma site, others through their relationship to people lost in the attacks, and still others via their work and volunteerism are exposed to the psychological sequelae of massive trauma.

Eddie—New York City Firefighter

"I had been raised to think that men don't cry. When I joined the department, I found that most of my fellow firefighters also subscribed to the same code. My image had to be one of being strong, reliable, brave, dedicated to taking care of others, and able to cope with tough situations. Then came September eleventh! Everything I thought I knew went out the window.

"As a professional I had been trained to keep my head under pressure. During the attacks I lost some of my closest friends. I thought I was losing it too. In past crisis situations, my buddies had always said, 'Keep a brave face,' 'Be strong for your families,' 'Leave your troubles at the job,' but I was unable to follow any of these beliefs. I felt guilty and ashamed that I couldn't 'be strong for my fallen brothers' and that I was selfish in putting my own feelings before my job. I was a mess.

"I felt depressed and couldn't control myself. I cried anytime I saw the TV reports of the events or read about it in the newspapers. My sleep was horrible, and I was ashamed to say that at several points I even thought about ending my life. I didn't know where to turn. Going for professional help was not in my vocabulary. I saw it as a sign of failure and of a flawed character. My religious beliefs were shaken by the tragedies I had witnessed. I felt completely lost.

"It wasn't until a 'shrink' came to our firehouse to talk about the normal reactions to stress and death that I realized that some of my 'weird' feelings were not only to be expected but were shared by fire department personnel all the way up the line to our chief. I date that as the turning point for me. I was better able to be more public about what I was going through and felt less isolated. It was extremely important to me not only that my buddies think less of me, but that many of them were experiencing similar reactions."

Wanda—New York City Police Department Paralegal

"The days immediately following the nine-eleven attacks were horrible. My boyfriend and I were in shock. We couldn't believe what had happened and what the consequences would be. In fact, we were so scared, we spent three consecutive days in bed having what we called 'apocalyptic sex'!

"We figured that since the world might be coming to an end, we might as well spend our remaining time on earth together and in the pursuit of happiness and pleasure.

"With the passage of time, I look back on those days and realize that we were so terrified, that we desperately clinging to each other out of fear and a need to be 'connected.'

"Although I had seen plenty of violence on the job, it was nothing compared to the terrorist attacks on the Trade Center. In my wildest dreams, I never could have imagined what I saw on September eleventh."

Duane—Second-Year Dental Student

"When I volunteered to help after nine-eleven, I wondered whether I had any useful skills that might make a contribution to the search-and-rescue effort. Little did I know that I would end up being assigned to the New York City morgue. My job was to assist in identifying victims of the World Trade Center disaster through their dental records.

"I wanted to be a part of the effort following the terrorist attacks, but I was not remotely prepared for what I saw. I had to sift through pieces of what were formerly living human beings in an attempt to put a name to the fragments assembled before me. I wanted to flee, but I felt too guilty to do so. Although I had seen physical trauma, surgical procedures, blood, and pain before, this was different. I felt that I had to find some way that would allow me to rise above my feelings of revulsion and depression.

"My consolation came from speaking with co-workers in the morgue who worked in the mortuary on a regular basis and had been through this ordeal many times before. I found comfort in knowing that what I had done might bring peace and closure to families and friends who were agonizing over the fate of loved ones unaccounted for in the Twin Towers disaster. That put many things in focus for me, including a change in my career plans; somehow being a wealthy dentist in the suburbs didn't seem relevant in the face of all the tragedy I witnessed."

"Spillover": The Ground Zero Neighborhood

Juan—Small Business Owner

"The morning of September eleventh, I was making breakfast for my regulars. Another steady customer came in and told me to turn on the news. For the next hour we all sat at the counter, watching in disbelief.

"When the towers came down, my store was in the line of fire. I had no idea the tragedy would actually reach us. I don't really remember much except being overwhelmed by the smoke and dust. Everybody was screaming—people were racing in off of the street, trying to outrun the cloud from the towers. I wanted to close the door because it was already so crowded inside, but instead I hid, hoping it would be over soon. It was dark as midnight; I couldn't see my own hand in front of my face. Some people were hyperventilating. I think we all thought we were going to die.

"I don't know how, but we got through that day. My business wasn't so lucky. I lost my plate glass window; everything inside was buried in dust. I had to throw away every last item in the kitchen.

"I got a small loan and insurance helped out a little bit, but it wasn't enough to start up again. I don't know if I want to, anyway."

The Employee Group

In the weeks following the 9/11 terrorist attacks, a group of co-workers came together in a conference room at one of the large financial centers adjacent to the World Trade Center to discuss their experiences in a group led by a pair of mental health professionals. Members continued to be depressed and shaken; most had recurring flashbacks and nightmares. In vivid detail, they all described aspects of the horror of hearing and watching the explosions as well as seeing people, some part of a handholding couple, jump to their deaths. All had been personally pessimistic about their own survival on that day. They had since all become aware that a maintenance crew had discovered the remains of a human scalp on their office roof.

Frank told of his twelve-hour ordeal extricating himself from the building, getting trapped in a basement down the block, then trudging through remains and body parts. Upon returning home, his hysterical wife removed bone fragments from his hair.

Helen, the child of Holocaust survivors, had to reach a ferry in order to return to her home. When she finally arrived at her destination, she was asked to remove her "favorite" suit jacket which was promptly sealed into a metal container marked "contaminated." She took her place in line to be "decontaminated" herself by being hosed down. This evoked visions and echoes of the Holocaust as well as the civil rights movement, which left Helen feeling all the more vulnerable and distraught.

Georgia complained that her supervisor had been out of town around the time of the incident. He was critical and angry that his department had not "pulled themselves together by now."

A few people voiced their reluctance to return to the office and the neighborhood, skeptical that they could ever resume their previous routine. Sarah was finding it difficult to get to work without crying or vomiting, and Bill said he no longer felt the way he earned his living was "meaningful," particularly as compared to firefighters and the police.

Families and Friends

Anna—Victim's Mother

"Every night at six-thirty I go into the yard and wait for Sandy to come home. Even though I saw her body, her beautiful face, after they recovered her, I still wait for my baby. God didn't mean for children to die before their parents. People say we're lucky in a sense because we had a real burial, but I think they're crazy when they talk about closure.

"There's no such thing as closure for my family or me. I can't imagine a day where I don't think about Sandy every single second of every minute of every hour. My youngest graduated from high school this summer and I don't remember it. All I remember is thinking how happy his big sister would have been to see him accomplish that goal.

"My son was angry with me because I couldn't stop crying to celebrate his milestone. My husband is angry with me because I won't go with him to the cemetery. Every day he goes and talks to Sandy; brushes the dirt away from her headstone; and lies against it. Me, I just wait for her. Six-thirty every night. I'll do it every day until it's my time. Maybe then I can see my beautiful girl again. Touch her face and call her my baby. I know that will happen. And I don't tell many people, but I can't wait for that day to come. Life here on earth just doesn't mean that much to me anymore."

Joanne

Joanne was home when the phone rang. It was her father, telling her to turn on the news. As soon as she saw the images from the World Trade Center she knew there was no hope for her husband of three years who worked on the ninety-eighth floor of the North Tower. Her fears were all but confirmed when she had not heard from him by that afternoon.

Then the wait began for the ultimate confirmation—the body. That came two weeks later. Workers on "the pile" uncovered remains, which, after DNA testing, proved to be Joanne's husband. Distraught but feeling some sense of closure, Joanne began making funeral preparations for what was left of his body. She buried him a week later and began to grieve in earnest.

Joanne expected the tears, the sleeplessness, and the anger—what she did not envision was another phone call from the city, telling her that additional body parts had been identified and were in the morgue. She had to decide whether to disinter the body to reunite the parts, bury them separately, or ask for them to be handled on her behalf.

Sadly, this would prove to be a decision Joanne would face again. Less than two months later, another call came in from the Fresh Kills site—more remains. In each instance, Joanne chose to reunite the remains, each time praying that it would be the last time she would have to disturb her beloved and re-open wounds she was so desperately trying to close.

Distant Repercussions

Often people assumed to be on the periphery of traumatic events and thus spared the emotional impact of them are, in fact, profoundly affected by catastrophic events.

Kevin—The Bagpiper

"The music hits the people you're playing to, but it also hits you while you're playing. It puts to music what you can't put into words. Sometimes the bagpipes fill a spot that's missing; you're playing a final farewell for someone who's passed on. I've been in the pipe-and-drum band in the department for fifteen years, playing the bagpipes since I was a kid. If you're Irish it's kind of tradition. Before September eleventh we'd perform at parades and ceremonies like promotions. Every once in a while there'd be a funeral. Now there's a funeral every weekend—sometimes two or three in one day. When I play 'Amazing Grace' or 'Danny Boy' I think my heart's going to pop. That's as bad as it gets during the ceremony though. I try not to think about it because the family's so sad. After all, I'm alive and playing and I'm thankful for that. It's worst when I get home, after I've put my pipes away.

"I just sit on my couch with the mass cards from church. Sometimes I cry; sometimes I pour a drink and toast the poor bastard looking back at me. I don't think I'll give up playing anytime soon, it's my way of answering the call. People who I've played for at some funerals come over and say 'thank you.' I should thank them. When it comes down to it I'm honored to be there, honored to play for the families of the city of New York."

Michael—Psychotherapist

"I have been a practicing therapist for many years, but never have I encountered tragedy on the scale of what I saw following the events of September eleventh. I volunteered my services at one of the major hospitals close to Ground Zero. For weeks after the disaster, I spent time counseling people who were dealing with the emotional impact of the terrorist attacks.

"Initially, I was using the critical incident stress debriefing format since it was the only trauma model I was familiar with. I wasn't quite sure whether or not what I was doing was appropriate given the controversy surrounding this method, but for me it was 'the only game in town.' Since I was appointed head of a post-nine-eleven drop-in center, I felt a particular need to appear composed and in control; after all, who would put any stock in what a nervous therapist was advising them to do?

"My style as a therapist is to be open and interactive, but I felt the need to let people unburden themselves as the initial priority early in their therapy, so I listened and absorbed more than I ordinarily would. On the surface, I was the anchor, the emotional rock for those shaken by the tragedy. Inwardly, however, I noticed that the work was having a dramatic effect on me.

"With so much input on the themes of loss, panic, despair, and the fears going forward related to the possibility of future biological, nuclear, and other terrorist actions, I felt overwhelmed. I began to feel emotionally numb and found myself 'tuning out' in therapy sessions. My sleep was disturbed; I felt frightened and depressed myself about the very same things my therapy clients were struggling with.

"I had become preoccupied with the safety of my family, particularly when they are not with me. This is exacerbated now that my children are late adolescents and increasingly independent and out of touch. I cannot seem to reassure myself, nor can my inner circle of colleagues including my own former therapist. I don't want to burden family or friends because I am usually in the role the caregiver and would be humiliated to admit to feeling this shaky.

"I worried that in the name of trying to be helpful to people, I might be inadvertently retraumatizing them. For the first time in my professional life, I questioned my ability to continue doing psychotherapy in general and psychological trauma work in particular."

Margo—New York City Elementary School Teacher

Another overlooked but vulnerable population are teachers—traumatized in their own right and also by the vicarious traumatization of their students. Margo is a thirty-one-year-old elementary school teacher who was reading to her first-grade class when she heard a loud noise. Although she was momentarily distracted by it, she did not change her behavior or let the children know of her curiosity. Less than an hour later, a colleague called Margo out into the hallway and told her that the Twin Towers had been struck by airplanes.

Reflexively, Margo organized thirty students for what she told them was a fire drill. When they got into the street, the air smelled like smoke and the streets were filled with people. Margo did not allow the children to look south, instead quickly shepherding them to a location several blocks uptown. They were safely evacuated when the buildings collapsed. Margo spent the rest of the day trying to reunite frightened children with even more frightened parents.

It was a week later when she saw her students again, this time in a temporary classroom. Many were tearful, had difficulty saying good-bye to their parents, and were easily agitated. Margo, with barely any time to process her own feelings, was faced with the daunting task of trying to answer questions that could not be answered and allay fears that were grounded in reality.

With the school's help Margo sought counseling. She returned to teaching the next fall but insisted upon being placed in a different school.

Daniel

Often overlooked in the aftermath of disaster is the group of people who are already psychologically vulnerable owing to their chronic mental illness. Daniel is a twenty-seven-year-old formerly homeless man who was discharged from the hospital on September 9, after an inpatient stay of ten days following a psychotic episode. He was stabilized on medication and in the phase of making his transition from hospitalization to a new living situation in the community.

When the events of 9/11 took place, Daniel was out walking in his neighborhood and noticed a crowd clustered around an electronics and appliances store. Upon closer inspection he realized that people were watching the news broadcasts on the TV sets on display in the storefront windows. He squeezed his way in among the viewers and was stunned by what he saw on the screens.

Daniel, whose emotional vulnerabilities often took the form of paranoid ideation, felt a sense of panic that perhaps his own "bad thoughts" were partly responsible for the tragic events that took place on September 11. He became restless, agitated, and unable to keep his focus. He stopped his medication due to the fear that it might be contaminated by the environmental fallout emanating from the burning buildings. He was unable to sleep because he felt so unsafe. He stopped eating and was found three days later wandering aimlessly through the city streets. The police brought him to the psychiatric emergency room where he was evaluated and rehospitalized.

Elizabeth—Television Reporter

"I've been called in for breaking news before, but never something like this. I remember getting dressed watching the live broadcast—watching the towers burn. One of my colleagues was live on the air when they started to collapse. You could hear the fear in her voice, terror, really, as she and her crew ran for safety. I kept thinking 'I can't believe I'm going *there*.' There were no taxis on the street; subways weren't running either. So I got on the bus. One man had a Walkman and was giving updates to everyone on board. Everyone was crying. I was thinking 'I can't believe I'm going there.'

"Ground Zero was a nightmare. Seeing in person was completely different from seeing it on TV. But my job is to gather information so, with a cameraman, I went toward the pile. I think about what we saw every day. Sometimes I even dream about it. The streets were knee-deep in ash—most likely remains of those who were incinerated when the planes hit. Beneath that, NYPD memo books singed at the edges. Invoices from companies whose offices were in the towers. Hundreds of shoes—Holocaust-like—left by people running for their lives. The worst, though, were the hands. Two of them—from different people—bodiless but still clutching a rail.

"I spent nearly three months down there working fourteen-hour shifts, mostly overnight—each time breathing the smoke and dust and trying not to think about the seven friends I'd lost 'in there.' That's how I came to call the pile and plume of smoke behind me. As a reporter, you have to be objective, calming, even. That took a lot out of me. For the first time in my career I felt completely overwhelmed. Then when I was cleared for the day I went home, only to find I couldn't sleep because my dreams were filled with images from the day.

"The more time I spent there the angrier I became. It was hard to talk to anyone who wasn't exposed to the same horrors I was. Friends asked 'can you get me in?' Like it was Disneyworld or a Yankees parade. I told them they didn't want to see it. *I* didn't want to see it but I didn't have the choice.

"It's been more than a year now and I still don't want to see it. But I do, both on the job and in my mind's eye. The visions are vibrant and they're haunting."

It should be noted that similar reactions are present in many who work in allied fields that put them in the eye of the disaster. Photojournalists, print reporters in combat arenas, Red Cross personnel, and other volunteer caregivers are all susceptible to the same trauma stimuli and reactions described in this vignette.

SUMMARY

Terrorism is an attempt to attack the human spirit. From the stories here, one can see the ways in which people from all walks of life have been touched by large-scale traumatic events. What we have observed in posttrauma and disaster work attests to the resiliency of the human spirit and the ability of individuals, families, and organizations to regain their equilibrium and restore a sense of emotional balance.

At the same time, the posttrauma and disaster period shows the creative and sensitive ways in which caregivers have designed programs and interventions to delimit the extent of psychological trauma.

The chapters that follow provide an in-depth study of both the phenomenology of trauma reactions and the cutting-edge techniques aimed at relief of the suffering caused in its wake.

Chapter 3

Mental Health Issues and Implications of Living Under Ongoing Terrorist Threats

Sara A. Freedman
Rivka Tuval-Mashiach

INTRODUCTION

I had just been shopping in the market, and was getting on the bus—there were loads of people, so I was being pushed from behind. Then—there was this enormous boom, and I flew to the other side of the street. There was blood everywhere, pools of it. I saw a woman—she wasn't whole—there were bits of bodies everywhere. Then I looked down at my leg, and I fainted—I thought that it wasn't there anymore. The next thing I remember was waking up in hospital.

I am always scared—I can't understand how anyone isn't. Even here [in the hospital] I keep thinking—someone is going to burst into the room with a gun. I can't watch television—and the news—all the time, more attacks.

I put my children to sleep at night, and then I stand there, crying. How will they cope without me? I know there will be another bomb, and this time I'll die.

Testimonies, Victims of Terrorist Attacks, Israel, 2000-2002

The intense fear resulting from witnessing or being injured in a terrorist attack is immense. When terrorist attacks continue, one after another, the cumulative effect is assumed to be exponentially greater. Terrorist attacks tend to target innocent victims, people who were in the wrong place at the wrong time. These attacks generally occur with no prior warning and take place in everyday surroundings: the café down the street, the market where you buy vegetables, your place of work.

Attacks need not actually occur for the effect to be felt. The threat that something *may* happen is sufficient to produce high levels of anxiety. This phenomenon was seen in the United States in the weeks following the anthrax deaths. Relatively few people were directly affected by this terrorism, but the effects were felt by millions (Susser, Herman, & Aaron, 2002).

Terrorism, and its threat, poses several conundrums for mental health care providers. It is generally concluded that the experience of terrorism is so horrific that *traumatization* follows. Accordingly, comprehensive intervention programs are needed. It is, therefore, essential to ascertain how these threats affect us and to develop appropriate and effective psychological interventions.

This chapter will explore what happens to people who live under the constant threat of terrorist attacks over a prolonged period. It will also examine the specific characteristics of clinical issues related to ongoing terrorist threats and review appropriate psychological interventions.

HISTORICAL PERSPECTIVE

Although terrorist attacks are not unusual, living in a society where they are frequent enough to become a daily threat is rarer. As a result, few studies have examined the impact of this type of threat. It is, however, worth examining the literature on ongoing conflict, because different situations of ongoing conflict have elements in common, whether the particular conflict is a regional war or a civil one; whether it is characterized by precision bombing or street fighting; or whether it involves actual or threatened danger to life. During every such conflict situation, individuals' lives change dramatically, as events cause modifications in day-to-day living, as their environment shifts from safe to dangerous, and as individuals realize the loss of control that they can exert over the course of events.

Such living conditions seem guaranteed to result in high levels of stress. Ongoing conflict involves both uncontrollability and unpredictability in daily life, two factors that have been shown to influence the likelihood of adverse reactions to potentially traumatic events (Foa, Zinbarg, & Rothbaum, 1992). Living under these conditions makes individuals more vulnerable to their circumstances. The fact that this type of environment inevitably involves loss of life also increases stress: Normal grief reactions are combined with the stresses of living under difficult conditions.

During World War II, England was subjected to a protracted period of bombing, known as the Blitz. It was assumed that this difficult and dangerous environment would result in many pathological reactions among civilians. Consequently, the government opened "neuroses centres" (Stokes,

1945) designed to provide treatment for those unable to cope emotionally with the nightly air raids, loss of life, and danger.

The literature shows that contrary to expectations, the majority of civilians coped remarkably well with horrifying circumstances, and consequently the government's centers treated more soldiers than civilians (Stokes, 1945). Treatment at these centers concentrated on rehabilitating patients as quickly as possible and getting them back to work and home.

One paper written during the war era describes the treatment of 100 patients hospitalized for traumatic reactions (Maclay & Whitby, 1942, p. 449). These patients had all been exposed to the bombing, and in 86 cases had "endured severe mental stress . . . loss of husband, wife, or children, loss of home or business, or being buried for hours." In fifty-seven cases, the precipitating traumatic incident was seen as the cause of the psychological symptoms. In eighteen cases the trauma was central but there were also predisposing personality characteristics, such as being "timid" and "irritable." The remaining cases were considered to be relapses of a previous condition. Their therapy included "persuasion, explanation, re-education, removal of symptoms by suggestion, occupational and physical therapy," and was considered successful in the majority of patients (21 percent recovered; 27 percent much improved; 28 percent improved).

These studies indicate that although the majority of civilians coped well with the strains of the ongoing bombing, a small minority developed psychological symptoms that required intervention. The majority of people requiring immediate treatment or displaying symptoms showed no long-term ill effects (Stokes, 1945). Risk factors for developing longer-term problems included being female, having premorbid psychological problems, and having immediate stressors such as financial worries.

Another example of an area of ongoing conflict is Northern Ireland, which has been subjected to violence for more than twenty years. Several studies have examined the effects of this conflict, and most researchers conclude that the violence has had no direct effect on psychiatric symptoms (Cairns & Wilson, 1985, 1989) or psychosomatic diseases (Clyde, Collins, Compton, Cooper, & Firel, 1975; Beare, Burrows, & Merrett, 1978). A study by Cairns and Wilson (1991) does show that people in Northern Ireland living in more violent areas, or perceiving a higher level of exposure to violence, report more physical symptoms than do people living in less violent areas. However, the authors stress that although the trend for this finding is statistically significant, they are less ready to assign cause: "No claim is being made that the violence plays a major role in influencing the physical health of the population of Northern Ireland" (p. 711).

Living in a more violent area of Northern Ireland, combined with perceiving it as more violent, was related to mild psychiatric morbidity as measured by the General Health Questionnaire. In addition, women were more likely to express symptoms than men (Cairns & Wilson, 1985). The authors conclude that the majority of the Northern Irish population coped well with the situation, with a small minority at risk. The majority, who lived in an objectively more dangerous area and were coping well, tended to perceive only low levels of danger in their areas. Thus, the study suggests that successful coping may be associated with a degree of denial.

Taken together, these studies, which are among the few empirically researched papers on traumatic stress during periods of ongoing threats, give a consistent picture of resiliency. The majority of people are able to cope with ongoing threats to their safety and major changes in their day-to-day lives. Those who do develop symptoms tend to recover quickly, and intervention programs for more chronic problems are successful in the majority of cases.

ONGOING TERRORIST THREATS: WHAT IS THE DIFFERENCE?

Certain aspects of ongoing terrorist attacks differentiate these experiences from other sources of trauma and ongoing stressors, including war. This is exemplified by comparing terrorist attacks with inner-city violence. The first salient difference is that ongoing terrorism may have a definable starting point, as is the case in Israel. This implies a prior period of stability, allowing individuals to compare their experiences "before" and "since." Terrorism creates a radical emotional climate change. Yet, despite this upheaval, the solid memory of a calm and stable "before" period encourages community members to have confidence in the restoration of normalcy, even if the threat of terrorism remains active for years. By contrast, inner-city violence may fluctuate from year to year, but it remains an established feature of daily life and is usually the only daily life its residents have ever known.

A second difference involves control and prediction. Residents in high crime areas know the geography of danger among their streets, behaviors, people, and other features of their daily lives, precisely because crime has been a constant. This is also true in war and in ongoing conflicts, where there are clear targets, such as the docks during the Blitz. Even when other "safe" places are targeted, there is still a feeling of predictability, however minimal. In contrast, the randomness of ongoing terrorist threats is germane to the terror that they cause.

A third factor is the grotesque result of terrorist attacks. The following examples are not easy to read, but they illustrate this aspect of terrorism.

People are exposed to horrific images, and their distress is compounded by the indelible memories these images create.

> Testimony of witnesses to suicide bomb attacks, Israel, 2000-2002:

> "I saw a hand, just a hand, with blood and nerves. And it was black—you know, it had been burnt."

> "It looked like a mask—but it was his head, and it was next to me."

> "I had flesh and blood all over my hands—it was like butter. And my feet—I was standing on flesh, his flesh."

The combination of a change in conditions in individual lives, the randomness of terrorist targets, and the grotesqueness of the aftermath of these events—along with the sheer frequency of actual and threatened attacks—differentiates ongoing terrorist attacks from other traumatizing situations. In single or site-specific terrorist attacks the predictability may be conserved (e.g., airplanes and tall buildings). But ongoing terrorism is also distinct from other ongoing stress-inducing situations, which may not be random, and other types of traumatic events not characterized by grotesque images.

ONGOING TERRORIST THREATS: CURRENT RESEARCH ON THE ISRAELI EXPERIENCE

Since its establishment in 1948, Israel has been a site of repeated terrorist acts against civilians. Since September 2000, however, there has been a sharp rise in terror threats and actual attacks against civilians (Figure 3.1). Although terrorists aim to attack both soldiers and civilians, 70 percent of Israeli victims have been civilians. Terrorist attacks include suicidal acts, bombings, and shootings, and take place throughout the country (some geographic sites, however, have been more exposed than others). The risk of being attacked and killed is the greatest in the Judea and Samaria settlements of the West Bank, compared with the rest of the country (Figure 3.2). Despite the continuous danger, the threat to security, and the acute disruption of many daily routines, people in Israel continue their everyday activities, including work, recreation, travel, schooling, and family life. This normalcy despite terror has also been experienced by other world communities, but its negative effects and its toll have not been adequately described.

Studies that have examined the consequences of living under ongoing terrorist threats over the past two years in Israel, many of which are still in

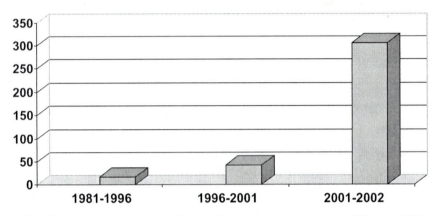

FIGURE 3.1. Yearly Average of Terror Victims in Israel Between 1981 and 2002

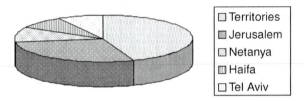

FIGURE 3.2. Israeli Victims of Terrorist Attacks, According to Residency (August 2002)

preparation, have shown conflicting results. Bleich, Gelkopf, & Solomon (2003) found that of 512 adults, only 9 percent suffered from posttraumatic symptoms. Most (84 percent) were not exposed or injured in a terror attack; 37.3 percent knew someone close who was exposed; and 15 percent reported they knew someone who was wounded and/or died in a terror attack. However, about 60 percent of these individuals reported feelings of sadness and dejection.

The numbers regarding children are higher. Laufer (Eldar, 2003) studied 3,000 adolescents (ages thirteen to fifteen) from areas differing in exposure to terror and found 42.7 percent to be suffering from PTSD (mild to severe). Of them, 15 percent reported moderate to very severe PTSD, and 9 percent reached a clinical level of PTSD. Seventy percent reported that terror attacks had some impact on their lives, 33 percent reported knowing at least one person who was injured in a terror act; 20 percent had a family member affected by such an act, and 2 percent were directly exposed to and injured in a terror attack.

On the other hand, Lavie (cited in Rotem, 2002) surveyed 1,300 children and found 30 percent PTSD in Israeli children in settlements, 50 percent in Arab–Israeli children, and 70 percent PTSD in Palestinian children. Another survey (Pat-Horenczyk et al., 2003) examined 1,029 adolescents (ages twelve to eighteen) in Jerusalem and nearby settlements. This study found 5.4 percent of the sample suffered from PTSD after twenty months of terrorist attacks, and 12.4 percent were suffering from partial PTSD. Girls showed higher rates of full and partial PTSD, more somatic complaints, more behavioral changes, and restrictions in activities. However, boys reported more functional impairment and more suicidal thoughts. Adolescents living in the settlements, although exposed to more terrorist attacks, exhibited less full and partial PTSD than those in Jerusalem. As described later, this difference may be accounted for by the greater community support in the more exposed settlements.

A professional hotline specializing in trauma reported an increase in calls from 600 between the years 1998 and 2001 to more than 800 in the first four months of 2002 (Rotem, 2002). Laor (in preparation) found that religious beliefs were positively correlated to better coping and to lower PTSD levels. In a highly exposed religious settlement the percentage of symptoms was the lowest (7 percent) as compared with Jerusalem (9 percent) and Beersheva (14 percent), the latter being the least exposed area among the three. This last piece of the data suggests that the appearance of PTSD symptoms may be correlated not only with the level of exposure to violence but also to cultural and religious methods of coping with trauma-related stress. Another study of 149 adults (Gidron, 2002b) showed that levels of PTSD were not necessarily related to direct exposure to terrorist attacks. Women who perceived themselves as having low personal control over the situation had higher levels of PTSD. Men, meanwhile, who perceived that the government had control, were less symptomatic.

In the summer of 2001, two civilian populations differing in their level of exposure to terror incidents and threats were compared (Shalev, Tuval-Mashiach, Frenkiel, & Hadar, 2003; Shalev, Tuval-Mashiach, & Hadar, 2004). One of the samples was drawn from Efrat, a suburb near Jerusalem, with approximately 1,500 families, which was continually exposed for a period of more than ten months (at the time of the study) to daily terrorist incidents such as stoning, shooting, ambushes, and bombings. The main road to the suburb was frequently closed for this reason, and residents were often delayed on their way to or from their homes. Several people were mortally shot, and several others were injured. Two children from a nearby suburb were kidnapped, stoned, and tortured to death. Given the effect of terror on the entire population in Israel, a control group was chosen, which was assumed to have experienced little or only indirect exposure. This "control" suburb, Ramat Beit Shemesh, was very similar in its demographic characteristics (Table 3.1), yet its roads were safe and no terrorist incident had ever occurred there (Figure 3.3).

The 271 subjects of this study included 177 from the highly exposed site (Efrat), and 94 from the less exposed site (Ramat Beit Shemesh). The residents were compared on a number of measures designed to assess exposure to danger, perception of this danger, disruption of daily activities (such as driving to work and seeing friends), coping strategies, and self-help activities (such as visiting a physician). Psychological symptoms were assessed in three ways: First were symptoms of PTSD, using the PTSD Symptom Scale (Foa & Tolin, 2000). Second, an abbreviated form (fifty-one items) of the SCL-90, Derogatis's Brief Symptom Inventory (BSI) (Derogatis &

TABLE 3.1. Sample Characteristics

Characteristics	Efrat (n = 177)	Ramat Beit Shemesh (n = 94)
Gender (Female)	68%	72%
Age	40.5 + 13.5	35.3 + 8.7
Marital Status		
Married	83%	92%
Divorced	2.3%	3.3%
Widowed	2.8%	0
Religious observance	95%	95%
College education	77%	78%
Income > Average	55%	70%
People in household	5.2	5.9
Average number of children	3.4	3.6

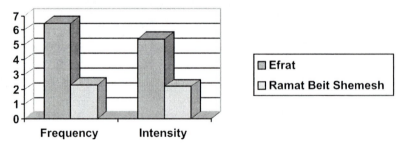

FIGURE 3.3. Differences on Exposure, by Suburb

Spencer, 1982), was used as a general measure of psychopathology with population norms for Israel. Finally, the General Well-Being questionnaire (Dupuy, 1975) was included, to measure selected aspects of subjective well-being and distress.

The picture found in comparing the two suburbs regarding symptomatology is unexpected. Efrat residents did not express significantly more PTSD (28 percent versus 19 percent in Ramat Beit Shemesh). In both communities PTSD percentage is higher then the aforementioned Israeli studies during the same period, in which the average percentage was about 10 percent. However, it is quite similar to the average percentage of PTSD reported by Gidron (2002a). Based on six studies that evaluated PTSD in reaction to direct exposure to terror attacks, Gidron (2002a) found the mean prevalence of PTSD to be 28.2 percent. Also, on the General Well-Being scale, there were no differences, and Ramat Beit Shemesh residents showed even *more* symptoms than Efrat residents when measured on the BSI.

Residents of both sites appraised the situation as more severe than it was ten months earlier, and their fears as increasing. The people in Efrat appraised their ability to cope with the situation as stable, while the people in the control group appraised it as slightly deteriorating. Another finding was that although subjective experiences of fear were higher in Efrat, those experiences were much more specific and related directly to realistic dangers. In the control group the fears were much more general or not related to realistic dangers.

Help-seeking behavior differed in the two populations on some measures. People in Ramat Beit Shemesh turned significantly more to marriage counselors (8.7 percent in Ramat Beit Shemesh versus 2.3 percent in Efrat, $p < 0.02$). In addition, 22.6 percent of the sample in Ramat Beit Shemesh turned to rabbis or other religious authorities as compared with 10.8 percent in Efrat, the more exposed suburb ($p < 0.01$). All other kinds of help did not vary significantly between the sites. Among adults who were diagnosed as suffering from PTSD, turning to a family doctor was much more common (25 percent) than it was for others (11 percent). Participants from Efrat found the situation to be more controllable than did participants from Ramat Beit Shemesh, and appraised the situation as less frightening and less severe than their counterparts in Ramat Beit Shemesh, despite the fact that Efrat was actually a far more dangerous place.

Participants of both areas rated their social support networks similarly, as well as their ability to function effectively with daily routines. In both places, the level of reported functioning was found to be intermediate.

This study shows that greater frequency and intensity of exposure to terrorist threats, together with a higher subjective (but reality-based) estimate of the level of disruption, were not consistently correlated with higher symptom levels. In fact, people living in the less exposed area (Ramat Beit Shemesh) were more symptomatic on one measure, suffered from more diffused anxiety, looked more for different kinds of professional help, felt less control over their situation, and rated it as *more* severe, than those living under objectively harder conditions.

The findings from this research are similar to research from Northern Ireland (i.e., living in an objectively more violent area, but perceiving it to be less frightening, is related to better functioning). It may be that people who share a sense of community and purpose, as indicated in the high community attendance at terror-related funerals, may be protected from some of the psychological effects of chronic violence (Hobfoll, 1989). According to Hobfoll's "conservation of resources" theory, the aftermath of a traumatic event is characterized by multiple losses of various levels, from the private and personal to the public and societal level. Better coping is related to fewer losses, and therefore less symptomatology. This study's results may show that the level of community resources, which was not harmed, but on the contrary strengthened, had a protective effect on the residents in Efrat.

These studies of Israel, Northern Ireland, and the London Blitz all demonstrate similar patterns in response to ongoing violence. The majority of people cope well. Even among those who suffer from psychological symptoms, many function at a high level. We conclude that symptom levels are not necessarily related to actual exposure dose. It remains a topic of investigation to find who among people exposed to chronic violence are at risk to develop psychological symptoms; the data suggest it may be individuals who feel more isolated from their community.

MENTAL HEALTH ISSUES

It is well documented that ongoing stress of any kind is associated with increased levels of psychosomatic illness, drug and alcohol use, and family problems (Kiecolt-Glaser, McGuire, Robles, & Glaser, 2002). But few data are available examining these issues in relation to ongoing terrorist threats. This section will describe the studies that have been carried out and also indicate those unexamined areas of research that may be important.

Physical Symptoms

Epidemiological research from France examining all victims of terrorist attacks between 1982 and 1987 (Abenhaim, Dab, & Salmi, 1992) shows that the long-term effects of terrorist attacks go beyond psychological symptoms. Many victims still suffered hearing loss three years after an attack, and 25 percent frequently used painkillers. Research from Lebanon indicates that death from cardiovascular disease is much higher among women exposed to human trauma and among men exposed to financial and work-related trauma during wartime (Sibai, Fletcher, & Armenian, 2001).

Following the September 11, 2001, terrorist attacks in New York, one study showed no difference in levels of fibromyalgia pain pre- and post-attacks (Raphael, Natelson, Janal, & Nayak, 2002). An increase in asthma attacks has been reported, but this has been linked with stressors predating September 11, combined with the effects of smoke and debris. A weak relationship was found between asthma attacks and symptoms of PTSD and depression (CDC, 2002b). An anecdotal report from the *Journal of American Dental Association* (Lund, 2002) suggests that more cases of teeth clenching were seen in the few months immediately following the attacks.

Medication, Alcohol, and Smoking

Two studies, one from France (Abenhaim et al., 1992) and one following the September 11 attacks (Vlahov et al., 2002), describe an increase in the use of painkillers following terrorist attacks. Again following the September 11 attacks (CDC, 2002a), an increase in alcohol use (3 percent of alcohol users, with men more likely to increase alcohol consumption than women) and cigarette use (21 percent increase in smokers, 1 percent of nonsmokers started smoking, and women were more likely to increase smoking than men) was observed. In another study, in the five to eight weeks following the September 11 attacks, an increase was found in drug taking, smoking, and alcohol use. This increase was linked to PTSD (Vlahov et al., 2002). Because stress has been shown to predict relapse in drug abuse, this trend seems logical.

Other Effects

In the study from France (Abenhaim et al., 1992), 25 percent reported relationship difficulties, and 10 percent had separated from their partners. More than half the victims reported financial difficulties related to the attack, including unemployment.

In a study of firefighters from New York, there was a seventeenfold increase in stress-related incidents in the year following September 11, compared to the previous year (CDC, 2002c).

Age, Gender, Ethnicity, and Culture

No research has explicitly examined the effects of these factors on the reactions postterror. Levels of symptoms among adolescents are no higher
than those seen in adults (see Pat-Horenczyk et al., 2003), although a California study of twelve- to sixteen-year-olds indicates that fear of dying from
general causes rose significantly post-September 11 (Halpern-Felsher &
Millstein, 2002).

Gender. In a study described earlier (compare Shalev et al., 2003), gender differences were found in rates of reported PTSD as measured by the
PSS-SR. This study showed that women, although being significantly less
exposed to adversities, were more likely to develop PTSD than men.
Women reported significantly more symptoms of the Brief Symptoms Inventory, as well as diminished energy and increased anxiety, reported by the
General Well-Being Questionnaire (GWB). Gender differences were also
found in coping strategies. Women used emotion-focused coping more frequently and in general used more different coping strategies than men. An
interaction between gender and exposure level was also noted, such that in
the high-exposure group, men used more problem-oriented coping, but no
such gender difference was observed in the low-exposure group.

Ethnicity and culture may play a crucial role in determining symptom levels
and coping. As stated previously, a high level of community cohesion may be
protective, and this may be related to certain cultures and ethnic backgrounds.

These data, suggesting that individuals respond physiologically to terror-related trauma, indicate that a *biopsychosocial* model of trauma is appropriate (see Shalev, Galai, & Eth, 1993). The interaction of all these factors—biological, psychological, and social—is part of the normal response
to trauma. Therefore, it seems logical to expect disruption in all areas of
people's lives, not just the psychological. Ongoing terrorist attacks tend to
be related to worsening economic conditions as well, which in turn have a
negative impact on health, relationships, and the ability to cope. The relative dearth of data and the complexity of the interaction among terrorism,
social conditions, and individual well-being mean that any conclusions regarding the effects of ongoing terrorist attacks on mental health issues such
as drug abuse and family relationships remain incomplete, awaiting further
research in this important area.

TREATMENT IMPLICATIONS

Although the general conclusions from the data presented suggest that
only a minority of people develop a psychiatric diagnosis as a result of liv-

ing under the threat of terrorist attacks, major mental health implications remain due to the large numbers involved.

Treatment issues are complex and cannot be fully described within the scope of this chapter. The aim is to provide a framework that is useful when making clinical decisions during a time of ongoing terrorist attacks, indicate the relevant literature, and describe in detail the specific elements of treatment that are unique to this situation. The model presented next is based on a continuum of care model, such that different levels of intervention are made available depending on need.

Table 3.2 illustrates the two main factors in the treatment model: level of exposure to terrorist attack, and the type of appropriate interventions for each group. For the whole model the following conceptual issues regarding treatment may need to be taken into account.

Concepts

"Caseness"

The documented responses of Israelis during terrorist attacks indicate that the definition of a clinical case is more complex in these circumstances of ongoing threat. Summerfield (1999, p. 1455) suggests that "the notion that war collapses down in the head of an individual survivor to a discrete mental entity, the 'trauma,' that can be meaningfully addressed by . . . counselling or other talk therapy is absurdly simplistic." He also argues that Western concepts of "trauma" and "posttraumatic stress" are often irrelevant when living during war. This conceptualization neatly applies to the experience of individuals living under ongoing terrorist threats. It is a misconception to regard every symptom as pathology. As illustrated in the two Israeli sites study reported (Shalev et al., 2003), many people had a diagnosis or symptoms of PTSD, but only a small minority were disabled by them. In spite of their "symptoms" and the challenges in daily life, most people were able to function well in all areas of their lives. Clinical assessments

TABLE 3.2. Continuum of Care Model for Intervention

Exposure/Time	ER	Outreach	Prevention	Intervention
Directly exposed	Psychological first aid	Telephone follow-up	CBT	CBT Medication Other
Indirectly exposed	Information helplines	Via media information and education		

must take this issue into account: having intrusive thoughts, avoiding certain areas, and jumping at every noise—all symptoms of PTSD—may constitute "normal" adaptive responses to the surroundings. This interpretation also explains the reason why therapy may be inappropriate for the majority of people who are somewhat symptomatic but who are still able to cope.

Our clinical experience does suggest the importance of acknowledging that an ongoing conflict is composed of numerous individual traumatic events. PTSD and other psychiatric disorders, such as depression, commonly found after discrete traumatic events, are the primary reasons individuals seek treatment during ongoing conflict. Most symptomatic conditions are a result of the ongoing stress, whereas most psychiatric referrals are a result of exposure to a specific event. Our experience is that ongoing symptoms are rarely the cause for seeking mental health interventions, in contrast to specific event exposure and its consequences. This can be explained in many ways: these individuals do not need any intervention. People who are living within a stressful environment of ongoing conflict but have not been directly exposed to a trauma-inducing event may feel that they have more pressing needs than therapy. The upheaval in their lives may make therapy impossible. They may assume (correctly) that their suffering is normal. However, it is possible that some people who are in need of help will not seek it; therefore, we maintain that active outreach should be an integral part of any intervention program, in order to identify individuals in need who are not receiving help.

Studies examining reactions to traumatic events have shown that the majority of people are symptomatic in the first days following exposure to a trauma, but that these symptoms quickly fade in a high percentage (Shalev, Freedman, Peri, Brandes, & Sahar, 1997; Shalev et al., 1998). In addition, 50 percent of trauma survivors who had a diagnosis of PTSD one month posttrauma had fully recovered five months later (Shalev et al., 1998). These data and our view of the literature regarding reactions to ongoing conflict and terrorist threats strongly suggest that the majority of people will not need psychological intervention. The data clearly show that symptoms following discrete traumatic events, and those that occur as a result of ongoing stress, are transient, do not adversely affect functioning, and are part of a normal coping mechanism. This suggests that any interventions should be carefully targeted and resources not squandered on those people who continue to function well.

Timing

It is generally assumed that early intervention following a traumatic event is useful and may prevent problems from becoming chronic (see

Everly & Mitchell, 1999). Some studies indicate that this may not be the case and that very early debriefing treatment programs may do more harm than good (Rose & Bisson, 1998). It may be that this type of intervention can in some cases interfere with the natural recovery generally seen. In addition, the original model as described by Everly and Mitchell is rarely used, and the negative results may reflect this. For instance, treatment groups are not homogenous, the debriefing is carried out individually, and ongoing care for those who need it is not provided. More research is needed to examine whether the mixed results reflect an issue of the timing of the intervention, its nature, or a combination of the two.

Patients usually present for treatment as a result of a specific traumatic event. This is important because patients under constant threat may best be understood as seeing the environment as both a constant traumatic event and as a series of discrete traumas. In our treatment of survivors of terrorist attacks, we have found it useful to use the specific attack as the starting point of any intervention, but the treatment also takes the ongoing nature of attacks into account.

Clinical Vignette

After shopping in a Jerusalem open-air market, G, a twenty-eight-year-old single man, suffered severe injuries when a suicide bomber exploded next to a bus he was boarding. He was thrown to the other side of the road, temporarily lost consciousness, and then witnessed grotesque images: rivers of blood, body parts, people dying next to him.

G presented for treatment three months after the trauma, with severe PTSD. He was unable to think about anything else and had horrific intrusive images. He was unable to return to work as a security guard, and his avoidance was so severe that he did not leave the house apart from his hospital appointments.

G was treated using prolonged exposure, a well-established intervention for PTSD (Foa & Rothbaum, 1998). He was adamant that many of the places he was avoiding—traveling on buses, crowded places, the open-air market—were still dangerous, and he was not prepared to approach them again. This was treated by first focusing on the avoidance that involved no objective danger. For instance, he was instructed to begin reading the newspaper and watching the television news. Although it was agreed that it was neither desirable nor necessary for him to resume traveling on buses or going to the market, his level of anxiety was such that we felt it essential that some direct anxiety reduction should be integrated into the treatment.

G was presented with reminders of these places, which were not in themselves dangerous. For instance, exposure to the open-air market began with looking at a map of the area, with the market clearly marked. Following this, he was presented with photographs of the market, photos of terrorist attacks that did not include the one he was involved in, and, last, pictures of the Jerusalem market after the terrorist attack. G's anxiety reduced relatively quickly to these images and he continued to look at them at home, between sessions. This exposure was extremely important to G's overall symptom improvement. It helped him to differentiate between the terrorist attack in which he was injured and the threat of another attack. G's initial decision to not go to the market was not changed by the end of therapy; however, his fear regarding the market changed dramatically.

In order to help G cope with future terrorist attacks, their possibility and likelihood were discussed, and natural responses to these were predicted— that is, G may have a temporary setback following an attack, but this was a reaction that everyone shared and would be transient. In addition, it was recommended to G that he avoid repeatedly viewing the grotesque images that are shown on television following an attack, and continue in his avoidance of "dangerous" places, such as the market, for as long as the threat continued. This avoidance was relabeled as a positive safety behavior that would help protect him both from potential danger (the market) and potential reactions to a terrorist attack (television). G completed treatment after eight sessions, with a 65 percent improvement in his PTSD symptoms.

Treatment Goals

Many interventions for PTSD assume that the end point of treatment is a return to normal everyday functioning (Foa, Keane, & Friedman, 2000). These goals may be inappropriate when heightened danger still exists and the behavior of the whole population has changed accordingly. For instance, many people in Efrat, described previously, stopped travelling along the main road to their suburb after dark. Any intervention aimed at helping the patient to resume such an activity would not only be likely to fail but might also make the patient susceptible to real danger. As one patient succinctly put it: "You can't tell me my fear is exaggerated—what is happening is terrifying, and I think most people aren't scared enough." As described in the clinical vignette, treatment goals need to be carefully designed to take into account the day-to-day surroundings. We have found it useful to explain to patients that some of their behavior is not avoidance per se but is a safety behavior. However, we stress that this is related to the situation: once it becomes safe again, at some undefined point in the future, behavior should change accordingly.

Resources

Given the sheer number of patients, it is likely that in any ongoing situation of terrorist threats, resources will be depleted, and the number of people in need of treatment will be far greater than the pool of therapists available. We suggest, through our own experience, two ways to address this issue. First, we have actively trained therapists in working with traumatized clients. Second, we have used group therapy whenever possible. We have found through anecdotal evidence that although this helps only about half of the participants substantially, it still reduces the necessity for individual intervention with some patients. The literature on group interventions suggests that they are mostly related to positive outcome, although it is not clear which treatment approach may be the most useful. No randomized controlled trials are available comparing the effectiveness of group interventions with individual. (See Foa et al., 2000, for further discussion of group treatments.)

Interventions

As stated previously, proposed interventions will be described briefly, since they have been extensively covered in the literature. The specific elements of treatment during ongoing terrorist attacks are outlined in detail.

Direct Exposure

Emergency Room

Once someone has been involved directly with a terrorist attack, the first stage of mental health intervention is often in the emergency room (ER), hospital ward, or home in the hours immediately following the attack. This is the point at which the greatest number of personnel is needed, because the number of presenting patients at one time is highest. Patients in an ER present with a wide range of symptoms which may be diverse. One young woman, for instance, who was brought to the ER after being exposed to (although not injured in) a suicide bomb attack in a bus, alternated between crying, shaking, and lack of ability to communicate effectively, and periods of calm, when she was able to chat with visitors about mundane events. The clinician's purpose at this point is to stop the trauma for the patient: for example, by providing information regarding the situation and preventing further exposure to the trauma (e.g., badly injured patients, reports on the television). The next goal is to help the patient regain control of her emotions. This may well need no intervention at all, but monitoring. In other cases, a wide range of techniques, including medication, relaxation exercises, and talking about the experience, may all be utilized.

Outreach

The second stage in any intervention program should be follow-up of all those initially seen in the ER, as well as other exposed individuals who did not require immediate treatment. Ideally, all exposed people should be followed up. Distributing leaflets describing common reactions and coping mechanisms, as well as how and when to seek help, are regularly used and have been shown to be helpful (Hartsough & Myers, 1985). In Israel, outreach is primarily carried out by the National Insurance Institute, which actively seeks out people caught up in each terrorist attack. The Institute distributes leaflets in the ER, invites all known exposed individuals to a debriefing session shortly after the attack, and may even provide home visits. Initial intervention and consequent follow-up are offered. When resources are scarce, which is often the case amid sustained terrorist threats and attacks, then follow-up should be concentrated on those known to be most vulnerable (e.g., previous psychiatric history, previous exposure to terrorist attacks, no social support).

Psychological Treatment

Many studies have examined the relative effectiveness of psychological treatments following discrete traumatic events. These will be briefly described (for a full discussion, see Foa et al., 2000). The way in which ongoing threats affect therapy will be discussed in detail.

In the first month following any traumatic event, the major aim of therapy is to prevent the development of chronic PTSD. Three major intervention approaches have been shown to be helpful at this stage:

1. Cognitive-behavioral therapy (CBT) has been successfully used to prevent chronic PTSD (Foa, Hearst-Ikeda, & Perry, 1995; Bryant, Harvey, Dang, Sackville, & Basten, 1998) and was more helpful than supportive counseling. Reprocessing therapies, such as eye movement desensitization reprocessing (EMDR), have also been effective (Carlson, Chemtob, Rusnak, Hedlund, & Murakoa, 1998), as has short-term psychodynamic therapy (Brom, Kleber, & Defares, 1989). No study has compared these three approaches.

2. The use of pharmacotherapy in the initial phase is widespread and seems to be directed primarily at symptom relief. It is helpful in bringing about a reduction of specific problems, such as sleep disturbance or irritability, but does not seem to provide an approach that can pre-

vent the development of PTSD (for a fuller discussion, see Friedman, Davidson, Mellman, & Southwick, 2000).

3. To minimize the impact of the social factors relevant in situations of ongoing terrorist violence, psychoeducation might play an important role. This can be accomplished either directly or by talking to family or employers. By suggesting to the people who have been traumatized that they explain common trauma reactions to those around them, or through less direct means, such as clinicians distributing printed brochures, a community dealing with ongoing violence can come to recognize PTSD symptoms. This psychoeducation may come about naturally, as reactions to threats are a topic of media interest. This was clearly seen in the weeks following the September 11 attacks, when information regarding normal reactions was made much more available via the Internet. In addition, mental health professionals may be active in distributing this information, both to other professionals and to members of the public.

Given that people directly exposed to a terrorist attack during a period of ongoing threats are highly likely to be reexposed, follow-up is essential. This follow-up allows clinicians to assess whether treatment gains are maintained in the face of subsequent attacks.

Chronic PTSD has been shown to respond well to different types of cognitive-behavioral therapy following exposure to a trauma-inducing event. In addition, other treatment modalities (group, family, psychotherapy) have some effectiveness (see Foa et al., 2000, for a full description of the literature).

Indirect Exposure

The majority of people indirectly exposed to ongoing terrorist attacks may express psychological symptoms, but this rarely affects functioning. These symptoms can be considered a normal reaction that requires no direct intervention. Providing information regarding common reactions and treatment when necessary are appropriate and useful strategies (Erikson, 1976). Susser et al. (2002) state that strong leadership can be a form of intervention, in that it helps people cope with situations such as terrorist attacks. The authors give as examples Winston Churchill boosting morale with his speeches during the Blitz. New York City's former mayor Rudolph Giuliani is another example of strong leadership following the September 11 attacks. Giuliani's willingness to show grief and terror while remaining calm and his immediate public appearances in the hours following the attacks in New

York made him the primary public leader during the period of anxiety following the attacks. Thus, intervention for the indirectly exposed group will probably not involve classical psychological intervention, but rather reports about the ongoing situation, information regarding symptoms, and where and when to go for help.

TREATMENT DURING ONGOING TERRORIST THREATS

Specific Challenges and Suggested Solutions

Various factors affect trauma treatment during ongoing conflict, and any intervention must take these into account. The first issue to be considered is avoidance. Avoiding reminders of a traumatic event, whether by actively not approaching memory-triggering people, places, or objects, not talking about the event, or emotional numbing, are symptoms of PTSD (DSM-IV, APA, 1994). As such, they are a focus of intervention in most treatment modalities, and any measure of treatment effectiveness would consider level of avoidance.

Avoidance, however, becomes an integral part of most peoples' lives during ongoing threat, and certain avoidance behaviors are an important part of staying safe in the face of real danger. These safety behaviors must be distinguished from avoidance that represents clinical characteristics of PTSD without affecting safety. For instance, people who wandered around during a nightly air raid in the Blitz instead of going to a shelter were considered to be taking unnecessary risks. The behavior—going for a walk at night—was not considered dangerous before the Blitz, but circumstances change, and behavior changes with them. Any ongoing threat involves such necessary change and is an important part of coping with the situation. People may feel they are better able to control the danger and to be safer by avoiding—when avoiding realistic dangers in their new environment, they can.

The situation of ongoing threat differs from other situations where traumatic events are the norm, for example, living in a dangerous inner city, where violent attacks are commonplace. This difference influences the nature of the intervention. An ongoing terrorist threat situation is usually a departure from the norm, so that people are able to remember when their lives were different. Inner-city violence, however high, is likely to be relatively stable. In the example of Israel given earlier (Shalev et al., 1997) there was a 300 percent increase in the number of terrorist attacks between 2000 and 2001. The change in actual levels of danger resulted in changed behavior and was not considered to be part of daily living.

How then does this affect treatment? Many people, including those with no direct exposure to a terrorist attack, change their behavior and begin avoiding dangerous areas. This avoiding is not part of a psychiatric problem; rather, it is an adaptive response to prevailing conditions. Any treatment that focuses on this avoiding will be asking patients to behave in a way not consistent with norms.

Consideration of a few examples of individuals' behavior from the Israeli context can shed light on how to distinguish normative, adaptive avoidance from avoidance that signifies PTSD symptomotology. Many people in Jerusalem avoided going to the center of town unless this was necessary (e.g., for work). A patient who had been involved in a terrorist attack downtown avoids going there. Should the therapist suggest he go downtown as part of a PTSD intervention? The answer we have found is both yes and no. This decision will be made with the patient, according to the importance to him or her. For example, a patient with PTSD following a terrorist attack in the center of Jerusalem is reluctant to go back to work, which is located next to where the attack took place. It is, however, essential for him to do so, since his business will not survive if he stays at home. Even though this may involve some objective danger, given the importance to the patient, it is encouraged. Conversely, the patient described earlier who felt that he had no reason to ever return to the place of the attack was not encouraged to do so.

Distinct from instances where behavior changes are part of adaptation when an environment becomes dangerous, we suggest that avoidance that clearly involves no objective danger, such as sleeping with the light on, should be challenged. Behaviors that involve a certain degree of danger, such as going into town, should be assessed regarding saliency to the individual. When these are important, the patient is helped to tackle this, but the consideration that another attack might take place is always discussed first.

It is assumed that the same avoidance in a person with PTSD and one without it actually differ in underlying perceptions. The person without PTSD understands that the danger is minimal but will not take any unnecessary risk. The person with PTSD overestimates the danger, believing that he or she will definitely be in another attack and is likely to die. As a result, any avoidant behavior that cannot be directly tackled by the patient, due to dangerous circumstances, is tackled indirectly in therapy. For instance, as described earlier, the patient who avoided the market was shown a map of the area and photographs of the market both before and during the attack. We have found that this helps patients to differentiate between the real danger they experienced and daily life, which—though more dangerous that it used to be—is not constantly dangerous.

Another factor adversely affecting treatment is reexposure. Almost all the patients we have seen were exposed to another attack during the course of treatment. Sometimes patients were directly involved; in other cases the exposure was indirect. For instance, one woman was near the city center when a bomb went off; she was not injured, nor did she witness anything. She personally knew people who had been injured in another bombing and was exposed to every attack via the media. Reexposure, even as indirect as this instance, was consistently associated with relapse among patients. For some patients, hearing about another attack was a sufficient trigger for symptoms of depression. This is substantiated by research on the impact of the media following terrorist attacks. After the Oklahoma City bomb, one set of studies found that children's exposure to television following the bomb was related to PTSD levels, both in the short- and long-term (Pfefferbaum et al., 1999, 2001). More recent work following the September 11 attacks has also shown that exposure to the television coverage among adults from Manhattan was related to higher levels of PTSD and depression in those who were directly affected by the attacks (Ahern et al., 2002).

We suggest to patients that though some reexposure is not under their control—such as being in the supermarket when a bomb explodes—the more indirect exposure is. The media reports of attacks in Israel tend to be graphic and prolonged. After a major attack, all regularly scheduled programming is replaced with live coverage of the attack, often including horrific images. We recommend that our patients stop watching these programs. Similarly, newspaper items about the terrorist attacks, and generally discussing the situation at length with friends and family, should be avoided or minimized. We explain this to patients in a positive way. Rather than seeing all avoidance as "pathological," some avoidance is helpful. If an individual feels bad watching horrible images on television, the best "cure" is simply to turn the television off. This strategy has been very helpful for patients, who become better able to withstand the impact of the next attack, and the next attack.

The final factor that affects such interventions is related to the goals of therapy described previously. The end of therapy never coincides with the end of the threat of terrorist attacks. The therapy therefore includes techniques to help patients cope in the future. This long-term planning includes relaxation, cognitive therapy, and relapse prevention. For instance, patients are asked to describe what will help them when another terrorist attack occurs. Strategies that have been helpful, such as relaxation, are encouraged. Cognitive techniques that challenge unhelpful thinking patterns are also indicated (e.g., "I can't cope when there is another bomb"). Relapse prevention helps patients understand that their reactions to the attacks are

natural and shared by most of the population; any reaction is likely to be transient; and there are techniques that they can utilize to ensure this, e.g., not watching the television and using social support.

Helping the Helpers: Implications for Mental Health Professionals

In any situation where those working in the mental health field are experiencing the same general stressors as their patients, there is a great toll on the clinicians. The therapists themselves may not be expressing PTSD, but the combination of the normal symptoms and listening empathetically to patients who have experienced terrible terrorist attacks may be too much to bear. One therapist living under the stressful conditions of Efrat, the suburb subjected to repeated attacks, refused to treat a particular patient. This patient had been in a terrorist attack in which the therapist's daughter's friend had been killed. Such overlap of roles is very stressful and may result in burnout if precautions are not taken.

The grotesque nature of terror attacks is also stressful, and many therapists will feel that they are unable to withstand descriptions of the carnage. If resources permit, therapists should be able to volunteer for this work, not have it forced upon them.

In addition, the lack of resources is likely to be an additional burden for mental health professionals. They may be working more hours than normal and treating patients who by definition present a constant indirect exposure to the terrorist attacks. We advocate supporting these clinicians by offering a choice if working with traumatized patients and recruiting as many people as possible to share the burden of care. For instance, working in an emergency room immediately following a terrorist attack was seen as the most stressful among our psychologists. Once we had recruited more people, and the work was more widely shared, the staff described relief and higher motivation for this work.

CONCLUSIONS

The material presented in this chapter clearly highlights the essential problem of mental health issues during ongoing terrorist threats: a paucity of data.

Although we know about PTSD in general, PTSD following isolated terrorist attacks, and the effects of living in a war situation, the specific effects of ongoing terrorist threats are only beginning to be described.

The preliminary data examined here, almost all from Israel in the past two years, indicate that coping with this situation may be somewhat similar to adjustment to the effects of war. Most people, even symptomatic, are able to adjust and cope to the ongoing stress. Those who do develop psychopathology can often be helped by treatment that is adapted to the situation.

Several questions remain: Do different types of terrorist threats affect people in different ways? How do these ongoing threats affect the family and social structure? Is there a process of burnout as terrorist threats continue? What are the cultural, ethnic, gender, and age differences in coping with ongoing threats? How does this situation impact on alcohol abuse, substance abuse, and crime?

Most important, the interventions that are commonly used, from psychological first aid in the ER, to outreach programs and more structured interventions, need to be evaluated.

Terrorist attacks are, by definition and intent, terrifying, and ongoing attacks even more so. The resiliency that we have encountered in people living in these conditions is humbling. The strength of those who do develop PTSD and still respond to treatment, despite daily reports of additional terrorist attacks is astounding. This resiliency and strength must also be studied, as it may provide direction for our efforts to understand mental health issues during ongoing terrorist threats and in planning effective and appropriate interventions.

REFERENCES

Abenhaim, L., Dab, W., & Salmi, L. R. (1992). Study of civilian victims of terrorist attacks (France 1982-1987). *Journal of Clinical Epidemiology, 45*(20), 103-109.

Ahern, J., Galea, S., Resnick, H., Kilpatrick, D., Bucuvalas, M., Gold, J., & Vlahov, D. (2002). Television images and psychological symptoms after the September 11 terrorist attacks. *Psychiatry, 65*(4), 289-300.

American Psychiatric Association. (1994). *Diagnostic and statistical manual of mental disorders* (4th ed.). Washington, DC: Author.

Beare, J. M., Burrows, D., & Merrett D. (1978). The effects of mental and physical stress on the incidence of skin disorders. *British Journal of Dermatology, 98,* 553-558.

Bleich, A., Gelkopf, M., & Solomon, Z. (2003). Exposure to terrorism, stress-related mental health symptoms, and coping behaviors among a nationally representative sample in Israel. *Journal of the American Medical Association, 290*(5), 612- 617.

Brom, D., Kleber, R. J., & Defares, P. B. (1989). Brief psychotherapy for post-traumatic stress disorders. *Journal of Consulting and Clinical Psychology, 57*(5), 607-612.

Bryant, R. A., Harvey, A. G., Dang, S. T., Sackville, T., & Basten, C. (1998). Treatment of acute stress disorder: A comparison of cognitive-behavioral and supportive counseling. *Journal of Consulting and Clinical Psychology, 66*(5), 862-866.

Cairns, E., & Wilson, R. (1985). Psychiatric aspects of violence in Northern Ireland. *Stress Medicine, 1,* 193-201.

Cairns, E., & Wilson, R. (1989). Mental health aspects of political violence in Northern Ireland. *International Journal of Mental Health, 18,* 38-56.

Cairns, E., & Wilson, R. (1991). Northern Ireland: Political violence and self-reported physical symptoms in a community sample. *Journal of Psychosomatic Research, 35*(6), 707-711.

Carlson, J. G., Chemtob, C. M., Rusnak, K., Hedlund, N. L., & Muraoka, M. Y. (1998). Eye movement desensitization and reprocessing (EMDR) treatment for combat-related posttraumatic stress disorder. *Journal of Traumatic Stress, 11,* 3-24.

Centers for Disease Control. (2002a, September 6). Psychological and emotional effects of the September 11 attacks on the World Trade Center—Connecticut, New Jersey, and New York, 2001. *Morbidity and Mortality Weekly Report.*

Centers for Disease Control. (2002b, September 6). Self-reported increase in asthma severity after the September 11 attacks on the World Trade Center—Manhattan, New York, 2001. *Morbidity and Mortality Weekly Report.*

Centers for Disease Control. (2002c, September 11). Stress-related illnesses during the 11 months after the attacks (September 11, 2001-August 22, 2002). *Morbidity and Mortality Weekly Report.*

Clyde, R., Collins, J. S. A., Compton, S. A., Cooper, N. K., & Firel, C. M. (1975). Peptic ulcer and civil disturbances. *Lancet, 1,* 1302.

Derogatis, L. R., & Spencer, P. M. (1982). *The Brief Symptom Inventory: Administration, scoring and procedures, Manual 1.* Baltimore, MD: Clinical Psychometrics Diagnostics.

Dupuy, D. F. (1975). "Utility of the National Center for Health Statistics, General Well-Being Schedule in the assessment of self representations of subjective well-being and distress." Paper presented at the National Conference on the Evaluation of Drug, Alcohol, and Mental Health Programs.

Eldar, E. (2003, April 24). 4 out of 10 Israeli children suffer from PTSD. *Ha'aretz Newspaper* (Tel Aviv, Israel).

Erikson, K. T. (1976). Loss of communality at Buffalo Creek. *American Journal of Psychiatry, 133*(3), 302-305.

Everly, J. S., & Mitchell, J. T. (1999). *Critical Incident Stress Management (CISM): A New Era and Standard of Care in Crisis Intervention.* Ellicot City, MD: Chevon.

Foa, E. B., Hearst-Ikeda, D., & Perry, K. J. (1995). Evaluation of a brief cognitive-behavioral program for the prevention of chronic PTSD in recent assault victims. *Journal of Consulting and Clinical Psychology, 63*(6), 948-955.

Foa, E. B., Keane, T. M., & Friedman, M. J. (Eds.). (2000). *Effective treatments for PTSD.* New York: Guilford Press.

Foa, E. B., & Rothbaum, B. O. (1998). *Treating the trauma of rape: Cognitive behavior therapy for PTSD.* New York: The Guilford Press.

Foa, E. B., & Tolin, D. (2000). Comparison of the PTSD Symptom Scale—Interview Version and the Clinician Administered PTSD Scale. *Journal of Traumatic Stress, 13*(2), 181-191.

Foa, E. B., Zinbarg, R., & Rothbaum, B. O. (1992). Uncontrollability and unpredictability in post-traumatic stress disorder: An animal model. *Psychological Bulletin, 112*(2), 218-238.

Friedman, M. J., Davidson, J. R. T., Mellman, T. A., & Southwick, S. M. (2000). Pharmacotherapy. In E. B. Foa, T. M. Keane, & M. J. Friedman (Eds.), *Effective treatments for PTSD* (pp. 84-105). New York: The Guilford Press.

Gidron, Y. (2002a). "Israelis' responses to terrorist threats." Presentation at Women's Mental Health Conference, Beersheva, Israel, October.

Gidron, Y. (2002b). Posttraumatic stress disorder after terrorist attacks: A review. *Journal of Nervous & Mental Disease, 190*(2), 118-120.

Halpern-Felsger, B. L., & Millstein, S. G. (2002). The effects of terrorism on teens' perceptions of dying: The new world is riskier than ever. *Journal of Adolescent Health, 30*(5), 308-311.

Hartsough, D. M., & Myers, D. G. (1985). *Disaster work and mental health: Prevention and control of stress among workers.* Rockville, MD: National Institute of Mental Health.

Hobfoll, S. E. (1989). Conservation of resources: A new attempt at conceptualizing stress. *American Psychologist, 44*(3), 513-524.

Kiecolt-Glaser, J. K., McGuire, L., Robles, T. F., & Glaser, R. (2002). Psychoneuroimmunology and psychosomatic medicine: Back to the future. *Psychosomatic Medicine, 64*(1), 15-28.

Laor, N., manuscript in preparation.

Lavie, T., manuscript in preparation

Lund, A. E. (2002). Question of the month. *Journal of the American Dental Association, 133,* 297.

Maclay, W. S., & Whitby, J. (1942). In-patient treatment of civilian neurotic casualties. *British Medical Journal, 502,* 449-451.

Pat-Horenczyk, R., Abramovitz, R., Brom, D., Horwitz, S., Baum, N., & Chemtob, C. (2003). Symptoms of PTSD and functional impairment among adolescents exposed to ongoing terror: A comparison between two Israeli contexts. Annual Conference, International Society of Traumatic Stress Studies, Chicago, November.

Pfefferbaum, B., Nixon, S., Tivis, R., Doughty, D., Pynoos, R., Gurwitch, R., & Foy, D. (2001). Television exposure in children after a terrorist incident. *Psychiatry, 64,* 202-211.

Pfefferbaum, B., Nixon, S. J., Tucker, P. M., Tivis, R. D., Moore, V. L., Gurwitch, R. H., Pynoos, R. S., & Geis, H. K. (1999). Posttraumatic stress responses in bereaved children after the Oklahoma City bombing. *Journal of the American Academy of Child Adolescent Psychiatry, 38*(11), 1372-1379.

Raphael, K. G., Natelson, B. H., Janal, M. N., & Nayak, S. (2002). A community-based survey of fibromyalgia-like pain complaints following the World Trade Center terrorist attacks. *Pain, 100,* 131-139.

Rose, S., & Bisson, J. I. (1998). Brief early psychological interventions following trauma: A systematic review of the literature. *Journal of Traumatic Stress, 11,* 697-710.

Rotem, T. (2002, April 8). What kind of person had she been if not the terror attack? *Haaretz Newspaper* (Tel Aviv, Israel).

Shalev, A. Y., Freedman, S., Peri, T., Brandes, D., & Sahar, T. (1997). Predicting PTSD in trauma survivors: Prospective evaluation of self-report and clinician-administered instruments. *British Journal of Psychiatry, 170,* 558-564.

Shalev, A. Y., Freedman, S., Peri, T., Brandes, D., Sahar, T., Orr., S. P., & Pitman, R. K. (1998). Prospective study of posttraumatic stress disorder and depression following trauma. *American Journal of Psychiatry, 155*(5), 630-637.

Shalev, A. Y., Galai, T., & Eth, S. (1993). Levels of trauma: A multidimensional approach to the treatment of PTSD. *Psychiatry, 56*(2), 166-177.

Shalev, A. Y., Tuval-Mashiach, R., Frenkiel, S., & Hadar, H. (2003). Psychological reactions to continuous terror. Submitted to *British Medical Journal,* July, 2003.

Shalev, A. Y., Tuval-Mashiach, R., & Hadar, H. (2004). Posttraumatic stress disorder as a result of a mass trauma. *Journal of Clinical Psychiatry, 65*(Supp.1), 4-10.

Sibai, A., Fletcher, A., & Armenian, H. (2001). Variations in the impact of long-term wartime stressors on mortality among the middle aged and older population in Beirut, Lebanon, 1983-1993. *American Journal of Epidemiology, 154,* 128-137.

Stokes, A. B. (1945). War strains and mental health. *Journal of Nervous and Mental Disease, 101,* 215-219.

Summerfield, D. (1999). A critique of seven assumptions behind psychological trauma programmes in war-affected areas. *Social Science & Medicine, 48,* 1449-1462.

Susser, E. S., Herman, D. B., & Aaron, B. (2002). Combating the terror of terrorism. *New Scientist, 287*(2), 54-62.

Vlahov, D., Galea, S., Resnick, H., Ahern, J., Boscarino, J. A., Bucuvalas, M., Gold, J., & Kilpatrick, D. (2002). Increased use of cigarettes, alcohol, and marijuana among Manhattan, New York, residents after the September 11th terrorist attacks. *American Journal Epidemiology, 155*(11), 988-996.

Chapter 4

Neurobiology of Trauma

Craig L. Katz
Rachel Yehuda

INTRODUCTION

The terrorist attacks of September 11, 2001, visited death and destruction upon New York City, Washington, DC, and Pennsylvania on a distressingly massive and sudden scale. Anguish, grief, and shock coalesced into what was a uniquely traumatic event for those people who were direct and indirect victims or witnesses to what transpired that day. With few surviving casualties and rapidly fading hope for the rescue efforts, considerable attention was paid toward ministering to the psychological fallout. Organizations devoted to providing mental health assistance in the wake of disasters, including the American Red Cross, mobilized as part of a mental health intervention that was unprecedented in its magnitude (McQuistion & Katz, 2001).

The involvement of mental health professionals in the relief efforts following 9/11 raises an important question about what constitutes "normal" reactions to trauma and disasters. If emotion and distress are expected elements of the human response to catastrophe, why involve professionals whose work traditionally encompasses "abnormal" reactions? In this chapter, we try to answer the fundamental question of the role of mental health professionals following disasters by examining the physical substrate for responses to stressful and traumatic events. Can our knowledge of the physiology and anatomy of human fear and anxiety inform if, when, and how mental health professionals should intervene in the aftermath of trauma?

IMPACT OF THE TRAUMATIC EVENT
ON MIND AND BRAIN

The time course of disasters can be thought of as spanning the following phases: impact, recoil, and postimpact (Kinston & Rosser, 1974; Tyhurst,

1951). *Impact* encompasses the period during which the threat is present, real, and active, followed by a time of *recoil* in which survivors become aware of what has happened to them and to the people and things around them. During *postimpact,* secondary stresses prevail, occurring as a consequence of the event as survivors attempt to recover.

When the first hijacked airplane hit the World Trade Center, those who bore direct witness experienced an initial visceral reaction. Whether fleeing down the stairs of the towers, staring from the nearby streets, or watching from afar by television, people naturally experienced fear and disbelief while engaging in a number of behaviors intended to ensure or maintain their safety. The initial response to disaster occurs largely because information from our senses floods into the region of the brain known as the amygdala (see Figure 4.1). The amygdala then initiates a cascade of physiological reactions characteristic of fear responses (Davis, 2000): heightened arousal and alertness; elevated heart rates; deepened breathing; widened pupils to permit receipt of as much visual information as possible; and a rise in blood sugar to fuel the muscles of flight and action. These are known collectively as the *fight-or-flight reaction.*

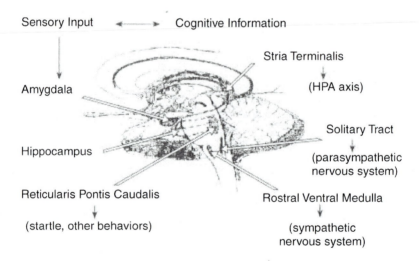

Note: Arrows indicate "activation" of specified system; HPA = hypothalamic-pituitary-adrenal.

FIGURE 4.1. Neuroanatomy of Fear (*Source:* Yehuda, R. "Biology of Posttraumatic Stress Disorder." *The Journal of Clinical Psychiatry, 61*(Suppl 7): 14-21, 2000. Copyright 2000, Physicians Postgraduate Press. Reprinted by permission.)

As will be discussed in greater detail later, the amygdala induces the release of many hormonal systems that coordinate the physiologic response to stress. These stress-activated responses help people to flee the disaster scene. This is accomplished immediately and, likely, before any true cognitive processing begins. The visceral feeling of fear and its concomitant neurobiological response signals people to run; the attendant, rapid physical changes provide the resources with which to act on this aversive feeling.

The physiological process of reacting to a fearsome situation appears to err on the side of safety and false alarms, reacting quickly and not taking any chances. As stated by LeDoux (1996): "It's better to mistake sticks for snakes than snakes for sticks." Other brain regions assist the amygdala in determining whether to let the fear response continue. On 9/11, for example, one could imagine that the person involved in running for his or her life needed not only to initiate a fear response but also to determine whether the information being received by the amygdala was accurate (Armony & LeDoux, 1997). Was that really an airplane flying into the World Trade Center? Cells in the higher-level brain region known as the sensory cortex, which is responsible for organizing and bringing into awareness information received through the senses, confirmed that it was an airplane. Meanwhile, another structure, the hippocampus, may have been called upon to perform an important check on that information by determining the context in which it arose (Armony & LeDoux, 1997; LeDoux, 1996). That is, although a plane indeed flew into the tower, was this *really* happening amid the otherwise normal flow of a September day or was it a stunt, a scene from a movie, or a matter of imagination? The hippocampus seems to play an important role in determining context, and context can further exacerbate or attenuate an initial response to fear.

Early theories about the fight-or-flight response emphasized its role in helping remove organisms from danger and therefore as part of an adaptive stress response mechanism shared by animals and humans (Cannon, 1914). The period of impact encompasses intense and focused action bent exclusively on survival. In the economy of feeling and behavior, this phase seems to permit little else in the way of thinking, feeling, or acting. In fact, LeDoux has argued that conscious awareness, including the consciousness of fear, may not be essential for engaging in survival behaviors when faced with what we nonetheless describe as fear-inducing events (LeDoux, 1996). These behaviors, and their neural basis, are remarkably similar between humans and a number of other species, including snails and fruit flies, none of which are understood to have any form of consciousness. Fight-or-flight responses in humans, in both their biological and behavioral manifestations, appear to be evolutionarily primitive responses preserved over time and species because of their adaptiveness. The immediate feeling of fear, and the thoughts which accompany it, may not cause humans to initiate a fear response but rather constitute a distinctly human manifestation of that otherwise biologically ancient and largely unconscious response (LeDoux, 1996). On the other hand, the feeling or consciousness of fear may sustain the fear response once it begins in humans, potentially sustaining a reaction akin to the fight-or-flight response beyond its period of usefulness.

Recoil

At some point during the process of fleeing from the Twin Towers on 9/11, those who escaped to safety began to take stock of what had happened. This period has been described as *recoil* (Kinston & Rosser, 1974). Once a tentative sense of safety was attained, emotions and cognitions surrounding the event could become a focus of attention. In the hours, days, and weeks that followed, survivors and witnesses to the event began trying to understand what had happened and whether they were still under threat. Anxiety, hypervigilance, and sleep and appetite disturbances are common during this period, reflecting a state of lingering fear of the unknown. People faced an important psychological challenge at this time, wondering if it would be safe to let down their guard. Until this was resolved, their physiological stress response might very well remain active as if still engaged in some form of fight or flight.

With the cessation of an immediate threat, the body's physiological task becomes that of *containing* the stress reaction. The neurotransmitter responsible for much of the initial stress response is norepinephrine (NE), which is released by the sympathetic nervous system. This state of alert in the nervous system is initiated by the amygdala. It has been suggested that NE is necessary for initiating and maintaining the fight-or-flight reaction, including elevating heart rate and blood sugar which permits enhanced delivery of energy to muscles engaged in strenuous exertion, as well as contributing to the formation of memories of emotionally charged events in the amygdala (Cahill, 1997). NE may not only lead to immediate action for the sake of survival but also promotes learning about the current danger for future use. Another stress hormone released by the adrenal gland is cortisol, and it is simultaneously released in response to stress. Cortisol is involved in diminishing the nervous system response, reducing levels of NE, and turning off the physiological state of alert (Yehuda, 2000). Failure to contain the machinery of the stress response may result in significant emotional and mental disequilibrium and eventual psychiatric disorder (Yehuda, 2002). This may include problems maintaining balance in mood, intellectual performance, and even basic functions such as sleeping and eating.

Secondary stressors may be defined as those stressful elements of survivors' lives that have been set in motion, or exacerbated, by the initial event. The ongoing burden on survivors makes recovery from the traumatic event itself additionally challenging. Secondary stressors may include grieving, displacement from one's home, unemployment, and many other problems. Following 9/11, there were the added stressors of the anthrax mystery, the fear of additional acts of terrorism, and the possibility of war, among others. The longer these stressors persist, the higher the probability that the initial fight-or-flight reaction will be perpetuated. This increases the probability of subsequent pathological consequences such as post-traumatic stress disorder (PTSD) (Yehuda, 2000). The obvious clinical implication is the crucial need to reduce the likelihood or impact of secondary stress in the aftermath of a disaster (Yehuda, 2002). This is where the efforts of mental health workers of various disciplines may affect both the mind and the physiology of the brain.

ROLE OF THE MENTAL HEALTH PROFESSIONAL DURING THE RECOIL PHASE

Mental health professionals can provide the support that promotes the attenuation of deleterious sympathetic nervous system activity. Disturbances in life routines and alterations in people's social support systems are all disruptions that can potentially be remedied or modulated. Often, trauma survivors are quite psychologically devastated and lack the energy to attend to the practical matters of daily life and survival. Mental health professionals can offer advice about the primacy of attending to these matters and become important sources of support and guidance. In this way, psychotherapeutic efforts have a favorable impact on the neurobiological milieu of those who simultaneously derive a psychological benefit from the intervention.

This effort requires that resources be marshaled to ensure the availability of mental health professionals in the immediate aftermath of disasters. During this period, a survivor may find himself or herself without family, friends, or other natural supports. In time of disaster, the traumatized individual may find that traditional sources of support, whether from individuals, groups, or even institutions, have become temporarily or even permanently unavailable. If not lost altogether, such support systems may have been rendered uncharacteristically preoccupied with their own survival or recovery, having a number of deleterious consequences, and the mental health professional can step into this physiologically and psychologically vital breach.

Adults and children are vulnerable to experience regression in the face of events such as 9/11 (Katz & Nathaniel, 2002). Regression consists of a temporary return to earlier, more childlike or even infantile stages of development that were previously mastered. It is a phenomenon that may cause disaster survivors to become more disorganized, needy, and dependent. Research into development and regression may further inform our understanding of the neurobiology of the recoil period, when regression may be especially prone to occur.

For example, it has been found that rat pups that are handled during infancy appear less "emotional" when compared with nonhandled rat pups exposed to an open field (Hess, Denenberg, Zarrow, & Pfeiffer, 1969; Levine, Haltmeyer, Karas, & Denenberg, 1967). In comparison to their counterparts, the "handled" rat pups are more active in the novelty of the open field; defecate less; and, intriguingly, have lower plasma levels of corticosterone (another adrenal hormone related to cortisol that typically elevates with stress). These findings have been interpreted to mean that handling in infancy has the beneficial effect of rendering rats less emotionally vulnerable and more organized in the face of new or frightening stimuli.

A human analogue of these situations may be found in the children of Holocaust survivors, who have been found to have elevated rates of PTSD compared to controls despite not having reported more events in their lives that meet strict definitions of trauma (Yehuda, Schmeidler, Wainberg, Binder-Brynes, & Duvdevani, 1998). On the other hand, they did report experiencing more events consistent with stress. This may be interpreted to mean that these individuals are more prone to experience what other people see as ordinary life stresses as more overwhelming traumas. Their vulnerability to this, like the rat pups, may in turn arise from the type of parenting they received. Their parents' PTSD may have altered the quality of care the children received and may be associated with emotional and physical neglect and even abuse (Yehuda, Halligan, & Bierer, 2002). Children of PTSD sufferers may receive less nurturance and "handling" than those individuals with healthier parents. Deficient parenting in the form of inadequate physical and/or emotional "handling" may render both rats and people ill-equipped to cope with any form of stress, traumatic or otherwise.

In the case of disaster, a community may be less available to support, nurture, and, at least metaphorically, "handle" its members precisely when they are most dependent and in need of support. This may engender the twin consequences of worsening people's emotional liability and fearfulness while exacerbating their physiological stress response. The proper functioning of cognitive faculties, maintenance of physiological homeostasis, and the regularity of circadian rhythms of the body rely at least in part on the steady and familiar social interactions of daily life (Hofer, 1984), and disasters disrupt social rhythms and customs. In the event of a man-made disaster such as 9/11, the need for extra nurturance may be especially important in restoring physiological and psychological equilibrium amid ongoing threats. Aware of these heightened needs and well trained in attending to the raw and often regressed emotions of their patients, mental health professionals constitute a unique reservoir of skilled nurturance for their community. In addition to education, advice, and support, mental health professionals may contribute in other ways to the neurobiological transition necessary to recover from stressful events. The panoply of emotional symptoms associated with the aftermath of traumas deserves attention not only from the perspective of reducing suffering but also because the symptoms may become secondary stressors in themselves.

Symptoms that appear consistent with PTSD are very common in the immediate period following a traumatic event. For example, a study of rape victims found that 94 percent of the women interviewed twelve days after the assault had PTSD symptoms (Rothbaum, Foa, Riggs, Murdock, & Walsh, 1992). Anxiousness, hypervigilance, loss of appetite, disturbed sleep, nightmares, and impaired concentration all make a difficult situation worse. These symptoms are likely the product of the same hormonal and neuronal changes necessary for the adaptiveness of the initial fear response, yet they may pose a burden in excess of the survival benefits that may have long-term consequences. Similarly, elevations in heart rate immediately posttrauma, although a potentially adaptive part of the fight-or-flight response, have been found to be a potential risk factor for the development of subsequent PTSD (Bryant, Harvey, Guthrie, & Moulds, 2000; Shalev et al., 1998).

Providing maximal comfort with regard to the symptomatic manifestations of stress and fear may thus aid in preventing the development of PTSD. Mental health professionals should consider the use of psychotherapeutic or pharmacological interventions toward accomplishing this end. Following 9/11, psychiatrists working with Disaster Psychiatry Outreach at the New York City Family Assistance Center provided symptomatic relief of people's anxiety and insomnia via the prescription of brief courses of antianxiety or sleep medications (Katz, 2002). The efficacy and appropriateness of this approach needs confirmation through randomized, double-blind placebo-controlled trials, as there is uncertainty in the literature whether the use of antianxiety medications immediately posttrauma reduces, exacerbates, or has no effect on the symptoms of, or risk for, PTSD (Gelpin, Bonne, Peri, Brandes, & Shalev, 1996; Marshall & Pierce, 2000; Mellman, Byers, & Augenstein, 1998). The limited number of studies on this subject handicaps arguments on all sides, although opponents of the use of medications point out their questionable benefit for symptoms of PTSD coupled with their potential for abuse in PTSD sufferers (Marshall & Pierce, 2000). Proponents, including psychiatrists who worked at the 9/11 Family Assistance Center, point to the immediate clinical benefits for some of the anxiety-related symptoms immediately posttrauma, before PTSD develops. Whether the use of these medications immediately posttrauma prevents the future onset of PTSD warrants further investigation (Gelpin et al., 1996; Mellman et al., 1998).

A NEUROENDOCRINE PERSPECTIVE ON THE POSTTRAUMATIC PERIOD

PTSD, other anxiety disorders, major depression, and possibly substance abuse disorders are potential psychiatric sequelae in the long-term aftermath of disasters (Katz, Pellegrino, Pandya, Ng, & DeLisi, 2002). Although PTSD is not the sole psychiatric phenomenon of the posttraumatic period following disasters, we focus on it here because the biology of it has been well characterized.

PTSD consists of intense reexperiencing of the inciting trauma, persistent avoidance of reminders of the event, and heightened arousal (American Psychiatric Association, 2000). It has an acute subtype, which is presumed to last less than three months. It has a chronic subtype that persists beyond that duration. PTSD may be a common or not infrequent manifestation in the aftermath of disasters and other traumas. Incidence of PTSD after a traumatic event ranges from as low as 6 percent to as high as 20 percent (Yehuda, 2002). Following a disaster the rates of PTSD vary among individuals based on a number of risk factors, including existence of a prior trauma/disaster history, degree of exposure to the event, and presence of psychosocial problems both before and after the disaster (see review by Katz et al., 2002).

From a phenomenological perspective, PTSD may be distinguished from initial stress responses as the continuation of the physical and emotional effects of the fear response despite cessation of the stressful event (Yehuda, 2000). Patients with PTSD appear to react to the world around them as though the event were still happening, reliving the past and misinterpreting cues from the world as indications of an ongoing or imminent threat (van der Kolk, 1997). PTSD symptoms typically diminish over the first three months. Persistence of the symptoms beyond three months suggests chronic PTSD (Simon, 1999). Even in chronic forms of the disorder there is gradual reduction in its rates and severity. This has been observed in the three-year period following several transportation disasters, including incidents involving air, rail, and ferry (Duggan & Gunn, 1995). Thus, the longer the symptoms persist, the more inappropriate and dysfunctional they become.

If 100 percent of people experience a disaster and yet only a small percentage develop PTSD, it is necessary to ask why initial fear responses precipitate the disorder in some people but not others. A number of psychosocial and psychological variables appear to bear on the likelihood of developing the disorder. There appear to be biological differences among those with chronic PTSD and those at greater risk for developing PTSD in the immediate aftermath of a trauma (van der Kolk, 1997; Yehuda, 1997, 2000, 2002).

Elevations of the adrenal stress hormone cortisol, or related adrenal hormones, have historically been associated with stress states in both animals and humans. For example, newborn rats exposed to either heat or electric shocks have been found to have elevations in plasma levels of corticosterone (the adrenal hormone related to cortisol) when compared to their nonstressed litter mates (Haltmeyer, Denenberg, Thatcher, & Zarrow, 1966). In the comparison of handled and nonhandled rat pups mentioned in the previous section, the handled pups appeared to be less stressed by being in an open field than their counterparts and indeed were found to have lower levels of corticosterone. So consistently has elevated cortisol been associated with stress states that its mere presence is often taken as de facto proof that an organism has undergone stress (Yehuda, 1997). The magnitude of the stress has often been estimated based on the magnitude of the increases in cortisol (Yehuda, 1997). These findings led researchers interested in the biology of PTSD to examine the role of cortisol in the development of the disorder, since PTSD has been conceptualized as an abnormal stress response. The levels of cortisol have often been found to be *diminished* in PTSD sufferers, unlike normal stress responses, where cortisol is elevated (van der Kolk, 1997; Yehuda, 1997, 2000, 2002). PTSD appears to be biologically distinct from normal stress reactions because the system regulating the release of the adrenal stress hormones functions at a threshold that maintains unexpectedly low cortisol levels (Yehuda & McFarlane, 1995). Not only do people diagnosed with PTSD have lower levels of cortisol, but recent trauma victims who have low cortisol levels are more likely to eventually develop PTSD than those with higher levels of the hormone in the immediate aftermath of a traumatic event (Resnick, Yehuda, Pittman, & Foy, 1995; Yehuda, 2000). Given that stress-induced release of cortisol is hypothesized to be vital to turning off the fear response fueled by the amygdala's activation of the nervous system, the possible failure of this mechanism in PTSD sufferers may permit the fear response to proceed unchecked. Reduced cortisol levels at the time of trauma facilitate sustained sympathetic arousal, which may ultimately lead to a pathological stress response such as PTSD (van der Kolk, 1997; Yehuda, 1997, 2000, 2002). Of course, low levels of adrenal stress hormones per se do not automatically lead to a pathological stress response, as the handled rats cited earlier suggest. Thus, the role of low cortisol levels in PTSD is complicated, and a fuller understanding of it will likely involve study of other factors, including the timing and duration of cortisol level fluctuations in the course of a stress response.

Norepinephrine, the hormone essential for maintaining sympathetic arousal, is observed to be central to the normal stress response, suggesting that NE also warrants investigation in PTSD. If, in PTSD, cortisol has failed to turn off the sympathetic nervous system activation underlying the fear response, can it be predicted that NE would be found in persistently elevated levels in PTSD sufferers? Research has shown that individuals with chronic PTSD have elevated urinary levels of NE (Southwick et al., 1997) and other evidence of existing in a state of heightened nervous system responses. These include a proneness to elevated heart rate and other physical effects of NE, compared with patients without PTSD, both in the initial aftermath of the trauma (Shalev et al., 1998) and in reexposure to reminders of the event later (Southwick et al., 1997).

Unfortunately, because not all studies have confirmed that trauma survivors who go on to develop PTSD demonstrate early evidence of either too much NE (Blanchard, Hickling, Galovski, & Veazey, 2002) or insufficient cortisol (Yehuda, Resnick, Schmeidler, Yang, & Pitman, 1998), this picture will probably become more complex as further studies are done. From what we already know about the relationship among NE, cortisol, and PTSD, however, it is possible to offer some clinical speculations. First and perhaps most important, the hormonal findings in individuals with PTSD suggest that immediate attempts to either reduce NE and its physiological consequences, or somehow strengthen the cortisol counter-stress response, could help prevent PTSD. This has been attempted by administering the antihypertensive medication propranolol to recent trauma victims (Pitman et al., 2002). Propranolol is known as a beta-blocker which works by preventing NE from attaching to the beta-receptors of the nervous system, thereby blocking the effects of NE release on the organs under the influence of the sympathetic nervous system (i.e., heart, lungs, and blood vessels). Interestingly, this study found no reduction in subsequent PTSD among those who received propranolol. However, recipients of the propranolol experienced less physiological reactivity to reminders of the trauma than did those who received placebo. This attempt at PTSD prophylaxis deserves further study. All measures that reduce stress-related activity of the nervous system could have great clinical merit. Such measures could include the administration of antianxiety medication, the teaching and practice of relaxation techniques, physical exercise, and support through psychotherapy. Although these measures are not known to act specifically at the level of NE receptors within the sympathetic nervous system, they may act either "upstream" or "downstream" from these receptors. They may reduce either the release of NE or the physiological consequences of the persistent activation of receptors by the NE that has been released. For example, the relaxation techniques of deep breathing and guided imagery help reduce the amount of anxiety felt by trauma victims, an effect that likely takes place in the distinctly human, higher-level areas of the brain known as the frontal cortex. The frontal cortex may then exert a moderating influence on the sympathetic nervous system.

Attempts at modulating the diminished cortisol response associated with PTSD are more complicated. Administering hydrocortisone, a steroid hormone related to cortisol, to patients admitted to an intensive care unit with severe sepsis, a life-threatening condition in which the blood becomes infected, reduces the possibility of developing PTSD from the trauma of being in an ICU (Schelling et al., 1999). Given how different the clinical application is in severe sepsis from trauma, prior to applying this treatment to trauma patients, studies would have to show similar benefits, with few side effects. Although cortisol and related adrenal steroids are necessary for the maintenance of human life, administration of exogenous steroids to humans can lead to a range of serious side effects, including diabetes, osteoporosis, and even psychiatric symptoms such as depression and psychosis. Although intriguing from a research perspective, the use of adrenal steroid hormones for prevention of PTSD in trauma survivors cannot yet be considered safe or effective.

The alterations in NE may explain why PTSD sufferers have such distressing and recurrent recollections, images, and nightmares of the event. It has been theorized that during the normal fear response described earlier, NE facilitates the "consolidation" of memories of the event in regions such as the amygdala (Cahill, 1997). Consolidation consists of the formation and storage of new memories. An abnormally prolonged NE response can lead to the *overconsolidation* of bad memories (Yehuda, 2000), meaning that PTSD sufferers may have such uncomfortably intense and vivid memories of the inciting event because of the unmitigated effects of NE. An appreciation of the proposed biological basis for these elements of PTSD can enhance psychotherapeutic and pharmacological treatments. Hence, prazosin, a medication that reduces the effect of NE by blocking some NE receptors, has been shown to reduce nightmares in PTSD sufferers (Raskind et al., 2003).

The fact that hormonal responses of NE and cortisol are sustained in those individuals who develop PTSD suggests that this disorder is not just a normal variant of stress or fear responses, but rather an entity distinct from normal stress responses (Yehuda & McFarlane, 1995). Scientific knowledge is currently insufficient concerning precisely where, when, and how otherwise common fear responses cross what might be conceived of as a biological threshold, to become PTSD. Based on what we know, testing of heart rate and cortisol levels in recent trauma victims would constitute an intriguing approach to physiological screening for pathological fear responses. However, until further scientific knowledge about these and other biological correlates of PTSD is gained, nonphysiological predictors of PTSD are necessary for helping to determine who might cross the biological threshold to PTSD. It is known that prior life experiences may constitute risk factors that increase the likelihood of developing PTSD. These risk factors include prior trauma/psychiatric history, psychosocial problems, predisaster, or the lack of social supports postdisaster, and extent of exposure to the event. Screening for the presence of these factors among recent survivors of traumata might permit the identification of those people at highest risk for developing PTSD from an otherwise superficially unremarkable fear response.

Other Neurochemicals and PTSD

A great many other compounds active within the central nervous system besides cortisol and NE may play a role in the human response to trauma. Research has suggested a role for neurochemicals other than NE and cortisol in the development and manifestations of PTSD. For example, both physiologically produced, or endogenous, opioids and exogenous opioids, such as the drugs morphine and heroin, act at various opioid receptors in the brain. Analgesia is among their major effects. This has led investigators to explore the role of opioids in the often observed phenomena of stress-induced analgesia. Veterans with and without PTSD were asked by investigators to view a combat videotape. Immediately afterward, those with PTSD were found to have decreased sensitivity to the pain of a heat stimulus compared to the non-PTSD sufferers (van der Kolk, Greenberg, Orr, & Pitman, 1989). The PTSD sufferers' increased pain threshold was eliminated by administration of naloxone, a drug that blocks opioid receptors. Observations that both animals and humans under stress feel less physical pain may thus be attributable to a surge of naturally produced opioids during traumatic events. This may underlie the high rates of heroin abuse among war veterans and PTSD victims, because heroin's effects are mediated via its actions at opioid receptors (Charney, Deutsch, Krystal, Southwick, & Davis, 1993). Further elucidation of this phenomenon and its underlying neurobiology is required before more clinical implications can be drawn.

Benzodiazepines represent a major class of antianxiety medications that bind at sites throughout the brain at what are known as GABA receptors. Gamma-aminobutyric acid binds to GABA receptors and thereby acts as the major inhibitory neurotransmitter of the brain, reducing the activation level of neurons. Benzodiazepines, such as lorazepam (Ativan), diazepam (Valium), and alprazoloam (Xanax), are thought to mitigate anxiety by enhancing GABA's effects. Given its role in fear and anxiety, the amygdala is thought to have high concentrations of these receptors (Janicak, Davis, Preskorn, & Ayd, 1997). Researchers have examined the connection between GABA receptors and stress. Mice exposed to the repeated stress of swimming exhibit reduced capacity to bind benzodiazepines in a number of brain regions (Weizman et al., 1989). In humans, investigators have found fewer and/or less-sensitive receptors for benzodiazepines in the frontal region of the brains of people with PTSD compared to those without the disorder (Bremner, Innis, Southwick, & Staib, 2000). Alterations in the brain's GABA receptors may thus underlie some of the anxiety symptoms in patients with PTSD and, despite recent controversy about whether this class of medications helps or possibly worsens PTSD (Marshall & Pierce, 2000), this finding suggests a possible role for benzodiazepines in the treatment of PTSD.

A NEUROANATOMICAL PERSPECTIVE ON THE POSTTRAUMATIC PERIOD

Some of the very portions of the brain involved in mounting an appropriate fear response have also been implicated in PTSD (see Figure 4.2). The hippocampus is essential for the formation of memories about information that can be verbalized and recalled, which is known as declarative memory (Grossman, Buchsbaum, & Yehuda, 2002). It plays a specific role in the formation of memories related to emotionally laden events; this includes giving context to stressful events. The sense of context enables the brain to distinguish, for instance, between a building collapse occurring in a movie and the collapse that occurred amid the shocking but real events of 9/11, and then to mount an appropriate response to either one (e.g., fleeing in response to the latter but not the former). Given its prominent role in memory and fear, the hippocampus has been studied in relation to abnormal fear responses. Postmortem studies in primates exposed to significant social stress in the form of aggression have found degeneration of the hippocampus in otherwise unremarkable brain specimens (Uno, Tarara, Else, Suleman, &

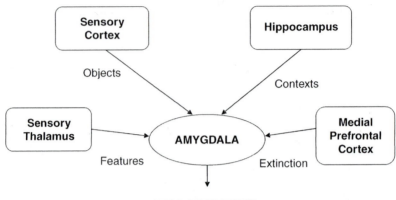

FEAR RESPONSES

FIGURE 4.2. The amygdala receives parallel inputs from a wide range of levels of sensory and cognitive processes. The thalamus conveys low-level stimulus features, whereas information about objects is provided by the cortex, especially during the late stages of cortical processing. The hippocampus plays a crucial role in setting the emotional context. The medial prefrontal cortex has been implicated in the process of extinction of fear responses through its inputs to the amygdala. An imbalance between these input channels may lead to powerful inappropriate fear responses, such as those seen in PTSD. (*Source:* Armony, J. L. & LeDoux, J. E. 1997. How the brain processes emotional information. *Annals of the New York Academy of Sciences, 821,* 259-270. Copyright © 1997 New York Academy of Sciences, USA. Reprinted with permission.)

Sapolsky, 1989). These changes are believed to be associated with elevated levels of cortisol amid stress (Sapolksy, Uno, Rebert, & Finch, 1990). Persistent exposure to cortisol during stress appears to cause cell damage within the hippocampus, suggesting an area of overlap between the neuroendocrine system and anatomical abnormalities of PTSD sufferers.

Investigators using MRI images of the brain found reduced right hippocampal volume in Vietnam veterans with PTSD compared to those veterans without the disorder (Bremner, Randall, Scott, & Bronen, 1995). Other imaging studies of patients with PTSD have found a reduction in the volume of the hippocampus across several different trauma populations (Tucker & Trautman, 2000; van der Kolk, 1997), although the consistency and accuracy of these findings has recently been questioned (Grossman et al., 2002). As with nonhuman primates, reduced hippocampal size in PTSD sufferers appears to result from the effects of cortisol on that brain region (van der Kolk, 1997). Because cortisol levels have consistently been found to be reduced in humans with PTSD, this effect may stem from the hypersensitivity of the hippocampus to whatever cortisol is present (Yehuda, 2001). Hippocampal changes likely account for these patients' difficulties forming and processing new experiences. The damaged hippocampus is less able to put these experiences into the current and appropriate context, leading patients to see otherwise nontraumatic events as though they were threats akin to the initial trauma.

It is impossible to study the anatomy of the brain using neuroimaging while the fear-producing agent is exerting its effects. We have this opportunity only in cases of prolonged fear responses. Functional MRI represents a form of MRI that detects areas of increased cerebral flow thought to reflect increased activation of those areas. A recent case study using functional MRI found that different regions of the brain become more active in PTSD sufferers depending upon whether the individual experiences intense memories and arousal or numbing in response to exposure to a script of their traumatic event (Lanius, Hopper, & Menon, 2003). Patients with PTSD have likewise undergone positron emission tomography (PET) scanning while being alternately exposed to audiotapes of traumatic and emotionally neutral scripts (Rauch et al., 1996). PET utilizes radioactively labeled biological substrates such as water and glucose to detect changes in blood flow or metabolism in areas of the brain. These changes reflect the overall level of activity of particular areas. Exposure to the person's trauma, unlike exposure to neutral scripts, is associated with increased cerebral blood flow to the right-sided structures associated with emotion, globally known as the limbic system and including the amygdala, and to the visual cortex of the brain.

The involvement of the amygdala and associated limbic structures is predictable based on the fact that fear and emotion are involved. The finding of enhanced flow to the visual cortex is notable since it supports the theory that posits PTSD as a defect in the storage of traumatic memories. Some believe that the flashbacks and nightmares of PTSD reflect a failure to encode these memories in semantic terms (van der Kolk & van der Hart, 1989). Instead, they "escape words" and are stored instead in visual terms, as underscored by the enhanced flow to the part of the brain responsible for vision. The detour around semantic memory, and the subsequent intrusion of traumatic memories in visual rather than verbal terms, is also consistent with the observation of a decrease in cerebral blood flow to the left inferior frontal cortex when subjects were exposed to the traumatic script. This portion of the brain contains Broca's area, the region responsible for the encoding of words, and this PET finding may explain PTSD patients' difficulty processing traumas.

The findings of reduced activation or size of structures involved in the verbalization and integration of memories and experience, the left inferior frontal cortex and the hippocampus, could underscore how important it is to assist traumatized patients with the verbal processing and evaluation of their experiences. This may be an argument in support of the promotion of verbalizing of experiences as soon as possible after a traumatic event, possibly through debriefing exercises (Kaplan, Iancu, & Bodner, 2001). We may also interpret these findings as providing an anatomical basis for the use of cognitive psychotherapy in the prevention of chronic PTSD by enhancing otherwise defective cognitive or verbal processing (Bryant, Harvey, Basten, Dang, & Sackville, 1998).

To the extent that the anatomical abnormalities found in PTSD are the result of traumatic events, prevention and early treatment of PTSD may prevent such damage from occurring. It remains to be determined whether these alterations or abnormalities in the amygdala, visual cortex, left inferior frontal cortex, and the hippocampus are already present in those who subsequently suffer from PTSD, or gradually develop or worsen following PTSD onset. This debate has been highlighted by two recent studies. In one, recent survivors of a trauma who did and did not go on to develop PTSD underwent MRI studies of their brains one week and six months after the event (Bonne et al., 2001). There was no difference in hippocampal volume between the two groups at either time, nor did those subjects with PTSD experience a reduction in hippocampal size over time. The investigators concluded that reduced hippocampal volume may not constitute a risk factor for the development of PTSD and may be associated with only chronic PTSD.

On the other hand, an MRI study of identical twins concluded that reduced hippocampal volume may increase the risk of PTSD following trauma exposure (Gilbertson et al., 2002). In this study, identical twins of combat-exposed Vietnam veterans with PTSD had reductions in hippocampal volume despite not having themselves been exposed to combat, and not having PTSD. Both of these groups had significantly smaller hippocampal volumes compared to Vietnam veterans without PTSD and their non-combat-exposed twins who also did not have PTSD. These findings were interpreted to suggest that genetic predisposition, rather than stress-induced atrophy, are the source of pretrauma reductions in hippocampal size which then predispose people to developing PTSD after a trauma. Further investigation of the hippocampal alterations associated with PTSD is thus warranted. Immense clinical benefit will ultimately come from knowing if and how we could apply this knowledge when following the course of people's emotional recovery from trauma and disaster.

Overall, the possibility exists that whatever physical vulnerabilities certain brain regions have to the effects of trauma may be reversible. Long-term treatment with antidepressant medication in animals appears not just to increase or decrease amounts of certain neurotransmitters in the brain but also to increase the amounts of substances necessary for promoting the growth and function of neurons themselves, such as brain-derived neurotrophic factor (BDNF) (Duman, Heninger, & Nestler, 1997). This has been particularly noted to occur in the hippocampi of animals that previously underwent experimentally induced stress (Duman et al., 1997). The adult human brain has historically been considered unable to generate new neurons once they are damaged or otherwise lost. However, one recent study has found that the hippocampus indeed retains the ability to grow new neurons throughout human life (Eriksson et al., 1998). Investigators found that the postmortem brains of cancer patients who had been given a chemically modified form of DNA, bromodeoxyuridine (BrdU), for diagnostic purposes still contained the BrdU. Only dividing cells could have incorporated the BrdU and maintained it in the brain, meaning that the hippocampus was still growing new neurons prior to these patients' deaths.

One study has examined the role of treatment in modulating the brain images of patients with PTSD (Levin, Lazrove, & van der Kolk, 1999). In this study, PTSD patients underwent single photon emission computed tomography (SPECT) prior to and following a three to four session treatment with eye movement desensitization and reprocessing (EMDR). SPECT creates images by tracing and compiling patterns of radiation emission from the human brain injected with radioactive materials. EMDR is a procedure consisting of prompted eye movements, recall of traumatic imagery, and cognitive evaluation of thoughts associated with these images, and is a treatment of uncertain benefit for PTSD (Shapiro, 1989). Following EMDR, the anterior cingulate gyrus of the brains of four of six patients demonstrated hyperactivity compared to pretreatment. Because this brain region is involved in discerning real from perceived threats, the investigators concluded that the treatment successfully enhanced the accuracy of patients' perceptions of threat. These results, while promising, must still be regarded with caution given the small number of subjects and lack of controls; and it certainly cannot be interpreted to provide neuroanatomical support for the practice of EMDR.

CONCLUSION

Neurobiological findings can inform when and how mental health professionals may intervene in the wake of trauma and disaster, but they have not done so yet because we know so little about the biological and psychological aftermath of trauma. Neurobiological information about both normal stress response and pathological response in the form of PTSD suggests the possibility of numerous avenues for intervention by mental health professionals. Without more research, we don't know exactly which avenues to pursue. Mental health clinicians may possibly help recent disaster victims to modulate and, where appropriate, terminate their stress responses, thereby avoiding pathological persistence and intensification of the state of fear induced by traumatic events.

Knowledge of the neurobiological and neuroanatomical changes of PTSD suggests that some prophylactic measures, immediate interventions, and long-term treatment options might be instituted during the acute aftermath of trauma and disaster. We do not yet know what these interventions would include. Hopefully, with further investigation, the negative impact on mental health of events such as 9/11 can be minimized.

Mental health professionals may have a role in helping those whose reactions to disaster and trauma become maladaptive, especially when such patients' other sources of support are exhausted or inadequate. Not all negative emotion is maladaptive, as demonstrated by the fear during the impact phase. Likewise, not all symptoms associated with highly emotional states are bad in and of themselves. The determination of what is adaptive biologically and psychologically may help us determine in the future where to draw the line between what is adaptive and what is maladaptive in individual responses to trauma, both long-term and short-term.

REFERENCES

American Psychiatric Association (2000). *Diagnostic and statistical manual of mental disorders, Fourth edition, Text revision.* Washington, DC: Author.

Armony, J., & LeDoux, J. (1997). How the brain processes emotional information. *Annals of the New York Academy of Sciences, 821,* 259-270.

Blanchard, E., Hickling, E., Galovski, T., & Veazey, C. (2002). Emergency room vital signs and PTSD in a treatment seeking sample of motor vehicle accident survivors. *Journal of Traumatic Stress, 15*(3), 199-204.

Bonne, O., Brandes, D., Gilboa, A., Gomori, J.M., Shenton, M., Pitman, R., & Shalev, A. (2001). Longitudinal MRI study of hippocampal volume in trauma survivors with PTSD. *American Journal of Psychiatry, 158*(8), 1248-1251.

Bremner, J.D., Innis, R., Southwick, S., & Staib, L. (2000). Decreased benzodiazepine receptor binding in prefrontal cortex in combat-related posttraumatic stress disorder. *American Journal of Psychiatry, 157*(7), 1120-1126.

Bremner, J.D., Randall, P., Scott, T.M., & Bronen, R.A. (1995). MRI-based measurement of hippocampal volume in patients with combat related post-traumatic stress disorder. *American Journal of Psychiatry, 152*(7), 973-981.

Bryant, R., Harvey, A., Basten, C., Dang, S., & Sackville, T. (1998). Treatment of acute stress disorder: A comparison of cognitive-behavioral therapy and supportive counseling. *Journal of Consulting and Clinical Psychology, 66,* 862-866.

Bryant, R., Harvey, A., Guthrie, R.M., & Moulds, M.L. (2000). A prospective study of psychophysiological arousal, acute stress disorder, and posttraumatic stress disorder. *Journal of Abnormal Psychology, 109,* 341-344.

Cahill, L. (1997). The neurobiology of emotionally influenced memory. *Annals of the New York Academy of Sciences, 821,* 238-246.

Cannon, W. (1914). Emergency function of adrenal medulla in pain and major emotions. *American Journal of Physiology, 3,* 356-372.

Charney, D., Deutch, A., Krystal, J., Southwick, S., & Davis, M. (1993). Psychobiologic mechanisms of posttraumatic stress disorder. *Archives of General Psychiatry, 50,* 294-305.

Davis, M. (2000). The role of the amygdala in conditioned and unconditioned fear and anxiety. In J.P. Aggleton (Ed.), *The amygdala* (Vol. 2, pp. 213-287). Oxford, UK: Oxford University Press.

Duggan, C. & Gunn, J. (1995). Medium-term course of disaster victims: A naturalistic follow-up. *British Journal of Psychiatry, 167,* 228-232.

Duman, R., Heninger, G., & Nestler, E. (1997). A molecular and cellular theory of depression. *Archives of General Psychiatry, 54,* 597-606.

Eriksson, P., Perfilieva, E., Bjork-Eriksson, T., Alborn, A.-M., Nordborg, C., Peterson, D., & Gage, F. (1998). Neurogenesis in the adult human hippocampus. *Nature Medicine, 4*(11), 1313-1317.

Gelpin, E., Bonne, O., Peri, T., Brandes, D., & Shalev, A. (1996). Treatment of recent trauma survivors with benzodiazepines: A prospective study. *Journal of Clinical Psychiatry, 7*(9), 390-394.

Gilbertson, M., Shenton, M., Ciszewski, A., Kasai, K., Lasko, N., Orr, S., & Pitman, R. (2002). Smaller hippocampal volume predicts pathologic vulnerability to psychological trauma. *Nature Neuroscience, 5*(11), 1242-1247.

Grossman, R., Buchsbaum, M., & Yehuda, R. (2002). Neuroimaging studies in post-traumatic stress disorder. *Psychiatric Clinics of North America, 25,* 317-340.

Haltmeyer, G., Denenberg, V., Thatcher, J., & Zarrow, M.X. (1966). Response of the adrenal cortex of the neonatal rat after subjection to stress. *Nature, 212,* 1371-1373.

Hess, J., Denenberg, V., Zarrow, M.X., & Pfeiffer, W.D. (1969). Modification of the corticosterone response curve as a function of handling in infancy. *Physiology and Behavior 4,* 109-111.

Hofer, M. (1984). Relationships as regulators: A psychobiologic perspective on bereavement. *Psychosomatic Medicine, 46,* 183-197.

Janicak, P., Davis, J., Preskorn, S., & Ayd, F. (1997). *Principles and practice of psychopharmacotherapy.* Philadelphia, PA: Williams and Wilkins.

Kaplan, Z., Iancu, I., & Bodner, E. (2001). A review of psychological debriefing after extreme stress. *Psychiatric Services, 52*(6), 824-827.

Katz, C. (2002). 9/11: The DPO perspective. Presented at the Second International Congress on Disaster Psychiatry, Mount Sinai School of Medicine, New York, April 19-20.

Katz, C., & Nathaniel, R. (2002). Disasters, psychiatry, and psychodynamics. *Journal of the American Academy of Psychoanalysis, 30*(4), 519-526.

Katz, C., Pellegrino, L., Pandya, A., Ng, A., & DeLisi, L. (in press). Research on psychiatric outcomes and interventions subsequent to disasters: A review of the literature. *Psychiatry Research, 110,* 201-217.

Kinston, W., & Rosser, R. (1974). Disaster: Effects on mental and physical state. *Journal of Psychosomatic Research, 18*(6), 437-456.

Lanius, R., Hopper, J., & Menon, R. (2003). Individual differences in a husband and wife who developed PTSD after a motor vehicle accident: A functional MRI case study. *American Journal of Psychiatry, 160*(4), 667-669.

LeDoux, J. (1996). *The emotional brain.* New York: Simon and Schuster.

Levin, P., Lazrove, S., and van der Kolk, B. (1999).What psychological testing and neuroimaging tell us about the treatment of posttraumatic stress disorder by eye movement densensitization and reprocessing. *Journal of Anxiety Disorders, 13*(1-2), 159-172.

Levine, S., Haltmeyer, G., Karas, G., & Denenberg, V. (1967) Physiological and be-havioral effects of infant stimulation. *Physiology and Behavior, 2,* 55-59.

Marshall, R., & Pierce, D. (2000). Implications of recent findings in posttraumatic stress disorder and the role of pharmacotherapy. *Harvard Review of Psychiatry, 7*(5), 247-256.

McQuistion, H., & Katz, C. (2001). The September 11, 2001 disaster: Some lessons learned in mental health preparedness. *Emergency Psychiatry, 7,* 61-64.

Mellman, T., Byers, P., & Augenstein, J. (1998). Pilot evaluation of hypnotic medi-cation during acute traumatic stress response. *Journal of Traumatic Stress, 11*(3), 563-569.

Pitman, R., Sanders, K., Zusman, R., Healy, A., Cheema, F., Lasko, N., Cahill, L., & Orr, S. (2002). Pilot study of secondary prevention of post-traumatic stress disor-der with propranolol. *Biological Psychiatry, 51*(2), 189-192.

Raskind, M.A., Peskind, E.R., Kanter, E.D., Petrie, E.C., Radant, A., Thompson, C.E., Dobie, D.J., Hoff, D., Rein, R.J., Straits-Troster, K., Thomas, R.G., & McFall, M.M. (2003). Reduction of nightmares and other PTSD symptoms in combat veterans by prazosin: A placebo-controlled study. *American Journal of Psychiatry, 160*(2), 371-373.

Rauch, S., van der Kolk, B., Fisler, R., Alpert, N., Orr, S., Savage, C., Fischman, A., Jenike, M., and Pitman, R. (1996). A symptom provocation study of posttraumatic stress disorder using positron emission tomography and script driven imagery. *Archives of General Psychiatry, 53,* 380-387.

Resnick, H., Yehuda, R., Pitman, R., & Foy, D. (1995). Effect of previous trauma on acute plasma cortisol level following rape. *American Journal of Psychiatry, 152*(11), 1675-1677.

Rothbaum, B., Foa, E., Riggs, D., Murdock, T., & Walsh, W. (1992). A prospective examination of post-traumatic stress disorder in rape victims. *Journal of Trau-matic Stress Studies, 5,* 455-475.

Sapolksy, R., Uno, H., Rebert, C., & Finch, C. (1990). Hippocampal damage associ-ated with prolonged glucocorticoid exposure in primates. *Journal of Neurosci-ence, 10*(9), 2897-2902.

Schelling, G., Stol, C., Kapfhammer, H.-P., Rothenhausler, H.-B., Krauseneck, T., Durst, K., Haller, M., & Briegel, J. (1999). The effect of stress doses of hydrocortisone during septic shock on posttraumatic stress disorder and health-related quality of life in survivors. *Critical Care Medicine, 27,* 2678-2683.

Shalev, A., Sahar, T., Freedman, S., Peri, T., Glick, N., Brandes, D., Orr, S., & Pitman, R. (1998). A prospective study of heart rate responses following trauma and the subsequent development of PTSD. *Archives of General Psychiatry, 55*(6), 553-559.

Shapiro, F. (1989). Eye movement desensitization: A new treatment for post-traumatic stress disorder. *Journal of Behavioral Therapeutics and Experimental Psychiatry, 20*(3), 211-217.

Simon, R. (1999). Chronic post-traumatic stress disorder: A review and checklist of factors influencing prognosis. *Harvard Review of Psychiatry, 6,* 304-312.

Southwick, S., Morgan, C.A., Bremner, A.D., Grillon, C., Krystal, J., Nagy, L., & Charney, D. (1997). Noradrenergic alterations in posttraumatic stress disorder. *Annals of the New York Academy of Sciences, 821,* 125-141.

Tucker, P., & Trautman, R. (2000). Understanding and treating PTSD: Past, present, and future. *Bulletin of the Menninger Clinic, 64*(Suppl. A), A37-A51.

Tyhurst, J.S. (1951). Individual reactions to community disasters. *American Journal of Psychiatry, 108,* 764-769.

Uno, H., Tarara, R., Else, J., Suleman, M., & Sapolsky, R. (1989). Hippocampal damage associated with prolonged and fatal stress in primates. *The Journal of Neuroscience, 9*(5), 1705-1711.

Van der Kolk, B. (1997). The psychobiology of posttraumatic stress disorder. *Journal of Clinical Psychiatry, 58*(Suppl. 9), 16-24.

van der Kolk, B., Greenberg, M., Orr, S., & Pitman, R. (1989). Endogeneous opiods, stress induced analgesia, and posttraumatic stress disorder. *Psychopharmacology Bulletin, 25*(3), 417-421.

van der Kolk, B., & van der Hart, O. (1989). Pierre Janet and the breakdown of adaptation in psychological trauma. *American Journal of Psychiatry, 146*(12), 1530-1540.

Weizman, R., Weizman, A., Kook, K., Vocci, F., Deutsch, S., & Paul, S. (1989). Repeated swim stress alters brain benzodiazepine receptors measured in vivo. *Journal of Pharmacology and Experimental Therapeutics, 249*(3), 701-707.

Yehuda, R. (1997). Sensitization of the hypothalamic-pituitary-adrenal axis in posttraumatic stress disorder. *Annals of the New York Academy of Sciences, 821,* 57-75.

Yehuda, R. (2000). Biology of posttraumatic stress disorder. *Journal of Clinical Psychiatry, 61*(Suppl. 7), 14-21.

Yehuda, R. (2001). Are glucocorticoids responsible for putative hippocampal damage in PTSD? How and when to decide. *Hippocampus, 11,* 85-89.

Yehuda, R. (2002). Post-traumatic stress disorder. *New England Journal of Medicine, 346*(2), 108-114.

Yehuda, R., Halligan, S., & Bierer, L. (2002). Cortisol levels in adult offspring of Holocaust survivors: Relation to PTSD symptom severity in the parent and child. *Psychoneuroendocrinology, 27,* 171-180.

Yehuda, R., & McFarlane, A. (1995). Conflict between current knowledge about posttraumatic stress disorder and its original conceptual basis. *American Journal of Psychiatry, 152*(12), 1705-1713.

Yehuda, R., Resnick, H., Schmeidler, J., Yang, R.-K., & Pitman, R. (1998). Predictors of cortisol and 3-methoxy-4-hydroxy-phenylglycol responses in the acute aftermath of rape. *Biological Psychiatry, 43,* 855-859.

Yehuda, R., Schmeidler, J., Wainberg, M., Binder-Brynes, K., & Duvdevani, T. (1998). Vulnerability to posttraumatic stress disorder in adult offspring of Holocaust survivors. *American Journal of Psychiatry, 155*(9), 1163-1171.

Chapter 5

Children and Families:
A New Framework
for Preparedness and Response
to Danger, Terrorism, and Trauma

Robert S. Pynoos
Alan M. Steinberg
Merritt D. Schreiber
Melissa J. Brymer

INTRODUCTION

Catastrophic disasters are nothing new to the human condition, deeply affecting populations across large regions, often dramatically altering the course of history, from earliest recorded time. Beyond the individual child, family, or community, researchers are beginning to perform studies that consider how the repercussions of catastrophic disasters may alter the social and political character of nations (Pynoos, 1992). After massive trauma, a large segment of the child and adolescent population may experience posttraumatic stress reactions and traumatic grief. As evidenced by exceedingly high rates of chronic psychiatric morbidity among children after the devastating 1988 earthquake in Armenia (Pynoos et al., 1993), the existence of thousands of traumatized children in different stages of recovery places special burdens on society. These may include widespread disturbances in moral development and conscience functioning (Goenjian et al., 1999), impairment in school, peer, family, and community functioning, and increased vulnerability to future stress. Changes in outlook on the future may not only affect the individual child, but on a massive scale permeate and transform cultural expectations, altering the social ecology of the next generation.

Disasters can lead to radical shifts in fundamental beliefs and philosophical perspectives. For example, Luke and Reeves (von Kleist, 1978, p. 16) reflected on how the devastating earthquake of 1755 "not only shattered Lisbon, but severely shook the optimistic theodocy of the Enlightenment."

The earliest known written personal account of the direct experience of a catastrophic disaster by an adolescent is from a letter by Pliny the Younger concerning his experience in Pompeii during the eruption of Mt. Vesuvius:

> My mother began to beseech, exhort and command me to escape as best I might: A young man could do it; She, burdened with age and corpulency, would die easily if only she had not caused my death. I replied, I would not be saved without her, and taking her hand I hurried her on. Even then however, my mother and I notwithstanding the danger we had passed, and that which still threatened us, had no thoughts of leaving the place, till we should receive some tiding of my uncle.
>
> I might have boasted that amidst dangers so appalling, not a sign or expression of fear escaped me, had not my support been founded in that miserable, but strong consolation, that all mankind were involved in the same calamity, and that I was perishing with the world itself.

In this passage, Pliny the Younger articulated his intense dilemma, pitting his own survival against that of his mother and uncle, Pliny the Elder. At one poignant moment, he describes no longer feeling any fear, resigned to the fatalistic expectation that the whole world was coming to an end.

September 11, 2001, brought to the forefront like no other recent national tragedy an awareness throughout communities in the United States of the psychological reverberations of traumatic experience and loss, and the challenge of living amid ongoing and new dangers and threats. In addition, one of the most important legacies of 9/11 should be an enduring commitment by our nation to plan for, and effectively respond to, the mental health impact of traumatic experience, loss, and a posttrauma ecology of danger on the lives of children, adolescents, their families, and their communities.

THE NATIONAL CHILD TRAUMATIC STRESS NETWORK

The National Child Traumatic Stress Network was funded by the U.S. Department of Health and Human Services, Center for Mental Health Services, Substance Abuse and Mental Health Services Administration, beginning October 1, 2001. Subsequently, federal funding was increased to support additional network partners who would be providing services for those areas most affected by the terrorist attacks. Funding was provided to establish a Terrorism and Disaster Branch to increase readiness, response, and capacity of federal, state, and local recovery efforts on behalf of the mental health needs of children and families after terrorism and disaster. The Network's three components include (1) a joint coordinating National Center for Child Traumatic Stress at UCLA and Duke University; (2) fifteen university-affiliated Intervention Development and Evaluation Centers primarily responsible for development, delivery, and evaluation of evidence-based prevention, assessment, early intervention, and treatment approaches, and service delivery models, and (3) thirty-eight Community Treatment and Service Centers primarily engaged in implementing community or specialty child service settings, model treatment interventions, and community services to increase availability of and access to care for children and their families who have experienced trauma. This unique federal mental health initiative leverages collaboration across the Network to meet a national mission of raising the standard of care and increasing access to services for traumatized children and their families across the United States.

Network partners include academic and community mental health leaders specializing in child and adolescent trauma, including neglect, sexual and physical abuse, sexual assault, school and community violence, family/domestic violence, war, political violence and torture, serious accidental injury, catastrophic medical illness (and its treatment), traumatic loss, disasters, and terrorism. A wide range of service sectors is represented, including emergency rooms, pediatric clinics and hospitals, child and adolescent mental health clinics, schools, home, child welfare and protection, juvenile justice and law enforcement, and residential care facilities. The Network also serves culturally diverse and special-needs populations in both urban and rural settings, and is dedicated to incorporating a sound developmental, cultural, and ecological framework in all of its collaborative activities.

SELECTED LITERATURE REVIEW ON CHILDREN AND TERRORISM

A growing body of literature has documented that major natural and technological disasters result in significant acute and long-term psychological distress for children and adolescents. Scant evidence is available regarding the impact of nuclear, biological, or chemical terrorism on youth. However, given the magnitude of potential terrorist events, a multitude of psychological consequences for children, adolescents, and families may be anticipated, with unprecedented psychological morbidity and enormous demands placed on health and mental health infrastructures.

Meta-analytic results combining numerous studies reveal that disaster exposure results in a 17 percent increase in the rate of adult psychological disorders (Rubonis & Bickman, 1991). There are no corresponding meta-analytic studies among children after terrorism. Findings regarding children's psychological morbidity after disasters (including hurricanes, earthquakes, fires, transportation disasters, the Oklahoma City bombing, and 9/11) indicate that school-age children are at risk for more severe and prolonged posttraumatic reactions (Norris et al., 2002). Studies also suggest the need to differentiate clinical psychiatric conditions from a range of distress reactions in children and adults who do not meet criteria for formal psychiatric diagnoses (North, Pfefferbaum, & Tucker, 2002).

Salter (2001) reported on an incident in which several young children in Brazil ingested discarded radioactive medical materials. As many as 125,000 individuals, including many children, were subsequently screened for radiation exposure. Among those screened, 5,000 individuals reported symptoms of acute radiological illness, but careful investigation found that *none* of these individuals had any exposure to radiological material. Long-term findings indicate that, compared to those with confirmed exposure, those who feared that they had been exposed experienced similar levels of long-term psychological consequences and chronic stress (Collins & De Carvalho, 1993).

Findings from nuclear and chemical accidents, such as the Three Mile Island (Bromet, Parkinson, & Dunn, 1990) and Chernobyl disasters (Havenaar et al., 1997), have suggested similar long-term risk for a subset of the exposed population. Estimates for these events indicate that the ratio of psychological to physical casualties can be 10 to 15:1 (Pastel, 2000). Following attacks with SCUD missiles in Israel, 43 percent of emergency department admissions were deemed to be psychological casualties (Bleich, Dycian, Koslowsky, Solomon, & Wiener, 1992). Bioterrorism, with exposure impossible to discern, confusing information on extended incubation, person-to-person transmission, limited or nonexistent treatments, quarantine, and horrific death further compounds pediatric risks. In Bhopal, India, following an industrial release of toxic chemicals in 1984, one in ten individuals in the surrounding community developed psychological disorders (Murthy, 2003). Reports have indicated that among the most striking effects of the Chernobyl nuclear incident was a marked increase in the suicide rate among the Estonian workers who cleaned up after the power plant accident (Vanchieri, 1997). Mothers of children under eighteen were also found to be at long-term risk for a range of psychological disorders (Havenaar et al., 1997).

Since 9/11, several major studies have appeared in the literature. A large-scale study looking at the psychological impact on children was conducted by the New York City Board of Education (2002) indicating that a broad range of mental health problems emerged among children in grades four to twelve at higher than expected prevalence, with an estimated 75,000 children meeting criteria for post-traumatic stress disorder (PTSD). Schuster et al. (2001) conducted a national sample random-digit dialing study approximately five days after 9/11. They found that 44 percent were bothered "quite a bit" by posttraumatic stress symptoms, and that 34 percent of children had at least one stress symptom following the attack. Galea and colleagues (2002) reported on 1,000 adults in the vicinity of the World Trade Center, also using the random-digit dialing method five to eight weeks post-9/11. Seven and one-half percent of adults living south of 110th Street reported posttraumatic stress symptoms, with 9.7 percent reporting depressive symptoms. Those living closer to the World Trade Center were more at risk for symptoms. Using the same sample of 1,000 adults, Stuber et al. (2002) interviewed 112 parents of children aged four to eighteen, five to eight weeks after the attack. They found that, among those residing below 110th Street, 22 percent of the children had received some type of mental health care. Schlenger et al. (2002), using a Web-based national sample several months after 9/11, found that the prevalence of PTSD was approximately three times higher in New York City than in Washington, DC, and other urban areas across the United States.

Although data on children's reactions to 9/11 are still emerging, following the 1993 terrorist attack on the World Trade Center, Koplewicz and colleagues (2002) found that children who had been trapped in an elevator and those on the observation deck during the explosion had long-term PTSD and disaster-related fears three and nine months later. In the most comprehensive series of studies on the impact of terrorism on children and adults, Pfefferbaum and colleagues (1999), following the 1995 Oklahoma City bombing, found in their studies of bereaved children that 63 percent continued to feel acutely worried about themselves or their families, with 15 percent reporting not feeling "safe at all" seven weeks after the bombing. In their studies of elementary school students, Gurwitch, Sitterle, Young, & Pfefferbaum (2002) found that close to one year after the bombing, 5 percent of their elementary age (grades three to five) sample of Oklahoma City public school students had clinically significant levels of posttraumatic stress symptoms, with nearly one-third having continued worries about family members. In their study among adults, North and colleagues (1999) found that approximately 34 percent of impacted adults met criteria for PTSD, with 22 percent having comorbid depression.

UNIQUE CHALLENGES REGARDING WEAPONS OF MASS DESTRUCTION

In unprecedented fashion, weapons of mass destruction place significant numbers of children at enormous risk for psychosocial morbidity for an extended period and place extreme demands on caring adults in a variety of settings, including families, schools, shelters, primary care offices, and health care facilities. The emerging risk of bioterrorism, the so-called silent agent of terror (Holloway, Norwood, Fullerton, Engel, & Ursano, 1997), is associated with particularly daunting challenges. The inherent fear created by a weapon composed of invisible microbes that kill silently, with extended incubation, and spread by strangers, friends, and family alike, sets the stage for catastrophic psychological causalities. At the time of this writing, there are no reported studies on the magnitude of psychological impact of biological terrorism. There are, however, reports of behavioral compliance with recommended antibiotic treatment among Brentwood postal workers after the Washington, DC, anthrax attacks of 2001, with only 47 percent complying with the recommended treatment guidelines (Altman, 2002), and with perceptions of differential treatment when congressional workers and postal employees received differing antibiotics and quantities.

Because many children are at developmental stages where they have not yet acquired the capacity to understand the concepts of "illness," "infection," "vaccination," "quarantine," or "contagion," the stress of biological events will be even more complicated. Similarly, the psychological effects of nuclear or radiological terrorism could produce significant levels of psychological morbidity. Worries about children will likely be experienced by families, first responders, and health care providers. Large-scale death and injury of children would portend serious psychological trauma for the nation.

Further compounding the impact is that, unlike conventional disasters with a clear demarcation of when the event is past and recovery has begun, bioterrorism and radiological attacks lack such fundamental delimitation. Bioterrorist attacks, as with radiological exposure, create difficulty in judging one's exposure over an extended time frame of expected effects, compounded by potentially confusing media coverage involving highly complex medical information. Mass panic may ensue in highly specific circumstances, such as where local resources are overwhelmed and governmental response is perceived as ineffective (DeMartino, 2001). Complicating the psychological impact is that specific differences in perceptions of medical risks accompany the many types of bioterror agents and methods of exposure/delivery (inhalation, ingestion, dermal) (Diamond & Schreiber, 2000).

Following the 1995 sarin attack in Tokyo, the local emergency medical system (EMS) was severely taxed with 640 walk-ins in the first hour, and 4,500 psychological visits in the first few days presenting with possible nerve agent exposure. It was later determined that these 4,500 admissions were primarily psychological casualties due to *perceived* exposure. This population has been characterized with expressions such as "multiple unexplained physical symptoms" (MUPS) (Richardson & Engel, 2004) and "the worried well." We strongly view this latter term as a misnomer; these individuals are likely not "well," in the psychological sense, as they are experiencing acute anxiety, somatic symptoms, and, as a result, significant untoward reactions likely resulting in functional impairment.

Should a weapons of mass destruction event involving nuclear, biological, or chemical materials ever occur, it is likely to result in unprecedented numbers of children with severe and prolonged posttrauma psychological reactions. No adequate protocol for their management yet exists. Large numbers of children and families with perceived risk will seek medical treatment, potentially overwhelming the health and mental health care systems. This, in turn, could induce panic in those who perceive the system as being overwhelmed (DeMartino, 2001).

THE ECOLOGY OF TERRORISM AND DANGER

Although there have been previous foreign and domestic acts of terrorism, including the bombings of the World Trade Center in 1993 and the Alfred P. Murrah Federal Building in Oklahoma City in 1995, the coordinated catastrophic attacks of September 11, 2001, and subsequent anthrax bioterrorism awakened the American people to enormous tragedy and put the nation on heightened alert to extreme danger and their first profound sense of national vulnerability. In response, mental health professionals have focused their attention on developing strategies to address an anticipated unprecedented incidence of post-traumatic stress disorder. Recent advances in characterizing risk factors for PTSD, improved methods of assessment and case identification, and the emergence of empirically tested treatments such as trauma-focused psychotherapy and cognitive behavioral therapy (American Academy of Child and Adolescent Psychiatry, 1998) provided a foundation for this convergence of mental health attention (Asarnow, Glynn, & Pynoos, 1999; Saltzman, Steinberg, Layne, Aisenberg, & Pynoos, 2001; Vernberg & Vogel, 1993; Goenjian et al., 1997).

However warranted, a narrow focus on PTSD does not provide a comprehensive public mental health framework to define the spectrum of needed strategies to serve the wide population of children, adults, and families affected by terrorism. PTSD must be considered within a broader context of appraisal and response to terrorist dangers, including a set of objective ecological features and associated subjective features, along with subsequent efforts directed at adaptation, prevention, and protection. In addition to the risks of PTSD, the deaths of several thousand victims in the 9/11 terrorist attacks multiplies to hundreds of thousands of directly grieving friends and family members across the United States and beyond, bringing substantial attention to the overlapping field of traumatic bereavement. Further, the enormity of the physical damage, environmental impact, community disruption, and economic repercussions invoke an entirely separate set of public mental health and public policy concerns related to postevent secondary stresses and adversities. As lives go on, intercurrent trauma and loss constitute additional components of a comprehensive conceptual framework. For all of these categories, monitoring and intervention strategies may be geared to the individual child or family, peer group or school, community and society, and may be designed to enhance support through parents, child-care providers, health and mental health professionals, schools, religious institutions, community and societal organizations and agencies, the media, and public policy. Exhibits 5.1 and 5.2 provide an overview of this expanded framework for conceptualizing the ecology of danger and adversity as a model for public mental health planning.

Terrorism has both personal and environmental impacts. As a priority, public mental health programs need to incorporate strategies to address the immediate, intermediate, and long-term psychological and functional impact on children and families directly exposed to traumatic situations and

---

EXHIBIT 5.1. The Ecology of Terrorism and Danger

1. *Terrorist Act*
   Personal Impact: Trauma Exposure, Injury, Loss of Loved Ones, Prolonged Separation
   Environmental Impact: Destruction, Damage to Infrastructure
2. *Consequence Management (Response and Recovery Phases)*
   Evacuation, Rescue, Decontamination, Mass Inoculation, Treatment
3. *Ongoing/New Threats*
4. *False Alarms/Hoaxes*
5. *Trauma and Loss Reminders*
6. *Information Management*
   Official Risk Communication, Media Coverage
   Personal Exchanges of Information
7. *Redefinition of High-Risk Professions*
8. *Heightened Security*
   Mobilization of Law Enforcement, National Guard
9. *Political Responses*
   Military Mobilization, War
10. *Attributions of Responsibility*
11. *Subsequent Societal, Community, and Personal Stress*

EXHIBIT 5.2. The Ecology of Postterrorism Adversity

1. *Dislocations and Physical Hardships*
2. *Disrupted or Reduced Family/Community Resources*
3. *Economic Hardships and Uncertainties*
   Loss of Family Income, Parental Unemployment
4. *Altered Peer, Family, School, and Community Interpersonal Functioning*
5. *Increased Work/School-Related Stress*
6. *Conflicting Demands of Personal, School, Family, Job Responsibilities*
7. *Increased Hassles of Daily Life*
8. *Intercurrent Trauma and Loss*

those in which family members have suffered serious physical injury (Pynoos, Goenjian, & Steinberg, 1998a). These groups will be at heightened risk for severe and persistent posttraumatic stress reactions (Pynoos et al., 1993) and behavioral disturbances (Durkin, Kahn, Davidson, Saman, & Stein, 1993). Families that lost loved ones under traumatic circumstances are also at risk for complicated grief reactions with comorbid depression (Goenjian et al., 1995). Postterrorism interventions need to include interventions and treatment strategies that are accessible and specifically address these different psychological consequences. Environmental impacts, depending on their nature and severity, can interfere significantly with coping and the course of recovery from posttraumatic stress and depressive reactions, and may be associated with their own set of psychological consequences (Goenjian et al., 1995).

Within the category of consequence management, immediate activities such as evacuation and decontamination may themselves be imbued with traumatic elements, pointing to the need for public mental health planning to minimize their traumatogenic aspects. Protective strategies for certain biological, chemical, and radiological risks, for example home sheltering, may be an appropriate step for reducing exposure to toxic agents, but such a strategy needs careful planning, especially where children may be attacked at school.

From several catastrophic violent events at schools, we have learned that evacuation plans need to incorporate protocols for tracking injured children who have been transferred to medical facilities, and for effectively and efficiently reuniting children with their families in a timely manner (Nader & Pynoos, 1993). Certain events may entail specific psychological issues for parents and children, requiring, for example, strategies to minimize the mental health impact on young children as they are separated from family members during decontamination, isolation, or evacuation procedures. The need to address specific communication issues for children, families, and schools has been recently highlighted by the Secretary of Emergency Public Information and Communications Advisory Board (United States Department of Health and Human Services, 2003). Such strategies might include use of various communication technologies and the media.

In the aftermath of a terrorist attack, children and families live in an atmosphere of ongoing threats and even false alarms. Recovery from false alarms should be an integral component of public mental health planning for families, schools, and communities. Principles of psychological first aid suggest that repeated clarifications of threat information are needed to help young children with confusions, misunderstandings, and distortions, and their tendency to bring fears closer to home and concretize them in their daily lives. School-age children are likely to develop incident-specific new fears that should not be overpathologized, for example as diagnosable phobias, without proper reference to their origins in terrorist threats and without provision of psychological support for ongoing fears of recurrence.

Under normal circumstances, children and parents are constantly renegotiating the capacity of children or adolescents to appraise danger and become self-reliant, with the dangers under discussion often changing with age and cultural circumstances. After terrorist attacks, parental concerns about safety in public places can run into conflict with adolescents' appraisals of risks and their unwillingness to compromise their developmental opportunities. Open discussions about these parental and adolescent differences, with adjustments seen as renewable but temporary, are important to maintaining parental guidance and adolescent progress toward independence.

Fears of terror recurrence, false alarms, heightened security, trauma/loss reminders, and media coverage can also be engendered by myths, rumors, and misconceptions. These fears are not bound by the same trauma and loss exposure parameters that typically predict posttraumatic stress and grief reactions; they can significantly disrupt children's behaviors at home and at school. Adolescents are prone to propagate rumors and prophesies that are then shared within their peer group and, potentially, their culture. For example, within the first week after 9/11, thirteenth-century writings of Nostradamus that predicted a catastrophic collapse of two giant towers in the twenty-first century were circulated via the Internet. It was weeks before the writings were exposed as fraudulent, but by then many individuals across the United States had fueled their own internal set of catastrophic expectations. Schools, families, and communities can put procedures in place to keep informed about these, help to debunk misattributions, confusions, or distortions, and mitigate their easy transmission.

In the aftermath of 9/11, several New York City schools experienced subsequent evacuations, including one school in lower Manhattan that was evacuated repeatedly because of bomb threats to a police station nearby. School and community postdisaster plans need to include protocols for responding to false alarms by providing appropriate information and support immediately afterward to help in the recovery from renewed fears of recurrence or increased anxiety and arousal.

Ongoing danger and threats also have reverberations through the family because they raise chronic worries about significant others that affect many circumstances of separation and reunion. These worries are not necessarily related to PTSD but are pervasive after catastrophic events (Goenjian et al., 1995), especially when family members had been separated and at varying distances from the danger. For example, in New York City, worry about significant others reached far beyond Ground Zero and its immediate zip code, as reflected in relatively high rates of separation anxiety across the five boroughs (New York City Board of Education, 2002). As we found after the 1994 Northridge, California, earthquake, separation anxiety is often experienced by parents as well as by children, in families responding to heightened concerns over members' welfare. Studies conducted after catastrophic events have found separation anxiety to be as pervasive among school-age children and adolescents as among younger children (Goenjian et al., 1995).

Trauma and loss reminders are often ubiquitous in the aftermath of catastrophic events. Such reminders can include both external and internal cues, including places, people, sights, sounds, situations, and feelings of anxiety that trigger thoughts and feelings about the traumatic or loss event. It is essential that public mental health programs for children, families, and schools promote awareness and skills for managing pervasive reminders and the reactivity they can elicit; this may be among the most effective public mental health interventions. The most recent neurobiological studies of recovery from acute fear would suggest that it is through continued context discrimination that fears subside (Bouton, 2002). Children of different ages require different sets of supports and skills to manage fear reminders (Pynoos, Steinberg, & Wraith, 1995; Pynoos, Steinberg, & Piacentini, 1999).

Seeking out and responding to information is an essential component of the ecology of danger, and modulating the media information that children are exposed to presents a challenge, especially to directly affected families where information gathering about missing family members, clarification of circumstances, and estimating the type of protective action are among the ongoing needs. At the same time, it can be aversive and debilitating to view repeated traumatic images, sounds, and other reminders of the tragedy.

Television, despite its usefulness, can constitute a major source of unnecessary secondary and repeated exposure to traumatic details. Studies conducted in Oklahoma City and in New York City have indicated that the amount of event-related television viewing among school-age children was correlated with increased severity of posttraumatic stress symptoms (Pfefferbaum et al., 2001; New York City Board of Education, 2002).

Children, as a function of their developmental status, require fundamental modifications in the content and process of providing emergency risk communication, including parameters such as who should provide it, when, and how often. It is not only through television coverage that children may receive unnecessary secondary exposure to traumatic details that elicit increased fear and anxiety responses; care must also be exercised in the ways in which children are engaged in discussions about the events. Clinical or school-based interventions with children need to be respectful of differing exposures and further perceived threats to extended family, and should always conclude any anxiety-provoking drawing or activity with focus on a constructive drawing or plan to address the child's fear and his or her need to reconstruct a sense of safety.

Children and parents face redefinitions of what constitutes a high-risk profession or working circumstance, as occurred with airline and post office personnel after 9/11. It may be hard for parents who themselves are now more worried in their work situation to have the emotional wherewithal to be aware of their children's worries and to provide the additional support they need. Terrorist attacks commonly lead to mobilization of first responders who may have to be away from their families for extended periods of time while working under difficult or dangerous circumstances. Military mobilization has a far-reaching impact for a large number of families.

Omnipresent signs of heightened security constitute another aspect of the ecology of terrorism and danger. After 9/11, these included National Guard officers at airports, police or military personnel at bridges and tunnels, and augmented security procedures in office buildings, schools, and train or subway stations. These measures are aimed at increasing safety. However, the social ecology of vigilance to danger can lead to incident-specific new fears: anxious, restrictive, aggressive, or reckless behavior among children and adolescents. The consequences of these behaviors differ by age, circumstance, and community. Among children, for example, reckless behavior may lead to minor injuries; among adolescents with access to cars, drugs, and firearms, lethal outcomes may result from their reckless behavior.

Danger typically invites greater group cohesion. Although the peer group constitutes an important source of social support among adolescents, over-reliance on peers may lead to misappraisals of threat and consequence. At the same time, it can also increase a sense of loneliness and isolation among teens, especially those who are not as successful at peer relationships and group affiliation.

From a public mental health standpoint, certain high-risk groups of children and parents are especially vulnerable to less adaptive responses to living with the danger and the threat of terrorism. Children and their caretakers with prior anxiety conditions may be less able to modulate information exposure, either pursuing information to excess or avoiding necessary survival information (Asarnow et al., 1999). They also may be less capable of being comforted by reassurance and support. Public mental health prevention and intervention strategies for this group of children and adults may include school and community programs that build resilience, and targeted school and community surveillance and screening that permits effective postevent case identification and specialized intervention. Other groups at increased risk for psychological morbidity include youth with a history of trauma, loss, or insecure attachment, youth whose parents are in newly redefined high-risk professions, youth whose group identity is appraised as dangerous, families with reduced resources due to the event, and those who had been or are currently living in dangerous environments, including communities with high levels of violence or potential terrorist targets.

For those most directly affected by terrorist events, subsequent societal, community, and personal stresses add to the severity and persistence of their posttraumatic stress responses (Pynoos, Goenjian, & Steinberg, 1995). Public mental health programs for children and families need to monitor these and advocate for public decision making that addresses these types of postevent adversities.

Terrorism, war, and other catastrophic events involve recognition of heroism and often enhance a sense of patriotism. There can also be much more reliance on spiritual comfort and support. However, continued acts of terror and real threats can lead to pessimistic changes in the spiritual schema that make children as well as adults less future oriented and less constructive in their current daily behavior. Further, as parents' confidence in their ability to protect their children erodes, parental demoralization may permeate the family environment and adversely affect the developmental course of their children (Baker, 1990). Finally, spiritual support, philosophical beliefs, and moral values are extremely important in contending with threats and finding meaning in the face of danger, trauma, and loss. At the same time, catastrophic events can challenge basic beliefs, as noted previously, and spiritual schemas may become pessimistic and even apocalyptic.

A public mental health approach to terrorism and its aftermath must recognize the cascade of severe and persistent postevent adversities that affect children, families, schools, workplaces, communities, and society. These adversities strongly affect mental health outcomes. In the immediate aftermath of a terror attack, the closing of schools, which is planned in response to certain terrorist events, can immediately interfere with peer support that is so critical to functioning and recovery.

As documented around the world, postdisaster adversities have many pernicious effects, including increased child abuse, domestic violence, delinquency, and substance abuse (Goenjian, 1993; Norris et al., 2002). These effects represent an important area of need for mental health intervention. Efforts at restoring a sense of community, increasing community resources, reducing unemployment, and promoting improvements in living circumstances are all essential components of a comprehensive public mental health recovery program after terrorism.

THE ECOLOGY OF RESPONSE

A terrorist event involving weapons of mass destruction, with chemical, biological, radiological or nuclear agents, or conventional explosives, triggers a cascade of federal responses involving multiple agencies carrying out specific emergency support functions. Currently the system of "crisis management" and "consequence management" is being condensed into a single national response plan and incident management system reflecting the centralized Department of Homeland Security and the changing face of mass casualty events still threatening the American public. In the system operating today, the Substance Abuse and Mental Health Services Administration/Center for Mental Health Services (SAMHSA), with a twenty-seven-year history of responding to and providing crisis counseling, is the lead federal agency handling mental health issues arising from disasters and terrorism. Complicating the disasters, terrorist events also result in active crime scenes. The uncertainty of future acts and the difficulty assessing exposure of the population further the potential of weapons of mass destruction to become weapons of mass *disruption*. Such events pose unique mental health risks, prompting the need for a fundamental paradigm shift in the way public mental health services are conceptualized and provided to children and families in the immediate aftermath.

Weapons of mass destruction terrorism may also result in mass casualties that have the potential to overwhelm both medical and mental health systems, further contributing to widespread disruption. The mental health impact on children will vary as a function of the specific type of agent, available treatments, morbidity and mortality (pediatric treatments, quarantine, isolation, disfiguring symptoms, etc.), mode of release (covert or with a clear epicenter), and aspects of the ensuing ecology of danger as described previously (dislocations, damage to schools, losses). Among mass casualty incidents, events with radiological, nuclear, and contagious biological agents create the greatest psychological morbidity and likely pose the greatest challenge.

Children have unique risks from weapons of mass destruction due to a variety of physiological and psychological risk factors, including the propensity to become hypothermic from mass decontamination, inadequate availability of pediatric emergency care and equipment, contraindications for pediatric use of several standard treatments, and possibly greater risk from biological agents themselves (American Academy of Pediatrics, 2000). Adolescents, in particular, have been especially prone to massive deleterious psychological reactions even in the absence of disease-causing agents (Jones et al., 2000), as compared with other age groups. Consequently, large numbers of adolescents may self-present in health care facilities (brought not by the typical EMS route, but by highly distressed and anxious parents or other caretakers to emergency departments or other types of health care setting). There will be the need to implement targeted risk communication strategies addressing the particular concerns of parents and the nation as it cares for its youth. These strategies should take advantage of a variety of communication mechanisms, including electronic and written media, specialized prompts, and community outreach (Sorensen & Mileti, 1991).

The capacities of public health and EMS to cope with a sudden surge of patients is of significant concern, as evidenced by the current focus of federal efforts to strengthen the public health infrastructure, following the passage of the Public Health Security and Bioterrorism Preparedness and Response Act (2002). This act also highlights the "special needs of children and other vulnerable populations."

With the exception of limited specialized pediatric facilities, most health care facilities still lack specialized staffing and equipment for the care of children, and would face critically ill children with fewer available treatments than for adults. Children can be at risk when separated from their parents during periods of acute stress, and weapons of mass destruction events could result in large numbers of unaccompanied children developmentally compromised in their ability to accurately gauge risk and immediate protective action. This scenario poses a psychological risk not only to children and their families, but also to the beleaguered health care providers and the general population. Procedures to handle child and parental anxiety before, during, and after emergency medical care, and those with perceived but medically unexplained symptoms, need to be developed for every health care facility and emergency provider in the United States. Based on findings for long-term outcomes, developing response protocols for the mass psychological casualties along with medical preparedness activities will be vitally important for the national recovery. Even in the absence of life threat, children experience significant stress from emergency medical procedures, and these procedures pose unique challenges to health care systems not trained specifically in pediatric emergencies (Institute of Medicine, 1993). Consider the distress to children inherent in one example, the hazmat decontamination procedures, in which faceless adults enveloped in voice-distorting masks and protective gear remove children's clothing and thoroughly wash away contaminants even before direct emergency care is provided. Emergency workers themselves are at high risk when faced with dead and dying children for whom there may be little care beyond supportive, or who may expire during decontamination procedures. With chemical agents in particular, time is of the essence, and the impact of agent intoxication overlaps the psychological symptoms of acute stress and anxiety reactions, complicating emergency triage efforts. Children who may lack verbal skills and be so fearful that they cannot aid those caring for them in rapid assessment and triage will be daunting emergency medical challenges.

Following terrorist events, two main exposure groups of children at highest risk may emerge: those who are at risk for developing PTSD based on proximity to the event and those who are at risk for traumatic grief reactions/depression due to loss of a loved one. Given that observational exposure may be a route to formal symptom and clinical reactions or to less severe but highly troubling distress, very large numbers of the entire American child population may be at risk for some level of distress, at least in the short-term. Many of these children may develop disabling symptoms without meeting full syndrome criteria in the acute aftermath. However, once these disorders become established, there is increased refraction to treatment and a requirement of longer, more costly treatment (Foa, Hearst-Ikeda, & Perry, 1995). The association of disaster exposure with depression is of particular concern due to the demonstrated link between depression and the third leading cause of death in the adolescent population, suicide (Centers for Disease Control and Prevention, 1997).

Evidence also suggests that disaster exposure can exacerbate existing psychological and learning disorders in children within the special health care need population (including those with premorbid psychological disorders such as anxiety, preexisting PTSD, school phobia, or separation anxiety). Left untreated, these symptoms can endure for years and lead to long-term impairment, result in missed developmental opportunities, and portend negative outcome in later life (Schreiber, 1999).

Services such as critical incident stress management have been developed and adapted for acute phase response in disasters with adults (Mitchell, 1983), but no outcome studies with children have been reported, and questions remain about its effectiveness, unintended side effects, and applicability beyond the immediate aftermath for first responders (Everly, 2001; Rose, Bisson & Wessely, 2001; Raphael, 2000). A recent NIH federal consensus concluded that the term *debriefing* should be used to refer only to operational, not psychological, debriefing, and expressed concerns about the impact of a single-session recital of events. Following recommendations of the consensus report, the Terrorism and Disaster Branch and the School Crisis and Intervention Unit of the National Child Traumatic Stress Network are currently operationalizing psychological first aid for the acute post-impact phase. It is also clear that the threat posed by weapons of mass destruction requires planning.

No national public health model, policy, or protocol is designed especially to mitigate the unique needs of children in mass casualty events and disaster to include the extended course of recovery that these weapons of mass destruction events portend. A broad model of public mental health care for children and families affected by weapons of mass destruction terrorism that results in mass casualties is needed. Our model proposes a *disaster system of care* that would integrate medical, public, and mental health services of multiple participants, including the prehospital, emergency departments, disaster relief settings such as American Red Cross service sites, schools, evacuation medical settings, first responders, and individual families themselves (Schreiber, 2002). It aims to develop a shared, common definition of psychological risk and collaboration to identify those at the highest risk and coordinate immediate and sustained services. Without a common model in place and integrated triage systems available, mass pediatric casualties might overwhelm existing systems.

It will be critical to intervene quickly for those at highest risk, to interrupt the trajectory of impairment and chronic distress. Because weapons of mass destruction events will involve many agencies and services and disrupt traditional community capabilities, a shared real-time rapid triage system with a common model of risk and triage is a needed first step of a multitiered response that links rapid triage to psychological first aid and more definitive screening and assessment, that is in turn keyed to providing evidence-based treatments. A model protocol could be distributed to states and communities and combined with disaster mitigation training in advance of disaster.

When disaster strikes, the protocol could also be deployed instantly with a simple-to-use Web-based tool. It could then be used to rapidly estimate individual and community mental health needs, and establish linkage and coordinated delivery of care, such as with schools. This electronic triage system could also be used to locate concentrations of at-risk children with geographic interface and potentially isolate outbreaks of mass psychological reactions. This model would bring together response capacities of schools, first responders, community crisis teams, emergency medical services, public mental health agencies, and American Red Cross disaster services to address the short- and long-term pediatric psychological risk with preset, mutually supportive response protocols and reach-back architecture. To address this need, a Web-based tracking and triage system, PsySTART, has been developed by the National Center for Child Traumatic Stress Terrorism and Disaster Branch (Schreiber, 2002).

PUBLIC MENTAL HEALTH STRATEGIES

Over the past decade, advances in the assessment of traumatized children and adolescents, their postdisaster reactions, and strategies of intervention have set the stage for a modern public mental health approach to child populations affected by disaster (Pynoos, Goenjian, & Steinberg, 1995, 1998b). The first step of this approach is to determine the extent of trauma and loss exposure, and the degree of current adverse psychosocial impact among affected youth. Our experience has been that schools, not surprisingly, provide the most effective setting for conducting this type of needs assessment. This effort should be conducted in conjunction with provision of immediate psychological first aid provided directly to children in their classrooms or other group settings (Pynoos & Nader, 1988; Brymer, Steinberg, McGlenn, & Pynoos, 2002). Periodic screenings will permit identification of affected individual children or groups at risk, as well as provide a mechanism for monitoring course of recovery. Depending on the magnitude of the event, sampling strategies should include schools from a spectrum of exposures, assessments made of life-threatening experience and injury, as well as extent of physical damage to school, homes, and community (Goenjian et al., 2001).

The initial screening should include basic exposure information about where the children were and what happened to them and those around them. This should be followed by specific questions about high-risk experiences for example, direct life threat, being trapped or injured, witnessing grotesque injury, hearing screams of distress, being separated from family members or caretakers, or injury or death of family members. Additional exposure screening questions should address the child's *subjective* appraisal of the event and associated emotional responses. These exposure questions should be complemented by a brief evaluation of prior trauma, posttraumatic stress, and depressive and grief reactions.

Such an approach can be carried out with school-age children and adolescents reliability and with high validity. The screening of preschool children remains more problematic, although methods to improve assessment procedures for this age group are under development (Gurwitch, Sullivan,

& Long, 1998). Such procedures have been implemented in areas of major disasters under extremely difficult circumstances, as in Armenia, where screenings were conducted in tents and makeshift schools (Pynoos et al., 1993), as well as situations of ongoing warfare, as in the UNICEF program in Bosnia-Hercegovina (Layne et al., 2001). In addition to providing useful triage information, screening, within the framework of psychological first aid, serves a psychoeducational function. Screening of children should be supplemented by information gathered through public health resources set up to assess disaster-related damage. Subsequent screenings should include additional assessments of the nature and frequency of trauma and loss reminders and secondary stresses and adversities faced by children and their families. Sharing of aggregate information derived from data collected by governmental and other public health agencies can help to promote more judicious allocation of resources and support for intervention efforts.

A clinical intervention team can use the screening information to better define the most at-risk populations. This information can then be shared with school administrators and used to guide immediate case finding and outreach efforts. It also provides a basis for estimating the amount of classroom, family, group, and individual work that may be required, including personnel requirements, planning and timing of interventions, and selection of appropriate treatment strategies and techniques (Saltzman, Pynoos, Layne, Steinberg, & Aisenberg, 2001). Ongoing monitoring contributes to effective evaluation of interventions and permits prompt recognition of exacerbations due to unanticipated occurrences or intercurrent traumas or adversities. Finally, information regarding a child's disaster-related exposure, reminders, and losses should be included as part of his or her health record. Such a record can help prevent mislabeling of behavioral or emotional responses to future reminders or disaster-related adversities and promote timely identification of renewed symptoms or subsequent academic, interpersonal, or developmental disturbances and appropriate intervention.

A Three-Tiered Public Health Model

This public health model has three tiers. Tier 1 provides general psychosocial support interventions to the affected child and community. These may include psychoeducation, information about communication and coping strategies, the use of community and social support resources, and risk communication. Tier 2 provides more specialized support and treatment for those with moderate-to-severe persisting distress and associated impairment which includes five focal areas of treatment: traumatic experience(s); trauma and loss reminders; traumatic bereavement; current stresses and adversities; and assistance in maintaining normal developmental progression. Tier 3 provides specialized psychiatric services for youths who need immediate and/or intensive intervention (Saltzman, Layne, Steinberg, Arslanagic, & Pynoos, 2003).

RESEARCH ISSUES FOR CHILDREN AND FAMILIES IN THE AFTERMATH OF WEAPONS OF MASS DESTRUCTION

Weapons of mass destruction terrorism pose the prospect of mental health consequences unparalleled in magnitude for children and families. Recent biological terrorism modeling on a hypothetical covert release of smallpox in three states projected 300,000 sick and 100,000 dead within two months, and 3,000,000 sick and 1,000,000 dead within three months (O'Toole, Mair, & Inglesby, 2002). Addressing the enormous public mental health consequences of such events is clearly integral to our homeland security agenda and is an especially urgent need.

In the aftermath of events involving chemical and radiological agents outside the United States, psychological casualties reach proportions likely to overwhelm the health care system, with resulting multiple occurrences of unexplained somatic symptoms (Salter, 2001). The potential for this to occur in the United States, based on worldwide events and recent United States exercises testing national response capacity to biological and radiological weapons of mass destruction, such as "Topoff I," "Topoff II," "Dark Winter," and "Silent Vector," remains significant. After the anthrax letters during the fall of 2001, some health care systems reported increased demands for evaluation, based on fear of exposure.

Mothers are among those at highest risk for severe and persistent posttrauma reactions. After a radiation accident at a nuclear power plant at Chernobyl, mothers whose children were under eighteen years of age showed persistent posttraumatic distress several years later (Havenaar et al., 1997). Mothers who were evacuated also rated their children's psychological functioning as significantly worse compared to age-matched controls from an unaffected area eleven year later (Bromet et al., 2000). Somatic symptoms were particularly elevated in the evacuated group.

At five years following the SCUD attacks in Israel, long-term risk status among children correlated with being displaced and having a mother with poor psychological functioning (Laor, Wolmer, & Cohen, 2001). Following the 9/11 terrorist attack in New York City, children with parents who developed PTSD were more often seen for mental health care (Galea et al., 2002).

Numerous issues also arise related to unique issues for children and mass vaccination, an especially complicated situations for children. The American Academy of Pediatrics is currently recommending limited vaccination for smallpox only after a case occurs, with no universal inoculations before an attack. They note that children may be at higher risk for complications, and that the smallpox vaccine has not yet been tested on children, nor currently licensed for use with children. Numerous issues also arise in pediatric testing of live-virus vaccines, further complicating compliance and risking communication. Parental compliance is problematic regarding some public health vaccination strategies, such as measles/mumps/rubella (MMR). Parental misperceptions about the MMR inoculation being linked to increased rates of autism not surprisingly results in decreased compliance with recommended practices (American Academy of Pediatrics, 2000). Management of children during health crises has historically been complicated. For example, in a polio outbreak in Long Island in 1916, some ill children were forcibly removed from their parents to isolation, resulting in conflict between parents and health authorities (Glass & Schoch-Spana, 2002).

In addition, medical management is more complicated for children in terms of diagnostics, therapeutics, and shortages of pediatric specialists and specialized facilities. There are additional complexities (both technical and psychological) for medical staff in biological/chemical/radiological protective clothing in performing emergency medical procedures on small children. Currently, a limited number of pathogens have countermeasures. However, several of these, such as ribavirin, are contraindicated for use in pediatric populations. Up until recently, there had been no pediatric dosing guidelines or standardized nerve agent antidote kits (Mark I) available for pediatric use in the United States after a chemical attack.

After a bioterrorism incident, compliance with public health strategies will be essential. For example, after the anthrax letters, approximately half of the postal workers completed the directed course of antibiotics (Altman, 2002). An important untoward consequence of biological terrorism is that schools may close after the detection of a contagious agent release as a containment/quarantine strategy. This could have a significant adverse consequence for children, as it blocks access to schools, an important source of support and site of early, intermediate, and long-term interventions.

There is limited research on the impact of biological, chemical, and radiological terrorism on psychological functioning, and no articulation of the unique national challenges presented by the mental health consequences to children and families. It is clear that a large population with physical symptoms can emerge. Yet, there is little, if any, research informing predictors of individual response and health care seeking during and after catastrophic emergencies. This research base, if available, could potentially guide preparedness, response, recovery, mitigation, and risk-communication strategies.

Research issues can be conceptualized along a continuum of preparation and response/recovery and by type of agent (biological, chemical, or radiological/nuclear), as well as direct versus indirect victim. Selected examples of needed research by disaster phase include the following.

Preparedness Phase

1. Research to develop evidence-based risk communication strategies (Fischhoff, 2003; Sorenson & Mileti, 1991) and to determine the most effective methods of risk communication to safeguard the unique mental health of children and families, including format, content, and quantity of information, dissemination strategies and timing, especially risk communication directed to parents, educators, religious leaders, health, and mental health professionals, children, adolescents, and special health care-needs populations;
2. Research to identify, develop, and evaluate methods of increasing pre-disaster parental knowledge of, and postdisaster compliance with, public health directives about vaccination, prophylaxis, and school disaster plans, including protocols for evacuation, reunion, and protection of unaccompanied minors;
3. Research to evaluate and improve school disaster and safety plans, especially the extent to which they include preparedness for mass casualty events, and emergency medical care, such as plans for dispensing potassium iodide (KI) or Prussian blue/DTPA at school in the event of a nuclear occurrence during school hours, and the extent to which school safety and disaster plans are integrated into local community, county, and state disaster plans; and
4. Research to develop effective methods and instrumentation for conducting rapid needs assessment and collection of triage information within and across specific types of disaster and terrorism events.

Response/Recovery Phase:

1. Research on the effectiveness of, and compliance with, public health and mental health efforts toward recovery, including parental reactions/behaviors, factors that affect compliance, and strategies to enhance compliance;
2. Research on the specific psychological and functional impact of disaster and terrorist events on the immediate and long-term recovery of children, parents, families, and schools;

3. Research among children with prior anxiety conditions, unaccompanied children, children with parents in high-risk professions, and special-needs child and adolescent populations to document the adverse psychosocial impact and the effectiveness of specialized interventions for these vulnerable groups;
4. Research to identify, develop, and evaluate strategies to reduce contagion of escalating fears, including interventions to improve functioning in school, home, workplace, and community; and
5. Research to determine psychological responses by *specific* disaster or terrorism agent (biological, chemical, nuclear, conventional, or radiological) and by specific subtype (e.g., Marburg, anthrax, ricin, VX), as medical risks and public health strategies vary quite specifically by type of agent.

Crosscutting Public Mental Health Issues

Public health research issues that cut across the preparedness and response/recovery phases include

1. research on the relationship between implementing the Federal Incident Management Plan and psychological functioning among children and families;
2. research on the effectiveness of the Federal Emergency Management Plan for high-risk groups, including children with prior anxiety conditions or other mental health conditions, unaccompanied minors, those with special health care needs or those in the juvenile justice or social service systems;
3. research on the effectiveness of acute, intermediate, and long-term trauma/grief focused mental health treatment strategies to reduce the psychological and functional impact on children and adolescents; and
4. research to identify and enhance protective factors that moderate and mediate disaster and weapons of mass destruction terrorism psychological consequences for children and families.

ETHICAL ISSUES IN MENTAL HEALTH RESEARCH AFTER TERRORISM

There is currently an increased interest in developing formal guidelines for the ethical conduct of mental health research in the aftermath of terrorism (Fleishman & Wood, 2002). This issue was highlighted at a conference cosponsored by the New York Academy of Medicine and the National Institute of Mental Health (2003). A major topic under discussion was the need for an ethically appropriate mechanism to coordinate and integrate research within an affected area so as not to overburden a research population. For example, after the bombing of the Murrah Federal Building in Oklahoma City, a process was put in place to review and coordinate all proposed research through the University of Oklahoma Health Sciences Center. Such coordination was directed at preventing redundancy and oversampling. Unfortunately, no such system was implemented in New York City after 9/11.

Another important question raised was whether, and in what respect, potential research subjects affected by disaster constitute a vulnerable population that needs special protections in regard to solicitation for participation in research and the informed consent process. Specific protocols need to be developed to assess the extent to which decision-making capacity may be impaired, either to understand and appreciate essential information about the risks and benefits of research participation, or the ability to make reasonable judgments based on weighing such risks and benefits. Competence to participate in research needs to be assessed prior to inviting subjects to participate. Mandatory reporting of child abuse is also an important issue for disaster researchers (Steinberg, Pynoos, Goenjian, Sossanabadi, & Sherr, 1999). It may also be necessary to carefully evaluate the extent to which individuals, after mass casualties, are especially vulnerable to being harmed by the research procedures.

REFERENCES

Altman, L. (2002, October 30). Many workers ignored anthrax pill regimen. *The New York Times*, p. A18.

American Academy of Child and Adolescent Psychiatry. (1998). Practice parameters for the assessment and treatment of children with posttraumatic stress disorder. *Journal of the American Academy of Child and Adolescent Psychiatry, 37,* 4S-26S.

American Academy of Pediatrics, Committee on Environmental Health and Committee on Infectious Disease. (2000). Chemical-biological terrorism and its impact on children: A subject review. *Pediatrics, 105,* 662-670.

Asarnow, J., Glynn, S., & Pynoos, R. S. (1999). When the earth stops shaking: Earthquake sequelae among children diagnosed for pre-earthquake psychopathology. *Journal of the American Academy of Child and Adolescent Psychiatry, 38,* 1016-1023.

Baker, A. M. (1990). The psychological impact of the intifada on Palestinian children in the occupied West Bank and Gaza: An exploratory study. *American Journal of Orthopsychiatry, 60,* 496-505.

Bleich, A., Dycian, A., Koslowsky, M., Solomon, Z., & Wiener, M. (1992). Psychiatric implications of missile attacks on a civilian population: Israeli lessons from the Persian Gulf War. *Journal of the American Medical Association, 268,* 613-615.

Bouton, M. E. (2002). Context, ambiguity, and unlearning: Sources of relapse after behavioral extinction. *Biological Psychiatry, 52,* 1-11.

Bromet, E. J., Goldgaber, D., Carlson, G., Panina, N., Golovakha, E., & Gluzman, S. F. (2000). Children's well-being 11 years after the Chernobyl catastrophe. *Archives of General Psychiatry, 57,* 563-571.

Bromet, E. J., Parkinson, D. K., & Dunn, L. O. (1990). Long term mental health consequences of the accident at Three Mile Island. *International Journal of Mental Health, 19,* 48-60.

Brymer, M. J., Steinberg, A. M., McGlenn, R., & Pynoos, R. S. (2002, November). Santana High School shooting: A public mental health approach. Presented at the 18th Annual International Society for Traumatic Stress Studies Conference, Baltimore, Maryland.

Centers for Disease Control and Prevention. (1997). Rates of homicide, suicide, and firearms related deaths among children. *Morbidity and Mortality Weekly Report, 46,* 101-105.

Collins, D. L., & De Carvalho, A. B. (1993). Chronic stress from the Goiania 137Cs radiation accident. *Behavioral Medicine, 18,* 149-157.

DeMartino, R. (2001, November). Behavioral health in a new era of bioterrorism. Paper presented at SAMHSA National Summit on Addressing Terrorism, New York City, New York.

Diamond, D., & Schreiber, M. (2000, November). Psychological implications of weapons of mass destruction for adults and children. Paper presented at Disaster Medical Assistance Team California 1 Mental Health Symposia, Western Medical Center, Santa Ana, California.

Durkin, M. S., Khan, N., Davidson, L. L., Saman, S. S., & Stein, Z. A. (1993). The effects of a natural disaster on child behavior: Evidence for posttraumatic stress. *American Journal of Public Health, 83,* 1549-1553.

Everly, G. (2001). Debriefing. Appendix F. In National Institute of Health, *Mental health and mass violence: Evidence based early psychological intervention for victims and survivors of mass violence: A workshop to reach consensus* (NIH Publication No. 02-5138, pp. 34-35). Washington, DC: U.S. Government Printing Office.

Fischhoff, B. (2003). Assessing and communicating the risks of terrorism. In A. H. Teich, S. D. Nelson, & S. J. Lita (Eds.), *Science and technology in a vulnerable world, Supplement to AAAS Science and Technology Policy Yearbook* (pp. 51-64). Washington, DC: American Association for the Advancement of Science.

Fleischman, A. R., & Wood, E. B. (2002). Ethical issues in research involving victims of terror. *Journal of Urban Health: Bulletin of the New York Academy of Medicine, 7,* 315-321.

Foa, E. B., Hearst-Ikeda, D., & Perry, K. J. (1995). Evaluation of a brief cognitive-behavioral program for the prevention of chronic PTSD in recent assault victims. *Journal of Consulting and Clinical Psychology, 63,* 948-955.

Galea, S., Ahern, J., Resnick, H., Kilpatrick, D., Bucuvalas, M., Gold, J., & Vlahov, D. (2002). Psychological sequelae of the September 11 terrorist attacks in New York City. *New England Journal of Medicine, 346,* 982-987.

Glass, T. A., & Schoch-Spana, M. (2002). Bioterrorism and the people: How to vaccinate a city against panic. *Clinical Infectious Diseases, 34,* 217-223.

Goenjian, A. K. (1993). A mental health relief program in Armenia after the 1988 earthquake: Implementation and clinical observation. *British Journal of Psychiatry, 163,* 230-239.

Goenjian, A. K., Karayan, I., Pynoos, R. S., Minassian, D., Najarian, L. M., Steinberg, A. M., & Fairbanks. L. A. (1997). Outcome of psychotherapy among pre-adolescents after the 1988 earthquake in Armenia. *American Journal of Psychiatry, 154,* 536-542.

Goenjian, A. K., Molina, L., Steinberg, A. M., Fairbanks, L. A., Alvarez, M. L., & Pynoos, R. S. (2001). Posttraumatic stress and depressive reactions among adolescents in Nicaragua after Hurricane Mitch. *American Journal of Psychiatry, 200,* 788-794.

Goenjian, A. K., Pynoos, R. S., Steinberg, A. M., Najarian, L. M., Asarnow, J. R., Karayan, I., Ghurabi, M., & Fairbanks, L. A. (1995). Psychiatric co-morbidity in children after the 1988 earthquake in Armenia. *Journal of the American Academy of Child and Adolescent Psychiatry, 34,* 1174-1184.

Goenjian, A. K., Stilwell, B. M., Steinberg, A. M., Fairbanks, L. A., Galvin, M., Karayan, I., & Pynoos, R. S. (1999). Moral development and psychopathological interference with conscience functioning among adolescents after trauma. *Journal of the American Academy of Child and Adolescent Psychiatry, 38,* 376-384.

Gurwitch, R. H., Sitterle, K. S., Young, B. H., & Pfefferbaum, B. (2002). The aftermath of terrorism. In A. LaGreca, W. Silverman, E. Vernberg, & M. Roberts (Eds.), *Helping children cope with disasters and terrorism* (pp. 327-357). Washington, DC: American Psychological Association.

Gurwitch, R. H., Sullivan, M. A., & Long, P. (1998). The impact of trauma and disaster on young children. *Psychiatric Clinics of North America, 7,* 19-32.

Havenaar, J. M., Rumyantzeva, G. M., van den Brink, W., Poelijoe, N. W., van den Bout, J., van Engeland, H., & Koeter, M. W. (1997). Long-term mental health effects of the Chernobyl disaster: An epidemiologic survey in two former Soviet regions. *American Journal of Psychiatry, 154,* 1605-1607.

Holloway, H. C., Norwood, A. E., Fullerton, C. S., Engel, C. C. Jr., & Ursano R. J. (1997). The threat of biological weapons: Prophylaxis and mitigation of psychological and social consequences. *Journal of the American Medical Association, 278,* 425-427.

Institute of Medicine, Division of Health Care Services. (1993). *Emergency medical services for children: Summary.* Washington, DC: National Academy Press.

Jones, T. F., Craig, A. S., Hoy, D., Gunter, E. W., Ashley, D. F., Barr, D. B., Brock, J. W., & Schaffner, W. (2000). Mass psychogenic illness attributed to toxic exposure at a high school. *New England Journal of Medicine, 342,* 96-100.

Koplewicz, H. S., Vogel, J. M., Solanto, M. V., Morrissey, R. F., Alonso, C. M., Abikoff, H., Gallagher, R., & Novick, R. M. (2002). Child and parent response to the 1993 World Trade Center bombing. *Journal of Traumatic Stress, 15,* 77-85.

Laor, N., Wolmer, L., & Cohen, D. J. (2001). Mothers' functioning and children's symptoms 5 years after a SCUD missile attack. *American Journal of Psychiatry, 158,* 1020-1026.

Layne, C. M., Pynoos R. S., Saltzman, W. R., Arslanagic, B., Black, M., Savjak, N., Popovic, T., Durakovic, E., Music, M., Campara, N., Djapo, N., & Houston R. (2001). Trauma/grief-focused group psychotherapy: School-based postwar intervention with traumatized Bosnian adolescents. *Group Dynamics: Theory, Research and Practice, 5,* 277-290.

Mitchell, J. T. (1983). When disaster strikes: The critical incident stress debriefing process. *Journal Medical Emergency Services, 8,* 26-39.

Murthy, R. S. (2003). Bhopal gas lead disaster: Impact on mental health. In J. M. Havenaar (Ed.), *Toxic turmoil: Psychological and societal consequences of ecological disasters* (pp. 129-145). New York: Kluwer Academic Press.

Nader, K., & Pynoos, R. S. (1993). School disaster: Planning and initial interventions in the handbook for post-disaster interventions. *Journal of Social Behavior and Personality, 8,* 299-320.

New York Academy of Medicine & the National Institute of Mental Health (2003). *Ethical issues pertaining to research in the aftermath of disaster.* Washington, DC: Authors.

New York City Board of Education. (2002). *Effects of the World Trade Center attack on NYC public school students: Initial report.* Applied Research and Consulting, LLC, Columbia University Mailman School of Public Health, New York State Psychiatric Institute.

Norris, F. H., Friedman, M. J., Watson, P. J., Byrne, C. M., Diaz, E., & Kaniasty, K. (2002). 60,000 disaster victims speak: Part I. An empirical review of the empirical literature 1981-2001. *Psychiatry, 65,* 207-239.

North, C. S., Nixon, S. J., Shariat, S., Mallonee, S., McMillen, J. C., Spitznagel, E. L., & Smith, E. M. (1999). Psychiatric disorders among survivors of the Oklahoma City bombing. *Journal of the American Medical Association, 282,* 755-762.

North, C. S., Pfefferbaum, B., & Tucker, P. (2002). Ethical and methodological issues in academic mental health research in populations affected by disasters: Oklahoma City experience relevant to 9/11. *CNS Spectrums, 7,* 580-584.

O'Toole, T., Mair, M., & Inglesby, T. V. (2002). Shining light on "Dark Winter." *Clinical Infectious Diseases, 34,* 972-983.

Pastel, R. (2000). Collective behaviors: Mass panic and outbreaks of multiple unexplained symptoms. *U.S. Military Medicine, 166*(Suppl. 2), 80-82.

Pfefferbaum, B., Nixon, S. J., Tivis, R., Doughty, D., Pynoos, R., Gurwitch, R., & Foy, D. (2001). Television exposure in children after a terrorist incident. *Psychiatry, 64,* 202-211.

Pfefferbaum, B., Nixon, S., Tucker, P., Tivis, R., Moore, V., Gurwitch, R. H., Pynoos, R. S., & Geis, H. (1999). Posttraumatic stress responses in bereaved children after the Oklahoma City bombing. *Journal of the American Academy of Child and Adolescent Psychiatry, 38,* 1372-1379.

Public Health Security and Bioterrorism Preparedness and Response Act of 2002, Pub. L. No. PL 107-188.

Pynoos, R. S. (1992). Violence, personality and politics. In A. Kales, C. M. Pierce, & M. Greenblatt (Eds.), *The mosaic of contemporary psychiatry in perspective* (pp. 53-65). New York: Springer Verlag.

Pynoos, R. S., Goenjian, A. K., & Steinberg, A. M. (1995). Strategies of disaster intervention for children and adolescents. In S. E. Hobfoll, & M. de Vries (Eds.), *Stress and communities* (pp. 445-471). Dordrecht, the Netherlands: M. Kluwer Academic Publishers Publications.

Pynoos, R. S., Goenjian, A. K., & Steinberg A. M. (1998a). Children and disasters: A developmental approach to posttraumatic stress disorder in children and adolescents. *Psychiatry and Clinical Neurosciences, 52*(Suppl.), S129-S138.

Pynoos, R. S., Goenjian, A. K., & Steinberg, A. M. (1998b). A public mental health approach to the post-disaster treatment of children and adolescents. *Psychiatric Clinics of North America, 7,* 195-210.

Pynoos, R. S., Goenjian, A., Tashjian, M., Karakashian, M., Manjikian, R., Manoukian, G., Steinberg, A. M., & Fairbanks, L. (1993). Posttraumatic stress reactions in children after the 1988 Armenian earthquake. *British Journal of Psychiatry, 163,* 239-247.

Pynoos, R. S., & Nader, K. (1988). Psychological first aid and treatment approach for children exposed to community violence: Research implications. *Journal of Traumatic Stress, 1,* 445-473.

Pynoos, R. S., Steinberg, A. M., & Piacentini, J. C. (1999). Developmental psychopathology of childhood traumatic stress and implications for associated anxiety disorders. *Biological Psychiatry, 46,* 1542-1554.

Pynoos, R. S., Steinberg, A. M., & Wraith, R. (1995). A developmental model of childhood traumatic stress. In D. Cicchetti and D. J. Cohen (Eds.), *Manual of developmental psychopathology* (pp. 72-93). New York: John Wiley & Sons.

Raphael, B. (2000). Conclusions: Debriefing-science, belief and wisdom. In B. Raphael & J. P. Wilson (Eds.), *Psychological debriefing theory, practice and evidence* (pp. 351-359). New York: Cambridge University Press.

Richardson R. D., & Engel C. C. Jr. (2004). Evaluation and management of medically unexplained physical symptoms. *Neurologist, 10,* 18-30.

Rose, S., Bisson, J., & Wessely, S. (2001). A systematic review of brief psychological interventions ("debriefing") for the treatment of immediate trauma related symptoms and the prevention of post-traumatic stress disorder. *The Cochrane Library* (3). Oxford: Update Software.

Rubonis, A. V., & Bickman, L. (1991). Psychological impairment in the wake of disaster: The disaster-psychopathology relationship. *Psychological Bulletin, 109,* 384-399.

Salter, C. (2001). Psychological effects of nuclear and radiological warfare. *Military Medicine, 166,* 17-18.

Saltzman, W. R., Layne, C. M., Steinberg, A. S., Arslanagic, B., & Pynoos, R. S. (2003). Developing a culturally and ecologically sound intervention program for youth exposed to war and terrorism. *Child and Adolescent Psychiatric Clinics of North America, 12,* 319-342.

Saltzman, W. R., Pynoos, R. S., Layne, C. M., Steinberg, A. M., & Aisenberg, E. (2001). Trauma- and grief-focused intervention for adolescents exposed to community violence: Results of a school-based screening and group treatment protocol. *Group Dynamics: Theory, Research and Practice, 5,* 291-303.

Saltzman, W. R., Steinberg, A. M., Layne, C. M., Aisenberg, E., & Pynoos, R. S. (2001). A developmental approach to school-based treatment of adolescents exposed to trauma and traumatic loss. *Journal of Child and Adolescent Group Therapy, 11,* 43-56.

Schlenger, W. E., Caddell, J. M., Ebert, L., Jordan, B. K., Rourke, K. M., Wilson, D., Thalji, L., Dennis, J. M., Fairbank, J. A., & Kulka, R. A. (2002). Psychological reactions to terrorist attacks: Findings from the national study of Americans' reactions to September 11. *Journal of the American Medical Association, 288,* 581-588.

Schreiber, M. D. (1999). Children's reactions to disasters. In L. M. Horowitz, M. D. Schreiber, I. Hare, V. I. Walker, & A. L. Talley (Eds.), *Psychological factors in emergency services for children* (pp. 41-48). Washington DC: American Psychological Association.

Schreiber, M. D. (2002). Children's emergencies in weapons of mass destruction and terrorism: Disaster system of care, rapid triage, and consequence management. Presented at the 16th Annual California Conference on Childhood Injury Control, Sacramento, California, September 23-25.

Schuster, M. A., Stein, B. D., Jaycox, L. H., Collins, R. L., Marshall, G. N., Elliott, M. N., Zhou, A. J., Kanouse, D. E., Morrison, J. L., & Berry, S. H. (2001). A national survey of stress reactions after the September 11, 2001, terrorist attacks. *New England Journal of Medicine, 345,* 1507-1512.

Sorensen, J., & Mileti, D. (1991). Risk communication in emergencies. In R. E. Kasperson & P. J. M. Stallen (Eds.), *Communicating risks to the public: International perspectives* (pp. 367-392). Dordrecht, the Netherlands: Kluwer Academic Publishers.

Steinberg, A. M., Pynoos, R. S., Goenjian A. K., Sossanabadi, H., & Sherr, L. (1999). Are researchers bound by child abuse reporting laws? *Child Abuse and Neglect, 23,* 771-777.

Stuber, J., Fairbrother, G., Galea, S., Pfefferbaum, B., Wilson-Genderson, M., & Vlahov, D. (2002). Determinants of counseling for children in Manhattan after the September 11 attacks. *Psychiatric Services, 3,* 815-822.

United States Department of Health and Human Services (2003). *Emergency Public Information and Communications Advisory Board final report.* Washington, DC: Author.

Vanchieri, C. (1997). Chernobyl liquidators show increased risk of suicide, not cancer. *Journal of the National Cancer Institute, 89,* 1750-1752.

Vernberg, E. M., & Vogel, J. (1993). Interventions with children following disasters. *Journal of Clinical Child Psychology, 22,* 485-498.
von Kleist, H. (1978). *The marquise of O* (D. Luke & N. Reeves, Trans.). London: Penguin Books.

Chapter 6

Traumatic Reactions to Terrorism: The Individual and Collective Experience

K. Chase Stovall-McClough
Marylene Cloitre

INTRODUCTION

Traumatic events produce extreme fear and horror and significantly disrupt feelings of security. Trauma has profound and enduring effects on both victims and their loved ones. A great deal is understood about the short- and long-term effects of trauma, especially with regard to experiences such as combat, rape, and accidents. The impact of terrorism as a traumatic experience, however, is significantly less well understood. In this chapter, we attempt to provide a comprehensive discussion and a review of the literature pertinent to the terrorism experience. We also provide a discussion of terrorism from an attachment theory perspective. Using this framework, we examine terrorism as a unique trauma in its ability to disrupt feelings of security on both an individual and a community level. This chapter also provides a comprehensive review of research on the psychological impact of terrorism, emphasizing both individual and community factors that are important in predicting risk and resilience. Included in this discussion will be the latest research emerging from the September 11 attacks. Finally, we will discuss the role of attachments and community support in repairing a sense of security and promoting recovery from terrorism.

TRAUMA AND ATTACHMENT THEORY

People need to feel safe to function in the world. To feel safe, we rely on our system of attachments (Bowlby, 1969/1982, 1973, 1980). Feeling safe is, in essence, an implicit confidence in one's ability to effectively handle dangerous situations that arise and, if not, that others can be relied upon for aid. When we are young our sense of security is first learned within the context of our primary relationships. When children are distressed they reach out for the comforting contact and protection of an adult. As adults, our need for safety and protection does not change, but we become vastly more sophisticated in our ability to obtain a sense of security. For instance, our feelings of safety become less organized around the physical availability of others and more organized around our mental representation of their availability; our overall sense of safety is based on our *expectations* of the availability of others in times of crisis. In addition, adults come to rely on a wider social network for emotional comfort and feelings of safety. Healthy individuals develop *systems* of attachment relationships that can be called upon to varying degrees when one's sense of safety is disrupted. In this way, our attachments allow us to feel safe and protected.

Under normal circumstances, our system of attachments works so well that we spend little energy making sure we are safe. In psychologically healthy adults, little time is spent worrying about our safety or assuring ourselves of the availability of others, while the majority of time is spent on other activities (e.g., education, work, and recreation). For example, most of us do not cry in protest when our spouse leaves for work in the morning, because we know he or she will remain psychologically available to us if we need them. We can go about our day with a confident sense of security in the availability of our attachments. However, when something occurs to threaten our sense of safety and overwhelms our own coping mechanisms, the equation remains the same as it did when we were children—our attachment system is activated and we seek the contact and comfort of others.

Traumatic events, in particular, lead to increased reliance on others. From an attachment theory perspective, trauma overwhelms our internal coping capacities leading to increased proximity seeking of those we can rely on, usually those we perceive to be stronger and wiser. Ideally the support of such people will help to reassure us, not only through words and touch but also through provision of physical resources needed to complete important tasks. Thus, our relationships help to reestablish our sense of security in the midst of trauma, allowing an eventual return to more typical levels of functioning.

THE NATURE AND IMPACT OF TRAUMATIC EVENTS

A traumatic event is understood as an experience that involves the direct or indirect threat to one's life or physical integrity and that causes extreme terror or horror. A traumatic event is often, but not necessarily, sudden and unpredictable. It is intense, may be dangerous, and is, by definition, overwhelming. A trauma may be an acute one-time event, as in the case of a rape or assault, or a sustained occurrence, as in the case of combat or terrorism. A range of experiences can potentially be defined as traumatic, including witnessing others involved in a trauma. Finally, traumas may be experienced alone (e.g., rape, assault) or with others (e.g., combat, natural disasters, terrorist attacks). The traumas most thoroughly studied include combat experience, violent crimes such as rape, childhood abuse, domestic violence, motor vehicle accidents, natural disasters, and, most recently, terrorism.

With the range of potential traumas being so broad, it should come as no surprise that experiencing a traumatic event is not uncommon. Epidemiological studies indicate that in the United States there is a 39 percent to 78 percent lifetime risk of being exposed to a traumatic event as defined by the DSM-IV (Breslau, Davis, Andreski, & Peterson, 1991; Breslau & Kessler, 2001; Norris, 1992). As many as 21 percent of the population encounter a trauma each year (Norris, 1992). For women, these rates are higher; 82 percent of women are expected to be exposed to trauma during their lifetime (Breslau & Kessler, 2001). This high percentage likely reflects the fact that women are the predominant victims of childhood sexual abuse, domestic violence, and rape. Moreover, married women still outlive their husbands by significant numbers, and losing one's partner even by natural causes is still a traumatic event in many cases.

From the careful study of individuals exposed to a wide variety of traumas, we have learned a great deal about the universality of the human response to such events. At least in the immediate response, there is usually disorganization on multiple levels. For instance, on a cognitive and behavioral level, a traumatic event may produce intense feelings of helplessness and/or terror which can temporarily overwhelm one's coping capacities. In addition, the most basic systems of belief about oneself, one's environment, and other people may suddenly be shattered, especially with regard to danger and vulnerability. During and immediately following a trauma, an individual must confront the reality of a new, more dangerous world and a new sense of one's own susceptibility to that danger. Sudden disorganization of emotions, behaviors, and belief systems is central to the experience of trauma. Also critical is the disruption of basic feelings of safety and protection which our attachment system normally provides. Continued trouble in these areas is at the heart of the symptom profile of post-traumatic stress disorder (PTSD).

ASSESSMENT OF POST-TRAUMATIC STRESS DISORDER

PTSD first became a formal diagnosis in the DSM-III in 1980. At that time, a diagnosis of PTSD was given when an individual had (1) experienced a traumatic event, defined as an event "outside the range of usual human experience" and (2) met criteria for two primary symptom clusters identified as being core to PTSD; intrusive/re-experiencing symptoms and avoidant/numbing symptoms (Horowitz, 1976). The criteria for PTSD have changed in several ways since 1980, with two important changes occurring in the current definition of traumatic events (Criterion A). First, the DSM-IV (APA, 1994) incorporated a more detailed description of trauma events which now requires that a person "experienced, witnessed, or was confronted with an event or events that involved actual or threatened death or serious injury, or a threat to the physical integrity of self or others." The new definition was a response to the controversy regarding the issue of whether traumatic events had to be uncommon to be traumatic (Weathers & Keane, 1999). In fact, research suggests that traumatic events such as rape, accidents, disasters, and assault are quite common (Kessler, Sonnega, Bromet, Hughes, & Nelson, 1995). Moreover, the revised DSM-IV definition requires that "the person's response involved intense fear, helplessness, or horror," thus formally incorporating the *subjective* experience as part of the definition of a trauma.*

The Criterion A Problem

A focal point of controversy among trauma experts is the definition of a trauma (Criterion A). The DSM-IV version is widely regarded as a more conservative characterization of a traumatic event than the DSM-III version, resulting in fewer experiences qualifying as traumatic (Weathers & Keane, 1999). This has made it difficult to draw general conclusions about the prevalence of traumatic events and rates of subsequent PTSD in studies conducted between the 1980s and the present. Some controversy also exists regarding the use of terms such as "confronted with" in the definition of a traumatic event, as it leaves room for interpretive ambiguity. Similarly, although many events are clear-cut with regard to meeting Criterion A, other events, such as chronic and/or life-threatening illnesses, are less clearly traumatic although extremely stressful. In addition to objectively determining whether an event is traumatic, there is also the issue of the accuracy of subjective reports. Problems with memory and report bias may complicate the task of accurately documenting the nature of an event. (For further discussion of the Criterion A question, see Corcoran, Green, Goodman, & Krinsley, 2000; Davidson & Foa, 1993; Resnick, Kilpatrick, & Lipovsky, 1991; Weathers & Keane, 1999).

---

*Reprinted with permission from the *Diagnostic and Statistical Manual of Mental Disorders,* Fourth Edition (Copyright 1996). American Psychiatric Association.

Assessment of PTSD

The DSM-IV includes three primary symptom clusters in addition to Criterion A that must be met for a diagnosis. An individual must have at least one Criterion B reexperiencing symptom (e.g., nightmares, flashbacks, heightened distress in response to trauma-related cues, intrusive thoughts/images); three or more Criterion C symptoms of numbing or avoidance (e.g., avoidance of reminders, emotional numbing, alienation from the community), and evidence of two or more Criterion D symptoms of general sympathetic hyperarousal (e.g., exaggerated startle response, hypervigilence, irritability, disturbed sleep). According to the DSM-IV, symptoms of PTSD may appear within the first three months following a traumatic event but may also appear up to several months, or even years, later (called "delayed onset"). In the latter case, significant life events are often the trigger for the onset of symptoms. Under these circumstances, it is critical that PTSD symptoms be linked directly and specifically to the original traumatic event for an accurate differential diagnosis to be made.

Currently more than thirty measures are available for assessing PTSD symptoms (Weathers & Keane, 1999). These measures include self-report questionnaires and structured clinical interviews and vary with respect to method of administration, the construction of the rating scales, and the time frame being assessed (past two weeks, past month). Some of the more well-validated clinical interviews include the Structured Clinical Interview for DSM-IV-TR (SCID) (Spitzer, Gibbon, & Williams, 1996); the Diagnostic Interview Schedule (DIS); and the Clinician-Administered PTSD Scale (CAPS) (Blake et al., 1990, 1995). Among the well-validated self-report measures are the PTSD Symptom Scale–Self-Report (PSS-SR) (Foa, Riggs, Dancu, & Rothbaum, 1993) and the PTSD Checklist (PCL) (Weathers, Litz, Herman, Huska, & Keane, 1993). Other commonly used self-report measures with high construct validity and reliability but that do *not* correspond to current DSM-IV diagnostic criteria include the Impact of Event Scale (IES) (Horowitz, Wilner, & Alvarez, 1979; Weiss & Marmar, 1997) and the Mississippi Scale for Combat-Related PTSD (Keane, Caddell, & Taylor, 1988).

Much has been written with regard to the procedures, methodology, and controversies surrounding the assessment of PTSD (for reviews see Davidson & Foa, 1991, 1993; Davidson, Smith, & Kudler, 1989; Keane, Wolfe, & Taylor, 1987; Litz & Weathers, 1994; Solomon & Canino, 1990; Weathers & Keane, 1999; Wilson & Keane, 1997). Some of the concerns highlighted in these discussions include the proliferation of PTSD measures in the field, linking symptoms explicitly to particular traumatic events, definitions of traumatic events, report bias resulting from secondary gain of having a PTSD diagnosis, and the use of self-reports in the absence of clinical interview data. As a result, the use of multiple measures for assessing PTSD has become the "gold standard" of PTSD assessment. Integrating information from several sources (self-report and interview data) reduces the likelihood of diagnostic error and report bias.

Terrorism As a Unique Trauma

Although we know quite a bit about the psychological sequelae of various traumas, we know less about the impact of large-scale and ongoing traumatic experiences such as terrorism. We expect that the psychological impact of terrorism might be similar to that of other traumas. An act of terrorism is likely to induce extreme fear, helplessness, and anger, and be experienced as arbitrary, violent, and unpredictable. However, as a trauma, terrorism has unique characteristics that we will review in depth in the following pages. These differences are important when considering the symptom picture associated with terrorism. The ongoing threat of future attacks creates a unique challenge both to researchers and clinicians in effectively assessing, treating, and studying the aftermath of terrorism. It is not clear if our current models of PTSD and its treatment will apply to terrorism survivors. Are there psychological sequelae unique to the experience of terrorism that might complicate the diagnosis and course of PTSD? For example, we do not know whether PTSD or other mood, anxiety, or substance use disorders dominate the symptom picture as is the case with most other traumas. In order to formulate the potential psychological impact of terrorism, we should first consider how terrorism differs from traditionally defined traumas.

Ostensibly, terrorism is intended to manipulate nations or political groups to change in accord with the terrorists' sociopolitical or religious objectives through the use of intimidation and fear (Ayalon, 1993; Fredrick, 1994). Acts of terror are often supported by governments or radical groups, although they may also include acts of individual terrorism. Importantly, terrorist acts carried out by groups are often more calculated and well planned than terrorist acts by individuals; and terrorist group members are less likely to suffer from an identifiable mental illness (Fredrick, 1994).

The trauma of terrorism differs in several critical ways from acute traumas such as rape, assault, disasters, and accidents. First, the impact of terrorism is chronic. Unlike single-incident traumas, the threat of terrorism is not a contained experience with a clear-cut beginning and end. This is, in part, because terrorist attacks are sustained efforts of random violence. The experience of continued risk and the associated heightened sense of ongoing fear is the intended effect of the terrorists. In addition, members of terrorist organizations are often extremely difficult to identify, with ill-defined boundaries and identities. These enigmatic characteristics can induce a generalized vigilance to the point of paranoia across a community. An attack could occur at anytime and anywhere, and be perpetrated by almost anyone.

Terrorism also involves a different type of perpetrator compared to other traumas in which there may be either no human perpetrator (e.g., natural disaster) or a single perpetrator (e.g., assault). Unlike accidents, assaults, or natural disasters, terrorism usually entails mass violence perpetrated *intentionally* by an *organized* human effort. Most terrorist attacks are well planned by several individuals and carefully designed to maximize fear and suffering. This characteristic of terrorism adds a layer of malevolence that dramatically increases the sense of violation in victims.

Terrorism As a Community Trauma

In addition to the characteristics described previously, perhaps the feature most unique to terrorism is the *collective* experience of the trauma within a community. Terrorism is a shared experience because attacks are most often directed toward *groups* of people rather than individuals. It is rare that only one person is injured or killed during a terrorist attack; rather, there are often large numbers of casualties, up to several thousands as was the case with the September 11 attack. In addition, the particular group of people attacked is often chosen based upon their symbolic membership in a larger racial, ethnic, or sociopolitical community. Thus, there is both a direct attack on a set of individuals and an indirect attack on an entire identified sociopolitical community.

The disruption to the collective sense of safety that results from terrorism occurs on at least four levels. First, terrorists primarily target places where large numbers of ordinary people congregate to conduct the ordinary activities of daily life (i.e., business, commuting, socializing, dining, etc.) that keep the society in a functional equilibrium. Attacking a random set of innocent civilians and disrupting a society's equilibrium is precisely the intent of terrorism. An attack on ordinary people, doing ordinary things, greatly intensifies the sense of danger across the community because it suggests that almost anyone could be next (Ayalon, 1993). The systematic targeting of civilians (versus combatants) by terrorists is illustrated by the fact that from 1996 to 2001 an average of fifteen government and eight military facilities worldwide were attacked by terrorists each year. In comparison, an average of 317 civilian businesses were attacked each year during the same time period (U.S. Office of the Coordinator for Counterterrorism, 2001). Such indiscriminate destruction of people and property, combined with disruption of daily routines, is designed to functionally paralyze a community and intimidate governments into compliance.

Second, terrorism disrupts the collective sense of safety as a result of the destruction and loss of physical resources. Because terrorists often target civilian businesses, attacks result in the loss of employment and financial instability. As a result of the terror acts of 9/11, thousands of individuals lost months of revenue and had their businesses destroyed. The economy of New York City, and even the country, was negatively affected. Many also temporarily lost access to their homes. The economic instability that results from these losses significantly disrupts the daily routines of communities, as people struggle to rebuild their lives and reestablish reliable sources of income.

Third, terrorism often results in large numbers of casualties at once, to be grieved by an even larger number of survivors. The potentially large number of sudden losses following a terrorist attack is particularly detrimental from an attachment perspective. This is because the loss of loved ones compromises the very network of support that is so critical to reestablishing a sense of security following an attack. New York City was particularly hard hit on the morning of September 11. To put this in perspective, between the years 1996 and 2001 over 2,000 persons died at the hands of terrorists in the Middle East (U.S. Office of the Coordinator for Counterterrorism, 2001). By contrast, over 3,000 persons were killed in just one day on September 11, leaving thousands of widows, parentless children, and childless parents. As a result, survivors and those connected to victims must negotiate not only their own fears and trauma, but also the loss of the very persons from whom they would otherwise seek comfort and support. The larger the number of casualties that occur at one time, the more the devastation to a community, its support network, and its capacity to respond effectively in recovery efforts.

Finally, the goals of terrorism are often global and political. Thus, to whom one looks for protection includes not only other people, but also larger governmental institutions organized to protect the community (e.g., police, firefighters, military, relief organizations). Because so many people can be affected at once by a terrorist attack, individuals must depend on a complex network of local, state, and national government agencies for financial, medical, and mental health relief. Although a great number of people are provided with significant aid, negotiating the complicated bureaucracy that is necessary to get help can sometimes further frustrate those in need.

Communities targeted by terrorist groups must also rely on governmental protection from further attacks. As such, disruption of safety following a terrorist attack can also come from the perceived failure of the government to protect its citizens. That one's government failed to protect the community from such an attack in the first place may bring about a new sense of vulnerability in the form of a loss of faith in our national and global security. That such destruction and suffering planned at the hands of other individuals could not be stopped by government defense systems may bring about a unique sense of disillusionment. Shattered beliefs in the limitations of others and the safety of one's global environment are a unique part of the terrorism experience and can significantly affect one's ability to function normally. In addition, because much government activity surrounding national security is often protected from the public, developing a sense of safety and protection often requires some leap of faith. Under these circumstances, we often look to our leader, much as we would a parent figure, to contain our fears and worries. Like a good parent, when such a leader is able to provide a sense of strength, confidence, and stability under conditions of threat and fear, individual citizens can find much solace.

From an attachment theory perspective these multiple levels of damage to a community are important because, in addition to our close relationships, we also derive a sense of safety through our membership in a larger community (Chemtob, 2002). Membership in a social network is evolutionarily adaptive because groups provide protection from danger and play an important reparative role following trauma. At the most basic level, communities protect individuals from danger by maximizing the likelihood that basic life-sustaining needs will be met. For instance, members of social groups provide physical protection to one another and ensure the provision of food and shelter. Communities can also mediate the psychological effect of trauma and disaster. Following such an event, communities act as powerful sources of comfort and support. For example, participating in healing community rituals following terrorism (e.g., religious gatherings, funerals, political rallies, etc.) helps to repair the psychological impact of loss and injury and can give one a sense of hope. In addition, one's membership in a larger community can influence the long-term psychological impact of trauma. A community that effectively responds to a trauma provides a sense of confidence in the ability to successfully handle future traumas and increases the confidence in the potential for future recovery and growth in the face of ongoing threat.

Circles of Vulnerability in a Community

The extent to which a particular individual experiences anxiety and fear following a terrorist attack can be predicted to a large degree by how much that individual identifies with the victims and the wider targeted socio-political group. In other words, the belief that "It could have been me" or "I could be next" will likely translate into feelings of vulnerability. As a result, terrorism is likely to have a stronger impact on a larger number of individuals compared to other types of traumas. This effect is maximized by targeting individuals who hold symbolic similarity to a wider community.

A useful model for understanding the uniquely far-reaching impact of terrorism is borrowed from Wright, Ursano, Bartone, & Ingraham (1990) and Tucker, Pfefferbaum, Nixon, & Dickson (2000). In this model, the psychological impact of terrorism can be seen as organized within concentric "circles of vulnerability" (Ayalon, 1993; see Figure 6.1). In the middle are those survivors most directly affected, followed by first-degree relatives

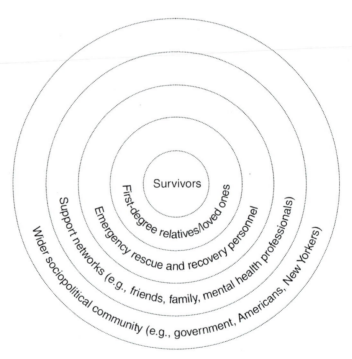

FIGURE 6.1. Circles of Vulnerability Following Terrorism

and loved ones of those killed or injured. Next are the service providers directly involved in the rescue and recovery efforts, followed by friends, family, mental health workers, and other individuals offering support, followed by the larger community psychologically linked to the victims. In the case of terrorism, this larger community is the sociopolitical group being targeted (e.g., Americans, Jews, African Americans, etc.). As one moves from the center to the outer circles, there is less emotional and physical proximity to the original attack.

The outermost circle, containing the wider national community, is unique to the experience of terrorism. The suffering and death of the individuals in the middle circle are meant to symbolize the threat to the wider community. The victims' suffering is a means of communicating a form of omnipotent destructive power. This message is not only communicated by targeting symbolic groups of individuals, but can also be communicated through the simultaneous destruction or desecration of sacred cultural symbols (Ayalon, 1993). The attack on the Alfred P. Murrah Federal Building was meant as a communication to the U.S. government and the IRS. The targeting of the World Trade Center and the Pentagon, as well as the use of commercial airliners carrying the names "United" and "American" held symbolic meaning to all Americans. Unique fears arise in such contexts where the psychological damage can include injury to the national group identity.

TERRORISM AND MENTAL HEALTH: RESEARCH FINDINGS

Prior to the Oklahoma City bombing in 1995, research examining the psychological impact of terrorist attacks was given relatively little attention or funding in the United States. Only a handful of studies were conducted prior to the Oklahoma City bombing and the events of 9/11. In addition, most of these studies were conducted outside the United States, in countries where terrorist activities have been significantly more common.

In an epidemiologic survey of 254 French civilians who survived terrorist attacks between 1982 and 1987, Abenhaim, Dab, & Salmi (1992) found that a large percentage of survivors met DSM-III criteria for PTSD and major depression. Survivors who were directly exposed to terrorist attacks, and found through police and hospital records, were asked to fill out a self-report questionnaire several years after the attacks. Based on DSM-III criteria, 18 percent of the overall sample had PTSD. Those who had suffered a serious physical injury in the attack had a 30 percent incidence of PTSD and a 22 percent incidence of major depressive disorder.

Curran et al. (1990) examined survivors of the 1987 Enniskillen IRA bombing, which claimed eleven lives and injured sixty. Six months following the bombing, 50 percent of the survivors had PTSD; however, there was no association between physical injury and overall symptoms found.

Arieh Shalev (1992) examined rates of PTSD in a sample of injured civilians who survived a 1989 terrorist attack in Israel when a local bus traveling to Jerusalem was attacked, killing fourteen and injuring twenty-three. Fifteen of the survivors who had been hospitalized following the attack were interviewed a few days later, and twelve of those were interviewed again eight to fourteen months later. Subjects answered a self-report questionnaire (translated into Hebrew) regarding the impact of the trauma (IES) and a DSM-III-R symptom checklist for PTSD. Results indicated that 33 percent had PTSD on follow-up, with most survivors reporting significant hyperarousal (92 percent) and reexperiencing (83 percent) symptoms.

Terrorism on U.S. Soil

Research resulting from the recent terrorist attacks on U.S. soil also reveals varying rates of PTSD in those directly exposed to the trauma. Approximately six months after the Oklahoma City bombing, North et al. (1999) surveyed 182 adult survivors who were drawn randomly from a confidential registry of those directly exposed based on their proximity to the bombing. Using the Diagnostic Interview Schedule for DSM-III-R (DIS), the authors found that almost all of those surveyed reported having had some trauma-related symptoms on the day of the attack (96 percent). At the time of the interview 45 percent of those surveyed met criteria for at least one DSM diagnosis. The most common diagnosis was PTSD (34 percent) followed by major depression (23 percent). The researchers also found that intrusive reexperiencing and hypervigilence were extremely common among subjects and, therefore, did not uniquely predict PTSD or other forms of psychopathology. On the other hand, avoidance and numbing symptoms were relatively uncommon. They found that when avoidance and numbing symptoms were present, a diagnosis of PTSD was more likely. In a follow-up survey on the same sample one year later, North (2001) reported little or no reduction in the rates of PTSD and no new cases of PTSD, indicating that almost all the cases originally detected were chronic. These findings are highly significant and may have important implications for the prognosis of recovery from terrorism.

The data emerging from these attacks suggest that large numbers of New Yorkers experienced PTSD. Using random-digit dialing, 1,008 adults living in Manhattan within six miles of the World Trade Center Towers were surveyed between October 14 and November 15, 2001 (Galea, Ahern, et al., 2002). Using a modified version of the DIS for PTSD, subjects were asked about symptoms of PTSD and major depression. Results indicated that 7.5 percent of those surveyed met DSM-IV criteria for PTSD and 9.7 percent met DSM-IV criteria for depression. These rates translated into significant numbers of individuals with PTSD (67,000) and with depression (87,000) five to eight weeks following the attacks. The rates for PTSD rose dramatically (20 percent) the closer the subjects lived to Ground Zero (Galea, Ahern, et al., 2002) in keeping with concept of "circles of vulnerability" discussed earlier.

Symptoms of PTSD were not exclusive to those living in Manhattan. In a national survey conducted within five days of the attack, 560 adults were randomly chosen from across the United States and interviewed about their reactions. Schuster et al. (2001) found 90 percent of those interviewed reported at least one stress symptom had bothered them "a little bit" and 44 percent reported at least one stress symptom had bothered them "extremely." Symptoms assessed included disturbing memories, lack of concentration, upset by reminders, irritability, trouble with sleep, and nightmares. Similarly, in an Internet-based survey of individuals living outside Manhattan, 17 percent reported some symptoms of PTSD two months following the 9/11 attacks (Silver, Holman, McIntosh, Poulin, & Gil-Rivas, 2002). These numbers decreased to 5.8 percent at six months.

Media Exposure and Psychological Symptoms

Modern communication technology allows for the transmission of important information almost instantaneously around the world. Media coverage during times of disaster is now especially rapid and extensive. It provides the needed information for those potentially in danger, explaining what has occurred and communicating methods for obtaining safety (location of police, shelters, hospitals, and other basic services). For many people, the media also serves the much broader function of shaping the nature of a terrorist attack itself. Especially for those not directly involved in a terrorist attack, the initial experience of an attack is filtered through the media reports as the event unfolds. For many in the United States and across the globe, watching the 9/11 tragedy unfold on television *was* their experience of the event.

In addition to shaping the experience of terrorism, the intensive and often unedited media coverage of terrorist acts, especially in the immediate aftermath, may be traumatizing for some (e.g., Schlenger et al., 2002). Several studies of local children were conducted following the Oklahoma City bombing. Overall, the findings suggest that PTSD symptoms were positively correlated with the amount of attack-related media exposure in children grades six to twelve (Pfefferbaum, Moore, et al., 1999; Pfefferbaum et al., 2001; Pfefferbaum, Nixon, et al., 1999). This was true both for children who lost a relative and for those who did not. Ahern et al. (2002) also identified an important relationship between 9/11 attack-related television viewing and symptoms. Rates of depression and PTSD were predicted in part by the frequency of viewing graphic images related to the World Trade Center attack in people living in Manhattan. In particular, 22.5 percent of those who said that they had repeatedly watched images of people falling or jumping from the towers had PTSD, compared to only 3.6 percent of those who said they had not watched such images. Outside of Manhattan, significant distress, but not necessarily PTSD, was associated with the amount of terrorism-related television viewing in a national sample of 2,773 adults (Schlenger et al., 2002). However, as will be discussed later, these findings should be interpreted cautiously, as many other variables that might influence the watching of television could potentially mediate this relationship.

Several studies have failed to find an association between television exposure and PTSD symptoms per se. In a sample of eighty-five adults seeking treatment six months after the Oklahoma City bombing, Tucker, Pfefferbaum, Nixon, & Dickson (2000) found that the amount of television exposure was unrelated to PTSD symptoms. Similarly, in a sample of over 1,000 adults living in Manhattan, Ahern et al. (2002) reported that during the first two months following the 9/11 attacks, rates of PTSD and depression were not related to television viewing of terrorism-related stories.

The research provides a mixed picture of the role of media exposure in the adaptation of individuals following terrorist attacks. One likely explanation for the mixed results could be the differences in the degree of vulnerability in each sample studied. For instance, it is likely that children (as identified in the Pfefferbaum studies) are more frightened and confused than adults by news reports and television images. In a similar way, adults who have lost a loved one or have directly survived an attack could be particularly vulnerable to the effects of repeated images of the events compared to those who have had no direct contact with the attack.

We believe that studies indicating a positive relationship between trauma-related symptoms and television exposure should be interpreted cautiously, at least in adult samples. The findings could reflect the fact that people high in symptoms may have a stronger need to watch attack-related news coverage. Most adults viewing television coverage of terrorism-related news are doing so voluntarily. Schuster et al. (2001) found that in a random national sample of adults, respondents watched an average of 8.1 hours of television the day of the September 11 attacks. Holloway and Fullerton (1994) point out that voluntary exposure to fearful stimuli may be an attempt to master feelings of helplessness among traumatized people. They argue that "vicarious exposure to terror may help us deny actual terrors—we can watch a horror movie from a distance and not be destroyed" (p. 36). Knowing what is happening during and immediately following an attack could perhaps increase a viewer's sense of control and, therefore, could be seen as an active coping technique for some.

Limitations of Assessment and Methodology

Drawing conclusions from these data on the psychological impact of terrorism as a whole is limited by the methodological disparity across studies. The current studies include a variety of samples from different regions, each with its own sociopolitical history of violence. The symptom measures, timing of data collection, and method of sampling are highly disparate. For instance, a variety of standard and nonstandard measures of PTSD were used. Some studies relied on self-report (Abenhaim et al., 1992; Curran et al., 1990), while others used clinical interviews (North et al., 1999; Galea, Ahern, et al., 2002; Galea, Resnick, et al., 2002). In addition, diagnostic decisions were made according to varying versions of the DSM, and several interviews were translated into native languages. The time between the terrorist event studied and collection of data ranged from two days to more than five years. Sampling methods included the use of random sampling across an affected community (Galea, Ahern, et al., 2002; Galea, Resnick, et al., 2002) to interviewing 100 percent of the attack survivors (Shalev, 1992), resulting in widely varying sample sizes and compositions. Unfortunately, these differences preclude one from drawing decisive conclusions about the general impact of terrorism.

Uniformity across studies that assess the psychological impact of terrorism is exceptionally difficult (Silove & Kinzie, 2001). Because of the sudden nature of these disasters, researchers must often conduct their work in the field with limited control over variables. Collection of information on acute reactions soon after an attack is particularly difficult, as researchers must mobilize quickly and navigate already chaotic disaster sites. Use of planned and careful sampling techniques is often constrained by the need for timely data collection and the difficulty of obtaining access to victims. As such, researchers must often rely on convenience sampling or preselected samples using hospital records or government registries. The study of terrorism in non-Western cultures is especially complicated by the lack of available resources, including standardized measures. Finally, as in the case of 9/11, information about survivors is often kept from the public and from researchers, in an effort to protect the privacy of people who have already lost so much (Ursano, Fullerton, & McCaughey, 1994).

In addition to the methodological and logistical challenges, there is also a lack of expert consensus about the definition of terrorist trauma and its psychological sequelae. Determining whether the experience of terrorism meets the Criterion A standard for trauma is complicated. For example, although it is clear that an individual who is directly exposed to a terrorist attack (e.g., witnesses the attack and/or injury of others, or is injured himself or herself) would meet Criterion A, it is not so clear that an individual who only watches the event unfold on television also meets this criterion. Second, there is the question of whether our current assessment instruments are adequate for measuring terrorism-related trauma. Green (1993) suggests that few data support the idea that different traumas predict different PTSD symptoms. Rather, the PTSD response profile seems to be common across traumas. On the other hand, it is possible that the unique features of terrorism may lead to unique symptom profiles. The intermingling of trauma symptoms and intense grieving has been noted by New York therapists, especially in patients with severe exposure to the attack.

Risk and Resilience

Although a variety of studies suggest that the number of individuals impaired by PTSD, depression, and traumatic grief are significant, not all individuals directly exposed to terrorism suffer from significant functional impairment. Bonanno (2004) argues that in fact most individuals display resilience following a trauma; that is, they are able to maintain "relatively stable, healthy levels of psychological and physical functioning" following a traumatic event (p. 20). He cites several studies highlighting such resilience such as the post-9/11 survey by Galea, Resnick, et al. (2002) which indicated more than 40 percent of those of those living in Manhattan on the morning of 9/11 reported no trauma symptoms when surveyed months later. In addition, Ozer, Best, Lipsey, & Weiss (2003) recently reported that despite high national rates of exposure to trauma, only 5 percent to 10 percent develop PTSD. Bonanno (2004) suggests that these data indicate most individuals are likely to remain quite stable following a trauma. In turn, he warns us against pathologizing normal reactions to traumatic stress which might lead to premature and potentially damaging psychological interventions.

Such wide variation in the prevalence of impairment following trauma suggests that mere exposure to terrorism does not account for the presence of PTSD symptoms. Understanding factors important to individual resilience following trauma or terrorism may teach us something about what can be done to help those who are suffering. Predicting those at highest risk also aids the efficient delivery of treatment. Unfortunately, few studies specifically examine risk in response to terrorism. On the other hand, extensive work has been conducted in the past ten years with regard to risk factors for PTSD following a broader range of traumas.

Gender, Race, and Age

Gender. Studies have been conducted over the past two decades examining the role of demographic characteristics, including gender, in the development of PTSD following a trauma. Epidemiologic surveys suggest that women are more vulnerable to PTSD following a trauma as compared to men (Breslau et al., 1991; Brewin, Andrews, & Valentine, 2000; Kessler et al., 1995). Drawing conclusions about the effects of gender, however, is cautioned, as many studies documenting gender differences involve rape and sexual assault, traumas associated with very high rates of PTSD (Breslau & Kessler, 2001; Wolf & Kimerling, 1997). Wolf and Kimerling also argue that effects seemingly attributable to gender could be explained by other variables, such as women's propensity toward greater symptom reporting than men. Interestingly, Breslau, Davis, Andreski, Peterson, & Schultz (1997) have found that when lifetime prevalence or frequency of traumatic events was the same in men and women, women still were more likely to develop PTSD.

Other interesting findings regarding gender have emerged from recent studies of U.S. terrorism. In a study of treatment-seeking adults in Oklahoma, Tucker et al. (2000) failed to find an effect for gender on rates of PTSD. However, the study population was primarily female (71 percent). In a more evenly distributed sample, North et al. (1999) found that among those directly exposed to the Oklahoma City bombing, women were twice as likely to be diagnosed with PTSD. Data emerging from the 9/11 attacks also suggest that female gender is predictive of poorer outcome. Galea, Ahern, et al. (2002) found that women were more likely to be diagnosed with PTSD and to be depressed in the months following the attack. These data as a whole support findings from other studies of mass violence, including natural disasters, where women have been identified as being at elevated risk for PTSD (e.g., Steinglass & Gerrity, 1990).

Race and ethnicity. Until the 9/11 attacks, few data were available on the impact of race in response to terrorism, perhaps because most groups targeted by terrorists are ethnically uniform. Even in the Oklahoma City bombing, no main effect for ethnicity was found in North et al.'s study since the sample was 89 percent white. But the 9/11 victims represented a wide variety of ethnocultural backgrounds, and Galea, Ahern, et al. (2002) found that respondents' ethnicity significantly predicted PTSD and depression following the attacks. In particular, Manhattan Hispanics were found to be almost three times more likely to be diagnosed with PTSD and over three times more likely to be depressed compared to whites. This finding is consistent with studies of natural disasters, which also hit ethnically heterogeneous regions. These studies suggest that membership in a minority group poses a unique risk to survivors of natural disasters (Fothergill, Maestas, & Darlington, 1999). Minority group members may be more vulnerable to the damage of natural disasters due to housing configurations, language barriers, lack of resources, and isolation from community resources (Fothergill et al., 1999).

Some evidence also suggests that gender and ethnicity may interact to predict vulnerability to psychopathology following a large-scale disaster. In a unique study of natural disaster survivors, Norris, Perilla, Ibanez, & Murphy (2001) examined gender differences in PTSD as a function of culturally defined roles. Specifically, they found that identification with the Mexican culture exaggerated gender differences. The authors argued that the fostering of traditional feminine and masculine roles in the Mexican culture resulted in females reporting more severe distress than males. This was not the case among African Americans studied, where the authors argued more egalitarian gender roles are fostered.

Cultural differences in definitions of trauma and in the willingness to report symptoms can also be noted. For example, among Israeli military, bravery and strength are seen as necessary to ensure the survival of the *sabra* (literally "new Jew," and meant to indicate Israeli-born Jews) (Witztum & Kotler, 2000). Admission of war-related symptoms would be directly contradictory to the ideal of resilience in the face of historical persecution. As such, some argue that the existence of war-related trauma in soldiers has been significantly downplayed if not completely denied by Israeli military authorities (Witztum & Kotler, 2000).

Age. The symptoms of PTSD are similar among children, young and middle-aged adults, as well as the elderly. Among adults, there is little evidence that the elderly are particularly vulnerable to trauma-related distress (Weintraub & Ruskin, 1999) compared with other adult age groups. This may be due to the fact that research on the elderly is most often concerned with the resurgence of symptoms years after trauma exposure. Studies of natural disaster survivors suggest that middle-aged adults show greater trauma-related impairment than other age groups (Norris, 2002). This is most likely due to the cumulative responsibilities of middle-age that is associated with the maintenance of a home, job, and family (Thompson, Norris, & Hanacek, 1993). Younger adulthood age is consistently associated with greater PTSD symptomology, although the relationship, overall, is a relatively weak one and varies with the population being studied (e.g., military versus civilian; Brewin et al., 2000).

Prior History Characteristics

A range of pretrauma risk factors have been examined in relation to PTSD. Two key characteristics which appear to predict outcome following a variety of traumas are a history of childhood trauma and psychiatric disorders. Both of these variables, although related themselves, are independently associated with increased risk for exposure to trauma as well as PTSD following trauma (Breslau, Davis, & Andreski, 1997; Breslau, Davis, et al., 1997; Koenen et al., 2002). Childhood abuse is an especially damaging type of trauma because it occurs early in life, during crucial developmental years. It is a risk factor for subsequent victimization in various forms, including sexual assault, physical assault, and domestic violence (Polusny & Follette, 1995). Among women, childhood abuse is associated with greater susceptibility to PTSD than traumas occurring later in life (Breslau, Davis, & Andreski, 1997). A generally accepted theory is that prior trauma sensitizes one to the effects of subsequent trauma.

In a national comorbidity study, preexisting mental disorders, especially anxiety disorders, increased the risk for PTSD (after controlling for type of trauma; Bromet, Sonnega, & Kessler, 1998). This is also true among military personnel exposed to combat (Kulka et al., 1990). In a twin study of the familial and individual risk factors for PTSD, Koenen et al. (2002) found that preexisting mood and anxiety disorders increased the risk for PTSD. In a study of survivors of the Oklahoma City bombing, North et al. (1999) found pretrauma psychiatric disorder to be a strong predictor of PTSD approximately six months following the blast. They found that 45 percent of those with a history of mental disorder, and only 26 percent of those without such a history, were PTSD positive. These findings suggest that perhaps the cognitive biases associated with previous psychiatric disorders may play a role in susceptibility to PTSD following trauma. Individuals cope with trauma primarily by relying on the coping skills they have used before while under stress. A person with a history of poor mood regulation or who is susceptible to catastrophic attributions of internal states is at particular risk for developing trauma-related psychopathology.

Acute Reactions

A commonly studied risk factor is an individual's immediate reaction to trauma. Acute reactions most often studied immediately following trauma include dissociation, reexperiencing, avoidance, and arousal reactions. Numerous studies have linked early dissociative reactions to the development of PTSD (Bremner & Brett, 1997; Koopman, Classen, & Spiegel, 1994; McFarlane, 1986; Shalev, Peri, Canetti, & Schreiber, 1996; Shalev et al., 1998; Solomon & Mikulincer, 1992). However, as Bryant and Harvey (2000) point out, "although there are initial indications that suggest a relationship between peritraumatic dissociation and PTSD, it appears that this relationship is not linear or uniform" (p. 26). Early intrusive reexperiencing symptoms are quite common and only weak predictors of PTSD (e.g., McFarlane, 1988; North et al., 1999). Early avoidance has also been studied as a predictor of PTSD (e.g., Creamer, Burgess, & Pattison, 1992; McFarlane, 1992; Bryant & Harvey, 1995); but overall, the evidence suggests that ongoing rather than immediate avoidance symptoms are more important to long-term prognosis (Bryant & Harvey, 2000).

In addition to dissociation, reexperiencing, and avoidance, one can also imagine other ways in which an individual might lose the capacity to self-regulate his or her emotions in the aftermath of a traumatic event. Impulsive and self-injurious behavior such as mutilation can be a response to the overwhelming stress of a trauma. There is little information about the prevalence of such behaviors following trauma and whether such behaviors are short-lived or evolve into more complicated conditions. It is likely that unless there is a history of such behavior under stress, as in the case of preexisting character disorder, such behaviors would likely dissipate after a brief period. Trauma tends to bring out habitual and instinctive coping strategies. If an individual has good coping techniques prior to the onset of trauma, these strategies will likely reemerge.

In one of the most important and comprehensive reviews of risk for PTSD, Brewin et al. (2000) conducted a meta-analysis of seventy-seven studies examining PTSD risk factors. Studies included in the meta-analysis covered traumatic experiences such as combat, assault, motor vehicle accidents, natural disasters, and serious medical conditions. Brewin et al. (2000) selected fourteen of the most commonly studied risk factors including education, gender, age, socioeconomic status (SES), intelligence, race, psychiatric history, childhood abuse, previous trauma, childhood adversity, family psychiatric history, trauma severity, life stress, and lack of social support. Results revealed three clusters of variables which emerged according to the magnitude of their impact (i.e., effect size). Minority status and age emerged as risk factors with the weakest, although still significant, effect sizes (-.38 to .28 for age and -.27 to .39 for minority status). Second, a group of demographic and prior history characteristics were identified, including gender, SES, education, intelligence, psychiatric history, history of abuse, and family psychiatric history. The authors found that a history of psychiatric illness was moderately associated with the development of PTSD, with an average effect size of .11. A history of childhood abuse showed effect sizes ranging from low (.07) to moderate (.30). Demographic variables, such as SES, education, gender, intelligence, race, and age were also significantly associated with PTSD with average effect sizes between .05 for race to .18 for intelligence. The third group of variables, which showed the strongest effect sizes, included characteristics of the trauma and circumstances following the trauma. Brewin et al. found that the severity of the trauma, life stress following the trauma, and lack of social support were the strongest predictors of PTSD, producing average effect sizes of .23 (range .14 to 76), .32 (range .26 to .54), and .40 (range .02 to .54), respectively. Overall, the lack of social support was the strongest and most consistently identified risk factor for the development of PTSD following a traumatic event.

Social Support and Resilience

It is a well-known fact that social contact is a powerful buffer of the effects of stress (e.g., Cohen & Willis, 1985). This is not only true for humans, but also for other species, such as nonhuman primates and rats. When primates or rats are put through a stressful experience, they will show a classic stress response (i.e., activation of the HPA axis and secretion of glucocorticoids). If the same animal is put through a stressful experience in the presence of "friends" then the stress response is either attenuated or disappears altogether. In humans, the physical and/or psychological presence of friends and loved ones during stressful times is an important factor predicting adaptation to stress (Cobb, 1976; Cohen & McKay, 1984; Holahan & Moos, 1981; Norris & Murrell, 1990). For instance, the availability of family support has been found to be especially important to the physical health of women (Holahan & Moos, 1982), and decreases in social support have been found to be related to psychological maladjustment (Holahan &

Moos, 1981; Norris & Murrell, 1990) and survival rates following myocardial infarction (Frasure-Smith, Lesperance, & Talajic, 2000).

With regard to traumatic stress, attachment theory predicts a sharp increase in social contact in the immediate aftermath of a trauma to reestablish a sense of safety and obtain security. According to attachment theory, there is not only a strong pull to be comforted by others, but an equally strong pull to help those in danger, especially loved ones. On the morning of 9/11, witnesses, victims, and others made gallant efforts to save, rescue, and comfort those in need, often without regard for their own safety. This pull to help and be helped under conditions of danger reflects one of our most primitive tools and instinctual skills for survival. Unlike non-social species that might flee to safety by running to a *place* (e.g., a den, cave, up a tree), when we are afraid we seek safety in *each other*. Think of your own reaction on the morning of 9/11 and you will understand the powerful role of attachments as the primary mechanism of obtaining safety. Upon hearing news of the attack, many people made immediate efforts to contact loved ones in an effort to reestablish a sense of safety and protection ("Are the people I depend on okay?") and to verify the safety of others ("Are those that depend on me okay?"). This was especially true for those in close proximity to the danger. The widespread efforts to contact loved ones was so immediate and intense around New York City that many of the ground and satellite phone services were shut down as a result of the deluge of phone calls (in addition to the technical difficulties brought about by the towers' destruction itself) (Schiesel & Hansell, 2001).

It is not surprising then that one of the most robust predictors of recovery from trauma, and even from everyday stress, is the perception of available support from others. More specifically, the study of social support and trauma suggests that it is a *lack of support* that is uniquely predictive of outcome. Evidence suggests that those who do not anticipate that friends and loved ones would be available if needed cope with stress and trauma far less well than those with high perceived social support (Brewin et al., 2000; Cohen & Willis, 1985; Norris & Murrell, 1990; Solomon, Mikulincer & Avitzur, 1988).

Although Brewin et al. (2000) did not include outcome studies of terrorism, data from natural disasters, which share some common features with large-scale acts of terrorism (massive destruction and loss; collective sense of trauma) indicate social support is critical to recovery. In a prospective longitudinal study of depression following a natural disaster, Kaniasty & Norris (1993) surveyed 222 individuals before and after a severe flood. They found that the degree of deterioration in social support mediated the impact trauma on mental health. Specifically, those who did not report deterioration in their perceived support following the flood were less likely to be depressed compared to those who did report such deterioration. Freedy, Shaw, Jarrell, & Masters (1992) found that the loss of resources, including social support, was a significant predictor of distress and psychopathology following a natural disaster. A loss of psychosocial resources was more important than individual and coping characteristics in predicting short-term adjustment to a natural disaster. Benight, Swift, Sanger, Smith, and Zeppelin (1999) separated loss of material resources (e.g., property, financial loss) from loss of social support to examine the role of each in trauma-related distress following Florida's 1995 Hurricane Opal. They found that a loss of material resources had a direct impact on loss of perceived social support and reported distress.

Data from the Oklahoma City bombing and the 9/11 attacks suggest that social support, and specifically reaching out to others, is important for recovery from such large-scale terrorist attacks. Tucker et al. (2000) assessed various arenas of social support used by a sample of eighty-five treatment-seeking adults exposed directly or indirectly to the Oklahoma City bombing, and examined the association with symptoms six months following the event. Those adults who reported finding support from the workplace were less likely to be symptomatic compared to those who reported finding support from counseling. These findings suggest that those who have strong social networks outside of counseling have a better prognosis. The authors note that the findings could also reflect the fact that those who found help from counseling may also have had more severe symptoms, necessitating treatment in the first place. North et al. (1999) examined the psychiatric impact of the Oklahoma City bombing in a sample of injured survivors who had registered with the State Health Commissioner. Although North et al. did not ask specifically about social support, they noted that "regardless of diagnostic status, turning to others for support was a nearly universal response" (p. 759). Those diagnosed with PTSD six months following the bombing reported significant worsening in the quality of relationships compared to those without PTSD. However, the cause and effect relationship here is not always clear.

Similar to North et al. (1999), Schuster et al. (2001) found that almost everyone interviewed three to five days after the 9/11 attacks coped in part by talking with others (98 percent). Additional evidence of the critical role of social networks in coping and recovery was the significant use of public or group activities such as memorializing the victims (60 percent), religious activities (90 percent), and efforts to help others by making donations (36 percent). Galea, Ahern, et al. (2002) found that among those living in Manhattan, low levels of perceived support in the six months prior to the attack were predictive of depression within one to two months following the attack, but not predictive of a PTSD diagnosis.

Although there is no research on the role of family relationships and/or the quality of marital relationships in recovery from terrorism, several studies with combat veterans suggest these relationships are critical to long-term functioning. Veterans with chronic PTSD are more likely to have marital problems compared to veterans without PTSD, including problems with self-disclosure, marital violence, and overall marital satisfaction (Carroll, Rueger, Foy, & Donahoe, 1986; Riggs, Byrne, Weathers, & Litz, 1997). Much of the data also suggest that traumatized individuals have difficulty relying on others and feel more distress about the quality of their relationships following the trauma (MacDonald, Chamberlain, Long, & Flett, 2000).

Taken together, the research on social support and trauma suggest that not only is social support important to resiliency following trauma, but that trauma-related distress can have deleterious effects on relationships, thus creating a vicious cycle of distress and loss of support. In most cases, not only are individuals who have limited social support networks more vulnerable to PTSD, the distress associated with PTSD can also have taxing effects on existing relationships. Problems in important relationships following trauma may stem in part from the tension created by individual differences in coping strategies. Under conditions of threat, most individuals rely rigidly on coping strategies that are most familiar to them. If two people handle trauma differently, each is more likely to feel unsupported by the other. For example, in a family that has lost a member following terrorism, one person may instinctually withdraw as a way to cope with the situation, while another person may need to talk about the traumatic loss. Such differences within a family or couple are likely to cause significant conflict, which is also likely to have detrimental effects on the individuals trying to recover from the trauma.

Long-Term Adjustment

It is important to clarify the differences between acute reactions to trauma and chronic responses. A large percentage of individuals directly exposed to a traumatic event such as terrorism will likely develop some symptoms of PTSD immediately following the event, including intrusive recollections and emotional numbing. During the first year after the trauma, however, these symptoms will substantially decrease in most individuals. For a subgroup of people, symptoms of PTSD may be unrelenting, lasting years or even decades. It is this chronic subgroup that is most likely to seek some type of professional help and will be the group most clinicians see.

It is not clear how many survivors of terrorist attacks will develop long-term problems. Some studies suggest up to one-third of PTSD cases in general do not remit after ten years without treatment (Kessler et al., 1995). In a study of the effects of natural disasters in the general population, time did not predict a decrease in PTSD severity (Briere & Elliott, 2000), suggesting that those who are diagnosed with PTSD following such a trauma remained PTSD positive. McFarlane and Papay (1992) found that forty-two months following a natural disaster, 18 percent of firefighters had significant symptoms of PTSD, with 12 percent meeting diagnostic criteria. In some cases following a natural disaster, PTSD remained chronic, while depression and other anxiety disorders abated over time (Shore, Tatum, & Vollmer, 1986). Goenjian et al. (2000) compared the course of PTSD between victims of a severe natural disaster and severe political violence. They found no differences in PTSD chronicity between subjects exposed to political violence and subjects exposed to natural disaster, with more than 80 percent in both groups reporting high PTSD scores both at baseline and at a 4.5-year follow-up. Unfortunately, the impact of treatment on the chronicity of PTSD was not addressed in these studies.

Studies of the long-term consequences of terrorism are limited. Desivilya, Gal, & Ayalon (1996) examined survivors of a terrorist attack seventeen years following the trauma. The surviving subjects were adolescents at the time of the attack and had been taken as hostages for sixteen hours by armed guerrillas, during which time twenty-two students had been killed. Sixty-one percent of those interviewed reported at least five symptoms of PTSD seventeen years later, with those that had suffered some type of injury during the attack having the most symptoms. Injury also predicted severity of PTSD symptoms four to five years following terrorist attacks in a study by Abenhaim et al. (1992). In a study of the Oklahoma City bombing, North (2001) noted that the onset of PTSD among victims was very quick, with 98 percent meeting criteria within the first month afterward. Approximately 34 percent of those interviewed six months and again 1.5 years following the bombing met criteria for PTSD. Data regarding the chronicity of PTSD and other trauma-related symptoms following the 9/11 attacks have not yet been published.

Although PTSD appears to be a relatively unrelenting disorder for some, particularly without treatment, a variety of factors are important in predicting chronic outcomes. Several studies suggest that the severity of the trauma is a powerful predictor of long-term recovery; injury and prolonged victimization (e.g., being held hostage, combat exposure) are associated with the highest rates of chronic PTSD (Yehuda, McFarlane, & Shalev, 1998). For instance, in a sample of survivors exposed to a factory disaster, the intensity of exposure (e.g., injury, proximity to disaster, direct involvement) predicted not only early rates of PTSD, but also rates of chronic PTSD (Weisaeth, 1996). Research on PTSD in the elderly is most often focused on the resurgence of symptoms years after trauma exposure. As a result, little is known about the course of PTSD in the elderly compared to young adults.

Few data support the association between early intrusive and dissociative symptoms and chronic PTSD (Freedman, Brandes, Peri, & Shalev, 1999). However, early comorbidity may be a marker for long-term problems. In a study of trauma survivors recruited from a hospital emergency room, Shalev et al. (1998) found the co-occurrence of major depression and PTSD in almost half the sample (44.5 percent). The authors also found that, although each disorder appeared to be a separate sequelae of traumatic events, their co-occurrence was associated with significantly increased impairment and distress. In addition, depressive symptoms in the early months following the trauma were found to be the best predictor of continuing PTSD symptoms one year after the event. These data have important implications for recovery from terrorism-related experiences following 9/11, given the high number of deaths and the high incidence of major depression and PTSD among survivors (Galea, Resnick, et al., 2002). It is also important to note that not all individuals with chronic difficulties will meet criteria for PTSD. In fact, the majority of survivors will likely suffer from a variety of symptoms that do not meet the full DSM criteria but are associated with significant decreases in functioning.

In the case of terrorism, social support and other external factors are also likely to promote recovery in those exposed. For instance, exposure to ongoing or periodic reminders of the trauma could likely exacerbate or rekindle trauma-related symptoms. Among the most persistent reminders are ongoing media coverage, anniversary dates, and the threat of additional attacks. In fact, exposure to bombing-related television coverage was found to play a role in sustaining PTSD symptoms in children living far away from the Oklahoma City attack two years afterward (Pfefferbaum et al., 2000). In addition, it is clinically accepted that anniversary dates serve as powerful reminders of past trauma; these dates are often anxiously anticipated by both client and clinician. In one study, Morgan, Kingham, Nicolaou, & Southwick (1998) found that a large number of Gulf War veterans reported significant anniversary reactions two years after the war. In a follow-up study six years after the war, the authors replicated their findings and showed these reactions to be corroborated by spouses (Morgan, Hill, Fox, Kingham, & Southwick, 1999).

It is certainly expected that traumatic reactions to terrorism will likely be chronic in a subset of people, whether the symptoms give rise to a diagnosis of PTSD or not. As discussed earlier, terrorism is an ongoing trauma having no clear end. Terrorism also affects entire communities, which must adapt to a new sense of vulnerability. The resulting changes (e.g., more stringent safety precautions, new attitudes, heightened vigilance) serve as ongoing reminders of one's own individual vulnerability, complicating the recovery process.

TRAUMATIC REACTIONS TO TERRORISM: SPECIAL CONSIDERATIONS

Rescue and Recovery Personnel

Firefighters, police officers, emergency medical technicians, volunteers, and other emergency personnel involved in rescue and recovery efforts following terrorist acts make up a particularly susceptible population. These workers are called upon to put themselves in extreme and prolonged danger while remaining calm and in control. Prolonged exposure to graphic scenes of mutilation and mass destruction while working exhausting hours put these workers at particular risk. In addition, rescue personnel are often working to recover victims they know and have lost. This was particularly true of the 9/11 firefighters and police force, both of whom lost many of their own.

Given these experiences, surprisingly low rates of PTSD and other psychiatric symptoms have been noted in several samples of rescue workers following disasters (Durham, McCammon, & Allison, 1985; McFarlane, 1988; Raphael, 1986; Raphael, Singh, Bradbury, & Lambert, 1983; Weiss, Marmar, Metzler, & Ronfeldt, 1995). Weiss et al. (1995) found as few as 10 percent of EMT workers displayed significant symptoms after exposure to disasters. Similarly, very low rates of PTSD were identified in a study of police officers in Ireland who were exposed to terrorist attacks (Wilson, Poole, & Trew, 1997). Officers who were interviewed between seven and ten months following exposure to a terrorist attack had only a 5 percent incidence of PTSD. The same relatively low rate (10 to 20 percent) of serious symptoms has been found in other samples of Australian firefighters (McFarlane, 1988), British police officers (Alexander & Wells, 1991), and American and Australian EMS workers (Durham et al., 1985; Raphael et al., 1983). Among male Oklahoma City firefighters, 13 percent were found to have PTSD almost three years following the bombing (North et al., 2002). Although this indicates a large number of chronic PTSD cases, this rate was significantly lower than rates found among other male survivors of the bombing (23 percent). Firefighters also reported little impairment and high job satisfaction.

Low rates of psychological distress in rescue workers could in part be the result of their well-defined role. A clear set of responsibilities combined with extensive training are likely to be significant buffers to distress during times of chaos. Some evidence for this is reflected in the higher levels of trauma-related symptomology that have been noted in untrained volunteer rescuers following a natural disaster (Johnsen, Eid, Lovstad, & Michelsen, 1997). In this study, "spontaneous" rescuers during a natural disaster were found to be equally as symptomatic as direct victims in the months immediately following the disaster.

Unique to the role of rescuer is the pressure to be courageous and strong while performing extremely dangerous and distressing tasks. Especially in the wake of 9/11, rescue personnel have been labeled "heroes." For some, such a label may be uplifting, but for others, such a stereotyped role may feel burdensome or oppressive, especially for those suffering from distressing symptoms. The "hero" label may be isolating to some rescue workers, as they are upheld as having a special status apart from the rest of the traumatized community. Although such a role could feel burdensome to some, this image may help account for the relatively low rates of self-reported trauma-related symptoms in rescue workers. It is very likely that the pressure to project a "macho" image precludes many rescue workers from admitting to trauma-related symptoms, even to themselves. They may feel that their "real" feelings or symptoms are shameful and must be hidden. This may lead to avoidance of talking or thinking about their experiences. As a result, many rescue personnel may be truly suffering but are not identified through self-report measures.

The notion that rescue workers may be significantly impaired but unlikely to talk about their symptoms comes from North et al. (2002), who found that among Oklahoma City firefighters with PTSD, most showed strongly avoidant coping behavior. Although many reported that they turned to friends (50 percent), the second most frequently reported coping method was increasing consumption of alcohol (19 percent). In addition, only half with PTSD sought professional mental health treatment. North et al. highlighted the presence of substance abuse among firefighters following disasters, half of whom will likely meet criteria for an alcohol use disorder at some point in their lifetime. Indeed, the most common preexisting psychiatric disorder among the firefighters interviewed was alcohol abuse/dependence (47 percent). Similarly, the most common postdisaster psychiatric disorder of those reported was not PTSD but alcohol abuse/ dependence (88 percent).

Children

A population especially vulnerable to the impact of terrorism is children. It is estimated that as many as 3 million children experience a traumatic event each year (Schwarz & Perry, 1994). Tremendous variability in rates of PTSD following traumas has been documented in children depending on type and severity of trauma, timing of assessment, and factors including prior traumas. In a review of epidemiological data regarding crime, abuse, war, and disaster-related traumas in children, Saigh, Yasik, Sack, & Koplewicz (1999) demonstrated PTSD prevalence rates ranging from 0.0 to 95 percent. Much of the variability seemed to be accounted for by the type and intensity of the stressor. Deliberate acts or events (e.g., combat, criminal victimization, and accidents) were associated with higher rates of PTSD compared to accidental stressors such as chemical contamination.

In a report to the New York City Board of Education, more than 8,000 children in grades four through twelve were randomly sampled about their experiences and symptoms six months following the 9/11 attacks (Board of Education of the City of New York, 2002). The children were drawn from schools near Ground Zero and across the city. Using the Diagnostic Predictive Scale as a screening tool for mental health problems, it was estimated that as many as 27 percent of children in the New York City public schools had at least one of the psychiatric disorders assessed. Specifically, 8 percent of children had symptoms consistent with major depression, 10.3 percent with generalized anxiety, 15 percent with agoraphobia, 12 percent with separation disorder, 11 percent with conduct disorder, and 11 percent with PTSD. Rates of PTSD were highest among those who had been directly exposed to the attack (injured, running from debris or smoke), had a family member who died or escaped from the WTC towers, had been exposed to traumatic events prior to 9/11, or who had spent "a lot" of their time watching television coverage of the attacks. Other demographic variables predictive of PTSD status were being female, younger (grades four to five compared to grades six to twelve), and Hispanic.

Other Traumatic Reactions to Consider

It is also important to emphasize that terrorism may lead to other psychological, psychosocial, and/or health problems beyond PTSD and depression. PTSD, although most often associated with comorbid depression, is also very often diagnosed alongside other anxiety disorders, including panic disorder, agoraphobia, obsessive-compulsive disorder, social phobia, and specific phobia (Breslau et al., 1991; Kessler et al., 1995).

Somatic preoccupation and somatization disorder have been most consistently noted in those with previous chronic traumatic experiences, such as childhood abuse, and those with complex PTSD, but it is also an important disorder to consider following terrorism. Substance-related disorders are commonly associated with PTSD, especially for men, and in specific populations such as rescue workers. In a national comorbidity study, Kesseler et al. (1995) found that as many as 48 percent of those with PTSD also had depression at some point in their lifetime, and 51.9 percent of males with PTSD abused alcohol.

Problems in physical health following terrorism are also a concern. In disaster settings, physical complaints are often the reason for initial help seeking (McFarlane, Atchison, Rafalowicz, & Papay, 1994). Not unlike natural disasters, large-scale terrorist attacks may include the physical destruction of sanitation systems, which may result in compromised air and water quality. With the additional threat of biological and chemical weapons, intense somatic preoccupation following terrorist attacks is likely, and hospitals and doctors may be the primary mechanism through which people first seek professional help. Exposure to trauma has been associated with increased somatic concerns and lowered health status particularly in women (Cloitre, Cohen, Edelman, & Han, 2001). Furthermore, somatization, physical injury, and comorbidity have been linked to PTSD chronicity (for review see Breslau, 2001).

Finally, terrorism can be associated with high numbers of casualties, deaths, and injuries which often occur under horrendous conditions. Given the number of casualties on 9/11, particular attention has been paid to the role of grief as a critical piece of the psychological aftermath of terrorism. Although this topic is covered more fully elsewhere in this book, it is important to mention grief as a significant part of the individual experience. Both the suddenness of the deaths and the manner in which they occur greatly complicate bereavement. As a result, many who experience the loss of a loved one under such conditions experience an intermingling of grief and trauma. Not only do individuals grieve a loss, but they can also find themselves traumatized by the manner of death. Classic traumatic reexperiencing symptoms (nightmares, intrusive images, etc.) related to the manner of death complicate the grieving process. Termed "traumatic grief," many argue that it captures a unique aspect of the psychological aftermath of disasters. Prigerson et al. (1999) suggest that traumatic grief is characterized by symptoms of intrusion and reexperiencing with regard to a specific death, intermixed with grief associated with the loss of an attachment. Such a combination of symptoms may result in a unique risk for functional impairment. Indeed, following the Oklahoma City bombing, the presence of grief for a loved one who had died combined with PTSD was a strong predictor of functional impairment and substance use in treatment-seeking individuals (Pfefferbaum et al., 2001, 2002).

CONCLUSIONS AND IMPLICATIONS FOR RECOVERY

In this chapter, we have attempted to provide a comprehensive discussion of the effects of terrorism on the individual and the community. We have also adopted an attachment theory framework for understanding the unique ability of terrorism to disrupt security on both an individual and community level. We have discussed terrorism as an experience that is prolonged, without boundaries, and multilayered in its effects. We have also attempted to examine terrorism as a collective experience that is strongly influenced by the community response. These characteristics separate terrorism from other traumas studied and, unfortunately, complicate the recovery process.

Literature reviewed in this chapter suggests that a moderate to high prevalence of PTSD among community members most directly exposed to terrorist attacks should be anticipated. However, because the fear of terrorism reaches far beyond the immediate setting of an attack, we must look for PTSD and other pathological reactions not only in those directly exposed but also among the wider community. This includes those who have lost loved ones, those whose homes or businesses are destroyed, those taking part in rescue and recovery efforts, and all of us who feel directly threatened by an attack or a future attack. In addition to PTSD, research reviewed here suggests that an increase in substance abuse, depression, and a variety of anxiety disorders are common following events of mass violence. Also of concern is the *combination* of depression, grief, substance use, and PTSD, which occur following traumas similar to terrorism. Research suggests that such comorbid conditions may be particularly recalcitrant, especially without treatment. PTSD rates should be highest in the months immediately following an attack but should then slowly drop off. Individuals who continue to meet criteria for PTSD up to a year after an attack are at greatest risk for long-term problems; PTSD has been found to be a relatively unrelenting disorder, especially without appropriate treatment. The unremitting nature of PTSD, combined with the ongoing threat inherent to terrorism, implies that a subset of people most severely affected may have an unusually difficult task ahead.

Although many demographic factors are important for the efficient targeting of treatment efforts (e.g., minority status, gender, the bereaved, rescue personnel), those who perceive a lack of support from their surrounding community are particularly at risk. This corresponds with an attachment theory perspective predicting that the absence of supportive relationships following a trauma places victims at significant risk for maladaptive responses. Research indicates that the relationship between social support and trauma-related distress is bidirectional, with low social support associated with more severe PTSD and PTSD contributing to conflict in relationships.

From this perspective, a critical element of recovery is the reestablishment of a feeling of security in the face of vulnerability. A sense of security comes from the combined effects of feeling confident in one's ability to cope with life's obstacles and, if these abilities are overwhelmed, feeling confident that one can rely on the support of others to overcome those obstacles. In the face of ongoing trauma such as terrorism, this translates into a recovery process that entails a gradual reestablishment of feelings of competency through the use of active coping and an increase in social ties through sharing with others.

Active coping may take a number of forms but must involve decisive behavioral changes to reduce distress. Examples include the use of stress-reduction techniques, community service, hobbies, changes in routine, and improvements in health and fitness. Although simple, such active changes turn feelings of passive victimization into feelings of mastery. Strengthening relationships as the second part of promoting security begins with the sharing of traumatic experience with important others, whether family, friends, or lovers. For some individuals, the establishment of new relationships is needed. The use of support groups is particularly effective in this case. Such groups allow for the discovery of common ground with others, thereby reducing feelings of marginalization and alienation associated with the experience of trauma. Most effective are relationships with others who have endured similar traumatic experiences. These efforts promote the development of an adaptive mental framework for organizing the experience of terrorism that allows one to continue functioning in the face of threats. The importance of active coping has been highlighted by recent findings that passive coping following 9/11 (e.g., "giving up," "denial," "self-distraction") was associated with higher levels of PTSD symptoms compared to active coping (Silver et al., 2002).

On a community level, security and recovery can be established in similar ways. Just as active coping on an individual level is important, the orchestration of effective community responses to terrorism is critical to increasing feelings of well-being (Chemtob, 2002). This involves, among other things, fortifying community resources to treat the traumatized as well as implementing safety measures. This involves ensuring that adequately trained medical and mental health personnel are available, that hospitals and other recovery organizations (e.g., Red Cross) are fully supplied and staffed, that protocols for disaster response are developed and implemented, and that funds are obtained to finance these efforts. Also central to an effective community response are competent and capable leaders and intelligence agencies that provide a sense of protection. In addition to active coping, the community must also increase relatedness and feelings of unity that do not further marginalize those who are traumatized. Opportunities to help others on a community level must be made available and encouraged. Such opportunities may include blood drives, fund-raising drives, participation in recovery efforts, donation of services, food, money, or clothing, and displays of unity. Participation in such community projects helps to reduce division and increase confidence in the capacity of a community to handle future threats. Most important, we believe such efforts will diminish the individual's fear in the face of terrorism's threat.

REFERENCES

Abenhaim, L., Dab, W., & Salmi, L.R. (1992). Study of civilian victims of terrorist attacks (France 1982-1987). *Journal of Clinical Epidemiology, 45,* 103-109.

Ahern, J., Galea, S., Resnick, H., Kilpatrick, D., Bucuvalas, M., Gold, J., & Vlahov, D. (2002). Television images and psychological symptoms after the September 11 terrorist attacks. Poster presented at the Society for Epidemiologic Research Annual Meeting, Palm Desert, California, June 19.

Alexander, D.A. & Wells, A. (1991). Reactions of police officers to body-handling after a major disaster: A before-and-after comparison. *British Journal of Psychiatry, 159,* 547-555.

American Psychiatric Association. (1994). *Diagnostic and statistical manual of mental disorders* (4th ed). Washington, DC: Author.

Ayalon, O. (1993). Posttraumatic stress recovery of terrorist survivors. In J.P. Wilson and B. Raphael (Eds.), *International handbook of traumatic stress syndromes* (pp. 855-866). New York: Plenum Press.

Benight, C.C., Swift, E., Sanger, J., Smith, A., & Zeppelin, D. (1999). Coping self-efficacy as a mediator of distress following a natural disaster. *Journal of Applied Social Psychology, 29,* 2443-2464.

Blake, D.D., Weathers, F.W., Nagy, L.M., Kaloupek, D.G., Gusman, F.D., Charney, D.S., & Keane, T.M. (1995). The development of a clinician-administered PTSD scale. *Journal of Traumatic Stress, 8,* 75-90.

Blake, D.D., Weathers, F.W., Nagy, L.M., Kaloupek, D.G., Klauminzer, G., Charney, D.S., & Keane, T.M. (1990). A clinician rating scale for assessing current and lifetime PTSD: The CAPS-1. *Behavior Therapist, 13,* 187-188.

Board of Education of the City of New York (2002, May 6). *Effects of the World Trade Center attack on NYC public school students: Initial report to the New York City Board of Education.* Prepared by Applied Research and Consulting, LLC, Columbia University Mailman School of Public Health, and New York State Psychiatric Institute.

Bonanno, G.A. (2004). Loss, trauma, and human resilience: Have we underestimated the human capacity to thrive after extremely aversive events? *American Psychologist, 59,* 20-28.

Bowlby, J. (1969/1982). *Attachment and loss:* Vol. 1, *Attachment.* New York: Basic.

Bowlby, J. (1973). *Attachment and loss:* Vol. 2, *Separation.* New York: Basic.

Bowlby, J. (1980). *Attachment and loss:* Vol. 3, *Loss, sadness and depression.* New York: Basic.

Bremner, J.D., & Brett, E. (1997). Trauma-related dissociative states and long-term psychopathology in posttraumatic stress disorder. *Journal of Traumatic Stress, 10,* 37-49.

Breslau, N. (2001). Outcomes of posttraumatic stress disorder. *Journal of Clinical Psychiatry, 62,* 55-59.

Breslau, N., Davis, G.C., & Andreski, P. (1997). Risk factors for PTSD-related traumatic events: A prospective analysis. *American Journal of Psychiatry, 152,* 529-535.

Breslau, N., Davis, G.C., Andreski P., & Peterson, E.L. (1991). Traumatic events and posttraumatic stress disorder in an urban population of young adults. *Archives of General Psychiatry, 48,* 216-222.

Breslau, N., Davis, G.C., Andreski, M.A., Peterson, E.L., & Schultz, L.R. (1997). Sex differences in posttraumatic stress disorder. *Archives of General Psychiatry, 54,* 1044-1048.

Breslau, N., & Kessler, R.C. (2001). The stressor criterion in DSM-IV posttraumatic stress disorder: An empirical investigation. *Biological Psychiatry, 50,* 699-704.

Brewin, C.R., Andrews, B., & Valentine, J.D. (2000). Meta-analysis of risk factors for posttraumatic stress disorder in trauma-exposed adults. *Journal of Consulting and Clinical Psychology, 68,* 748-766.

Briere, J., & Elliott, D. (2000). Prevalence, characteristics, and long-term sequelae of natural disaster exposure in the general population. *Journal of Traumatic Stress, 13,* 661-679.

Bromet, E., Sonnega, A., & Kessler, R.C. (1998). Risk factors for DSM-III-R posttraumatic stress disorder: Findings from the National Comorbidity Survey. *American Journal of Epidemiology, 147*(4), 353-361.

Bryant, R., & Harvey, A.G. (1995). Avoidant copying style and post-traumatic stress following motor vehicle accidents. *Behaviour Research and Therapy, 33,* 631-635.

Bryant, R., & Harvey, A.G. (2000). *Acute stress disorder: a handbook of theory, assessment, and treatment.* Washington, DC: American Psychological Association.

Carroll, E.M., Rueger, D.B., Foy, D.W., & Donahoe, C.P. (1986). Vietnam combat veterans with posttraumatic stress disorder: Analysis of marital and cohabitating adjustment. *Journal of Abnormal Psychology, 94,* 329-337.

Chemtob, C. (2002). Keynote address presented at the Annual Jewish Board of Family and Children's Services, New York, June 30.

Cloitre, M., Cohen, L.R., Edelman, R.E., & Han, H. (2001). Posttraumatic stress disorder and extent of trauma exposure as correlates of medical problems and perceived health among women with childhood abuse. *Women & Health, 34,* 1-17.

Cobb, J. (1976). Social support as a moderator of life stress. *Psychosomatic Medicine, 38,* 300-314.

Cohen, S., & McKay, G. (1984). Interpersonal relationships as buffers of the impact of psychological stress on health. In A. Baum, J.E. Singer, & S.E. Taylor (Eds.), *Handbook of psychology and health* (pp. 253-267). Hillsdale, NJ: Erlbaum.

Cohen, S., & Willis, T.A. (1985). Stress, social support, and the buffering hypothesis. *Psychological Bulletin, 98,* 310-357.

Corcoran, C.B., Green, B.L., Goodman, L.A., & Krinsley, K.E. (2000). Conceptual and methodological issues in trauma history assessment. In A. Shalev & R. Yehuda, (Eds.), *International handbook of human response to trauma* (pp. 223-232). The Plenum series on stress and coping. New York: Kluwer/Plenum.

Creamer, M., Burgess, P., & Pattison, P. (1992). Reaction to trauma: A cognitive processing model. *Journal of Abnormal Psychology, 101,* 452-459.

Curran, P.S., Bell, P., Murray, A., Loughrey, G., Roddy, R., & Rocke, L.G. (1990). Psychological consequences of the Enniskillen bombing. *The British Journal of Psychiatry, 156,* 479-482.

Davidson, J.R.T., & Foa, E.B. (1991). Diagnostic issues in posttraumatic stress disorder: Considerations for DSM-IV. *Journal of Abnormal Psychology, 100,* 346-355.

Davidson, J.R.T., & Foa, E.B. (1993). *Posttraumatic stress disorder: DSM-IV and beyond.* Washington, DC: American Psychiatric Press.

Davidson, J.R.T., Smith, R.D., & Kudler, H.S. (1989). Validity and reliability of the DSM-III criteria for posttraumatic stress disorder: An epidemiological study. *Psychological Medicine, 21,* 1-9.

Desivilya, H.S., Gal, R., & Ayalon, O. (1996). Extent of victimization, traumatic stress symptoms, and adjustment of terrorist assault survivors: A long-term follow-up. *Journal of Traumatic Stress, 9,* 881-889.

Durham, T.W., McCammon, S.L., & Allison, E.J. (1985). The psychological impact of disaster on rescue personnel. *Annals of Emergency Medicine, 14,* 664-668.

Foa, E.B., Riggs, D., Dancu, C.V., & Rothbaum, B.O. (1993). Reliability and validity of a brief instrument for assessing post-traumatic stress disorder. *Journal of Traumatic Stress, 6,* 459-473.

Fothergill, A., Maestas, E.G., & Darlington, J.D. (1999). Race, ethnicity and disasters in the United States: A review of the literature. *Disasters, 23,* 156-173.

Frasure-Smith, N., Lesperance, F., & Talajic, M. (2000). The prognostic importance of depression, anxiety, anger, and social support following myocardial infarction: Opportunities for improving survival. In P.M. McCabe & N. Schneiderman (Eds.), *Stress, coping, and cardiovascular disease* (pp. 203-228). Mahwah, NJ: Lawrence Erlbaum Associates.

Fredrick, C.J. (1994). The psychology of terrorism and torture in war and peace: Diagnosis and treatment of victims. In R.P. Liberman & J. Yager (Eds.), *Stress in psychiatric disorders* (pp. 140-159). New York: Springer Publishing.

Freedman, S.A., Brandes, D., Peri, T., & Shalev, A. (1999). Predictors of chronic post-traumatic stress disorder: A prospective study. *British Journal of Psychiatry, 174,* 353-359.

Freedy, J.R., Shaw, D.L., Jarrell, M.P., & Masters, C.R. (1992). Toward an understanding of the psychological impact of natural disasters: An application of the conservation resources stress model. *Journal of Traumatic Stress, 5,* 441-454.

Galea, S., Ahern, J., Resnick H., Kilpatrick, D., Bucuvalas, M., Gold, J., & Vlahov, D. (2002). Psychological sequelae of the September 11 terrorist attacks in New York City. *New England Journal of Medicine, 346,* 982-987.

Galea, S., Resnick, H., Ahern, J., Gold, J., Bucuvalas, M., & Kilpatrick, D. (2002). Posttraumatic stress disorder in Manhattan, New York City, after the September 11th terrorist attacks. *Journal of Urban Health Studies, 79,* 340-353.

Goenjian, A.K., Steinberg, A.M., Najarian, L.M., Fairbanks, L.A., Tashjian, M., & Pynoos, R.S. (2000). Prospective study of posttraumatic stress, anxiety, and depressive reactions after earthquake and political violence. *American Journal of Psychiatry, 157,* 911-916.

Green, B.L. (1993). Disasters and posttraumatic stress disorder. In J.R.T. Davidson & E.B. Foa (Eds.), *Posttraumatic stress disorder: DSM-IV and beyond.* Washington, DC: American Psychiatric Press.

Holahan, C.J., & Moos, R.H. (1981). Social support and psychological distress: A longitudinal analysis. *Journal of Abnormal Psychology, 90,* 365-370.

Holahan, C.J., & Moos, R.H. (1982). Social support and adjustment: Predictive benefits of social climate indices. *American Journal of Community Psychology, 10,* 403-415.

Holloway, H.C., & Fullerton, C.S. (1994). The psychology of terror and its aftermath. In R.J. Ursano, B.G. McCaughey, & C.S. Fullerton (Eds.), *Individual and community responses to trauma and disaster: The structure of human chaos* (pp. 31-45). Cambridge, UK: Cambridge University Press.

Horowitz, M.J. (1976). *Stress response syndromes.* Northvale, NJ: Jason Aronson.

Horowitz, M.J., Wilner, N., & Alvarez, W. (1979). The Impact of Event Scale: A measure of subjective stress. *Psychosomatic Medicine, 41,* 209-218.

Johnsen, B.H., Eid, J., Lovstad, T., & Michelsen, L.T. (1997). Posttraumatic stress symptoms in non-exposed, victims, spontaneous rescuers after an avalanche. *Journal of Traumatic Stress, 10,* 133-140.

Kaniasty, K., & Norris, F.H. (1993). A test of the social support deterioration model in the context of natural disaster. *Journal of Personality and Social Psychology, 64,* 395-408.

Keane, T.M., Caddell, J.M., & Taylor, K.L. (1988). Mississippi Scale for Combat-Related Posttraumatic Stress Disorder: Three studies in reliability and validity. *Journal of Consulting and Clinical Psychology, 56,* 85-90.

Keane, T.M., Wolfe, J., & Taylor, K.L. (1987). Post-traumatic stress disorder: Evidence for diagnostic validity and methods of psychological assessment. *Journal of Clinical Psychology, 43,* 32-43.

Kessler, R.C., Sonnega, A., Bromet, E., Hughes, M., & Nelson, C.B. (1995). Posttraumatic stress disorder in the National Comorbidity Survey. *Archives of General Psychiatry, 52,* 1048-1060.

Koenen, K.C., Harley, R., Lyons, M., Wolfe, J., Simpson, J.C., Goldberg, J., Eisen, S.A., & Tsuang, M. (2002). A twin registry study of familial and individual risk factors for trauma exposure and posttraumatic stress disorder. *The Journal of Nervous and Mental Disease, 190,* 209-218.

Koopman, C., Classen, C., & Spiegel, D. (1994). Predictors of posttraumatic stress symptoms among survivors of the Oakland/Berkeley, Calif., firestorm. *American Journal of Psychiatry, 151,* 888-894.

Kulka, R.A., Schlenger, W.E., Fairbandk, J.A., Hough, R.L., Jordan, K.B., Marmar, C.R., & Weiss, D.S. (1990). *Trauma and the Vietnam War generation: Report of findings from the National Vietnam Veterans Readjustment Study.* Psychosocial Stress Series, No. 18. New York: Brunner/Mazel.

Litz, B.T., & Weathers, F.W. (1994). The diagnosis and assessment of post-traumatic stress disorder in adults. In M.B. Williams & J.F. Sommer Jr. (Eds.), *Handbook of post-traumatic therapy* (pp. 19-37). Westport, CT: Greenwood Press.

MacDonald, C., Chamberlain, K., Long, N., & Flett, R. (2000). Posttraumatic stress disorder and interpersonal functioning in Vietnam War veterans: A mediational model. *Journal of Traumatic Stress, 12,* 701-707.

McFarlane, A.C. (1986). Posttraumatic morbidity of disaster. *Journal of Nervous and Mental Disease, 174,* 4-14.

McFarlane, A.C. (1988). The longitudinal course of posttraumatic morbidity: The range of outcomes and their predictors. *Journal of Nervous & Mental Disease, 176*(1), 30-39.

McFarlane, A.C. (1992). Avoidance and intrusion in posttraumatic stress disorder. *Journal of Nervous and Mental Disease, 180,* 439-445.

McFarlane, A.C., Atchison, M., Rafalowicz, E., & Papay, P. (1994). Physical symptoms in post-traumatic stress disorder. *Journal of Psychosomatic Research, 38,* 715-726.

McFarlane, A.C., & Papay, P. (1992). Multiple diagnoses in posttraumatic stress disorder in the victims of a natural disaster. *The Journal of Nervous and Mental Disease, 180,* 498-504.

Morgan, C.A., Hill, S., Fox, P., Kingham, P., & Southwick, S.M. (1999). Anniversary reactions in Gulf War veterans: A follow-up inquiry 6 years after the war. *American Journal of Psychiatry, 156*(7), 1075-1079.

Morgan, C.A., Kingham, P., Nicolaou, A., & Southwick, S.M. (1998). Anniversary reactions in Gulf War veterans: A naturalistic inquiry 2 years after the Gulf War. *Journal of Traumatic Stress, 11*(1), 165-171.

Norris, F.H. (1992). Epidemiology of trauma: Frequency and impact of different potentially traumatic events on different demographic groups. *Journal of Consulting and Clinical Psychology, 60,* 409-418.

Norris, F.H. (2002). Psychosocial consequences of disasters. *PTSD Research Quarterly, 13,* 1-7.

Norris, F.H., & Murrell, S. A. (1990). Social support, life events, and stress as modifiers of adjustment to bereavement by older adults. *Psychology and Aging, 5,* 429-436.

Norris, F.H., Perilla, J.L, Ibanez, G.E., & Murphy, A.D. (2001). Sex differences in symptoms of posttraumatic stress: Does culture play a role? *Journal of Traumatic Stress, 14,* 7-28.

North, C.S. (2001). The course of post-traumatic stress disorder after the Oklahoma City bombing. *Military Medicine, 166,* 51-52.

North, C.S., Nixon, S.J., Shariat, S., Mallonee, S., McMillen, J.C., Spitznagel, E.L., & Smith, E.M. (1999). Psychiatric disorders among survivors of the Oklahoma City bombing. *Journal of the American Medical Association, 282,* 755-763.

North, C.S., Tivis, L., McMillen, J.C., Pfefferbaum, B., Spitznagel, E.L., Cox, J., Nixon, S., Bunch, K.P., & Smith, E.M. (2002). Psychiatric disorders in rescue workers after the Oklahoma City bombing. *American Journal of Psychiatry, 159,* 857-859.

Ozer, E.J., Best, S.R., Lipsey K.T.L., & Weiss, D.S. (2003). Predictors of posttraumatic stress disorder and symptoms in adults: A meta-analysis. *Psychological Bulletin, 129,* 52-71.

Pfefferbaum, B., Moore, V., McDonald, N., Maynard, B., Gurwitch, R., & Nixon, S. (1999). The role of exposure in posttraumatic stress in youths following the 1995 bombing. *Journal of the State Medical Association, 92,* 164-167.

Pfefferbaum, B., Nixon, S., Tivis, R., Doughty, D., Pynoos, R., Gurwitch, R., & Foy, D. (2001). Television exposure in children after a terrorist incident. *Psychiatry, 64,* 202-211.

Pfefferbaum, B., Nixon, S., Tucker, P., Tivis, R., Morre, V., Gurwitch, R., Pynoos, R., & Geis, H. (1999). Posttraumatic stress response in bereaved children after the Oklahoma City bombing. *Journal of the American Academy of Child and Adolescent Psychiatry, 38,* 1372-1379.

Pfefferbaum, B., Seale, T.W., McDonald, N.B., Brandt, E.N., Rainwater, S.M., Maynard, B.T., Meierhoefer, B., & Miller, P.D. (2000). Posttraumatic stress two years after the Oklahoma City bombing in youths geographically distant from the explosion. *Psychiatry, 63,* 358-370.

Pfefferbaum, B., Vinekar, S., Trautman, R.P., Lensgraf, S.J., Reddy, C., Patel, N., & Ford, A.L. (2002). The effect of loss and trauma on substance use behavior in individuals seeking support services after the 1995 Oklahoma City bombing. *Annals of Clinical Psychiatry, 14,* 89-95.

Polusney, M.A., & Follette, V.M. (1995). Long-term correlates of child sexual abuse: Theory and review of the empirical literature. *Applied and Preventive Psychology: Current Scientific Perspective, 4,* 143-166.

Prigerson, H.G., Bridge, J., Maciejewski, P.K., Beery, L.C., Rosenheck, R.A., Jacobs, S.C., Bierhals, A.J., Kupfer, D.J., & Brent, D.A. (1999). Influence of traumatic grief on suicidal ideation among young adults. *American Journal of Psychiatry, 156*(12), 1994-1995.

Raphael, B. (1986). Victims and helpers. In B. Raphael (Ed.), *When disaster strikes: How individuals and communities cope with catastrophe* (pp. 222-244). New York: Basic Books, Inc.

Raphael, B., Singh, B., Bradbury, L., & Lambert, F. (1983). Who helps the helpers? The effects of a disaster on the rescue workers. *Omega—Journal of Death & Dying, 14*(1), 9-20.

Resnick, H.S., Kilpatrick, D.G., & Lipovsky, J.A. (1991). Assessment of rape-related posttraumatic stress disorder: Stressor and symptom dimensions. *Journal of Consulting and Clinical Psychology, 3,* 561-572.

Riggs, D.S., Byrne, C.A., Weathers, F.W., & Litz, B.T. (1997). The quality of the intimate relationships of male Vietnam veterans: Problems associated with posttraumatic stress disorder. *Journal of Traumatic Stress, 11,* 87-101.

Saigh, P.A., Yasik, A.E., Sack, W.H., & Koplewicz, H.S. (1999). Child-adolescent posttraumatic stress disorder: Prevalence, risk factors, and comorbidity. In P.A. Saigh & J.D. Bremner (Eds.), *Posttraumatic stress disorder: A comprehensive text* (pp. 18-43). Boston: Allyn & Bacon.

Schiesel, S., & Hansell, S. (2001, September 12). Communications: A flood of anxious calls clog phone lines and TV channels go off the air. *The New York Times,* p. A8.

Schlenger, W.E., Caddell, J.M., Ebert, L., Jordan, B.K., Rourke, K.M., Wislon, D., Thalji, L., Dennis, J.M., Fairbank, J.A., & Kulka, R.A. (2002). Psychological reactions to terrorist attacks: Findings from the National Study of Americans' Reactions to September 11. *Journal of American Medical Association, 288,* 581-588.

Schuster, M.A., Stein, B.D., Jaycox, L.H., Collins, R.L., Marshall, G.N., Elliott, M.N., Zhou, A.J., Kanouse, D.E., Morrison, J.L., & Berry, S.H. (2001). A national survey of stress reactions after the September 11, 2001, terrorist attacks. *New England Journal of Medicine, 345,* 1507-1512.

Schwarz, E., & Perry, B.D. (1994). The posttraumatic response in children and adolescents. *Psychiatric Clinics of North America, 17,* 311-326.

Shalev, A. (1992). Posttraumatic stress disorder among injured survivors of a terrorist attack: Predictive value of early intrusion and avoidance symptoms. *The Journal of Nervous and Mental Disease, 180,* 505-600.

Shalev, A.Y., Freedman, S., Peri, T., Brandes, D., Sahar, T., Orr, S.P., & Pitman, R.K. (1998). Prospective study of posttraumatic stress disorder and depression following trauma. *American Journal of Psychiatry, 155,* 630-637.

Shalev, A.Y., Peri, T., Canetti, L., & Schreiber, S. (1996). Predictors of PTSD in injured trauma survivors: A prospective study. *American Journal of Psychiatry, 153,* 219-225.

Shore, J.H., Tatum, E.L., & Vollmer, W.M. (1986). Psychiatric reactions to disaster: The Mount St. Helens experience. *American Journal of Psychiatry, 143,* 590-595.

Silove, D., & Kinzie, J.D. (2001). Survivors of war trauma, mass violence, and civilian terror. In E. Gerrity, T.M. Keane, & F. Tuma (Eds.), *The mental health consequences of torture* (pp. 159-174). New York: Kluwer Academic/Plenum Publishers.

Silver, R.C., Holman, E.A., McIntosh, D.N., Poulin, M., & Gil-Rivas, V. (2002). Nationwide longitudinal study of psychological responses to September 11. *JAMA, 288,* 1235-1244.

Solomon, S.D., & Canino, G.J. (1990). Appropriateness of DSM-III-R criteria for posttraumatic stress disorder. *Comprehensive Psychiatry, 31,* 1-11.

Solomon, Z., & Mikulincer, M. (1992). Aftermath of combat stress reactions: A three-year study. *British Journal of Clinical Psychology, 31,* 21-32.

Solomon, Z., Mikulincer, M., & Avitzur, E. (1988). Coping, locus of control, social support, and combat-related posttraumatic stress disorder: A prospective study. *Journal of Personality and Social Psychology, 55,* 279-285.

Spitzer, R.L., Gibbon, M., & Williams, J.B.W. (1996). *Structured Clinical Interview for DSM-IV Axis I disorders.* New York: New York State Psychiatric Institute, Biometrics Research Department.

Steinglass, P., & Gerrity, E. (1990). Natural disasters and posttraumatic stress disorder: Short-term vs. long-term recovery in two disaster-affected communities. *Journal of Applied Social Psychology, 20,* 1746-1765.

Thompson, M.P., Norris, F.H., & Hanacek, B. (1993). Age differences in the psychological consequences of Hurricane Hugo. *Psychology and Aging, 8,* 606-616.

Tucker, P., Pfefferbaum, B., Nixon, S.J., & Dickson, W. (2000). Predictors of posttraumatic stress symptoms in Oklahoma City: Exposure, social support, peritraumatic response. *Journal of Behavioral Health Services and Research, 27,* 406-416.

United States Office of the Coordinator for Counterterrorism (2001). *Patterns of global terrorism.* Washington, DC: U.S. Department of State.

Ursano, R.J., Fullerton, C.S., & McCaughey, B.G. (1994). Trauma and disaster. In R.J. Ursano, B.G. McCaughey, & C.S. Fullerton (Eds.), *Individual and community responses to trauma and disaster: The structure of human chaos.* Boston: Cambridge University Press.

Weathers, F.W., & Keane, T.M. (1999). Psychological assessment of traumatized adults. In P.A. Saigh & J.D. Bremner (Eds.), *Posttraumatic stress disorder: A comprehensive text.* Boston: Allyn and Bacon.

Weathers, F.W., Litz, B.T., Herman, D.S., Huska, J.A., & Keane, T.M. (1993). The PTSD Checklist (PCL): Reliability, validity, and diagnostic utility. Paper presented at the annual meeting of the International Society for Traumatic Stress Studies, San Antonio, California.

Weintraub, D., & Ruskin, P.E. (1999). Posttraumatic stress disorder in the elderly: A review. *Harvard Review of Psychiatry, 7,* 144-152.

Weisaeth, L. (1996). PTSD: The stressor response relationship. *Bailliere's Clinical Psychiatry, 2,* 217-228.

Weiss, D.S., & Marmar, C.R. (1997). The Impact of Event Scale—Revised. In J.P. Wilson & T.M. Keane (Eds.), *Assessing psychological trauma and PTSD* (pp. 399-411). New York: Guilford Press.

Weiss, D.S., Marmar, C.R., Metzler, T.J., & Ronfeldt, H.M. (1995). Predicting symptomatic distress in emergency services personnel. *Journal of Consulting and Clinical Psychology, 63,* 361-368.

Wilson, F.C., Poole, A.D., & Trew, K. (1997). Psychological distress in police officers following critical incidents. *Irish Journal of Psychology, 18*(3), 321-340.

Wilson, J.P., & Keane, T.M. (1997). *Assessing psychological trauma and PTSD.* New York: The Guilford Press.

Witztum, E., & Kotler, M. (2000). Historical and cultural construction of PTSD in Israel. In A.Y. Shalev, R. Yehuda, & A.C. McFarlane (Eds.), *International handbook of human response to trauma* (pp. 103-114). New York: Kluwer Academic/Plenum Publishers.

Wolf, J., & Kimerling, R. (1997). Gender issues in the assessment of posttraumatic stress disorder. In J.P. Wilson & T.M. Keane (Eds.), *Assessing psychological trauma and PTSD* (pp. 192-238). New York: Guilford Press.

Wright, K.M., Ursano, R.J., Bartone, P.T., & Ingraham, L.H. (1990). The shared experience of catastrophe: An expanded classification of the disaster community. *American Journal of Orthospychiatry, 60,* 35-42.

Yehuda, R., McFarlane, A.C., & Shalev, A.Y. (1998). Predicting the development of posttraumatic stress disorder from the acute response to a traumatic event. *Biological Psychiatry, 44,* 1305-1313.

Chapter 7

Clinical Issues
in the Psychopharmacology of PTSD

Steven B. Rudin
Serena Y. Volpp
Randall D. Marshall

INTRODUCTION

Post-traumatic stress disorder (PTSD) is a complex and challenging condition to treat. Individuals with PTSD may suffer from considerable morbidity, including chronic emotional distress, disruption of interpersonal relationships, inability to work, and high rates of health problems. Although PTSD does seem to remit spontaneously for some individuals, for many symptoms may continue for years to decades (Kessler, Sonnega, Bromet, Hughes, & Nelson, 1995; Breslau & Davis, 1992).

Over the past twenty years, researchers and clinicians have developed a growing interest in the psychological and neurobiological underpinnings of PTSD. Research on the treatment of PTSD has focused on psychosocial treatments (mainly cognitive-behavioral approaches) as well as pharmacotherapy. Both have shown clinical efficacy. In clinical practice, they are often combined. However, there have been neither head-to-head trials of psychosocial therapy versus psychotropic medication treatment nor published trials of combined psychosocial/pharmacotherapy approaches. Case reports and sequential trials do suggest that combination treatment, in particular the addition of exposure therapy to medication treatment, can improve treatment response (Marshall, Carcamo, Blanco, & Liebowitz, 2003; Foa et al., unpublished trial).

The writing of this chapter was supported by NIMH Grant MH01412 (R.M.), the New York Times Foundation (R.M., S.R.), and the Atlantic Philanthropies Foundation (R.M., S.R.)

Residual symptoms following completion of controlled trials of both medication and psychotherapy demonstrate that many patients continue to have symptoms despite receiving treatment (Marshall & Cloitre, 2000). This may be due to a variety of limitations such as relative intractability in a severe, chronic disorder, mixed efficacy of existing treatments, clinicians' lack of special training required to deliver effective treatments, patients' inconsistent compliance with treatment, or other factors.

Psychosocial treatment and pharmacotherapy have been shown to be effective in all three symptom clusters of PTSD: reexperiencing, avoidance/numbing, and hyperarousal. Experts in PTSD advocate considering both treatment approaches when evaluating any patient with PTSD. This still leaves many clinicians uncertain about when and with whom to initiate which treatment.

This chapter will provide the clinician with an overview and guidance in the use of medication for patients with PTSD. To this end, several areas will be explored. First, a rationale for using psychotropic medications will be presented. We will review the clinical issues that may influence treatment recommendations, including comorbidity, compliance, the patient-therapist relationship, patients' attitudes toward medication, and therapists' attitudes toward medication. Next, a general review of the literature of pharmacotherapy in PTSD will be presented. This section will explain the applicability of the current research; symptom cluster response; and the issues of continuation, maintenance, and termination of medication. Finally, the evidence for the use of each class of medication will be critically summarized.

We focus here on the pharmacotherapy of PTSD in adults, as the majority of research has been conducted in this population. Research in children will be presented where there have been significant findings. The pharmacotherapy of acute stress disorder (ASD) is an area of substantial controversy that is beyond the scope of this chapter. In short, psychotropic agents might theoretically facilitate the normal recovery process that takes place after a traumatic event or might mitigate the neurobiological processes that precede the development of chronic PTSD. However, the data from which to draw these conclusions are insufficient at this time. Therefore, this chapter focuses primarily on PTSD in adults.

RATIONALE FOR PHARMACOTHERAPY

PTSD is the only psychiatric condition other than adjustment disorder that requires a particular stressor to precede the appearance of symptoms

(American Psychiatric Association, 1994). To some it may seem counter-intuitive to prescribe medication for a disorder that is triggered by an identifiable psychosocial stressor; however, many reasons and indications for using a biological approach to treatment exist. Some evidence indicates that patients with PTSD have neurobiological alterations in the central nervous system (Yehuda, 2002). Alterations have been noted in the autonomic nervous system, in the hypothalamic-pituitary-adrenal axis, in endogenous opioid regulation and function, in catecholaminergic and serotonergic systems, and in various aspects of the immune system (Marshall & Garakani, 2002). In addition, some have shown structural changes in the brain that may correlate with the symptoms of PTSD (Katz & Yehuda, Chapter 4 in this volume; Vermetten & Bremner, 2002). Although psychosocial therapies are assumed to modulate the neurobiology of PTSD, the existence of neurobiological alterations in PTSD is also a logical rationale for the use of psychotropic agents.

Other reasons to consider medication in patients with PTSD are more practical. Psychosocial treatments for PTSD, such as prolonged exposure, can be too psychologically and emotionally taxing for some patients and can temporarily worsen symptoms. Medication can be used to ameliorate severe symptoms prior to the initiation of, or concurrently with, psychosocial treatments. This may facilitate a patient's ability to engage in psychosocial treatments and may improve outcome. Medications may also be prescribed for patients who try psychosocial therapies but who (1) terminate such therapies because they cannot tolerate aspects of the treatment, (2) remain symptomatic after a course of psychotherapy, or (3) relapse after a successful course of psychotherapy. In addition, some patients will opt for medication instead of psychosocial treatments for personal or cultural reasons.

Medication should be considered more strongly in high-risk patients such as those with suicidal or homicidal ideation, or those with medication-responsive comorbidity such as psychotic features or severe panic disorder. Severe pretreatment PTSD or depressive symptoms, prominent levels of anger, great difficulties in interpersonal relationships, and problems tolerating anxiety have been associated with higher drop-out rates from psychosocial treatment (Chemtob, Novaco, Hamada, Gross, & Smith, 1997; Resick, Nishith, & Weaver, 1999; Riggs, Cancu, Gershuny, Greenberg, & Foa, 1992; Jaycox & Foa, 1996; Scott & Stradling, 1997; Marshall & Cloitre, 2000). These risk factors might be reduced via concomitant pharmacological treatment.

CLINICAL ISSUES RELATED
TO PHARMACOTHERAPY OF PTSD

Comorbidity

The presence of psychiatric comorbidity is an important factor to consider when deciding whether to use medication. Comorbidity should guide the class of medication chosen. The majority of patients with PTSD suffer from one or more comorbid psychiatric disorders. Breslau, Davis, Andreski, & Peterson (1991) reported that 83 percent of subjects in an urban population also met criteria for one or more additional DSM diagnoses. Kessler et al. (1995) found in the National Comorbidity Survey that a lifetime history of at least one other psychiatric disorder was present in 88.3 percent of men and 79 percent of women with lifetime PTSD. The most common comorbid psychiatric diagnoses include major depressive disorder, panic disorder, generalized anxiety disorder, somatization disorder, substance abuse, and personality disorders (Brady, Sonne, & Roberts, 1995; Breslau et al., 1991; Kessler et al., 1995).

It is important for the clinician to investigate the possibility of additional psychiatric disorders in the traumatized patient. Furthermore, an understanding of what is most problematic for the patient should be identified and addressed first. For example, disabling panic disorder or severe depression may need to be treated with medication before a patient can engage in a psychosocial treatment for PTSD. For severe personality disorders, some experts advocate using medication during a preliminary "stabilization" phase that focuses on fostering higher-level coping mechanisms, reducing impulsive and self-destructive behaviors, and teaching self-soothing techniques. Clearly, medication should be prescribed in patients with a comorbid psychotic or bipolar disorder.

Compliance

The use of medication in PTSD presents several unique clinical challenges. Owing to the almost universal issues of safety, trust, control, and intimacy involved in trauma-related psychopathology (Resick, 2001), PTSD patients are often difficult to engage and maintain in treatment. Avoidance as a defense against the anxiety produced by confrontation of the traumatic event may interfere with the traumatized patient's ability to initiate and stay in treatment. Other factors, such as secondary gain (i.e., material and psychological benefits derived from being in the sick role) as well as pessimism based on prior experiences with unsuccessful therapy may lead to reluctance to engage in a therapeutic alliance or may lead to noncompliance.

It is unclear how drop-out rates for patients with PTSD compare to drop-out rates of patients with other severe chronic disorders. It is also unclear how the drop-out rates differ between psychosocial treatment and medication treatment, primarily because of methodologic differences across sites and trials as to how "dropout" is defined and reported. There is also little research investigating patients' reasons for leaving treatment.

Reasons for dropping out of psychosocial treatment versus medication treatment may differ. For example, the challenges inherent in prolonged exposure (PE) therapy may deter the avoidant patient. The homework and consistency necessary for other cognitive-behavioral techniques may be too cumbersome for the patient who is very depressed, cannot concentrate, or who is actively using substances. Reasons for stopping medication treatment may be related to side effects, attitudes and fears about taking medication that have not been addressed by the physician, problems in the patient-therapist relationship, or the therapist's conscious and/or unconscious attitudes toward medication.

The Patient-Therapist Relationship

Successful psychopharmacotherapy involves active collaboration between the patient/psychiatrist pair, the patient/therapist pair, and the therapist/psychiatrist pair (in the case of split treatment). The therapeutic alliance is essential when prescribing medications to the patient with PTSD, just as it is with any other health-related intervention. The clinician should foster an atmosphere of *collaborative empiricism*. This phrase refers to a treatment situation in which the patient assumes an active part in the decision-making process. The patient perceives his or her role as that of observer of treatment response and adverse effects, along with the physician. It is important to recognize that patients with PTSD have often been violated in some way, whether by assault, rape, or terrorist attack. The patient with PTSD has often felt misunderstood and alone. In many cases, he or she has perceived others to be judgmental, critical, or blaming. This perception is unfortunately often accurate. Perceived helplessness and fear of being harmed or dominated are features of the disorder and may complicate the therapist-patient relationship. In addition, patients with chronic or refractory symptoms may have become disillusioned by mental health treatment. They may have tried various medications in the past without success, or may have been coerced by loved ones to see a mental health professional. Some patients are mandated to treatment as a requirement for receiving disability benefits. Patients with PTSD need to feel understood and safe in order to trust the clinician. By the same token, patients with PTSD often have powerful ideas about medication related to their views of safety, trust, control, and intimacy. If not discussed explicitly, these ideas, beliefs, or fears may manifest themselves in an increase of perceived adverse effects, noncompliance, or termination of the treatment.

The therapist's attitudes about medication may also influence the treatment. Clinicians who recommend a pharmacological treatment and those who prescribe medication should be aware of their own biases. Subtly negative views about medication may encourage the patient to give up on it. Overly optimistic views about medication may be received with disbelief and mistrust, especially when the patient has already tried medication without success, or in cases in which symptom remission is improbable. These biases are especially important in split treatments, in which one clinician is providing psychotherapy and the other medication.

Patients' Attitudes About Medication

For many patients, the idea of a medication that will help them to feel less tense, sleep better, and cope better is very appealing. Psychoeducation about the neurobiology of PTSD may be comforting, helping patients make sense of an experience that is often confusing and frightening. These patients are often reassured by factual explanations of the effects of stress, depression, and anxiety on the brain. Depending on previous experiences and attitudes, they may grasp the idea of correcting a "chemical imbalance" with optimism and relief.

Unfortunately, not all individuals feel reassured when offered a biological treatment for a psychological disorder. Conflicted or entirely negative views about medication are common. Such patients may feel a strong sense of injustice about taking a medication for a problem that is a result of their victimization. The idea of being "labeled" with a disorder may reinforce existing ideas that the world is critical and blaming. Therapists need to reflect on their attitudes to ensure they do not share the view that diagnoses represent labeling. This attitude has its roots in the long-standing stigma associated with psychiatric illness (Marshall, Spitzer, & Liebowitz, 1999). The suggestion of having a disorder that can respond to medication may have powerful negative effects on self-esteem. Medication may signify personal defectiveness or inability to cope. It is important for the prescribing clinician to listen to these concerns empathically, but to gently confront the patient's distortions with nonjudgmental, fact-based optimism.

Fear is another common attitude toward medication. The patient may fear ingesting an exogenous substance into a body that already feels out of control. Common fears include being oversedated, made more anxious, or being "turned into a zombie." Given that hypervigilance is a way to cope with the perception of perpetual threat, patients may equate relaxation via medication with being dangerously off guard. For example, one patient refused medication because he did not want to sleep more than four hours a night, saying he "needed to know what was going on around him." This may be especially frightening for patients who live in dangerous neighborhoods, need to care for others, or are involved in litigation. Fear of adverse effects may signify loss of bodily integrity or self-boundaries. This is especially salient if the adverse effects of the medication (e.g., nausea, dizziness, numbness) resemble the feelings that the patient experienced during the traumatic event.

Patients who are referred for medication evaluation after a psychosocial treatment has not been optimally effective may feel a sense of personal failure and shame about needing medication. Others may interpret the recommendation as the therapist's wish to suppress the patient's complaints, especially if medication is viewed as drugging or numbing. On the other hand, patients may hope that medication will erase memories of the traumatic event, especially those who wish to forget it ever happened. Very avoidant patients may be relieved to be in pharmacotherapy only, as the idea of talking specifically about their trauma may be seen as more threatening than taking medication.

Therapists' Attitudes Toward Medication

Historically, American psychiatrists and other mental health providers have been divided in their attitudes toward the value of drugs in the treatment of mental conditions, whether alone or in combination with psychotherapy (Klerman, 1991). With the proliferation of new therapies over the past thirty years, both psychosocial and biological, the debate on how to integrate psychotherapy with medication has become ever more important. This debate is relevant to PTSD, for which both psychosocial and biological treatments have demonstrated comparable efficacy. The dialogue articulated by Klerman (1991) has focused on the potential effects of drug therapy on psychotherapy and, to a lesser degree, vice versa.

Klerman (1991) describes several factors that may influence a therapist's negative view of medication. For example, drug-induced reduction of symptoms as motives for discontinuing psychotherapy may be a concern among therapists who have the bias that severe symptoms are needed as motivation to make character changes. Another concern is that pharmaco-therapy undercuts defenses and that symptom substitution or other compensatory symptom formation may ensue. Some therapists view drug therapy as a "crutch" that fosters dependence. This ironically resembles the belief among some biologically oriented mental health providers that psychotherapy fosters dependence and perpetuates symptoms. In addition, many therapists may doubt the emotional costs of psychotherapy, challenging the "no pain, no gain" underpinnings of some forms of psychotherapy. Although this has been considered a maxim in both behavioral and psychodynamic schools, no evidence indicates that moderating levels of anxiety and fear will impede treatment response. Research with phobias, for example, has suggested that reducing anxiety with diazepam may facilitate exposure to the phobic stimulus and does not reduce the efficacy of exposure therapy (Schneier, Marshall, Street, Heimberg, & Juster, 1996).

Although not proven in the case of PTSD, psychotherapy and drug therapy are thought to act through different yet complementary mechanisms, e.g., learning mechanisms versus neurotransmitter modulation. Drugs may facilitate psychotherapy for many patients, who may be better able to work through difficult issues without being distracted by intense emotional distress. Reciprocally, psychotherapy may facilitate drug treatment, by fortifying the patient's strengths, building trust, and encouraging compliance with medication.

OVERVIEW OF THE LITERATURE ON PHARMACOTHERAPY

Almost every class of medication has been tried in the treatment of PTSD. Multiple case reports have noted improvement in various combinations of specific and nonspecific symptoms associated with PTSD such as insomnia, nightmares, and anxiety. Since the operationalization of the diagnosis of PTSD in the early 1980s, investigators have been able to develop more refined diagnosis and assessment tools, e.g., the Clinician Administered PTSD Scale (CAPS) and the Davidson Trauma Scale (DTS), to identify and monitor improvement in patients with PTSD. Very recent controlled trials have shown that selective serotonin reuptake inhibitors (SSRIs) can be effective for the three core symptom clusters of PTSD. This contrasts the prior long-standing view that medications were helpful only as adjuncts to psychotherapy (Charney, Deutch, Krystal, Southwick, & Davis, 1993). In an examination of worldwide randomized controlled trials for the years 1966 through 1999, Stein, Zungu-Dirwayi, van der Linder, & Seedat (2002) concluded that no clear evidence shows that any particular class of medication is more effective or better tolerated than any other. However, the largest trials to date demonstrating efficacy have been with SSRIs. The only two medications with FDA approval for the treatment of PTSD are the SSRIs sertraline (Zoloft) and paroxetine (Paxil).

The pharmacotherapy literature in PTSD for medication classes other than the SSRIs is limited to a paucity of controlled trials in diverse populations and a small number of replication studies. In addition, no published controlled studies combine or compare pharmacologic and psychosocial treatments, in spite of the fact that the two modalities are often combined in clinical practice, particularly with chronic PTSD (i.e., meeting DSM-IV-TR criteria for PTSD for more than three months). Furthermore, most trials have studied only acute treatment response. This leaves clinical issues such as optimum length of treatment, long-term treatment response, relapse rates after a treatment is discontinued, and clinical approaches to relapse largely unanswered (Marshall, Davidson, & Yehuda, 1998).

Applicability of Research Findings to General Practice

Until recently, most pharmacotherapy studies in PTSD were conducted in adult male war veterans with PTSD. Recent trials with other populations, such as victims of rape, assault, accidents, natural disasters, and terrorist acts, are larger, but fewer of these studies have been undertaken. Research in children and in adults meeting criteria for acute stress disorder remains rare. Important questions remain as to the generalizability of research findings across veteran versus non–combat-related PTSD populations, from PTSD studies to persons with acute stress reactions, and from the adult literature to children with PTSD. Given that substantial gender differences in the brain may exist (Vermetten & Bremner, 2002), findings from studies done almost exclusively in males may not be applicable to women, and vice versa.

Veteran populations studied have tended to be treatment-refractory individuals. For example, many subjects were recruited from treatment centers where multimodal treatment was already available and where patients had been ill for many years. Many persons had already tried and failed to benefit from existing treatments. Issues such as comorbidity and secondary gain (i.e., threat of losing benefits) may have also affected results. Some studies have suggested that veteran status may be a predictor of poorer outcome, compared to civilians (Marshall, Davidson, & Yehuda, 1998). In summary, research findings among veterans who may be relatively treatment refractory might have been too hastily generalized to the entire disorder of PTSD. Because most males in PTSD studies had been war veterans and most women were survivors of other types of trauma, it was unclear whether the positive findings seen in civilian studies of PTSD were related to type of trauma, to gender, or to another variable. A large multicenter trial with paroxetine recently showed that treatment response was equivalent in men and women (Marshall, Beebe, Oldham, & Zaninelli, 2001).

Symptom Cluster Response

PTSD is characterized by three symptom clusters: reexperiencing, avoidance/numbing, and hyperarousal. The disorder is usually accompanied by other anxiety and depressive symptoms as well as by serious interpersonal, functional, and intrapsychic problems and deficits. The symptoms themselves can lead to social isolation and an experience of the self and the body as out of control or defective. Clinicians have long agreed that anxiety and depressive symptoms can respond dramatically to medication. Hence, the finding that these symptoms respond to medication in PTSD is not surprising. Historically, however, symptoms such as behavioral avoidance and negative self-appraisals were considered less responsive to pharmacotherapy without concurrent psychotherapy. This was based on the notion that these negative associations needed to be made conscious and then unlearned in various ways. Studies of SSRIs have unequivocally shown that behavioral avoidance does improve with pharmacotherapy alone. Whether these gains are sustained, even after the discontinuation of medication, has not been studied. However, it has been shown that SSRI medications treat all symptom clusters in PTSD (Marshall et al., 2001; Davidson, Rothbaum, van der Kolk, Sikes, & Farfel, 2001). Although not specifically reported in the early studies, monoamine oxidase inhibitors (MAOIs) and tricyclic antidepressants (TCAs) may treat all symptom clusters as well. At present, it is not known whether different categories of drugs have different specificities of action across the symptom spectrum of PTSD.

Rate and Course of Response

Despite the evidence that medication can provide relief of PTSD symptoms for many patients and has been proven to be superior to placebo in a multitude of studies, several questions regarding pharmacotherapy still remain. One such fundamental question is when to expect improvement of symptoms. The most comprehensive analysis of response rates in PTSD was conducted by Davidson, Malik, & Sutherland (1997). Although this was performed prior to the large trials of SSRIs, some interesting findings emerge in the analysis. Specifically, the time to achieve response for antidepressant and placebo is similar for tricyclics, irreversible MAOIs, and SSRIs. Placebo responders and drug responders will show improvement on the Clinical Global Inventory (CGI) relatively quickly (i.e., two to four weeks) and at around the same time. This is in contrast to the pattern in major depression, in which early responders may be more likely to be placebo rather than true responders (Quitkin, 1999). The magnitude of response in PTSD appears to be different between placebo and drug groups. Responders given active drug were more likely to be "very much improved" as compared to the rating of "improved" for placebo responders. Furthermore, the magnitude of the improvement for those on the drug predicted the overall improvement at the end of the studies. This may suggest that those patients with PTSD who are going to improve on a medication will improve early (two to four weeks) and in a robust manner. Whether this also means that patients who have not achieved response by two to four weeks should have their medications increased, switched, or augmented is still unclear. A multicenter paroxetine trial in adults with PTSD found that response rates were equivalent for 20 mg and 40 mg (Marshall et al., 2001), suggesting that response is not directly related to dose for groups of patients.

Continuation, Maintenance, and Termination of Medication

Anxiety and depressive disorders are generally chronic and/or recurrent and often require long-term medication treatment (Davidson, Pearlstein, et al., 2001). An extensive literature shows that medication reduces the likelihood of relapse and recurrence in depression, obsessive-compulsive disorder, panic disorder, and social phobia. At the time of this writing (2003), one study is looking at relapse after discontinuation of medication in patients with PTSD. In a double-blind placebo-controlled study of ninety-two patients with chronic, severe PTSD, Davidson, Pearlstein, et al. (2001) showed that those subjects maintained on sertraline for six months were 6.5 times less likely to experience relapse or clinical deterioration compared to those switched to placebo in double-blind fashion after six months of open treatment. Clinical deterioration among the placebo group occurred most frequently within the first two months after the switch to placebo and was equivalent across all three symptom clusters. Although limited by its small sample size, less comorbidly ill population, and exclusion of psychosocial treatment, this study may indicate that guidelines for maintenance treatment could be similar to those for depression and other anxiety disorders.

Still, there is little evidence in the literature regarding when to consider tapering medication in a patient who has achieved a therapeutic response. The Expert Consensus Guidelines (Foa, Davidson, & Frances, 1999) suggest considering tapering medication at the following time periods: six to twelve months for acute PTSD, twelve to twenty-four months for chronic PTSD with excellent response, and at least twenty-four months for chronic PTSD with residual symptoms. Factors to consider include presence of effective concomitant psychosocial treatment, duration of illness, time and course of recovery, and any substantial change in psychosocial factors. Current life stressors, suicide risk (past or present), history of violence, and comorbid conditions may be additional indications to continue treatment beyond the suggested guidelines (Foa et al., 1999).

Pharmacotherapy in Children

Psychotherapy, including play therapy, continues to be the standard practice for children with PTSD. Children who have disabling symptoms or comorbid disorders should be referred to a psychiatrist for a medication evaluation (AACAP, 1998). There are no published controlled studies of medication in children with PTSD. Based on evidence in the adult literature, SSRIs are considered first-line agents for children with PTSD in need of medication (Murphy, Bengtson, Tan, Carbonell, & Levin, 2000).

Pharmacotherapy in Acute Stress Disorder

There are no controlled medication trials to date in acute stress disorder; in fact, its utility as a diagnosis has been called into question (Marshall et al., 1999). The empirical literature suggests that the diagnosis fails to capture about one-third of those who will develop PTSD because it requires the presence of dissociative symptoms. Dissociation is present in only a subgroup of PTSD patients.

In a study of children, Robert, Blakeney, Villareal, Rosenberg, & Meyer (1999) found imipramine to be more effective than chloral hydrate in reducing all three acute stress symptom clusters in twenty-five pediatric burn patients with acute stress disorder, pain, and anxiety. A single controlled trial of propranolol in adults suffering severe trauma found minimal benefit for propranolol as a protective agent for the development of chronic PTSD (Pitman et al., 2002). Theoretically, prevention of central and perhaps peripheral catecholaminergic activation in the acute setting might decrease risk for developing a chronic disorder, and future research will likely focus on this promising hypothesis.

At present only a form of brief cognitive-behavioral therapy has been shown to reduce risk for developing PTSD. Medication may be used for individual symptoms (e.g., insomnia) but should not be considered a preventative treatment. However, patients with severe acute stress symptoms should be followed closely.

CLASSES OF MEDICATION: ANTIDEPRESSANTS

Selective Serotonin Reuptake Inhibitors

Selective serotonin reuptake inhibitors, also known as SSRIs, are a popular class of medications. This popularity derives from their indications in a wide array of psychiatric disorders, their relative safety, and their tolerability. Currently marketed SSRIs in the United States include fluoxetine (Prozac), sertraline (Zoloft), paroxetine (Paxil, Paxil CR), fluvoxamine (Luvox), citalopram (Celexa), and escitalopram (Lexapro).

SSRIs are considered first-line agents for the treatment of PTSD because they have come closest to targeting the entire PTSD syndrome (Yehuda, Marshall, Penkower, & Wong, 2002). They are the most extensively and systematically studied class of medications for PTSD. Studies of SSRI medications have benefited from the use of standardized clinical assessments (i.e., CAPS, DTS), superior design (double-blind and cross-over), and broadening of subject populations to non-combat-related PTSD (i.e., less chronic, refractory subjects) (Yehuda et al., 2002). Several large, controlled trials have demonstrated efficacy for sertraline (Brady et al., 1995; Davidson, Rothbaum, et al., 2001) and paroxetine (Marshall, Schneier, et al., 1998; Marshall et al., 2001; Tucker et al., 2001). Three controlled trials with negative results have also been published: two with sertraline (Medical Economics Company, 2002) and one with fluoxetine (Hertzberg, Feldman, Beckham, Kudler, & Davidson, 2002). Some explanations for this disparity are that the negative studies involved primarily war veteran subjects, who, as discussed previously, may be different from other populations in terms of type of trauma, length of illness, or severity of trauma (Marshall & Klein, 1995).

Mechanism of Action

SSRIs act by inhibiting reuptake of the neurotransmitter serotonin at the presynaptic terminal of the neuron. The immediate effect is to increase the availability of serotonin to certain regions in the brain. Symptoms associated with alterations in the serotonergic system (e.g., anxiety, depression, aggressive acting out, alcohol-related syndromes, and impulsivity) are frequently found in patients with PTSD, suggesting that serotonin has a role in the pathophysiology of the disorder (Vermetten & Bremner, 2002). That many patients' symptoms improve with SSRIs validates this hypothesis. Still, serotonin's exact role in PTSD is not entirely clear.

Adverse Effects

The adverse effects commonly attributed to SSRIs may be directly related to the excess serotonin they make available at various synaptic sites in the central and peripheral nervous systems as well as target organs with neurological innervation (i.e., the gastrointestinal system). Common adverse effects of SSRIs include gastrointestinal discomfort (e.g., nausea, vomiting), restlessness/internal agitation, insomnia, somnolence, weight gain, weight loss, apathy, sweating, and sexual dysfunction.

Most studies of SSRIs have used the same dosing strategies as those in studies of major depressive and panic disorder. For example, paroxetine should be started at 10 mg in the evening. Even lower doses (i.e., paroxetine 5 mg) can be appropriate starting doses for patients very sensitive to side effects. Given traumatized patients' tendency to be hypervigilant and somatically focused, lower starting doses and cautious titration would be advised. If patients discontinue an SSRI suddenly, they may suffer from an uncomfortable, but not dangerous, withdrawal syndrome.

Tricyclic Antidepressants

Tricyclic antidepressants were used much more frequently prior to the development of SSRIs. Their antidepressant properties were discovered fortuitously by clinicians who were testing them as antipsychotics in the 1950s and 1960s (Stahl, 2000). Although equally effective to SSRIs, TCAs have been prescribed less often over the past twenty years due to their difficult-to-tolerate adverse effects and their potential lethality in overdose. Because they were in common use prior to the advent of SSRIs, they were until recently the most commonly studied class of medication for PTSD.

TCAs all share a common baseline chemical structure of at least two joined benzene rings. They are split into two groups: tertiary amines and secondary amines. The tertiary amines include amitriptyline (Elavil), clomipramine (Anafranil), doxepin (Sinequan), imipramine (Tofranil), and trimipramine (Surmontil). Secondary amines include desipramine (Norpramin), nortriptyline (Pamelor), amoxapine (Asendin), and protriptyline (Vivactil). In the same family, maprotiline (Ludiomil) is a tetracylic compound (four benzene rings).

Three randomized clinical trials of TCAs have been conducted (Davidson et al., 1990; Kosten, Frank, Dan, McDougle, & Giller, 1991; Reist et al.,

1989), in addition to a number of open studies, retrospective studies, and case reports in adults with PTSD, as well as one controlled study of acute stress disorder in pediatric burn patients (Robert et al., 1999). Most have shown modest efficacy in reducing symptoms, although retrospective chart reviews have reported more mixed results (Yehuda et al., 2002).

Davidson et al. (1990) studied sixty-two veterans with PTSD in an eight-week double-blind trial of amitriptyline (doses 150 to 300 mg) and placebo (forty-six patients completed eight weeks of treatment). They found significant differences between amitriptyline and placebo on the Hamilton Anxiety and Hamilton Depression Scales, as well as on the CGI severity measure. Characteristics associated with a better outcome with amitriptyline included low combat exposure, good social/interpersonal adjustment, absence of comorbid disorders, low depression score, low PTSD symptom scores, lack of panic attacks, low anxiety score, and low avoidance of trauma cues.

Desipramine (doses up to 200 mg) was not found to be significantly different from placebo in a four-week crossover trial by Reist et al. (1989). The short length of this trial may have contributed to its negative results. In an eight-week trial by Kosten et al. (1991), however, imipramine (doses 150 to 300 mg) was shown to be more effective than placebo (and less effective than phenelzine) in global symptom improvement and improvement in intrusive symptoms.

Mechanism of Action

TCAs vary in their mechanism of action. Most block reuptake of both serotonin and norepinephrine, as well as, to a lesser extent, dopamine (Stahl, 2000). They have varying levels of serotonergic activity, with clomipramine notably being the most serotonergic. All tricyclics block muscarinic cholinergic receptors, histamine receptors (H-1 subtype), and alpha-1 adrenergic receptors; these three actions account for many of the adverse effects associated with TCAs. It is possible to use blood levels to guide treatment with some of the TCAs, including imipramine, nortriptyline, and desipramine, although blood level ranges have not been established specifically as indicators for effective treatment of PTSD.

Adverse Effects

As mentioned, TCAs have become less popular due to their troublesome side effects. In general, secondary amines tend to be associated with fewer adverse side effects than tertiary amines. The main types of adverse effects include those that are (1) anticholinergic, causing dry mouth, blurred vision, constipation, urinary hesitancy, and tachycardia; (2) antihistaminergic, causing sedation, carbohydrate craving, and weight gain; (3) anti-alpha-adrenergic, causing orthostatic hypotension (a drop in blood pressure when shifting from a lying or sitting position to a standing position); and (4) calcium channel effects, including slowed conduction through the atrioventricular node.

TCAs are considered antiarrhythmic drugs. They should not be prescribed to patients with preexisting cardiac conduction problems and should be used with great caution in combination with other antiarrhythmic drugs (Israel & Fava, 2000). With very rare exceptions, TCAs should not be combined with MAOIs, due to the possibility of serotonin syndrome.

Toxicity

The potential toxicity of TCAs deserves special mention. The gap between a therapeutic dose and a lethal dose of a TCA is relatively narrow. Thus, even a one-week supply of a TCA can be lethal, if taken all at once. In overdose, patients may suffer from arrhythmias, hypotension, and coma, as well as from anticholinergic toxicity (Israel & Fava, 2000), signs of which include confusion, delirium, seizures, dilated pupils, dry skin, blurred vision, fever, and the slowing down of bowel and bladder functions. Anyone suspected of overdosing on a TCA should receive immediate medical evaluation.

Monoamine Oxidase Inhibitors

Monoamine oxidase inhibitors are named for their ability to inactivate monoamine oxidase (MAO), an enzyme found in the brain and the gastrointestinal lining. MAO breaks down the monoamine neurotransmitters serotonin, norepinephrine, and dopamine, as well as tyramine and phenylethylamine. Medications that inhibit MAO (i.e., MAOIs) effectively increase the amount of these neurotransmitters in the nervous system. Currently, only the *irreversible, nonselective* MAOIs phenelzine (Nardil), tranylcypromine (Parnate), and isocarboxazid (Marplan) are available in the United States. However, the newer *reversible* and *selective* MAOI agents meclobemide and brofaromine are being studied and used outside the United States.

The nonselective MAOIs have a long tradition in psychiatric medicine and have been studied extensively. They are effective for a variety of psychiatric conditions, including panic disorder with agoraphobia, atypical depression, social phobia, bulimia, and borderline personality disorder (Liebowitz et al., 1990), as well as PTSD. Indeed, phenelzine, an irreversible MAOI, was the first medication considered as a possible treatment for PTSD. However, the use of MAOIs is currently reserved for refractory populations because, if not used properly, this class of antidepressants can have life-threatening food-drug and drug-drug interactions.

The first published study using MAOIs for PTSD (Hogben & Cornfield, 1981) reported that five patients with "traumatic war neurosis" who had not responded to multiple previous trials with antipsychotics, tricyclic antidepressants, and psychotherapy achieved favorable therapeutic responses with phenelzine. Specifically, they reported that patients felt calmer and experienced fewer nightmares, flashbacks, startle reactions, and aggressive, violent outbursts.

Since the initial observation that phenelzine could be helpful in treating the core symptoms of PTSD, several studies examining the efficacy of MAOIs in PTSD have been conducted (Stein et al., 2002). As mentioned previously, Kosten et al. (1991) compared phenelzine, imipramine, and placebo in sixty male war veterans with PTSD in an eight-week randomized trial. They showed that treatment with either phenelzine (mean dose 60 mg/day) or imipramine (150 to 300 mg/day) significantly reduced intrusion, but not avoidance or mild to moderate depressive symptoms. The study also showed a greater improvement, as assessed by the Impact of Events Scale (IES), in the phenelzine group (44 percent reduction) compared to the imipramine group (25 percent). The reader should note that the IES does not measure hyperarousal symptoms, which may limit interpretation of the results. Other randomized trials (Lerer et al., 1987; Shestatzky, Greenberg, & Lerer, 1988) reported no improvement with phenelzine compared with placebo. However, these studies have been criticized for their short length of treatment, small sample size, and comparably less globally symptomatic subjects (Marshall, Davidson, & Yehuda, 1998; Yehuda et al., 2002).

Studies with the *reversible, selective* MAOIs, brofaromine (Baker et al., 1995; Katz et al., 1995) and meclobemide (Neal, Shapland, & Fox, 1997), have shown moderate efficacy in PTSD but warrant further investigation.

Mechanism of Action

The enzyme monoamine oxidase is found in two forms, MAO A and MAO B. The MAO A form breaks down serotonin and norepinephrine. The MAO B form degrades phenylethylamine. Both MAO A and B break down dopamine and tyramine (a protein commonly found in food) (Marangell,

Silver, & Yudofsky, 1999). Compounds that inhibit MAO A are currently thought to be more relevant to psychiatric disorders. MAO B inhibitors (e.g., selegiline), helpful for the treatment of Parkinson's disease, are not currently used in mood or anxiety disorders.

Tranylcypromine, phenelzine, and isocarboxazid inhibit both MAO A and B, and are thus called nonselective. They are also called *irreversible* because once bound, the MAO enzymes are permanently inactivated. New protein must be regenerated for MAO enzymatic activity to be restored. Because these medications also inhibit MAO oxidase in the gastrointestinal lining, patients are required to adhere to special diets free of tyramine (e.g., aged cheese, wine) as well as to avoid certain medications (e.g., meperidine [Demerol], pseudoephedrine, SSRIs). If these substances enter the bloodstream in a patient on an MAOI, they can mimic neurotransmitters and overload the sympathetic nervous system, resulting in hypertensive crises (i.e., dangerously high blood pressure). Once an irreversible MAO inhibitor is discontinued, the body can produce new MAO enzyme. However, patients must continue on the special diet for two weeks after an MAOI is discontinued, as this time is required for the resynthesis of the MAO enzymes (Kennedy, McKenna, & Baker, 2000).

Selective MAOIs such as meclobemide preferentially bind to MAO A. Given that MAO B activity is preserved, patients do not need to adhere to the strict precautions required with nonselective agents. Because reversible MAOIs do not permanently deactivate MAOI, patients also may not require the same dietary restrictions (Marshall, Davidson, & Yehuda, 1998).

Adverse Effects

Irreversible, nonselective MAOIs. The most common adverse effects of irreversible, nonselective MAOIs are orthostatic hypotension (a drop in blood pressure when going from a lying or seated position to a standing position), headache, insomnia, weight gain, sexual dysfunction, peripheral edema (i.e., swelling of the extremities), afternoon somnolence, and dry mouth. As noted, the risk of hypertensive crisis can occur if a patient ingests food or medications that are contraindicated. Although there have been reports of spontaneous hypertensive crisis (Marangell et al., 1999), this is rare. The symptoms of a hypertensive crisis include severely painful headache, confusion, and sometimes neurological signs (e.g., paralysis, visual loss). Because hypertensive crises can lead to hemorrhagic stroke, patients complaining of these symptoms must have their blood pressure checked and be referred to the emergency room immediately. Another possible complication is serotonin syndrome, which may occur if a patient ingests a serotonin-enhancing medication (most commonly another antidepressant). Symptoms of serotonin syndrome include lethargy, restlessness, confusion, flushing, sweating, tremor, and jerking movements. Again, these symptoms warrant immediate medical attention.

Reversible and selective MAOIs. Meclobemide (not available in the United States) and brofaromine (discontinued in 1993) are selective and reversible inhibitors of monoamine oxidase A. Studies of meclobemide have shown no problematic interactions with normal dietary amounts of tyramine, no serious drug interactions, and few adverse side effects (Freeman, 1993; Neal et al., 1997). In a study of meclobemide (600 mg/day in divided doses) conducted by Neal et al. (1997), the most common adverse effects were mild headache and mild transient gastrointestinal upset including reflux esophagitis, increased bowel movement frequency, and loss of appetite. One participant developed a transient eczematous rash, but no participants discontinued treatment secondary to adverse effects. Brofaromine, which combines serotonin uptake inhibition with selective, reversible, MAO A inhibition, is not associated with the same safety and tolerability problems as phenelzine. In a study conducted by Katz et al. (1995), adverse effects of brofaromine included insomnia, headache, and dry mouth.

Other Antidepressants

Nefazodone

Nefazodone (Serzone) works as both a serotonin 5HT2A-receptor antagonist and as a serotonin and norepinephrine reuptake inhibitor. It is used for depression and anxiety. Although there have been no randomized controlled trials of nefazodone for PTSD, open-label trials have been positive. The Expert Consensus Guidelines on the treatment of PTSD suggest switching to nefazodone or venlafaxine if a patient has no response to an SSRI after eight weeks (Foa et al., 1999). Pooled results of six open-label trials of nefazodone for treatment of PTSD showed a 31.3 percent reduction of symptoms overall, with similar reductions in each symptom cluster (Hidalgo et al., 1999). Subsequent open-label trials have yielded similar results (Davis, Nugent, Murray, Kramer, & Petty, 2000; Garfield, Fichtner, Leveroni, & Mahableshwarkar, 2001; Zisook et al., 2000).

Although nefazodone often causes sedation, and can cause nausea, constipation, dizziness, headache, and blurred vision, it is less likely than SSRIs to cause sexual dysfunction. In 2002, the FDA issued a "black box warning" for nefazodone due to an extremely rare risk of liver failure in patients taking the drug (MEC, 2002). In January 2003, it was announced that nefazodone would be pulled off the market in Europe due to the risk of liver damage. Nefazodone is a strong inhibitor of the cytochrome P450 3A4, a liver enzyme responsible for drug metabolism, and as such it has some significant drug-drug interactions.

Venlafaxine

Like nefazodone, venlafaxine (Effexor) has been recommended for use as an alternative to SSRIs in patients with PTSD (Foa et al., 1999). It is FDA indicated for depression and generalized anxiety disorder. Venlafaxine inhibits the reuptake of serotonin and, at high doses, norepinephrine, and, at very high doses, dopamine. Almost no studies in the literature address using venlafaxine in the treatment of PTSD. One case report noted improvement in depression, anxiety, nightmares, panic attacks, sleep, irritability, and avoidance in a veteran who had not responded to several other antidepressants (Hamner & Frueh, 1998). Smajkic et al. (2001) reported that five Bosnian refugees with PTSD and depression treated with venlafaxine for six weeks showed statistically significant improvement in PTSD symptom severity but not in depressive symptoms. Eight additional patients started on venlafaxine discontinued the medication due to adverse affects.

Venlafaxine's most common adverse effects are nausea, insomnia, sedation, and dizziness. Blood pressure should be measured before initiation of the medication, as it can cause increased diastolic blood pressure. If patients discontinue venlafaxine suddenly, they may suffer from an uncomfortable, but not dangerous, withdrawal syndrome. The symptoms of withdrawal from venlafaxine include nausea, headache, and malaise. Thus, venlafaxine should be tapered slowly.

Mirtazapine

Mirtazapine (Remeron) increases the release and availability of serotonin and norepinephrine. Its main action is as an antagonist of presynaptic alpha-2 noradrenergic receptors. As with nefazodone and venlafaxine, very few studies of mirtazapine have specifically investigated its potential in PTSD. One open-label trial of six chronic PTSD patients, all with comorbid Axis I disorders, found global improvement in 50 percent of the patients (Connor, Davidson, Weisler, & Ahearn, 1999).

Patients on mirtazapine commonly complain of sedation and weight gain. Patients also may suffer from dizziness, dry mouth, constipation, and orthostatic hypotension.

Bupropion

Bupropion (Wellbutrin) is unique among antidepressants in its mechanism of action. It minimally increases dopamine and norepinephrine availability in the central nervous system. One open-label trial of bupropion for seventeen patients with chronic PTSD, many of whom had comorbid depression, found improvement in hyperarousal and depressive symptoms at the end of six weeks (Canive, Clark, Calais, Qualls, & Tuason, 1998).

Bupropion is known for its slight increase in a patient's risk of seizures. It is thus contraindicated for people at elevated risk for seizures, including those with bulimia or alcohol dependence. The seizure risk increases from one in 1,000 patients to three in 1,000 patients in doses above 450 mg/day. More common adverse effects are insomnia, agitation, tremor, headache, and dry mouth. Bupropion is rarely associated with sexual adverse effects and is sometimes used as an adjunct antidepressant for patients with SSRI-induced sexual dysfunction.

Buspirone

Buspirone (BuSpar) is a partial serotonin 5HT1A agonist that has FDA approval for generalized anxiety. Little research has been conducted with PTSD and buspirone. However, in case studies, it has been reported effective in reducing flashbacks, insomnia, and associated anxiety and depression (Fichtner & Crayton, 1994; LaPorta & Ware, 1992; Wells et al., 1991). Dizziness, headache, drowsiness, and nausea are the most commonly reported adverse effects.

Trazodone

Trazodone (Desyrel) inhibits serotonin uptake and blocks serotonin 5HT2 receptors. As it is very sedating even at low doses (25 mg), trazodone is most commonly used to treat insomnia related to another psychiatric condition. Patients generally cannot tolerate the high doses required for full antidepressant effects (400 to 600 mg/day). Hertzberg, Feldman, Beckham, & Davidson (1996) reported improvement of all PTSD symptom clusters in an open-label trial of trazodone (flexible dosing schedule 50 to 400 mg/day) in six combat veterans. Not surprisingly, sleep also improved; changes in depression and level of functioning were minimal. Men taking trazadone should be warned about the risk of priapism (1/1,000 to 1/10,000).

Cyproheptadine

Cyproheptadine (Periactin) is an antihistamine and partial serotonin antagonist; it is mainly used for rhinitis and allergic reactions. In the PTSD literature, there are a few reports of benefit in reducing PTSD-associated nightmares. Three case reports (Brophy, 1991; Harsch, 1986; Gupta, Austin, Cali, & Bhatara, 1998) and one retrospective review (Gupta, Popli, et al., 1998) report some benefit for nightmares in PTSD. Gupta, Austin, et al. (1998) noted improvement in sleep, reduction in nightmares, and better school performance in one nine-year-old boy with PTSD, attention deficit hyperactivity disorder, and obsessive-compulsive disorder. Clark, Canive, Calais, Qualls, Brugger, & Vosburgh (1999) reported no improvement in the nightmares of male combat veterans after a four-week trial of cyproheptadine.

CLASSES OF MEDICATION: SEDATIVES/HYPNOTICS

Benzodiazepines

Benzodiazepines are a class of commonly prescribed medications known for their anxiety-reducing and sedative properties. Commonly used benzodiazepines include alprazolam (Xanax), clonazepam (Klonopin), diazepam (Valium), and lorazepam (Ativan). In a variety of conditions, benzodiazepines can dramatically reduce anxiety levels, usually in less than an hour. Some are available in intramuscular and intravenous forms as well as oral pills. There is also a diazepam suppository. Owing to these qualities, the use of this class of medication has been extended to almost every field of medicine.

Benzodiazepines are extremely useful in psychiatric disorders. They are most often used for acute anxiety or agitation, or as "bridges" to other medications. Another common use is for transient insomnia. Given that patients sometimes require higher doses for desired effect (tolerance) and can experience life-threatening symptoms during sudden discontinuation (withdrawal), benzodiazepines are usually not preferred for long-term management of anxiety.

Historically, PTSD has been thought of as a disorder of memory and integration of affects. Early neurophysiologic models of memory consolidation supported the use of benzodiazepines shortly after a trauma, to prevent or diminish the intense, intrusive recollections of the event. However, this model has become controversial, as benzodiazepines may interfere with new learning (anterograde memory), independent of their sedative properties (Vermetten & Bremner, 2002). Theoretically, preventing new learning may preclude the adaptation, reappraisal, and learning that are intrinsic to success of the psychosocial treatments for PTSD (e.g., prolonged exposure and cognitive restructuring).

At present, no good studies substantiate the efficacy of benzodiazepines in the treatment of the core symptoms of PTSD. Some authors have observed positive effects in nonspecific PTSD symptoms (Dunner, Ishiki, Avery, Wilson, & Hyde, 1986; Lowenstein, Hornstein, & Farber, 1988; Mellman, Byers, & Augenstein, 1998). However, a randomized, double-blind trial found that alprazolam (Xanax) was not significantly better than placebo in patients with chronic PTSD (Braun, Greenberg, Dasberg, & Lerer, 1990). In addition, the use of benzodiazepines in PTSD can be problematic. For example, Risse et al. (1990) described eight patients who developed prominent rage with homicidal and suicidal ideation upon withdrawal from alprazolam (dose range 2 to 9 mg/day), even with tapering over three to eighteen weeks. Another study (Gelpin, Bonne, Peri, Brandes, & Shalev, 1996) showed that benzodiazepines given acutely after a traumatic event (i.e., two to eighteen days) appeared to increase, rather than decrease, risk of development of PTSD after six months.

Adverse Effects

Common adverse effects of benzodiazepines include daytime sedation, ataxia, memory disturbance, dependency, and withdrawal. Given high rates of comorbid substance abuse in patients with PTSD, clinicians need to be particularly aware of the potential for dependency. Benzodiazepines should be used with caution in the elderly, who are prone to falls. In addition, benzodiazepines can cause paradoxical disinhibition in patients with compromised cognitive status (e.g., from structural brain damage, mental retardation, or dementia).

CLASSES OF MEDICATION: MOOD STABILIZERS

The class of medications called mood stabilizers falls into two groups: lithium and anticonvulsants. Mood stabilizers are used in psychiatry mainly for patients with bipolar disorder to, as the name suggests, promote a stable mood state. At times they are prescribed for people who do not have bipolar disorder, but rather individual symptoms such as irritability, anger, and lack of impulse control. Published trials of mood stabilizers for treating PTSD have all been conducted among treatment-refractory combat veterans with problems of impulsivity and affect dysregulation (Yehuda et al., 2002). The Expert Consensus Guidelines on the treatment of PTSD suggest adding a mood stabilizer if the patient has had only a partial response after eight weeks of treatment with an SSRI (Foa et al., 1999). Please note that patients with comorbid bipolar disorder will most likely need to be on a mood stabilizer for the mood disorder, regardless of their treatment for PTSD.

Lithium

Lithium (Eskalith, Lithobid, Lithonate) was the first medication found to be effective in mania. Its effects were first reported by John Cade in the 1940s; it was approved for this use by the FDA in 1970 (Ghaemi, 2000). There are very few reports of lithium use in PTSD. Two open trials (Kitchner & Greenstein, 1985; van der Kolk, 1983) of lithium among combat veterans showed improvement in symptoms such as feeling out of control, explosiveness, feeling emotionally cut off, and preoccupation with guilt. Neither study used structured scales to assess comorbidity or severity of symptoms, or to measure improvement.

Mechanism of Action, Adverse Effects, and Toxicity

Lithium is a naturally occurring element, a salt. It is thought to act not directly on neurotransmitters at the synapse, but rather postsynaptically, at the level of second messengers such as G-proteins and phosphatidylinositol phosphate. The usual dose range is 900 to 1,200 mg/day (range 600 to 1,500 mg/day). Although quite effective in mania, lithium is associated with many adverse effects, including sedation, tremor, cognitive difficulties, increased appetite and weight gain, increased fluid intake, and frequent urination. It can cause nausea, diarrhea, and exacerbation of psoriasis and acne. Lithium can depress the pacemaking activity of the heart and, over time, can adversely affect thyroid function. Blood levels of lithium are followed in treatment, and toxic levels (generally above 1.2 mEq/L) may be associated with tremor, nausea, diarrhea, ataxia, seizures, and, at higher levels, acute renal failure, coma, and death (Ghaemi, 2000). Despite the plethora of possible adverse effects, patients often can be maintained on lithium safely for decades.

Anticonvulsants

Anticonvulsants, also called antiepileptic drugs, are a group of medications used to prevent seizures. Although only valproate (Depakote) has FDA approval for acute mania, in clinical practice many of the anticonvulsants are used by psychiatrists. At present, only lamotrigine (Lamictal) (Hertzberg et al., 1999) has been studied in a randomized control trial for PTSD.

Valproate

Valproate (Depakote) is probably the most commonly used anticonvulsant in psychiatry. There are two published open-label trials of valproate for PTSD, both in male combat veterans, most of whom were taking other medications concomitantly. While Fesler (1991) found improvement of hyperarousal and avoidant/numbing symptoms in his subjects, Clark, Canive, Calais, Qualls, & Tuason (1999) found improvement of hyperarousal and intrusive symptoms, but not in avoidant/numbing symptoms. Two case reports also cite valproate as helping reduce irritability, temper outbursts, and mood disturbance in patients also taking carbamazepine (Tegretol) (Ford, 1996; Tohen, Castillo, Pope, & Herbstein, 1994).

Patients on valproate may experience nausea, vomiting, sedation, dizziness, ataxia, tremor, weight gain, and hair loss. Blood levels can be monitored (therapeutic levels for epilepsy are 50 to 100 mcg/ml). Therapeutic levels are not well established for PTSD and mania, although a typical dose range is 750 to 1,500 mg/d. Complete blood counts and liver function tests should be monitored routinely, as the medication can sometimes (15 to 30 percent of cases) cause a mild, usually temporary, increase in liver enzymes as well as (less frequently) a decrease in platelet count. Valproate can be associated with increased ammonia levels, which may correspond clinically with cognitive deficits. Valproate is associated with approximately 4 percent incidence of neural tube defects and neonatal liver disease in exposed infants; thus, it should be avoided in pregnancy (Savitz & Sachs, 2000).

Carbamazepine

Although carbamazepine is not FDA indicated for acute mania, several placebo-controlled studies have indicated effectiveness (Savitz & Sachs, 2000). It is not prescribed as often as valproate by psychiatrists, because of its adverse effects, which include multiple drug-drug interactions. There are a few studies (open-label studies and case reports) of carbamazepine for

PTSD. Open-label studies by Lipper et al. (1986) and Wolf, Alavi, & Mosnaim (1988), each looking at ten veterans, found carbamazepine helpful for impulse control, anger, and hostility. Lipper et al. (1986) also noted improvement in intrusive recollections and sleep impairment. Davidson (1992) reported the case of a single patient with seizure disorder whose PTSD symptoms improved on carbamazepine. Looff, Grimley, Kuller, Martin, & Shoonfield (1995) reported on the use of carbamazepine in twenty-eight children ages eight to seventeen who had PTSD stemming from abuse. All were hospitalized at a state facility; the majority had comorbid disorders, and several were taking other medications concomitantly. All improved significantly, both globally and in regard to all three PTSD symptom clusters. By the end of their hospitalization (range seventeen to ninety-two days), twenty-two of the twenty-eight were reportedly asymptomatic.

Carbamazepine is an active inducer of hepatic enzymes. It induces its own metabolism, meaning that after a few weeks on the drug a patient will need to increase the dose in order to maintain the same blood level. Dosage usually starts at 200 mg a day and is increased gradually until a therapeutic blood level is reached or until the patient has intolerable side effects. Blood levels can be followed; therapeutic level for epilepsy is 4 to 12 mcg/ml; for bipolar disorder, it is 8 to 12 mcg/ml. The enzyme induction also leads to lower blood levels of many other medications taken concomitantly, including oral contraceptives, warfarin, benzodiazepines, and some antipsychotics and antidepressants. Rarely, carbamazepine can lead to dangerously low white blood cell counts, Stevens-Johnson syndrome, and liver failure. Thus, as with valproate, complete blood counts and liver function tests should be monitored. More common adverse effects include dizziness, clumsiness, sedation, nausea, and rash (Savitz & Sachs, 2000).

Lamotrigine

Lamotrigine (Lamictal), has recently been approved by the FDA for use in bipolar disorder. Typical dosing starts at 25 mg/day and is increased by 25 to 50 mg/day every one to two weeks; the usual maintenance dose is 75 to 250 mg. As noted previously, lamotrigine is the only anticonvulsant that has been studied for PTSD in a randomized, placebo-controlled, double-blind study. Hertzberg et al. (1999) studied fifteen veterans (male and female) over twelve weeks. Five of the ten subjects in the lamotrigine group (maximum dose 500 mg) showed improvements in intrusive and avoidance/numbing symptoms, compared to one of four subjects on placebo.

Lamotrigine can cause a rash that may progress to erythema multiforme major (Stevens-Johnson syndrome). This is an inflammatory skin disease involving other mucosal surfaces (including oral and conjunctival) that are fatal in the most severe cases. More common adverse effects are headache, blurred vision, dizziness, nausea, and fatigue. Valproate inhibits the metabolism of lamotrigine. Consequently, concurrent use of valproate and lamotrigine can increase the risk of fatal skin reactions, and thus this combination should be used with caution (Savitz & Sachs, 2000).

Gabapentin

Gabapentin (Neurontin) is another anticonvulsant that has gained popularity in psychiatry, most commonly for the treatment of anxiety. Its advantages include relatively few adverse effects and its lack of drug-drug interactions. The dosage ranges widely, from 900 to 3,000 mg/day, usually divided into two to three doses per day. One case study of gabapentin in the treatment of PTSD reported improvement in nightmares and anxiety at the end of eight weeks (Brannon, Labbate, & Huber, 2000). Adverse effects include somnolence, dizziness, and dry mouth.

Topiramate

Topiramate (Topamax), like the other newer antiepileptic drugs, has approval as an adjunct anticonvulsant, but not as a mood stabilizer. Typical doses range from 200 to 400 mg/day. Berlant has reported positive results with topiramate in patients with PTSD (Berlant, 2001; Berlant & van Kammen, 2002). In an open-label study of thirty-five civilian patients with chronic PTSD, his group found that topiramate as either add-on therapy or monotherapy (in doses of 12.5 to 500 mg) resulted in reduction of nightmares (79 percent of patients) and flashbacks (86 percent of patients), with about half reporting full suppression of both symptoms. In the subset (seventeen patients) that completed the PTSD Checklist Civilian Version (PCL-C), patients reported a similar reduction in all three symptom clusters. Notably, improvement in intrusive symptoms occurred rapidly, at a median of four days (Berlant & van Kammen, 2002).

Common adverse effects with topiramate include cognitive problems such as poor attention, concentration, and memory. Nausea, fatigue, and anorexia are also common. Weight loss associated with topiramate has been seen as a potentially beneficial side effect for some patients. Topiramate is approximately 80 percent excreted by the kidneys, but as the remainder is metabolized by the liver, levels can be affected by enzyme-inducing drugs, and topiramate may decrease effectiveness of oral contraceptives (Savitz & Sachs, 2000).

CLASSES OF MEDICATION: ANTIPSYCHOTICS

The advent of antipsychotic medications in the 1950s was revolutionary in the history of psychiatry. Prior to the discovery of chlorpromazine (Thorazine), biological treatments for mental illness were crude and only mildly effective. Older antipsychotic medications, including haloperidol (Haldol), chlorpromazine (Thorazine), fluphenazine (Prolixin), and perphenazine (Trilafon), are called typical, or first-generation, antipsychotics. The newer agents, including clozapine (Clozaril), risperidone (Risperdal), olanzapine (Zyprexa), quetiapine (Seroquel), ziprasidone (Geodon), and aripiprazole (Abilify), are commonly called atypical, or second-generation, antipsychotics. The second-generation antipsychotics are less likely to cause the classical adverse effects than the first-generation antipsychotics. In addition, the mechanism of action of first-generation and second-generation antipsychotics is different. Antipsychotics have been tried in PTSD, including in patients without comorbid schizophrenia or other psychotic disorders. The topic of psychotic symptoms as core components in PTSD is beyond the scope of this chapter. However, many authors have sought to differentiate severe dissociation and the phenomenon of reliving (i.e., flashbacks, intrusive thoughts) from psychosis. First-generation and second-generation antipsychotics will be reviewed separately.

First-Generation Antipsychotics

No formal studies have looked at the use of first-generation antipsychotics in PTSD. Nonetheless, expert opinion is that first-generation antipsychotic medications are not helpful in PTSD without comorbid psychosis and should be avoided. But they are used not uncommonly (i.e., around 10 percent in one study), at least in veteran populations (Sernyak, Kosten, Fontana, & Rosenheck, 2001).

Investigators at Veteran Administration hospitals affiliated with Yale University School of Medicine (Sernyak et al., 2001) performed a secondary analysis of more than 1,385 males (both inpatient and outpatient) receiving treatment for combat-related PTSD. Patients were excluded from the study if they had a comorbid diagnosis of schizophrenia or schizoaffective disorder. Researchers found that approximately 9 percent of inpatients and 10 percent of outpatients had received antipsychotic medications within the past twelve months. Although the authors did not look at which antipsychotic medications were used (i.e., typical or atypical), the treatments were conducted from 1990 to 1994, before the widespread availability of atypical agents. Predictors of neuroleptic use included younger age, non-white race, unmarried status, more intense service connection for PTSD (higher likelihood that the PTSD was deemed service related), and higher disability rating (i.e., less likely to be employed or well remunerated, or higher disability due to other comorbid psychiatric disorder). The neuroleptic-exposed group was also significantly more likely to have been hospitalized previously and more likely to be taking concurrent anxiolytic or antidepressant medications. This group also reported more hallucinations, violent behavior, and suicidality. The outpatient subgroup reported more intrusive, numbing and avoidance, hyperarousal, and total symptoms. Interestingly, the neuroleptic-exposed group was no different from the non-neuroleptic-exposed group in rates of comorbid alcohol abuse/dependence or drug abuse/dependence, in affective, dissociative, bipolar, adjustment, or sleep disorders, or in level of combat. Outcomes after one year for the group treated with neuroleptics were not significantly different from the other group. The authors concluded that neuroleptic use is common in this population, targeted toward the more seriously ill but not associated with substantial improvement.

Mechanism of Action and Adverse Effects

Typical antipsychotics have been shown to block the postsynaptic dopamine receptors in all regions of the brain. This global blockade is also responsible for many of the adverse effects. Dopamine is involved in many other brain processes, most notably movement, cognition, and hormonal regulation. Hence, the most common adverse effects associated with this class of medications can be attributed to the blockade of dopamine in pathways not associated with psychosis. These are the so-called extrapyramidal adverse effects, including Parkinsonian features (slow movements, constricted affect), dystonia (stiffness), and akathisia (sense of internal restlessness). Long-term use of first-generation antipsychotics may result in tardive dyskinesia (involuntary movements of the face, extremities, and trunk). Other common adverse effects include sedation, weight gain, amenorrhea, and galactorrhea (due to elevation of the hormone prolactin), orthostatic hypotension, and prolonged QT interval. In addition, use of first-generation antipsychotics may lead to neuroleptic malignant syndrome (NMS), a life-threatening condition characterized by fever, confusion, and muscle rigidity.

Second-Generation Antipsychotics

The second-generation (atypical) antipsychotics have several benefits over the typical antipsychotics. In general, they are less likely to cause extrapyramidal adverse effects, tardive dyskinesia, elevation of prolactin, and neuroleptic malignant syndrome. They are also believed to be helpful in mood disorders (both depression and bipolar mania), negative symptoms (social withdrawal, alogia, amotivation), and refractory psychosis. For these reasons, perhaps, they are increasingly used for nonpsychotic disorders.

Several case studies (Burton & Marshall, 1999; Krashin & Oates, 1999; Leyba & Wampler, 1998; Monnelly & Ciraulo, 1999) have reported improvement with adjunctive olanzapine and risperidone for symptoms such as insomnia, hyperarousal, nightmares, flashbacks, intrusive thoughts, anger, and aggression. For monotherapy with olanzapine there has been one positive open trial (Petty et al., 2001) and one negative trial (Butterfield et al., 2001); the dosages were titrated to clinical effect and varied for both medications in these studies.

Mechanism of Action and Adverse Effects

Second-generation antipsychotics block dopamine receptors but also block serotonin 2A receptors. For this reason, they are sometimes referred to as dopamine-serotonin 2A receptor antagonists. Preferential blockade of dopamine in the mesolimbic pathway, which is thought to be dysregulated in psychosis, is a purported explanation for the more favorable side effect profile of these drugs (Stahl, 2000).

The second-generation antipsychotics, however, are not without adverse effects. Most serious is the possibility of agranulocytosis (decimation of white blood cells) that can occur in 1 percent of those treated with clozapine. In addition, clozapine and olanzapine are associated with significant weight gain, placing patients at risk for diabetes and cardiac disease. Sedation, drooling, constipation, and seizure are common adverse effects with clozapine. Also, although much lower, the possibility of extrapyramidal symptoms, tardive dyskinesia, neuroleptic malignant syndrome, and prolactin elevation still exists. Some of these medications are sedating. Other adverse effects associated with individual second-generation antipsychotics include dizziness, hypotension, tachycardia, headache, liver enzyme abnormalities, nausea, vomiting, anxiety, rhinitis, coughing, and prolonged QT interval (Stahl, 2000). In addition, they are considerably more expensive than first-generation antipsychotics.

CLASSES OF MEDICATION:
AUTONOMIC NERVOUS SYSTEM MODULATORS

Several authors have suggested that autonomic dysregulation, particularly increased sympathetic discharge (neurohormonal fight-or-flight response), is present in patients with PTSD (Marshall & Garakani, 2002). Sympathetic overdrive has been hypothesized to contribute to the physiologic hyperarousal symptoms (e.g., increased startle) in PTSD. This is based on the observation of symptomatic relief with autonomic nervous system modulators (Harmon & Riggs, 1996; Horrigan, 1996; Kinzie & Leung, 1989; Kolb, Burris, & Griffiths, 1984), clinically significant differences in heart rates in PTSD and ASD patients (Shalev et al., 1998; Bryant, Harvey, Guthrie, & Moulds, 2000), and measurable increased catecholamine metabolites in serum and urine samples of patients with PTSD (Kosten, Mason, Giller, Ostroff, & Harkness, 1987; Perry, Southwick, Yehuda, & Giller, 1990; Yehuda, Southwick, & Giller, 1992). Findings are notably mixed, however, in that some investigators have found no differences (Mellman, Kumar, Kulick-Bell, Kumar, & Nolan, 1995), and others, reduced catecholamines and their metabolites (Murburg, McFall, Lewis, & Beith, 1995; Marshall et al., 2002; for a review see Charney & Bremner, 1999).

Propranolol

Propranolol (Inderal) is a nonselective beta-adrenergic receptor blocking agent (also called a beta-blocker) possessing no other autonomic nervous system activity (MEC, 2002). It has FDA approval for the treatment of cardiac arrhythmias, hypertension, pheochromocytoma, essential tremor, and migraine (MEC, 2002). Although there are few controlled trials for psychiatric disorders, it has also been widely reported as useful for panic disorder, simple phobia, behavioral manifestations of attention-deficit hyperactivity disorder (ADHD), and neuroleptic-induced akathisia. Two studies have suggested the utility of propranolol in chronic PTSD (Famularo, Kinscherff, & Fenton, 1998; Kolb et al., 1984), and one in acute PTSD (Pitman et al., 2002).

Kolb et al. (1984) found that open treatment with propranolol (dose range 120 to 160 mg/day in divided doses) was associated with improvement in explosiveness, nightmares, intrusive recollections, startle response, hyperalertness, impaired sleep, self-esteem, and psychosocial function in twelve Vietnam veterans. Famularo et al. (1998) reported eleven cases of PTSD in physically and/or sexually abused children who presented in an agitated, hyperaroused state using a B-A-B (off-on-off) medication design in a clinical setting. Scores on an inventory of PTSD symptoms indicated that patients exhibited significantly fewer symptoms while receiving propranolol than either before or afterward.

Adverse effects of beta-blockers include fatigue, mild gastrointestinal distress, and insomnia. The more lipophilic beta-blockers, such as propranolol, that enter the central nervous system have been associated with depression. Propranolol treatment is also associated with forgetfulness, sexual impairment, bradycardia, and hypotension (Yehuda et al., 2002). Beta-blockers must be used with caution in patients with cardiac problems and asthma, because these medications can decrease the strength of contractions of the heart and cause the lungs to constrict airflow.

Clonidine

Clonidine (Catapres) stimulates alpha-2 noradrenergic receptors in the central nervous system. This stimulation causes a decrease in firing of sympathetic nervous system activity which leads to a decrease in heart rate and blood pressure. Clonidine's main indication is for the treatment of hypertension. However, relaxing sympathetic tone has also been useful in a variety of psychiatric conditions characterized by hyperactivity of the autonomic nervous system. The most common use of clonidine in psychiatry is in the treatment of opiate withdrawal. However, clonidine has been reported useful in uncontrolled trials for the treatment of ADHD, Tourette's syndrome, smoking cessation, alcohol withdrawal, school phobia, narcolepsy, and PTSD (Kinzie & Leung, 1989).

The observation of similarities between the arousal seen in opiate withdrawal states and that in PTSD has led to several small studies examining the efficacy of clonidine in PTSD. Kolb et al. (1984) reported improvement in flashbacks and hyperarousal. In addition, Kinzie and Leung (1989) reported overall improvement as well as specific improvement in sleep, frequency of nightmares, and startle reaction when clonidine (0.1 to 0.6 mg/day) was added to imipramine. Harmon and Riggs (1996) reported that clonidine was helpful among children ages three to six in controlling aggression, impulsivity, emotional outbursts, mood lability, hyperarousal, insomnia, nightmares, and generalized anxiety.

Clonidine's main adverse effects include sedation, dry mouth, nausea, and photophobia. In high doses, it may cause hypotension and dizziness (Cozza & Dulcan, 1999). If stopped abruptly after protracted use (particularly with high doses, e.g., greater than 1 mg/day), clonidine can lead to dangerous rebound hypertension.

Guanfacine

Guanfacine (Tenex) is another alpha-2 agonist of central noradrenergic receptors which is indicated for the treatment of hypertension. For the same reasons as clonidine, guanfacine has been tried in PTSD. Horrigan (1996) used guanfacine in a seven-year-old girl who had not responded to psychotherapy and had experienced worsening of nightmares with clonidine. With guanfacine, the patient reported improvement in nightmares. Guanfacine is also longer acting and possibly less sedating than clonidine (Cozza & Dulcan, 1999).

Opiate Antagonists

Behavioral theorists have sought to make a connection between the stress-induced analgesia found in animals exposed to inescapable conditions and the emotional numbing observed in PTSD (Pitman, van der Kolk, Orr, & Greenberg, 1990). The animal condition of stress-induced analgesia is thought to be mediated by the endogenous opioid system because it can be reversed by the administration of opioid antagonist medications (naltrexone [ReVia, Depade], naloxone [Narcan], nalmefene [Revex]) (Glover, 1992). Furthermore, naloxone-reversible analgesia has been reported in humans under stress such as in response to foot shock and after long-distance running (Pitman et al., 1990). Opioid antagonist-reversible analgesia has also been observed in schizophrenia and schizoaffective disorder, conditions where there is a high threshold for pain.

Naloxone and nalmefene are given parenterally (intravenous or intramuscular), most commonly for opiate overdose. Naltrexone, available orally, is used in the treatment of alcohol and opiate dependence. Opiate antagonists must all be administered with great caution, as they immediately block all opiate receptors. Therefore, a patient with a hidden opiate dependence will go into acute withdrawal with a single injection or ingestion. At present, opioid antagonists for the treatment of emotional numbing associated with PTSD is still theoretical (Yehuda et al., 2002). Pitman et al. (1990) reported a 30 percent reduction in pain ratings in Vietnam veterans exposed to a fifteen-minute videotape of dramatized combat compared to a 0 percent reduction in those subjects given naloxone. Clinically, Glover (1993) reported improvement in numbing and other symptoms in eight of eighteen combat veterans using nalmefene (dose 200 to 400 mg/day). In addition, two case reports (Bills & Kreisler, 1993; Ibarra et al., 1994) showed some benefit of naltrexone in three complicated patients.

Adverse effects of opioid antagonists include insomnia, anxiety, abdominal pain, headache, muscle pain, low energy, and nausea and vomiting (MEC, 2002). At high doses, naltrexone can cause hepatic damage and is contraindicated in patients with acute hepatitis or liver failure.

ALTERNATIVE MEDICATIONS

Despite the lack of scientific validation for many alternative medications, a growing number of people are using these substances for a variety of conditions, both medical and psychiatric. Many patients do not volunteer information about their use of alternative medications for many reasons. Some may fear that the clinician will disapprove; some may not believe that herbs or nutritional supplements are medications. However, many of these substances have been associated with toxicity as well as drug interactions. Given this reality, clinicians should routinely ask patients about the use of any alternative medications, including herbal remedies.

Published literature on the use of herbs and nutrients for PTSD is scarce. The use of inositol will be discussed, as there is a small controlled trial in PTSD. Although there are no published PTSD-specific studies of kava and valerian, they are briefly discussed since they are commonly used for anxiety, and patients with PTSD may have tried them.

Inositol

Inositol is a glucoselike molecule that occurs naturally in the diet. It functions as a precursor in second-messenger systems used by noradrenergic and serotonin receptors. It has shown beneficial effects in small trials for depression, panic disorder, and obsessive-compulsive disorder in doses of twelve to eighteen grams a day (Mischoulon & Nierenberg, 2000). One double-blind, placebo-controlled crossover study of seventeen PTSD patients with mixed trauma types found no significant difference between inositol and placebo (Kaplan, Amir, Swartz, & Levine, 1996). Inositol appears to have mild adverse effects and no associated toxicity. It can cause gastrointestinal effects, including gas and loose bowels.

Kava

The use of kava *(Piper methysticum)* originated in the South Pacific, where it is used as a ritual drink as well as for analgesia. Kava is thought to have a calming effect via kavapyrones, a type of muscle relaxant. In animal models, kava also has anticonvulsant properties; whether this is via GABA receptor binding is unknown (Brown & Gerbarg, 2000). Although there have been no double-blind, placebo-controlled studies of kava for PTSD, five such studies found that kava is effective for anxiety (Mischoulon & Nierenberg, 2000). Connor and Davidson (2002), however, in a rigorous placebo-controlled trial, recently reported no effect for kava versus placebo in thirty-seven adults with generalized anxiety disorder. Kava is usually dosed at 60 to 120 mg/day. Toxic reactions to kava have been reported with high doses and/or prolonged use; symptoms include ataxia, hair loss, visual problems, respiratory problems, and rash. In March 2002, the FDA issued a consumer advisory on kava because of a number of case reports of liver failure leading to liver transplant (Centers for Disease Control, 2002).

Valerian

Valerian *(Valeriana officinalis)* is an herb usually used at doses of 450 to 600 mg before bedtime as a sedative. Trials, which have been few in number and small in size, have shown a decrease in sleep onset (less time to fall asleep) and improvement in sleep quality. However, valerian may require two to four weeks of use before having a clinical effect. Reported adverse effects include blurry vision, dystonias, and hepatotoxicity, but it is unknown whether these effects are from valerian or from other ingredients in valerian preparations. Valerian tea and tablets do, however, often have a distinctive foul odor and taste, leading both to problems with patient compliance and with executing a successful double-blind trial (Brown & Gerbarg, 2000; Mischoulon & Nierenberg, 2000).

SUMMARY

Pharmacotherapy is effective in the treatment of PTSD and should be considered in all patients presenting with this condition. Medication may be helpful alone or in combination with psychosocial treatment. Growing knowledge about the biological alterations present in PTSD underscores the rationale for considering medication. Pharmacological treatment should be considered more strongly in patients with severe and/or chronic symptoms, patients who continue to have symptoms despite psychosocial treatments, and patients with comorbid psychiatric disorders that are responsive to medication. Patients with PTSD suffer from high rates of comorbid mental illness which may dictate the need for medication. Certain patients may even prefer pharmacotherapy.

People with PTSD are invariably fearful, avoidant, and hypervigilant—these are among symptoms of the disorder—and many have been victimized. Accordingly, patients may be difficult to engage in treatment. Compliance with recommendations can be strengthened by a special attention to the working alliance, making all efforts to include the patient in a collaborative decision-making process. Clinicians should be aware of their own biases about pharmacotherapy, as well as those of cotherapists (as in the case of split treatment). In addition, patients often have fears and fantasies about medication that need to be explored.

Historically, most pharmacotherapy studies in PTSD were conducted among adult male war veterans with chronic PTSD. Drug trials with other populations such as victims of rape, assault, accidents, natural disasters, and terrorist acts are fewer in number. In addition, research in children, and in adults meeting criteria for acute stress disorder, is rare. Important questions remain as to the generalizability of research findings from veteran to non–combat-related PTSD populations, from PTSD studies to persons with acute stress reactions, and from adult PTSD patients to children with PTSD.

SSRIs are considered first-line medications for PTSD, due to their demonstrated efficacy in all three symptom clusters in several large randomized controlled trials. Sertraline and paroxetine, both SSRIs, are currently the only two medications with FDA approval for the treatment of PTSD. Tricyclic antidepressants and monoamine oxidase inhibitors have also been shown to be helpful but are used less frequently due to their potential for adverse effects. Other antidepressants, such as venlafaxine and mirtazapine, are reasonable alternatives, but more studies are required to demonstrate their efficacy.

Mood stabilizers, such as lithium, valproate, and other antiepileptic drugs, can be useful for augmentation in patients with residual symptoms such as irritability, explosiveness, and impulsivity. Autonomic nervous system (ANS) modulators, such as propranolol and clonidine, may be useful adjuncts for nightmares, insomnia, and hyperarousal. Theoretically, ANS modulators may have some role in the treatment of acute stress disorder.

Benzodiapines should, in general, be avoided in PTSD, secondary to their effects on memory and their potential for abuse. First-generation antipsychotics (e.g., haloperidol) may worsen affective blunting and are not recommended for patients without a comorbid psychotic disorder. Owing to their more favorable side effect profile and more broad spectrum action, second-generation antipsychotics (e.g., olanzapine, quetiapine) may be useful adjuncts for their sedative properties. The utility of opioid antagonists for the numbing symptoms associated with PTSD remains theoretical. Although some alternative medications, such as kava and valerian, may be helpful for other anxiety disorders, there is no evidence for their efficacy in PTSD.

Important questions regarding optimal length of treatment, relapse rate after discontinuation of medication, and standardized approaches to refractory symptoms remain. Studies combining pharmacotherapy with psychosocial treatment in PTSD have not yet been published. However, recently developed diagnostic and assessment tools have given way to more sophisticated clinical trials. In addition, the neurobiological alterations underlying the disorder continue to be an active area of investigation. As this chapter has shown, there are a variety of efficacious pharmacological treatments for PTSD. The challenge is to combine research findings with creativity and patience, to tailor an appropriate and effective medication regimen for each patient with PTSD.

APPENDIX: INDEX OF MEDICATIONS

Medications are listed as the generic name, followed by any brand names within parentheses.

Antidepressants

Selective Serotonin Reuptake Inhibitors (SSRIs)

>   Citalopram (Celexa)
>   Escitalopram (Lexapro)
>   Fluoxetine (Prozac)
>   Fluvoxamine (Luvox)
>   Paroxetine (Paxil, Paxil CR)
>   Sertraline (Zoloft)

Tricyclic Antidepressants (TCAs)

>   Amitriptyline (Elavil)
>   Amoxapine (Asendin)
>   Clomipramine (Anafranil)
>   Desipramine (Norpramin)
>   Doxepin (Sinequan)
>   Imipramine (Tofranil)
>   Maprotiline (Ludiomil)
>   Nortriptyline (Pamelor)
>   Protriptyline (Vivactil)
>   Trimipramine (Surmountil)

Monoamine Oxidase Inhibitors (MAOIs)

Brofaromine (not available in the United States)
Isocarboxazid (Marplan)
Tranylcypromine (Parnate)
Meclobemide (not available in the United States)
Phenelzine (Nardil)

Other Antidepressants/Antianxiety Agents

Bupropion (Wellbutrin, Wellbutrin SR,
   Wellbutrin XL)
Buspirone (BuSpar)
Mirtazapine (Remeron)
Nefazodone (Serzone)
Trazodone (Desyrel)
Venlafaxine (Effexor, Effexor XR)

Antihistamine

Cyproheptadine (Periactin)

Benzodiazepines

Alprazolam (Xanax)
Clonazepam (Klonopin)
Diazepam (Valium)
Lorazepam (Ativan)

Mood Stabilizers

Carbamazepine (Tegretol)
Gabapentin (Neurontin)
Lamotrigine (Lamictal)
Lithium (Eskalith, Eskalith CR, Lithobid,
   Lithonate)
Topiramate (Topamax)
Valproate (Depakote)

Antipsychotics

First-Generation ("Typical")

Chlorpromazine (Thorazine)
Fluphenazine (Prolixin)
Haloperidol (Haldol)
Perphenazine (Trilafon)

Second-Generation ("Atypical")

Aripiprazole (Abilify)
Clozapine (Clozaril)
Olanzapine (Zyprexa)
Quetiapine (Seroquel)
Risperidone (Risperdal)
Ziprasidone (Geodon)

Autonomic Nervous System Modulators

Clonidine (Catapres)
Guanfacine (Tenex)
Prazosin (Minipress)
Propranolol (Inderal)

Opiate Antagonists

Nalmefene (Revex)
Naloxone (Narcan)
Naltrexone (ReVia, Depade)

REFERENCES

American Academy of Child and Adolescent Psychiatry (AACAP). (1998). AACAP practice parameters for the assessment and treatment of children and adolescents with post-traumatic stress disorder. *Journal of the American Academy of Child and Adolescent Psychiatry, 37*(Suppl 10), 4S-26S.

American Psychiatric Association. (1994). *DSM-IV: Diagnostic and statistical manual of mental disorders* (4th ed.). Washington, DC: Author.

Baker, D.G., Diamond, B.I., Gillette, G., Hamner, M., Katzelnick, D., Keller, T., Mellman, T.A., Pontius, E., Rosenthal, M., Tucker, P., et al. (1995). A double-blind randomized placebo-controlled multi-center study of brofaromine in the treatment of post-traumatic stress disorder. *Psychopharmacology, 122,* 386-389.

Berlant, J.L. (2001). Topiramate in posttraumatic stress disorder: Preliminary clinical observations. *Journal of Clinical Psychiatry, 62*(Suppl 17), 60-63.

Berlant, J., & van Kammen, D.P. (2002). Open-label topiramate as primary or adjunctive therapy in chronic civilian posttraumatic stress disorder: A preliminary report. *Journal of Clinical Psychiatry, 63*(1), 15-20.

Bills, L.J., & Kreisler, K. (1993). Treatment of flashbacks with naltrexone. *American Journal of Psychiatry, 150,* 1430.

Brady, K.T., Sonne, S.C., & Roberts, J.M. (1995). Sertraline treatment of comorbid posttraumatic stress disorder and alcohol dependence. *Journal of Clinical Psychiatry, 56*(11), 502-505.

Brannon, N., Labbate, L., & Huber, M. (2000). Gabapentin treatment for posttraumatic stress disorder. *Canadian Journal of Psychiatry, 45,* 84.

Braun, P., Greenberg, D., Dasberg, H., & Lerer, B. (1990). Core symptoms of posttraumatic stress disorder unimproved by alprazolam treatment. *Journal of Clinical Psychiatry, 51,* 236-238.

Breslau, N., & Davis, G.C. (1992). Posttraumatic stress disorder in an urban population of young adults: Risk factors for chronicity. *American Journal of Psychiatry, 149,* 671-675.

Breslau, N., Davis, G.C., Andreski, P., & Peterson, E. (1991). Traumatic events and posttraumatic stress disorder in an urban population of young adults. *Archives of General Psychiatry, 48*(3), 216-222.

Brophy, M.H. (1991). Cyproheptadine for combat nightmares in post-traumatic stress disorder and dream anxiety disorder. *Military Medicine, 156,* 100-101.

Brown, R.P., & Gerbarg, P.L. (2000). Integrative psychopharmacology: A practical approach to herbs and nutrients in psychiatry. In P.R. Muskin (Ed.), *Complementary and alternative medicine and psychiatry* (pp. 1-66) (Review of psychiatry series, Vol. 19, No. 1; J.M. Oldham & M.B. Ribas, series eds.). Washington, DC: American Psychiatric Press.

Bryant, R.A., Harvey, A.G., Guthrie, R.M., & Moulds, M.L. (2000). A prospective study of psychophysiological arousal, acute stress disorder, and posttraumatic stress disorder. *Journal of Abnormal Psychology, 109,* 341-344.

Burton, J.K., & Marshall, R.D. (1999). Categorizing fear: The role of trauma in clinical formulation. *American Journal of Psychiatry, 156*(5), 761-766.

Butterfield, M.I., Becker, M.E., Connor, K.M., Sutherland, S., Churchill, L.E., & Davidson, J.R. (2001). Olanzapine in the treatment of post-traumatic stress disorder: A pilot study. *International Clinical Psychopharmacology, 16*(4), 197-203.

Canive, J.M., Clark, R.D., Calais, L.A., Qualls, C., & Tuason, V.B. (1998). Bupropion treatment in veterans with posttraumatic stress disorder: An open study. *Journal of Clinical Psychopharmacology, 18*(5), 379-383.

Centers for Disease Control. (2002). Hepatic toxicity possibly associated with kava-containing products—United States, Germany, and Switzerland, 1999-2002. *MMWR, 51*(47), 1065-1067.

Charney, D.S., & Bremner, J.D. (1999). The neurobiology of anxiety disorders. In D.S. Charney, E.J. Nestler, & B.S. Bunney, (Eds.), *Neurobiology of mental illness* (pp. 495-517). New York: Oxford University Press.

Charney, D.S., Deutch, A.Y., Krystal, J.H., Southwick, S.M., & Davis, M. (1993). Psychobiologic mechanisms of posttraumatic stress disorder. *Archives of General Psychiatry, 50*, 294-306.

Chemtob, C.M., Novaco, R.W., Hamada, R.S., Gross, D.M., & Smith, G. (1997). Anger regulation deficits in combat-related posttraumatic stress disorder. *Journal of Traumatic Stress, 10*, 17-35.

Clark, R.D., Canive, J.M., Calais, L.A., Qualls, C., Brugger, R.D., & Vosburgh, T.B. (1999). Cyproheptadine treatment of nightmares associated with posttraumatic stress disorder. *Journal of Clinical Psychopharmacology, 19*(5), 486-487.

Clark, R.D., Canive, J.M., Calais, L.A., Qualls, C.R., & Tuason, V.B. (1999). Divalproex in posttraumatic stress disorder: An open-label clinical trial. *Journal of Traumatic Stress, 12*(2), 395-401.

Connor, K.M., & Davidson, J.R. (2002). A placebo-controlled study of kava kava in generalized anxiety disorder. *International Clinical Psychopharmacology, 17*(4), 185-188.

Connor, K.M., Davidson, J.R., Weisler, R.H., & Ahearn, E. (1999). A pilot study of mirtazapine in post-traumatic stress disorder. *International Clinical Psychopharmacology, 14*, 29-31.

Cozza, S.J., & Dulcan, M.K. (1999). Treatment of children and adolescents. In R.E. Hales & S.C. Yudofsky (Eds.), *Essentials of clinical psychiatry* (pp. 851-852). Washington, DC: American Psychiatric Press.

Davidson, J. (1992). Drug therapy of post-traumatic stress disorder. *British Journal of Psychiatry, 160*, 309-314.

Davidson, J., Kudler, H., Smith, R., Marhorney, S.L., Lipper, S., Hammet, E., Saunders, W.B., & Cavenar, J.L. (1990). Treatment of posttraumatic stress disorder with amitriptyline and placebo. *Archives of General Psychiatry, 47*, 259-266.

Davidson, J.R., Malik, M.L., & Sutherland, S.N. (1997). Response characteristics to antidepressants and placebo in post-traumatic stress disorder. *International Clinical Psychopharmacology, 12*(6), 291-296.

Davidson, J., Pearlstein, T., Londborg, P., Brady, K.T., Rothbaum, B., Bell, J., Maddock, R., Hegel, M.T., & Farfel, G. (2001). Efficacy of sertraline in preventing relapse of posttraumatic stress disorder: Results of a 28-week double-blind, placebo-controlled study. *American Journal of Psychiatry, 158*(12), 1974-1981.

Davidson, J.R., Rothbaum, B.O., van der Kolk, B., Sikes, C.R., & Farfel, G.M. (2001). Multi-center double-blind comparison of sertraline and placebo in the treatment of posttraumatic stress disorder. *Archives of General Psychiatry, 58*, 485-492.

Davis, L.L., Nugent, A.L., Murray, J., Kramer, G.L., & Petty, F. (2000). Nefazodone treatment for chronic posttraumatic stress disorder: An open trial. *Journal of Clinical Psychopharmacology, 20*(2), 159-164.

Dunner, D.L., Ishiki, D., Avery, D.H., Wilson, L.G., & Hyde, T.S. (1986). Effects of alprazolam and diazepam on anxiety and panic attacks in panic disorder: A controlled study. *Journal of Clinical Psychiatry, 47,* 458-460.

Famularo, R., Kinscherff, R., & Fenton, T. (1998). Propranolol treatment for childhood post-traumatic stress disorder, acute type: A pilot study. *American Journal of Diseases of Children, 142,* 1244-1247.

Fesler, F.A. (1991). Valproate in combat-related posttraumatic stress disorder. *Journal of Clinical Psychiatry, 52,* 362-364.

Fichtner, C.G., & Crayton, J.W. (1994). Buspirone in combat-related posttraumatic stress disorder. *Journal of Clinical Psychopharmacology, 14,* 79-81.

Foa, E.B., Davidson, J.R., & Frances, A. (1999). The expert consensus guideline series: Treatment of posttraumatic stress disorder. *Journal of Clinical Psychiatry, 60*(Suppl 16), 1-76.

Ford, N. (1996). The use of anticonvulsants in posttraumatic stress disorder: Case study and overview. *Journal of Traumatic Stress, 9*(4), 857-863.

Freeman, H. (1993). Drug profile: Meclobemide. *Lancet, 342,* 1528-1531.

Garfield, D.A., Fichtner, C.G., Leveroni, C., & Mahableshwarkar, A. (2001). Open trial of nefazodone for combat veterans with posttraumatic stress disorder. *Journal of Traumatic Stress, 14*(3), 453-460.

Gelpin, E., Bonne, O., Peri, T., Brandes, D., & Shalev, A.Y. (1996). Treatment of recent trauma survivors with benzodiazepines: A prospective study. *Journal of Clinical Psychiatry, 57*(9), 390-394.

Ghaemi, S.N. (2000). Lithium. In T.A. Stern & J.B. Herman (Eds.), *Psychiatry: Update and board preparation* (pp. 365-368). New York: McGraw-Hill.

Glover, H. (1992). Emotional numbing: A possible endorphin-mediated phenomenon associated with post-traumatic stress disorders and other allied psychopathologic states. *Journal of Traumatic Stress, 5*(4), 643-675.

Glover, H. (1993). A preliminary trial of nalmefene for the treatment of emotional numbing in combat veterans with post-traumatic stress disorder. *Israel Journal of Psychiatry & Related Sciences, 30*(4), 255-263.

Gupta, S., Austin, R., Cali, L.A., & Bhatara, V. (1998). Nightmares treated with cyproheptadine. *Journal of the American Academy of Child and Adolescent Psychiatry, 37*(6), 570-572.

Gupta, S., Popli, A., Bathhurst, E., Hennig, L., Droney, T., & Keller, P. (1998). Efficacy of cyproheptadine for nightmares associated with posttraumatic stress disorder. *Comprehensive Psychiatry, 39*(3), 160-164.

Hamner, M.B., & Frueh, B.C. (1998). Response to venlafaxine in a previously antidepressant treatment-resistant combat veteran with post-traumatic stress disorder. *International Journal of Clinical Psychopharmacology, 13*(5), 233-234.

Harmon, R.J., & Riggs, P.D. (1996). Clonidine for posttraumatic stress disorder in preschool children. *Journal of the American Academy of Child and Adolescent Psychiatry, 35,* 1247-1249.

Harsch, H.H. (1986). Cyproheptadine for recurrent nightmares. *American Journal of Psychiatry, 143,* 1491-1492.

Hertzberg, M.A., Butterfield, M.I., Feldman, M.E., Beckham, J.C., Sutherland, S.M., Connor, K.M., & Davidson, J.R. (1999). A preliminary study of lamotrigine for the treatment of posttraumatic stress disorder. *Biological Psychiatry, 45,* 1226-1229.

Hertzberg, M.A., Feldman, M.E., Beckham, J.C., & Davidson, J.R. (1996). Trial of trazodone for post-traumatic stress disorder using a multiple baseline group design. *Journal of Clinical Psychopharmacology, 16*(4), 294-298.

Hertzberg, M.A., Feldman, M.E., Beckham, J.C., Kudler, H.S., & Davidson, J.R. (2002). Lack of efficacy for fluoxetine in PTSD: A placebo controlled trial in combat veterans. *Annals of Clinical Psychiatry, 12*(2), 101-105.

Hidalgo, R., Hertzberg, M.A., Mellman, T., Petty, F., Tucker, P., Weisler, R., Zinook, S., Chen, S., Churchill, E., & Davidson, J. (1999). Nefazodone in posttraumatic stress disorder: Results from six open-label trials. *International Journal of Clinical Psychopharmacology, 14*(2), 61-68.

Hogben, G.L., & Cornfield, R.B. (1981). Treatment of neurotic war neurosis with phenelzine. *Archives of General Psychiatry, 38,* 440-445.

Horrigan, J.P. (1996). Guanfacine for PTSD nightmares. *Journal of the American Academy of Child and Adolescent Psychiatry, 35,* 975-976.

Ibarra, P., Bruehl, S.P., McCubbin, J.A., Carlson, C.R., Wilson, J.F., Norton, J.A., & Montgomery, T.B. (1994). An unusual reaction to opioid blockade with naltrexone in a case of post-traumatic stress disorder. *Journal of Traumatic Stress, 7*(2), 303-309.

Israel, J., & Fava, M. (2000). Antidepressants and somatic therapies. In T.A. Stern & J.B. Herman (Eds.), *Psychiatry: Update and board preparation* (pp. 353-358). New York: McGraw-Hill.

Jaycox, L.H., & Foa, E.B. (1996). Obstacles in implementing exposure therapy for PTSD: Case discussions and practical solutions. *Clinical Psychology and Psychotherapy, 3,* 176-184.

Kaplan, Z., Amir, M., Swartz, M., & Levine, J. (1996). Inositol treatment in posttraumatic stress disorder. *Anxiety, 2,* 51-52.

Katz, R.J., Lott, M.H., Arbus, P., Crocq, L., Lingjaerde, O., Lopez, G., Loughrey, G.C., MacFarlane, D.J., Nugent, D., Turner, S.W., Weisath, L., & Yule, W. (1995). Pharmacotherapy of post-traumatic stress disorder with a novel psychotropic. *Anxiety, 1,* 169-174.

Kennedy, S.H., McKenna, K.F., & Baker, G.B. (2000). Monoamine oxidase inhibitors. In B.J. Sadock & V.A. Sadock (Eds.), *Comprehensive textbook of psychiatry* (pp. 2397-2407). Philadelphia: Lippincott, Williams and Wilkins.

Kessler, R.C., Sonnega, A., Bromet, E., Hughes, M., & Nelson, C.B. (1995). Posttraumatic stress disorder in the National Comorbidity Survey. *Archives of General Psychiatry, 52,* 1048-1060.

Kinzie, J.D., & Leung, P. (1989). Clonidine in Cambodian patients with post traumatic stress disorder. *Journal of Nervous and Mental Disease, 177,* 546-550.

Kitchner, L., & Greenstein, R. (1985). Low dose lithium carbonate in the treatment of post traumatic stress disorder: Brief communication. *Military Medicine, 150,* 378-381.

Klerman, G. (1991). Ideological conflicts in integrating pharmacotherapy and psychotherapy. In B. Beitman & G. Klerman (Eds.), *Integrating pharmacotherapy and Psychotherapy* (pp. 3-19). Washington, DC: American Psychiatric Association Press.

Kolb, L.C., Burris, B.C., & Griffiths, S. (1984). Propranolol and clonidine in the treatment of the chronic post-traumatic stress disorders of war. In B.A. van der Kolk (Ed.), *Post-Traumatic Stress disorder: Psychological and biological sequelae* (pp. 98-105). Washington, DC: American Psychiatric Press.

Kosten, T.R., Frank, J.B., Dan, E., McDougle, C.J., & Giller, E.L. (1991). Pharmacotherapy for posttraumatic stress disorder using phenelzine or imipramine. *Journal of Nervous and Mental Disease, 179,* 366-370.

Kosten, T.R., Mason, J.W., Giller, E.L., Ostroff, R.B., & Harkness, L. (1987). Sustained urinary norepinephrine and epinephrine elevation in posttraumatic stress disorder. *Psychoneuroendocrinology, 12,* 13-20.

Krashin, D., & Oates, E.W. (1999). Risperidone as an adjunct therapy for post-traumatic stress disorder. *Military Medicine, 164*(8), 605-606.

LaPorta, L.D., & Ware, M.R. (1992). Buspirone in the treatment of posttraumatic stress disorder. *Journal of Clinical Psychopharmacology, 12,* 133-134.

Lerer, B., Bleich, A., Kotler, M., Garb, R., Hertzberg, M., & Levin, B. (1987). Posttraumatic stress disorder in Israeli combat veterans: Effect of phenelzine treatment. *Archives of General Psychiatry, 44,* 976-981.

Leyba, C.M., & Wampler, T.P. (1998). Risperidone in PTSD. *Psychiatric Service, 49*(2), 245-246.

Liebowitz, M.R., Hollander, E., Schneier, F., Campeas, R., Welkowitz, L., Hatterer, J., & Fallon, B. (1990). Reversible and irreversible monoamine oxidase inhibitors in other psychiatric disorders. *Acta Psychiatrica Scandinavica, 360*(Suppl), 29-34.

Lipper, S., Davidson, J.R.T., Grady, T.A., Edinger, J.D., Hammett, E.B., Mahorney, S.L., & Cavenar, J.O. (1986). Preliminary study of carbamazepine in post-traumatic stress disorder. *Psychosomatics, 27,* 849-854.

Looff, D., Grimley, P., Kuller, F., Martin, A., & Shoonfield, L. (1995). Carbamazepine for PTSD. *Journal of the American Academy of Child and Adolescent Psychiatry, 34*(6), 703-704.

Lowenstein, R.J., Hornstein, N., & Farber, B. (1988). Open trial of clonazepam in treatment of post-traumatic stress symptoms in multiple personality disorder. *Dissociation, 1,* 3-12.

Marangell, L.B., Silver, J.M., & Yudofsky, S.C. (1999). Psychopharmacology and electroconvulsive therapy. In R.E. Hales & S.C. Yudofsky (Eds.), *Essentials of clinical psychiatry* (pp. 722-725). Washington, DC: American Psychiatric Press.

Marshall, R.D., Beebe, K.L., Oldham, M., & Zaninelli, R. (2001). Efficacy and safety of paroxetine treatment for chronic PTSD: A fixed-dose, placebo-controlled study. *American Journal of Psychiatry, 158*(12), 1982-1988.

Marshall, R.D., Blanco, C., Printz, D., Liebowitz, M.R., Klein, D.F., & Coplan, J. (2002). Noradrenergic and HPA axis functioning in PTSD vs. panic disorder, *Psychiatry Research, 110,* 219-230.

Marshall, R.D., Carcamo, J.H., Blanco, C., & Liebowitz, M. (2003). Trauma-focused psychotherapy after a trial of medication for chronic PTSD: Pilot observations. *American Journal of Psychotherapy, 57*(3), 374-383.

Marshall, R.D., & Cloitre, M. (2000). Maximizing treatment outcome in post-trau-matic stress disorder by combining psychotherapy with pharmacotherapy. *Current Psychiatry Reports, 2,* 335-340.

Marshall, R.D., Davidson, J.R.T., & Yehuda, R. (1998). Pharmacotherapy in the treatment of posttraumatic stress disorder and other trauma-related syndromes. In R. Yehuda (Ed.), *Psychological trauma* (pp. 133-177). Washington, DC: American Psychiatric Press.

Marshall, R.D., & Garakani, A. (2002). Psychobiology of the acute stress response and its relationship to the psychobiology of post-traumatic stress disorder. *Psychiatric Clinics of North America, 25,* 385-395.

Marshall, R.D., & Klein, D.F. (1995). Pharmacotherapy in the treatment of post-traumatic stress disorder. *Psychiatric Annals, 25,* 588-597.

Marshall, R.D., Schneier, F.R., Fallon, B.A., Knight, C.B., Abbate, L.A., Goetz, D., Campeas, R., & Liebowitz, M.R. (1998). An open trial of paroxetine in patients with noncombat-related, chronic posttraumatic stress disorder. *Journal of Clinical Psychopharmacology, 18*(1), 10-18.

Marshall, R.D., Spitzer, R., & Liebowitz, M.R. (1999). A review and critique of the new DSM IV diagnosis of acute stress disorder. *American Journal of Psychiatry, 156,* 1677-1685.

Medical Economics Company (MEC). (2002). *Physicians desk reference 2003* (57th ed.). Montvale, NJ: Thomson Healthcare.

Mellman, T.A., Byers, P.M., & Augenstein, J.S. (1998). Pilot evaluation of hyp-notic medication during acute traumatic stress response. *Journal of Traumatic Stress, 11,* 563-569.

Mellman, T.A., Kumar, A., Kulick-Bell, R., Kumar, M., & Nolan, B. (1995). Noc-turnal/daytime urine noradrenergic measures and sleep in combat-related PTSD. *Biological Psychiatry, 38,* 174-179.

Mischoulon, D., & Nierenberg, A.A. (2000). Natural medications in psychiatry. In T.A. Stern & J.B. Herman (Eds.), *Psychiatry: Update and board preparation* (pp. 399-408). New York: McGraw-Hill.

Monnelly, E.P., & Ciraulo, D.A. (1999). Risperidone effects on irritable aggression in posttraumatic stress disorder. *Journal of Clinical Psychopharmacology, 19*(4), 377-378.

Murburg, M.M., McFall, M.E., Lewis, N., & Beith, R.C. (1995). Plasma norepine-phrine kinetics in patients with posttraumatic stress disorder. *Biological Psychiatry, 38,* 819-825.

Murphy, T.K., Bengtson, M.A., Tan, J.Y., Carbonell, E., & Levin, G.M. (2000). Se-lective serotonin reuptake inhibitors in the treatment of paediatric anxiety disor-ders: A review. *International Clinical Psychopharmacology, 15*(Suppl 2), S47-63.

Neal, L.A., Shapland, W., & Fox, C. (1997). An open trial of meclobemide in the treatment of post-traumatic stress disorder. *International Clinical Psychophar-macology, 12,* 231-237.

Perry, B.D., Southwick, S.M., Yehuda, R., & Giller, E.L. (1990). Adrenergic recep-tor regulation in posttraumatic stress disorder. In E.L. Giller (Ed.), *Biological assessment and treatment of posttraumatic stress disorder* (pp. 87-114). Wash-ington, DC: American Psychiatric Press.

Petty, F., Brannan, S., Casada, J., Davis, L.L., Gajewski, V., Kramer, G.L., Stone, R.C., Teten, A.L., Worchel, J., & Young, K.A. (2001). Olanzapine treatment for post-traumatic stress disorder: An open-label study. *International Clinical Psychopharmacology, 16*(6), 331-337.

Pitman, R.K., Sanders, K.M., Zusman, R.M., Healy, A.R., Cheema, F., Lasko, N.B., Cahill, L., & Orr, S.P. (2002). Pilot study of secondary prevention of posttraumatic stress disorder with propanolol. *Biological Psychiatry, 51,* 189-192.

Pitman, R.K., van der Kolk, B.A., Orr, S.P., & Greenberg, M.S. (1990). Naloxone-reversible analgesic response to combat-related stimuli in posttraumatic stress disorder. *Archives of General Psychiatry, 47,* 541-544.

Quitkin, F.M. (1999). Placebos, drug effects, and study design: A clinician's guide. *American Journal of Psychiatry, 156,* 829-836.

Resick, P. (2001). Cognitive reprocessing therapy. In Workshop series. New York City Consortium for Trauma Treatment, New York.

Resick, P., Nishith, P., & Weaver, T. (1999). Predictors of CPT and PE outcome: Demographics and initial distress. In Symposium "Predictors of Treatment Outcome" (Chair: Resick P). International Society for Traumatic Stress Studies, Miami, Florida.

Resit, C., Kauffman, C.D., Haier, R.J., Sangdahl, C., DeMet, E.M., Chicz-DeMet, A., & Nelson, J.N. (1989). A controlled trial of desipramine in 18 men with posttraumatic stress disorder. *American Journal of Psychiatry, 146,* 513-516.

Riggs, D.S., Cancu, C.V., Gershuny, B.S., Greenberg, D., & Foa, E.B. (1992). Anger and posttraumatic stress disorder in female crime victims. *Journal of Traumatic Stress, 5,* 613-625.

Risse, S.C., Whitters, A., Burke, J., Chen, S., Scurfield, R.M., & Raskind, M.A. (1990). Severe withdrawal symptoms after discontinuation of alprazolam in eight patients with combat-induced posttraumatic stress disorder. *Journal of Clinical Psychiatry, 51*(5), 206-209.

Robert, R., Blakeney, P.E., Villareal, C., Rosenberg, L., & Meyer, W.J. (1999). Imipramine treatment in pediatric burn patients with symptoms of acute stress disorder: A pilot study. *Journal of the American Academy of Child and Adolescent Psychiatry, 38*(7), 873-882.

Savitz, A., & Sachs, G. (2000). Anticonvulsants. In T.A. Stern & J.B. Herman (Eds.), *Psychiatry: Update and board preparation* (pp. 369-374). New York: McGraw-Hill.

Schneier, F.R., Marshall, R.D., Street, L., Heimberg, R.G., & Juster, H.R. (1996). Social phobia and specific phobias. In G.O. Gabbard (Ed.), *Synopsis of treatments of psychiatric disorders* (2nd ed.) (pp. 617-625). Washington, DC: American Psychiatric Press.

Scott, M.J., & Stradling, S.G. (1997). Client compliance with exposure treatments for posttraumatic stress disorder. *Journal of Traumatic Stress, 10,* 523-526.

Sernyak, M.J., Kosten, T.R., Fontana, A., & Rosenheck, R. (2001). Neuroleptic use in the treatment of post-traumatic stress disorder. *Psychiatric Quarterly, 72*(3), 192-213.

Shalev, A.Y., Sahar, T., Freedman, S., Peri, T., Glick, N., Brandes, D., Orr, P., & Pitman, R.K. (1998). A prospective study of the heart rate response following trauma and the subsequent development of posttraumatic stress disorder. *Archives of General Psychiatry, 55,* 553-559.

Shestatzky, M., Greenberg, D., & Lerer, B. (1988). A controlled trial of phenelzine in posttraumatic stress disorder. *Psychiatric Research, 24,* 149-155.

Smajkic, A., Weine, S., Djuric-Bijedic, Z., Boskailo, E., Lewis, J., & Pavkovic, I. (2001). Sertraline, paroxetine, and venlafaxine in refugee posttraumatic stress disorder with depression symptoms. *Journal of Traumatic Stress, 14*(3), 445-452.

Stahl, S.M. (2000). *Essential psychopharmacology: Neuroscientific basis and practical applications* (2nd ed.). Cambridge, UK: Cambridge University Press.

Stein, D.J., Zungu-Dirwayi, N., van der Linder, G.J.H., & Seedat, S. (2002). Pharmacotherapy for posttraumatic stress disorder (Cochrane review). In The Cochrane Library, Issue 4. Oxford: Update Software.

Tohen, M., Castillo, J., Pope, H.G., & Herbstein, J. (1994). Concomitant use of valproate and carbamazepine in bipolar and schizoaffective disorders. *Journal of Clinical Psychopharmacology, 14*(1), 67-70.

Tucker, P., Zaninelli, R., Yehuda, R., Ruggiero, L., Dillingham, K., & Pitts, C.D. (2001). Paroxetine in the treatment of chronic posttraumatic stress disorder: Results of a placebo-controlled, flexible-dosage trial. *Journal of Clinical Psychiatry, 62*(11), 860-868.

van der Kolk, B.A. (1983). Psychopharmacological issues in posttraumatic stress disorder. *Hospital and Community Psychiatry, 34,* 683-691.

Vermetten, E., & Bremner, J.D. (2002). Circuit and systems in stress: II. Applications to neurobiology and treatment in posttraumatic stress disorder. *Depression and Anxiety, 16,* 14-38.

Wells, G.B., Chu, C., Johnson, R., Nasdahl, C., Ayubi, M.A., Sewell, E., & Statham, P. (1991). Buspirone in the treatment of posttraumatic stress disorder and dream anxiety disorder. *Military Medicine, 11,* 340-343.

Wolf, M.E., Alavi, A., & Mosnaim, A.D. (1988). Posttraumatic stress disorder in Vietnam veterans clinical and EEG findings: Possible therapeutic effects of carbamazepine. *Biological Psychiatry, 23,* 642-644.

Yehuda, R. (2002). Current concepts: Post-traumatic stress disorder. *New England Journal of Medicine, 346*(2), 108-114.

Yehuda, R., Marshall, R.M., Penkower, A., & Wong, C.M. (2002). Pharmacological treatments for posttraumatic stress disorder. In P.E. Nathan & J.M. Gorman (Eds.), *A guide to treatments that work* (2nd ed, pp. 411-445). New York: Oxford University Press.

Yehuda, R., Southwick, S.M., & Giller, E.L. (1992). Urinary catecholamine excretion and severity of PTSD symptoms in Vietnam combat veterans. *Journal of Nervous and Mental Disease, 180,* 321-325.

Zisook, S., Chentsova-Dutton, Y.E., Smith-Vaniz, A., Kline, N.A., Ellenor, G.L., Kodsi, A.B., & Gillin, J.C. (2000). Nefazodone in patients with treatment-refractory posttraumatic stress disorder. *Journal of Clinical Psychiatry, 61*(3), 203-208.

Chapter 8

Respecting Cultural, Religious, and Ethnic Differences in the Prevention and Treatment of Psychological Sequelae

Yael Danieli
Kathleen Nader

Contemporary habits of travel and communication have placed much of the world's population in "complex connectedness." Paradigms of cultural competence based only on recognizing cultural difference are not sufficient to take into account the subtleties and importance of the cross-cultural interactions between refugees and mental health services. Instead, paradigms from the "new ethnography" that reflect how cultures engage and influence one another must be incorporated into the delivery of mental health services. (Weine, 2001, p. 1214)

INTRODUCTION

Cultural differences are reflected in the majority of a civilization's beliefs and attitudes, behaviors, and customs; values; implicit rules of conduct (including the expression of emotions); patterns of family and social organization; and taboos and sanctions that may affect group participation. In addition to their histories, languages, cultures, traditions, and religions, civilizations are differentiated by their views on the relationships between God and humankind; the individual and the group; the citizen and the state; parent and child; husband and wife; rights and responsibilities; liberty and authority; equality and hierarchy; tolerance for an unequal distribu-

The authors would like to acknowledge the contribution of Anand Pandya early in this chapter's preparation. We would like to thank Patricia Saunders and Joe Sills for their excellent review and feedback throughout the writing process.

tion of power; the need for predictability; and time orientation (Hodgetts, 1993; Hofstede, 1980; Huntington, 1993). Symptoms (e.g., anxiety) and syndromes must be understood in terms of their social meanings, personal roles, and the beliefs and situations that may engender them and influence their management (Boehnlein, 2001; Kirmayer, Young, & Hayton, 1995).

An individual's identity involves a complex interplay of multiple systems including the personal (biological and intrapsychic), the interpersonal (familial, social, and communal), the cultural (ethnic, ethical, religious, and spiritual), the educational and occupational, the material (economic, environmental), and the national and international (cultural, political, and legal). For example, although there were no reported differences in functional impairment following 9/11, Pole, Best, Metzler, & Marmar (2005) found that culturally sanctioned practices (e.g., dissociation and somatization) combined with a religious predisposition for self-punitive coping were factors in Hispanic police officers' higher symptom levels compared to non-Hispanics. Individual and intermixed cultural issues may contribute to the emergence of traumatic events such as terrorism, the reactions to these events, and recovery from them. Thus, culture may serve as a transmitter, buffer, and healer of trauma (Danieli, 1998; Nader & Danieli, 2004).

Integration of a traumatic experience must take place in all of life's relevant systems and cannot be accomplished by the individual alone. One or more systems can change and recover independently of other systems. Repair of trauma's rupture to life and psyche, however, may be needed in all systems of the survivor, in his or her community and nation, and in their place in the international community. The need to repair cultural systems and meanings is often paramount to healing.

Many clinicians consider it important to incorporate elements of traditional culture into the healing process through the development and usage of culturally appropriate therapies. They also recognize the need to combine the modern (i.e., psychotherapy) with traditional methods (Duran, Duran, Brave Heart, & Yellow Horse-Davis, 1998; Raphael, Swan, & Martinek, 1998). Posttrauma healing may require restoring the cultural context as well as using culturally appropriate therapies. Deep appreciation of a culture, its historical roots, and the way it has shaped indigenous concepts of mental health and healing requires an ongoing commitment to learning. Consulting with individuals from other cultures makes clear the clinician's or consultant's need for adding to an existing knowledge base.

LEARNING ABOUT A CULTURE

> . . . whatever are our therapeutic models, listening and questioning in and of themselves are not quite good enough, and . . . special "knowledges" are helpful as long as we hold them tentatively. (Laird, 1998, p. 22)

Unearthing an individual's cultural story through, for example, reading, listening, and inquiring may be necessary but not sufficient for working with groups or individuals from other cultures (Dubrow & Nader, 1999; Laird, 1998; Nader, 2003b). In addition to knowing what to ask and what to notice, it is essential to recognize how our own cultural lenses and biases motivate and blind us to the unfamiliar and unrecognizable (Laird, 1998). Westermeyer (1987) suggests, in addition to an awareness of personal attitudes, biases, traditions, and norms, that cross-cultural treatment requires training, supervision, reading, and work within a culture for a minimum of one to two years.

Establishing a relationship with community members and leaders is an essential part of learning about and preparing to assist a cultural group (Stamm, & Stamm, 1999; P. Yee, personal interview, August 7, 2003). The sanctions by community leaders (e.g., religious, business, familial) of the training programs, assessments, and interventions of experts and researchers from inside and outside of a cultural community are beneficial. Following September 11, involving local leaders in outreach and treatment programs has enhanced their effectiveness (Waizer, Dorin, Stoller, & Laird, 2004; P. Yee, personal interview, August 7, 2003). Working with area schools, newspapers, and radio stations has also contributed to program success (Fang & Chen, 2004).

Clues to a culture's history and ways of thinking can be found in written works, local humor, folklore, and habits (McGoldrick, 1998; Nader, 2003b; Nader, Dubrow, & Stamm, 1999; van Suntum, 2001). Language, largely unconsciously but constantly, expresses historical identity (Ogundele, 2002). According to Bird (2002), shared cultural narratives (e.g., folklore) serve to construct a sense of place and cultural identity that includes some individuals while excluding others. Folklore (e.g., legends, magical tales, religious stories) becomes a part of the preservation of cultural identity (van Suntum, 2001). Oral memory is audience adaptable and may change during the performance as well as when it is handed down to the next generation. Each generation modifies, adds, and deletes according to its own needs (Ogundele, 2002). It is sometimes in that manner that repeatedly told tales influence and, to varying extents, represent cultural beliefs, expectations, and behaviors (Nader, 2003b; van Suntum, 2001).

The widespread telling of cultural tales across cultures underscores the need to understand their specific history and meaning for the group. For example, told in many cultures in the southern United States, the Uncle Remus tales include stories of playing dumb or helpless in order to escape harm or to outwit an adversary. A European-American salesman explained that because of his southern accent people often assumed that he was dumb; he used that expectation to his advantage. The true stories of Solomon Northup and Harriet Tubman show African-American slaves' very real need to play dumb or helpless for their survival, safety, or well-being.

IDENTIFYING THE COMPONENTS OF CULTURE

Cultures emerge nationally, regionally, locally, at churches/mosques/temples, and in organizations, schools, or workplaces. Each of these cultural groups influences individuals' and groups' characteristic reactions (i.e., thoughts, emotions, behaviors) (Nader, 2004). When the values of a person's subcultures conflict, additional difficulties may occur. For example, Korean-American psychologists reported distress as a result of struggling between their parents' traditional culture and the values of their American-trained psychology group (Park, 1996).

National Cultures

Although overall cultural statements about Western, Asian, African, Hispanic, or other societies do not automatically apply to individuals within or from those cultures (Hofstede, 1980), specific characteristics have been found to be common among members of a nation and individuals who share its heritage. Studying employees of multinational corporations, Hofstede defined *culture* as the collective mental programming that people of a group, tribe, geographical region, national minority, or nation have in common.

Hofstede identified four dimensions defining the values associated with national cultures: *power distance* (how much a society accepts the fact that power is distributed unequally in institutions and organizations), *individualism/collectivism* (people are supposed to take care of themselves and of their immediate families only versus a tight social framework in which people expect their relatives, clan, or organizations to look after them), *masculinity/femininity* (the extent to which a society's dominant values are "masculine"—assertiveness, acquisition of resources, not caring for others or the quality of life), and *uncertainty avoidance* (desire to avoid uncertainty or desire for predictability)

(Hodgetts, 1993; Hofstede, 1980). Michael Bond (see Hodgetts, 1993) added a fifth dimension: *long-term versus short-term time orientation.*

According to Hofstede, people of a nation become programmed to have particular expectations as well as beliefs and behaviors. For example, in nations with high levels of uncertainty avoidance (e.g., Latin, Catholic cultures; Soeters, 1996), citizens are accustomed to greater career stability, more formal rules, belief in absolute truths, the attainment of expertise from others, and intolerance of deviant ideas and behaviors (Hofstede, 1980). These countries are also characterized by higher levels of anxiety and aggressiveness that, among other outcomes, induce people to work hard.

Hofstede's (1980) dimensions occur in all combinations except one: no countries value both low power distance and high levels of collectivism. Nations with high levels of individualism (e.g., Australia, Great Britain, United States) most often have small power distance. Soeters (1996) suggests that nations with high levels of collectivism and high levels of uncertainty avoidance tend to handle intergroup conflicts in an inflexible, often violent manner (e.g., minorities are forcibly repressed or assimilated). Hofstede suggests that organizational cultures are a more superficial phenomenon than national cultures which reside mainly in deeply rooted values. He advocates that statements about a person's culture can be used only as working hypotheses.

CULTURAL ISSUES AND REACTIONS TO TRAUMA AND TREATMENT

Aspects of national and regional cultures can influence mental health assessments and interventions. Knowing the cultural differences in the ways that families function, distribute power, respond to and express stress, and view mental health interventions equips clinicians and researchers to provide accurate assessments and treatment.

Understanding culture is integral to understanding the predicament of trauma survivors and their families, whether or not cultural identity played a role in their victimization—perhaps more so if it did. Cultural beliefs and attitudes affect interpretations of events, symptoms, interviewers' questions, expectations (e.g., of behaviors after a death or disaster), issues of trust, establishing a time frame, and the admission and expression of emotions (Nader, 2003a). They shape expectations; transference and countertransference (see Danieli, Chapter 21 in this volume); perceptions and expressions of illness and bodily sensations, and beliefs about their causation; the social meaning of and response to symptoms, disorders, and those affected by trauma or illness; and the factors that increase stress (Boehnlein, 2001; de Silva, 1999).

Family and Hierarchy

The Importance of Family and Other Support

Family and other support systems are important to recovery in all cultures. Higher levels of social support have been associated with lower levels of traumatic response for both children and adults (Boehnlein, 2001; Compas & Epping, 1993; de Silva, 1999; La Greca, Silverman, Vernberg, & Prinstein, 1996; Nader, 2003a; Rabalais, Ruggiero, & Scotti, 2002). A good support system has been associated with reduced trauma and hostility. For example, Laotian immigrant women who have been observed to remain outside mainstream American society had higher hostility scores (Westermeyer & Ueker, 1997). For immigrants, joining a structured religion and thus gaining a supportive community has been linked to lower hostility and PTSD scores (Mollica, Cui, McInnes, & Massagli, 2002; Westermeyer & Ueker, 1997).

Individualism versus Collectivism

While families are important, societies and religions differ in their emphasis on the individual versus the collective. Individuals in different regions of the world define themselves and parent their children in relationship to independence or to connectedness-interdependence. For example, in independence-minded nations, such as the United States and countries of northern Europe, the good, moral self is highly individualistic and autonomous and seeks to conquer new frontiers; in interdependent nations, such as many found in Asia and Africa, the good, moral self puts the good of the group before individual needs (Markus, Kitayama, & Heiman, 1996; Shiang, 2000; Shiang, Kjellander, Huang, & Bogumill, 1998; Triandis, Kashima, Shimada, & Villareal, 1986). Therefore, Asians and Hispanics are more likely to emphasize group harmony and the needs of the group, while Europeans and Americans are likely to emphasize the needs of the individual and personal autonomy (Hofstede, 1980; McDermott, 1991). Pole et al. (2005) suggest that a greater value on collectivism may place greater emphasis on the need for social support. For example, studies of traumatized adults have found that Hispanic patients with poor familial and social relationships had higher and more intense PTSD symptoms (Pole et al., 2005).

Although Western societies tend to be individualistic rather than collectivistic, Americans are influenced to some degree by their cultures of heritage. Many American subcultures include values of interdependence. For example, the strength of extended family kinship networks in many American families (e.g., African American, Italian, Jewish) has been well documented (Boyd-Franklin & Franklin, 1998; McGoldrick, 1996; Watson, 1998).

Cultural values of interdependence or independence are reflected in religious ceremonies (e.g., those dealing with guilt) and interpretations of the meaning of traumatic events. As one American child of Nazi Holocaust survivors put it, "On Yom Kippur Jews say 'we have sinned, we have done. . . .' Nowhere does it say 'I have done . . . .'" (Danieli, 1993, p. 891). In contrast, for the Roman Catholic, atonement is dealt with individually in the confessional and through priest-prescribed personal tasks to make amends. When asked about September 11, a rabbi in Texas explained that it was God's message to all Americans that we must return to a more spiritual way of life. Similarly, a Hindu may ascribe such events to collective or personal karmas (i.e., justice for a nation's or a person's past actions).

Hierarchy

In individualistic cultures, the family leadership structure is based in the marital couple and depends upon reaching agreement. In collectivistic cultures, leadership and authority are invested in the older generation; throughout life, vertical relationships and age hierarchies are stressed (Falicov, 1998). In some cultures (e.g., Asian), the power hierarchy is age first, then males over females (McGoldrick, 1998). Thus, a therapist's youthful appearance may hinder credibility in some cultures (e.g., Asian, Native American) (Stamm, & Stamm, 1999). Moreover, advising a patient from an interdependent culture to go against a grandparent's or parent's wishes may be divisive to the extended family and may undermine treatment (Danieli, 1993; Waldegrave, 1998).

Some nations and groups (e.g., Asian, Native American, African) include ancestors in the family hierarchy (Peddle, Monteiro, Guluma, & Macaulay, 1999). Rites and rituals (e.g., funerals, feeding dead souls) may include ceremonies for contacting them (Peddle et al., 1999). The influence of elders thus extends beyond their deaths. A third-generation Chinese-American psychologist lamented that he would have to end his relationship with the woman with whom he felt mutual love. Many years before her death, his grandmother had told him that if he did not marry a Chinese girl she would come back to haunt him after her death. He was certain that if there was a way to do so, his grandmother would find it.

Roles of Men, Women, and Children

Gender Issues

In most societies (masculine or feminine), the majority of men have masculine values (i.e., directed toward ego goals, achievement, careers, and high salaries) (Hofstede, 1980; Lloyd, 1999; Soeters, 1996). Feminine societies (e.g., Scandinavian countries), however, value social goals: taking care of the poor, weak, and needy and the environment (Soeters, 1996); relationships are more important than money, and quality of life more important than performance (Hofstede, 1980). In masculine nations (e.g., Japan, Austria, Venezuela), men's values differ more from women's and older people's more from younger people's than is seen in feminine nations (Hofstede, 1980; Lloyd, 1999).

In 1993, Samuel P. Huntington purported that the dividing lines that would create conflict in the twenty-first century were likely to be between civilizations (Nader & Danieli, 2004). Inglehart and Norris (2003) suggest that the cultural fault line is primarily about gender equality and sexual liberation. Awwad (1999) points out that Palestinian women are among the most educated in the Arab world, yet they struggle for equal rights as do other women of the world. According to Meleis (2003), however, Middle Eastern men of Arab descent fear that, under American influence, their women would adopt Western values that separate them from established cultural, family, and religious values (see also "Clash of Civilizations?," 2003; Huntington, 1993; Inglehart & Norris, 2003). Women's sexual liberation depicted by mass media and prevalent in many Western nations are contrasted to expectations of women in other nations. For example, Latina girls are expected to be obedient, respectful, and feminine (Garcia-Preto, 1998). In some cultures (e.g., Latino, Middle Eastern), virgins and martyrs are revered (Garcia-Preto, 1998). In these cultures, virginity may be necessary for a safe future in the hands of an honorable man (Garcia-Preto, 1998) or even to physical survival (in some countries women who do not remain virgins until marriage are killed). Boyd-Franklin and Franklin (1998) quote the 1961 film *A Raisin in the Sun* in describing how some cultures (e.g., African American; Middle Eastern) "raise their daughters but love their sons." Stories of the killing of infant girls in some Asian cultures have been well publicized.

In many cultures, gender roles are clearly delineated, and any deviation may result in ostracism. Gender roles (as well as personality styles) may indicate an increased need for therapeutic privacy and the inadvisability of raising emotional or potentially shaming issues in a group. For example, in some cultures (e.g., Asian Indian, Hispanic), it is not acceptable for men to complain of distress (Canive, Castillo, & Tuason, 2001; de Silva, 1999). Prescribed roles and gender expectations may inhibit communication. For example, a Hispanic male may not be willing to describe some of his distressing experiences to a female because of the need to protect her. Moreover, cultural differences must be taken into account when determining what are normal differences between the genders. For example, in a study of six- to seven-year-old children in the People's Republic of China (PRC) and the United States, gender differences in activity levels, inhibitory control, perceptual sensitivity, and smiling rates were opposite (Ahadi, Rothbart, & Ye, 1993).

Role Reversals

Relocation can disrupt or challenge normally expected cultural roles. The expectation that age and gender define the hierarchy for Asians or Native Americans can be undermined by grandparents depending on English-speaking children as interpreters and cultural brokers (Gerber, Nguyen, & Bounkeua, 1999). This may also occur between spouses or between parents and children when one becomes more acculturated than the other. Men from some cultures lose power when they move to Western nations and their wives are able to work and bring home a salary.

Children

Cultural differences are the best predictors of temperament in early life (McDermott, 1991). Children's temperaments and behaviors reflect their parents' training. For example, different cultures value different kinds of intelligence (e.g., musical, bodily-kinesthetic, logical-mathematic, spatial, linguistic, interpersonal, intrapersonal) (Gardner, 1983; Watkins & Williams, 1992). Culture determines whether an adolescent transition occurs between childhood and adulthood. Culture affects the difficulties that occur for children as they grow up. For example, Puerto Rican children have fewer sleep problems and more discipline problems than American children. McDermott (1991) explained this finding with the Puerto Rican habit of allowing children to take a bottle to bed until age five—a practice almost never done in the United States (see also Rousseau & Drapeau, 1998).

Attitudes Toward Mental Health and Practitioners

Mental Health Issues

Most psychological theories were developed in Western Europe and white North America (hereafter referred to as the West or Western) and include a primary goal of individual self-worth. Individual concepts such as destiny, responsibility, legitimacy, and even human rights, along with concepts of self, individual assertiveness, and fulfillment, are central to most of these therapies (Waldegrave, 1998). In contrast, for people who come from communal or extended-family cultures, questions about the self, self-exposure, and self-assertion are confusing, alienating, or can be considered intrusive and rude (Waldegrave, 1998).

In some cultures (e.g., Cambodian, Chinese, Arabic), expressing mental health problems is associated with shame or stigma (Kinzie, 1993; Shiang, 2000). For example, a Kuwaiti adolescent seeking assistance for his war trauma explained that if one family member was diagnosed with mental health problems, the whole family was tainted. Similarly, when asked why they resisted therapy, a group of African Americans responded that they felt a stigma ("there is always something wrong with us") and a feeling of powerlessness. Some cultures (e.g., Asian, Hispanic) may express emotional problems in the physical realm. Complaining of physical symptoms allows the elicitation of social support without the stigmatization and shame of a mental problem (Kinzie, Boehnlein, & Sack, 1998; Shiang, 2000).

The taboo against having mental problems may be greater for males (e.g., Latino). For example, a Nicaraguan refugee father felt he should be tough with his traumatized nine-year-old son so that he would not whine (Bevin, 1999). The length of time (weeks, months, or years) it takes for a person to reveal the extent of personal traumatic reactions varies by culture (Kinzie, 1993).

Stress Behaviors

The DSM-IV provides a list of culture-bound syndromes. Although it is valuable to familiarize oneself with such distinctive clinical phenomena, it is important to remember that the influence of culture is often far more subtle than what is contained in any list of syndromes. In the DSM-IV-TR (American Psychiatric Association, 2000), a list of cultural issues are found in the appendix. Laria and Lewis-Fernández (2001) and Nelson (2002) suggest that incorporating "cultural-bound syndromes" into the main body of the DSM system would make it more cross-culturally relevant. For example, empirical evidence suggests a strong correlation between dissociative and somatic symptoms. Unlike the DSM-IV, the International Classification of Disorders (ICD-10) (World Health Organization, 1992) includes somatoform disturbances in the definition and categories of dissociation (Laria & Lewis-Fernández, 2001; Lee, Lei, & Sue, 2001). Some cultures more readily present with physical rather than emotional complaints when distressed.

Culture-Bound Syndromes

Research suggests that the same basic patterns and signs of psychopathology (e.g., insomnia, worry, crying spells, weakness, reduced energy, suicidal ideation, hallucinations) exist worldwide. Families and friends report, for example, inappropriate or purposeless behaviors, social withdrawal, assaultiveness, incomprehensible speech, or damage to property. The prevalence and presentation of syndromes, however, may differ by culture (Kirmayer et al., 1995; Laria & Lewis-Fernández, 2001; Westermeyer, 1987; Westermeyer et al., 2002). For example, a study of Cambodian combat veterans with PTSD revealed significantly higher levels of physiological responses to viewing trauma scenes than found in American combat veterans (Boehnlein, 2001; Kinzie, Denney, et al., 1998). Boehnlein (2001) suggests that cultural attitudes may be a trigger for some symptoms (e.g., panic symptoms).

Although nonpsychotic disorders that include a combination of emotional distress, behavioral abnormality, transient cognitive disturbances, and crises or situational problems have been observed across cultures (Westermeyer, 1987), the outward manifestation of universally found disorders can be influenced by cultural factors (de Silva, 1999). For example, *koro,* a disorder associated with threat to survival in South and East Asia, includes the intense fear that the penis or nipples/breasts are shrinking with the anticipation of loss of fertility and death.

Some culture-bound syndromes are similar to conditions described in Western cultures (with differing names in different cultures) or are correlated with a number of different disorders in the United States (Kirmayer et al., 1995). For example, *Agahinda gakabije* (i.e., sadness, poor relationships, lack of self-care, loss of mental ability, inability to work, isolation, and feeling that life is meaningless), the Rwandan (Africa) illness that is most similar to depression, is a more general disorder that includes depression (Bolton, 2001). Similarly, the Cambodian *kyol goeu* ("wind illness") includes orthostatic panic and associated somatic changes (Boehnlein, 2001). In addition to these similarities and variations, culture-bound syndrome labels have been used by the community to categorize deviant or so-cially problematic behaviors that vary among cultures. Moreover, differ-ences (e.g., biomedical, dietary, environmental conditions, traumatic exposure rates) between cultures and the distress of migration can influence rates and types of psychopathology (Westermeyer, 1987).

Cultural and social factors shape the expression and nature of emotional and attentional states. For example, after overwhelming emotionally dis-tressing events, the Hispanic (e.g., Puerto Rican) cultures sanction *ataques de nervios,* brief outbursts of intense emotionality and undercontrolled be-havior (e.g., intense fear, anger, grief, lashing out, crying) (Laria & Lewis-Fernández, 2001; Velez-Ibanez & Parra, 1999). Sanctioned alterations in states of consciousness may affect the expression of specific symptoms (e.g., dissociation, somatization). For example, for Westerners, dissociative identity disorder alters (multiple personality manifestations) are individual; for Indians, the same spirit or identity may possess different family or vil-lage members. Peritraumatic dissociation (the tendency to experience al-tered states of consciousness at the time of trauma) has been a robust predic-tor of PTSD. After 9/11, it was among the factors associated with higher symptom levels for Hispanic police officers than for non-Hispanics (Pole et al., 2005). The use of altered states as coping mechanisms varies among cultures. For example, meditative training may have assisted some impris-oned Tibetan monks in coping with torture by their Chinese captors (Laria & Lewis-Fernández, 2001). For Asian and American meditators, adapted meditation techniques may be used in conjunction with or to replace anxiety management techniques (e.g., relaxation methods).

Some variations in the prevalence of traumatic reactions among ethnic and cultural groups have been attributed to cultural reporting biases or the interaction of culture and other variables (Ahadi et al., 1993; de Silva, 1999; Mash & Dozois, 2003). For example, following World War II, Williams suggested that one reason British combat personnel had three and a half times the postwar psychiatric illness as Indians may have been that showing anxiety resulted in a great loss of face for Indians (de Silva, 1999).

Cultural backgrounds may contribute to the risk factors (e.g., history of discrimination stresses; acculturation stress; lack of access to services) and protective factors (e.g., a supportive community; culturally acceptable out-lets for emotional responses) following traumatic events (Rabalais et al., 2002). Ethnic or cultural experiences (e.g., discrimination, mass murder) can influence vulnerabilities or sensitivities. For example, the Nazi Holo-caust has added to the vulnerabilities of the offspring of its victims (Danieli, 1995, 1998). Similarly, during the Vietnam War, African Americans were more distressed by the victimization of villagers than were other soldiers (Boehnlein & Kinzie, 1997). In addition to community and family support, higher socioeconomic status (SES) and better access to resources serve as protective factors. Following September 11, DeVoe and Klein (2003) found no significant differences in rates of PTSD for highly exposed white, bicultural, and minority families with high SES. In this study, the white parents of white children reported the most negative shift in worldview. DeVoe and Klein (2003) suggested that greater experience serving as buffers against racism for their children may have resulted in minority parents reporting the most positive parenting changes.

Stress and Migrant Groups

Most refugees have had high levels of exposure to traumatic experiences (Hollifield et al., 2002). Most migrant groups (i.e., immigrants, refugees, international adoptees) have higher rates of psychiatric disorders than non-migrant groups (Hollifield et al., 2002; Westermeyer, 1990). In addition, terrorist events such as 9/11 have been especially difficult for those who left war conditions to find safety in the United States (Danieli, Engdahl, & Schlenger, 2003; Kinzie, 2004; P. Yee, personal interview, August 7, 2003). A large number of migrants and refugees exhibit a condition marked predominantly by depressive symptoms but not severe enough for a diagnosis of a major depressive disorder. It is long-lasting (too long for an adjustment disorder) and also includes mistrust or suspiciousness, mild-to-moderate anxiety, psychophysiological symptoms, posttraumatic stress symptoms, and social withdrawal (Westermeyer, 1987). Kirmayer et al. (1995) suggest that, rather than a discrete disorder (e.g., PTSD), the refugee experience includes a kind of cultural bereavement. Paranoia and depression are common among refugees. Social isolation appears to be a factor in increased hostility (i.e., reportedly more annoyance and irritation, outbursts of temper, urges to harm someone or break things, arguing, shouting, or throwing things) among migrants to the United States, which persists for years after migration (Westermeyer & Ueker, 1997).

Leadership roles and the consequent increased self-esteem, prestige, and influence (despite the associated increased stresses and work hours), religious practice, social support, and engagement in work/economic activity have all served as significant protective factors. In many countries (e.g., United States, Croatia), refugees have been forbidden to work, thus increasing their risk. Lack of work and the related want of resources also limit the ability to engage in spiritual practices that involve fees or donations to those who perform the spiritual ceremonies (Mollica et al., 2002).

Diagnosis

It is essential that diagnostic assessment and treatment planning consider the individual within his or her sociocultural context (e.g., family, local community, work/school community, health care system) (Canive et al., 2001). Difficulties in cross-cultural diagnosis have led to inappropriate treatment plans and therapeutic failures. A number of factors may contribute to misdiagnosing an individual whose culture is different from that of the mental health professional (Westermeyer, 1987, 1990). Among them are underassessment of the severity of a disorder (e.g., by attributing the behavior to cultural differences), overassessment of severity (e.g., because of cultural beliefs such as communication with the dead or other culturally valued preternatural experiences), misdiagnosis (e.g., failure to discover a physiological cause for an ailment frequent among members of the patient's culture), unwillingness to share information (e.g., due to mistrust, lack of a shared language or values), and poor understanding or interpretation of what the patient says (Kirmayer et al., 1995; Phan & Silove, 1997; Westermeyer, 1987, 1990; P. Yee, personal interview, August 7, 2003). For example, what an individual says when abruptly switching into a native language may be of diagnostic import (Canive et al., 2001).

Specific behaviors that are interpreted one way in a Western culture may mean something entirely different in another culture. For example, silence may be considered a sign of denial or resistance in some clinical settings. In the context of PTSD for Southeast Asian cultures, however, it may represent one of the following: (1) shame, (2) lack of trust, (3) fatalistic acceptance or resignation, or (4) repression or suppression of intimate issues such as "emotional pain" (Ton-That, 1998). Ton-That suggests that prolonged and frequent silence may indicate a severe illness with a poor prognosis for some Asian Americans. Kirmayer et al. (1995) suggest that locating the source of anxiety or distress in the spirit world, the social world, or the individual's existential predicament are all cultural coping strategies (e.g., for some African, Asian, and Hispanic cultures) for constructing and living in a coherent world. Cultural differences may either mask or be mistaken for disorders (Canive et al., 2001; Kirmayer et al., 1995). For example, culturally expected behaviors may include repetitive religious rituals (e.g., for Asians, Africans, Hispanics, Jews, Catholics, Muslims, Hindus), being homebound (e.g., for women from Qatar or Saudi Arabia), belief in magic or witchcraft (e.g., for Carribean African Americans or traditional Hispanics), adults living with or preadolescents sleeping with parents (e.g., in Hispanic or Asian cultures). When culturally valued, the experiences and behaviors are generally supported by family and community; preceded and followed by socially appropriate, productive, and coping behaviors (not psychological, behavioral, or social deterioration); do not diminish self-esteem or prestige; do not include pathological signs and symptoms; and are culturally congruent (Westermeyer, 1987).

War, terrorism, and inner-city violence in many nations and locations may make anxiety endemic and difficult to measure (Kirmayer et al., 1995). Individuals may continue to experience deleterious effects after becoming consciously oblivious or habituated to anxiety-provoking experiences and situations or after symptoms transform into patterns of behavior that are no longer linked to the traumatic experience (Kirmayer et al., 1995).

Assessment. Empirically developed and tested instruments covering the complete range of migrants' and refugees' experiences and responses have not been readily available (Hollifield et al., 2002). A few measures have been designed specifically for use with migrants and refugees (e.g., Harvard Trauma Questionnaire; Vietnamese Depression Scale), a few have been adapted for use with refugees (e.g., Hopkins Symptom Checklist-25; Beck Depression Inventory); and many scales have been translated into other languages (see also Wilson and Keane, 2004). When local professionals are not available to validate instruments for a culture, Bolton (2001) suggests comparing an instrument to the locally recognized illness that most closely resembles the disorder in question. Having multiple possible responses (on a rating scale) may be difficult for the respondents in some cultures (Hollifield et al., 2002). Using the language of the interviewee and establishing rapport are essential (P. Yee, personal interview, August 7, 2003).

Cultural Issues and Group Leader or Therapist

There are pros and cons to choosing a clinician of the same cultural or ethnic background as the participant/patient. A therapist/leader of the same background may more readily establish rapport, intuitively behave with the appropriate level of formality, and understand the client's language, word meanings, beliefs, and expectations (Canive et al., 2001; Westermeyer, 1989). On the other hand, a therapist/leader of the same background may overlook important social and educational differences, be overconfident about understanding/knowing, be easily sidetracked by the prominence of cultural symptoms, fail to explore deep underlying issues, or overly identify or compete with the patient (see Danieli, Chapter 21 in this volume).

In addition to rapport with group leaders or therapists, self-disclosure may be affected by cultural attitudes. For example, in some cultures (e.g., Hispanic, Asian), intimate emotional problems are not to be shared with people outside of the immediate family. Perhaps especially when shame is associated with admitting emotional difficulties (P. Yee, personal interview, August 7, 2003), individuals may be more likely to share information with someone who understands and is accepted in their culture. Group leaders may need to take the time to establish rapport in the community as well as with individual group members (Waizer et al., 2004).

The attitudes, feelings, and beliefs toward a patient's culture have an impact on the clinical encounter (Canive et al., 2001). In cross-cultural as well as trauma treatment, the patient's raw or disguised anger, protective defenses, and covered-up emotions may constantly test the group leader's own emotions and skills. Cultural countertransference, therefore, is important to explore for both same-culture and other-culture therapists/group leaders (Canive et al., 2001; Spiegel, 1976; Westermeyer, 1989). Gender may be an issue in the choice of a therapist/group leader or in group composition. For example, traditional Hispanic males may be extremely concerned about the fragility of females and may feel it is their duty to protect them from hearing about grotesque, inhuman, or disgusting events (Canive et al., 2001).

Religious, Cultural, and Spiritual Beliefs Relevant to Group Practice

Spiritual Beliefs

Beliefs are clearly intertwined with emotional and behavioral functioning. Our belief systems are at the heart of who we are; they help us to organize and make coherent meaning of the onslaught of stimuli, ideas, emotions, and memories (Hines, 1998). Many of our most basic beliefs are founded in religion—organized belief systems that include shared moral values, patterns for living, beliefs about God, and involvement in a religious or spiritual community (Walsh, 1998). Spirituality can be experienced either within or outside of formal religions (Tully, 1999). Spirituality may include personal beliefs (e.g., about an ultimate human condition, values, a supreme being, or a unity with nature and the universe), experiences (e.g., holy or mystical), a sense of meaning, inner wholeness, and connection with others. "It invites an expansion of awareness, with personal responsibility for and beyond oneself" (Walsh, 1998, p. 72).

Religious or spiritual beliefs may influence or dictate responses to crisis (e.g., culpability, view of therapy, coping strategies). In a crisis, beliefs may either function as a source of comfort and as anchors or may promote a sense of hopelessness and helplessness (Hines, 1998; Tully, 1999). Congruence between beliefs and practices may generate an overall sense of well-being and wholeness; incongruence may induce shame or guilt (Walsh,

1998). Validating these beliefs and practices will increase treatment compliance for many (Bibb & Casimir, 1996; Ton-That, 1998). This may be especially true when spiritual beliefs are central to group members. For example, for some Irish Americans, the Catholic Church is more important than the family (McGoldrick, 1996).

Some theorists and cultures believe that the soul, which is deep within the personality, is the key to change. Taking care not to attribute failure to a lack of spiritual purity, therapeutic efforts therefore must take into account the human spirit (Seligman, 1990; Walsh, 1998). Medical studies have confirmed that faith, prayer, and rituals have strengthened health and healing presumably by triggering emotions that influence the immune and cardiovascular systems (Walsh, 1998). Viewed as crises in most cultures, suicides or sexual assaults take on additional consequence in religions and cultures that deem them a severe source of shame for a family or a cause of unmarriageability in children (Sandoval & Lewis, 2002).

Religious or spiritual beliefs may place the responsibility for change on the external world and external forces (e.g., supernatural), minimizing individual responsibility and the ability of self-determination (Bibb & Casimir, 1996). In contrast, beliefs may emphasize the culpability of the individual or society. For example, the Irish Catholic Church purports the evil and untrustworthy aspects of human nature; people deserve to suffer for their sins (McGoldrick, 1996). The sense of personal culpability may add to feelings of guilt, isolation, and helplessness.

Spiritual distress can impede coping and the sense that life has meaning. Spiritual renewal can be found in a variety of ways such as specific rituals, communion with nature, attunement to spiritual teachers or teachings, attending places believed to have high levels of spiritual energy (e.g., healing waters, pilgrimages, sacred shrines and temples), beauty, music or dance, love, and transporting imaginations (Tully, 1999; Walsh, 1998). The injustice or senselessness of trauma and suffering are spiritual issues. Traumatic events can precipitate a questioning of long-held spiritual beliefs or a quest for something new that can be sustaining (Walsh, 1998). Following traumatic experiences, the feeling that God has betrayed, abandoned, or punished may increase symptoms or risk factors (e.g., a sense of isolation).

Basic spiritual principals permeate an entire culture no matter what organized religion a person may belong to. For example, although in Haiti 80 percent of the population identify themselves as Catholics and 20 percent as Protestants, voodoo rituals are integrated into Christian theology and there is a strong belief in voodoo's curative capabilities (Bibb & Casimir, 1996). Introducing the issue of voodoo when clients have not raised it themselves, however, will increase guardedness and may be interpreted as an underhanded method of assessing class status or "family secrets." Similarly, no matter what the formal religion (Christian, Hindu, Jewish, Muslim), karma (destiny, the results of one's actions), caste (a hierarchical organization), the connectedness of all living things, reincarnation (after death, the soul is born again into another human being or animal), and dharma (living life in accordance with the principle that orders the universe and in accordance with one's personal truth) are all essential to understanding the worldview of Asian Indian families (Almeida, 1996) and others whose spiritual practices or belief systems are largely based in eastern or New Age philosophies.

Rituals

Rituals are group methods that serve to maintain a culture's social structure and its norms, strengthen the bonds of individuals to their communities, assist adaptation (to changes or crises), manage fear and anxiety, and ward off threats (Johnson, Feldman, Lubin, & Southwick, 1995). In addition to supporting the existing social order, rituals have sometimes served to ward off unconscious or distressing emotions or to suppress individual expression. In order to be therapeutic, rituals also must help individuals to manage stressful emotions and may provide symbolic enactments that serve as metaphors for transformation (e.g., of emotions, injured relationships, disturbed social connection). To do so, Johnson et al. (1995) suggest that the ritual situations must contain rather than suppress emotions, permit more individual expression of emotions, designate specific times for spontaneous comments or actions, and allow for greater arousal of the disturbing experience toward greater catharsis.

Even when informal, specific cultural and religious rituals become important when interacting with communities (e.g., hospitality rituals). In some cultures (e.g., Arabic, African American), the offering of food has special significance; acceptance of the food may be an essential communication. Tully (1999) suggests that accepting food or refreshments in an African-American home is polite and honors the host. An American visiting a relative in Lebanon tells the story of how he enthusiastically admired a carved figure on the mantle. The relative responded by giving it to him. Realizing how much the item was prized by his relative's wife, he tried to return it, but she told him that her husband would beat her if he did not keep it. Some cultures expect a specific kind of reciprocation after giving a gift.

Formal cultural and religious rituals (e.g., protection, responding to a sudden death, funerals) may be easier to learn about than informal ones. Certain cultures (e.g., Liberian, Native American, Southeast Asian; Catholic, Baptist, Hindu) have rituals (e.g., lighting candles; burning incense, plants, or resins; placing bird feathers, garlic, or talismans; saying specific prayers or mantras) for protection or cleansing (e.g., to dispel evil spirits, ghosts, and negative thought forms or to invoke good or holy spirits) (Brown, 1989; Gerber et al., 1999; Nader et al., 1999; P. Yee, personal interview, August 7, 2003). For example, after a sniper attack at a school in Stockton, California, in 1989, Cambodian and Vietnamese Buddhist monks, a Vietnamese Catholic priest, and Protestant ministers performed a blessing ceremony that included the exorcism of spirits (e.g., the bad spirit of the sniper and the spirits of dead children who might grab other children and take them into the next world) (Dubrow & Nader, 1999). Rituals performed to please or appease God or supernatural forces (e.g., Catholic prayers to specific saints, Haitian "spiritual reshaping," Hindu puja) may be used to alleviate depressive and anxiety symptoms and feelings of vulnerability. Amulets or medals may be worn for self-protection and restoration of self-confidence (Bibb & Casimir, 1996).

Funerals, having or viewing a deceased body, and grieving vary across and within cultures. African-American funerals range from somber and silent events to loud celebrations or parades (Tully, 1999). For Liberian mourners, distant nieces and nephews are expected to distract the grievers from the death by dressing up as clowns, making funny jokes, singing funny songs, and dancing (Peddle et al., 1999). In contrast, Bosnian Muslim funeral mourners are expected to be quiet so as not to upset the soul of the dead (Mooren & Kleber, 1999). In some cultures (Liberian, African), children are forbidden to see dead bodies (Dubrow & Nader, 1999).

The terrorist attacks of September 11, 2001, left many surviving families without bodies to bury. The lack of a body to assist with closure and memorialization was distressing across cultures. Some families retrieved dust from the World Trade Center site to use in memorial services, as a symbol of the deceased, or for burial (P. Yee, personal interview, August 7, 2003). Other families included pictures of the deceased in ceremonies. When a person of Southeast Asian descent dies tragically, it is the duty of family members to perform appropriate spiritual rituals in order to prevent the person's soul or spirit from wandering around unable to live in peace or reach heaven (Gerber et al., 1999). Tending to the graves is especially important for Vietnamese who have a Confucian emphasis on ancestor worship (Gerber et al., 1999).

A number of Western clinicians have incorporated specific ceremonies or rituals into the treatment process. Treatment used following wars are often relevant after terrorist attacks as well. Ceremonies can compartmentalize the process of reviewing the traumatic experience, set it aside from everyday life, and recontextualize it, thereby lessening traumatic alienation (Johnson et al., 1995). Following wartime experiences, rituals have been used to reaffirm membership and assist reintegration of warriors into their culture. For example, Native American sweat lodge, homecoming, tree planting, and ceremonial fire rituals have been used with war veterans for purification, to assist transitions home and begin anew, and to free energy and restore hope (Wilson, 1989; Wilson, Walker, & Webster, 1989). Similarly, Canive et al. (2001) have included a veteran's pilgrimage to a nearby war memorial with the lighting of candles, reading emotional letters about traumas, and burning the letters to symbolize the cleansing of their inner beings. In addition to individual treatment with veterans, Johnson et al. (1995) have designed and used a series of ceremonies that address departure and homecoming, forgiveness of postwar behaviors and misunderstandings, return, and survival guilt.

Role of the Mental Health Practitioner

In addition to Westermeyer's (1987) recommendation earlier that cross-cultural treatment requires a year or two of training, supervision, reading, and working within the culture, Canive et al. (2001) suggest that clinicians engage in cultural sensitivity exercises (e.g., see Pinderhughes, 1989). No matter how well the clinician has learned about and within a culture, it is essential to assure good communication between therapist and group members.

Working with a Translator

Language plays a significant role in the expression and perception of identity, emotions, and psychopathological conditions (Ogundele, 2002; Westermeyer, 1990). The greater severity of the stress or disorder, the greater the undermining of a second language (Westermeyer, 1987). In treatment, patients may abruptly switch from English to a native language when anxious or experiencing strong emotions (Canive et al., 2001). Consequently, even when both patient and clinician speak a common language, it is frequently desirable to conduct an interview in the patient's native language or to have group leaders who are proficient in the languages of group members. Poor communication in an interview can lead to a variety of clinical misadventures, including preventable suicides (Nader, 2003a; Swiss & Giller, 1993; Westermeyer, 1990). Even if the patient insists on using the clinician's language, it is advisable to have (at least until the patient regains second language skills with treatment) an interpreter for clarification.

Although using the translator as "a word unscrambler" may work well in diplomatic or some business situations, it is inadvisable for therapeutic treatment. Unless an emergency requires the use of whoever is at hand, a trained interpreter is best (Westermeyer, 1990). Words have different meanings in different cultures. For example, the Vietnamese word that literally translates "terribly sad" refers to many worries rather than to depressive symptoms (Phan & Silove, 1997). Interpreters must be able to communicate fluently in both languages, exchange words from one language to the next without losing meaning, present the connotative as well as the denotative meaning, understand trauma and psychological terminology, appreciate the importance of nonverbal communication, communicate questions that do not come up in normal conversation (e.g., suicidal plans, hallucinations, conflicts), understand interviewing techniques with distraught patients, and take into account the power dynamics (e.g., victim versus perpetrator) that may be replayed in the group or other therapeutic sessions (Kirmayer et al., 1995; Phan & Silove, 1997; Westermeyer, 1990). The interview or treatment may be obstructed by an interpreter who is not trusted by the patient (e.g., because of politics, relationships with members of the patient's or conflicting faction's social/political group) who might share confidential information with others or not share with the clinician all that the patient has said (e.g., to prevent the patient looking crazy, to normalize, or to protect the patient from the authority figure), or who would not elicit all the information sought (e.g., to avoid embarrassing the patient or to prevent crying).

The Clinical Interview

Establishing rapport is a key step in eliciting relevant information and providing treatment for intracultural and intercultural groups (Westermeyer, 1987; P. Yee, personal interview, August 7, 2003). Those who have experienced racial, economic, legal, or political prejudice in the United States (Westermeyer, 1987) as well as those who have concerns about American values differing from their own (Meleis, 2003) may not trust a representative of a social institution or someone from a different culture. In addition to gaining basic knowledge of the ways a patient may respond to and in treatment because of cultural and subcultural membership, the clinician may enhance rapport by conducting a skilled interview, demonstrating the commitment to understanding the patient's point of view, and beginning with open-ended nonthreatening questions (e.g., regarding sleep, appetite, energy level) (Westermeyer, 1987). A longer period may be needed to establish trust with a nonnative clinician (Canive et al., 2001; Kinzie, 1993).

Thought patterns (e.g., orientation to time) and behaviors (e.g., body language) commonly assessed or referred to in therapeutic sessions vary across cultures. What is valued in one culture (e.g., shyness) can be the cause of concern in another (Chen, Rubin, & Li, 1995; Mash & Dozois, 2003). Behaviors (e.g., direct eye contact) that signify one thing in one culture (e.g., honesty or openness) may signify something very different in another (e.g., anger or disrespect) (Westermeyer, 1987). Cultural discontinuities as well as apparent contradictions, excessive emotional reactions, or unlikely behaviors indicate the need for clarification. Because cultural differences exist in physical signs of stress or illness (e.g., lower tension levels in Japanese than in Westerners), study and an educated and trained translator from the culture in question can clarify whether behaviors are culturally congruent or idiosyncratic (Westermeyer, 1987).

Choice of Methods

Culture may dictate to some extent the methods a group leader uses or the beliefs that must be addressed. It is essential to be cognizant of cultural or religious views (e.g., of treatment, assistance, authority figures) that may affect reactions to group interventions as well as to leaders. For example, when traditional Irish Catholics seek help, they may view therapy as similar to Catholic confession—a place to tell their "sins" and receive "penance" (McGoldrick, 1996). Canive et al. (2001) suggest that Hispanics may be more amenable to advice and counsel than to insight-oriented methods. Several methods have proven useful in treating Asian Americans under specific circumstances (Ton-That, 1998): psychopharmacology (temporarily); cognitive-behavioral treatment; supportive therapy; spiritual practices; and metaphysical practices.

Group Composition and Culture

In the aftermath of massive trauma that has affected multicultural populations, consideration must be given to the composition of therapy groups and the therapeutic availability of subgroups (Nader, 2004). Assessment of group membership must take into account multiple factors. In a therapy group, culture and religion in and of themselves may or may not be an issue for children. Their importance is likely to be affected by their age, heritage, location, local politics, and the nature of the traumatic event (e.g., whether culture was an issue in the event). For adult groups, however, questions must be asked routinely about how differences affect the group's ability to meet its goals (Nader, 2004).

After September 11, planned group meetings including individuals from various religious backgrounds following specific guidelines have helped individuals to use religion and one another as support systems (Drescher,

Chapter 11 in this volume). In contrast, if not addressed effectively, culture or religion may cause divisiveness in or ineffectiveness of groups. For example, following the 1992 Los Angeles riots, several all-female teacher groups that were established consisted of 50 percent Latina and 50 percent African-American members. The African-American teachers expressed rage and desires for retaliation. The Latino teachers were afraid and desired protection. Their fears resulted in reluctance to share anything that might upset the African-American teachers (Nader, 2004). Latinas may also tend to worry about *"el que diran"* (what people will say) (Garcia-Preto, 1998).

A main goal with cultural diversity is seeking balance between validating the differences while appreciating the sources of our common humanity. Dividing lines between cultural groups may prevent people from defining themselves in all their complexity. They may emphasize their differences from outsiders rather than their affiliation with group members. Some group members thus might feel pressure to suppress parts of self in order to pass for "normal" or not to offend. Group boundaries may also define covert power hierarchies, which remain invisible and therefore have a pernicious effect, as when ethnic differences are described in such a way that the status differences between groups become prominent.

Other issues may also be important for the effectiveness of group composition. Depending on the goals of the group, intellectual (e.g., education) or experiential (e.g., type or impact of the trauma) compatibility may assist group cohesion. Groups may be contraindicated for individuals who exhibit serious psychiatric disturbances, suicidal behavior, nonsuicidal self-mutilation, severe mood or thought disturbances, the capability to hurt others without remorse, or who were recently severely traumatized in a group (Homeyer, 1999). Moreover, experiencing "failure" in the group setting can exacerbate low self-esteem, feelings of powerlessness, victimization, and guilt. Therefore, recognizing and addressing "failure experiences" and reconsidering group composition may be necessary.

Attention must be given to the subcultures that arise from trauma. For example, parents of the traumatized, the dead, or the injured may develop tightly knit groups (Nader, 2004). The boundaries of preexisting groups may become less permeable (e.g., between parents of bereaved and parents of traumatized children). Following a tornado, it was important to include part-time teachers in teachers' group meetings in order to minimize their already existing division with the full-timers. Some adult and child groups benefited from including unexposed, nontraumatized members to provide support and mitigate the traumatic state of mind.

We have much to learn about the influences of culture, religion, and personality on group effectiveness. For example, what is the best method of combining individuals with conflicting resolution styles? Some personality or cultural styles value strong discussions (even "arguing things out") of what is or is not right/best in the process of finding a solution. Others value agreement or its appearance. Some are goal oriented; others are process oriented. Special attention is needed to find the combinations of these styles and the methods within a mixed group that facilitate recovery (Danieli, 2002; Nader, 2003a).

Emotion-Laden Issues in Diverse Cultural Groups

A number of culture-related and emotion-laden issues may arise in mixed-culture group sessions. Because of the clinician's own as well as group attendees' reactions to these issues, some preliminary knowledge of them (from multiple perspectives) is important. Among the issues are fear, prejudice, ongoing conflict, betrayal, humiliation/shame, guilt, rights, and more (see Danieli, 1998; Nader, 2003b).

Following the terrorist attacks of September 11, 2001, one of the ways U.S. Americans bonded was with an affirmation of American ideals. Despite America's pride in and demand for freedom of speech and belief, however, a number of Americans responded with violence toward Muslims or Arabs in general and, later, toward French individuals after France disagreed with going to war with Iraq. Although most Americans did not behave rudely or violently, the *behaviors* of a few demonstrate what happens when commonly experienced emotions (e.g., the need to fight back or express rage in response to horror and helplessness, the need to reduce dissonance by demanding agreement, and the desire to be free of fear) are unleashed. Group members and leaders may need or want to discuss their reactions (e.g., to Arab looking persons getting on a plane or of being an Arab looking person getting on a plane) and ways to contend with these reactions (e.g., fear, helplessness). For example, discussing how to recognize real danger, how to respond to terrorist attacks or attempts, and how to examine personal prejudices may help to contend with some postterrorism emotions.

Scheff (1997) suggests that when the source of unacknowledged emotions (anger) is feelings of rejection or inadequacy, rage and aggression may mask the resulting shame. The composite of shame and anger may result in a rage that generates violence or severe conflict. Retzinger and Scheff (2000) suggest that the fact that the shame is hidden prevents connection between disputing parties and leads to more alienation. In a dyad or group conflict, they propose that it is important to identify the patterns of alienation and dysfunctional communication and to interrupt the cycle. This can be done by disrupting humiliating communication; identifying and expressing (reflecting back) unacknowledged emotions (e.g., shame and hurt) without taking sides and without additional humiliation (and by permitting saving face); reframing behaviors or requests; and building support for a mediated solution.

CONCLUSIONS

Cultural, religious, and ethnic differences must be recognized by caregivers, healers, and therapists as important elements in the assessment, prevention, and treatment of the psychological sequelae of trauma in general and terrorism in particular. However, this recognition must go significantly beyond acknowledgment and understanding of the need. There has to be an ongoing commitment to learning (e.g., reading, studying, learning from and among them) in order to appreciate culture, its historical roots, and the way it has shaped indigenous concepts of mental health. Our "cultural lenses and biases" affect how we organize our thinking. An assumption that the central points of attention are the individual and the relatively isolated nuclear family is inadequate or even irrelevant in interdependence-based cultures. If adhered to, it will hinder the development and application of culturally appropriate therapies.

Therapists and other caregivers must receive professional instruction in how both to understand and appreciate cultural elements and how to incorporate them into their work. Cultural beliefs may interact with traumatic reactions. Training that includes cultural issues is necessary to ensure appropriate forms of therapeutic intervention, thus preventing harm in the process of assessing and treating trauma. Incorporating cultural elements into the work will facilitate understanding of how cultural differences can affect the patient's receptivity to therapy. It is also essential to appreciate the complexity and many nuances and variations in understanding cultural differences.

This work is best done in the context of the multidimensional framework that identifies the broad range of potential effects, including the psychological, social, economic, and political, on victims, their families, their helpers, and the societies and the world in which they live (Danieli, 1998). All interventions, even short-term, should be thought through from a long-term perspective, and each specific intervention component should be considered in its multidimensional context (Danieli et al., 2003). Accepting, honoring, and understanding our cultural, religious, and ethnic differences must go hand in hand with appreciating our common humanity.

REFERENCES

Ahadi, S. A., Rothbart, M. K., & Ye, R. (1993). Children's temperament in the US and China: Similarities and differences. *European Journal of Personality, 7,* 359-377.

Almeida, R. (1996). Hindu, Christian, and Muslim families. In M. McGoldrick, J. Giordano, & J. Pearce (Eds.), *Ethnicity and family therapy* (pp. 395-423). New York: Guilford Press.

American Psychiatric Association. (2000). *Diagnostic and statistical manual of mental disorders* (4th ed., text rev.). Washington, DC: Author.

Awwad, E. (1999). Between trauma and recovery: Some perspectives on Palestinian's vulnerability and adaptation. In K. Nader, N. Dubrow, & B. Stamm (Eds.), *Honoring differences: Cultural issues in the treatment of trauma and loss* (pp. 234-256). Philadelphia: Taylor & Francis.

Bevin, T. (1999). Multiple traumas of refugees—near drowning and witnessing of maternal rape. In N.B. Webb (Ed.), *Play therapy with children in crisis* (pp. 164-182). New York: Guilford.

Bibb, A., & Casimir, G.J. (1996). Haitian families. In M. McGoldrick, J. Giordano, & J. Pearce (Eds.), *Ethnicity and family therapy* (pp. 97-111). New York: Guilford Press.

Bird, S. E. (2002). It makes sense to us: Cultural identity in local legends of place. *Journal of Contemporary Ethnography, 31*(5), 519-547.

Boehnlein, J. K. (2001). Cultural interpretations of physiological processes in post-traumatic stress disorder and panic disorder. *Transcultural Psychiatry, 38*(4), 461-467.

Boehnlein, J. K., & Kinzie, J. D. (1997). Cultural perspectives on posttraumatic stress disorder. In T. W. Miller (Ed.), *Clinical disorders and stressful life events* (pp. 19-43). Madison, CT: International Universities Press.

Bolton, P. (2001). Cross-cultural validity and reliability testing of a standard psychiatric assessment instrument without a gold standard. *The Journal of Nervous and Mental Disease, 189*(4), 238-242.

Boyd-Franklin, N., & Franklin, A. J. (1998). African American couples in therapy. In M. McGoldrick (Ed.), *Re-visioning family therapy* (pp. 268-281). New York: Guilford.

Brown, J. E. (Ed.). (1989). *The sacred pipe: Black Elk's account of the seven rites of the Oglala Sioux*. Norman: University of Oklahoma Press.

Canive, J. M., Castillo, D. T., & Tuason, V. B. (2001). The Hispanic veteran. In W. Tseng & J. Streltzer (Ed.), *Culture and psychotherapy: A guide to clinical practice* (pp. 157-172). Washington, DC: American Psychiatric Press.

Chen, X., Rubin, K. H., & Li, Z. Y. (1995). Social functioning and adjustment in Chinese children: A longitudinal study. *Developmental Psychology, 31*, 531-539.

Clash of Civilizations? (2003). *Canada & the World Backgrounder, 68*(5), 17-21.

Compas, B. E., & Epping, J. E. (1993). Stress and coping, in children and families: Implications for children coping with disaster. In C. F. Saylor (Ed.), *Children and disasters* (pp. 11-28). New York: Plenum Press.

Danieli, Y. (1993). The diagnostic and therapeutic use of the multi-generational family tree in working with survivors and children of survivors of the Nazi Holocaust. In J. P. Wilson & B. Raphael (Eds.), *International handbook of traumatic stress syndromes* [Stress and Coping Series, Donald Meichenbaum, Series Editor] (pp. 889-898). New York: Plenum Publishing.

Danieli, Y. (1995). The treatment and prevention of long-term effects and intergene intergenerational transmission of victimization: A lesson from Holocaust survivors and their children. In C. R. Figley (Ed.), *Trauma and its wake* (pp. 295-313). New York: Brunner/Mazel.

Danieli, Y. (Ed.) (1998). *An international handbook of multigenerational legacies of trauma*. New York: Plenum Press.

Danieli, Y. (2002). *Sharing the frontline and the back hills: International protectors and providers, peacekeepers, humanitarian aid workers, and the media in the midst of crisis*. Amityville, NY: Baywood Publishing Company, Inc.

Danieli, Y., Engdahl, B., & Schlenger, W. E. (2003). The psychological aftermath of terrorism. In F. M. Moghaddam & A. J. Marsella, (Eds.), *Understanding terrorism: Psychological roots, consequences, and interventions* (pp. 223-246). Washington, DC: American Psychological Association.

de Silva, P. (1999). Cultural aspects of post-traumatic stress disorder. In W. Yule (Ed.), *Post-traumatic stress disorders: Concepts and therapy* (pp. 116-138). Chichester, England: John Wiley and Sons.

DeVoe, E. R., & Klein, T. P. (2003). Ethnic variation in PTSD in young children and their parents after 9/11. Presented at the International Society for Traumatic Stress Studies in Chicago, October 31.

Dubrow, N., & Nader, K. (1999). Consultations amidst trauma and loss: Recognizing and honoring differences. In K. Nader, N. Dubrow, & B. Stamm (Eds.), *Honoring differences: Cultural issues in the treatment of trauma and loss* (pp. 1-19). Philadelphia: Taylor & Francis.

Duran, E., Duran, B., Brave Heart, M., & Yellow Horse-Davis, S. (1998). Healing the American Indian soul wound. In Y. Danieli, (Ed.), *An international handbook of multigenerational legacies of trauma* (pp. 341-354). New York: Plenum Press.

Falicov, C. J. (1998). The cultural meaning of family triangles. In M. McGoldrick (Ed.), *Re-visioning family therapy* (pp. 37-49). New York: Guilford.

Fang, L., & Chen, T. (2004). Community outreach and education to deal with cultural resistance to mental health services. In N. B. Webb (Ed.), *Mass trauma, stress, and loss: Helping children and families cope* (pp. 234-255). New York: Guilford Press.

Garcia-Preto, N. (1998). Latinas in the United States. In M. McGoldrick (Ed.), *Re-visioning family therapy* (pp. 330-344). New York: Guilford.

Gardner, H. (1983). *Frames of mind*. Cambridge, MA: MIT Bradford Press.

Gerber, L., Nguyen, Q., & Bounkeua, P. K. (1999). Working with Southeast Asian people who have migrated to the United States. In K. Nader, N. Dubrow, & B. Stamm (Eds.), *Honoring differences: Cultural issues in the treatment of trauma and loss* (pp. 98-118). Philadelphia: Taylor & Francis.

Hines, P. M. (1998). Climbing up the rough side of the mountain: Hope, culture, and therapy. In M. McGoldrick (Ed.), *Re-visioning family therapy* (pp. 78-89). New York: Guilford.

Hodgetts, R. (1993). A conversation with Geert Hofstede. *Organizational Dynamics, 21,* 53-61.

Hofstede, G. (1980). Motivation, leadership, and organization: Do American theories apply abroad? *Organizational Dynamics, 9,* 42-61.

Hollifield, M., Warner, T. D., Lian, N., Krakow, B., Jenkins, J. H., Kesler, J., Stevenson, J., & Westermeyer, J. (2002). Measuring trauma and health status in refugees. *Journal of the American Medical Association, 288*(5), 611-621.

Homeyer, L. (1999). Group play with sexually abused children. In D. S. Sweeney & L. Homeyer (Eds.), *The handbook of group play therapy* (pp. 299-318). San Francisco: Jossey-Bass.

Huntington, S. P. (1993). The clash of civilizations? *Foreign Affairs, 72*(3), 22-50.

Inglehart, R., & Norris, P. (2003). The true clash of civilizations. *Foreign Policy, 135,* 62-66.

Johnson, D., Feldman, S., Lubin, H., & Southwick, S. (1995). The therapeutic use of ritual and ceremony in the treatment of post-traumatic stress disorder. *Journal of Traumatic Stress, 8*(2), 283-298.

Kinzie, J. D. (1993). Posttraumatic effects and their treatment among southeast Asian refugees. In J. Wilson & B. Raphael (Eds.), *The international handbook of traumatic stress syndromes* (pp. 311-319). New York: Plenum Press.

Kinzie, J. D. (2004). Some of the effects of terrorism on refugees. In Y. Danieli, D. Brom, & J.B. Sills (Eds.), *The trauma of terrorism: Sharing knowledge and shared care* (pp. 411-420). Binghamton, NY: The Haworth Press Inc.

Kinzie, J. D., Boehnlein, J., & Sack, W. H. (1998). The effects of massive trauma on Cambodian parents and children. In Y. Danieli (Ed.), *An international handbook of multigenerational legacies of trauma* (pp. 211-221). New York: Plenum Press.

Kinzie, J. D., Denney, D., Riley, C., Boehnlein, J. K., McFarland, B., & Leung, P. (1998). A cross-cultural study of reactivation of posttraumatic stress disorder symptoms: American and Cambodian psychophysiological response to viewing traumatic video scenes. *Journal of Nervous and Mental Disease, 186,* 670-676.

Kirmayer, L. J., Young, A., & Hayton, B. C. (1995). The cultural context of anxiety disorders. *The Psychiatric Clinics of North America, 18*(3), 503-521.

La Greca, A. M., Silverman, W. K., Vernberg, E. M., & Prinstein, M. J. (1996). Symptoms of posttraumatic stress in children after Hurricane Andrew: A prospective study. *Journal of Consulting and Clinical Psychology, 64,* 712-723.

Laird, J. (1998). Theorizing culture: Narrative ideas and practice principles. In M. McGoldrick (Ed.), *Re-visioning family therapy* (pp. 20-36). New York: Guilford.

Laria, A. J., & Lewis-Fernández, R. (2001). The professional fragmentation of experience in the study of dissociation, somatization, and culture. *Journal of Trauma & Dissociation, 2*(3), 17-47.

Lee, J., Lei, A., & Sue, S. (2001). The current state of mental health research on Asian Americans. *Journal of Human Behavior in the Social Environment, 3*(3/4), 159-178.

Lloyd, B. (1999). Book review essay: Tough love/tough theory. *Psychology, Evolution and Gender, 1*(3), 321-327.

Markus, H. R., Kitayama, S., & Heiman, R. J. (1996). Culture and basic psychological principles. In E. T. Higgins & A. W. Kruglanski (Eds.), *Social psychology: Handbook of basic principles* (pp. 857-913). New York: Guilford Press.

Mash, E. J., & Dozois, D. (2003). Child psychopathology: A developmental systems perspective. In E. J. Mash & R. A. Barkley (Eds.), *Child psychopathology,* (2nd ed., pp. 3-71). New York: Guilford.

McDermott, J. F. (1991). The effects of ethnicity on child and adolescent development. In M. Lewis (Ed.), *Child and adolescent psychiatry: A comprehensive textbook* (pp. 145-159). Baltimore, MD: Williams & Wilkins.

McGoldrick, M. (1996). Irish families. In M. McGoldrick, J. Giordano, & J. Pearce (Eds.), *Ethnicity and family therapy* (pp. 544-566). New York: Guilford Press.

McGoldrick, M. (1998). A framework for re-visioning family therapy. In M. McGoldrick (Ed.), *Re-visioning family therapy* (pp. 3-19). New York: Guilford.

Meleis, A. (2003). Reflections on September 11, 2001. *Health Care for Women International, 24*, 1-4.

Mollica, R. F., Cui, X., McInnes, K., & Massagli, M. P. (2002). Science-based policy for psychosocial interventions in refugee camps: A Cambodian example. *Journal of Nervous and Mental Disease, 190*(3), 158-166.

Mooren, G., & Kleber, R. (1999). War, trauma, and society: Consequences of the disintegration of former Yugoslavia. In K. Nader, N. Dubrow, & B. Stamm (Eds.), *Honoring differences: Cultural issues in the treatment of trauma and loss* (pp. 76-97). Philadelphia: Taylor & Francis.

Nader, K. (2003a). Assessing traumatic experiences in children and adolescents. In J. P. Wilson & T. M. Keane (eds.), *Assessing psychological trauma and PTSD* (2nd ed., pp. 513-537). New York: The Guilford Press.

Nader, K. (2003b). Walking among us: Culture and trauma. *The Child Survivor of Traumatic Stress, 4*, 1-8.

Nader, K. (2004). Treating traumatized children and adolescents: Treatment issues, modality, timing, and method. In N. B. Webb (Ed.), *Mass trauma and violence: Helping families and children cope* (pp. 50-73). New York: Guilford Press.

Nader, K., & Danieli, Y. (2004). Cultural issues in terrorism and in response to terrorism. In Y. Danieli, D. Brom, & J. Waizer, (Eds.), *The trauma of terrorism: Sharing knowledge and shared care* (pp. 399-410). Binghamton, NY: The Haworth Press.

Nader, K., Dubrow, N., & Stamm, B. (Eds.) (1999). *Honoring differences: Cultural issues in the treatment of trauma and loss.* Philadelphia: Taylor & Francis.

Nelson, J. (2002). Diversity as an influence on clients with anxiety and depressive disorders: What the responsible social worker should know. *Families in Society: The Journal of Contemporary Human Services, 83*, 45-53.

Ogundele, W. (2002). Devices of evasion: The mythic versus the historical imagination in the postcolonial African novel. *Research in African Literatures, 33*(3), 125-140.

Park, J. (1996). The universality of the MBTI, cultural ideal types and falsification issues in Korea. A paper presented at Psychological Type and Culture-East and West: A Multicultural Research Symposium, University of Hawaii at Manoa, January 5-7.

Peddle, N., Monteiro, C., Guluma V., & Macaulay, T. (1999). Trauma, loss, and resilience in Africa: A psychosocial community based approach to culturally sensitive healing. In K. Nader, N. Dubrow, & B. Stamm (Eds.), *Honoring differences: Cultural issues in the treatment of traumatic and loss* (pp. 121-149). Philadelphia: Taylor & Francis.

Phan, T., & Silove, D. M. (1997). The influence of culture on psychiatric assessment: The Vietnamese refugee. *Psychiatric Services, 48*(1), 86-90.

Pinderhughes, E. (1989). *Understanding race, ethnicity and power.* New York: Free Press.

Pole, N., Best, S. R., Metzler, T., & Marmar, C. R. (2005). Why are Hispanics at greater risk for PTSD? *Cultural Diversity and Ethnic Minority Psychology 11*(2), 144-161.

Rabalais, A. E., Ruggiero, J. K., & Scotti, J. R. (2002). Multicultural issues in the response of children to disasters. In A. M. La Greca, W. K. Silverman, E. M. Vernberg, & M. C. Roberts (Eds.), *Helping children cope with disasters and terrorism* (pp. 73-99). Washington, DC: APA Press.

Raphael, B., Swan, P., & Martinek, N. (1998). Intergenerational aspects of trauma for Australian Aboriginal people. In Y. Danieli (Ed.), *An international handbook of multigenerational legacies of trauma* (pp. 327-340). New York: Plenum Press.

Retzinger, S., & Scheff, T. (2000). Emotion, alienation, and narratives: Resolving intractable conflict. *Mediation Quarterly, 18*(2), 71-86.

Rousseau, C., & Drapeau, A. (1998). The impact of culture on the transmission of trauma: Refugees' stories and silence embodied in their children's lives. In Y. Danieli (Ed.), *An international handbook of multigenerational legacies of trauma* (pp. 465-486). New York: Plenum Press.

Sandoval, J., & Lewis, S. (2002). Cultural considerations in crisis intervention. In S. Brock & P. Lazarus (Eds.), *Best practices in crisis prevention and intervention in the schools* (pp. 293-308). Bethesda, MD: National Association of School Psychologists.

Scheff, T. (1997). Deconstructing rage. Available online at <http://www.soc.ucsb.edu/faculty/scheff/7.html>, accessed September 17, 2003.

Seligman, M. (1990). *Learned optimism.* New York: Random House.

Shiang, J. (2000). Considering cultural beliefs and behaviors in the study of suicide. In R. Maris, S. Canetto, J. McIntosh, & M. Silverman (Eds.), *Review of suicidology* (pp. 226-241). New York: Guilford.

Shiang, J., Kjellander, C., Huang, K., & Bogumill, S. (1998). Developing cultural competency in clinical practice: Treatment considerations for Chinese cultural groups in the U.S. *Clinical Psychology: Science and Practice, 5,* 182-209.

Soeters, J. L. (1996). Culture and conflict: An application of Hofstede's theory to the conflict in the former Yugoslavia. *Journal of Peace Psychology, 2*(3), 233-244.

Spiegel, J. P. (1976). Cultural aspects of transference and countertransference revisited. *Journal of the American Academy of Psychoanalysis, 4,* 447-467.

Stamm, B. H., & Stamm, H. E. (1999). Trauma and loss in Native North America: An ethnocultural perspective. In K. Nader, N. Dubrow, & B. Stamm (Eds.), *Honoring differences: Cultural issues in the treatment of trauma and loss* (pp. 49-75). Philadelphia: Taylor & Francis.

Swiss, S., & Giller, J. E. (1993). Rape as a crime of war: A medical perspective. *Journal of the American Medical Association, 270*(5), 612-615.

Ton-That, N. (1998). Post-traumatic stress disorder in Asian refugees. *Psychiatry and Clinical Neurosciences, 52*(Suppl), S377-S379.

Triandis, H., Kashima, Y., Shimada, E., & Villareal, M. (1986). Acculturation indices as a means of confirming cultural differences. *International Journal of Psychology, 21,* 43-70.

Tully, M. (1999). Lifting our voices: African American cultural responses to trauma and loss. In K. Nader, N. Dubrow, & B. Stamm (Eds.), *Honoring differences: Cultural issues in the treatment of trauma and loss* (pp. 23-48). Philadelphia: Taylor and Francis.

van Suntum, L. A. R. (2001). Creating Jewish identity through storytelling: The tragedy of Jacob Bendixen. *Scandinavian Studies, 73*(3), 375-398.

Velez-Ibanez, C. G., & Parra, C. G. (1999). Trauma issues and social modalities concerning mental health concepts and practices among Mexicans of the southwest United States with reference to other Latino groups. In K. Nader, N. Dubrow, & B. Stamm (Eds.), *Honoring differences: Cultural issues in the treatment of trauma and loss* (pp. 76-97). Philadelphia: Taylor and Francis.

Waizer, J., Dorin, A., Stoller, E., & Laird, R. (2004). Community-based interventions in New York City after 9/11: A provider's perspective. In Y. Danieli, D. Brom, & J. B. Sills (Eds.), *The trauma of terrorism: Sharing knowledge and shared care* (pp. 499-512). Binghamton, NY: The Haworth Press, Inc.

Waldegrave, C. (1998). The challenges of culture to psychology and postmodern thinking. In M. McGoldrick (Ed.), *Re-visioning family therapy* (pp. 404-413). New York: Guilford.

Walsh, F. (1998). Beliefs, spirituality, and transcendence: Keys to family resilience. In M. McGoldrick (Ed.), *Re-visioning family therapy* (pp. 62-77). New York: Guilford.

Watkins, J. M., & Williams, M. E. (1992). Cognitive neuroscience and adolescent development. In E. McAnarney, R. Kreipe, D. Orr, & G.D. Comerci (Eds.), *The textbook of adolescent medicine* (pp. 99-106). Philadelphia: WB Saunders.

Watson, M. F. (1998). African American sibling relationships. In M. McGoldrick (Ed.), *Re-visioning family therapy* (pp. 282-294). New York: Guilford.

Weine, S. (2001). From war zone to contact zone: Culture and refugee mental health services. *Journal of the American Medical Association, 285*(9), 1214.

Westermeyer, J. (1987). Cultural factors in clinical assessment. *Journal of Consulting and Clinical Psychology, 55*(4), 471-478.

Westermeyer, J. (1989). Research of stigmatized conditions: Dilemma for the sociocultural psychiatrist. *American Indian and Alaska Native Mental Health Research, 2*(3), 41-45.

Westermeyer, J. (1990). Working with an interpreter in psychiatric assessment and treatment. *Journal of Nervous and Mental Disease, 178*(12), 745-749.

Westermeyer, J., Canive, J. M., Garrard, J., Padilla, E., Crosby, R., & Thuras, P. (2002). Perceived barriers to mental health care for American Indian and Hispanic veterans: Report by 100 VA staff. *Transcultural Psychiatry, 39*(4), 516-530.

Westermeyer, J., & Uecker, J. (1997). Predictors of hostility in a group of relocated refugees. *Cultural Diversity and Mental Health, 3*(1), 53-60.

Wilson, J. P. (1989). *Trauma, transformation, and healing: An integrative approach to theory, research, and post-traumatic therapy.* New York: Brunner/Mazel.

Wilson, J., & Keane, T. (Eds.) (2004). *Assessing psychological trauma & PTSD* (2nd ed.). New York: The Guilford Press.

Wilson, J. P., Walker, A., & Webster, B. (1989). Reconnecting: Stress recovery in the wilderness. In J. P. Wilson (Ed.), *Trauma, transformation, and healing: An integrative approach to theory, research, and post-traumatic therapy* (pp. 159-195). New York: Brunner/Mazel.

World Health Organization. (1992). Classification of diseases and health related problems (ICD-10). Geneva: World Health Organization.

Chapter 9

Contextual Influences on Posttraumatic Adjustment: Retraumatization and the Roles of Revictimization, Posttraumatic Adversities, and Distressing Reminders

Christopher M. Layne
Jared S. Warren
William R. Saltzman
John B. Fulton
Alan M. Steinberg
Robert S. Pynoos

INTRODUCTION

Critical questions facing front-line clinicians, administrators, and policymakers in the immediate aftermath of disaster include the following: Who should be treated? Many individuals have been exposed to a variety of traumatic experiences, and many are exhibiting varying types and degrees of distress. How can we discriminate between who will, and who will not, recover without specialized intervention? What information can we gather *now,* in the immediate postdisaster phase, that will assist us in determining

Support for this chapter was provided by UNICEF Bosnia and Hercegovina; the Family Studies Center, Kennedy International Studies Center, and the College of Family, Home, and Social Sciences of Brigham Young University; and the UCLA Trauma Psychiatry Bing Fund.

The authors would like to express their gratitude to Russell T. Jones, PhD, and Laura Kay Murray, PhD, for their helpful comments on an early draft; to Miriam Iosupovici, PhD, for her helpful suggestions relating to the topic of retraumatization, and to Janine G. S. Shelby, PhD, for her helpful suggestions relating to the integration of developmental issues and posttraumatic recovery.

which types of treatments should be provided, to whom, by whom, when, and for how long, to decrease the likelihood of maladaptive outcomes?

Responses to these questions must draw upon a growing knowledge base concerning the pre-, peri-, and posttraumatic ecologies and their capacities to facilitate, or interfere with, posttrauma recovery processes. In particular, the identification of reliable predictors and determinants of long-term posttraumatic adjustment carries important implications for accurate risk screening and case identification, the creation of treatment triage algorithms, case conceptualization, treatment monitoring and evaluation, resource allocation, and theory development (McFarlane & de Girolamo, 1996; Pynoos, Goenjian, & Steinberg, 1995; Saltzman, Layne, Steinberg, Arslanagic, & Pynoos, 2003). Although it is in a comparatively early phase of development, this knowledge base helps to address additional questions: Which factors within trauma-exposed individuals' personal histories, physical environments, interpersonal relationships, and psychological and physiological constitutions can we assess to stratify individuals and groups according to who is at highest risk and for which types of maladaptive outcomes? Which variables can we reliably measure to predict the likelihood of a positive treatment response? What naturally existing coping resources should we include in our risk assessments to increase our predictive accuracy or directly target in our intervention programs to promote adaptive outcomes?

Our approach in this chapter is predicated on the assumption that the many variables typically available for assessment in the aftermath of disaster vary in their accessibility, function, psychometric properties, and overall value for theory building and intervention planning. In particular, the identification of reliable *markers of risk* (e.g., O'Connor & Rutter, 1996) in a given trauma-exposed population is useful because it facilitates case identification, risk stratification, and triage efforts. Identifying risk markers thus helps to address the predictive question, Who is at greatest risk for which types of maladaptive outcomes? For example, information concerning whether certain children exposed to a school bus accident are, due to a history of parental divorce or concurrent separation anxiety, at greater risk for specific maladaptive outcomes, may assist in the identification of those for whom in-depth assessment and follow-up are indicated. Notably, when evaluating the desirability of including specific variables in assessment activities, it is important to distinguish between *passive* markers of risk in contrast to actively operating processes that contribute to the perpetuation of distress over time. Passive markers of risk typically include demographic characteristics, historical events, or preexisting and/or concurrent conditions that contribute to accurate identification of higher risk status but which do not have any demonstrable active influence, either direct or indirect, on targeted outcomes at the time of assessment. Thus, passive markers contain little intrinsic information concerning *what* processes intervention programs should specifically target and in which order.

Of greater comparative value, then, is the task of identifying ongoing processes that actively contribute to maladaptive outcomes over time. More specifically, *etiological* and *maintenance factors* (see Schnurr, Lunney, & Sengupta, 2004; Simon, 1999) serve to initiate or perpetuate distress over time, respectively. Etiological and maintenance factors thus not only may assist in risk identification but also yield additional valuable information by constituting potential targets for direct intervention. Identifying either factor thus assists in answering the questions, Which treatment foci will reduce the likelihood of maladaptive outcomes and/or increase the likelihood of positive adaptation, and how should these therapeutic objectives be prioritized and sequenced? For example, such etiological variables as who was injured during the school bus accident, who witnessed the fatal injury of a classmate, and who was a close friend to the deceased, assist with the tasks of risk stratification, treatment triage, case conceptualization, identification of treatment objectives, and the selection of assessment instruments to monitor and evaluate treatment (see Pynoos & Nader, 1988; Pynoos, Steinberg, & Wraith, 1995). Similarly, a focus on maintenance factors, including family conflict, poor social support, peer rejection following the accident, and commuting on a school bus route that passes daily by the site of the accident, may yield similarly valuable information in their roles as active contributors to ongoing distress and maladaptation. This information may be used to supplement trauma-specific work with a concomitant focus on family, peer, and environmental factors, thus increasing the comprehensiveness, flexibility, overall therapeutic impact, and perceived personal relevance of the intervention.

In this chapter, we explore four fundamental issues relating to the effective conceptualization, assessment, and treatment of trauma-exposed populations in the aftermath of disaster. The first issue explores problems concerning the *lack of terminological*—and by extension, *conceptual—clarity* with which the existing literature defines and describes the perpetuation of trauma-related distress over time. The second issue addresses the larger *theoretical framework* through which processes theorized to influence posttraumatic adjustment are conceptualized (see Pynoos, Steinberg, & Wraith, 1995). In this section, we argue that posttraumatic adjustment is best viewed more comprehensively, encompassing both vulnerability and protective processes as they influence short-term, intermediate, and long-term posttraumatic adjustment. Indeed, as such large-scale traumata as the September 11, 2001, terrorist attacks recede in time, it is reasonable to propose that temporally proximal factors—including both vulnerability and protective processes embedded within the posttraumatic ecology—will explain an increasingly larger proportion of the variance in posttraumatic adjustment as compared to pre- or peri-traumatic factors. The third issue focuses on introducing three *intervening variables* theorized to mediate the link between initial trauma exposure and subsequent posttraumatic distress, and thereby influence the longitudinal course of posttraumatic adjustment. The last issue involves conducting two *empirical tests* of the proposed intervening variables, and discussing their implications for the conceptualization, assessment, and treatment of trauma-exposed populations.

THE NEED FOR TERMINOLOGICAL AND CONCEPTUAL CLARITY

Retraumatization: An Oft-Used but Ill-Defined Construct

Predicting the longitudinal course of posttraumatic adjustment is fraught with imprecision due to the complex and highly variable nature of traumatic exposure, posttraumatic adversities, and associated processes of post-traumatic recovery (e.g., McFarlane & Yehuda, 1996; Pynoos, Steinberg, & Wraith, 1995). Compounding this challenge is an additional imprecision introduced by the disparate meanings and uses given to the term *retraumatization* within the trauma literature. In preparation for this chapter, we located many sources that used the term *retraumatization,* often as a core subject; however, we were unable to identify any published sources that provided a definitive conceptual definition or specific operational definition for the term. Indeed, some studies (e.g., Dietrich, 2000) appear to use the terms *revictimization* and *retraumatization* interchangeably. One exception was the Sidran Foundation's Web site glossary (Sidran Institute, 2002), which posted the following definition: "Re-traumatizing: Re-enacting or reinforcing a traumatic experience or belief." Although it constitutes a useful beginning, this definition is vague, internally inconsistent ("re-enact" and "reinforce" have different meanings and mechanisms of action), and insufficiently broad to comprehensively cover all the varied meanings of the term as currently used in the trauma literature. As a second step, we sought to clarify the meaning of the term *retraumatization* by examining its contextual use within a number of diverse sectors within the trauma literature, including studies examining correlates of childhood physical and sexual abuse, combat trauma, and adult sexual assault.

Common Usage 1: Retraumatization As Revictimization

One primary use of the term *retraumatization* is to describe exposure to one or more traumatic events subsequent to an initial trauma. Published examples falling within this category of common use come from a diversity of populations, including investigations of the reactions of Holocaust survivors to subsequent trauma, such as a missile attack on Israel (Robinson, Hemmendinger, Rapaport, Zilberman, & Gal, 1994). Other examples are found within the sexual abuse literature, wherein such factors as previous trauma, typically childhood sexual or physical abuse, and PTSD status are used to predict exposure to subsequent traumatic life events (e.g., Cloitre, 1998; Dietrich, 2000). For example, retraumatization has served as an experimental grouping variable. Cloitre, Scarvalone, & Difede (1997) examined levels of PTSD and self/interpersonal psychosocial functioning in three groups of women: victims of adult sexual assault, retraumatized victims, and a nonassault control group. The retraumatized group was operationally defined as those with a childhood sexual abuse history who were subsequently reexposed to trauma in adulthood. The authors found that retraumatized women scored significantly higher on measures of alexithymia (i.e., difficulty labeling emotional states) and dissociation, had lower levels of interpersonal functioning, and had a higher risk of suicide than women in the other two groups. Similarly, using a prospective methodology, Orcutt, Erickson, & Wolfe (2002) tested the hypothesis that PTSD symptomatology mediated the relationship between previous trauma and subsequent trauma exposure. Notably, the authors used the term *retraumatization* in reference to trauma exposure following the Gulf War. Strong support for mediation was found, in that PTSD symptomatology explained an additional 48 percent of the variance in later trauma exposure beyond that of combat.

Interestingly, studies interchanging the meaning of the terms *retraumatization* and *revictimization* vary widely in the duration of time intervening between the initial and subsequent trauma, ranging from months to decades (e.g., Wyatt, Gunthrie, & Notgrass, 1992), and in the degree of similarity between the initial and subsequent trauma. For example, Cloitre, Tardiff, Marzuk, Leon, & Portera (1996) found higher levels of risk for adult sexual assault were associated with childhood physical abuse alone (36 percent), or a combination of physical and sexual abuse (51 percent), than with childhood sexual abuse alone (CSA) (13 percent). Further, using a longitudinal design with 174 women, Baynard, Williams, & Siegel (2001) reported that women with self-reported histories of CSA reported higher rates of exposure to nonsexual childhood trauma, adult sexual assault, and adult trauma exposure compared to non-CSA controls.

Other examples in which *retraumatization* is apparently used interchangeably with *revictimization* include studies linking diagnoses of antisocial or borderline personality disorder and risk for exposure to later traumatic events (Lauterbach & Vrana, 2001). Notably, the authors use the term *multiple traumatization* in reference to subsequent trauma exposure, and the term *retraumatization* in reference to "further abuse." In a paper that underscores the need to discriminate between the terms *retraumatization* and *revictimization,* Dietrich (2000) reports that child sexual abuse is a predictor of revictimization (i.e., exposure to subsequent trauma) and that these individuals often experience acute distress reactions evoked by cues that symbolize or resemble an aspect of the initial traumatic event (a condition the author refers to as "retraumatizing"). As a final example underscoring the need for conceptual clarity, Silove, McIntosh, & Becker (1993) expressed concern that Australia's immigration policies may place mentally fragile refugees at increased risk for retraumatization. Notably, the authors do not clarify the meaning of this term as it specifically applies to either shielding refugees from additional traumatic experiences, or from cues that evoke distress associated with past traumatic experiences. Clearly, these two potential uses of the term *retraumatization* have different meanings, intervention objectives, and associated intervention strategies.

Common Usage 2: Retraumatization As Exacerbation
of Trauma-Related Distress

In contrast, a second common use of the term *retraumatization* found within the literature is in reference to acute exacerbations in trauma-related distress evoked by cues that symbolize or resemble aspects of a past traumatic experience. For example, Smith (1995) described the use of physical restraints in inpatient settings as potentially "retraumatizing" to patients, citing as a case example a female client who reexperienced distressing memories of a previous sexual assault while being manually restrained. Further, Doob (1992) discusses the mental health system's response to sexual abuse victims, suggesting that current methods for diagnosis and treatment may mimic the denial and the disconfirmation of a client's childhood abuse experiences. As a consequence, the author asserts that symptomatology may be reevoked and the client may present as very ill and untreatable. Similarly, Rosenberg et al. (2001) note that the health care system may create triggers that reevoke traumatic memories and advocate for greater awareness of trauma-related problems on the part of care providers. Other researchers (e.g., Koss, 2000) have discussed the potential of the judicial system to retraumatize sexual assault victims through judicial proceedings. For instance, the experience of testifying in a judicial hearing emerged as one of four positive predictors of PTSD in a study of sexually victimized women (Epstein, Saunders, & Kilpatrick, 1997).

Other researchers have identified reexposure to additional threats as a factor that leads to traumatic distress reactivation. For example, Toren, Wolmer, Weizman, Magal-Vardi, & Laor (2002) reported distress symptom reactivation in individuals with previous Gulf War trauma exposure as a function of the reemergence of threat in the region. Further examples of distress-evoking reminders are found within the psychotherapy literature. Schamess, Streider, & Connors (1997) discussed the risk of retraumatization to children participating in group therapy and proposed specific therapeutic interventions designed to increase the effectiveness of group treatment while reducing the likelihood of retraumatization among the members. In addition, Pearlman and Saakvitne (1995) discuss the risk of vicarious traumatization to the therapist via exposure to clients' traumatic material. The authors assert that the vicarious traumatization of the therapist may lead to impaired clinical judgment, placing the client at greater risk for retraumatization.

The dilemma: Is retraumatization an intervening mechanism or an outcome variable?

Unfortunately, the varieties in usage of, and especially the discrepancies in meanings accorded to, the term *retraumatization* decrease both the precision of the term and the clarity of research findings relating to posttraumatic adjustment. Essentially, the duality in terminology with which the term *retraumatization* is currently used in the literature boils down to the circular and confusing explanation that *trauma survivors who are subsequently retraumatized become retraumatized as a result.* This confusion, caused by using the same term (retraumatization) with two very different phenomena ("retraumatizing" life events versus "retraumatization" as an outcome), is illustrated in Figure 9.1. Such terminological ambiguity invokes a counterproductive tautology, blurs the distinction between an intervening mechanism and an outcome variable, and thereby divests itself of explanatory power or any legitimate claim to predictive accuracy. Stated simply, the claim that a client has *become* retraumatized because she has *been* retraumatized has little or no meaning.

Etiological Factor   Intervening Socioenvironmental Mechanism   Outcome Variable

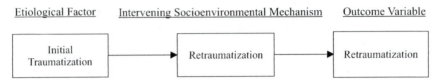

FIGURE 9.1. Circularity Found in the Literature's Current Use of the Term "Retraumatization."

Suggested Terminological Clarifications

To increase conceptual precision, we propose that the term "retraumatization" be discontinued in the professional literature due to its ambiguity and imprecision. Instead, we recommend that the term "trauma reexposure" be used to refer to *exposure to a traumatic event following exposure to one or more antecedent traumatic events.* The appendix at the end of this chapter contains a glossary of suggested definitions of concepts relating to trauma reexposure and its sequelae. The term *revictimization* may also be profitably employed, but we suggest that its meaning be broadened to encompass trauma reexposure within the larger context of its accompanying social consequences. Specifically, victimization refers to the abrogation, either within or outside the family system, of expectations implied by the social contract. The status of victim is defined by society to carry special sanctions to (1) assume a social role that warrants restitution and special treatment and (2) expect society to bring the perpetrator to judicial accountability. As proposed by Pynoos, Sorenson, & Steinberg (1993), because many forms of trauma involve failures to abide by the social contract, the resolution of accountability is intrinsically involved in the recovery process for many victims. This accountability requires that assailants be properly identified, arrested, prosecuted, and punished in the interests of promoting security, providing accountability to the public, and assuaging fears of recurrence.

We also propose three additional terms relating to trauma reexposure and its consequences in the interests of increasing terminological precision where needed. *Serial trauma* is defined as repeated exposure to the same type of trauma, either within the same developmental period or across successive developmental periods. Examples of serial trauma include repeated instances of sexual abuse during preadolescence or repeated instances of physical abuse from early childhood through school-age years. *Sequential trauma* is defined as repeated exposure to different types of trauma, either within the same developmental period or across different developmental periods. For example, sequential trauma may involve exposure to domestic violence during early childhood, followed by sexual abuse during middle adolescence, followed by exposure to community violence during young adulthood. Clearly, serial trauma and sequential trauma are not mutually exclusive circumstances and thus may overlap within the same individual. Last, we recommend that the term *traumatic distress reactivation* be used in reference to acute exacerbations in posttraumatic distress reactions among individuals with a history of exposure to one or more traumatic events. These reactions may assume a variety of forms, including posttraumatic stress reactions, depressive reactions, anxiety, somatic distress, and acute grief reactions. This reactivation is presumably evoked by a variety of mechanisms, including exposure to trauma reminders, loss reminders, posttraumatic adversities, serial trauma, or sequential trauma.

In summary, it is hoped that the use of these terms will assist in clarifying the important distinctions between antecedent, intervening, and outcome variables, such as in the case of early childhood sexual abuse, sexual abuse by one's dating partner during adolescence, and reactivation of posttraumatic distress reactions in adolescence, respectively.

Having established these working conceptual definitions, we now turn to our second issue, introducing a theoretical framework of factors contributing to reactivations in posttraumatic distress.

A THEORETICAL FRAMEWORK OF POSTTRAUMATIC ADJUSTMENT

A review of the traumatic stress literature reveals a generally consistent dose-response relationship between exposure to objective threat and posttraumatic stress reactions in a wide variety of trauma-exposed populations (see Foy, Madvig, Pynoos, & Camilleri, 1996). However, recent reviews of the extant literature report that the prevalence rate of PTSD and other disorders, such as depression, is quite variable across trauma-exposed samples (Foy et al., 1996; Saigh & Bremner, 1999), and that the effect size between trauma exposure and subsequent distress is often comparatively modest (Fletcher, 2003; see also Brewin, Andrews, & Valentine, 2000, for an application with adults). Further, studies generally reveal that posttraumatic distress is a phasic, rather than static, phenomenon that tends to decrease over time, accompanied by intermittent and typically temporary upsurges in distress. Importantly, a minority of trauma-exposed subgroups do not show general reductions in posttraumatic distress over time but rather evidence a chronic and severe course (Saigh & Bremner, 1999).

These disparate findings have stimulated a movement away from a generally static "main effects" approach to the study of posttraumatic adjustment that emphasizes the identification of exposure-related risk variables and their associated odds ratios for increased risk, to a conceptually more sophisticated approach emphasizing the dynamic and interactive influences of other vulnerability and protective mechanisms and processes over time. In particular, recent efforts have sought to clarify the influence of factors embedded within the pre-, peri-, and posttraumatic contexts that serve to mediate and/or moderate the trauma exposure-posttraumatic distress relationship (e.g., Yehuda, 1999; Yehuda, McFarlane, & Shalev, 1998; see Holmbeck, 1997, for an explication of mediator and moderator variables). This emerging literature is dedicated to identifying processes—in the form of antecedents, correlates, cause-effect relationships, moderating influences, and mediating mechanisms—that lead to traumatic distress reactivation.

An essential part of this movement, both with respect to research and intervention efforts, is the development of etiological theories of posttraumatic stress (e.g., Foy, Osato, Houskamp, & Neumann, 1992; Jones & Barlow, 1990; Pynoos, Steinberg, & Wraith, 1995; Wilson, 1994; Yehuda, 2002). These theories generally posit that exposure to traumatic events leads to acute, and potentially chronic, posttraumatic stress and other distress reactions. These theories also propose that the pathways linking trauma exposure, acute distress reactions, and chronic distress are mediated and/or moderated by vulnerability- and resilience-enhancing variables embedded within the traumatic event, the individual, and the surrounding physical and social ecology, which serve to increase or decrease risk, respectively, for adverse proximal and distal outcomes.

Building on these theoretical developments, we now present a conceptual framework for considering the influences of a number of mechanisms theorized to mediate the trauma exposure-distress relationship, and by extension contribute to the reactivation of posttraumatic distress. In so doing, we note that although the literature on resilience has enjoyed comparatively limited direct emphasis within the trauma field, it shows considerable promise for clarifying why only a minority of trauma-exposed individuals generally develop clinically significant posttraumatic outcomes (e.g., Engelhard, Van Dan Hout, & Arntz, 2001; de-Jong et al., 2001; Koren, Arnon, & Klein, 2001), and for developing early intervention programs designed to minimize the risk for long-term maladaptation (e.g., Luthar & Cicchetti, 2000; Masten, 1999; Rolf & Johnson, 1999). As noted by Harvey (1996), the field of traumatic stress has overemphasized the study of risk factors and maladaptive outcomes at the expense of delineating the nature of adaptive processes and outcomes.

Overview of the Conceptual Model: Therapeutic Foci, Resilience, and Vulnerability Factors

The conceptual framework used in this chapter is based on an ecological/developmental psychopathology model of traumatic stress developed by Pynoos, Steinberg, & Wraith (1995). This model posits that the course of posttraumatic adjustment in children and adolescents is influenced by vulnerability and protective processes present within the pretrauma, peritrauma, and posttraumatic ecological contexts. Further, Pynoos and his colleagues propose that five critical domains should be systematically addressed to build theory, predict posttraumatic adjustment, guide intervention efforts, and develop effective public policy.

The first therapeutic focus, *trauma exposure,* is directed toward the identification of both objective and subjective elements of a focal traumatic experience. These exposure characteristics, in turn, are used to stratify trauma-exposed individuals according to their risk for severe posttraumatic distress reactions and need for specialized intervention. This domain of focus also includes consideration of risk for, and potentially interactive effects of, other traumatic experiences that have taken place either prior or subsequent to the focal trauma. These include careful assessment of the presence and effects of prior trauma, as well as that of reexposure to traumatic events. The second domain, *trauma and loss reminders,* focuses on identifying posttraumatic reminders of trauma and loss, their frequency, and the degree to which they evoke distress and disrupt significant life activities. Pynoos, Steinberg, & Wraith (1995) propose that reminders serve as a primary mediating mechanism that links trauma exposure and posttrauma distress, and thus serve as important targets for assessment and intervention. The third domain, *posttrauma adversities,* addresses the effects of losses, life changes, and other stresses generated or exacerbated by traumatic events. These stressful events and conditions often serve as significant sources of stress in themselves, may persist long after the traumatic incident, and may significantly interfere with posttraumatic adjustment by diverting coping resources or by inducing maladaptive coping.

The fourth domain of therapeutic focus, *interplay of trauma and grief,* addresses the dual stresses of traumatization and bereavement imposed on survivors of traumatic loss that significantly increase the risk for both complicated grief and posttraumatic distress reactions (Pynoos, 1992; Rando, 1993). The concurrent presence of posttraumatic distress and grief reactions may indeed increase the risk for the chronicity of each disorder (Nader, Pynoos, Fairbanks, & Frederick, 1990). Last, the fifth domain addresses the *developmental impact* of trauma. This impact may be manifest in the form of altered core beliefs regarding the self, the world, and the future. It may also be manifest in the interruption, premature initiation, or delay of important developmental tasks and activities; and in the form of disruptions within interpersonal relationships, including disturbances in family functioning, peer relationships, and romantic attachments (see Cloitre, Davis, Mirvis, & Levitt, 2002; Layne, Pynoos, & Cardenas, 2001; Layne, Saltzman, Steinberg, & Pynoos, 2003, for examples of therapeutic interventions that explicitly focus on these issues).

Although not explicitly addressed in the model, an indispensable sixth domain of focus consists of identifying *other vulnerability and protective processes* as part of risk assessment, treatment triage, and treatment planning. These include systematically assessing and addressing such demographic variables as gender, age at the time of trauma, socioeconomic status, education level, and ethnicity/race. Other pertinent variables include history of childhood physical abuse, sexual abuse, or neglect; age at first trauma; psychiatric history; intelligence; family psychiatric history; perceived social support; use of coping resources; and stressful life events and circumstances (see Brewin et al., 2000; Yehuda, 1999). Similarly, research from the resilience literature has identified a number of protective factors linked to adaptive outcomes in individuals exposed to various forms of stress or high-risk conditions. Some of these factors include an easy temperament, high intellectual ability, a positive family environment, internal locus of control, socioeconomic advantage, and supportive relationships with peers, family members, and other adults (Cowen, Wyman, & Work, 1996; Garmezy, 1985; Grossman et al., 1992; Luthar, 1991; Luthar & Zigler, 1991; Masten, Best, & Garmezy, 1990; Masten et al., 1988; O'Grady & Metz, 1987; Parker, Cowen, Work, & Wyman, 1990; Rutter, 1987; Weist, Freedman, Paskewitz, Proescher, & Flaherty, 1995; Werner, 1989). Incorporation of resilience research into the assessment and treatment of trauma-exposed individuals holds considerable promise for improved therapeutic outcomes.

An understanding of these therapeutic foci and the processes by which they exert their influence requires a basic understanding of how resistance, resilience, and vulnerability factors are conceptually defined and presumed to operate. Drawing on the general developmental psychopathology literature, resistance denotes a capacity to maintain relatively normal functioning upon exposure to stress or adversity (Pynoos, Steinberg, & Wraith, 1995). On the other hand, resilience has been defined as a dynamic process encompassing early effective positive adaptation within the context of significant stress or adversity (Luthar, Cicchetti, & Becker, 2000; Masten, 2001; Masten & Coatsworth, 1998). Inherent in the definition of resilience are two essential components: (1) exposure to stress or disadvantage of a magnitude sufficient to constitute a significant challenge to well-being and (2) demonstration of early effective positive adaptation within this context (Luthar & Zigler, 1991; Masten et al., 1990; Werner & Smith, 1982). Conversely, vulnerability encompasses conditions or circumstances that make successful adaptation more difficult (Cicchetti, Rogosh, & Toth, 1997). In contrast to resilience, vulnerability denotes an impaired capacity to withstand stress and adversity. Fundamental to the study of resilience has been the identification and exploration of adaptive or maladaptive adjustment processes— that is, characteristics or conditions that moderate (by attenuating or exacerbating, respectively) the influences of stressful life events or other disadvantages which, in turn, place individuals at increased risk for negative outcomes (Luthar et al., 2000; Masten & Coatsworth, 1998; Masten & Garmezy, 1985).

A more conceptually rich and clinically useful perspective on stress and adaptation has been advanced by Pynoos and his colleagues (Pynoos, Steinberg, & Wraith, 1995; Pynoos, Rodriguez, Steinberg, Stuber, & Fredericks, 1999). In their developmental psychopathology framework of child traumatic stress, the authors introduce the dimension of time in their distinction between resistance and resilience processes. Specifically, the authors posit that, upon exposure to trauma or adversity, *resistance factors or processes* act to mitigate their adverse influence, thereby promoting the *maintenance* of adaptive functioning without the development of significant distress, impaired role functioning, or developmental disturbance. These resistance factors or processes interface in complex ways with *vulnerability factors or processes,* which conversely increase the likelihood of developing these same adverse outcomes. In the advent that clinically significant distress develops, Pynoos and his colleagues propose that *resilience* comes into play. In particular, the authors define *resilience* as the capacity to recover more efficiently and effectively from distress once it develops (thus "springing back," in keeping with the metallurgical origins of the concept). Resilience processes can be characterized along a continuum of adjustment-maladjustment, in which those factors demonstrated to be most effective in restoring well-being are viewed as possessing greater resilience-promoting properties.

Adding to this definitional complexity is the observation that resilience (and presumably resistance) processes often have domain-specific properties (Luthar & Zelazo, 2003). For example, a lack of significant externalizing behavior problems in response to major adversities does not preclude the presence of other problems such as depression, anxiety, or academic failure. This point is underscored by Hetherington and Elmore (2003), who observe that in families led by divorced mothers, a subset of daughters who appear socially responsible and overtly well adjusted also experience considerable levels of internalizing symptoms such as anxiety and depression. Comparable results have been reported by Zucker, Wong, Puttler, & Fitzgerald (2003) in longitudinal studies of the children of alcoholic parents. Zucker et al. noted that many children of alcoholics who appeared to demonstrate minimal levels of externalizing behavior problems from childhood through adolescence later developed severe internalizing behavior problems. These findings serve to illustrate the temporal and domain-specific features of resilience processes, as well as the need for utilizing a sufficiently broad range of constructs and associated measures with which to measure resilient outcomes (Luthar & Zelazo, 2003).

A simplified hypothetical example involving three youths who are exposed to equivalent levels of chronic domestic violence may illustrate the distinctions between stress resistance, vulnerability, and resilience. Youth A exhibits resistance to the adverse effects of domestic violence exposure to the extent that significant distress, impaired role functioning, and developmental disturbance do not develop. In contrast, Youth B exhibits *vulnerability* to domestic violence-related stress and develops stress-related pathology. Nevertheless, he also exhibits *resilience* in the form of quickly and effectively recovering to a relatively normal and adaptive level of functioning. Youth C, conversely, is vulnerable to the effects of stress, lacks resilience, and consequently develops persisting pathology and exhibits signs of impaired role functioning and developmental disturbance (see Compas, Hinden, & Gerhardt, 1995; Layne, Warren, Shalev, & Watson, in press).

Using these conceptual definitions as a background template, we now address our third issue, proposing three vulnerability processes theorized to reactivate posttraumatic distress as an outcome.

Vulnerability Variables Theorized to Reactivate Posttraumatic Distress

To promote a clearer understanding of the processes through which posttraumatic distress may become reactivated, we theorize that at least three factors may influence the onset, duration, intensity, and/or periodicity of distress reactivation. As diagrammed in Figure 9.2, we propose that distress reactivation may be evoked by any of three intervening variables: (1) revictimization, (2) posttraumatic adversities that may be consequent to, or independent of, a prior traumatic event; and (3) exposure to distressing reminders, also termed "cues" or "triggers," that symbolize or resemble aspects of a past traumatic experience. Each of these pathways of influence will be reviewed in turn.

Etiological Factor   Intervening Socioenvironmental Mechanism   Outcome Variable

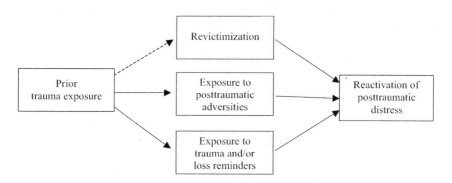

FIGURE 9.2. Proposed Alternative Model: Factors Theorized to Reactivate Posttraumatic Distress. Solid lines denote causal pathways theorized to characterize all types of traumatic events. Broken line indicates that certain types of traumatic events/circumstances increase the risk for subsequent exposure to certain types of trauma but not others (e.g., childhood sexual abuse may increase the risk for sexual assault, but not exposure to natural disasters, in adulthood).

Pathway 1: Risk for Posttraumatic Distress Reactivation via Revictimization

Many studies have identified early trauma exposure as a significant marker of risk for subsequent trauma exposure and associated distress reactivation (e.g., Baynard, Williams, & Siegel, 2001; Chu, 1992; Cloitre, 1998; Cloitre et al., 1996; Dietrich, 2002; Glodich, 1998; Orcutt et al., 2002; Simon, 1999; Wyatt et al., 1992). For example, using a longitudinal design with 174 women with self-reported histories of childhood trauma, Baynard, Williams, & Siegel (2001) reported that women with self-reported histories of childhood sexual abuse reported higher rates of exposure to nonsexual childhood trauma, adult sexual assault, and adult trauma exposure than non-CSA controls. Exposure to traumatic events in childhood and adulthood were both significantly correlated with poorer mental health functioning. Notably, total exposure to traumatic life events fully mediated the relationship between CSA and adult mental health outcomes.

Pathway 2: Risk for Posttraumatic Distress Reactivation
via Posttraumatic Adversities

A second mechanism by which distress reactivation is theorized to occur is through adverse life events or circumstances existing in the aftermath of a given traumatic event. These adversities may be either consequential to, or independent of, the focal trauma, and thus may precede, coincide with, or arise following the trauma. One set of studies addresses the adversities experienced by children of trauma survivors, whom the evidence suggests are at increased risk for aggression, anxiety, psychosomatic difficulties, guilt, and depression (Felsen, 1998). Indeed, as noted by Nelson and Schwerdtfeger (2002), family systems may incorporate trauma-related themes into their interaction patterns, becoming "trauma-organized systems" (Baynard, Englund, & Rozelle, 2001). Nelson and Schwerdtfeger (2002) also note that children may be adversely affected by parental exposure to trauma via exposure to a parent's posttraumatic stress symptoms, including flashbacks, startle responses, and mistrust of others (Steinberg, 1998). In addition, parenting may be compromised in traumatized adults in such forms as minimizing the perceived effects of trauma on children (Applebaum & Burns, 1991; Freeman, Shaffer, & Smith, 1996; Marans, Berkman, & Cohen; 1996).

Other examples of increased risk for posttraumatic distress reactivation imparted by posttraumatic adversities include participation in stressful judicial proceedings following violent victimization (Epstein et al., 1997), and participation in physical rehabilitation following severe injury. Kutlac et al. (2002) have studied the adversities consequent to living as a war refugee following violent eviction from one's home, and Djapo et al. (2000) have studied the mental health correlates of widespread unemployment, poverty, and the physical destruction of infrastructure in the aftermath of war. Both authors reported that posttraumatic stress symptomatology reported two years postwar is positively correlated with the number of postwar adversities reported by Bosnian youths.

Pathway 3: Risk for Posttraumatic Distress Reactivation
via Exposure to Distressing Reminders

As noted in our review, a primary factor in the reactivation of posttraumatic distress reactions is exposure to reminders that symbolize or resemble aspects of a previous traumatic experience. For example, Solomon (1990) reported that rape victims often report a reactivation in trauma-related distress following discussions of their sexual assault. A second example is found in Christenson, Walker, Ross, & Maltbie's (1981) clinical case description of Mr. "A," a fifty-five-year-old World War II veteran who shot a boy, under orders, while guarding a military base. The patient experienced breakdowns both directly after the war (at which time he was hospitalized for "nerves") and later at age fifty-two, when he was asked to care for the remains of a dead boy at the emergency room where he worked as an attendant. At this time, the patient experienced a resurgence of trauma-related nightmares and depressive reactions. The authors suggest that posttraumatic distress can be reactivated by reminders or triggers of previous traumatic events. Another example involving distressing reminders is found in Chairamonte's (1992) discussion of the reactivation of distress in Vietnam veterans as a result of extensive media coverage of the Gulf War.

Because they are comparatively newer and less familiar constructs within the trauma literature, we will devote comparatively more space in this chapter to a review of the theoretical underpinnings, conceptual distinction, and clinical utility of trauma and loss reminders.

Theoretical and Clinical Perspectives on Trauma and Loss Reminders

Trauma and loss reminders are theorized to act as mediators through which traumatic experiences and losses exert a persisting influence over the longitudinal course of posttraumatic adjustment. In particular, the frequency with which survivors are exposed to trauma and loss reminders, the subjective intensity of the distress reactions they elicit, and the availability and specific use of coping resources to contend with distressing reminders over time, are theorized to act as primary determinants of the course of posttraumatic adjustment (Pynoos, Steinberg, & Wraith, 1995). *Trauma reminders* are defined as cues, either physically internal or external to survivors of trauma exposure, which resemble or symbolize aspects of traumatic events and thereby elicit distressing reactivity (including emotional, cognitive, physiological, and/or behavioral reactions) associated with one or more past traumatic experiences. These reactions are subjectively experienced as distressing, unbidden, intrusive, and involuntary. The usefulness of trauma reminders is underscored by their broad clinical utility. These include establishing a therapeutic alliance based on recognition of the ongoing impact of past trauma; psychoeducation regarding distress reactions and the cues that elicit them; developing coping strategies that pair identification of "high risk" reminder-laden situations with effective coping strategies to contend with them; and therapeutic reprocessing of traumatic experiences, especially "worst moments." Trauma reminders are theorized to serve as therapeutic links to the traumatic past, and as such assist in identifying and reprocessing the horrific, terrifying, overwhelming, and/or tragic circumstances within which traumatic events or losses originally took place (Layne, Saltzman, et al., 2003; Saltzman, Layne, Steinberg, & Pynoos, Chapter 18 in this volume). The therapeutic importance of working explicitly with trauma reminders, including interventions focusing on psychoeducation, increased coping effectiveness, prolonged therapeutic exposure, and relapse prevention, is underscored by observations that these reminders may evoke intense emotional reactivity associated with the original trauma without an accompanying conscious awareness of its historical reference (Janoff-Bulman, 1992).

In accordance with their definition as stimuli that resemble or symbolize aspects of traumatic events, trauma reminders are integrally related to DSM-IV-TR diagnostic criteria for post-traumatic stress disorder (PTSD) (American Psychiatric Association, 2000). In particular, trauma reminders consist of either objective events or circumstances, or such internal states as physiological or kinesthetic sensations, mood states, or mental images that trigger posttraumatic stress reactions. These stress reactions include, but are not limited to, such DSM-IV-TR diagnostic symptoms as psychological distress at exposure to internal or external cues that symbolize or resemble an aspect of the traumatic event (Symptom B4); physiological reactivity on exposure to internal or external cues that symbolize or resemble an aspect of the traumatic event (Symptom B5); efforts to avoid thoughts, feelings, or conversations associated with the trauma (Symptom C1); and efforts to avoid activities, situations, or people who arouse recollections of the trauma (Symptom C2).

Exposure to trauma reminders is presumed to occur via two primary channels: external cues and internal cues. *External trauma cues* consist of things that are seen, heard, smelled, touched, and/or tasted within the external environment. Examples of external trauma reminders cited by youths and adults exposed to trauma include the following.

1. *Visual:* Ambulances; jets flying overhead; seeing blood; worried faces; young men with long hair. In an interview, a withdrawn Bosnian student described being flooded with complicated grief reactions each day after his school bus drove past the tree where his older brother was brutally hanged by paramilitary troops during the war. He reported that it often took him hours each morning to calm down sufficiently to attend to his schoolwork.
2. *Auditory:* Sirens; people shouting; telephones ringing; every time my father says good-bye; the song playing on the radio at that moment.

3. *Olfactory:* The smell of smoke; the smell of the cologne the rapist was wearing.
4. *Taste:* Canned food; the taste of the food I had in my mouth when a bomb exploded nearby.
5. *Tactile:* Putting my hands in warm water reminds me of lifting my fallen friend's head and then feeling his warm blood spilling over my hands.
6. *Seasonal/cyclical:* Thursdays; my birthday, when my dad called from the front lines to wish me happy birthday and to tell me he hoped to be home soon. We learned a week later that he had been killed the day after.
7. *Complex/multisensory:* My school; an emotionally exhausted high school teacher resigns from her tenured post in an inner-city Chicago school to escape its chronic gang violence and moves to a small town in Oregon. Two years later, she nearly collapses in shock in front of her fifth-grade classroom when a televised educational program on youth violence unexpectedly displays an image of the violence-ridden public park where a favorite student had been brutally murdered years before. She tearfully describes herself as "really out of it" and "sucker punched" for the next three days.

Internal trauma cues consist of internal phenomena that resemble or symbolize an aspect of a given traumatic event, including cognitions, mental images, emotions, dreams, or bodily/kinesthetic sensations. Examples of internal cues identified by trauma survivors include the following:

1. *Emotions:* When I start feeling lonely.
2. *Cognitions:* I worry about being molested, and that reminds me of times during the war when I feared for the safety of our family; when I am suddenly surprised.
3. *Dreams:* I have frequent dreams about my dead friend, and these remind me of when he was stabbed to death.
4. *Bodily/kinesthetic sensations:* Sitting down to have lunch; feeling tense; feeling my heart beat quickly; feeling sick or queasy in my stomach; sexual arousal. A former kidnap victim suffers debilitating flashbacks after allowing her child to tie up her hands in play. In treatment, she recognizes that this constituted a reenactment of a "worst moment" of her abduction when she felt completely helpless and vulnerable.

In contrast, *loss reminders* are defined as cues, either internal or external, that evoke acute grief-related reactions in an individual with a history of significant loss. They are presumed to do so by directing attention either to the absence of the lost object or to life changes consequential to the loss. Similar to the therapeutic role of trauma reminders, loss reminders may serve as therapeutic tools for understanding the subjective meaning and impact of the ongoing and anticipated consequences of a given loss. More specifically, the loss of a loved one's physical presence is known to precipitate a cascade of ensuing losses, given that the cessation of the physical relationship brings with it the loss of all of the present and future supportive provisions, hopes, plans, expectations, activities, and modes of behavior that the relationship formerly provided (Parkes, 1998). For example, a cousin's wedding may evoke a grief pang linked to the felt absence of a cherished grandmother, a bittersweet recollection of past times spent with her discussing weddings and other life plans, and a distressing anticipation of what one's own future wedding will be like without her. Loss reminders are theorized to stem from the loss of any cherished entity or object, including family members, friends, teachers, neighbors, or cherished pets who are deceased, missing, or from whom one is separated or estranged. Loss reminders may also be linked to material possessions, including homes, villages, money, and other personal valuables, or to the perceived loss of such abstract yet cherished "possessions" as one's previous way of life, self-concept, childhood, moral conscience, innocence, culture, or nationality (e.g., Apfel & Simon, 1996; see also Rando, 1993, for a description of cyclical and linear cues theorized to evoke "temporary upsurges in grief"). Like trauma reminders, exposure to loss reminders is theorized to occur via external and internal cues. Examples of *external loss reminders* cited by youth and adult survivors of loss include these:

1. *Visual:* Whenever I see a happy, normal teenager; looking at pictures from my old life; seeing other girls out walking with their fathers; seeing groups of friends out having a good time; my uncle's resemblance to my dad reminds me of him.
2. *Auditory:* Songs of my birthplace; hearing my old friends' names; favorite sayings and mode of speech from my old town; hearing the crying of our neighbors on the anniversary of their son's death.
3. *Olfactory:* The smell of my uncle's favorite cologne.
4. *Taste:* When our family eats my uncle's favorite dish; red wine (my uncle's favorite drink); my uncle's favorite chewing gum.
5. *Tactile:* Feeling the rain dripping on us from our leaky roof where we now live.
6. *Seasonal/cyclical:* The months of autumn, when we were driven from our home; my father's birthday; Christmas holiday.

7. *Complex/multisensory:* Our dirty basement lodgings reminds me of our family's home before we were driven out to become war refugees who have nothing; family celebrations; being without my parents; being in the company of lonely people who have lost loved ones.

Internal loss reminders consist of internal cues that call attention to the loss, and may include cognitions, mental images, emotions, dreams, or bodily/kinesthetic sensations. Examples of internal cues identified by survivors of loss include these:

1. *Emotions:* When I feel lonesome; when I feel sad.
2. *Cognitions:* Thoughts of how my life would have been if ugly things had not happened; when I see that I have no money; thinking of the financial hardships my parents face; when I study hard twelve hours a day but in vain; when I see children and think about their lives.
3. *Dreams:* When I dream about my dead father.
4. *Bodily/kinesthetic sensations:* When I wake up in the morning all alone in my bed.

Clinical Case Example

A poignant account involving two Bosnian youths helps to make concrete the distinction between trauma and loss reminders and to underscore the idiosyncratic functions that a given cue may assume. Two students participating in the same school-based trauma/grief-focused psychotherapy group (Layne, Neibauer, et al., 2003) were invited, as part of a psychoeducational group activity, to identify current events or circumstances that evoked distressing feelings or memories related to the war. Both described the commonplace act of walking past a butcher shop as particularly distressing. When invited to explore the connection between this setting and their war-related experiences, one student recounted witnessing a horrific marketplace massacre and the role played by butcher shops as a graphic trauma reminder:

It was a nice day, so my friend and I went out to do some shopping downtown. Suddenly, we heard a muffled sound that we recognized—it was the sound that a grenade launcher makes. We were waiting for a big bang. Suddenly, there was a huge flash and everything around us was leveled. A spot 30 meters away became a river of blood and a heap of flesh. Only afterwards did I begin to feel "happiness" that I had survived. Nowadays, walking by a butcher shop reminds me of all those mutilated bodies and the screams and crying of the wounded and makes me feel shaky and sick to my stomach.

The second youth, whose father had been killed in the war, linked the setting of a butcher shop to experiencing the father's continued absence. This student's narrative underscores the role of butcher shops as a potent loss reminder for her:

My father was a butcher before the war, and I used to visit him while he was at work. Nowadays, when I walk by a butcher shop, I miss him terribly and become very sad, remembering the old times and wishing the war had never happened. Life without Father is getting harder and harder. My mother is often depressed and very stressed, working long hours at low-paying jobs to keep food on the table and a roof over our heads, so I have to do most of the housework and look after my younger brother and sister. It is a big responsibility that leaves me little time for my schoolwork and friends. I miss my old life before the war, when Father was here.

The same cue serves two distinct roles for the youths as a result of their individual histories. For the first, butcher shops elicited posttraumatic stress symptoms consisting of distressing, intrusive images associated with witnessing a horrific marketplace massacre. For the second, butcher shops evoked acute grief pangs associated with her father's permanent physical absence from her family's life. These were reflected in her sorrowful descriptions of the "empty situation" of the shops and of the pressing and persistent adversities generated by his absence.

Clinical Utility of Trauma and Loss Reminders

Interventions with traumatically bereaved individuals often entail therapeutic work with both trauma and loss reminders (e.g., Figley, Bride, & Mazza, 1997; Jacobs, 1999; Rando, 1993; Rynearson, 2001). Because each category of reminder is associated with distinct treatment objectives and intervention strategies—requiring trauma- and grief-focused work, respectively—discriminating between trauma and loss reminders and the traumatic distress and/or grief reactions that they evoke constitutes an important therapeutic objective. This work may be particularly relevant to such therapeutic interventions as psychoeducation, building a coping skills tool kit, developing a relapse prevention plan, and processing trauma and loss-related experiences and associated distress reactions (e.g., Cohen, Greenberg, Padlo, Shipley, Mannarino, & Deblinger, 2001; Layne, Saltzman, et al., 2003; see Cohen, Berliner, & March, 2000, for general treatment recommendations).

Understanding the distinction between trauma and loss reminders and the roles that each plays in eliciting distress reactions also aids in clarifying the dynamics underlying traumatic bereavement. These dynamics are manifest in cases in which normal grieving processes that promote adaptive accommodation to the permanent physical absence of loved ones are disrupted or inhibited by posttraumatic distress reactions linked to the traumatic circumstances of their death. These distress reactions may include (1) intrusive, distressing thoughts, mental images, and/or emotions linked to the violent, tragic circumstances of the death; (2) avoidance of, or escape from, distressing cues linked to the death or to the deceased; (3) psychological numbing; and/or (4) social estrangement, mistrust, or withdrawal. Adaptive grief processes theorized to be vulnerable to trauma-related intrusions and avoidance include reminiscing about and remembering the deceased; learning more about the deceased; processing painful emotions associated with accepting and adapting to the loss; participating in grief rituals; and forming new, gratifying relationships and activities to partially compensate for the loss. Other vulnerable grief processes may include accepting the reality of the death; arriving at an explanation of why the loved one died; making meaning of the death; transforming one's relationship with the deceased from a physically based to a psychologically based relationship; and adapting and accommodating to the permanent physical absence of the deceased (Cohen, Mannarino, & Greenberg, 2002; Nader, 1997; Pynoos, 1992; Rando, 1993; Stubbenbort & Cohen, Chapter 16 in this volume; Worden, 1996).

Among "doubly distressed" individuals experiencing both posttraumatic distress reactions associated with a loved one's traumatic death and bereavement reactions associated with their permanent physical absence, trauma reminders are theorized to reflect the nature of, and subjective meanings attached to, the *traumatic circumstances* of the death. Conversely, loss reminders are presumed to reflect the subjective impact and deeply personal meanings ascribed to the *ongoing effects* of that loss. For example, the ringing of the telephone may function as a trauma reminder by eliciting distressing memories of the night family members were notified of a traumatic death. In contrast, being served the deceased loved one's favorite dish, hearing his favorite song on the radio, seeing his former girlfriend, or playing in a school basketball game without seeing his face in the bleachers may evoke acute grief reactions in their role as loss reminders.

Loss reminders often evoke differential, and sometimes dissonant and incompatible, grief reactions within families or other systems that may constitute significant secondary adversities to system members. For example, a Bosnian youth recounted that she and her mother sometimes argued over her refusal to practice the piano. When asked why, she described how playing the piano brought up sad memories of past times when she had played for her father, who loved music and had been killed in the war. During these present-day arguments, her mother maintained that playing music was an important way to remember her father and to honor his wish that she enrich her life through music. From a therapeutic perspective, playing the piano became for the mother a means for memorializing her husband and carrying on his legacy in their daughter's life. For the daughter, it contrastingly symbolized the painful emptiness created by his ongoing absence and the irrevocable and unwanted change it brought to their lives.

Figure 9.3 is a proposed typology of reminders theorized to follow in the aftermath of the traumatic death of a loved one. The first circle, *reminders of the circumstances of the death,* is comprised of trauma reminders related to the traumatic circumstances within which the death took place. The second and third circles, *reminders of the subsequent changes in one's life* and *reminders of the loss or absence of the deceased,* are both reminders of loss and its aftermath. These reminders are linked to life changes consequential

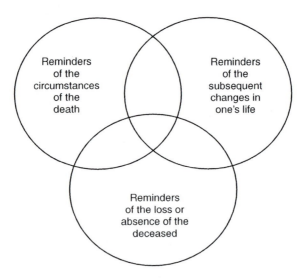

FIGURE 9.3. Typology of Distressing Reminders Associated with Violent Death

to the loss (commonly termed *change reminders*) or to the ongoing absence of the lost object (commonly termed *loss reminders*), respectively.

Notably, the three categories of reminders partially overlap in the figure, signifying that a given reminder may simultaneously fall within up to three functional categories. For example, a family holiday gathering may serve as a trauma reminder (many of the same relatives who attended a father's funeral are gathered together in a manner reminiscent of the funeral; grandparents in attendance conveyed the terrible news about his death to other family members), as a loss reminder (his absence is keenly felt within the gathering; several relatives resemble him in appearance or behavior; family members speak about him), and as a change reminder (the loss of father's income necessitated a move to a smaller home within which the gathering is held; fewer and less expensive presents are given to the children; there is no Christmas bonus to take the family on their traditional holiday vacation). Moreover, ostensibly positive life changes, such as the material improvements brought about by donations to families of emergency personnel who perished in the September 11, 2001, disaster, may constitute distressing loss reminders. Affected youths have been observed to complain, "I don't want a new computer or a nicer TV. I want my dad back!"

The presence of loss reminders should not be presumed to indicate that the loss was traumatic in nature. Rather, survivors of both traumatic and nontraumatic losses report encountering loss reminders (Rando, 1993). For example, families of loved ones who have died a natural death often report feeling lonely or sad in the presence of reminders of the loved ones' absences, such as an empty place at the dinner table or a loved one's empty bedroom (Parkes, 1998). Conversely, the presence of trauma reminders does not necessarily indicate that actual loss occurred, such as in the case in which a loud bang brings to mind a shooting incident in which no one was killed or seriously injured. In summary, although trauma and loss reminders share some similarities in their function and conceptual underpinnings, they should be assessed, interpreted, and clinically utilized as distinct constructs.

EMPIRICAL TESTS OF THE PROPOSED INTERVENING VARIABLES

In the remaining portion of this chapter, we report the results of two preliminary empirical studies designed to address two questions pertaining to the conceptualization and correlates of posttraumatic distress reactivation, respectively. Because of the comparatively new status in the field and controversy regarding the distinction between posttraumatic stress reactions and distressing reminders, the first study addresses the question, Are post-traumatic stress symptoms, trauma reminders, and loss reminders empirically distinguishable constructs? This exploratory factor-analytic study tested two hypotheses: (1) Self-report measures of posttraumatic stress symptoms, frequency of exposure to loss reminders, and frequency of exposure to trauma reminders will load, respectively, on three moderately correlated but empirically and conceptually distinct factors; and (2) factor scores created from the trauma and loss reminder variables will correlate strongly with measures of posttraumatic stress and grief reactions, respectively, and to a moderate degree with measures of other distress reactions.

The second study addresses, on a broader level, the theoretical and methodological relevance of the mechanisms underlying posttraumatic distress reactivation proposed in this chapter. This study utilized structural equation modeling to test the hypothesis that exposure to postwar adverse life events, trauma reminders, and subsequent traumatic events mediates the relationship between war trauma and long-term postwar posttraumatic stress symptoms.

Data for both studies were collected as part of the UNICEF Bosnia and Hercegovnia School-Based Psychosocial Program for War-Exposed Adolescents (Layne, Pynoos, Saltzman, et al., 2001). In fall 1997, following the program protocol, trained school counselors selected classrooms within their respective schools that, based on school records and their professional knowledge, contained high overall concentrations of students exposed to significant war-related trauma. After making presentations introducing the program and its components to school personnel, parents, and the selected classrooms (see Saltzman et al., 2003), and after securing the consent of the school director and participating teachers, parents, and students, the counselors administered a risk screening survey (hereafter termed "Time 1") in the selected classrooms in the presence of the teacher. Students who reported or showed signs of significant distress were, according to the protocol, individually debriefed by a counselor and evaluated as appropriate for specialized services or a community referral. The screening survey contained measures of exposure to war trauma (Layne, Stuvland, Saltzman, Steinberg, & Pynoos, 1999), trauma reminders (Layne, Steinberg, Saltzman, Wood, & Pynoos, 1998), postwar adverse life events (Layne, Djapo, & Pynoos, 1998), and posttraumatic stress symptomatology (Pynoos, Rodriguez, et al., 1999), in addition to measures of depression (Birleson, Hudson, Buchanan, & Wolff, 1987) and grief reactions (Layne, Pynoos, Savjak, & Steinberg, 1998). Using standardized selection criteria, six to eight students per school who were deemed at highest risk for severe, persisting posttraumatic distress reactions and developmental difficulties were selected to participate in trauma/grief-focused group treatment (Layne, Saltzman, et al., 2003).

Approximately six months later, in spring 1998, a follow-up survey (hereafter, "Time 2") was administered to a subset of the same students. This survey contained the same distress, trauma reminder, and life adversities measures contained in the first survey, in addition to several other measures not used in the study. Approximately six months later, in fall/winter 1998, a third survey was administered (hereafter, "Time 3") to a subset of students who participated in Time 2. Time 3 contained the same measures used in Time 2, in addition to several measures not used in the study, a newly created loss reminders inventory (Layne, Savjak, Steinberg, & Pynoos, 1999) and measures of parenting behaviors. In summary, Times 1 through 3 covered the period of approximately two, two and one half, and three years after the Bosnian conflict, respectively.

Study 1

The first study, reported by Layne, Neibauer, et al. (2003), utilized data collected during Time 3 ($N = 85$, 70.4 percent girls and 29.4 percent boys; mean age = 17.34). Data were subjected to exploratory factor analysis Principal Axis factor extraction with oblique rotation to measure the dimensionality of a set of three measures (posttraumatic stress, reported frequency of exposure to trauma reminders, and reported frequency of exposure to loss reminders) completed by a sample of war-exposed youth attending public schools in the Republika Srpska, Bosnia.

The authors reported that the factor analysis results were consistent with the hypothesis that trauma reminders, loss reminders, and posttraumatic stress reactions are empirically distinct constructs. Each of the three factors appeared to be well-defined, possessing six or more items with strong factor loadings. Based on the patterns of factor loadings and the guiding theoretical model, the factors were labeled *loss reminders, posttraumatic stress reactions,* and *trauma reminders,* respectively. Consistent with the hypothesized distinctness and theoretical relatedness of the factors, correlations between the factor scores fell within the moderate range.

Finally, Layne, Neibauer, et al. (2003) calculated factor scores for the loss reminder and trauma reminder factors and correlated them with total-scale scores from measures of posttraumatic stress, depression, grief, postwar interpersonal adversities, and postwar trauma subscales. As predicted, the trauma reminder factor score correlated with a broad range of outcomes, including posttraumatic stress symptoms, the loss reminder factor score, and measures of depressive symptoms, grief reactions, postwar interpersonal adversities, and postwar trauma. Contrary to prediction, the magnitude of correlation between trauma reminders and posttraumatic stress symptoms was not significantly larger in comparison to those between trauma reminders and other distress measures. As hypothesized, loss reminders correlated highly with grief and to a moderate degree with measures of posttraumatic stress, depression, interpersonal adversities, and postwar trauma.

Study 2

The second study, reported by Layne, Legerski, et al. (2003), used data collected from Times 1, 2, and 3. The sample was comprised of 166 Bosnian secondary school students, consisting of 63 percent girls and 37 percent boys; mean age = 16.11. This study tested the hypothesized role of three sets of variables—postwar adversities, frequency of exposure to trauma reminders after the war, and postwar revictimization—as mediators of the relationship between wartime trauma and postwar distress reactivation. Figure 9.4 presents an overview of the basic model, which was tested at each of the three time points. Specifically, *postwar revictimization* (e.g., confirmation of the death of a loved one disappeared during the war), *postwar adversities,* and *postwar trauma reminders* were hypothesized to serve as intervening variables that mediated the relationship between *war trauma* and *postwar posttraumatic stress symptoms,* the hypothesized outcome variable as reported at Times 1, 2, and 3. Given this chapter's emphasis on posttraumatic distress reactivation as an outcome produced by intervening mechanisms, *posttraumatic stress symptoms* was selected as the outcome of interest and was modeled as a latent variable. All remaining variables were modeled as measured variables.

The structural equation model tested the hypothesis that the intervening variables postwar revictimization, postwar adversities, and postwar trauma reminders would *partially mediate* the relationship between war trauma exposure and postwar posttraumatic stress symptoms measured at Time 1. This hypothesized relationship took the form of direct pathways from the two war variables to posttraumatic stress symptoms, and via indirect pathways through the mediating variables. After making minor modifications to the originally proposed model, the resulting final model indicated a good fit, accounting for over 50 percent of the variance in posttraumatic stress symptoms measured at each of the three times. The model provided generally consistent support for the hypothesized roles of postwar adversities, frequency of exposure to trauma reminders, and postwar revictimization as mediators of the relationship between wartime trauma exposure and long-

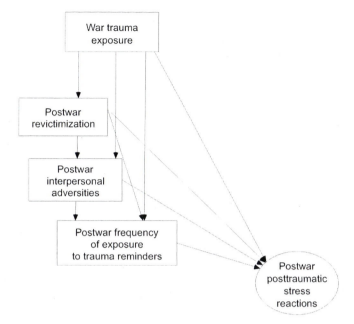

FIGURE 9.4. Overview of Multiply Mediated Model of Postwar Adjustment

term postwar posttraumatic stress reactions. Notably, predictive relationships between the variables extended over consecutive times, such that that the war trauma exposure variables measured at Time 1 had significant paths to the mediating mechanisms *postwar revictimization* and *interpersonal adversities* measured at Time 2.

Discussion of Studies and Preliminary Conclusions

We conclude this chapter with an interpretation of the two preliminary studies described above, followed by a discussion of the clinical and research implications of issues surrounding traumatic distress reactivation. The results of the first study provided preliminary empirical support for the hypothesis that trauma reminders and loss reminders are conceptually and empirically distinct constructs. As hypothesized, frequency of exposure to loss reminders was strongly related with grief reactions and moderately related to measures of posttraumatic stress and depression. Conversely, frequency of exposure to trauma reminders was moderately related to all distress measures and thus did not demonstrate the same specificity as loss reminders. These findings are generally consistent with the assertion that frequency of exposure to distressing reminders may serve as important mediators of the long-term relationship between trauma exposure and/or loss and a range of posttraumatic distress outcomes, including posttraumatic stress reactions, grief reactions, depression, and other anxiety reactions.

The results of the pilot longitudinal study generally supported the proposed role of reminders, adversities, and revictimization as mechanisms that mediate the relationship between trauma exposure and posttraumatic psychosocial adjustment. Of particular note, wartime trauma exposure variables measured retrospectively at Time 1 shared significant paths to the mediating mechanisms postwar adversities and postwar revictimization measured at Time 2. These findings point to the long and reverberating reach of large-scale traumata across time as they impregnate the posttraumatic ecology with distressing reminders, generate or exacerbate posttraumatic adversities, and increase the risk for revictimization. In turn, these factors may contribute to an increased likelihood of persisting distress reactivation, developmental disturbance, and impaired role functioning.

These preliminary empirical studies clearly contain a number of significant methodological limitations, including a reliance on self-report methodology, utilization of brief clinical screening measures, retrospective reports of war-related exposure, small sample size, and the use of a select population (war-exposed Bosnian youths) which, although understudied in the general trauma literature, may not generalize well to other populations. Notwithstanding these weaknesses, it is hoped that this theoretical framework and the methodology described herein will stimulate efforts to build theory, to improve assessment and other intervention technologies, and to empirically replicate and extend these findings in a variety of populations exposed to trauma and disaster.

Clinical Implications

Taken as a whole, these preliminary studies suggest that a range of intervention-related and theory-building tasks may benefit from the systematic inclusion of intervening variables (specifically vulnerability and protective mechanisms and their associated processes), which explain a significant proportion of the variance in short- and/or long-term posttraumatic adjustment. More generally, findings reported herein that posttraumatic reminders, adversities, and revictimization may function as primary mediators through which the effects of initial trauma exposure are transmitted over time hold important implications for developing more sophisticated and comprehensive theory, assessment, and intervention technologies (see McFarlane & Yehuda, 2000). For example, such knowledge, by identifying critical variables, mechanisms, and pathways of influence, can inform efforts to increase the accuracy of a spectrum of intervention-related tasks, including situation analysis, needs assessment, surveillance, program development, risk screening, treatment triage, case conceptualization, treatment planning, treatment monitoring, treatment/program evaluation, the development of effective public policy, and judicious resource allocation.

Further, identifying the specific pathways and processes through which intervening variables transmit their influence can greatly increase the range of therapeutic options open to interventionists. Possession of this knowledge base will provide interventionists with a firm empirical and theoretical foundation needed for accurate case conceptualization and for developing systematic, theoretically coherent, and evidence-based algorithms with which to form and prioritize treatment objectives. Informed interventionists can utilize these algorithms to intervene initially at the level of an antecedent causal variable, at the level of mediating or moderating variables, at the level of outcome variables, or at some combination thereof. For example, treatment might profitably be directed toward prolonged therapeutic exposure to memories of a focal traumatic experience (an etiological variable), toward reducing day-to-day exposure to nontherapeutic trauma reminders (a putative mediating variable), or toward enhancing support-seeking skills (a putative moderating variable) to cope with distressing reminders and adversities generated by the trauma. Alternatively, treatment may focus exclusively on symptom management (an outcome variable) via psychoeducation, abdominal breathing, distraction, self-soothing, and so forth.

The development of guiding theory that is sufficiently powerful to elucidate the linkages between "start point" etiological factors such as traumatic loss, and intervening variables such as persisting familial disruption secondary to the loss, may be particularly helpful in conducting risk assessment, case conceptualization, and treatment planning/monitoring of complex clinical cases. Under these conditions, intervening variables may gain an independent momentum that will not remediate even if the initiating trauma is directly therapeutically addressed. For example, interventionists may use a theory-based algorithm to determine whether, and in which order, to therapeutically treat a child suffering from distressing images relating to the traumatic death of a parent; complicated bereavement reactions in the surviving parent; and/or severe and persisting disturbances in family interaction patterns. Interventionists may also rely on theory to revise the initial treatment plan if clients prove refractory to the initial treatment protocol. If judiciously applied, the results of such theory development hold great potential for developing and refining evidence-based interventions by improving case conceptualization, accurately targeting therapeutic foci across multiple ecological levels, clarifying intervention objectives, and improving intervention techniques and strategies.

One example of targeting these mediating variables directly is found in an adolescent trauma/grief-focused group treatment protocol used in post-war Bosnia, the post–September 11, New York region, and in violence-ridden communities in Southern California (see Layne, Saltzman, et al., 2003; Saltzman, Layne, Steinberg, & Pynoos, Chapter 18 in this volume; see Cohen et al., 2001, for an application with younger children). This program addresses trauma and loss reminders in the context of psychoeducational activities, then focuses on developing coping skills with which to contend with distressing reminder-laden situations. These activities are in part dedicated to developing a coping plan to predict, where possible, and effectively cope with reminder-laden situations such as holidays and graduation dates. The program also contains skills-building and problem-solving exercises in which group members identify and cope with secondary adversities using such skills as support seeking, cognitive restructuring, problem solving, and giving appropriate support to others. The program concludes with an emphasis on resuming normal developmental progression, avoiding high-risk situations that increase the possibility of revictimization and developmental derailment, and building a coping plan with which to contend with developmentally linked adversities likely to emerge in the future.

SUMMARY AND DIRECTIONS FOR FUTURE RESEARCH

We have reviewed the varied meanings conferred on the term *retraumatization* by the existing traumatic stress literature and proposed a modification in definitions that we hope will increase terminological precision and utility. We paired the emphasis on conceptual clarity with a proposed conceptual framework that specifies a number of mechanisms theorized to lead to the reactivation of trauma-related distress. These mechanisms include posttraumatic adversities, exposure to distressing reminders (also varyingly termed *cues* or *triggers*), and trauma reexposure or revictimization. We have given special emphasis to the definition and distinction of trauma versus loss reminders and to their utility for both clinical and research efforts. We paired this conceptual explication with a set of two studies designed to test the empirical viability of trauma and loss reminders as distinct entities, and the roles of posttraumatic adversities, distressing reminders, and revictimization as mediators of the long-term link between trauma exposure and distress reactivation, the latter construct being assessed by a measure of posttraumatic stress symptomatology. Notably, both studies' results were generally consistent with hypotheses that trauma and loss reminders are empirically, as well as conceptually, distinct phenomena, and that distressing reminders, posttraumatic adversities, and revictimization act as significant mediators that link catastrophic trauma exposure to long-term posttraumatic adjustment.

Proposed Research Agenda

We hope that this conceptual framework, in combination with the empirical results presented herein, will contribute to the growing literature on determinants and correlates of long-term posttraumatic adjustment. In this section, we highlight five lines of inquiry that show promise for developing theoretically sound, evidence-based intervention strategies for individuals and communities exposed to disaster.

Developing Dynamic, Process-Oriented Models of Short-Term, Intermediate, and Long-Term Adaptation Following Traumatic Stress

One promising recent development within the field of traumatic stress is directed toward looking beyond the comparatively static approaches to the study of posttraumatic adaptation that have thus far been advanced. Such static approaches often involve the development of typologies of trauma exposure (e.g., acute versus chronic/repeated versus "crossover" traumatic events, see Terr, 1995) and an exclusive focus on the relative risks for various adverse outcomes conferred by specific (and often comparatively distal) etiological or marker variables. In contrast, more dynamic, process-based models are likely to yield much more information concerning *how* and *why* various types and degrees of risk, vulnerability, resistance, and resilience are imparted through the ecologies of trauma-exposed individuals over time. Such models are not only more likely to increase the predictive accuracy of risk identification and triage methods, but just as importantly will facilitate theory development and spotlight specific mechanisms and processes that trauma interventions can directly assess and therapeutically target.

Such a focus is consistent with recent calls to "unpack" traumatic and other stressful events and circumstances—and by extension the concatenated sequences that *interlink* events and circumstances over time and across ecological levels—into their constituent components (see Hobfoll, Dunahoo, & Monnier, 1995; Monnier & Hobfoll, 2000). The information gleaned thereby will likely assist in answering such questions as, What etiological agents are most important in eliciting posttraumatic distress reactions, and why? What types of events and processes do traumatic events set in motion that transmit the adverse effects of trauma over time? What adverse factors, though independent of traumatic events, interfere with positive posttraumatic adjustment and thus should be assessed and/or therapeutically targeted? What naturally occurring protective processes should be included in risk assessment or enhanced therapeutically to prevent long-term dysfunction and to facilitate recovery? What specific types of coping skills and supportive resources will help trauma survivors to recover most effectively over time? These variables must be systematically considered within the context of cultural and regional differences in recognition of the fact that sociocultural factors will influence both short- and long-term disaster responses (Adachi et al., 2002).

Investigating the Influence of Temporal Spacing
Between Stressors on Posttraumatic Recovery

An important area for future research relating to the consequences of serial and sequential trauma focuses on the influence of temporal spacing between repeated trauma and/or posttraumatic adversities. Some evidence suggests that traumata that are repeated in close temporal proximity (e.g., the second following only weeks or months after the initial trauma) may exert a more deleterious effect on adjustment than traumata that are repeated at longer intervals (e.g., the second following years later). For example, in a follow-up study of children exposed to a schoolyard shooting, Nader et al. (1990) found that new exposures or losses within the first year after the shooting were associated with higher distress and a longer recovery time compared to children who reported no new trauma exposure or losses. Further, in their population-based study of postearthquake Armenia, Goenjian et al. (1994) found evidence of an additive effect of exposure to earthquake subsequent to exposure to political violence. The authors suggest that the second trauma occurred before sufficient recovery could take place from the first.

These findings point to important research questions concerning the role played by the temporal spacing between traumata, the duration and development impact of serial trauma, and the interplay between serial and sequential trauma. Pynoos, Steinberg, & Piacentini (1999) propose that posttraumatic adjustment can be conceptualized along two major axes: distress and contextual discrimination. An implication of this model is that an excess accumulation of posttraumatic adversities will load onto the distress axis and thereby exacerbate and/or prolong the course of recovery. Moreover, subsequent dangers or threats, risk information, and/or trauma can complicate contextual discrimination, thereby making recovery more difficult, independent of the adverse effects of stressful life events and circumstances. Thus, exposure to distressing cues may both exacerbate distress relating to the initial traumatic event and complicate or challenge contextual discrimination, thereby affecting the course of recovery in a dual fashion. These findings point to the need for researchers and clinicians to differentiate between overwhelming traumatic stressors on the one hand, and other types of stressors that can be reframed as challenges and thereby used to promote positive posttraumatic adjustment (e.g., Lazarus & Folkman, 1991).

Examining Predictors and Determinants of Stress Resistance and Resilience in Posttraumatic Adaptation

Although the constructs of resistance and resilience were addressed only briefly in this chapter, the intersection of the traumatic stress and resilience literatures constitutes a particularly promising area for future research. The observed robustness and generalizability of several of the protective factors reviewed above across a wide variety of populations and outcomes suggests that these concepts may prove useful in both theory building and predicting the course of posttraumatic recovery. Regrettably, relatively little is yet known concerning how these factors function in relation to posttraumatic adjustment and recovery, given that most studies have focused almost exclusively on the adverse sequelae of trauma (Witmer & Culver, 2001). However, researchers are increasingly recognizing the importance of examining protective processes in conjunction with risk factors (Garbarino, 2001; Masten, 2001; McFarlane & Yehuda, 1996; Shakoor & Fister, 2000; Witmer & Culver, 2001), and some studies have made attempts to elucidate the relation between putative protective factors, exposure to trauma, and adaptive outcomes (e.g., Ferren, 1999; King, King, Fairbank, Keane, & Adams, 1998; Llabre & Hadi, 1997). As reviewed by Yehuda (1999), the majority of trauma-exposed individuals do not develop clinically significant long-term symptomatology and functional impairment, suggesting that chronic, severe maladjustment following trauma is the exception rather than the rule.

In addition to the need for increased attention to correlates of vulnerability versus resistance, and resilience versus maladaptation in trauma-exposed populations, there is a particular need for the more precise examination of protective models for trauma exposure and recovery. For example, additional studies are needed to explore risk, mediator, and moderator models of protective factors in trauma-exposed populations, as well as longitudinal studies that explicate the role of protective factors as they operate within the various phases of trauma exposure and its sequelae. These methods of operation include reducing the initial likelihood of exposure to trauma, promoting resistance to the adverse effects of exposure once it occurs, and buttressing resilience in the forms of expediting recovery and/or promoting adaptive role functioning in the presence of persisting distress. Such studies may also profitably examine the role of protective processes in promoting posttraumatic growth (e.g., Calhoun & Tedeschi, 2001).

Taken as a whole, the body of research on stress resistance, resilience, and protective processes suggests that individual factors by themselves typically do not account for large amounts of variance in broad adaptive outcomes. However, researchers have argued that targeting multiple protective processes simultaneously, especially those that span multiple systems—such as the individual, family, peer group, school, and community—may yield substantial therapeutic benefits (Masten, 1999). As such, it behooves professionals to incorporate the assessment, facilitation, and monitoring of relevant protective factors into therapeutic approaches rather than relying exclusively, or even primarily, on symptom-focused treatments. Although the empirical evaluation of resilience-based approaches to prevention and intervention as a field of research is currently in its infancy, the available literature suggests that clinicians are likely to observe improved outcomes when these factors are systematically incorporated into existing treatment approaches (Luthar & Cicchetti, 2000; Masten, 1999; Rolf & Johnson, 1999).

Although not limited to trauma-focused resistance and resilience research, the need persists for empirical studies and theoretical models that shed light on *how* protective factors work—that is, the specific mechanisms by which they promote stress resistance and resilience, and thereby reduce risk for adverse outcomes. For example, given the multidimensional nature of social support as a protective factor, are specific dimensions of supportive provisions, such as emotional support, instrumental aid, and companionship support, more important in promoting adaptive outcomes for certain types of traumatic experiences? The importance of understanding the workings of social support is underscored by findings that a deficit in support constitutes one of the leading indicators of risk for developing posttraumatic distress disorder in trauma-exposed adults (Brewin et al., 2000). More generally, these types of questions concerning protective processes and correlates must be answered in order to enhance the effectiveness of interventions designed to promote resilience, particularly those that rely on enhancing the effectiveness of naturally existing support networks (Gottlieb, 1996, 2000). Clearly, resilience research holds important implications for understanding and facilitating posttraumatic adjustment and recovery.

Likewise, studies conducted with populations exposed to significant trauma, because they involve the study of such "high magnitude" events exerting effects over a protracted period, constitute a fertile domain for the ongoing exploration of resistance and resilience and the method of operation and relative potency of their respective mechanisms.

Integrating Development into Trauma-Focused Theory, Assessment Technologies, and Intervention Protocols

In this chapter we have emphasized the need to adopt dynamic, process-oriented conceptual frameworks for explicating the mechanisms through which focal traumatic events interface with pre-, peri-, and posttraumatic contextual variables to influence posttraumatic adaptation. We nevertheless acknowledge that such models will be incomplete and inaccurate if they do not include development as an integral and pervasively influential component. As noted by a variety of developmentally savvy theoreticians and clinicians (e.g., Nader, 2001; Saywitz, Mannarino, Berliner, & Cohen, 2000; Shelby, 2000; Vernberg, 2002; Vernberg & Johnston, 2001), developmental factors are inextricably interwoven into the matrix of variables that influence risk for various forms of trauma exposure, appraisal and initial response processes, risk and protective resources, stress resistance, vulnerability, resilience, and response to various forms of intervention. Developmental factors also influence the timing, duration, developmental tasks, vulnerabilities, and competencies associated with developmentally sensitive periods. As advocated by Shelby (2000), the careful matching of intervention strategies to the developmental challenges, vulnerabilities, and competencies associated with these sensitive periods may be critical to ensuring positive therapeutic outcomes.

Examining Pathways Between Types of Trauma and Specific Posttraumatic Adversities

A last promising area for future applied clinical research centers on explicating the mechanisms and processes through which risk and vulnerability are transmitted over time and across levels of the ecology. More specifically, such assessment and intervention-related tasks as situation analysis, needs assessment, risk identification, treatment triage, and case conceptualization may benefit from models that address such questions as (1) Do specific types of traumatic events (e.g., violent expulsion from one's home during war; death of a parent) differentially increase the risk for specific types of secondary adversities (e.g., postwar status as a displaced person; chronic economic strain; family disruption)? (2) Do specific forms of postwar adversities differentially contribute to adverse outcomes, such as post-traumatic stress, depression, anxiety, somatization, or grief reactions? and (3) Which forms of traumatic events confer the highest degree of risk for specific posttraumatic adversities, and in turn for adverse psychosocial outcomes?

Such efforts to map out the concatenated pathways of influence by which specific types of wartime traumata and stressful life circumstances lead to specific forms of posttraumatic adversities, which in turn contribute to specific adverse outcomes, will facilitate two important intervention-related tasks. First, the sophisticated models that will be developed will greatly assist in risk assessment, case identification, treatment triage, and surveillance by assisting with the accurate identification of which subgroups are at highest risk for specified adverse outcomes. Second, these models will also assist with case conceptualization, treatment planning, and treatment monitoring/evaluation by identifying specific risk and protective mechanisms and processes that impart, convey, or buffer against the effects of risk. These processes, in turn, can then be assessed and therapeutically targeted to increase the effectiveness, efficiency, and sustainability of treatment gains.

These observations lead to six predictions relating to the construction of theoretical and empirical models of posttraumatic adjustment and their associated implications for risk assessment.

1. Due to their nature, the links between traumatic events and psychosocial adjustment outcomes are complexly determined due to a "ripple effect" of spreading influence through the surrounding ecology. This ripple effect is manifest via mediated and/or moderated pathways linking the traumatic event to current psychosocial adjustment. For example, witnessing a fatal school shooting will lead directly to acute distress reactions in its immediate aftermath. In the weeks and months that follow, the effects of the shooting will also be indirectly transmitted via distressing trauma reminders (e.g., the murder occurred at the school's outdoor basketball courts), loss reminders (e.g., passing the student's empty locker), secondary adversities (e.g., the adoption of stringent "zero tolerance" school policies designed to promote safety, intensive monitoring and behavioral control by concerned parents leading to increased teen/parent conflict), via social adversities (e.g., highly affected students are increasingly irritable and withdrawn, leading to increased peer rejection and social alienation), and similar mechanisms.

2. As a focal traumatic event recedes in time, the pathways of *direct effect* between the traumatic event and indices of current psychosocial adjustment will tend to diminish, both in their relative number and magnitudes of influence. Conversely, the number and magnitudes of influence of *indirect*

or mediated pathways linking traumatic events to posttraumatic adversities, and in turn to psychosocial adjustment as an outcome, will tend to increase. For example, preliminary longitudinal data indicate that direct paths linking war exposure variables with postwar psychosocial adjustment outcomes in Bosnian youth are almost entirely mediated via the vulnerability variables *postwar adversities, distressing reminders,* and *revictimization.* Notably, only extremely traumatogenic forms of war exposure, including personal injury and traumatic death, appear to share direct, nonmediated paths of influence with postwar psychosocial outcomes (Legerski et al., 2003).

3. As a focal traumatic event recedes in time, the total proportion of variance in indices of current psychosocial adjustment that is explained by pre- and peritraumatic variables will tend to diminish, whereas the total proportion of variance that is explained by posttraumatic contextual variables—especially risk and protective processes—will tend to increase.

4. In the acute aftermath of trauma, case identification of groups at highest risk will benefit primarily from a careful focus on assessing (1) the types and degree of exposure, (2) known vulnerability variables, such as previous history of trauma or preexisting psychiatric history, and (3) acute distress responses. In contrast, as the traumatic event recedes in time, accurate case identification will benefit from the inclusion of additional posttraumatic contextual variables, including proximal and distal posttraumatic adversities, frequency of exposure to proximal and distal distressing reminders, and revictimization experiences (see Pynoos, Steinberg, & Piacentini, 1999).

5. In general, as the criterion variables targeted by risk identification procedures become increasingly distal (e.g., shifting the prediction of psychosocial adjustment from one month to one year posttrauma), the assessment battery must include increasingly reliable, valid, sensitive, and specific tests of an increasingly broad array of posttraumatic contextual and outcome variables in order to both facilitate accurate case identification and to explain equivalent amounts of criterion variance. Of particular importance is the inclusion of putative mediating and moderating mechanisms, especially risk and protective processes.

6. As the magnitude and scope of a focal traumatic event increases, by definition an increasing number of surrounding levels in the ecology will be impacted, and a greater degree of impact will be manifest within each affected level due to a massive depletion of resources (Hobfol et al., 1995). As a consequence, the pathways of influence that mediate the links between peritraumatic variables and current psychosocial adjustment will assume increasingly complex forms. These forms may include multiple mediators, sequenced chains of multiple mediators, bidirectional causal relationships between variables, positive feedback loops within concatenated sequences

of variables, and interactive effects between and within levels of the ecology. As an example of one potential pathway, wartime violent expulsion from one's home may lead to persisting financial strain during and after the war, which is positively related to persisting parental psychological distress, which is inversely related to perceived parenting self-efficacy, which is positively related to parental role functioning, which is positively related to adolescent psychosocial adjustment. Further, adolescent psychosocial adjustment may be related, via a positive feedback loop, to both parental psychological distress and perceived parenting self-efficacy.

CONCLUSION

In an era characterized by increased calls for methodological consistency across studies to enhance the comparability of research findings, we propose that one of the most straightforward but often neglected tools for deriving clear meaning and, by extension, less ambiguous conclusions from existing and future studies is to standardize the conceptual terminology. In the absence of a commonly accepted lexicon of critical concepts and constructs, encounters with dissimilar terminologies across studies will inevitably generate such perplexing questions as, Are these constructs different and therefore not directly comparable, or are these the same constructs clothed in different guises? Ironically, such unintentional collective obfuscation will hamper efforts to build clear lines of systematic inquiry within the literature; unnecessarily compartmentalize the literature based on superficial terminological distinctions; needlessly confuse concepts that *should* be kept distinct due to similarities in their terminological references; blur distinctions between antecedent variables, intervening mechanisms, and outcomes; encumber the development of precise and parsimonious theory; and introduce unnecessary subjectivity into literature reviews and meta-analyses by imposing a heavy reliance on individual judgments regarding which constructs and associated measures are which.

Unfortunately, the varied usage of the term *retraumatization* in the current literature renders it difficult to pinpoint a definitive conceptual meaning for the term. In response, we have proposed a series of conceptual definitions that we hope will promote the clarity with which studies are able to identify predictors, determinants, correlates, and sequelae of this clinically and theoretically important phenomenon. A more sophisticated understanding of the precise mechanisms through which traumatization is initially produced, and by which distress reactivation subsequently occurs, is expected to promote theory development, increase the accuracy of assessment tools and practices, and enhance the effectiveness of therapeutic interventions undertaken in the aftermath of disasters.

APPENDIX: GLOSSARY OF PROPOSED TERMS RELATING TO POSTTRAUMATIC DISTRESS REACTIVATION

loss reminder: A cue, either physically internal or external to individuals with a history of significant loss, that evokes grief reactions by directing attention either to the ongoing absence of the lost object or to life changes consequential to the loss. These grief reactions may include emotional, cognitive, physiological, and/or behavioral responses and may have a negative, neutral, and/or positive subjective valence.

maintenance factor for posttraumatic distress: A general term describing any variable that significantly contributes to the perpetuation of posttraumatic distress reactions, developmental disturbance, and/or functional impairment in the aftermath of a traumatic event.

mediating variable: An intervening variable that transmits the effects of a focal causally prior independent variable to a causally subsequent dependent variable (see Baron & Kenny, 1986; Holmbeck, 1997).

moderator variable: A moderator variable partitions a focal independent variable into subgroups according to its range of maximal effectiveness with respect to a given dependent variable. Moderator variables are typically introduced when the relationship between an independent variable and a given dependent variable is unexpectedly weak, inconsistent, or otherwise appears to depend on (i.e., interact with) levels of a third variable. For example, prior history versus no prior history of trauma will moderate the effect of a given quasi-independent variable, trauma exposure versus no trauma exposure, on a given dependent variable, posttraumatic stress reactions, if a Trauma Exposure × Prior Trauma History interaction effect indicates that individuals with a prior trauma history evidence higher distress rates after being exposed to the current trauma than similarly exposed individuals with no prior trauma history. "True" protective and vulnerability factors function as moderator variables, given that their effects become manifest only in the presence of a significant stressor (see Baron & Kenny, 1986; Holmbeck, 1997; Rutter, 1987, 1991).

posttraumatic adversity: An adverse life event or circumstance existing in the aftermath of a focal traumatic event. These adversities may include secondary adversities, which are caused by the trauma itself, as well as other preexisting, cooccurring, or subsequently occurring stressful life events and circumstances. Posttraumatic adversities are theorized to impede post-traumatic recovery by contributing to posttraumatic distress, either directly and/or by exacerbating the impact of the trauma; by eliciting maladaptive coping, and/or by diverting resources needed for adaptive coping and recovery.

posttraumatic distress reactivation: An acute exacerbation in posttraumatic distress reactions in individuals with a history of exposure to one or more prior traumatic events. Posttraumatic distress may assume a variety of forms, including posttraumatic stress reactions, depressive reactions, anxiety, somatic distress, acute grief reactions, or some combination thereof. This reactivation is presumably evoked by a variety of mechanisms, including exposure to trauma reminders, loss reminders, posttraumatic adversities, serial trauma, or sequential trauma.

retraumatization: This ambiguous and often confusing term is not recommended for use in the professional or forensic literature.

revictimization: Exposure to a traumatic event following exposure to one or more antecedent traumatic events, which constitutes an abrogation of expectations implied by the social contract. The status of victim carries special sanctions to assume a social role that warrants restitution and special treatment, and to expect society to bring the perpetrator to judicial accountability.

secondary adversity: An adverse life event or circumstance caused by, and persisting in the aftermath of, one or more traumatic events. Examples include undergoing reconstructive surgery and/or intensive physical therapy following an automobile accident, or participating in stressful judicial proceedings stemming from childhood sexual abuse.

sequential trauma: Repeated exposure to different types of trauma, either within the same developmental period or across different developmental periods. For example, sequential trauma may involve exposure to domestic violence during early childhood, followed by sexual abuse during middle adolescence, followed by exposure to community violence during young adulthood.

serial trauma: Repeated exposure to the same type of trauma, either within the same developmental period or across successive developmental periods. Examples include repeated instances of sexual abuse during preadolescence or repeated instances of physical abuse from early childhood through school-age years.

trauma reminder: A cue, either physically internal or external to individuals with a history of exposure to one or more traumatic events, that resembles or symbolizes aspects of one or more past traumatic experiences and thereby evokes distressing reactivity. This reactivity may include emotional, cognitive, physiological, and/or behavioral responses and is subjectively experienced as distressing, intrusive, unbidden, and/or involuntary.

REFERENCES

Adachi, K., Bertman, S., Corr, C., Cory, J., Doka, K., Gilbert, K., Gjertsen, E., Glassock, G., Hall, C., Hansson, R., Jaramillo, I., Kallenberg, K., Lattanzi-Licht, M., Lickiss, N., Oechsle, P., Oltjenbruns, K., Papadatou, D., Parkes, C. M., Schuurman, D., & Worden, W. (2002). Assumptions and principles about psychosocial aspects of disasters. *Death Studies, 26,* 449-462.

American Psychiatric Association (2000). *Diagnostic and statistical manual of mental disorders* (4th ed., text rev.). Washington, DC: American Psychiatric Association.

Apfel, R. J., & Simon, B. (Eds.) (1996). *Minefields in their hearts: The mental health of children in war and community violence.* New Haven, CT: Yale University Press.

Applebaum, D. R., & Burns, G. L. (1991). Unexpected childhood death: Posttraumatic stress disorder in surviving siblings and parents. *Journal of Clinical Child Psychology, 20,* 114-120.

Baron, R. M., & Kenny, D. A. (1986). The moderator-mediator variable distinction in social psychological research: Conceptual, strategic, and statistical considerations. *Journal of Personality & Social Psychology, 51,* 1173-1182.

Baynard, V. L., Englund, D. W., & Rozelle, D. (2001). Parenting the traumatized child: Attending to the needs of nonoffending caregivers of traumatized children. *Psychotherapy, 38,* 74-87.

Baynard, V. L., Williams, L. M., & Siegel, J. A. (2001). The long-term mental health consequences of child sex abuse: An exploratory study of the impact of multiple traumas in a sample of women. *Journal of Traumatic Stress, 14,* 697-715.

Birleson, P., Hudson, I., Buchanan, D. G., & Wolff, S. (1987). Clinical evaluation of a self-rating scale for depressive disorder in childhood (Depression Self-Rating Scale). *Journal of Child Psychology and Psychiatry and Allied Disciplines, 28,* 43-60.

Brewin, C. R., Andrews, B., & Valentine, J. D. (2000). Meta-analysis of risk factors for posttraumatic stress disorder in trauma-exposed adults. *Journal of Consulting and Clinical Psychology, 68,* 748-766.

Calhoun, L. G., & Tedeschi, R. G. (2001). Posttraumatic growth: The positive lessons of loss. In R. A. Neimeyer (Ed.), *Meaning reconstruction & the experience of loss* (pp. 157-172). Washington, DC: American Psychological Association.

Chairamonte, J. A. (1992). And the war goes on. *Social Work, 37,* 469-470.

Christenson, R. M., Walker, J. I., Ross, D. R., & Maltbie, A. A. (1981). Reactivation of traumatic conflicts. *American Journal of Psychiatry, 138,* 984-985.

Chu, J. A. (1992). The revictimization of adult women with histories of childhood abuse. *Journal of Psychotherapy Practice and Research, 1,* 259-269.

Cicchetti, D., Rogosh, F. A., & Toth, S. L. (1997). Ontogenesis, depressotypic organization, and the depressive spectrum. In S. S. Luthar, J. A. Burack, D. Cicchetti, & J. R. Weisz (Eds.), *Developmental psychopathology: Perspectives on adjustment, risk, and disorder* (pp. 273-313). Cambridge, UK: Cambridge University Press.

Cloitre, M. (1998). Sexual revictimization risk factors and prevention. In V. M. Follete & J. I. Ruzek (Eds.), *Cognitive and behavioral therapies for trauma* (pp. 278-304). New York: Guilford.

Cloitre, M., Davis, L., Mirvis, S., & Levitt, J. (2002). STAIR-NST: A phase-based treatment for traumatized adolescents. Unpublished treatment manual, New York University Child Study Center, New York University School of Medicine.

Cloitre, M., Scarvalone, P., & Difede, J. (1997). Posttraumatic stress disorder, self and interpersonal dysfunction among sexually retraumatized women. *Journal of Traumatic Stress, 10,* 437-452.

Cloitre, M., Tardiff, K., Marzuk, P. M., Leon, A. C., & Portera, L. (1996). Childhood abuse and subsequent sexual assault among female inpatients. *Journal of Traumatic Stress, 9,* 473-482.

Cohen, J. A., Berliner, L., & March, J. S. (2000). Treatment of children and adolescents. In E. B. Foa, & T. M. Keane (Eds.), *Effective treatments for PTSD: Practice guidelines from the International Society for Traumatic Stress Studies* (pp. 106-138). New York: Guilford.

Cohen, J. A., Greenberg, T., Padlo, S., Shipley, C., Mannarino, A. P., & Deblinger, E. (2001). Cognitive behavioral therapy for childhood traumatic grief. Unpublished Treatment Manual, Drexel University College of Medicine, Pittsburgh, Pennsylvania.

Cohen, J. A., Mannarino, A. P., & Greenberg, T. (2002). Childhood traumatic grief: Concepts and controversies. *Trauma Violence & Abuse, 3,* 307-327.

Compas, B. E., Hinden, B. R., & Gerhardt, C. A. (1995). Adolescent development: Pathways and processes of risk and resilience. *Annual Review of Psychology, 46,* 265-293.

Cowen, E. L., Wyman, P. A., & Work, W. C. (1996). Resilience in highly stressed urban children: Concepts and findings. *Bulletin of the New York Academy of Medicine, 73,* 267-284.

de-Jong, J. T. V. M., Komproe, I. H., Van-Ommeren, M., El-Masri, M., Araya, M., Khaled, N., van-de-Put, W., & Somasundaram, D. (2001). Lifetime events and posttraumatic stress disorder in 4 postconflict settings. *Journal of the American Medical Association, 286,* 555-562.

Dietrich, A. M. (2000). As the pendulum swings: The etiology of PTSD. *Traumatology, 6*(1), Article 3. Available online at <http://www.fsu.edu/~trauma/v6i1a4.html>.

Djapo, N., Katalinski, R., Pasalic, H., Layne, C. M., Arslanagic, B., Saltzman, W. S., & Pynoos, R. S. (2000, July). Long-term post-war adjustment in war-exposed Bosnian adolescents, parents, and teachers: In search of risk and protective mechanisms. In R. Stuvland & M. Black (Chairs), UNICEF Psychosocial Projects in Bosnia & Hercegovina 1993-1999. (Paper symposium conducted at the conference Psychosocial Consequences of War: Results of Empirical Research from the Territory of Former Yugoslavia.) University of Sarajevo, Bosnia.

Doob, D. (1992). Female sexual abuse survivors as patients: Avoiding retraumatization. *Archives of Psychiatric Nursing, 6,* 245-251.

Engelhard, I. M., Van Den Hout, M. A., & Arntz, A. (2001). Posttraumatic stress disorder after pregnancy loss. *General Hospital Psychiatry, 23,* 62-66.

Epstein, J. N., Saunders, B. E., & Kilpatrick D. G. (1997). Predicting PTSD in women with a history of childhood rape. *Journal of Traumatic Stress, 10,* 573-587.

Felsen, I. (1998). Transgenerational transmission of effects of the Holocaust: The North American research perspective. In Y. Danieli (Ed.), *International handbook of multigenerational legacies of trauma* (pp. 43-68). New York: Plenum.

Ferren, P. M. (1999). Comparing perceived self-efficacy among adolescent Bosnian and Croatian refugees with and without posttraumatic stress disorder. *Journal of Traumatic Stress, 12,* 405-420.

Figley, C. R., Bride, B. E., & Mazza, N. (Eds.) (1997). *Death and trauma: The traumatology of grieving.* Washington, DC: Taylor and Francis.

Fletcher, K. E. (2003). Childhood posttraumatic stress disorder. In E. J. Mash & R. A. Barkley (Eds.), *Child psychopathology* (2nd ed.) (pp. 330-371). New York: Guilford.

Foy, D. W., Madvig, B. T., Pynoos, R. S., & Camilleri, A. J. (1996). Etiologic factors in the development of posttraumatic stress disorder in children and adolescents. *Journal of School Psychology, 34,* 133-146.

Foy, D. W., Osato, S. S., Houskamp, B. M., & Neumann, D. A. (1992). Etiology of posttraumatic stress disorder. In P. A. Saigh, (Ed.), *Post-traumatic stress disorder: A behavioral approach to assessment and treatment* (pp. 28-49). New York: Pergamon Press.

Freeman, L. N., Shaffer, D., & Smith, H. (1996). Neglected victims of homicide: The needs of young siblings of murder victims. *American Journal of Orthopsychiatry, 66,* 337-345.

Garbarino, J. (2001). An ecological perspective on the effects of violence on children. *Journal of Community Psychology, 29,* 361-378.

Garmezy, N. (1985). Stress-resistant children: The search for protective factors. In J. E. Stevenson (Ed.), *Recent research in developmental psychopathology, Journal of Child Psychology and Psychiatry (Book Supplement No. 4)* (pp. 213-233). Oxford: Pergamon Press.

Glodich, A. (1998). Trauma exposure to violence: A comprehensive review of the child and adolescent literature. *Smith College Studies in Social Work, 68,* 321-345.

Goenjian, A., Najarian, L. M., Pynoos, R. S., Steinberg, A. M., Petrosian, P., Sterakyan, S., & Fairbanks, L. A. (1994). Posttraumatic stress reactions after single and double trauma. *Acta Psychiatrica Scandanavia, 90,* 214-221.

Gottlieb, B. H. (1996). Theories and practices of mobilizing support in stressful circumstances. In C. L. Cooper (Ed.), *Handbook of stress, medicine, and health* (pp. 339-356). Boca Raton, FL: CRC Press.

Gottlieb, B. H. (2000). Selecting and planning support interventions. In S. Cohen and L. G. Underwood (Eds.), *Social support measurement and intervention: A guide for health and social scientists* (pp. 195-220). London: Oxford University Press.

Grossman, F. K., Beinashowitz, J., Anderson, L., Sakurai, M., Finnin, L., & Flaherty, M. (1992). Risk and resilience in young adolescents. *Journal of Youth and Adolescence, 21,* 529-550.

Harvey, M. R. (1996). An ecological view of psychological trauma and trauma recovery. *Journal of Traumatic Stress, 9,* 3-23.

Hetherington, E. M., & Elmore, A. M. (2003). Risk and resilience in children coping with their parents' divorce and remarriage. In S. S. Luthar (Ed.), *Resilience and vulnerability: Adaptation in the context of childhood adversities* (pp. 182-212). Cambridge, UK: Cambridge University Press.

Hobfoll, S. E., Dunahoo, C. A., & Monnier, J. (1995). Conservation of resources and traumatic stress. In J. R. Freedy & S. E. Hobfoll (Eds.), *Traumatic stress: From theory to practice* (pp. 29-47). New York: Plenum Press

Holmbeck, G. N. (1997). Toward terminological, conceptual, and statistical clarity in the study of mediators and moderators: Examples from the child-clinical and pediatric psychology literatures. *Journal of Consulting and Clinical Psychology, 65,* 599-610.

Jacobs, S. C. (1999). *Traumatic grief: Diagnosis, treatment, and prevention.* Philadelphia: Brunner/Mazel.

Janoff-Bulman, R. J. (1992). *Shattered assumptions: Toward a new psychology of trauma.* New York: Macmillan.

Jones, J. C., & Barlow, D. H. (1990). The etiology of posttraumatic stress disorder. *Clinical Psychology Review, 10,* 299-328.

King, L. A., King, D. W., Fairbank, J. A., Keane, T. M., & Adams, G. A. (1998). Resilience-recovery factors in post-traumatic stress disorder among female and male Vietnam veterans: Hardiness, postwar social support, and additional stressful life events. *Journal of Personality and Social Psychology, 74,* 420-434.

Koren, D., Arnon, I., & Klein, E. (2001). Long term course of chronic posttraumatic stress disorder in traffic accident victims: A three-year prospective follow-up study. *Behaviour Research and Therapy, 39,* 1449-1458.

Koss, M. P. (2000). Blame, shame, and community: Justice responses to violence against women. *American Psychologist, 55,* 1332-1343.

Kutlac, M., Layne, C. M., Wood, J., Saltzman, W. S., Stuvland, R., & Pynoos, R. S. (2002). Psychological adjustment in war-exposed secondary school students two years after the war: Results of a large-scale risk screening survey. In S. Powell & E. Durakoviæ-Belko (Eds.), *Sarajevo 2000: The psychosocial consequences of war. Results of empirical research from the territory of former Yugoslavia.* Sarajevo: UNICEF. Available online at <http://www.psih.org>.

Lauterbach, D., & Vrana, S. (2001). The relationship among personality variables, exposure to traumatic events, and severity of posttraumatic stress symptoms. *Journal of Traumatic Stress, 14*, 29-45.

Layne, C. M., Djapo, N., & Pynoos, R. S. (1998). Post-war Adversities Scale. Unpublished psychological test, University of California, Los Angeles.

Layne, C. M., Legerski, J. P., Pasalic, A., Pasalic, H., Katalinski, R., Saltzman, W. R., & Pynoos, R. S. (2003, November). Parent-adolescent relationship variables in the aftermath of war. In M. Armsworth (Chair), "Fostering Intergenerational Resilience from War and Genocide." Symposium presented at the International Society for Traumatic Stress Studies, Chicago, Illinois.

Layne, C. M., Neibauer, N., Manwaring, A., Money, K., Arslanagic, B., Saltzman, W. R., & Pynoos, R. S. (2003, November). Treating complicated bereavement in adolescents. In S. Ley (Chair), "Treating Childhood Traumatic Grief: A Developmental Perspective." Pre-convention institute presented at the International Society for Traumatic Stress Studies, Chicago, Illinois.

Layne, C. M., Pynoos, R. S., & Cardenas, J. (2001). Wounded adolescence: School-based group psychotherapy for adolescents who have sustained or witnessed violent interpersonal injury. In M. Shafii & S. Shafii (Eds.), *School Violence: Contributing factors, management, and prevention* (pp. 163-186). Washington, DC: American Psychiatric Press.

Layne, C. M., Pynoos, R. S., Saltzman, W. R., Arslanagic, B., Black, M., Savjak, N., Popovic, T., Durakovic, E., Music, M., Campara, N., Djapo, N., & Houston, R. (2001). Trauma/grief-focused group psychotherapy: School-based post-war intervention with traumatized Bosnian adolescents. *Group Dynamics: Theory, Research, and Practice, 5*, 277-290.

Layne, C. M., Pynoos, R. S., Savjak, N., & Steinberg, A. M. (1998). Grief Screening Scale. Unpublished psychological test, University of California, Los Angeles.

Layne, C. M., Saltzman, W. R., Steinberg, A. M., & Pynoos, R. S. (2003). Trauma/grief-focused group psychotherapy manual for adolescents. Unpublished treatment manual, University of California, Los Angeles.

Layne, C. M., Savjak, N., Steinberg, A. M., & Pynoos, R. S. (1999). Loss Reminders Screening Scale. Unpublished psychological test, University of California, Los Angeles.

Layne, C. M., Steinberg, A. M., Saltzman, W. R., Wood, J., & Pynoos, R. S. (1998). War Trauma Reminders Screening Scale. Unpublished psychological test, University of California, Los Angeles.

Layne, C. M., Stuvland, R., Saltzman, W. R., Steinberg, A. M., & Pynoos, R. S. (1999). *War Trauma Exposure Scale.* Sarajevo, Bosnia: UNICEF Bosnia and Hercegovina.

Layne, C. M., Warren, J., Shalev, A., & Watson, P. (in press). Risk, vulnerability, resistance and resilience: Towards an integrative conceptualization of posttraumatic adaptation. In M. J. Friedman, T. M. Kean, & P. A. Resnick (eds.), *PTSD: Science & practice—A comprehensive handbook* (pp. _____). New York: Guilford.

Lazarus, R. S., & Folkman, S. (1991). The concept of coping. In A. Monat & R. S. Lazarus (Eds.), *Stress and coping: An anthology* (3rd ed.) (pp. 189-206). New York: Columbia University Press.

Legerski, J. P., Layne, C. M., Isakson, B., Saltzman, W. S., Djapo, N., & Kutlac, M. (2003, October). Post-war psychosocial adjustment in war-exposed Bosnian youths: A SEM analysis. Poster presentation at the 19th Annual Meeting of the International Society for Traumatic Stress Studies, Chicago, Illinois.

Llabre, M. M., & Hadi, F. (1997). Social support and psychological distress in Kuwaiti boys and girls exposed to Gulf Crisis. *Journal of Clinical Child Psychology, 26,* 247-255.

Luthar, S. S. (1991). Vulnerability and resilience: A study of high-risk adolescents. *Child Development, 62,* 600-616.

Luthar, S. S., & Cicchetti, D. (2000). The construct of resilience: Implications for interventions and social policies. *Development and Psychopathology, 12,* 857-885.

Luthar, S. S., Cicchetti, D., & Becker, B. (2000). The construct of resilience: A critical evaluation and guidelines for future work. *Child Development, 71,* 543-562.

Luthar, S. S., & Zelazo, L. B. (2003). Research on resilience: An integrative review. In S. S. Luthar (Ed.), *Resilience and vulnerability: Adaptation in the context of childhood adversities* (pp. 510-549). Cambridge, UK: Cambridge University Press.

Luthar, S. S., & Zigler, E. (1991). Vulnerability and competence: A review of research on resilience in childhood. *American Journal of Orthopsychiatry, 61,* 6-22.

Marans, S., Berkman, M., & Cohen, D. (1996). Child development and adaptation to catastrophic circumstances. In R. J. Apfel & B. Simon (Eds.), *Minefields in their hearts: The mental health of children in war and communal violence* (pp. 104-127). New Haven, CT: Yale University Press.

Masten, A. (1999). Commentary: The promise and perils of resilience research as a guide to preventive interventions. In M. D. Glantz & J. L. Johnson (Eds.), *Resilience and development: Positive life adaptations* (pp. 251-257). New York: Plenum Press.

Masten, A. S. (2001). Ordinary magic: Resilience processes in development. *American Psychologist, 56,* 227-238.

Masten, A. S., Best, K. M., & Garmezy, N. (1990). Resilience and development: Contributions from the study of children who overcome adversity. *Development and Psychopathology, 2,* 425-444.

Masten, A. S., & Coatsworth, J. D. (1998). The development of competence in favorable and unfavorable environments: Lessons from research on successful children. *American Psychologist, 53,* 205-220.

Masten, A. S., & Garmezy, N. (1985). Risk, vulnerability, and protective factors in developmental psychopathology. In B. Lahey & A. Kazdin (Eds.), *Advances in clinical child psychology* (Vol. 8, pp. 1-52). New York: Plenum Press.

Masten, A. S., Garmezy, N., Tellegen, A., Pellegrini, D. S., Larkin, K., & Larsen, A. (1988). Competence and stress in school children: The moderating effects of individual and family qualities. *Journal of Child Psychology and Psychiatry, 29,* 745-764.

McFarlane, A. C., & de Girolamo, G. (1996). The nature of traumatic stressors and the epidemiology of posttraumatic reactions. In B. A. van der Kolk, A. C. McFarlane, and L. Weisaeth (Eds.) *Traumatic stress: The effect of overwhelming experience on mind, body, and society* (pp. 129-154). New York: The Guilford Press.

McFarlane, A. C., & Yehuda, R. (1996). Resilience, vulnerability, and the course of posttraumatic reactions. In B. A. van der Kolk, A. C. McFarlane, & L. Weisaeth (Eds.), *Traumatic stress: The effects of overwhelming experience of mind, body, and society* (pp. 155-181). New York: Guilford.

McFarlane, A. C., & Yehuda, R. (2000). Clinical treatment of posttraumatic stress disorder: Conceptual challenges raised by recent research. *Australian and New Zealand Journal of Psychiatry, 34*, 940-953.

Monnier, J., & Hobfoll, S. E. (2000). Conservation of resources in individual and community reactions to traumatic stress. In A. Y. Shalev & R. Yehuda (Eds.), *International handbook of human response to trauma* (pp. 325-336). Dordrecht, the Netherlands: Kluwer Academic Publishers.

Nader, K. O. (1997). Childhood traumatic loss: The interaction of trauma and grief. In C. R. Figley and B. E. Bride (Eds.), *Death and trauma: The traumatology of grieving* (pp. 17-41). Washington, DC: Taylor and Francis.

Nader, K. (2001). Treatment methods for childhood trauma. In J. P. Wilson & M. J. Friedman (Eds.), *Treating psychological trauma and PTSD* (pp. 278-334). New York: Guilford Press.

Nader, K., Pynoos, R., Fairbanks, L., & Frederick, C. (1990). Children's PTSD reactions one year after a sniper at their school. *American Journal of Psychiatry, 147*, 1526-1530.

Nelson, B. S., & Schwerdtfeger, K. L. (2002). Trauma to one family member affects entire family. *Traumatic Stress Points, 26*(3), 7.

O'Connor, T. G., & Rutter, M. (1996). Risk mechanisms in development: Some conceptual and methodological considerations. *Developmental Psychology, 32*, 787-795.

O'Grady, D., & Metz, J. R. (1987). Resilience in children at high risk for psychological disorder. *Journal of Pediatric Psychology, 12*, 3-23.

Orcutt, H. K., Erickson, D. J., & Wolfe, J. (2002). A prospective analysis of trauma exposure: The mediating role of PTSD symptomatology. *Journal of Traumatic Stress, 15*, 259-266.

Parker, G. R., Cowen, E. L., Work, W. C., & Wyman, P. A. (1990). Test correlates of stress resilience among urban school children. *Journal of Primary Prevention, 11*, 19-35.

Parkes, C. M. (1998). *Bereavement: Studies of grief in adult life* (3rd ed.). Madison, CT: International Universities Press.

Pearlman, L. A., & Saakvitne, K. W. (1995). *Trauma and the therapist: Countertransferance and vicarious traumatization in psychotherapy with incest survivors*. New York: W.W. Norton.

Pynoos, R. S. (1992). Grief and trauma in children & adolescents. *Bereavement Care, 11*, 2-10.

Pynoos, R. S., Goenjian, A., & Steinberg, A. M. (1995). Strategies of disaster intervention for children and adolescents. In S. E. Hobfoll and M. W. deVries (Eds.), *Extreme stress and communities: Impact and intervention* (pp. 445-471). NATO ASI series. New York: Kluwer Academic/Plenum Publishers.

Pynoos, R. S., & Nader, K. (1988). Psychological first aid and treatment approach to children exposed to community violence: Research implications. *Journal of Traumatic Stress, 1,* 445-473.

Pynoos, R. S., Rodriguez, N., Steinberg, A. M., Stuber, M., & Fredericks, C. (1999). Reaction Index—Revised. Unpublished psychological test, University of California, Los Angeles.

Pynoos, R. S., Sorenson, S. B., & Steinberg, A. M. (1993). Interpersonal violence and traumatic stress reactions. In L. Goldberger & S. Breznitz (Eds.), *Handbook of stress: Theoretical and clinical aspects* (2nd ed., pp. 573-590). New York: Free Press.

Pynoos, R. S., Steinberg, A. M., & Piacentini, J. C. (1999). A developmental psychopathology model of childhood traumatic stress and intersection with anxiety disorders. *Biological Psychiatry, 46,* 1542-1554.

Pynoos, R. S., Steinberg, A., & Wraith, R. (1995). A developmental model of childhood traumatic stress. In D. Cichetti & D. J. Cohen (Eds.), *Manual of developmental psychopathology* (Vol. 2, pp. 72-95). New York: Wiley.

Rando, T. A. (1993). *Treatment of complicated mourning.* Champaign, IL: Research Press.

Robinson, S., Hemmendinger, J., Rapaport, M., Zilberman, L., & Gal, A. (1994). Retraumatization of Holocaust survivors during the Gulf War and SCUD missile attacks on Israel. *British Journal of Medical Psychology, 67,* 353-362.

Rolf, J. E., & Johnson, J. L. (1999). Opening doors to resilience intervention for prevention research. In M. D. Glantz & J. L. Johnson (Eds.), *Resilience and development: Positive life adaptations* (pp. 229-249). New York: Plenum Press.

Rosenberg, S. D., Mueser, K. D., Freidman, M. J., Gorman, P. G., Drake, R. E., Vivader, R. M., Torrey, W. C., & Jankowski, M. K. (2001). Developing effective treatments for posttraumatic stress disorders among people with severe mental illness. *Psychiatric Services, 52,* 1453-1461.

Rutter, M. (1987). Psychosocial resilience and protective mechanisms. *American Journal of Orthopsychiatry, 53,* 316-331.

Rutter, M. (1991). Protective factors: Independent or interactive? *Journal of the American Academy of Child and Adolescent Psychiatry, 30,* 151-152.

Rynearson, E. K. (2001). *Retelling violent death.* New York: Brunner-Routledge.

Saigh, P. A., & Bremner, J. D. (Eds.) (1999). *Posttraumatic stress disorder: A comprehensive text.* Boston: Allyn and Bacon.

Saltzman, W. R., Layne, C. M., Steinberg, A. M., Arslanagic, B., & Pynoos, R. S. (2003). Developing a culturally and ecologically sound intervention program for youth exposed to war and terrorism. *Child and Adolescent Psychiatric Clinics of North America, 12,* 319-342.

Saywitz, K., Mannarino, A. P., Berliner, L., & Cohen, J. A. (2000). Treatment for sexually abused children and adolescents. *American Psychologist, 55,* 1040-1049.

Schamess, G., Streider, F. H., & Connors, K. M. (1997). Supervision and staff training for children's group psychotherapy: General principles and applications with cumulatively traumatized, inner-city children. *International Journal of Group Psychotherapy, 47*, 399-425.

Schnurr, P. P., Lunney, C. A., & Sengupta, A. (2004). Risk factors for the development versus maintenance of posttraumatic stress disorder. *Journal of Traumatic Stress, 17*, 85-95.

Shakoor, M., & Fister, D. L. (2000). Finding hope in Bosnia: Fostering resilience through group process intervention. *Journal for Specialists in Group Work, 25*, 269-287.

Shelby, J. S. (2000). Brief therapy with traumatized children: A developmental perspective. In H. Kaduson & C. Schaefer (Eds.), *Short-term play interventions* (pp. 69-104). New York: Guilford.

Sidran Institute (2002). Sidran Foundation trauma disorders glossary. Available online at <http://www.sidran.org/glossary.html#retraumatizing>.

Silove, D., McIntosh, P., & Becker, R. (1993). Risk of retraumatisation of asylum-seekers in Australia. *Australian and New Zealand Journal of Psychiatry, 27*, 606-612.

Simon, R. I. (1999). Chronic posttraumatic stress disorder: A review and checklist of factors influencing prognosis. *Harvard Review of Psychiatry, 6*, 304-310.

Smith, S. B. (1995). Restraints: Retraumatization for rape victims? *Journal of Psychosocial Nursing, 33*, 23-28.

Solomon, Z. (1990). Does the war end when the shooting stops? The psychological toll of war. *Journal of Applied Social Psychology, 20*, 1733-1745.

Steinberg, A. (1998). Understanding the secondary traumatic stress of children. In C. R. Figley (Ed.), *Burnout in families: The systemic costs of caring* (pp. 29-46). Boca Raton, FL: CRC Press.

Terr, L. C. (1995). Childhood traumas: An outline and overview. In G. S. Everly Jr. & J. M. Lating (Eds.), *Psychotraumatology: Key papers and core concepts in post-traumatic stress* (pp. 301-320). New York: Plenum Press.

Toren, P. T., Wolmer, L., Weizman, M., Magal-Vardi, O., & Laor, N. L. (2002). Retraumatization of Israeli civilians during a reactivation of the Gulf War threat. *Journal of Nervous and Mental Disease, 190*, 43-45.

Vernberg, E. M. (2002). Intervention approaches following disasters. In A. M. La Greca, W. K. Silverman, E. M. Vernberg, & M. C. Roberts (Eds.), *Helping children cope with disasters and terrorism* (pp. 55-72). Washington, DC: American Psychological Association.

Vernberg, E. M., & Johnston, C. (2001). Developmental considerations in the use of cognitive therapy for PTSD. *Journal of Cognitive Psychotherapy, 15*, 223-237.

Weist, M. D., Freedman, A. H., Paskewitz, D. A., Proescher, E. J., & Flaherty, L. T. (1995). Urban youth under stress: Empirical identification of protective factors. *Journal of Youth and Adolescence, 24*, 705-721.

Werner, E. E. (1989). High-risk children in young adulthood: A longitudinal study from birth to 32 years. *American Journal of Orthopsychiatry, 59*, 72-81.

Werner, E. E., & Smith, R. S. (1982). *Vulnerable but invincible: A study of resilient children.* New York: McGraw-Hill.

Wilson, J. P. (1994). The need for an integrative theory of post-traumatic stress disorder. In M. B. Williams & J. F. Sommer Jr. (Eds.), *Handbook of post-traumatic therapy.* Westport, CT: Greenwood Press.

Witmer, T. A. P., & Culver, S. M. (2001). Trauma and resilience among Bosnian refugee families: A critical review of the literature. *Journal of Social Work Research, 2,* 173-187.

Worden, J. W. (1996). *Children and grief: When a parent dies.* New York: Guilford.

Wyatt, G. E., Gunthrie, D., & Notgrass, C. M. (1992). Differential effects of women's child sexual abuse and subsequent sexual revictimization. *Journal of Consulting and Clinical Psychology, 60,* 167-173.

Yehuda, R. (Ed.) (1999). *Risk factors for posttraumatic stress disorder.* Washington, DC: American Psychiatric Association.

Yehuda, R. (2002). Clinical relevance of biologic findings in PTSD. *Psychiatric Quarterly, 73,* 123-133.

Yehuda, R., McFarlane, A. C., & Shalev, A. Y. (1998). Predicting the development of posttraumatic stress disorder from the acute response to a traumatic event. *Biological Psychiatry, 44,* 1305-1313.

Zucker, R. A., Wong, M. M., Puttler, L. I., & Fitzgerald, H. E. (2003). Resilience and vulnerability among sons of alcoholics: Relationship to developmental outcomes between early childhood and adolescence. In S. S. Luthar (Ed.), *Resilience and vulnerability: Adaptation in the context of childhood adversities* (pp. 76-103). Cambridge, UK: Cambridge University Press.

Chapter 10

The Syndrome of Traumatic Grief and Its Treatment

M. Katherine Shear
Allan Zuckoff
Nadine Melhem
Bonnie J. Gorscak

INTRODUCTION

Loss of a loved one is a universal experience. An estimated 16 to 20 million people, including spouses, parents, children, siblings and close friends, are bereaved in the United States each year (Rando, 1993, p. 5). Although death is a normal part of life, bereavement is widely understood to be a major life stressor, perhaps the most severe. Grief is the reaction to bereavement that may be thought of as the mind's natural healing process, activated by a significant loss.

When the grief process is working well, it occurs in two phases. Initially there is a sense of disbelief, recurrent pangs of anguish, social withdrawal, and preoccupation with thoughts and images of the person who died. During this acute phase, the bereaved person must find a way to accept the death. This entails a shift in the relationship to the deceased, as well as psychological and behavioral accommodations to life without the deceased. Over time, occasionally as quickly as hours, usually over weeks or months, disbelief is superseded by acceptance of the finality of the death. The intensity and frequency of painful emotions, including sadness, anxiety, guilt, anger, and shame, subside; they are gradually replaced by positive, comforting feelings when thoughts turn to the deceased person. Interest is regained in daily activities and in other relationships, and practical problems that arise from the death are able to be solved. Thoughts of the deceased, although still present, no longer dominate the bereaved person's mind.

This chapter is supported by NIMH grants MH 60783 MH 30915 and MH 52247.

Nonetheless, because of our close psychological ties to the people we love, bereavement and grief are enduring states. In most cases, it is the attenuated form of grief that is permanent. Though its manifestations differ from one person to another, grief commonly resurfaces at the anniversary of a death, during holiday periods, and at other important times such as birthdays and wedding anniversaries. Grief may also intensify at the time of another loss or during periods of stress. When an increase in painful emotions, preoccupation, and social withdrawal occurs though, it is short-lived, and the bereaved person is often able to draw some comfort or other benefit from new thoughts of the deceased. This is illustrated well by Robert Neimeyer when he says of his father's suicide,

> Many of the subsequent emotional, relational and occupational choices made by my mother, my brother, my little sister, and me can be read as responses to my father's fateful decision, although their meaning continues to be clarified, ambiguated, and reformulated across the years. (Neimeyer, 2001, p. xi)

The period of time that is required to make the adjustments associated with grief is not uniform, but when grief is working, it waxes and wanes in intensity during the first months or years after a death, while gradually attenuating overall. Although there is not yet full agreement about the precise boundaries of normal grief, it is clear that the process does not always progress smoothly. Bereavement may trigger symptoms of a DSM-IV depressive or anxiety disorder, most commonly major depressive disorder (MDD), post-traumatic stress disorder (PTSD), or generalized anxiety disorder (GAD) (Bonnano & Kaltman, 2001). However, some bereaved individuals experience clinically significant symptomatology that is distinct from these disorders. Such a condition has been described in the literature and may be reliably identified in clinical settings. Recently, two sets of similar diagnostic criteria have been proposed, naming the disorder as either "complicated" or "traumatic" grief.

This chapter outlines strategies and procedures for assessment and treatment of the pathological grief syndrome that we shall refer to, for reasons made clear below, as traumatic grief (TG). We provide definitions of this condition and of other terminology that has thus far remained ambiguous, for the sake of clarity, but without making any special claim for the validity of these specific terms. We review the history of psychological approaches to the understanding of bereavement and its pathologies, as well as what more recent empirical investigations have discovered about the specific nature and effects of TG. We describe, as plainly as possible, the goals, strategies, and techniques we have used in individual treatment of patients with

TG, as well as our ideas for adapting our approach to a group treatment setting. And we conclude with brief consideration of the empirical status of our treatment approach, and of the further work that remains to be done.

Definitions

First, as in Stroebe, Hansson, Stroebe, & Schut (2001, p. 5), we define *bereavement* as "the objective condition of having lost someone significant," and *mourning* as "the actions and manner of expressing grief, which often reflect the mourning practices of one's culture." The definition of *grief* is somewhat more problematic. As Bonanno and Kaltman (2001) point out, "The bereavement field has yet to agree on a clear, empirically defensible definition of grief, or its normal and abnormal course and manifestations." Denoting a state of sorrow, heartache, anguish, pain, or unhappiness, grief is often thought of, as Stroebe et al. (2001) suggest, as an emotional reaction. However, Bonanno (2001) makes a strong case for differentiating grief from an emotional response, contending instead that grief is a "complex molar experience" encompassing, but not fully described by, "a range of specific emotions." Bonanno's definition emphasizes the multiplicity of emotions during grief and guides us to identify these and other components. We favor this approach and use it in our work with patients who have traumatic grief.

HISTORICAL PERSPECTIVES

In *Mourning and Melancholia,* Freud (1917) provided the first modern theoretical conceptualization of grief and its pathology. Freud understood *mourning* as a necessarily painful and intense process of disinvestment from the person who died, necessary to allow the bereaved to return to psychological health. Although disruptive to normal functioning, mourning was not considered a condition requiring intervention, due to its typically transient nature. However, Freud observed that in some cases the death of a loved one induced a state of pathological mourning which he called *melancholia,* in which the loss was attributed not to the world but to the ego itself, resulting in self-blame and other feelings and cognitions deleterious to self-esteem. The primary source of melancholia, Freud believed, was a strong but ambivalently held fixation to the lost object.

Freud's views have considerably influenced the bereavement literature over the past century (Bonnano & Kaltman, 1999). Several of his ideas, however, are not supported by current research. First, as we will discuss below, pathological grief and depression are best understood as separate, if often overlapping, disorders. Second, many individuals do not seem to need the experience of an intense grieving process before returning to normal functioning. Third, there is evidence to suggest that ambivalence is related to *less* intense grieving, and that stronger "mutual affiliation" between the griever and the deceased is predictive of more intense grief (Piper et al., 2001).

In 1944, Lindemann provided an observational report of acute grief following accidental death. He described a syndrome of grief that included psychological and somatic symptomatology, and defined six characteristics of an "unresolved grief reaction": somatic distress, preoccupation with the deceased, guilt, hostile reactions, changes in behavioral patterns, and identification with the deceased. Lindemann believed that onset of this syndrome in bereaved individuals could be predicted by delayed grief, a history of depression, or a relationship with the deceased characterized by hostile interaction.

Like Freud's theoretical speculations, Lindemann's ideas have been influential but, until recently, untested. Data now confirm that a syndrome with the features of *unresolved* (now referred to as *complicated* or *traumatic*) grief can be reliably identified, and that a past history of major depression does seem to exist with a high frequency among these patients. However, the syndrome is not related to delayed grief or to a hostile or ambivalent relationship to the deceased.

Engel (1961) argued that grief, though "normal" in being the common response to loss, is "abnormal" in the sense of total health, and thus should be considered a disease. Engel described what he termed *uncomplicated* grief as having a consistent, predictable course: an initial phase of shock and disbelief; a period of awareness of the loss accompanied by sadness, guilt, shame, hopelessness, crying, emptiness, sleep disturbance, somatic complaints, loss of interest in usual activities, and functional impairment; and recovery, wherein the loss is overcome and a state of health and well-being is reestablished. However, in more recent research grief has not been found to progress through predictable stages, and there is no period in which "recovery" could be said to be complete, since loss of an important relationship has reverberating effects throughout the bereaved person's life.

Bowlby (1961, 1980) viewed grief as an expression of attachment to the deceased person—regardless of how painful, unrealistic, and dysfunctional it may at times be—and linked loss of a loved one to clinical states of depression and anxiety. Bowlby described a separation response occurring in three phases: *protest,* focused on trying to recover the lost object, and manifested by anxiety, weeping, anger, and disequilibrium; *despair* when these efforts failed, manifested by disorganization and depression; and *adaptation,* in the form of reorganization and direction toward new attachment objects. Pathological mourning and depressive illness were seen as arising from failure to tolerate disorganization and/or remaining oriented toward the lost object.

Parkes's (1965, 1970, 1987) conceptualization of grief was consistent with Bowlby's; both defined the psychological distress following bereavement in terms of separation anxiety that included pangs of grief and searching behavior. Parkes published some of the earliest and most influential research describing grief and its course. However, as noted above, clearly defined phases of grief have not been confirmed, and separation distress does not necessarily take the form of anxiety. Instead, there may be anger or bitterness, a refusal or inability to accept the death, and accompanying strong feelings of yearning and longing.

Clayton and colleagues (Clayton, Desmarais, & Winokur, 1968; Clayton, Herjanic, Murphy, & Woodruff, 1974) were among the first to conduct empirical studies related to Freud's model of grief as a depressive reaction. Several studies indicated that psychological symptoms indistinguishable from depressive illness frequently occurred following bereavement. These symptoms were accepted by the bereaved and those in their environment as "normal," however, in contrast to symptoms of primary affective disorder, which were experienced as a change that made persons "not their usual self" (Clayton, 1974). Unlike clinical depression, the condition they referred to as the "normal depression of widowhood" was not found by Clayton and colleagues to be associated with a personal or family history of depression (Clayton, Halikas, & Maurice, 1972), or with psychomotor retardation, worthlessness, or suicidal thoughts (Clayton, Halikas, & Maurice, 1971).

DIAGNOSIS OF BEREAVEMENT-RELATED SYNDROMES

Past and Current Diagnostic Approaches

Influenced by the depressive model of bereavement, the DSM-III (American Psychiatric Association, 1980) and DSM-III-R (APA, 1987) considered a major depressive episode to be a normal reaction to the death of a loved one, instructing clinicians to consider such symptoms uncomplicated bereavement. However, later researchers' reports of similarity between bereavement-related depression and other forms of depression led to recommendations for revisions in DSM-III-R criteria and terminology.

Zisook and Shuchter (1991, 1993) reported feelings of worthlessness and suicidal ideation among a sample of bereaved spouses at two months, one year, and two years after the loss. However, these symptoms were found in a minority of grieving individuals and were associated with continued states of depression at least two years following the death; personal history of depression predicted depression following bereavement. Based on these findings, Zisook and Shuchter (1993) recommended use of the term *uncomplicated bereavement* only for bereaved who never meet the criteria of a DSM-III-R depressive episode. They argued that, as with other event-triggered episodes of major depression, the diagnosis should be made only if criteria are met for two or more weeks.

Likewise, Brent, Peters, & Weller (1994) argued that depressive reactions observed among adolescents exposed to a peer's suicide can be truly termed "major depression," as they are indistinguishable from depression occurring in any other context in terms of duration, recurrence, and associated functional impairment. Consistent with the findings of Zisook and Schuchter, Brent found that previous psychiatric history and family history of affective disorders were risk factors for depression in this population. Similar findings were reported by Karam (1994) in a community-based study in Beirut; in this sample of 658 subjects, eighteen to sixty-five years old, the length of bereavement-related depressive episodes, the risk of recurrence, associated dysfunction, and the frequency of seeking treatment were similar to those of other forms of depression.

Influenced by these data, the DSM-IV (APA, 1994) directs clinicians to diagnose major depressive disorder if criteria are met more than two months following the loss, and/or if there is marked functional impairment, morbid preoccupation with worthlessness, suicidal ideation, psychotic symptoms, or psychomotor retardation. A separate category of complicated or pathological grief was rejected based on a judgment of insufficient empirical evidence for its existence and lack of a standardized diagnostic criteria set (Horowitz et al., 1997).

The focus on depression as the pathology of bereavement, however, has been associated with inattention to other clinically significant symptoms that may appear after the death of a loved one (Bruce, Kim, Leaf, & Jacobs, 1990; Kim & Jacobs; 1991; Zisook & Schuchter, 1991). Emotional numbness, disbelief, longing, yearning for the lost person, sighing, crying, dreams, illusions, hallucinations, searching for the deceased, and disorganization of behavior are manifestations of grief that are unique and not accounted for in existing DSM diagnostic categories (Jacobs, 1999). Contemporary grief researchers have argued that recognition of unique grief-related symptoms is necessary so that bereaved individuals in need of treatment are not denied access to the health care system (Stroebe, van Son, Stroebe, Kleber Schut, & van den Bout, 2000), and that standardized diagnostic criteria are needed to help clinicians accurately identify and treat bereaved persons with complications (Prigerson, Shear, et al., 1999).

Proposed Diagnostic Criteria for Grief Disorders

Substantial data support the existence of a pathological grief syndrome that can be reliably identified, is distinct from major depression, and does not respond to standard treatment for depression. Despite agreement on relevant symptoms, researchers have not agreed upon a name for this condition, and criteria vary slightly.

Complicated Grief Disorder

Horowitz (1974) described a stress model of bereavement incorporating traumatic distress as a main component of the response. In a more recent study, his group has proposed three sets of symptoms that characterize prolonged grief reactions among bereaved individuals seeking therapy: intrusion, avoidance, and failure to adapt (Horowitz et al., 1997). The diagnostic criteria for what they refer to as complicated grief disorder are presented in Exhibit 10.1.

Traumatic Grief

A similar criteria set was proposed by a panel convened in Pittsburgh in 1997. This group, with broad expertise in bereavement and trauma and in the formulation of psychiatric diagnostic criteria, developed the consensus criteria for traumatic grief (Prigerson, Shear, et al., 1999) shown in Exhibit 10.2. These criteria include core components of separation distress and traumatic distress. Separation distress is characterized by symptoms of yearning and searching for the deceased, inability to accept the death, and severe emptiness and loneliness (Chen et al., 1999). The traumatic distress component consists of symptoms such as intrusive thoughts about the deceased, emotional numbness, sustained disbelief, and shock about the death.

The proposed consensus criteria were empirically tested on a sample of widows and widowers, mean age sixty-one years, who were interviewed seven weeks and seven months after the death of their spouse. Endorsing at least three of the four items in criterion A2 (Exhibit 10.2) as at least "sometimes true" yielded a sensitivity and specificity of .93 and .81. For criterion B, endorsing four of the eight items as "mostly true" yielded a sensitivity and specificity of .89 and .81. Combining both criteria A2 and B, as would occur in clinical practice, resulted in the highest sensitivity (.93) and specificity (.93) (Prigerson, Shear, et al., 1999).

EXHIBIT 10.1. Proposed Diagnostic Criteria for Complicated Grief Disorder

A. Event Criterion/Prolonged Response Criterion

Bereavement (loss of a spouse, other relative, or intimate partner) at least fourteen months ago (twelve months is avoided because of possible intense turbulence from an anniversary reaction)

B. Signs and Symptoms Criteria

In the last month, any three of the following seven symptoms with a severity that interferes with daily functioning

Intrusive Symptoms

1. Unbidden memories or intrusive fantasies related to the lost relationship
2. Strong spells or pangs of severe emotion related to the lost relationship
3. Distressingly strong yearnings or wishes that the deceased were there

Signs of Avoidance and Failure to Adapt

4. Feelings of being far too much alone or personally empty
5. Excessively staying away from people, places, or activities that remind the subject of the deceased
6. Unusual levels of sleep interference
7. Loss of interest in work, social, caretaking, or recreational activities to a maladaptive degree

Source: From Horowitz et al. (1997). Diagnostic criteria for complicated grief disorder. *American Journal of Psychiatry, 154*: 904-10. Reprinted with permission from the *American Journal of Psychiatry,* Copyright 1997. American Psychiatric Association.

Comparison of the Proposed Criteria

All of the signs and symptoms included in the complicated grief criteria proposed by Horowitz et al. (1997) are also included in the traumatic grief criteria, with the exception of the symptom of unusual levels of sleep interference. Omission of the sleep criterion was based on evidence from a study that found that hyperaroused sleep did not occur in subjects with syndromal levels of traumatic grief (McDermott et al., 1997). On the other hand, some

EXHIBIT 10.2. Consensus Criteria for Traumatic Grief

Criterion A

1. Person has experienced the death of a significant other
2. Response involves three of the four symptoms below experienced at least sometimes:
    (a) Intrusive thoughts about the deceased
    (b) Yearning for the deceased
    (c) Searching for the deceased
    (d) Loneliness as result of the death

Criterion B

In response to the death, the four of the eight following symptoms experienced as mostly true:

1. Purposelessness or feelings of futility about the future
2. Subjective sense of numbness, detachment, or absence of emotional responsiveness
3. Difficulty acknowledging the death (e.g., disbelief)
4. Feeling that life is empty or meaningless
5. Feeling that part of oneself has died
6. Shattered worldview (e.g., lost sense of security, trust, control)
7. Assumes symptoms or harmful behaviors of, or related to, the deceased person
8. Excessive irritability, bitterness, or anger related to the death

Criterion C

Duration of disturbance (symptoms listed) is at least two months

Criterion D

The disturbance causes clinically significant impairment in social, occupational, or other important areas of functioning

---

Source: © 1999 The Royal College of Psychiatrists. Reproduced with permission from Prigerson, Shear, et al. (1999). Consensus criteria for Traumatic Grief: A preliminary empirical test. *British Journal of Psychiatry,* 174: 67-73.

symptoms included in the traumatic grief criteria are not found in Horowitz's criteria for complicated grief. These include feelings of futility about the future, detachment and absence of emotional responsiveness, difficulty acknowledging the death, feeling that part of oneself has died, shattered worldview, and assuming symptoms or harmful behaviors related to the deceased.

The two definitions differ in the duration criterion as well. According to Horowitz et al. (1997), the diagnosis is not to be made before fourteen months postloss. The consensus panel preferred to follow the DSM-IV two-month criterion required to diagnose major depression following bereavement, even though previous studies have shown that a six-month assessment of complicated grief is superior to two- or three-month assessments in predicting long-term dysfunction and depressed mood at eighteen months (Prigerson, Bridge et al., 1999; Prigerson, Frank, et al., 1995; Prigerson, Shear, et al., 1996).* Some researchers have expressed concern that the two-month criterion may overlap normal grief (Fox, Reid, Salmon, McKillop-Duffy, & Doyle, 1999); however, the panel felt that it would be inhumane for bereaved individuals to have to suffer for six months before a diagnosis can be made, when they could benefit from earlier intervention. Thus, the duration criterion is still in need of validation.

Validation of the Syndrome of Traumatic Grief

Prigerson and colleagues have conducted extensive empirical research to establish the validity of an independent grief disorder. They initially adopted Horowitz's term *complicated grief,* and the self-report measure they developed to assess for the presence of this syndrome, the Inventory of Complicated Grief (ICG) (Prigerson, Frank, et al., 1995), also reflects this nomenclature. Later, as the similarity between some of the symptoms of the syndrome—i.e., disbelief, anger, shock, intrusive thoughts, numbness, detachment, sense of futility about the future, and irritability—and those of PTSD was recognized, they renamed it *traumatic grief.* This term is used in the consensus criteria and in the clinical work that we describe below.

The symptoms of TG were found to cluster together among a subgroup of bereaved persons, distinct from clusters of depression and anxiety symptoms (Prigerson et al., 1997; Prigerson, Frank, et al., 1995; Prigerson, Shear, et al., 1999); presence of this grief symptom cluster in turn predicted global functioning, depressed mood, sleep quality, and self-esteem eighteen months after the loss of a spouse. Also of interest, the existence of these

---

*For this reason, we have chosen in our studies of traumatic grief treatment to maintain the six-month postloss duration criterion.

symptoms six months following a loss was more significantly associated with long-term dysfunction than was their existence at three months. Grief symptoms also proved to be distinct from depression and anxiety in clinical and nonclinical samples of the bereaved (Prigerson, Bierhals, et al., 1996; Prigerson, Frank, et al., 1995; Prigerson, Shear, et al., 1996) and in men and women separately (Chen et al., 1999); moreover, a recent study (Boelen, van den Bout, & de Keisjer, 2003) confirmed the presence of a grief symptom cluster distinct from depression and anxiety in a group of bereaved midlife Dutch psychiatric outpatients, supporting its generalizability across cultures.

Further evidence of the distinctiveness of TG from other disorders is provided by studies examining the effects of antidepressant medication on TG symptoms, sleep patterns among TG patients, and mental and physical health outcomes associated with TG. In a study of eighty-two bereaved subjects, sixty years or older and with no previous psychiatric history, subjects depressed at the time of entry into the study were treated, with nortriptyline. Depression and anxiety declined among those treated, but no changes were observed in grief symptoms (Prigerson, Frank, et al., 1995; Prigerson, Shear, et al., 1996). A sixteen-week open trial of paroxetine, given to patients receiving an early form of the psychotherapy for TG that will be described later in this chapter, found that subjects scoring at least 20 on the ICG reported greater beneficial effects for depressive than for grief symptoms (Zygmont et al., 1998). Similar findings were observed in a double-blind placebo controlled trial of nortriptyline for the treatment of bereavement-related depressions (Pasternak et al., 1991), and in another report of open treatment with bupropion for early bereavement-related depression (Zisook, Shuchter, Pedrelli, Sable, & Deaciuc, 2001).

An electroencephalographic (EEG) sleep study (McDermott et al., 1997) found that subjects with TG only were similar to healthy nonbereaved controls in their EEG sleep continuity measures, and these subjects were also able to begin sleep and easily descend into deeper levels of sleep. On the contrary, increasing levels of depression were associated with worsening in sleep efficiency and sleep maintenance. Thus, TG does not entail the same changes in EEG sleep physiology as those observed in depression.

In a community-based sample of 150 subjects between the ages of forty and eighty years, interviewed before and after the death of a hospitalized spouse, TG at six months after spousal death predicted suicidal ideation after twenty-five months when controlling for depression and anxiety (Prigerson et al., 1997); Szanto, Prigerson, Houck, Ehrenpreis, & Reynolds (1997) reported similar findings among bereaved elderly. In a sample of young adults exposed to a peer's suicide, syndromal levels of TG were associated with a four times higher risk of suicidal ideation, after controlling for depression (Prigerson, Bridge, et al., 1999). TG is also associated with physical health outcomes such as heart trouble, cancer, headaches, high blood pressure, stroke, hospitalization, and changes in sleep and eating habits (Prigerson et al., 1997).

As noted, many symptoms of TG are similar to those of PTSD. However, TG and PTSD also differ in important ways. Traumatic grief occurs after a loss and entails an emotional response dominated by sadness; in contrast, PTSD occurs in response to a life-threatening event, and the primary emotional response is fear. Also, separation distress plays a prominent role in TG, manifested by yearning, longing, and searching for the person who died, which is not seen with PTSD. In addition, in TG there is often preoccupation with memories of the deceased and a tendency to engage in periods of pleasurable reveries and pleasant dreams of the person who died that are often a source of comfort as well as distress (Jacobs, 1993; Raphael, Martinic, & Wooding, 1997). These symptoms do not occur in PTSD. Finally, hypervigilance, common in PTSD, is less frequent in TG. When present, it often involves searching for signs of the deceased rather than monitoring potential threats (Jacobs, 1999).

In her doctoral dissertation, Melhem (2001) was the first to examine TG among adolescents exposed to a peer's suicide and provide empirical evidence of the distinctiveness of TG from PTSD as well as depression among adolescents. A cluster of grief symptoms similar to that described by Prigerson, Frank, et al. (1995) was identified, including numbness, crying, yearning, poor functioning, and preoccupation with thoughts of the deceased. Traumatic grief at six and twelve to eighteen months following a suicide predicted depression twelve to eighteen and thirty-six months later, respectively. Depressed and nondepressed adolescents were equally likely to develop TG, and the likelihood of developing TG was independent of the incidence of PTSD.

EPIDEMIOLOGY

Prevalence

Several studies have examined the prevalence of major depressive disorder (MDD) following bereavement, but few have focused on the prevalence and course of TG or on grief symptoms. Zisook and Devaul (1984) reported the prevalence of "unresolved" grief as measured by three items: (1) "I feel I have grieved for the person who died," (2) "Now I can talk about the person without discomfort," and (3) "I feel I have adjusted well to the loss." Fourteen percent of the 211 individuals bereaved of a first-degree relative, spouse, or close friend experienced unresolved grief, and grief symptoms did not change over time.

Studies of the time course of grief have not used consistent criteria and have not included the concept that grief is a permanent state, with an acute and chronic phase. From the perspective of the work reported in this chapter, most of the literature referring to *grief resolution* refers to subsiding of the acute phase reaction of disbelief, painful emotions, preoccupation with the deceased, and social withdrawal. Thus, according to Clayton (1990), by four months after the death, 85 percent of bereaved individuals experience a reduction in symptoms *without any intervention*. Nevertheless, in support of the occurrence of a chronic phase response, Clayton also notes that sleep may remain disturbed for a year and mood may be affected intermittently, especially on holidays, anniversaries, and special occasions. The time course of normal "grief resolution" was found by Parkes (1972) to be, on the average, six months to one year, while Jacobs and Ostfeld (1980) reported a duration of normal grief of four to six months. Zisook and colleagues (Zisook, Devaul, & Click, 1982; Zisook & Devaul, 1984; Zisook & Schuchter, 1985) make the important point that complete resolution is the exception rather than the rule. Nevertheless, the presence of TG after six months puts the bereaved at increased risk for psychological, physical, and social impairment (Prigerson, Frank, et al., 1995) thirteen and twenty-five months after the death, with symptoms present at six months and tending to remain high at twenty-five months and beyond (Prigerson et al., 1997).

Because of the absence of consistent definitions, the reported prevalence of pathologic grief varies widely across studies. In a sample of twenty-five bereaved spouses, average age fifty-one years, 64 percent were found to meet criteria for pathologic grief (Kim & Jacobs, 1991). Among this sample, 48 percent met criteria for one or more of subtypes: 12 percent had delayed separation distress, defined as onset two weeks or more after the loss; 69 percent had severe separation distress; and 94 percent had prolonged distress, defined as no change or worsening six months after the loss. Jacobs (1993) reported that approximately 20 percent of acutely bereaved individuals developed clinical complications, which included not only pathologic grief but also depression and anxiety disorders. Middleton, Burnett, Raphael, & Martinek (1996) found 9.2 percent of their sample of 120 bereaved spouses, children, and parents to have chronic grief, as defined by Jacobs (1993), over the twelve-month period following bereavement; the intensity of grief among parents who lost children was more likely to be higher than that of widows or widowers, who in turn had more intense grief reactions than adult children losing a parent. In a sample of seventy bereaved subjects interviewed six and fourteen months after the death of a spouse, 41 percent were found to meet the criteria for complicated grief proposed by Horowitz et al. (1997). Prigerson, Frank, et al. (1995) reported 26 percent of elderly bereaved spouses, sixty years or older, had complicated grief. McDermott et al. (1997) found that forty-one of sixty-five subjects (63.1 percent), who had lost their spouse within the past twelve months and had originally been recruited to study EEG patterns following bereavement, had high levels of TG. The prevalence of TG among adolescents exposed to a peer's suicide was 25 percent six months following the suicide (Melhem, 2001). Thus, problematic grief reactions have been found to range from 14 to 64 percent of bereaved persons. Different sample selection criteria used in these studies and the lack of standardized criteria seem to account for the wide range of estimates.

Comorbidity

Although TG is distinct from depression and PTSD, a high rate of comorbidity of TG with these DSM-IV Axis I disorders has been reported. Among twenty-three participants in our pilot study of traumatic grief treatment (TGT), 52.2 percent of patients had a current diagnosis of MDD and 30 percent had a current diagnosis of PTSD (Melhem et al., 2001). Among adolescents exposed to a peer's suicide, 44.8 percent, 40 percent, and 33.3 percent of those who had TG were also diagnosed with MDD at six, twelve to eighteen, and thirty-six months, respectively, following the suicide; 41.7 percent, 50 percent, and 22.2 percent also had PTSD at six, twelve to eighteen, and thirty-six months, respectively (Melhem et al., 2001).

Risk Factors

Gender

Prigerson, Maciejewski, et al. (1995) found female gender to predict traumatic grief, and Chen et al. (1999) found widows to have higher levels of TG at all time points following bereavement. Jacobs, Kasl, Ostfeld, Berkman, & Charpentier (1986) found widows to score higher than widowers on numbness and disbelief. However, no differences were found in separation anxiety and pangs of grief. Bierhals, Prigerson, Fasiczka, Frank, & Miller (1996) did not find any difference in TG between widows and widowers.

Age

Zisook and Devaul (1984) found that subjects with unresolved grief were younger than those without unresolved grief; Prigerson, Maciejewski, et al.

(1995) found the same result. However, other studies did not find age to be a significant predictor (Jacobs et al., 1986; Kim & Jacobs, 1991).

Previous Psychiatric History

Horowitz et al. (1997) assessed bereaved subjects for previous history of depression and anxiety using the Structured Clinical Interview for DSM-IV-TR–Non-Patient Edition (SCID-NP). Previous history of depression was found to be associated with complicated grief: 79 percent of subjects with complicated grief had a previous history of depression, compared to 41 percent of those without complicated grief. However, previous history of anxiety disorders was not associated with complicated grief. By contrast, Kim and Jacobs (1991) did not find past personal or family history of depression or anxiety disorders to be associated with pathological grief. However, subjects with pathological grief were significantly more likely to be currently depressed than those without pathological grief (94 percent versus 33 percent, P < .01). No difference in the rate of current anxiety disorder was found between the two groups. Zisook and Devaul (1984) also found subjects with unresolved grief to be more depressed. Rates of traumatic grief appear to be higher in treatment-seeking samples than in the general population; Piper et al. (2001) estimated that 33 percent of those seeking outpatient psychiatric treatment in local community settings met criteria for moderate to severe pathological grief.

Melhem et al. (2001) found TG patients to have high rates of previous psychiatric history. Fifty-seven percent of patients in our TGT pilot study also had a previous psychiatric history; the remaining 43 percent had the onset of a DSM-IV diagnosis after the loss. Of the patients with a previous psychiatric history, 83 percent were ill at the time of the loss. There was no group of bereaved individuals without TG with which to compare rates of previous psychiatric history in this study; nevertheless, this finding has led us to hypothesize that prior psychiatric illness is a likely risk factor for TG, and that this may be especially true when the death of a loved one occurs during an episode of illness. Similarly, onset of a psychiatric disorder may be a consequence of untreated TG. These hypotheses need to be tested in future studies.

Type of Death

Deaths that occur suddenly and violently (an accident, homicide, or suicide) have been thought to increase the risk of TG (Jacobs, 1993). However, chronic terminal illness has also been associated with a worse grief outcome (Jacobs et al., 1986; Prigerson et al., 1997).

Bereavement due to suicide is widely believed to be an especially distressing form of grief, although Reed and Greenwald (1991) found survivors of adults' suicides to experience less emotional distress, often welcoming the suicide as a relief from years of anguish caused by the victim's mental illness. Among adolescents exposed to a peer's suicide, Melhem (2001) found TG to be associated with previous history of depression, anxiety disorders, and family history of anxiety disorders. Previous history of depression was a significant predictor of TG; however, TG tended to occur among peers of certain suicide victims or specific "networks" of suicide victims. Once the network effect was controlled for, previous history of depression was no longer statistically significant. Looking more closely at the characteristics of networks with high rates of TG, these networks were more likely to have PTSD than networks with lower rates of TG. They were also more likely to have a physical or psychiatric illness themselves or in their relatives, although the difference was only marginally significant. These results indicate a confounding effect between previous psychiatric history and the network effect. The network effect in TG could be attributed to the fact that vulnerable youths may be more likely to be friends with other vulnerable youths. Personal predisposition or vulnerability may place individuals in situations and social environments or networks that are more likely to increase their risk of exposure to traumatic events and to increase their risk of subsequent psychological consequences.

Characteristics of the Relationship

Psychoanalytic authors have suggested that ambivalent relationships with the deceased are likely to be associated with poor grief outcomes (Anderson, 1949; Freud, 1917), but recent research findings challenge this association. Van Doorn, Kasl, Beery, Jacobs, & Prigerson (1998) found loss of supportive or "security-increasing" marriages, characterized by positive interactions and a close, confiding relationship, as well as the bereaved being possessed of an insecure attachment style, to be associated with higher levels of TG. The study by Piper et al. (2001) noted earlier, which reported ambivalence to be related to less intense grief and stronger affiliation to be related to more intense grief, echoes this finding, as do our clinical observations: individuals presenting for treatment of TG regularly describe the loss of a very close, identity-defining relationship.

CLINICAL FEATURES OF TRAUMATIC GRIEF

TG is a debilitating disorder that can wreak havoc upon the life of the bereaved. Inability to work, socialize, and parent effectively are common and can be exacerbated by demoralizing effects of a loss of purpose and direction, uncertainty as to one's current and future identity, and anxiety in the face of one's own mortality.

Individuals who meet criteria for TG consistently report that the death seemed to them to be sudden and unexpected. This occurs even when the person was clearly terminally ill. Following the death, there is typically an intense emotional response and/or feelings of numbness. Patients frequently report that they "went onto autopilot" and took care of the practicalities while feeling almost nothing. The core symptoms of TG described previously—the persistent, anguished sense of longing and disbelief, as well as the recurrent intrusive images and/or reveries of the longed-for past with the loved one—follow soon after. In addition, these individuals demonstrate persistently diminished interest in daily activities and other people. Feelings of emptiness and loneliness are prevalent, as efforts to maintain a sense of proximity alternate with avoidance of reminders of the deceased. Often there are guilty ruminations about alleged failings in relation to the lost loved one. For example, persons with TG may feel they have been insufficiently loving while the person was alive, or that they made inadequate efforts to prevent the death or ameliorate the suffering of the deceased. They may fear relinquishing their preoccupation and/or resist accepting the death because they feel that this would be disloyal, or because they are afraid that if they were to grieve less or accept the death they would lose the person forever.

Case Vignette: Traumatic Grief
Following a Violent Death

To illustrate the clinical presentation of TG and its treatment, we present a composite case example. Tom W. presented for treatment five years after the death of his nineteen-year-old son, Donald, in a bus accident. Forty-eight years old and married to his high school sweetheart for twenty-eight years, Tom was also the father of an eleven-year-old son and two daughters, seventeen and twenty-seven.

The relationship between Tom and his deceased son Donald had been especially warm and close. Tom made it a point to be actively involved in all of his children's lives and did his best to show no favoritism. Still, as his first son, Donald held a special place in his heart. Donald was a popular and outgoing person. He was a good student and the editor of his high school year-book. Shortly before the accident, he had begun his first year of college, intending to pursue a career in business. Tom himself was quiet and intro-verted, and he had taken pride in seeing his son excel not only in his studies, but socially as well. Nevertheless, when, in the beginning of his freshman year, Donald had been focusing on his social life to the detriment of his stud-ies, this introduced some mild tension between father and son.

Donald was killed when a bus in which he was riding was hit head-on by a truck driver who had fallen asleep at the wheel. He had been on his way home from college for a visit. Tom had fallen asleep when a call came in from the state police, informing the family of what had happened. Tom described a sensation of searing pain, in which "everything that mattered disappeared," followed by a feeling of empty numbness that had been with him since. He recalled the days that followed as a kind of a blur, and the events continued to feel unreal to him. However, images from that period—identifying his son's body in the morgue, seeing him in his casket at the funeral home—came un-bidden into his mind at least several times each day, accompanied by in-tense anguish and sorrow.

Soon after the accident Tom quit his job of twenty years and took a part-time position with minimal responsibility. Withdrawing from friends and rela-tives, he attended no social or family gatherings, and rarely left the house except when it seemed unavoidable. Tom felt bitter about his son's death and found himself envious of parents who had not lost children. He was furi-ous with God for allowing Donald's death to happen, and had largely avoided church since the funeral, though he had previously been an active member. Compounding his distress, he was ashamed of these distasteful feelings but could not control them.

Although Tom continued to spend time with his other children, they did not talk freely to him as they had before Donald's death. Tom's wife, Sally, complained that he was distant and uninvolved, and she worried about his health. He had hypertension and diabetes that he had neglected since the death. Their marriage had been a lively and satisfying one, but during the past five years their sex life had dissipated, and days could go by without their saying anything to each other that did not involve practical necessities. Sally, too, was devastated by the death of her son, but unlike Tom, she had come to a sense of acceptance. She devoted much of her time to raising awareness of the dangers of sleep deprivation in professional truckers and trying to get legislation passed limiting the number of hours that truckers can drive consecutively or in a given week.

Tom spent most of his free time alone in his workshop, though he could not say what he did there, as he no longer tinkered and built things as he had in the past. He had placed photos of his son on display, and he would day-dream for hours, with longing and sadness, about the events and occasions when the pictures had been taken. He also pursued the cul-de-sac of "what ifs": he often found himself ruminating over Donald 's decision to visit home, the timing of which Tom had felt was inopportune. He chastised himself for not arguing more forcefully against it. He frequently imaged about Donald 's last moments: had he known what was about to happen or was he taken by surprise? Had he suffered in the accident or died instantly? Some of the answers were contained in the police report, but Tom had never been able to bring himself to read it, or to discuss it with his wife, who had. Tom slept with a copy of the yearbook Donald edited under his pillow, but had not opened or read it since his son's death.

TRAUMATIC GRIEF TREATMENT

TGT is a structured, time-limited intervention in which the primary goal is to relieve the symptoms of TG and facilitate a more natural progression of grief. The treatment is based on the idea that TG occurs when natural grief is stuck, or not working. The goal is to remove impediments so that grief can work. TGT targets the two major components of grief: adjusting to the loss and reengaging in life without the deceased. To accomplish this requires reducing the intensity of painful emotions, ameliorating disturbing thoughts and images related to the death itself, decreasing avoidance of daily activities and social withdrawal, developing personal goals, and optimizing current relationships. An important initial objective is acceptance of the death so that the bereaved person is able to think of it as a part of the deceased's life. A further goal is helping the bereaved patient find a way at the same time to keep the deceased close by and to have comforting memories, and meanwhile to renew interest in and involvement with life, with a sense of meaning and purpose, resumption of self-care, and reengagement with significant others.

To achieve these goals, TGT utilizes a two-track approach that targets emotions, thoughts, and behaviors related to the loss, and works on achieving personal goals beyond grief, enhancing self-care and satisfying daily activities, and optimizing ongoing relationships with loved ones. TGT incorporates an interpersonal psychotherapy (Klerman, Weissman, Rounsaville, & Chevron, 1996) framework that conceptualizes TG as a medical illness, caused when the biobehavioral response to bereavement is blocked from its natural progression. Interpersonal psychotherapy (IPT) strategies are employed in giving the person a "sick role," focusing on emotions related to grief, and working to optimize relationships with others; IPT strategies also provide ways of helping to resolve problems in ongoing relationships that may interfere with the resumption of satisfying interpersonal interactions.

TGT employs cognitive-behavioral strategies in working with the reaction to the death and in addressing avoidance of daily life activities. TGT techniques of revisiting the death and creating daily activity exercises are derived from imaginal and in vivo exposure strategies devised by Foa and Rothbaum (1998) for the treatment of PTSD in rape victims. To further work with the loss, TGT employs structured forms for recording memories and utilizes pictures and other cues to evoke positive comforting memories of the deceased. TGT also uses a technique modified from motivational enhancement therapy (Miller, Zweben, DiClemente, & Rychtarik, 1992) to identify and work on personal goals beyond grief. When needed, elements of motivational interviewing are used to promote resolution of ambivalence about change and active treatment participation.

TGT is administered in three phases, in weekly sessions lasting sixty to ninety minutes each. The "introductory" phase, typically completed in three sessions, focuses on initiation of assessment and monitoring procedures, information about grief that is working, introduction to the treatment, goal setting, and engagement of a supportive other. The active "grief treatment" phase, typically completed in six to seven sessions, focuses on use of techniques to help with adjustment to the loss and with reengagement in life. The "consolidation" phase, also completed in six to seven sessions, focuses on residual symptoms, relationship problems, personal goals, preparation for termination, and maintenance of gains. We describe the elements of each phase in detail below.

Introductory Phase

*Getting Started, Eliciting the Story, Assessing
and Monitoring Patients' Grief*

Friends and family of people suffering from TG often respond with increasing confusion, impatience, and frustration to their continued preoccupation with the deceased and unrelenting anguish over the death. Given their heightened emotional sensitivity, feelings of vulnerability, and their inclination to isolate themselves interpersonally, people with TG often find these responses particularly painful. Even the most well-intentioned efforts to convince persons with TG to "get over" or "let go of" their grief are likely to be met with helplessness, resentment, and resistance.

Thus it is critical for the TGT therapist to employ an actively empathic, collaborative therapeutic stance. The therapist spends time listening to the story of the death and to the way patients have been feeling, making clear in word and deed the intention to be as "present" and supportive as possible, and to work together in a cooperative effort. As the treatment begins, patients' difficulties confronting and working to change the experience of the death is acknowledged, and the therapist inquires about any mixed feelings patients have about doing so. Patients are told that the therapist will describe each step of the treatment, attempt to explain the rationale, and obtain their willingness to do what is recommended.

After presenting a brief overview of the treatment structure, the therapist elicits the story of the relationship with the deceased, the death, and its aftermath. Patients are asked about their perceptions of their current difficulties. The therapist reviews the intensity of feelings of grief, typical beliefs (e.g., guilt, hopelessness, mistrust, self-denigration, and/or separation-anxious ideas) that contribute to the suffering, and avoidance of situations or activities associated with the death or the deceased. Current relationships are queried, with close attention paid to potential sources of support during the treatment, as well as to anyone who might be unsupportive and thus pose an obstacle to the treatment. Times of the year when patients are likely to experience intensified grief ("difficult times") are identified and anticipated so that the therapist may help patients prepare for these as they approach.

From the first session of the treatment, patients are asked to monitor the intensity of their grief using a diary in which they note the average daily level of grief, as well as its highest and lowest intensity. Monitoring negative affect has been found to be therapeutic, as it helps people focus on the feelings instead of avoiding them and helps them see that fluctuations occur in the level of grief they are feeling. Often, at the beginning of treatment, patients feel as Tom did, that grief is all there is in their lives. This is one of the concepts TGT aims to challenge. Grief monitoring is a simple, powerful tool in achieving this goal. Becoming aware of variations in grief intensity helps therapists and patients identify grief triggers, and recognize when patients are engaged in activities that allow them to set the grief aside. Both are important in this treatment. In Tom's case, grief intensity was very high during the first week of treatment, with an average intensity of 8 on a 1 to 10 scale, and a low of 7 and high of 10 for several days. However, by the second week, there were several days where the lowest level was 4, a noticeably lower rating. Tom and the therapist explored this and discovered that the lower levels occurred every morning after Tom read the sports page of the newspaper. Playing softball with his buddies was something Tom had always valued. Since Donald's death, he felt too guilty to do this and had quit the team he had played on for fifteen years. Since Donald, too, had been an athlete, Tom felt it was unfair for him to enjoy playing when his son had been unfairly deprived of this pleasure. In their discussions, the therapist helped Tom to see that playing on a softball team could actually be a way of honoring Donald. Tom ultimately agreed that this was something Donald would have wanted him to do.

Introducing the Model of TG and TGT

Psychoeducation about the nature of TG and TGT are key components of the introductory phase. The patient is given a handout, "Traumatic Grief and its Treatment: A Handout for Patients, Friends, and Family Members," and asked to review it. An audiotaped version of the handout is available for patients who prefer it. The therapist works to ensure that patients understand this model of TG, emphasizing that it is viewed as an illness resulting from the loss of a very special person, the circumstances of the death, and the state of the person who sustained the loss.

The therapist presents the rationale for TGT as a set of techniques that have been successfully used with people with TG in order to help them get their grief back on a natural progressive track. Patients are told that the therapist will work with them on two tracks concurrently: accepting the loss and going on with life. Patients are helped to understand how these techniques work by creating a balance between engagement with painful feelings and the ability to set them aside.

Imaginal revisiting of the death and memory work are introduced as key techniques for working with the loss. Revisiting is described as helpful in reducing the intensity of painful emotions through the increasing familiarity that comes from repetition. Revisiting also helps to identify the ways in which this death was particularly difficult. Patients are told that *revisiting* the death means going back to it, but only for a visit; this process allows the person to bring current perspectives to bear on the events of the past. This is how people typically solve a difficult problem: they think about the problem, try to understand what they can do about it, and then put it aside and do something else until later, when they revisit the problem again. The back-and-forth process of revisiting the problem and then putting it aside eventually leads to progress in solving it. Revisiting is also meant to be a "re-visitation" in the sense of honoring the person who died and providing comfort to the bereaved person. Working with memories of the deceased's life, both positive and negative, is described as a way to help the patient access memories in a way that is comforting and to put the memory of the death in perspective, as only one moment in a rich and varied relationship with the deceased.

Work on personal goals, involvement of a significant other, and working on daily activities are described as core techniques for helping with adjustment to life without the deceased. Patients are told that the therapist will ask them to think about their goals beyond the lessening of their grief and that the therapist will work with them to start taking steps to actualize these goals. They are also told that they will be asked to invite a significant other to attend one of their sessions, in order to elicit that person's active support during the treatment. The intention to collaboratively plan daily activities that help the patient to stop avoiding important activities or areas of their lives is also introduced. The potential to address unresolved interpersonal problems such as role transitions or role disputes later in the treatment is also noted, as is the importance of patients more actively taking care of, being kind to, and comforting themselves.

Information is provided using a Socratic approach rather than a didactic style. The therapist elicits patients' understanding of the phenomena under discussion and, using skillfully targeted questioning, helps patients to arrive at important understandings. Socratic work allows the therapist to discover surprising ideas that patients may have about their struggles and the therapeutic efforts at ameliorating them.

Patients' receptiveness to information varies. Most understand and accept it; some find certain aspects helpful but reject others, and some show little interest. Lack of understanding or receptiveness may be an indicator of ambivalence. TGT is most effective when patients can be full collaborators in the treatment; this means that they know what they are going to be doing, and why, and that they concur with the plan. It is therefore worth working with patients to help them accept and understand the model and rationale as best as possible. Even if this is unsuccessful, though, patients can still benefit from TGT if they are willing to make a commitment to accept the treatment procedures. Often, after some gains are made, the therapist can return to the rationale, which may now have become more comprehensible and acceptable to the patient. Patients who do grief monitoring and revisiting exercises often show substantial gains, and it may be only at this point that they begin to trust enough to profit from the information.

Personal Goals

An important focus of introductory phase work is the development of personal goals. The idea is to encourage patients to look beyond their grief to their hopes and dreams, and to find something practical they can begin to work toward again. The technique used for this is drawn from motivational enhancement therapy (Miller et al., 1992). The therapist asks patients to try to imagine that their grief is much less intense and prominent, and to consider what they would want for themselves if this were the case. Once goals are identified, the therapist asks patients how they would know that they were making some progress in accomplishing their goals. Patients are then asked how committed they are to doing so, what might stand in their way, and who might be helpful to them in accomplishing the goals they have specified. Questions about how patients will know they are progressing toward their goals provide markers and targets for ongoing work, and inquiries about potential obstacles and how support from others might be able to help overcome them connect the treatment process to the concrete challenges patients are likely to face as they attempt to restore themselves to a full and meaningful life.

Perhaps surprisingly, once they understand this idea, most patients are able to articulate goals and hopes for the future and what they need to do to address these. Most endorse a strong commitment to achieving their goals, and this can provide a powerful and salient incentive to begin work on them, even before their anguish related to the loss lessens. In TGT it is important that the work on goals proceed hand-in-hand with the work on the loss. When this does not occur, it can be much more difficult to work on the loss, since the person may feel there is little to live for. In fact, 65 percent of 154 patients who have completed initial assessments in our TGT project reported that they did not care if they lived or died, and most also felt only a mild to moderate desire to live. For the most part, those who do not wish to die are held back only by guilt about the effect of their death on someone else. The process of identifying personal life goals is an important first step to enhance the desire to live.

A patient who had a lifelong dream of starting an antique business with her husband provides an example of this process. After he died, she abandoned this idea. In the second TGT session the therapist identified this as one of her goals and began discussion of how the patient would know she was making progress and who could help her. The patient readily answered these questions and by the end of the treatment had successfully rented a space, fixed it up, and scheduled a grand opening.

In Tom's treatment, the goal of putting athletics back into his life was identified. He proposed as a first step finding what teams he could possibly join the next summer. He thought an obstacle would be that he would not be home for dinner two nights a week and his other children might be upset by this. He decided his wife could help him in this by being there for the children and helping them enjoy his return to his normal self.

Optimizing Support

Optimization of support from others during treatment while beginning the process of rebuilding relationships is a key component of TGT. People suffering from TG are often loving, passionate people who have formed close ties with others over the course of their lives. Yet once they develop this syndrome they feel very estranged from people who have previously been close friends and family members. For this reason we ask patients to invite someone, typically to the third treatment session, who might be willing to learn more about TG and its treatment.

Supportive others who attend this session are asked to provide their own perspective on the patient's experience and behavior since the death, and to be available to assist and support the patient, especially during the more demanding phase of the treatment. Patients are often surprised by the unambiguous support provided by the person who comes and by the views expressed by this person. For example, an elderly woman had not realized that she had been ignoring her beloved grandchildren until her daughter pointed this out in such a session. Because of her mother's obvious distress, this daughter had been reluctant to bring this to her mother's attention. However, when the patient heard this, she immediately responded, changing this behavior, though it was several weeks before she began to make real treatment gains.

Occasionally, important relationship problems may become clearer in the conjoint session, and these often become a focus of the treatment. A typical example of this is a couple who has lost their child. Though only one parent is identified as having TG, the other may also suffering from this condition. However, very often the natural tendency of the parental pair is opposite, with one person favoring extensive avoidance and the other favoring intense preoccupation. When this happens, a rift in the relationship often results, with each person feeling misunderstood. Tom and Sally exemplified this problem. While Tom was immobilized by survivor guilt, and thus unable to engage in any satisfying activities, Sally felt pressured to work to change the world so that tragedies like their son's death would be less likely to happen. Tom, usually very supportive of Sally's volunteer activities, thought this one was useless and futile. He even mildly resented the fact that she seemed to get satisfaction from indications that her work was making progress in getting legislation passed. Sally, for her part, was frustrated and annoyed with Tom for what she saw as his "wallowing in grief." Meeting with the couple, it was clear that they had previously had a loving and mutually supportive relationship and that, in their separate grief, they had become estranged and hostile to each other. Helping Tom to repair this breach would thus become an important component of Tom's treatment.

Grief Treatment Phase

Avoidance

The initial step in addressing patients' grief more directly entails discussions and exercises targeting avoidance of reminders of the death. Using a

Socratic approach, the therapist addresses the nature of avoidance. The following points are made:

1. People tend to avoid situations and thoughts that feel threatening or trigger painful feelings, especially when they fear that the intensity will be overwhelming, or that they may feel embarrassed or out of control.
2. Although avoidance is effective in the short term, it blocks the process of readjustment to life and reduces the chance that the intensity of painful feelings will decrease.
3. Avoidance prevents the bereaved from thinking about the death in the broader context of their own ongoing lives, so that shared experiences can be integrated into their lives.
4. Avoidance interferes with a process of bringing their full range of psychological resources to bear on the problem of how to make sense of and accept the death.
5. Avoidance gradually increases the difficulty of engaging in daily activities. In other words, in the long run, it has the opposite effect to what patients want: avoidance leads to a persistent, tense apprehension that painful feelings could be unpredictably triggered and results in progressively growing restrictions in what a person can think, remember, or do.
6. What is needed is to gradually and systematically reduce avoidance and to repeatedly engage in activities in order to reverse the detrimental effects.

Imaginal Revisiting

Imaginal revisiting of the immediate period surrounding the death, and sometimes of the funeral or memorial service, is the main in-session technique used to overcome avoidance and its effects on patients' lives. Imaginal revisiting is usually done for the first time in the fourth session and is repeated in each of the next three to four sessions.

When beginning imaginal revisiting, the therapist first reminds patients of the overall goal of TGT—to help patients find a way to accept the death so that they can go on to live a fulfilling life, keeping their loved one in their heart—and of the multifaceted rationale for using revisiting to help accomplish this. Acknowledging that this exercise is difficult, the therapist emphasizes that it has been found to be very helpful and well worth the price of the briefly intensified pain it evokes. The therapist also openly affirms and supports the patient's courage in doing this work.

Because imaginal revisiting is emotionally evocative, it is done early in the session, in order to allow sufficient time to facilitate return to baseline levels of distress. Patients are asked to close their eyes and to tell the story as though it were happening in the present, beginning when the death was imminent or, in the case of sudden death, when the patient first learned of the death, and ending after leaving the body or by the conclusion of that day, if there is no encounter with the body. During the imaginal revisiting, the therapist intermittently asks patients to rate the intensity of their feelings from 0 to 100 on a Subjective Units of Distress Scale (SUDS), and may also ask for brief clarification of thoughts or feelings that are not immediately obvious. This is done not only to ensure that the therapist knows what is happening and to document changes in grief intensity, but to let patients know that the therapist is there with them. Imaginal revisiting exercises are tape recorded, and patients are asked to listen to the tapes daily throughout the week. SUDS during each episode of tape listening are tracked by the patient on a form designed for that purpose.

To revisit the death of a loved one in imagination is an emotionally painful process. During the acute phase of grief that is working, thoughts of the death are very prominent, and it is common for the bereaved person to tell the story of the death repeatedly to friends and relations. Repeated retelling of an emotional story usually results in gradual attenuation of the intensity of the emotions, and in this process the person finds a way to accept the death and to think about it without enormous pain. People with TG, however, continue to have disturbing thoughts related to the death for many months, and often years after a loved one dies. A sense of disbelief remains strong, and acceptance of the death remains difficult. It is futile to resist acknowledgment of this event, and to do so blocks the natural healing progress of grief. Yet many people with TG are frightened to think or talk about the death, and though they have intrusive thoughts about it, they try to push these away. They may fear the experience of the deep anguish associated with the death, thinking they are barely keeping it at bay, and that if they did not fight it, the intense emotions would be more than they could bear and/or would last forever. Of course neither actually happens in the telling of the story, and it is very helpful for the bereaved person to see this. The revisiting exercise thus permits the bereaved person to engage this event in the supportive, empathic atmosphere of the therapy and to discover the easing of their pain that comes from repetition.

It is important to note, however, that after the first few sessions in which this exercise is done, it is common for distressing thoughts and feelings to increase. For this reason, the therapist must work with patients from the beginning to find ways to engage both in thinking about the deceased and also in setting these feelings aside. This is the ultimate goal of working with grief: adjustment to the death of a loved one is achieved optimally when the bereaved person can comfortably think about the deceased and can comfortably put these thoughts aside. TGT works toward this goal by conducting revisiting for about ten to fifteen minutes in the first third of the session, spending at least as long debriefing and discussing self-care, and being very explicit about the ways in which healthy, healing grief works by pushing us toward and also away from thoughts about the person who died. To assist patients in setting the story of the death aside, the therapist may instruct them in mindfulness breathing exercises, visual imagery (e.g., "rewinding" the "tape" of the story, imagining being someplace peaceful), or using contemplation to gradually shift attention, or distraction to shift attention more abruptly.

Following imaginal revisiting, the therapist engages patients in discussion of the exercise, focusing on confirming emotional engagement, empathic reflection of distress, direct congratulations for their courage in confronting this painful event, and inquiry about what it was like. The therapist then directs patients' attention to self-care and rewards. Many people with TG have difficulty accepting care from others. TGT fosters such acceptance. One way to present this is to say that when we lose someone we love, we have lost someone who we take care of and who takes care of us. Both being taken care of and being a caretaker are important to most adults. Many people feel somewhat uncomfortable taking care of themselves, but this is a very good solution, at least for an interim period, to the problem of losing a loved one. Taking good care of oneself is one way of honoring the person who died, of keeping his or her memory alive. It is also a very good idea during a period of active grief that bereaved individuals allow others to take care of them. Most cultures prescribe a period of caretaking for bereaved people, in which things are done for the grieving person that are not usually done. This caregiving allows the bereaved person space and time to adjust to the death.

In TGT, therapists make themselves as available as possible outside of the session, bringing themselves into patients' lives. This is important because individuals with TG often feel isolated and mistrustful of others. They feel pressure to "get over it" and sense that others are uncomfortable and impatient with them for being preoccupied with thoughts and longings for a person who died so long ago. They also feel ashamed of themselves and often think that others believe they are weak. Since the therapy targets relief of the grief, it is especially important for the therapist to display acceptance and understanding in every way possible. It is important to be as "real" as possible for these patients. One way to do this is for the therapist to call patients during the week, after they have listened to the tape of the revisiting exercise. Calls are typically more frequent in the first part of the revisiting phase. They should be brief and, like the postrevisiting discussion in the session, focused on empathizing with the difficulty of the treatment, applauding patients directly for their courage, assisting them in setting the thoughts and feelings aside, and encouraging plans for self-care.

After several weeks of revisiting exercises, patients sometimes request a break, most often because something else is problematic in their lives and they want to talk about this. It is a good idea for the therapist to permit patients to take such breaks. It is especially important to do this if there is a current life problem. To progress, grief must address both the adjustment to the loss and the adjustment to life without the person who died. If work is not done on both tracks, the patient will remain stuck: problems in dealing with loss will interfere with adjustment to life without the deceased, and vice versa.

Daily Activities

In-session revisiting exercises are complemented by weekly plans to engage in activities that patients have been reluctant to do. Most often these entail situations that are reminders of the person who died, reminders of the death itself, or reminders that the bereaved are now alone. These exercises can be particularly helpful, as they have the dual purpose of helping to cope with the loss and promoting adjustment to life without the deceased.

To decide the order in which activities will be done, a hierarchy is developed, with patients identifying situations that are difficult or that they simply avoid, and rating the distress associated with such situations or activities on the SUDS scale. The therapist then explains that she or he wants the patient to start doing these things. The plan is to engage in distressing activities in a gradual way. Patients are told that the therapist wishes to treat these tasks as exercises which, in addition to gradually decreasing the intensity of emotion, will also build patients' confidence in their ability to directly confront and handle whatever situations or activities arise, rather than feeling a need to avoid them. The therapist explains that the activity should thus be "challenging, but doable."

An activity that is expected to produce moderate levels of distress is then chosen from the hierarchy. It is best to begin with things that patients really want to do but have not been doing. For example, sometimes people with TG cannot bring themselves to visit the cemetery or final burial place of the person who died. They may feel very badly about this, thinking they are being disrespectful and/or depriving themselves of a kind of comfort. If the expected SUDS rating is not too high, visiting the cemetery would be a good choice of activity for this exercise. Another example might be going to a place frequented by the person who died. A patient may have enjoyed this place (e.g., a favorite restaurant, a movie theater, a park), and may want to revisit the place both to get pleasure and to feel connected to the deceased, but this may seem too sad. In addition, many people with TG have belongings from the person who died and/or pictures or condolence cards that they would like to spend some time with, but they are afraid to do so because they anticipate too much distress and/or triggering of time-consuming reveries that will interfere with their lives. These, too, can be good early choices for daily activities. It is also important in this process to ask patients to engage in activities that were once enjoyed but have been neglected since the death. Socializing, attending public events, or even regularly reading the newspaper may be scheduled, not only to reduce the intensity of grief, but also to help patients reengage in life and begin again to experience pleasure or satisfaction.

Patients are asked to plan to engage in the selected activity repeatedly during the week between sessions. Each activity, like the between-session listening to the imaginal revisiting tape, is tracked on a form designed for that purpose. This form is reviewed at the start of the following session, allowing therapists and patients to see the progress of reduction in distress levels, and/or to determine if an agreed-upon activity was too challenging and needs to be rethought. Week by week, over the course of the treatment, patients work their way up the hierarchy until they have successfully tackled their most emotionally difficult situations.

Memories and Pictures

Following several sessions of imaginal revisiting, the therapist introduces the first of a series of five "memories forms." These forms are given to patients in five consecutive sessions. The first asks patients to record positive memories of the person who died. The second asks for more positive memories, and the third for favorite memories. The fourth asks about the annoying or difficult side of the deceased, things that patients really do not miss, and the fifth form asks about both positive and not-so-positive memories.

Patients are asked to recall less-than-positive memories not because ambivalence plays an important role in TG, but this exercise encourages patients to begin thinking about the deceased as that person really was. Patients are told that positive memories will almost certainly far outweigh the negative, but that no relationship is perfect. This helps to give patients permission to acknowledge negative aspects of the relationship. Nonetheless, occasionally patients have difficulty thinking of the more negative side of the person who died. Most people with TG do not idealize the deceased, so when this does occur, it is a good idea to look for problems with guilt.

In addition to the memories forms, the therapist invites patients to bring in pictures of the person who died. Therapist and patient look at the pictures together, and patients are asked to talk about what they portray and to tell stories of the deceased related to the pictures. The therapist responds as he or she would to a friend showing pictures of other friends and family.

Most people find both the sharing of pictures and the completion of memories forms very comforting, when done after several successful revisiting sessions. The two activities are used similarly, to help free patients to talk and think about the deceased without fear of their emotions, as well as to help patient and therapist come together in memorializing the deceased. The sharing of pictures is often intimate and frequently moving, as patients lovingly describe the persons and situations who appear in the photographs. If pictures and/or memories forms are started too early, however, they tend to evoke tremendous sadness and are not as useful.

Imaginal Conversation

The next exercise used to work with the loss is to ask the patient to conduct an imaginal conversation with the person who died. Again, it is important that this is not done too early in the treatment and that some progress has been made in relieving the intense emotions associated with the death and developing comfort in thinking about positive memories. When these have occurred, the therapist tells patients that she or he would like them to have a conversation in their imagination, talking out loud to the deceased and also answering for the deceased. Patients are asked to put themselves at the bedside, or beside the casket, at the gravesite, or sometimes somewhere else, and to imagine that the person has died but is still able to hear and respond. The therapist reiterates the idea that this is simply an imaginary exercise that has been found to be very helpful, and asks for patients' agreement to do this.

The exercise is then conducted in a manner similar to that used for revisiting. Patients are asked to close their eyes and to imagine themselves beside their loved one after the person has died. The therapist then invites patients to ask and/or tell the deceased anything they like. The therapist asks for SUDS levels at regular intervals during this exercise. At a natural stopping point, the therapist asks patients to take the role of the deceased and to answer. The conversation may end at that point or it may continue, with patient and deceased having a discussion. Again at a natural stopping point, the therapist tells patients that when they are ready, they should stop and open their eyes.

Most often patients respond that this has been a very powerful and comforting exercise. The therapist guides patients in a discussion of the experience and then redirects them, as following the revisiting exercise, to focus on daily activities and plans for the remainder of the day and the upcoming week. These conversations may be repeated through several sessions, if patients discover previously unarticulated thoughts and feelings that they wish to "discuss" with the loved one.

Work on Personal Goals

Work on personal goals progresses side-by-side with the series of strategies used to cope with the loss. As previously noted, a typical misconception is that people must "grieve a loss" in order to detach from a loved one, put that relationship behind them, before they can move forward in their own lives. In fact, it is very difficult to adjust to the loss if there is not a simultaneous growing interest in daily life activities and relationships. However, people with TG report little or no interest in their own lives, often continuing to feel they are on "autopilot" or just fulfilling a duty, waiting until they can join the person who died.

Thus, an important component of TGT is to work to revitalize interest in their current life activities. Patients are helped to formulate specific plans that are steps on the way to achieving the personal goals articulated at the beginning of treatment. The therapist discusses progress on goals at the end of each TGT session and does not wait for grief to "be resolved" before attending to this important aspect of the process. Given time constraints, the discussion of goals is often more brief during the earlier active grief treatment sessions, but it is important that an emphasis be placed on this work nevertheless. The therapist helps patients track their progress in working toward identified goals, and problems that arise are addressed together. Patients are offered strong support and affirmation for their efforts.

Difficult Times

In TGT, the therapist needs to be aware of the occurrence of difficult times during which there is an upsurge in distress and preoccupation with the deceased. These are usually times such as the anniversary of the death, birthdays, holidays, or other special occasions. An important reason to attend to these times is to normalize the increase in distress that occurs. Even when grief is working well, and the bereaved is comforted by thoughts of the deceased and well engaged in life, the vulnerability to difficult times remains. This is not an indication that there is more "grief work" to do, or a signal that the bereaved has not really dealt with the loss. Instead, it is a normal fluctuation in the accessibility of grief-related thoughts and emotions.

For both patients with TG and people whose grief has worked in a healing way, difficult times are self-limited. They do not trigger a new ongoing level of more intense distress, but rather represent a temporary increase in pain. These times provide an opportunity to honor the deceased and a call to increase self-care activities. This is a time to be alone, if that is what is really most comfortable, or a time to seek support if that is what helps. The therapist uses the discussion of difficult times to introduce the expectation that grief is never entirely gone and that people find their own personal ways of managing the difficult times. The person with TG can expect them to come, and also to go. A goal of the treatment is to ameliorate the fear of such times and to encourage active planning.

Consolidation Phase

After approximately nine treatment sessions, the therapist reviews patients' progress with them and considers how to plan the remaining sessions. In this treatment therapists provide, assessment instruments used to rate the components of grief are readministered and compared with baseline ratings. Personal goals are reviewed and progress on that front is acknowledged. The status of relationships with others is discussed. A brief interpersonal inventory may be administered in order to determine the extent to which patients' relationships are satisfactory in terms of intimacy and shared activities, and whether any significant conflicts remain to be addressed.

On the basis of this reevaluation, patient and therapist decide how to proceed and develop a plan for the remaining sessions. In some cases it is clear that revisiting has been effective but that more remains to be done; an agreement may thus be made to return to revisiting or to an imaginal conversation. In some cases it becomes apparent that transitions in interpersonal roles brought on by the death still remain to be negotiated, or that lingering interpersonal disputes are interfering with patients' ability to enjoy the people who are still present. In these cases, a brief module of IPT may be conducted once a specific focus has been identified. This may include clarification of nonreciprocal relationship expectations or communication difficulties, various forms of education, problem solving, role-playing and rehearsal of new behaviors, and guidance in decision making where relationship problems are concerned. Whichever focus is chosen, work on personal goals continues to be a component of the treatment.

As the consolidation phase proceeds, the therapist begins to prepare patients for termination. Patients are reminded of the number of remaining sessions and are asked how they feel about the treatment drawing to a close. Patients have different responses to termination, depending in part on how comfortable they are feeling with their progress. For people who have successfully transitioned to grief that is working, the treatment termination is usually comfortable and a positive experience. These patients have obtained a new model for adjustment to loss that seems to be easily transferred to the therapy. Their response to discussions of termination is to feel sad, but grateful, and, at the same time, to describe a sense of confidence that they are ready to move on. About a third of the people we have treated manifest this response.

For those who have benefited but are not yet completely adjusted to the loss, termination is more difficult. The therapist needs to make a judgment about whether more treatment is indicated or if, instead, a period of time to continue to work with new ideas and skills is the best next step. More extensive discussion of this work is indicated, as is more extensive discussion of feelings about ending the treatment. The end of the treatment is used in these cases as a model for the approach to loss that is provided in TGT. The therapist encourages open confrontation with the loss and attendant feelings, and also discusses ways patients can engage in their own lives. A sense of continuity is fostered, and the therapist may even invite patients to call in a few months and report how they are doing. Patients' concerns about how they will maintain or continue their progress are discussed. Beliefs that there will be work yet to be done after the treatment has ended are normalized, and plans are made for helping patients to address these remaining pockets of difficulty on their own.

In the final session, changes in ratings of grief components and social support are again reviewed, and progress toward achievement of personal goals is assessed and affirmed. Patients are asked about their perceptions of what about the treatment has been most important to their progress, and after adding his or her own perspective, the therapist encourages them to continue these activities, and supports their ability to do so. The therapist warmly expresses his or her genuine feelings of pleasure in and admiration for patients' hard work and accomplishments, and patients are invited to contact the therapist in the future with further questions or to report ongoing efforts and progress.

Case Vignette Revisited: Traumatic Grief Treatment of Tom W.

During the introductory phase of his treatment, Tom W. describes his relationship with his deceased son, Donald, and recounts, at the therapist's request, the story of the death and its immediate aftermath. Tom emphasizes the mixed sense of unreality and horror that he felt upon receiving the call from the state police, and the extent to which he felt his life to be shattered in that moment. He describes his life in the five years since the death, as the therapist listens empathically and elicits information about his efforts at coping with the death through other forms of counseling. Tom describes his participation in a grief support group from which he found some temporary comfort, but not lasting relief. The therapist reviews Tom's self-reports of the intensity of his grief since the death, which Tom reports as never having dipped below a 6 on a scale from 1 to 10, and his self-described "difficult times," which include Donald's birthday, the anniversary of the death, and the Christmas season. Tom's sources of support include only his wife, Sally. The therapist engages Tom in a Socratic discussion of the model of grief that is working and grief that is not working. They review self-reports of avoided situations: the grave-site, social occasions of all kinds, funerals, talking about Donald, and going to sports events (a favorite activity that he and Donald shared). TG-related beliefs Tom endorses include, "You could have done something to prevent his death," "You should have made a greater effort to overcome your differences," "Life will never be good again," "You cannot cope with your feelings since the death," "There is little justice in the world," and "If your grief diminishes, you are abandoning your loved one."

The therapist reviews key elements and goals of the planned treatment and explores Tom's personal goals. Among these, Tom identifies a goal of feeling the sense of loving closeness with his family that he shared with them prior to Donald's death—in his words, "That Donald's death not also be the death of our family." This provides the underlying theme of a conjoint session with Sally, who describes her perspective on Tom's difficulties and her own struggles to come to terms with their son's death. In this session, Sally is helped to understand TG, the model of treatment, and how she can support Tom's participation—at Tom's request, to be available to him as he works through exercises planned for each week.

As the grief treatment phase begins, the therapist reviews the imaginal revisiting procedure and rationale. Tom closes his eyes and recounts the evening of the death: from the moment the phone rang and he had a premonition that something was wrong, through the visit to the morgue to identify the body. The revisiting exercise is audiotaped. The therapist asks him, every few minutes, to provide a SUDS rating; when Tom pauses, obviously experiencing intense emotion, she tells him he is doing an excellent job and invites him to continue the narrative.

When the story is concluded, the therapist asks Tom to open his eyes and "put the story away for now." They discuss what this experience was like for him and whether there was anything that surprised him. Tom says he was surprised by how intense his emotions still were, especially when they peaked at the most emotional time. He asks if he could ever expect to feel better. The therapist reassures him that it is very likely that he will feel much better than he does now and that the story will become far less emotional for him, at the same time explaining that of course there will always be a great deal of sadness and anguish related to this loss. The discussion then turns to Tom's plans for finding something comforting to do after the session. He says that he will go for a walk in the park as it is springtime and he loves to look at the budding trees. He will pay close attention to sounds and smells in the park, using a kind of mindfulness technique. During this discussion, the therapist periodically checks Tom's SUDS level, and finds by the session's close that it has come down to a 6, slightly higher than at the start of the session, but also noticeably lower than the 9 that he reported at the end of the revisiting exercise.

Prior to the first imaginal revisiting, the therapist had worked with Tom to formulate a hierarchy of avoided situations, which looked, in part, like this:

SUDS	Situation
100	Visiting the grave
95	Reading the police report
90	Going to a baseball game (shared favorite with Donald)
85	Funeral home
80	Reading the yearbook Donald edited
80	Talking about Donald to his other children
70	Attending annual family gathering
70	Going to church
60	Contemporary movies
50	Talking about Donald to his wife
40	Being out and around strangers

Now they agree upon an activity for the upcoming week. They choose one of the least difficult situations as the first exercise: Tom makes a commitment to walk around the local mall each evening for about twenty minutes. He also agrees to listen to the imaginal revisiting tape each day.

Later that week the therapist calls Tom at home, and he tells her that he was so exhausted following their session that he'd gone home and slept for several hours. Each evening since he had walked around the mall, which he noticed was somewhat easier the second time than the first. Listening to the imaginal revisiting tape, he'd been struck that, while still very painful, the experience was less intense than it had been when he told the story in the session. The therapist tells him this is quite typical and strongly affirms him for engaging so faithfully in the planned activities.

During further repetitions of the story in the ensuing sessions, Tom's SUDS level gradually drops, except when he revisits the moment when he first sees his son's body in the morgue, which he continues to report as a SUDS of 100. The therapist has him repeat this very brief part of the story, which lasts less than a minute, six times consecutively. His SUDS level drops to 90 after two repetitions, but they don't budge after that. When the therapist notices that, during the last repetition, Tom seemed to pause for a moment when he described seeing the body, she asks him to tell her what he is thinking. For the first time, Tom shouts, in an anguished voice, "It should be me lying there and not you!"

At this point, the therapist asks Tom to have an imaginal conversation with his son. She suggests that he imagine standing there at the morgue and that he can talk with his son, who can hear and respond. Tom agrees and pours out thoughts that he had not previously described in detail: that he should have stopped Donald from taking the bus home that night, that he felt as though every day he went on living was an affront to Donald, and that he was afraid that the gradual decrease in the intensity of his grief meant that he did not love Donald as much as he had claimed. The therapist then asks Tom to speak for Donald. As Tom speaks for Donald, his son tells him that he knows how much Tom loves him, that his family meant more to him than anything else in his life, and that Tom must stop pulling away from his mother and siblings and help them with their own feelings about Donald's death. He strongly refutes the idea that feeling less pain now would suggest that his love was not strong. Instead, Donald describes his own pain in seeing his father so anguished and says that it would be a gesture of love to let his grief take the natural course toward lessening of pain.

In the next session, Tom reports that he felt tremendously relieved after the imaginal conversation, and that he has had other thoughts about things he would like to say to Donald. The therapist agrees that further use of imaginal conversations is a good idea. This time Tom recalls an episode when Donald was ten and he himself was under more than the usual stress at work. He remembers being irritable and chastising Donald unfairly for something. This had weighed on his mind for years, even before the death, but he has found that since the death, the remorse over this episode is intense. In the conversation with Donald, his son tells him that he can barely remember the episode, and that Tom was one of the best fathers he can imagine having. He says he long ago forgave Tom, as he understood his father loved him so deeply.

After these sessions Tom begins to experience a new feeling of comfort and closeness to Donald. He now finds himself imagining their laughing together over the foibles of other family members. Tom brings pictures of his son, which he and the therapist admire as he describes the events they portray. He works on memories forms, recalling on the fourth form how his son's sloppiness and forgetfulness could drive him crazy. Each week Tom progresses up the hierarchy of daily activities, one week bringing in Donald's yearbook to look at with the therapist, and later bringing a copy of the police report to read in the session and discovering, to his relief, that they believed Donald had died instantly and painlessly in the moment the bus was hit.

At each session, Tom is also encouraged to talk about the future. In addition to his goal of improving his relationship with his wife and children, he has also identified a goal of finding a meaningful activity beyond his part-time job. He has no interest in returning to his old line of work and instead is considering a change to a career in teaching. He has always loved learning and had a lifelong dream of being a teacher, but had not chosen this career because he was concerned that he would not be able to provide well enough for his family. However, he finds himself drawn to children and loves to watch the science channel on TV. He decides he wants to investigate training to become a middle school science teacher. During the first ten sessions of the treatment this idea has been increasingly elaborated, and he has discovered that the needed classes will be available at a local college beginning in a few weeks. The idea of taking these courses both stimulates and frightens him, and the therapist agrees to focus part of the remaining sessions on exploring what this new career will mean for him and his family.

As the treatment ends, Tom continues to experience feelings of sadness when he thinks of Donald, but no longer reports intrusive memories. Instead, he has bittersweet memories of time spent together, that provide as much comfort as pain, and occasional humorous interludes imagining how Donald would react to a current situation. After going alone to the local ballpark and alternating between cheering and crying throughout the game, he has taken his younger son to a game and discovered that he, too, shares the sense of excitement that Tom and Donald had enjoyed together. They decided they would go again soon, buying a third ticket and leaving the seat next to them empty. Tom still spends time in his workshop and has returned to tinkering, occasionally describing out loud an idea that pleases him while looking at one of Donald's pictures. He has joined a softball team and still sometimes daydreams of Donald as he waits for a ball in the outfield. In discussions with Sally, Tom has decided that their financial needs make pursuing a full-time teaching job untenable, and instead he plans to sign up to become a tutor, a volunteer position he could fill on evenings or weekends. He has reconsidered his reluctance to return to a promising career in accounting and has applied for a job at a firm that was a competitor of the one where he previously worked. He and Sally are beginning to spend more time together and have made love for the first time in years. Both were slightly taken aback at the intensity of mixed emotions they felt, and they decided to spend more time together before attempting sexual intimacy again. In their last session, he tells the therapist with a smile that he's sad that he will no longer be meeting with her, but even more relieved that he's done with all of her forms and exercises.

COMPONENTS OF A POTENTIAL
GROUP TREATMENT APPROACH OF TGT

Many aspects of TGT lend themselves naturally to group work. In fact, it could be desirable to incorporate group components into this treatment, as they have the potential to contribute both to its effectiveness and to the ease of its dissemination. Another clinical research group led by William Piper (Piper, McCallum, & Azim, 1992) has developed and tested a psychodynamically oriented, short-term group therapy for adaptation to loss, with promising results. A colleague in our institution has been leading multifamily groups for families of suicide completers, and has done this work successfully for more than a decade. The suggestions made here thus derive from our own thoughts, reports in the literature, and observation of the suicide groups, and are provided in the hope that others may use them as a springboard for further development of group interventions.

Two Potential Types of TGT Groups

Group treatment for traumatic grief could be designed either as an adjunct to TGT conducted individually or as a stand-alone intervention. In an adjunct group, information about grief and traumatic grief (grief that is or is not working) is provided and discussed. Personal goals are identified and the group works together to encourage pursuit of these. Group members are instructed in completing grief monitoring, and the group provides support and encouragement for the revisiting exercises and imaginal conversation. These more emotional and personal components of the treatment are conducted during individual sessions, as is currently done. Members are prepared for revisiting via psychoeducation. Ambivalence is explored in the group, and strategies for enhancement of motivation are employed to encourage participation in these difficult exercises. Such strategies include elements of role induction (Walitzer, Dermen, & Connors, 1999) and motivational interviewing (Zweben & Zuckoff, 2002). Once revisiting has begun, giving members the opportunity to discuss their reactions could have salutary effect. Mutual support is given, and ways of tolerating the distress that accompanies revisiting exercises is shared. People who have already undertaken revisiting could describe the experience and report on its benefits, providing a sense of hope for those who have not yet begun.

In a stand-alone group intervention, provision of information about grief and traumatic grief, identification and support of work toward achievement of personal goals, and grief monitoring would most likely be augmented by a revisiting process of reduced intensity. Experience with attempts at prolonged exposure in groups of patients with PTSD has not been promising, and it is difficult to encourage intensely emotional exercises in a group format. However, repeated retelling of the story of the death, in a more muted way, with open eyes and without SUDS ratings, would certainly be possible. This is something that is often done in grief support groups in the community. It is possible that this would be fully beneficial to at least some participants. In this case members have the chance to tell and retell the story of the death to a caring and attentive group of individuals who share similar experiences, and in doing so gain some of the benefits of the more formal imaginal revisiting procedure. Similarly, members could develop individualized hierarchies of avoided situations with the assistance of the leader and each other. Participants are encouraged to undertake planned weekly exercises in a similar fashion as is done in the individual treatment. Pictures and memories forms are used as they are in the individual treatment.

Common Elements

Whether group therapy is employed as an adjunct to individual treatment or as a stand-alone treatment in its own right, group applications offer some clear advantages. They provide excellent opportunities for offering psychoeducation, using a Socratic style to generate a guided discussion about grief and how it works, and discussion of the symptoms of TG could arise naturally through shared awareness of its manifestations. This might occur even more readily in a group than in individual treatment. Patients can facilitate shared identification of symptoms, as they describe their own experiences. Telling of one's own and listening to other group members' accounts of personal experiences since the death might help to specify the more general symptom description offered by the therapist. Hearing others endorse many of the same symptoms could help people feel understood and to feel less strange and abnormal. The shared group experiences could help people see their TG reactions as a common pathway rather than as idiosyncratic reactions to the death produced by personal defect.

In a comparable manner, explanations of the principles and procedures of TGT may prove to be at least as effective when delivered in group as in individual settings. Group discussion and illustration could facilitate understanding of such key concepts as the two-track model of grief, the principles of repeatedly revisiting the death, and the goals of being comfortable thinking about the person who died and also not thinking about that person. This might help patients to make a commitment to the treatment process and to resolve any problems or obstacles as they arise.

Group psychoeducation may be useful not only for patients but also for members of their families and others who are supportive. Realizing that their family member is neither simply unwilling to move forward in their grief, nor uniquely debilitated, may help significant others to be more tolerant of their loved ones' struggles and more supportive of their efforts at using treatment to overcome them. Family members and supportive others may also benefit from feeling less alone in facing the challenges of helping and caring about a loved one with TG.

People with TG feel isolated and estranged from others, and this often increases as time passes following the death. The experience of isolation is intensely painful, yet the person with TG often feels out of control of the sense of difference they feel. The sense of isolation that results from the loss of interest in the world without the loved one, and the experience of frustration felt by others with the bereaved person's continued suffering, can be intense. This isolation intensifies TG symptoms and further interferes with natural processes of healing.

A group in which others are experiencing similar struggles can be validating and affirming and can foster the process of reconnecting with others. In a traumatic grief treatment group, members may find the kind of understanding and safety that is often missing from their natural social network. At the same time, they would also find mutual understanding of the difficulty of engaging in the treatment tasks of TGT and emotional support and comfort in their efforts at meeting these important challenges. Patients could also find valuable support for their struggles in working toward and achieving their personal goals "beyond grief." Although friends and family members of course want to see their loved ones freed from the pain of acute grief, they may in some cases be less welcoming of the changes that a person recovering from TG might make in the service of accomplishing a transition to a new social role or life structure. Encouragement from others with TG, who may more easily appreciate the need for striking out in new directions or seeing oneself in new ways, could be important in helping patients make these fledgling efforts come to fruition.

Another benefit of working in a group derives from sharing of memories and pictures of the deceased. The opportunity to share mementos with other group members can bring deceased loved ones "into the room" and make their presence palpable. This helps to enhance an atmosphere of closeness and mutual support, while also serving to enhance the desired effect of evoking previously inaccessible memories and warm and comforting feelings associated with them. The telling of stories about the lives of the deceased, and their effects on the bereaved person, would naturally follow from such sharing and would further amplify, through laughter and tears, the loving memories recalled.

A group could be beneficial in providing an opportunity for discussion of interpersonal problem solving. Interpersonal psychotherapy has been successfully employed in a group format (Wilfley, Ayres, Welch, Weissman, & MacKenzie, 2000). This might occur through focus on interpersonal processes within the group and also through group members serving as resources for one another in efforts to resolve conflicts with significant others outside the group. This could include not only the sharing of ideas and experiences but also the chance, via role-plays and rehearsals, to practice and refine new styles of communication or interpersonal conflict-resolution skills before trying them out in relationships with family members and friends.

An additional benefit of group work is facilitation of discussions aimed at helping patients to develop new understandings of the meaning of death in their lives. Loss of an important person confronts people with the need to come to terms not only with that person's death but also with their own mortality. Although coming to terms with this existential reality and finding a way to make it an acceptable part of life is a universal developmental task, it takes on greater salience when a loved one dies. Persons with TG sometimes struggle to accept not only the finality of the death of their loved one but also the inevitability of their own. Opportunities to explore their feelings with others who share the same salience of this question could prove valuable for patients wrestling with these concerns.

RESEARCH EVIDENCE FOR TREATMENT OF TRAUMATIC GRIEF

We have begun the process of confirming the efficacy of TGT with a publication of very promising pilot results (Shear et al., 2001). In this open treatment study, twenty-one patients who scored >25 on the Inventory of Complicated Grief (ICG) (Prigerson, Maciejewski, et al., 1995) and were at least three months postloss (m = 2.9 years, range = three months to nine years), began treatment. Thirteen of twenty-one completed four months of treatment, while eight were dropouts (mean number of sessions = 5.75, SD 3.37, median = 7, range 1 to 10). For both the completer and intent-to-treat groups, we found significant pretreatment-posttreatment differences in the main outcome measure, the ICG. Effect sizes for the changes were 2.19 and 1.45, respectively. Furthermore, the mean decrease in ICG scores for the intent-to-treat group was nearly twice that found in a prior study of depressed patients whose scores on the ICG indicated that they were suffering from TG and who received standard IPT. We have now completed treatments with fifty-four individuals and the results remain comparable. Of note, given the specific focus on grief symptoms in TGT, cooccurring anxiety, as measured by Beck Anxiety Inventory scores, and depression, as measured by Beck Depression Inventory scores, showed concomitant marked reductions in both completer and intent-to-treat groups, with large effect sizes (BAI, 2.04 and 1.08; BDI, 1.8 and 1.16). We are currently conducting a randomized controlled trial of TGT versus IPT, in which we are treating individuals at least six months after the death with an ICG score of ≥30, in order to see whether we can confirm our findings thus far.

CONCLUSION

Traumatic grief is a chronic, debilitating condition that has recently been clearly differentiated from other sequelae of bereavement. Several research groups are currently working to encourage the inclusion of diagnostic criteria in official nosologies. In this chapter we have provided an introduction to the understanding and treatment of this prevalent and debilitating condition, reviewing both research and clinical perspectives on its nature and describing essential principles and concrete strategies for its successful treatment.

REFERENCES

American Psychiatric Association. (1980). *Diagnostic and statistical manual of mental disorders* (3rd ed.). Washington, DC: Author.

American Psychiatric Association. (1987). *Diagnostic and statistical manual of mental disorders* (3rd ed., text rev.). Washington, DC: Author.

American Psychiatric Association. (1994). *Diagnostic and statistical manual of mental disorders* (4th ed.). Washington, DC: Author.

Anderson, C. (1949). Aspects of pathological mourning. *International Journal of Psycho-Analysis, 30,* 48-55.

Bierhals, A.J., Prigerson, H.G., Fasiczka, A., Frank, E., & Miller, M. (1996). Gender differences in complicated grief among the elderly. *Omega: Journal of Death and Dying, 32,* 303-317.

Boelen, P., van den Bout, J., & de Keisjer, J. (2003). Traumatic grief as a disorder distinct from bereavement-related depression and anxiety: A replication study with bereaved mental health care patients. *American Journal of Psychiatry, 160,* 1339-1341.

Bonanno, G.A. (2001). Grief and emotion: A social-functional perspective. In M.S. Stroebe, R.O. Hansson, W. Stroebe, & H. Schut (Eds.), *Handbook of bereave-*

ment research: Consequences, coping and care (pp. 493-515). Washington, DC: American Psychological Association.

Bonanno, G.A., & Kaltman, S. (1999). Toward an integrative perspective on bereavement. *Psychological Bulletin, 125,* 760-766.

Bonanno, G.A., & Kaltman, S. (2001). The varieties of grief experience. *Clinical Psychology Review, 21,* 705-734.

Bowlby, J. (1961). Processes of mourning. *International Journal of Psycho-Analysis, 42,* 317-340.

Bowlby, J. (1980). *Attachment and loss,* Vol. 1: *Loss.* New York: Basic Books.

Brent, D.A., Peters, M.J., & Weller, E. (1994). Resolved: Several weeks of depressive symptoms after a exposure to a friend's suicide is "major depressive disorder." *Journal of the American Academy of Child and Adolescent Psychiatry, 33*(4), 582-587.

Bruce, M.L., Kim, K., Leaf, P.J., & Jacobs, S. (1990). Depressive episodes and dysphoria resulting from conjugal bereavement in a prospective community sample. *American Journal of Psychiatry, 147,* 608-611.

Chen, J.H., Bierhals, A.J., Prigerson, H.G., Kasl, S.V., Mazure, C.M., Reynolds, C.F., Shear, M.K., Day, N., and Jacobs, SC. (1999). Gender differences in the effects of bereavement-related psychological distress on health outcomes. *Psychological Medicine, 29,* 367-380.

Clayton, P.J. (1974). Mortality and morbidity in the first year of widowhood. *Archives of General Psychiatry, 30,* 747-750.

Clayton, P.J. (1990). Bereavement and depression. *Journal of Clinical Psychiatry, 51*(Suppl), 34-38.

Clayton, P., Desmarais, L., & Winokur, G. (1968). A study of normal bereavement. *American Journal of Psychiatry, 125,* 168-178.

Clayton, P.J., Halikas, J.A., & Maurice, W.L. (1971). The bereavement of the widowed. *Diseases of the Nervous System, 32,* 597-604.

Clayton, P.J., Halikas, J.A., & Maurice, W.L. (1972). The depression of widowhood. *British Journal of Psychiatry, 120,* 71-77.

Clayton, P.J., Herjanic, M., Murphy, G.E., & Woodruff, R. (1974). Mourning and depression: Their similarities and differences. *Canadian Psychiatric Association Journal, 19,* 309-312.

Engel, G.L. (1961). Is grief a disease? *Psychosomatic Medicine, 23,* 18-22.

Foa, E.B., & Rothbaum, B.O. (1998). *Treating the trauma of rape: Cognitive behavioral therapy for PTSD.* New York: The Guilford Press.

Fox, G.C., Reid, G.E., Salmon, A., Mckillop-Duffy, P., & Doyle, C. (1999). Criteria for traumatic grief and PTSD. *The British Journal of Psychiatry, 174,* 560-561.

Freud, S. (1917). Mourning and melancholia. *Internationale Zeitschrift fur arzliche Psychoanalyse, 4,* 288-301.

Horowitz, M.J. (1974). Stress response syndromes: Character style and dynamic psychotherapy. *Archives of General Psychiatry, 31,* 768-781.

Horowitz, M.J., Siegel, B., Holen, A., Bonanno, G.A., Milbrath, C., & Stinson, C.H. (1997). Diagnostic criteria for complicated grief disorder. *American Journal of Psychiatry, 154,* 904-910.

Jacobs, S.C. (1993). *Pathologic grief: Maladaptation to loss.* Washington, DC: American Psychiatric Press.

Jacobs, S. (1999). *Traumatic grief: Diagnosis, treatment and prevention.* Philadelphia: Brunner/Mazel.

Jacobs, S.C., Kasl, S., Ostfeld, A.M., Berkman, L., & Charpentier, P. (1986). The measurement of grief: Age and sex variation. *British Journal of Medical Psychology, 59,* 305-310.

Jacobs, S., & Ostfeld, A. (1980). The clinical management of grief. *Journal of the American Geriatrics Society, 28,* 331-335.

Karam, E.G. (1994). The nosological status of bereavement-related depressions. *British Journal of Psychiatry, 165,* 48-52.

Kim, K., & Jacobs, S.C. (1991). Pathologic grief and its relationship to other psychiatric disorders. *Journal of Affective Disorders, 21,* 257-263.

Klerman, G.L., Weissman, M.M., Rounsaville, B.J., & Chevron, E.S. (1996). Interpersonal psychotherapy for depression. In J.E. Groves (Ed.), *Essential Papers on short-term dynamic therapy* (pp. 134-148). Essential Papers in Psychoanalysis. New York: New York University Press.

Lindemann, E. (1944). Symptomatology and management of acute grief. *American Journal of Psychiatry, 101,* 141-148.

McDermott, O.D., Prigerson, H.G., Reynolds, C.F., Houck, P.R., Dew, M.A., Hall, M., Mazumdar, S., Buysse, D.J., Hoch, C.C., & Kupfer, D.J. (1997). Sleep in the wake of complicated grief symptoms: An exploratory study. *Biological Psychiatry, 41,* 710-716.

Melhem, N.M. (2001). Traumatic grief among adolescents exposed to their peer's suicide. PhD dissertation, University of Pittsburgh, Graduate School of Public Health.

Melhem, N.M., Rosales, C., Karageorge, J., Reynolds, C.F. III, Frank, E., & Shear, M.K. (2001). Comorbidity of Axis I disorders in patients with traumatic grief. *Journal of Clinical Psychiatry, 62,* 884-887.

Middleton, W., Burnett, P., Raphael, B., & Martinek, N. (1996). The bereavement response: A cluster analysis. *British Journal of Psychiatry, 169,* 167-171.

Miller, W.R., Zweben, A., DiClemente, C.C., & Rychtarik, R.G. (1992). *Motivational enhancement therapy manual.* Bethesda, MD: National Institute on Alcohol Abuse and Alcoholism.

Neimeyer, R.A. (2001). The language of loss: Grief therapy as a process of meaning reconstruction. In R.A. Neimeyer (Ed.), *Meaning reconstruction & the experience of loss* (pp. 261-292). Washington, DC: American Psychological Association.

Parkes, C.M. (1965). Bereavement and mental illness, part I: A clinical study of the grief of bereaved psychiatric patients. *British Journal of Medical Psychology, 38,* 1-12.

Parkes, C.M. (1970). The first year of bereavement: A longitudinal study of the reaction of London widows to the death of their husbands. *Psychiatry, 33,* 44.

Parkes, C.M. (1972). *Bereavement: Studies of grief in adult life.* New York: International Universities Press.

Parkes, C.M. (1987). *Bereavement: Studies of grief in adult life* (2nd ed.). Madison, CT: International Universities Press, Inc.

Pasternak, R.E., Reynolds, C.F. III, Schlernitzauer, M., Hoch, C.C., Buysse, D.J., & Houck, P.R. (1991). Acute open-trial nortriptyline therapy of bereavement-related depression in late life. *Journal of Clinical Psychiatry, 52,* 307-310.

Piper, W.E., Ogrodniczuk, J.S., Joyce, A.S., McCallum, M., Weideman, R., & Azim, H.F. (2001). Ambivalence and other relationship predictors of grief in psychiatric outpatients. *The Journal of Nervous and Mental Disease, 189,* 781-787.

Piper, W.E., McCallum, M., & Azim, H.F.A. (1992). *Adaptation to loss through short-term group psychotherapy.* New York: Guilford.

Prigerson, H.G., Bierhals, A.J., Kasl, S.V., Reynolds, C.F. III, Shear, M.K., Day, N., Beery, L.C., Newsom, J.T., & Jacobs, S. (1997). Traumatic grief as a risk factor for mental and physical morbidity. *American Journal of Psychiatry, 154,* 616-623.

Prigerson, H.G., Bierhals, A.J., Kasl, S.V., Reynolds, C.F. III, Shear, M.K., Newsom, J.T., & Jacobs, S. (1996). Complicated grief as a disorder distinct from bereavement-related depression and anxiety: A replication study. *American Journal of Psychiatry, 153,* 1484-1486.

Prigerson, H.G., Bridge, J., Maciejewski, P.K., Beery, L.C., Rosenheck, R.A., Jacobs, S.C., Bierhals, A.J., Kupfer, D.J., & Brent, D.A. (1999). The influence of traumatic grief on suicidal ideation among young adults. *American Journal of Psychiatry, 156,* 1994-1995.

Prigerson, H.G., Frank, E., Kasl, S.V., Reynolds, C.F. III, Anderson, B., Zubenko, G.S., Houck, P.R., George, C.J., & Kupfer, D.J. (1995). Complicated grief and bereavement-related depression as distinct disorders: Preliminary empirical validation in elderly bereaved spouses. *American Journal of Psychiatry, 152,* 22-30.

Prigerson, H.G., Maciejewski, P.K., Reynolds, C.F. III, Bierhals, A.J., Newsom, J.T., Fasiczka, A., Frank, E., Doman, J., & Miller, M. (1995). Inventory of complicated grief: A scale to measure maladaptive symptoms of loss. *Psychiatry Research, 59,* 65-79.

Prigerson, H.G., Shear, M.K., Jacobs, S.C., Reynolds, C.F. III, Maciejewski, P.K., Davidson, J.R., Rosenheck, R., Pilkonis, P.A., Wortman, C.B., Williams, J.B., et al. (1999). Consensus criteria for traumatic grief: A rationale and preliminary empirical test. *British Journal of Psychiatry, 174,* 67-73.

Prigerson, H.G., Shear, M.K., Newsom, J.T., Frank, E., Reynolds, C.F. III, Maciejewski, P.K., Houck, P.R., Bierhals, A.J., & Kupfer, D.J. (1996). Anxiety among widowed elders: Is it distinct from depression and grief? *Anxiety, 2,* 1-12.

Rando, T.A. (1993). *Treatment of complicated mourning.* Champaign, IL: Research Press.

Raphael, B., Martinek, N., & Wooding, S. (1997). Assessing traumatic bereavement. In J.P. Wilson & T.M. Keanes (Eds.), *Assessing psychological trauma and PTSD* (pp. 492-512). New York: Guilford Press.

Reed, M.D., & Greenwald, J.Y. (1991). Survivor-victim status, attachment, and sudden death bereavement. *Suicide and Life Threatening Behavior, 21,* 385-401.

Shear, M.K., Frank, E., Foa, E., Cherry, C., Reynolds, C.F. III, Vander Bilt, J., & Masters, S. (2001). Traumatic grief treatment: A pilot study. *American Journal of Psychiatry, 158,* 1506-1508.

Stroebe, M.S., Hansson, R.O., Stroebe, W., & Schut, H. (2001). Introduction: Concepts and issues in contemporary research on bereavement. In M.S. Stroebe, R.O. Hansson, W. Stroebe, & H. Schut (Eds.), *Handbook of bereavement research: Consequences, coping and care* (pp. 4-22). Washington, DC: American Psychological Association.

Stroebe, M., van Son, M., Stroebe, W., Kleber, R., Schut, H., & van den Bout, J. (2000). On the classification and diagnosis of pathological grief. *Clinical Psychology Review, 20,* 57-75.

Szanto, K., Prigerson, H.G., Houck, P.R., Ehrenpreis, L., & Reynolds, C.F. (1997). Suicidal ideation in elderly bereaved: The role of complicated grief. *Suicide & Life-Threatening Behavior, 27,* 194-207.

Van Doorn, C., Kasl, S., Beery, L.C., Jacobs, S.C., & Prigerson, H.G. (1998). The influence of marital quality and attachment styles on traumatic grief and depressive symptoms. *The Journal of Nervous and Mental Disease, 186,* 566-573.

Walitzer, K.S., Dermen, K.H., & Connors, G.J. (1999). Strategies for preparing clients for treatment. *Behavior Modification, 23,* 129-151.

Wilfley, D., Ayres, V.E., Welch, R.R., Weissman, M.M., & MacKenzie, K.R. (2000). *Interpersonal psychotherapy for group.* New York: Basic Books.

Zisook, S., & Devaul, R. (1984). Measuring acute grief. *Psychiatric Medicine, 2,* 169-176.

Zisook, S., Devaul, R., & Click, M. (1982). Measuring symptoms of grief and bereavement. *American Journal of Psychiatry, 139,* 1590-1593.

Zisook, S., & Shuchter, S.R. (1985). Time course of spousal bereavement. *General Hospital Psychiatry, 7,* 95-100.

Zisook, S., & Shuchter, S.R. (1991). Depression through the first year after the death of a spouse [see comments]. *American Journal of Psychiatry, 148,* 1346-1352.

Zisook, S., & Shuchter, S.R. (1993). Uncomplicated bereavement. *Journal of Clinical Psychiatry, 54,* 365-372.

Zisook, S., Shuchter, S.R., Pedrelli, P., Sable, J., & Deaciuc, S.C. (2001). Bupropion sustained release for bereavement: Results of an open trial. *Journal of Clinical Psychiatry, 62,* 227-230.

Zweben, A., & Zuckoff, A. (2002). Motivational interviewing and treatment adherence. In W. Miller & S. Rollick (Eds.), *Motivational interviewing: Preparing people for change* (pp. 299-319). New York: Guilford.

Zygmont, M., Prigerson, H.G., Houck, P.R., Miller, M.D., Shear, M.K., Jacobs, S., & Reynolds, C.F. III. (1998). A post hoc comparison of paroxetine and nortriptyline for symptoms of traumatic grief. *Journal of Clinical Psychiatry, 59,* 241-245.

Chapter 11

Spirituality in the Face of Terrorist Disasters

Kent D. Drescher

INTRODUCTION

The primary purpose of this chapter is to describe a group therapy model for addressing spiritual issues that arise following terrorist disasters. The aim is to help survivors utilize their spiritual resources in their healing process. The problem of *theodicy,* a key spiritual issue for trauma survivors, will be introduced, and current research on links between spirituality and health will be reviewed. A rationale for incorporating spiritual themes and exercises into other existing group therapies for terrorist disaster survivors is also provided. Finally, several session vignettes are offered to illustrate various approaches for dealing with key spiritual themes in recovery from trauma.

The terrorist attacks that occurred on September 11, 2001, were significant for the people of the United States as well as the rest of the world. That series of events has had major repercussions on mental health systems as it has in other life domains. The impact that the disaster had on the spirituality and/or religious perspectives of many people is a major focal point of exploration as well. A 2001 study conducted after the September 11 terrorist events indicated that 90 percent of respondents in this nationally representative sample reported coping with the terrorist event by "turning to religion" (Schuster et al., 2001). This is not surprising in a society such as the United States where high numbers of people report interest and participation in religion or spirituality. For example in a recent Harris survey (Harris

The author would like to acknowledge the insights, wisdom, and creativity of several individuals without whose effort this chapter would not be possible. Helena Young, PhD, and Gilbert Ramirez, MS, collaborated with the author in the creation of a therapeutic group similar to that described in this chapter, and cofacilitated it for the past two years in the PTSD residential rehabilitation program of the National Center for PTSD. In addition, the author would like to thank David Foy, PhD, Patricia Chuo, BA, and Sherry Riney, LCSW, for their thoughtful and helpful comments about this manuscript.

2000 National Issues Survey, study no. 12851, 2000), 93 percent of respondents indicated belief in God, while only 4 percent stated they did not believe in God.

This presents a unique challenge to the mental health profession, as religion is one area in which clinicians differ markedly from their clients (Bergin & Jensen, 1990; Maugans & Wadland, 1991; Shafranske, 1996). One study (Ragan, Malony, & Beit-Hallahmi, 1980) indicated that only between 40 to 45 percent of psychologists and psychiatrists report a "belief in God." In addition, though many providers agree that spiritual well-being is an important component of health, few actually address the topic with their patients. Nearly 60 percent of family physicians indicate they have inadequate training even to inquire about a patient's spiritual history (Ellis, Vinson, & Ewigman, 1999). Many patients, however, express a wish that their doctor/provider would address their spiritual needs and concerns as a part of treatment (D. E. King & Bushwick, 1994). This desire may help explain the finding of one study that four of ten individuals with mental health disorders seek counseling from clergy, which is a percentage equal to or greater than the number seeking treatment from psychologists or psychiatrists (Weaver et al., 1997).

Terrorist disasters, because they are sometimes perpetrated by individuals of one religion against members of other religious groups, may uniquely and powerfully raise religious and spiritual concerns and questions among survivors. All of recorded history, the past several decades providing no exception, contains gruesome examples of violence spawned by religious intolerance directed toward other religious or cultural groups. Since World War II alone, religious intolerance has been responsible for Nazi genocide perpetrated against some 6 million Jews; violence between Protestants and Catholics in Northern Ireland; "ethnic cleansing" of Muslims in Bosnia and Croatia; fighting between Muslims and Hindus in the border dispute between India and Pakistan in the Kashmir region; and suicide bombings, retaliatory violence, and assassinations in the Palestinian–Israeli conflict. All these events show the powerful interconnection between religious perspectives and political objectives in many societies. It seems logical that when faced with the consequences of intense religious hatred directed at their society, individuals may seek to better understand and make sense of these events through examination of their own spirituality.

There is increasing emphasis in the literature on the connection between traumatic events and existential concerns and questions among survivors. For example, Janoff-Bulman (1992) suggests that traumatic events shatter the naive core beliefs that most people hold, such as the safety of the world, the meaningfulness of life, and one's personal self-worth. These existential core values are spiritual in nature and are areas directly affected by spiritual beliefs and practices for most people.

DEFINITION OF TERMS

Religion versus Spirituality

For the purposes of this chapter we will define *religion* as "a system of beliefs, values, rituals, and practices shared in common by a social community as a means of experiencing and connecting with the sacred or divine." And we will define *spirituality* quite broadly as "an individual's understanding of, experience with, and connection to that which transcends the self." The object of that understanding, experience, and connection may be God, nature, a universal energy, or something else unique to a particular individual. A person's spirituality may be realized in a religious context, or it may be entirely separate and distinct from religion of any sort. In most cases, however, religion can be understood as a spiritual experience, with spirituality a more broad, generic way of describing the experience. Kenneth Pargament (1997, p. 32) has suggested a brief but powerful definition of religion as a process, "a search for significance in ways related to the sacred." Spirituality has been defined perhaps most broadly as "multidimensional space in which everyone can be located" (Larson, Swyers, & McCullough, 1997). Miller and Thoresen (1999) have provided a review, citing a number of useful definitions of and distinctions between these two terms. One of the most common distinctions calls attention to the individual versus corporate nature of the two terms. From a mental health perspective, both these constructs can be observed and measured to some degree as practices, beliefs, and experiences. At times in this chapter the two terms may be used somewhat interchangeably.

Natural disasters, because of their relatively random nature, are frequently called "acts of God." Although terrible in their effect for some victims, these events rarely receive the attributions of malevolence or "evil" that are assigned to disasters intentionally perpetrated by humans. Similarly, traumatic events intentionally perpetrated by people pose particular challenges among survivors in coming to terms with the strong feelings of rage and desire for revenge associated with these events. Developing mental health approaches that address spiritual and existential questions and marshal spiritual resources and support for survivors of such events is especially important.

Theodicy

Another context in which to address issues of spirituality in the face of terrorist disasters is a theological idea first articulated in the seventeenth century called *theodicy*. From the Latin *théos díe,* meaning justification of God, the term was coined by Leibniz, who in 1710 wrote an essay attempting to show that the existence of evil in the world does not conflict with belief in the goodness of God (Leibniz, 1890). Frequently called "the problem of evil," the logic behind theodicy is in answering the question it poses: If God is all-powerful, and God is all-good, how does God allow evil to exist in the world? Historically, varied solutions have been proposed to the theodical problem, including philosophical solutions that diminish God (e.g., God is not all-powerful, God is not all-good, God does not exist), or that diminish evil (i.e., it is a punishment for sin, it may bring about some greater good), and perhaps those nonphilosophical solutions that diminish the self (e.g., self-blame, rage, or loss of meaning, purpose, or hope).

From a psychological perspective, Festinger's (1957) cognitive dissonance theory posits that individuals tend to seek consistency among their cognitions and experiences. When there is inconsistency between cognitions and experience there is strong motivation for change to eliminate the dissonance. In the case of a traumatic event, the event cannot be changed; survivors must struggle to adapt their beliefs and attitudes to accommodate their experience in order to resolve the dissonance. Many trauma survivors, along with their families and friends, begin a lifelong journey toward making sense out of their experiences. Dr. David Blumenthal, professor of Judaic studies at Emory University, has written about theodicy and its relation to trauma. He sees theodicy as a healing journey that is neither simple nor linear. Rather, he describes the healing process that occurs after trauma as follows:

> Healing itself is a seriatim process, a tacking into the wind, an alternation between empowerment and desire for revenge, between acceptance and protest, between love and rage. How could it be otherwise? The past cannot be erased or ignored (at least not for any length of time). It must be coped with by mourning and empowerment, and by protest. Further, this must be done, not simultaneously, not linearly, but in an alternating rhythm. This healing-by-tacking is not unethical; it is not dis-integrative; it is not a miring down in a cyclic process. Rather, it is a moving forward by alternating directions. It is sewing with a backstitch, repeatedly. It is integrative—more integrative than healing procedures that urge survivors to "forgive and go beyond," to "be healed once and for all. (Blumenthal, 1998, p. 98)

RESEARCH ON SPIRITUALITY AND HEALTH

It should be noted that over the years, mental health clinicians and researchers have postulated and investigated various aspects of religious experience and spirituality as both healthy and unhealthy. The literature on the relationship between trauma and spirituality is part of a larger body of literature looking at the general relationship between spirituality and health (both physical and mental). In 1996 the National Institute for Healthcare Research convened three panels to examine the relationship between spirituality and physical health, mental health, and substance abuse (W. R. Miller & Thoresen, 1999). All three panels' reviews of the literature suggest that spirituality is generally related positively to health and inversely related to disorders of various sorts (Larson et al., 1997). Several other published reviews of the same literature confirm this finding for physical health (Levin, 1994), mental health (Bergin, 1983; Larson, Pattison, Blazer, Omran, & Kaplan, 1986; Larson et al., 1992), and substance abuse (Gorsuch, 1995).

In addition to research that is PTSD-specific, the literature on medical health outcomes and on stress and coping have relevant findings related to spirituality that do not utilize a PTSD model. Thus there is research examining the effect of spirituality on the course and outcome of life-threatening illnesses such as HIV and cancer (Powell, Shahabi, & Thoresen, 2003; Remle & Koenig, 2001). There is also research about the beneficial effects of religious coping with survivors of severe negative life events such as homicide (Murphy, Johnson, & Lohan, 2003; Rynearson, 1995), and the sudden, unexpected death of a child (e.g., SIDS) (McIntosh, Silver, & Wortman, 1993).

The relationship between traumatic experience and spirituality is complex, and the literature is as yet relatively undeveloped. The literature examining the impact of traumatic events is complicated by the many types of trauma and the difficulty in directly comparing the intensity of trauma exposure across types. Thus, literature tends to cluster by trauma type. Alongside PTSD literature is that of the effects of negative life events. The stress literature tends to see trauma as the high end of a continuum that includes other more mundane (nontraumatic) life stressors with which individuals must cope.

Much of the literature on religion is even more ambiguous. Religion and spirituality are often poorly defined, and few measures are standardized. One study reviewed seven major APA journals from 1991 to 1994 and found that only 2.7 percent of articles utilized any measure of religion, and 79 percent of those utilized a single-item measure (Weaver et al., 1998). Many aspects of religious or spiritual experience have measures with established psychometric characteristics, i.e., motivation, practices/behaviors, beliefs/values, and coping. A frequent limitation of the literature, however, is created by the use of single-item and unvalidated measures.

As we begin to examine the research on trauma and spirituality we find that the research is *bidirectional*. It examines (1) associations between religious and spiritual variables and known outcomes of trauma (i.e., PTSD symptoms, physical/mental health) and (2) the direct effect of trauma on spirituality—both positive and negative.

There are findings suggestive of religion's positive effect upon trauma survivors. For example, two studies found that religious commitment was associated with lower levels of PTSD symptoms among battered women (Astin, Lawrence, & Foy, 1993), and among family members of individuals killed by drunk drivers (Sprang & McNeil, 1998). Subjective religiousness was associated with better overall mental health in a sample of 3,443 female U.S. military veterans, some of whom had experienced sexual assault. Religious attendance was associated with better mental health and lower incidence of depression among the survivors of sexual assault in this same study, a finding that might be explained as resulting from increased social support (Chang, Skinner, & Boehmer, 2001). A recent study by Witvliet (2004) indicates that lack of forgiveness and negative religious coping are associated with more severe PTSD and depression among veterans with PTSD. Another recent study (Fontana & Rosenheck, 2004) shows negative war zone events associated with weakened religious faith, and then weakened faith associated with greater current VA service utilization.

In studies examining medical outcomes from life-threatening illnesses, religious coping has been associated with better recovery from kidney transplant surgery (Tix & Frazier, 1998), and with lower levels of depression among HIV patients (Woods, Antoni, Ironson, & Kling, 1999). In a study examining the relationship between religious practice and affective and immune status of 106 HIV-seropositive mildly symptomatic gay men, religious behavior (e.g., service attendance, prayer, spiritual discussion, reading religious literature) was significantly associated with higher T-helper-inducer cell (CD4+) counts and higher CD4+ percentages, but was not associated with depression (Woods et al., 1999). This finding was not mediated by self-efficacy or ability to utilize positive health coping, and was independent of HIV symptom status.

Mixed results were obtained in a sample of forty-nine lower socioeconomic status (SES) Hispanic women in treatment for early-stage breast cancer; levels of distress were different between Evangelical and Catholic women. Among Catholic women, church attendance at six months was associated with greater distress at twelve months; among Evangelical women, obtaining emotional support from church members at six months was associated with less distress at twelve months (Alferi, Culver, Carver, Arena, & Antoni, 1999). Religious attendance was positively related to perceived social support and greater meaning found in the loss among parents coping with the sudden death of a child. Importance of religion was positively correlated with both cognitive processing and ability to find meaning in a child's death (McIntosh et al., 1993).

Other evidence suggests that not all forms of religious experience are positive. A study by Plante and Manuel (1992) examined the stress responses among college students (psychology and religious studies students) at the beginning of the Persian Gulf War. Results indicated greater subjective distress among Catholic students as opposed to non-Catholics, while strong religious faith was associated with higher levels of intrusive symptoms.

A number of studies have indicated a positive relationship between both religion and religious coping and an improved ability to deal with several types of negative life events, including natural disasters (Gillard & Paton, 1999), war stress among college students (Pargament et al., 1994; Park, Cohen, & Herb, 1990), negative life events in general (Pargament et al., 1990; Pargament, Olsen, Reilly, Falgout, & Ensing, 1992), and effective management of life crisis (Hall, 1986). Conversely, in a study of adults who had lost a family member to homicide, a lack of religious faith (and history of sexual abuse) was associated with greater levels of treatment seeking (which was associated with greater psychiatric symptoms) (Rynearson, 1995).

Studies also indicate that traumatic events may be associated with an increase in spirituality following the event. Carmil and Breznitz (1991), for example, found higher levels of belief in God and belief in a better future among Holocaust survivors ($N = 125$) and their children ($N = 189$) than among a control group of adults of European Jewish descent ($N = 219$). In two studies, women with breast cancer had deeper religious or spiritual satisfaction than women with benign breast problems (Andrykowski et al., 1996), and greater appreciation for life and spiritual growth than healthy controls (Cordova, Cunningham, Carlson, & Andrykowski, 2001). In these studies the authors suggest that certain traumatic experiences can push survivors toward growth in several areas, including spirituality. In a study of 266 women working in religion-related jobs, those who had experienced sexual abuse in childhood (31 percent) were more likely to have turned to faith and spirituality for support that those who had not been abused (Reinert & Smith, 1997).

However, several studies contradict these findings and indicate that traumatic life events may have a negative relationship to spirituality in survivors, particularly among those exposed to human-perpetrated (especially family-perpetrated) trauma, such as child abuse. Several studies have found child sexual abuse associated with decreased religiousness. In one study of 2,964 professional women, religious activity as an adult was mediated by the religious orientation of the family of origin and by whether the abuse occurred within the immediate family. In another nationally representative sample of 2,626 adults, victims of childhood sexual abuse involving penetration were more likely to report a tendency to not participate in religion (Finkelhor, Hotaling, Lewis, & Smith, 1989). In a study of seventy-five Christian women, those who had experienced childhood abuse had lower spiritual functioning than psychiatric (nonabused) and normal controls (Hall, 1995). Incest survivors ($N = 33$) in another study had more anger at God and a stronger feeling that God is distant than did nonabused controls ($N = 33$) (Kane, Cheston, & Greer, 1993), and negative perceptions of God were found in greater percentage among the abused group (Pritt, 1998).

Though not all these areas have been researched, several important clinical themes have been noted among trauma survivors involving religion and spirituality. For example, anger, rage, and a desire for revenge may be tempered by spiritual beliefs or practices. Feelings of isolation, loneliness, and depression related to grief and loss may be lessened by the social support of religious participation (McIntosh et al., 1993). Recovery of meaning in life may be achieved through changed ways of thinking and involvement in meaningful, caring activities brought about by religious or other interpersonal involvement. In addition, such disappointment may be used as a starting point for discussing the many ways in which group members define what it is to have "faith."

There have been several examinations of the avenues by which involvement in religion or spirituality may be beneficial. One is social support, which is known to be beneficial. In other words, religion as it is frequently experienced in social settings places people in proximity of caring individuals that may provide encouragement, emotional support, as well as possible instrumental support in the form of physical or even financial assistance in times of trouble.

A second way that religion and spirituality may benefit the trauma survivor is by facilitating new ways of thinking about the event, such as spiritual schemas that may allow for improved cognitive processing of the trauma. Spirituality and religion also may provide ways to create meaning related to an event (and in so doing, ameliorate some cognitive dissonance), and may involve the individual in a variety of volunteer activities which may also help the survivor to feel useful, actively making a difference, rather than being only a passive victim (McIntosh et al., 1993).

Finally, religion and spirituality may be associated with beliefs about healthy lifestyles and may deter people from unhealthy coping behaviors. This may, for instance, decrease substance abuse and isolation in the trauma aftermath. It also may provide stress reduction through practices such as prayer and meditation.

Although frequent positive associations have been identified between various aspects of religion and overall health, including mental health, several aspects of religious experience have been associated with negative outcomes as well. For example, holding a concept of God as *wrathful* and *punitive* is linked to increased risk of substance abuse (Gorsuch, 1995). Likewise, while collaborative religious coping (i.e., working with God as partner) and religious support coping (i.e., seeking help from clergy) are associated with positive health outcomes, anger at God and punitive religious appraisal (e.g., "God was punishing me") are related to poorer outcomes (Pargament & Brandt, 1998). Identifying the negative aspects of religious beliefs experienced by members within a group can provide important background information as facilitators target the group activities and processes to best meet the actual needs of group members.

NON–MENTAL HEALTH RESOURCES FOR SPIRITUAL SUPPORT

Though the fundamental intent of this chapter is to describe therapeutic techniques for addressing spiritual issues that arise following a traumatic event such as a terrorist disaster and for helping survivors utilize spiritual resources as part of their healing process, it must be recognized that a wide variety of additional resources exist within the local community that can also support these goals. Frequently forgotten by mental health providers is the fact that clergy within a local community provide "front-line" resources for many people, particularly in times of grief and loss. Clergy members are often the first helping professionals consulted during a period of life crisis (Weaver et al., 1997). It can be extremely helpful for trauma therapists to develop close ties with clergy members within the local community to facilitate the sharing of ideas, and as resources and for mutual referrals.

Churches and synagogues have a long history of providing caring, supportive communities and small groups to their members in need. In many Christian churches there has been an emphasis on creative use of small support groups for the past fifty years or more. These group programs come in many forms, providing a variety of formats focusing on faith development, experiential community, and frequently need-specific support (e.g., grief/loss groups). Those involved in promoting this type of ministry activity suggest that these small groups provide opportunities for personal sharing, a sense of connectedness, and spiritual renewal for which many people yearn (Kirkpatrick, 1995; Leslie, 1971).

Peer-to-peer pastoral care programs found within many congregations of all faiths are another potential means of accessing spiritual support for survivors of terrorist disasters and other traumatic events. A prime example of this type of program is the Stephen Ministries (*Media Fact Sheet,* 2001). This program, which began in 1975, provides fifty hours of training to laypersons who then begin to provide structured, supervised, confidential care and listening services to other church members experiencing stressful life circumstances. Over 300,000 laypeople in 7,800 churches in more than twenty countries have been trained and have provided these care and listening services to more than 1 million people in formalized one-to-one caring relationships, and to millions of other individuals through informal caring and listening.

Given the empirical evidence that trauma recovery and spirituality may intersect, and the fact that spirituality is reported to be a primary domain of life experience among many people in many cultures, the remainder of this chapter describes the rationale and methods for conducting group treatment sessions that address spiritual issues. Much of the thinking and experience that led to the group procedures described in this chapter occurred in the implementation of a treatment group in the PTSD residential rehabilitation program of the National Center for PTSD in Menlo Park, California (Drescher, Loew, & Young, 2001; Drescher, Young, & Loew, 2002).

RATIONALE FOR INCORPORATING SPIRITUAL THEMES INTO GROUP-BASED TRAUMA TREATMENT

Several factors provide a primary basis for the use of spiritual themes in treatment for the effect of traumatic events such as terrorist disasters. First, strong evidence indicates that individuals in this society value and appreciate personal spirituality and tend to seek out spiritual resources following traumatic events (Schuster et al., 2001). In addition, trauma theorists have suggested that traumatic events frequently call into question existential and spiritual issues related to the meaningfulness of life, personal self-worth, and the safety of life (Janoff-Bulman, 1992). As seen in the earlier review, empirical support is generally growing for a link between healthy spirituality and positive physical and mental health outcomes. Some empirical support also exists for a link between negative religious coping (e.g., attributions that God is punishing me, or feeling angry at God) and poorer health outcomes. It may be that survivors of traumatic life events are at higher risk for forming these negative beliefs and could benefit from some assistance in reconsidering these and possibly modifying these attributions. This all argues for the importance of addressing these spiritual and existential issues in treatment for survivors of traumatic events.

One rationale for utilizing group-based treatments is the fact that one primary pathway by which spirituality may be associated with positive health outcomes is through social support (McIntosh et al., 1993). The social support that comes from spiritual involvement with other like-minded individuals may help explain positive outcomes found in some research. Group-based therapies offer the possibility of increased social support for survivors of trauma as well as other individuals with similar experiences as they wrestle with these existential questions. This is especially important because avoidance and isolation are primary features of PTSD. Group-based trauma treatment that allows for processing of spiritual questions and concerns and integrates these with standard mental health approaches would seem to be an optimal clinical approach.

The Treatment Model and Theory of Change

The treatment model presented in this chapter shares some characteristics with more traditional present-centered treatments. Specifically, clients are asked to keep their focus on here-and-now concerns. As a result of this present-centered focus, much therapeutic work may be client directed. Group sessions will not be focused on the details of past traumatic events, but rather how the impact of the events is being experienced in the present. The model views an individual's spiritual background and history as aspects of their life that may affect and be affected by traumatic events and the direct symptoms of trauma. The tension or *dissonance* created between the experience of trauma and the individual's pretrauma beliefs, values, and expectations generate the motivation to pursue change in the aftermath of the events.

These group sessions differ slightly from many present-centered therapeutic approaches in that they address very specific topics. They devote focused attention on particular spiritual and existential issues that might or might not arise naturally in a present-centered group. These sessions are specifically designed to address concerns that clients may not have resources to voice and that clinicians frequently do not have background, experience, or training to address.

Goals and Purposes of the Group Exercises

The primary focus of this chapter is to describe therapeutic group interventions for survivors, including families of victims of terrorist disasters, that focus on the interrelationship of spirituality and the impact of a terrorist disaster. Individuals exposed to such events may experience a variety of symptoms with a range of intensity up to and including diagnosis of psychiatric disorders. These issues may include PTSD symptoms, traumatic bereavement issues, major depression, adjustment disorder with depressed or anxious mood, and religious or spiritual disillusionment and doubt.

The group sessions and spirituality-enhancing interventions that follow are designed to be used in one of two ways. They may be used as a series of group sessions for trauma survivors interested in pursuing these issues who also have other primary means of clinical support; these sessions *are not* designed as a primary treatment for PTSD or other psychiatric disorders but as an adjunct to other types of intervention in the aftermath of a terrorist disaster. All or part of these sessions may also be utilized as individual modules that could be incorporated into other group treatments for PTSD to enhance their effectiveness. In either situation timing in the presentation of these interventions is important. Though there is no empirical evidence to inform the optimal time to address these issues, logic would suggest that it should be at a time when the participants have sufficient attentional resources to direct toward consideration of existential issues. Ideally these interventions should be presented after the immediate "crisis" phase of the disaster when emphasis is appropriately placed on providing immediate resources (i.e., shelter, food, clothing) and facilitating access to other direct disaster relief services.

A number of primary goals exist for interventions focused on the intersection of spirituality and traumatic events such as a terrorist attack. One such goal is to encourage development of a healthy vital spirituality that might serve as a healing resource in coping with traumatic events. This involves strengthening and deepening the group members' present spiritual or religious understandings and practices. It involves reconnecting with their religious or spiritual roots and traditions from childhood. It may also involve searching out and exploring new avenues of spiritual experience and expression that are more immediately relevant to the members' recent experiences. Exposure to and direct experience of a variety of spiritual practices from a number of traditions need to be incorporated into the interventions. Group activities are selected to express both diversity and the inherent value of a wide variety of spiritual experiences.

Another goal of this intervention is to help facilitate appropriate cognitive processing of the existential meaning associated with the disaster events and the personal significance individuals might attach to them. This includes identification of cognitive distortions (e.g., inappropriate survivor guilt, including self-blame). It certainly includes helping individuals begin to process and seek personal answers to the difficult existential questions of "why" and "how" these events have occurred, as well as what impact such events should and will have on the course of each individual's life in the future. It facilitates the shared feedback and reflections from other group members about how they are coming to terms with these things.

A third primary goal of the group interventions is to increase perceived social support and encourage enhanced development of a healthy family and community support system. Communities usually rally and provide additional support for survivors of disasters such as a terrorist attack in their immediate aftermath. Over time, however, as people continue with their own day-to-day struggles of life, this enhanced sense of community begins to fade. Also, individual and cohort differences in the degree of support experienced in the aftermath of terrorist events are likely. Some individuals and groups may feel quite isolated in the midst of the many rescue and reconstruction activities. A good network of support may allow an individual to feel less alone and more capable of obtaining necessary resources.

The intended outcome of these interventions and group sessions is to increase hope, to decrease anger and hostility, and to decrease the intensity of feelings of grief and loss, while enhancing the individual's sense of purpose and the meaningfulness of life.

CONSIDERATIONS PRIOR TO IMPLEMENTING ACTIVITIES

Criteria for Selection of Members

Of primary concern in any group intervention is identifying the appropriate target population. Identifying individuals that might be expected to benefit from these group sessions, and those for whom the sessions may not be beneficial or who might disrupt the group in some way, is a key initial task for the facilitators. Methods of assessing and evaluating individual participants as well as the overall clinical outcomes of the group must be identified. Some areas to be considered in assessment of participants for this treatment should include

1. an understanding of each individual's unique spiritual history, not for purpose of exclusion but for insight into how life experiences have affected it;
2. each individual's unique history of exposure to traumatic and other negative life events (including degree of exposure to the specific terrorist disaster);

3. screening for developing or worsening psychiatric problems (e.g., PTSD, depression, substance use); and
4. general states (i.e., life satisfaction, well-being, or happiness) indicative of quality of life.

Assessment of spirituality can be performed in a variety of ways ranging from engaging in unstructured clinical interviews to distributing highly specific questionnaires. A good overview of helpful strategies and useful instruments can be found in the book *Integrating Spirituality into Treatment* (Gorsuch & Miller, 1999). Of key importance for the function of the group is identifying potentially deleterious belief systems and the individual participants' levels of motivation in examining their spiritual history, current attitudes and values, and spiritual practices.

Measurement During Assessment

In recent years, a collaborative effort between the Fetzer Institute and the National Institutes of Health (NIH) created a working group that gathered together many of the key researchers investigating spirituality. One result was a new instrument, the Brief Multidimensional Measure of Religion/Spirituality (BMMRS) (Fetzer Institute, 1999). This scale measures many aspects of spirituality, including daily spiritual experiences, spiritual practices, spiritual coping, forgiveness, religious support, and commitment and meaning. The brief scales were created using the best items from a number of well-validated measures. Long versions to assess each of the domains are included in the manual and may be most appropriate when more reliable and detailed assessment of a particular area is required. The BMMRS was incorporated into the 1998 General Social Survey (GSS), a random national representative survey examining attitudes about a wide variety of social topics (Davis, Smith, & Marsden, 2001). Findings supported the psychometric properties of the instrument as well as the utility of the multidimensional approach to measurement (Idler et al., 2003).

Assessment of the exposure to and the effects of trauma is another important task in attempting to understand the participants in the group process. Many useful and well-validated instruments exist. For the assessment of the lifetime experience of trauma there are only a few instruments with well-demonstrated psychometric properties. One of these is the Traumatic Life Events Questionnaire (Kubany et al., 2000). This scale is a twenty-five-item self-report instrument that assesses PTSD Criterion A for a number of possible traumas that may occur in the course of a person's lifetime. The scale takes approximately ten to fifteen minutes to complete. In a group formed after a terrorist disaster, most group members will have been exposed in some way to the disaster. However, it is also important for facilitators to know what other types of traumatic experiences to which members may have been exposed. These traumas may be exacerbated by the immediate disaster and contribute to coping difficulties that may arise during the group process.

Another important area for assessment is the extent of PTSD symptoms present. Many brief self-report scales are adequate for these purposes. One that is both brief and which has been used across many different trauma populations including terrorist disasters, is the Los Angeles Symptom Checklist (Foy, Wood, King, King, & Resnick, 1997; King, King, Leskin, & Foy, 1995). The scale has good psychometric properties and normative scores for several trauma types.

Understanding how an individual thinks about a traumatic event and identifying ways in which this thinking may be distorted is an important clinical activity. One method of assessing this is the Posttraumatic Cognitions Inventory (Foa, Ehlers, Clark, Tolin, & Orsillo, 1999). This scale has good psychometric properties and factors that measure negative cognitions about self, negative cognitions about the world, and self-blame. Group member responses to specific items can also provide a launching point for discussing ways that the disaster has affected each member's view of the world and their place in it. Although most trauma-focused scales measure the negative consequences of traumatic experiences, in recent years growing attention has been paid to the potential positive effects of traumatic events. The Posttraumatic Growth Inventory (Tedeschi & Calhoun, 1996) is one of the first scales to attempt to measure these changes. Factor analysis has identified several areas of potential growth, including relating to others, new possibilities, personal strength, spiritual change, and appreciation of life that individuals may report as positive consequences of experiencing traumatic circumstances.

Inclusion/Exclusion Criteria

Inclusion criteria for participation in group sessions that address spirituality in the context of a terrorist disaster are relatively straightforward. First, a participant must be willing to begin an exploration of how spirituality might be a resource in attempts to better cope with aftereffects of trauma. This is not to say that individuals need to feel entirely positive about this idea. It is acceptable for participants to be hostile to religion or be extremely angry at God, or angry at some particular religious group for "allowing" the events to occur. It is sufficient for individuals to be willing to participate actively in the group process, to express themselves honestly, and to be willing to receive support and honest feedback from other group members. Second, there must be some significant exposure to the particular disaster; this may be direct exposure through actually surviving the events or it may be exposure to the event through loss of or injury to a friend or other loved one. There may be community exposure by living or working in the vicinity of the event. Exposure may also occur by helping with the disaster recovery efforts. In some cases individuals may be exposed to the event in several of these ways.

The primary exclusion criteria for this group would be the presence of psychiatric symptoms of such intensity that in the clinical judgment of the facilitators an individual's participation would be harmful to the participant, would keep him or her from receiving more appropriate treatment, or would be frightening or harmful to other group members. An example of this might be evidence of spirituality that has been harmfully modified or created by a psychiatric process, such as psychotic religious or spiritual delusions, or a manic episode characterized by extreme forms or religious activity. In such instances, group activities designed to enhance spirituality might actually worsen or intensify the psychiatric problems rather than helping.

Group Composition

As is true for all group processes, the size of the group may influence the effectiveness of the interventions. Extremely large groups make it difficult to utilize direct interpersonal processing unless the group subdivides to allow more personal sharing and ensure some members are not left out. Small groups limit the variety of resources available from other members and may be vulnerable to losing the character of a group if the drop-out rate is high. For group sessions like these that involve some amount of interpersonal connection and sharing, a group size of six to perhaps fifteen members with two facilitators would be ideal. All members should have some degree of exposure to the terrorist event, either as direct survivors, family members of victims, or perhaps rescue workers. It is not necessary for all members to have the same type of exposure. To the degree possible, it is also helpful to provide for as much diversity as possible among group members and facilitators. A group that includes a healthy variety of spiritual and cultural traditions, and experiences can provide a wider array of resources and ideas for coping with the new challenges posed by terror's aftermath. It may also provide expanded ways of thinking about other life problems and challenges. Obviously, such diversity may not be possible in instances in which trauma impacts a large number of people in a particular ethnic or cultural community.

Recruitment of members for specialized group sessions such as these can potentially come from clinicians and case managers familiar with their clients' individual needs and concerns that may overlap with the goals of the group. In-service presentations and brochures describing group sessions and goals can alert providers about the clinical opportunity for their clients. In addition, posters or flyers advertising the start of a group can be posted in places where survivors are likely to be (e.g., Red Cross stations, FEMA offices, hospitals, and clinics).

Preparation of prospective group members prior to the start of treatment which includes a one-on-one meeting between each group member and a facilitator that addresses spirituality in the context of trauma can be very useful to prevent early drop out by members who do not quite understand the aims of the group. The screening interview for individuals interested in participating in such a group is an opportunity for group facilitators to review their clinical and social history and evaluate their suitability for treatment. Included in this evaluation should be a review of prior trauma history, including exposure to terrorist disasters and screening for possible Axis I or II disorders, including PTSD, depression, and substance use. This is also the time when motivation for treatment and openness toward the exploration of spiritual beliefs, values, and practices are evaluated. During this one-on-one meeting, the goals, rules, and timing of the group are thoroughly reviewed to ensure that each client has a clear understanding of the purpose of the group sessions and what they might expect from participation. Finally, a formal treatment contract is signed by all members that includes important clinical understandings such as the limits of confidentiality within group treatment, as well as particulars such as the meeting dates/times and duration of the group.

The one-on-one meeting with the client is also an opportunity to ensure that he or she is adequately prepared for group participation. Because these group sessions are conceptualized as adjunctive treatment, they do not contain all necessary components for treatment of all disorders which might occur subsequent to a terrorist disaster. All members must have adequate case management available and as clinically necessary to be participating in other forms of treatment. In conjunction with the client, group facilitators must identify specific family, church, and community supports that may help them better manage day-to-day stressors. Finally, it is important for all group members to have a clear understanding that this treatment group is a mental health intervention and not a church or religious service.

Role of the Therapist

Therapist style and modes of intervention can have a tremendous effect on treatment outcomes. As in other group treatment settings, in a group addressing trauma and spirituality it is beneficial to have two therapists, particularly for larger groups of ten to fifteen members. A second therapist can provide diversity of style and personality and possibly different viewpoints that can be helpful. In addition, because some issues addressed in this type of group may generate disagreements and tension given the depth of feelings connected with these issues, a second facilitator can be very helpful in observing and helping process underlying issues. Facilitators provide initial leadership for the group by introducing topics and contributing to the overall flow of the group. They do this by observing interactions, keeping their finger on the pulse of the group, and providing interpersonal process comments about what seems to be happening in a particular interaction. They also encourage the group to provide feedback to individual members, and then model this process themselves. Therapists should be comfortable with some level of personal self-disclosure as a means of modeling this for the group but should also maintain clear boundaries and a focus on clients' experiences as central to the treatment.

An initial task of the facilitators along with the group is to create a working alliance as in any treatment context. One element of this alliance is a shared understanding of the group's limits and boundaries which could be called the "group culture." With survivors of a terrorist disaster as with any group with potential PTSD symptoms, safety is a key issue. It is important for members to feel that group is a safe place to discuss fears and express strong emotions, and that the facilitators and other group members will be able to tolerate and even benefit from their expressions. In the early sessions, facilitators will be required to play a more active role by describing, modeling, and encouraging behaviors that are characteristic of an effective working alliance. This will be done by telling group members how the group will work, gently reminding them when they stray from the guidelines, and processing the difficulties that may arise.

Another initial task of the facilitators is to help shape a group culture that sustains itself across sessions. At the heart of this culture are the expectations of the members as to what types of behaviors and forms of expression are appropriate and tolerable within the group. This is especially important in a group addressing issues of religion and spirituality, as many group members may have strong expectations instilled from childhood about what kinds of behaviors or statements are appropriate in a religious or spiritual context. These childhood expectations may differ across group members and may be very different from what the facilitators intend. These expectations may differ in important ways even in groups in which all members share a similar religious, ethnic, or cultural background. Facilitators will likely be aware of some of these issues of religious background and expectations from the pre-group assessment of members' spiritual history. Of key importance is to create a group environment in which it is acceptable to express religious doubts or struggles, and to be free to describe ways in which individuals may feel let down or disappointed by God or their spiritual community. In addition, it can be useful for the group to begin to accept the expression of strong negative feelings—even those directed toward God.

Facilitators need to pay close attention to their use of language during group sessions. They must use clear and understandable definitions of key words and ideas presented in the group. Frequent repetition and review is also important. Words such as *religion, spirituality,* and *forgiveness* should be carefully and clearly defined so that all members may articulate their meanings. An example of how to do this is presented in the sample clinical vignettes later in this chapter. These definitions should be reviewed from time to time throughout the course of the group to ensure members continue to share the same definitions with facilitators and one another. Facilitators should carefully observe member reactions during presentation of new ideas for nonverbal indicators of discomfort, disagreement, or simply emotionality. It is important to check in with members about their reactions and to resolve the issues that arise to allow questions and disagreements to be processed effectively.

Core Group Values

A number of core group values should be described and modeled by facilitators to members in the earliest sessions and repeated as necessary throughout the duration of the group. The first of these is openness to new ways of thinking, new behaviors, and new experiences. Members should be encouraged to withhold judgment about the experiences of others and to value (or at least tolerate) differences within the group. One of the benefits of group therapy that should be acknowledged and used to its fullest advantage is the opportunity to receive insights and ideas from the coping strengths of other trauma survivors as well as from the training and experience of the facilitators. Differences expressed among members should be viewed as opportunities for learning and growth. This value is particularly important in a group that addresses spirituality in the context of trauma. It is important for members to understand how important religious and spiritual perspectives may be to people and how strongly they may react if their beliefs or values are challenged, threatened, or demeaned by others. As a consequence, one group rule that should be strictly adhered to is the use of "I" statements in discussing personal beliefs, theological perspectives, thoughts, and values. Facilitators should intervene early and utilize frequent gentle reminders when members fall into global statements that sound as though the person is speaking for the group or is expounding the "right" way of looking at an issue. At the same time, "I" statements should not be used as a "semantic mask" from behind which to do those very things (Hammer, personal communication, 2003).

Self-Disclosure

Self-disclosure is a key aspect of any group treatment as facilitators and members interact with one another. Group members may need to be coached on occasion throughout the sessions as to the type and appropriate amount of self-disclosure that should occur during treatment sessions. These group sessions are present centered and focus on group members' current attitudes, thoughts, beliefs, and values that have been affected by trauma. As such, self-disclosure on the part of members should generally be encouraged, particularly as it relates to the members' spiritual experiences, beliefs, and practices and how these are related to current life functioning and recent life events. Group sharing of the spiritual or religious issues that may arise from traumatic experiences is completely appropriate, as they are a core focus of the group sessions. However, disclosure of detailed trauma descriptions beyond what is necessary to help others understand their current experience should be discouraged. Members should be encouraged to be sensitive to the fact that others in the group may have PTSD symptoms or other clinical issues that might be triggered or exacerbated by graphic disclosures of traumatic events. It is easy for members to begin sharing unnecessarily explicit trauma details as a natural part of a narrative when they are talking about the religious impact of the events. In such cases, a gentle reminder by the facilitator that specific details of the experience are not necessary may be required to keep members from getting sidetracked into recounting too many specifics.

COMPONENTS OF THERAPY

Several components are involved to varying degrees during treatment. These are discussion and interpersonal processing, learning about and direct experience of several spiritual practices common in many spiritual traditions (e.g., meditation, prayer, service), and homework that allows for individuals to continue to explore spirituality outside of the group and report their experiences back to the group. During discussion portions of a session, a facilitator may introduce a topic and ask members to share their relevant thoughts, feelings, and experiences. The goal of discussion at one level is to broaden the knowledge base of the group and to stretch members' thinking in a particular direction, but it is also to facilitate the next component of the group, which is interpersonal processing. It is important that members encounter one another through self-disclosure and by giving and receiving feedback about how they experience other members. A goal of interpersonal processing within the group is to broaden social support networks and to enhance trust. Depending on the size of the group, small-group sharing in dyads or groups of three to four people allows for periods of more intimate sharing that may make interacting easier for more reticent members of the larger group.

In addition, most sessions should include an experiential component in which the group actually participates in spiritual practices or rituals such as meditation, prayer, or special readings. These experiential elements should be selected with a view toward exposing members to a broad and diverse spectrum of spiritual experience. When possible, several experiential options can be made available and group members encouraged to select an experience based on their needs and interests.

Finally, homework should be assigned at each session and reviewed during the next meeting. Homework assignments should be designed to be enjoyable and should contain a mix of introspection (e.g., writing in a journal) and direct experience (e.g., attendance at spiritual activities, practice of rituals). The use of homework is an element of many forms of individual and group treatments. Homework provides opportunities for generalization of therapeutic experience into members' actual lives and experiences. It also provides additional time during the week when members are focused on the goals of therapy. This group utilizes homework as a means of extending and broadening the group experience through the use of both cognitive activities (i.e., thinking, reflection) and behaviors (i.e., spiritual practices, attending services, reading, writing). Homework also affords members other opportunities to participate in religious and community activities that may help shift destructive meanings associated with the disaster and increase social support. Homework that involves activities may encompass helping other victims and families, volunteering in the community, getting involved in rebuilding efforts, and participating in fund-raising activities. Whenever homework of any kind is given, time must be set aside in the next group session to process members' experiences. This communicates that the activities are important and holds members accountable. It also enhances motivation for the group as a whole to acknowledge that the activities of members outside the group are important to group process and to each individual's healing. Facilitators should attempt to combine a mix of all these components during each session. Facilitators should also endeavor to share leadership among group members in selection and sharing of various readings or prayers.

One important factor in facilitating effective rapport among group members is encouraging constructive engagement. Managing the level of tension within the group is one way of doing this. Given the elevated level of stress that may exist within the traumatized community, facilitators should not be surprised if minor conflicts and tensions emerge within the group. Strong emotions are frequently close to the surface with survivors, and issues of blame and responsibility abound—particularly in a human-perpetrated disaster, such as a terrorist attack. Although it is important to allow for emotional expression, it is quite possible for the group to become sidetracked by political or theological arguments when members have different viewpoints. Effective facilitators walk a fine line by encouraging both expression and tolerance, by leading discussions about core beliefs and values that do not degenerate into divisive battles that detract from group goals. It is also important to help the group manage individual members' levels of participation and sharing. This includes gently limiting the role of individuals who may dominate the group process and encouraging those who may find it more difficult to share their thoughts and feelings.

Skill Development

Trauma and spirituality group treatment should help members develop a number of therapeutically useful skills. First would be the ability to seek and give support to others with regard to their spiritual journeys. The ability to tolerate, accept, and even celebrate differences in beliefs, values, and viewpoints allows members to consider the benefits they may receive from engaging in dialogue with others. Speaking with others about spiritual perspectives and experiences can provide members with the ability to enlarge their supportive network in the community. These group sessions also help members reflect and reexamine issues of meaning in the context of disastrous events. Finally, such a group should provide members with experience in the exploration of a number of forms of spiritual practice which may be useful in managing stress and coping more effectively with the psychological and emotional impact of a terrorist disaster.

Spirituality-Enhancing Exercises for Group Therapy

A number of exercises are available to support the overall treatment goals of increasing social support, enhancing cognitive processing of the existential issues arising from the terrorist event, and enhancing a sense of personal meaning and purpose. This section will explain some of these exercises and their relation to these goals.

Spiritual Autobiography

The spiritual autobiography is an exercise that provides opportunity for personal reflection and sharing among group members. It is designed to enhance the third primary goal of group sessions by increasing the sense of social support among members. Because trauma frequently isolates survivors and leaves them thinking that no one else has experienced what they have or could possibly understand their experience, sharing spiritual history in the context of various life events reveals to members frequently how alike they are. Prior to exploring the activity, members are given an autobiography worksheet as in Figure 11.1.

Each member is asked to reflect on important events of his or her life, both positive and negative. Using the symbol codes at the bottom of the worksheet, members are asked to place each event on the timeline, using the symbol for the type of event and an arrow to indicate whether the event was positive or negative. Each symbol should be placed vertically somewhere between the High and Low ends of the spectrum, indicating the importance that spirituality or religion played in their life during the time of each event. After each member has drawn the personal timeline, each is given the opportunity to briefly tell his or her story by explaining the graph to the group. If the group is large (i.e., more than twelve members), it may be subdivided into two or three smaller groups and each person can share the story there. Members should be instructed to think about what they learned that was new from each person's story, as well as what they learned about themselves. The group should reflect together about common themes that emerge from the exercise and any revelations they have about how various types of life events (especially trauma) affect spirituality.

Values Exploration

Exploration of values is an exercise that contributes to the members' growth and skill development and facilitates the group goal of increasing their cognitive processing of the traumatic events. One method of exploring values is through the use of a values assessment measure such as described in the BMMRS manual (Fetzer Institute, 1999). After introducing the concept of values, facilitators should explain to members the idea that values may change in the aftermath of traumatic events. The values inventory can

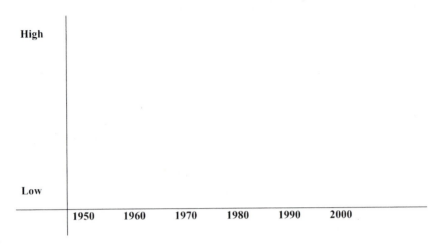

Instructions: Place symbols on the timeline to correspond with the time(s) in your life that events like these occurred. Use an arrow next to the symbol to indicate whether these were positive or negative life events. Place the symbol within the range from Low to High to indicate the importance, intensity, or value of religion or spirituality in your life at the time the event occurred. Draw a line which connects the various events as an indicator of the role of religion/spirituality in your life over time.

✗ = work/financial events (new job, moving, promotion, job loss, bankruptcy)
♥ = relationship events/family events (birth, death, marriage, divorce)
∿ = traumatic life events (fires, accidents, crime victim)

♉ = religious events (baptisms, confirmation, bar mitzvah)
♒ = military events (induction, combat, discharge)
📖 = educational events (high school, college, graduation)

↑ = positive events
↓ = negative events

FIGURE 11.1. Spiritual Autobiography

be assigned as homework two sessions prior, so that it can be returned, scored, and summarized prior to the values session. This activity helps members identify and clarify the things that are most (and least) important to them now and contrast it with their priorities before the traumatic event. During the session dedicated to the discussion of values, members will have an opportunity to reflect on things that surprise them about the group's patterns of responses, and ways their responses are alike and different from those of other members. Members should share how they see their values changing from before the disaster and what if any changes in behavior might be necessary to have their behavior and lifestyle more consistent with their new values.

Spiritual Practices

Exercises involving participation in various spiritual practices can be effectively utilized both within group sessions and in homework assignments. These types of exercises address all three primary goals of the group sessions. Utilized within the group sessions, they enhance relationships among members, improving perceived social support. Most of the exercises are reflective in nature, allowing for healthy cognitive processing of important existential themes, and perhaps providing a sense of meaning directly.

The reading of written prayers when utilized during group sessions should reflect a broad spectrum of religious traditions over time so that group members can begin to widen their appreciation of the wisdom and contributions of people who differ from themselves. Members should be involved as much as possible in the selection and reading of prayers used within group. Outside of group, members should be encouraged to develop styles and modes of spiritual practice that are comfortable and consistent with their own beliefs and values.

Spiritual practices such as meditation can be useful in providing an avenue for both relaxation and developing insight. Nearly all religious traditions include occasions for meditation. Research on the impact of several forms of meditation has been done, and results have been generally positive (Alexander, Robinson, & Rainforth, 1995; Gelderloos, Walton, Orme-Johnson, & Alexander, 1991; Harris, Thoresen, McCullough, & Larson, 1999; J.J. Miller, Fletcher, & Kabat-Zinn, 1995). In a helpful chapter on meditation, Marlatt and Kristeller (1999) note that many reviewers of the literature on meditation cluster various techniques into two main types: concentrative meditation and mindfulness meditation. They describe concentration techniques as those in which attentional focus is narrowly concentrated on an experience (such as breathing), an object (such as a candle), or a particular word or sound. The goal of these techniques is to observe one's physical and mental processes, and to redirect one's attention whenever it wanders, back to the area of focus.

Mindfulness meditation techniques have the goal of insight, by being fully aware of the range of experiences available to the individual's senses from moment to moment (Marlatt & Kristeller, 1999). It involves awareness of one's immediate experience with acceptance rather than judgment. One issue to be aware of in the use of mindfulness techniques, in particular with trauma survivors, is that an "opening up" of the mind may provide an opportunity to experience distressing intrusive memories about the trauma, such that meditation becomes an exercise of painful rumination. If this happens consistently it may be useful to redirect and train the individual in the use of more concentrative techniques that encourage the person to focus attention and pull away from distracting thoughts back to the object, experience, or sound being utilized. Meditation exercises properly practiced allow a person to develop the ability to be the detached observer of his or her experience which can help people make informed intentional choices about their behavior and goals.

Rituals are a part of spiritual practice in every spiritual tradition and provide a way of connecting the inner (cognitive, spiritual) experience with the outer (behavioral and experiential) life. A ritual is defined as "a ceremonial or formal solemn act or observance" (Neufeldt & Guralnik, 1988). Simple rituals selected and enacted regularly by the group can provide members a meaningful way of connecting with one another that transcends simply verbal interaction. Rituals can be as simple as joining hands in a moment of silence as a way of closing the group every week, or they may be more elaborate.

In addition, many other forms of spiritual practice within Western Judeo-Christian traditions and in other religious and cultural traditions can be explored. One possible homework assignment for group members is to search out and bring together resources to create a library of materials to support the group's spiritual exploration. Books such as *Celebration of Discipline* (Foster, 1988) provide a good resource about traditional spiritual practices or "disciplines" within the Christian tradition. The author identifies and describes a number of less familiar traditional disciplines that can assist in a person's spiritual journey. Exposing group members to new ideas and potentially beneficial spiritual practices addresses one of the primary goals of the group sessions.

Forgiveness Exercises

Exercises centered on forgiveness can be potentially important in working with victims of terrorist disasters. Because of the attributions of evil and malevolence attached to intentionally perpetrated traumatic events such as terrorist disasters, survivors and family members of victims frequently struggle with feelings of hatred, rage, and vengeance that are difficult to get rid of and may interfere with functioning. Forgiveness exercises strongly support the second primary goal of the group: cognitive processing of the meanings associated with the traumatic events. Group members may feel "stuck" with these feelings and unable to move forward. Research has begun to show positive benefits from forgiveness interventions in a number of contexts, including marital problems (DiBlasio, 2000; Gordon, Baucom, & Snyder, 2000), incest (Freedman & Enright, 1996), and men struggling with women's pro-choice decisions (Coyle & Enright, 1997). Forgiveness interventions can focus on forgiving oneself, others (possibly including perpetrators of terror), and even God. Clear definitions of forgiveness are important, as confusion and even anger may result from misunderstanding the term. Members must understand that forgiveness does not mean condoning an act of terror or forgetting the victims. Moreover, there is no requirement that *reconciliation* with the perpetrator be part of the process. The primary purpose of a forgiveness intervention is to allow the survivors to loosen the hold that the event and the related emotions have on them and to begin to move forward. The goal is to help people get "unstuck."

Several key elements are part of a forgiveness exercise (Worthington, Sandage, & Berry, 2000). A first element is clarification of responsibility. One problem with conceptualizations of responsibility in the midst of or shortly after a traumatic event is that they may contain significant distortions or logical or even factual errors. People may blame themselves or others for injury or death of loved ones for nonsensical or illogical reasons. People may blame themselves for simply surviving when others did not. Allowing and encouraging group members to speak about their beliefs concerning blame and culpability and to receive corrective feedback from other members and facilitators is very important before the concept of forgiveness is addressed.

These distortions may also be theological in nature (e.g., "God is punishing me"). One difficult problem is that these distortions may be part of strongly and long-held theological ideas that are actually part of the members' religious upbringing. It is important to create a group environment that can gently challenge members to reexamine those beliefs and allow other members to share different understandings that are part of their own upbringing and tradition. Facilitators may share the fact that sometimes people change their beliefs and ideas, and solicit from the group examples of that. It is important not to engage in theological arguments, but rather to keep people focused on and allow for feedback about how their ideas, beliefs, and traditions either help or hinder their own healing process.

Another element of the forgiveness process is helping group members move in the direction of making a *decision* to forgive. It should be emphasized throughout the group process that forgiveness is a personal choice. This should be reiterated when members introduce objections or reasons why forgiveness is inappropriate in a given situation. It should also be emphasized that this choice is a reflection of the members' desires to pursue their own health, not a decision to be made because outside influence (e.g., church, family and friends, or even group leaders) is being exerted. It should be a decision made after weighing carefully the costs and benefits of such a decision.

A final step in the forgiveness process has to do with reinforcing the decision to forgive. This may involve a reframing of the event or a retelling of the story within its larger context. It should also be suggested that it is normal for old ways of thinking about the events, and old feelings of anger, betrayal, and pain to reemerge at times. Members should be encouraged to see this as a normal part of the healing process and not as a failure in forgiveness. They should be encouraged to bring to mind the reasons for their decision and to recommit themselves to their pursuit of health.

Research and clinical practice on forgiveness are not without controversy. Particularly with female survivors of malevolent male-perpetrated traumas such as incest, in which victim and perpetrator have differing amounts of power on several levels (including physical, social, and relational), serious questions have been raised as to whether forgiveness is appropriate (Lamb, 2002). These authors propose that maintaining healthy levels of both anger and compassion is a more mature, healthier goal for these survivors. Clearly, it is important to recognize that clinical goals are unique for any given individual. Forgiving may be the prescription for release and moving forward for one person; not forgiving, and permitting oneself to be angry, may be the best prescription for regaining the power taken away in the course of the offense for another survivor.

Situations and Members Experienced As Difficult

Several types of situations and behaviors exhibited by members may be difficult for facilitators to manage. Some of these are anger and hostility that may manifest as conflict among members, aggression directed toward a facilitator, or overt hostility directed at a particular religious or spiritual tradition. Several techniques aid the management of these emotions within the group. The first is to remind group members that the group values honest and appropriate expression of strong emotions of all types. If anger seems so strong as to feel threatening or intimidating to others, those reactions should be identified by members; the angry member is then gently confronted as to whether he or she intended to create that type of reaction. In addition, it may be helpful to ask the angry member to reflect on what other emotional cues may have elicited the anger, such as feeling sad, hurt, discounted, or betrayed in some significant way.

Premature termination by a group member is another issue that must be managed by facilitators. How this is handled will depend on whether the member simply stops attending, has to drop out due to some unplanned change in availability, or has a specific problem with the group and chooses to drop out. A general principle should be adhered to regardless of the reason for termination. It is important to process the loss of a member; in any group dealing with trauma, loss is a sensitive issue—even the "loss" of a group member—that may stimulate emotions related to other traumatic losses. Giving members an opportunity to express these emotions and to separate the new loss from other losses is therapeutically useful. If possible, members leaving the group should be allowed to say good-bye and to express feelings with other members prior to leaving. If members give indications of potentially dropping out, such as anger at the group, spotty attendance, or diminished participation in the group process, and are unwilling or unavailable to express their experience and feelings within the group, it may be appropriate for the facilitator to check in with and perhaps meet briefly with the member separately to see if the problem can be identified and a solution formulated, with a goal of retaining the member as an active group participant.

VIGNETTES DEMONSTRATING INTERVENTIONS AT DIFFERENT STAGES

Session Vignette: Introduction and Explanation of Guidelines

This session is designed to be the first of several sessions that address trauma and spirituality issues. It sets the tone for subsequent sessions by providing clients information about group ground rules, expectations about how the group will function, and definitions of important terms used throughout. This first session focuses on introducing all members to one another and beginning to model together the format the group will take in weeks ahead. Members introduce themselves and share what brought them to consider participating in this group, as well as hopes they might have for their group experience. Next, group rules and guidelines are reviewed as facilitators remind members of things they spoke about in the screening session. Mutual respect for the beliefs and feelings of other members should be emphasized. The importance of using "I" statements rather than more global language as referred to earlier in this chapter should be emphasized, including the caveat about "semantic masking." Facilitators introduce a discussion of the terms "religion" and "spirituality," and have members share their concepts of any distinguishing or overlapping characteristics for each term while facilitators list responses on a blackboard or large poster where the group can see it and amend it together. The goal is to allow members to begin working (i.e., to listen, share, and think) together. Facilitators should be ready if necessary to make suggestions or ask questions that will help shape the emerging definition. Of particular importance is the introduction of the idea that spirituality and religion are not static but may change and evolve through all life's experiences, both positive and negative, to become something both meaningful and intensely personal. The remainder of the session should be spent utilizing the spiritual autobiography exercise described earlier. This exercise provides a structure for some personal sharing about life history and the role that religion or spirituality has played for each member over time in the context of significant personal events. Facilitators should attempt to make the process interactive by politely intervening and redirecting the sharing when necessary to allow others to volunteer their own similar experiences and to elicit feedback, allowing members to experience elements of commonality.

LEADER: Now that we've all had a chance to meet one another and start the getting-acquainted process, I'd like to talk a little about the way this group will operate. By design this group is breaking one of our society's taboos. Anyone remember what the two topics are that people aren't supposed to talk about in public?

MEMBER: Politics?

LEADER: And . . . what else?

MEMBER: Religion.

LEADER: And why don't people talk about those things?

MEMBER: Because people get mad. They have strong feelings!

LEADER: Yes, and our group will end up talking a lot about one of those topics during our time together. Religious ideas are very personal. No one wants their personal thoughts and feelings to be trampled on. We want this group to be a place where everyone feels safe and comfortable, and able to open up. In order to help that happen, we need some guidelines that we all follow. I'd like to just lay out for you some of these guidelines, and then we can talk about them and see if everyone feels comfortable with them.

Really there are only two main guidelines for this group. The first one is respect. This is so important that I don't think a group like this can function without it. We all need to respect one another and feel respected in return.

So what do I mean by respect? The first way I express respect is to respect the privacy of others. We will be talking about personal things in this group, and one of the cornerstones of any therapy is confidentiality. Now if I am in individual therapy—there are laws that protect my confidentiality. With very few exceptions my therapist can't share my private words and thoughts with anyone without my permission. Group therapy is a little different. Those laws only apply to licensed therapists. It will be up to each group member to protect each other's confidence. We do that by only talking about the group inside the group. We respect one another by not talking about what happens in the group with people outside the group.

Another way we respect one another is to allow for each of us to be different and to value the ways that we are different. You know, if everyone in this group were exactly the same, there would be no point in our meeting. No one could learn anything from anyone else. The differences among us provide avenues for us to learn and grow and change. It's important not to make assumptions about other people and the way they think and feel. Because of that, when people share I'm going to ask us all to try very hard to use "I" statements. Let me just give an example. Let's say I have had some painful experiences in churches in the past. If I express my feelings by saying, "All church people are hypocrites!" what might that feel like to someone else in the group who is a church member?

MEMBER: I would feel a little put down!

[The leader can continue to elicit other reactions here.]

LEADER: What might be a better way for someone to express something about their past history in churches using "I" statements?

MEMBER: Maybe something like, "I've had some difficult experiences in my past where I've felt like some of the church people I knew were hypocrites."

LEADER: Good—what was the difference? How did that feel to people?

[The leader can continue to process with the group until members have some practice using I statements.]

Finally, everyone here in this group has one thing in common. Everyone here has had their lives impacted by the recent terrorist events. You are all in very different places about this. I am going to ask that out of respect for one another that we not go into explicit descriptions of the actual traumatic events that we saw or experienced. That is so that my experiences won't cause someone else in the group to relive or re-experience events they are trying to learn to deal with. That doesn't mean in any way that we won't talk about how the events affect us emotionally, what they meant to us, or the anger or confusion that the events have brought into our lives.

It also doesn't mean we can't talk about important things. It doesn't mean we won't get upset or angry and feel misunderstood or hurt at times. We probably will, and we'll need to work together to process and understand those feelings. It also doesn't mean we will always agree about things.

[At this point the leader should introduce the second group guideline—involvement—and process it in a similar way with the group.]

Involvement means regular group attendance and active participation. The leader should acknowledge differences in individual comfort levels with speaking—that some members will have to push themselves to speak, while others will have to self-monitor to be sure they don't dominate the group. Aspects of involvement to be explored are willingness to try new behaviors, to actively listen, to try to learn from others, and attempt to explore and expand the scope of spiritual experience.

[Leaders should process with members definitions of important terms which will be used throughout subsequent group sessions.]

LEADER: Religion and spirituality will be things we talk about as we meet together over these next weeks. It is important that we are all on the same page as to what we are talking about. Let's brainstorm together. How would you define these two words? What do they mean to you? How are they alike and different? Let's start with the word *religion*. How would you define it?

MEMBER: Connecting with God?

LEADER: Good—someone else?

MEMBER: I think of rules and rituals!

LEADER: Okay, someone else.

MEMBER: It's a bunch of people with the same beliefs and values about God.

LEADER: Okay. So I'm hearing that religion has something to do with God, and beliefs, and that it involves other people as well. Anything else?

MEMBER: I think religion is a bunch of hypocrites trying to tell me how to live my life!

LEADER: So, it sounds like for you religion also has a downside—that it feels pushy to you.

MEMBER: Yeah.

LEADER: Okay. What about spirituality? How is that word different from religion. What does it mean to you?

MEMBER: Well for me spirituality seems more individual. I don't have to believe the same as anyone else.

MEMBER: Yeah, more individual, more personal.

MEMBER: I don't think you have to believe in God to be spiritual. I am spiritual when I feel close to nature . . . like when I'm hiking.

LEADER: Okay, so spirituality is more individual—less bound by other people's traditions, and it might or might not be about God. Is that what I'm hearing?

LEADER: Does this picture [show them Figure 11.2] sort of capture what we're saying? That religion and spirituality deal with some of the same themes—that of connection with something divine or transcendent, and that for some people the domains may overlap or be the same, while for others they may be different? In other words some individuals may experience spirituality within religion—while others may not. What about these definitions? How about a definition of religion as *a system of beliefs, values, rituals, and practices shared in common by a social community as a means of experiencing and connecting with the sacred or divine?* Does that capture what you've been saying? There is a community—a shared rather than individual—aspect to religion, isn't there?

MEMBER: Yes, and that can be either good or bad—can't it? I mean relationship with others can be both good and bad?

LEADER: That's right. What do others think?

[Continue to process as necessary.]

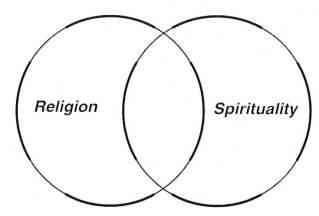

FIGURE 11.2. The Relationship Between Religion and Spirituality

LEADER: How about a definition for spirituality? How about this? *Spirituality is an individual's understanding of, experience with, and connection to that which transcends the self.* So both definitions have to do with the sacred, and the primary difference seems to be the shared nature of religion—where beliefs, values, traditions, practices are held together in common.

MEMBER: To me spirituality is personal. It's what I think, not what someone tells me to believe.

LEADER: So this group we are involved in is an attempt for us to share together our own individual spirituality and to develop it as a resource in recovering from and making sense of the terrorist events that have affected each one of us so powerfully.

Now I'd like each of you to take out of your client workbook a page called "Spiritual Autobiography." I explained this exercise to you when I met with you individually, and I asked you to fill it out and bring it to this first session. Does everyone have their copy? What I'd like to do is have us break into smaller groups of four people. [*Leader helps divide the small groups.*] I'd like to have each of you take about two minutes to briefly share with your subgroup the highs and lows of your spiritual journey, from birth until now, and to illustrate what you are saying by showing others your graph. Does everyone understand? Okay—we have about fifteen minutes to share with one another, and then we'll come back and talk about this experience.

[*Small groups share their graphs and stories, and then reform as one group.*]

LEADER: Okay as we regroup—I want you to respond to two questions. First, what did you learn about yourself from this exercise? Second, what did you learn about someone else by listening to them tell their story?

MEMBER: I learned that we're all very different.

MEMBER: I realized that the terrorism had affected someone else like it did me. I thought I was the only one that felt like I do.

MEMBER: It seemed like positive life experiences seemed to have a positive impact on faith, while bad experiences sometimes dragged our faith down.

The remainder of the session can be spent sharing reflections about the past, hope for the future, and for the group. Leader(s) should have people check in about what the group experience was like, making sure that everyone is heard from.

Session Vignette: Values and Meaning

This session deals with recovery of meaning and purpose in life. It should occur midway throughout a series of sessions addressing trauma and spirituality issues. A values questionnaire previously completed as homework and returned to facilitators at the previous session should be scored and summarized by facilitators prior to this session. This questionnaire should be returned to group members at the start of the session to remind them of the responses they made. This questionnaire's values lists are rated on a scale from 1 (opposed to) to 7 (of supreme importance). Facilitators should review with the group the most and least important values expressed by the group and elicit feedback about their selections. Members should look for values held in common, and individuals should be allowed to express the reasons for their ratings. Then the group should reflect together how they feel they might have rated values differently before the terrorist disaster. Members should be asked to reflect on changes in behavior that they might have already made or might consider making that would reflect changes in their values. Another topic for discussion during this session is the sources of meaning that members find in life. What individuals or experiences are associated with meaning and purpose in life? Have these changed since the traumatic event? If so, what actions or changes in behavior might make present life experience more meaningful?

LEADER: Today's session is about values and meaning. Trauma has a powerful effect on the way that we look at life. When a person is exposed to extreme loss, pain, or horror the emotional load feels intense and overwhelming; everyone is changed by that. Sometimes those changes leave us confused. It feels like we're in a storm and our compass is broken. Sometimes we lose the sense that life has meaning and purpose, and it's hard to get our bearings again.

A few years ago there was a movie that perhaps some of you saw called *City Slickers*. It's a comedy with Billy Crystal who plays Mitch, a man from New York City that is going through sort of a midlife crisis. Mitch decides his crisis solution is to go on vacation with some of his friends to a ranch out west that lets visitors take part in a cattle drive. Jack Palance plays Curly, the old grizzled cowboy that is directing the cattle drive. At any rate, at one point in the movie Curly is talking to Mitch and says, "None of you get it. . . Do you know what the secret of life is . . . one thing, just one thing. You stick to that and everything else don't mean shit." Mitch answers, "That's great, but what's the one thing?" Curley replies, "That's what you've got to figure out."

When life settles down after a trauma, and the immediate emergency is over, survivors have the task of recovering what is left of their lives and figuring out what is important to them. How can they again have a life that matters? Sometimes after trauma, ordinary daily life may begin to feel small and unimportant. The sense of the meaningfulness and the purpose of life may diminish.

In last week's homework you filled out a values questionnaire. We reviewed the group's responses over the past week, and I am handing back your answer sheet as well as a summary of the group's responses. Take a minute to look at the summary and your answers. What do you see?

MEMBER: I notice that a couple people put "financial gain" as the least important value on their values questionnaire. I also did that, and it seems that my values have changed since the disaster.

LEADER: Tell us more about that.

MEMBER: Just things like cars and possessions that I cared a lot about before seem less important. People seem more important to me now.

LEADER: Someone else. What do you notice about the group's responses?

[Continue processing. Allow people an opportunity to share what they listed as the most important and least important values and why.]

LEADER: Now I'd like to do a little different exercise. I'd like people to think back to some period of time during your childhood. It might be grade school, middle school, or high school. At that period of time, what did you want to be when you grew up, and why? What did that mean to you? Let's hear from people.

[Allow members time to reminisce, to get better acquainted. Press for what happened and how they changed their minds about their future career, if they did. What events influenced the outcome?]

LEADER: Now let's move back into the present. You've had some time to think about this. Has anything changed about your values since the terrorist attack? How are you seeing things differently? Are there things that are more or less important to you now?

[Listen for changes that might be viewed as healthy or positive. Listen for thoughts that might reflect cognitive distortions, and gently inquire and elicit feedback about what other members think.]

LEADER: Let's talk about meaning. What is that? What is life meaning? How would you define that? Why is it important? Is it important?

MEMBER: For me meaning is importance or significance.

MEMBER: Yeah, it's what makes life matter.

LEADER: That's right. But sometimes tragedy and loss makes us feel like life doesn't matter anymore—that the important things have been taken away. Survivors sometimes have to rediscover old meaning or find new meaning in order to move forward. Let's hear from people about meaning in your life right now since the disaster.

MEMBER: It's been a struggle for me, trying to put the attack out of my mind. Trying to focus, I feel like I'm in a daze all the time.

MEMBER: My family has gotten closer since the attack. That's been really important to me.

MEMBER: It really helped me to work with my church at some fund-raising events for the relief fund.

LEADER: That's right. I think that often we get meaning ourselves by being meaningful to others. We can get a lot of benefit by serving others. How many of you have been more involved in service activities since the attack?

Continue to process this idea of meaning and purpose. Brainstorm individual ways members could be involved in service as a means to meaning.

Session Vignette: Forgiveness

This session involves forgiveness. This can be very helpful for those who have experienced significant trauma or loss and who are stuck in their anger. Sessions related to forgiveness should occur toward the end of the trauma and spirituality sessions. This allows for the development of maximum rapport, peer support, and trust within the group prior to addressing these very personal issues.

From a PTSD perspective, fear is the core conditioning element of trauma (Foa, Riggs, Massie, & Yarczower, 1995), in the same way that sadness is the primary element of grief following a significant loss. Anger in each of these situations functions as a secondary emotion that may prevent the effective processing of the primary emotion. Anger following trauma may provide survivors with a greater sense of control, which is not experienced with sadness, fear, and grief. To the degree that forgiveness allows people to move beyond anger, it may facilitate the processing of sadness and fear, and allow healing to occur.

The initial portion of the session may involve introduction and definition of the term "forgiveness" and a presentation of the possible benefits therein. It is important to allow time for people to express themselves fully during the early part of this session. Members may have strong reactions to examining this topic with regard to a terrorist act. It should be expected that some members will feel strongly that they could never forgive this type of intentional, malevolent act. It is important to allow for the expression of these views, since if they remain unspoken they will hinder the group process. The session should not move forward until all reactions have been heard. Facilitators should be accepting of such views as natural reactions and reinforce the idea that forgiveness is entirely a personal choice which members are free to consider, or not, for their personal situations. At some point, it will be helpful to proceed to a more nuanced discussion of the term, aided by the suggestions from group members. The group should define together what forgiveness is and what it is not. Some key elements to be highlighted include that forgiveness is a choice, a personal decision done to help oneself, involving acceptance of a hurtful event, letting go of rage toward the event and the perpetrator, moving forward, or getting unstuck. In examining what forgiveness is not, it should be emphasized that it is never synonymous with or predicated on forgetting or condoning evil actions, nor does it necessarily involve reconciliation with the perpetrators. Allow members to understand that forgiveness may be directed toward oneself, others, or even toward God. For those not ready to explore the construct in the context of the disaster, they may be encouraged to practice the process by using a less intense hurt, betrayal, or disappointment they have experienced.

LEADER: Today we are going to address a difficult topic, but one that is perhaps the most relevant for the things you all experienced with the terrorist attack. I'm going to start by introducing a word and asking you to respond to the word. What does it mean to you? What thoughts and feelings come up when we start to talk about it? The word today is *forgiveness*. What comes up for you when I write this word on the board?

MEMBER: Forgiveness is getting away with something, not being punished.

LEADER: Okay, what else?

MEMBER: I guess it means getting past something, moving on.

MEMBER: I have this image of my priest telling me I need to forgive everyone.

MEMBER: I'm never going to forgive those terrorists. I don't think they should be forgiven. What they did was evil!

LEADER: So it sounds like you have a strong reaction to this word.

MEMBER: Yes, I do.

LEADER: That's understandable! It's why I want to talk about this. But again, I want to make sure we are all talking about the same thing. So let's define *forgiveness*. Okay? I'd like to put some ideas on the board about what forgiveness is, and what it isn't, so that we are all thinking about the same thing and so that we don't misunderstand one another. That's important because there are very strong feelings about these terrorist events in all of us.

Let's talk about what it *is* first. When I think about forgiveness, I think about three things that are very important. First, forgiveness is a personal choice. I really want everyone to agree on this that we are not pushing people to forgive. It's okay not to choose forgiveness. It's okay to not be ready to even think about it. It is choosing to let go of bitterness, hatred—negative emotions that cloud my present experience and prevent experience of positive hope and joy.

There is another thing that forgiveness is. For me, forgiveness is an opportunity to begin to change my current experience, to get unstuck, to start moving again. Sometimes, when something terrible happens we feel stuck. We can't move backward and make the event go away, but somehow we can't move forward either. We live every day remembering, reliving, being stuck with the trauma. It's hard to live like that! Forgiveness is something I choose to do for me—not for someone else. It is a process that I begin to take back control over my life and my emotions. One of the inherent aspects of trauma is the lack of control we feel. Life has taken us on a ride without our permission. The sense that our lives are under our control is exposed as an illusion. The only real control we have is over our behavior: what will we choose to do after the trauma.

Now, I'd like to talk about what forgiveness is *not*. This is really important, because these are the things that get in the way of even considering this. We've all heard the phrase "forgive and forget," so the first thing I want to say is that forgiving does not involve forgetting. That's important, because for many of us forgetting an event like this would dishonor those that died. So forgiving is not forgetting.

Another thing that forgiving is *not:* Forgiving is not excusing or condoning evil! It does not mean that the evil act was okay! We can acknowledge the hatred that drove the event, the innocence of the victims, and it's not okay. Evil must be confronted and prevented; we can't be passive.

Finally, forgiving does not always mean reconciliation. It does not always assume that people will be friends, that everything is all better. Reconciliation might happen, but it might not. Sometimes it is impossible; the gulf between the parties is too big to be bridged.

So how would I define it? How about this: *Forgiveness is the decision to reduce negative thoughts, affects and behaviors . . . toward the offender . . . to begin to gain better understanding of the offense and the of-*

fender [Thoresen, Harris, & Luskin, 2000]. So, I've been talking for a while. What did you hear? Tell me what forgiveness is.

MEMBER: You said a choice.

MEMBER: Letting go of anger?

LEADER: Good, someone else?

MEMBER: I'm still not willing to forgive what happened! I don't want to talk about this! It makes me mad!

LEADER: That's okay! I'm sure that there is a lot of anger and hurt and pain in this room, with good reason. Let's take a step backward for a minute. Let's separate our discussion for the moment from the terrorist attack and think about forgiveness in more general terms. Even if everyone here in this room were to decide that forgiveness was not appropriate for the terrorist event, there might still be some personal benefit in talking about forgiveness for other aspects of our lives. Can anyone think of any way that forgiveness in general might be helpful? How might my life be different if I were generally more forgiving?

[It is important to take time to build a case for forgiveness and get members' buy in for the general idea before taking on the most difficult case. Try to get the group to withhold judgment about how forgiveness might relate to the trauma and examine its benefits and costs related to more mundane hurts and disappointments.]

MEMBER: I don't know—maybe relationships. Maybe you'd hold fewer grudges.

LEADER: Okay, good. What other benefits might there be?

MEMBER: Less angry all the time. Maybe I wouldn't feel so stressed on the freeway!

LEADER: Great. That would be good wouldn't it? What else?

[Continue processing—allow time for all members to contribute if possible. Check in with silent members to see what they are experiencing.]

LEADER: Okay. Let's switch gears a minute and start to talk about barriers to forgiveness. What might make it a bad idea? What might make me choose to not forgive?

[The therapist here wants to walk a tightrope between eliciting the perceived barriers to forgiveness that exist in the group and gently challenging and reframing the concerns. The goal is to begin to address some of the fears and concerns without shutting down their expression.]

MEMBER: I might be afraid that if I let it go—it might happen again. I will get hurt again.

LEADER: That's right. The prospect of more hurt is scary. But maybe I also allow for the possibility of more joy, or more hope too! What do others think?

MEMBER: I'd feel weak. I don't like that.

LEADER: There's that fear of losing control, of being vulnerable to being hurt again. And that's a real concern. But remember, control is somewhat of an illusion. The control I really have over what happens to me is only minimal—unless I lock myself in a steel box, but there are some pretty lonely consequences to that too!

MEMBER: The person won't get what they deserve!

LEADER: That may be true. Justice might not happen. Evil might not change. But remember, forgiveness is a personal choice. If I choose to forgive, it is for me, for what I might get out of it—so that I don't have to continue to be the victim. I can take control of my own destiny—not over what happens to me, but over how I choose to respond to it.

[It's important not to rush this processing. Barriers to considering forgiveness are real. The therapist doesn't want to appear to not take them seriously. The leader could move to discussion of self-forgiveness here, particularly the need to look for distortions related to guilt and self-blame that are likely to occur as attempts to maintain the illusion of control. "If I had some control, maybe I could have done something to prevent the event." People tell themselves "If only I hadn't. . ."]

SUMMARY

Spirituality is a primary human dimension and a potential healing resource for survivors of traumatic events. Traumatic events such as terrorism frequently elicit from survivors important core life questions that must be worked through. Frequently these questions are spiritual or existential in nature. Why did this happen to me? Does life still have meaning? How did God let this happen? How can I ever feel safe again? The struggle to come to terms with questions such as these is a spiritual struggle sometimes called "theodicy."

Growing evidence indicates that healthy spirituality is associated with both physical and mental health. However, other evidence shows that certain negative spiritual attributions (e.g., God is punishing me, anger at God) are sometimes associated with negative outcomes, such as poorer health. Trauma victims may be particularly vulnerable to making attributions such as these as they attempt to cope with and make sense of horrible events.

This chapter has presented some suggestions for integrating spirituality into group treatment for survivors of terrorist attacks. Though designed as an adjunct to other trauma-related treatments, the purpose of these sessions is to address these spiritual, existential issues in ways not addressed by other therapies. The hope is that the exercises utilized in these specialized group treatment sessions will provide a forum for survivors to begin to speak about, reflect upon, and perhaps even change the way they think about the spiritual implications of these events. In addition, it is hoped that as survivors engage with other members that they can benefit from the social support and discover together new pathways to meaning and purpose in the wake of trauma and tragedy.

REFERENCES

Alexander, C. N., Robinson, P., & Rainforth, M. (1995). "Treating and preventing alcohol, nicotine, and drug abuse through transcendental meditation: A review and statistical meta-analysis": Errata. *Alcoholism Treatment Quarterly, 13*(4), 97.

Alferi, S. M., Culver, J. L., Carver, C. S., Arena, P. L., & Antoni, M. H. (1999). Religiosity, religious coping, and distress: A prospective study of Catholic and Evangelical Hispanic women in treatment for early-stage breast cancer. *Journal of Health Psychology, 4*(3), 343-356.

Andrykowski, M. A., Curran, S. L., Studts, J. L., Cunningham, L., Carpenter, J. S., McGrath, P. C., Sloan, D. A., & Kenady, D. E. (1996). Psychosocial adjustment and quality of life in women with breast cancer and benign breast problems: A controlled comparison. *Journal of Clinical Epidemiology, 49*(8), 827-834.

Astin, M. C., Lawrence, K. J., & Foy, D. W. (1993). Posttraumatic stress disorder among battered women: Risk and resiliency factors. *Violence & Victims, 8*(1), 17-28.

Bergin, A. E. (1983). Religiosity and mental health: A critical reevaluation and meta-analysis. *Professional Psychology-Research & Practice, 14*(2), 170-184.

Bergin, A. E., & Jensen, J. P. (1990). Religiosity of psychotherapists: A national survey. *Psychotherapy, 27*(1), 3-7.

Blumenthal, D. R. (1998). Theodicy: Dissonance in theory and praxis. *Concilium, 1*, 95-106.

Carmil, D., & Breznitz, S. (1991). Personal trauma and world view: Are extremely stressful experiences related to political attitudes, religious beliefs, and future orientation? *Journal of Traumatic Stress, 4*(3), 393-405.

Chang, B.-H., Skinner, K. M., & Boehmer, U. (2001). Religion and mental health among women veterans with sexual assault experience. *International Journal of Psychiatry in Medicine, 31*(1), 77-95.

Cordova, M. J., Cunningham, L. L., Carlson, C. R., & Andrykowski, M. A. (2001). Posttraumatic growth following breast cancer: A controlled comparison study. *Health Psychology, 20*(3), 176-185.

Coyle, C. T., & Enright, R. D. (1997). Forgiveness intervention with postabortion men. *Journal of Consulting and Clinical Psychology, 65*(6), 1042-1046.

Davis, J. A., Smith, T. W., & Marsden, P. V. (2001). *General Social Surveys, 1972-2000: 3rd version.* Retrieved 8/27/02, 2002 Cumulative Codebook. Chicago: NORC.

DiBlasio, F. A. (2000). Decision-based forgiveness treatment in cases of marital infidelity. *Psychotherapy, 37*(2), 149-158.

Drescher, K. D., Loew, D., & Young, H. E. (2001). Spirituality and PTSD: Making sense of trauma. Paper presented at the Seventeenth Annual Meeting of the International Society for Traumatic Stress Studies, New Orleans, Louisiana, March 31.

Drescher, K. D., Young, H., & Loew, D. (2002). Spirituality and complex trauma: Group intervention with combat veterans. Paper presented at the Eighteenth Annual Meeting of the International Society for Traumatic Stress Studies, Baltimore, Maryland.

Ellis, M. R., Vinson, D. C., & Ewigman, B. (1999). Addressing spiritual concerns of patients: Family physicians' attitudes and practices. *Journal of Family Practice, 48*(2), 105-109.

Festinger, L. (1957). *A theory of cognitive dissonance.* Stanford, CA: Stanford University Press.

Fetzer Institute. (1999). *Multidimensional Measurement of Religiousness/Spirituality for Use in Health Research.* National Institute on Aging Working Group. Kalamazoo, Michigan: Author.

Finkelhor, D., Hotaling, G. T., Lewis, I., & Smith, C. (1989). Sexual abuse and its relationship to later sexual satisfaction, marital status, religion, and attitudes. *Journal of Interpersonal Violence, 4*(4), 379-399.

Foa, E. B., Ehlers, A., Clark, D. M., Tolin, D. F., & Orsillo, S. M. (1999). The Posttraumatic Cognitions Inventory (PTCI): Development and validation. *Psychological Assessment, 11*(3), 303-314.

Foa, E. B., Riggs, D. S., Massie, E. D., & Yarczower, M. (1995). The impact of fear activation and anger on the efficacy of exposure treatment for posttraumatic stress disorder. *Behavior Therapy, 26*(3), 487-499.

Fontana, A., & Rosenheck, R. (2004). Trauma, change in strength of religious faith, and mental health service use among veterans treated for PTSD. *Journal of Nervous Mental Disorders, 192*(9), 579-584.

Foster, R. J. (1988). *Celebration of discipline.* New York: Harper/Collins.

Foy, D. W., Wood, J. L., King, D. W., King, L. A., & Resnick, H. S. (1997). Los Angeles Symptom Checklist: Psychometric evidence with an adolescent sample. *Assessment, 4*(4), 377-384.

Freedman, S. R., & Enright, R. D. (1996). Forgiveness as an intervention goal with incest survivors. *Journal of Consulting & Clinical Psychology, 64*(5), 983-992.

Gelderloos, P., Walton, K. G., Orme-Johnson, D. W., & Alexander, C. N. (1991). Effectiveness of the transcendental meditation program in preventing and treating substance misuse: A review. *International Journal of the Addictions, 26*(3), 293-325.

Gillard, M., & Paton, D. (1999). Disaster stress following a hurricane: The role of religious differences in the Fijian Islands. *Australasian Journal of Disaster & Trauma Studies, 3*(2), NP.

Gordon, K. C., Baucom, D. H., & Snyder, D. K. (2000). The use of forgiveness in marital therapy. In M. E. McCullough, K. I. Pargament, & C. E. Thorsen (Eds.), *Forgiveness: Theory, research, and practice* (pp. 254-280). New York: Guilford Press.

Gorsuch, R. L. (1995). Religious aspects of substance abuse and recovery. *Journal of Social Issues, 51*(2), 65-83.

Gorsuch, R. L., & Miller, W. R. (1999). Assessing spirituality. In W. R. Miller (Ed.), *Integrating spirituality into treatment* (pp. 47-64). Washington, DC: American Psychological Association.

Hall, C. (1986). Crisis as opportunity for spiritual growth. *Journal of Religion & Health, 25*(1), 8-17.

Hall, T. A. (1995). Spiritual effects of childhood sexual abuse in adult Christian women. *Journal of Psychology & Theology, 23*(2), 129-134.

Harris 2000 National Issues Survey, study no. 12851. (2000). New York: Harris Interactive, Inc.

Harris, A. H., Thoresen, C. E., McCullough, M. E., & Larson, D. B. (1999). Spiritually and religiously oriented health interventions. *Journal of Health Psychology, 4*(3), 413-433.

Idler, E., Musick, M., Ellison, C. G., George, L. K., Krause, N., Levin, J., Ory, M., Pargament, K. I., Powell, L. H., Williams, D. R., & Underwood-Gordon, L. (2003). NIA/Fetzer Measure of Religiousness and Spirituality: Conceptual Background and Findings from the 1998 General Social Survey. *Research on Aging, 25,* 327.

Janoff-Bulman, R. (1992). *Shattered assumptions: Towards a new psychology of trauma.* New York: Free Press.

Kane, D., Cheston, S. E., & Greer, J. (1993). Perceptions of God by survivors of childhood sexual abuse: An exploratory study in an underresearched area. *Journal of Psychology & Theology, 21*(3), 228-237.

King, D. E., & Bushwick, B. (1994). Beliefs and attitudes of hospital inpatients about faith healing and prayer. *Journal of Family Practice, 39*(4), 349-352.

King, L. A., King, D. W., Leskin, G., & Foy, D. W. (1995). The Los Angeles Symptom Checklist: A self-report measure of posttraumatic stress disorder. *Assessment, 2*(1), 1-17.

Kirkpatrick, T. G. (1995). *Small groups in the church: A handbook for creating community.* Bethesda, MD: Alban Institute.

Kubany, E. S., Haynes, S. N., Leisen, M. B., Owens, J. A., Kaplan, A. S., Watson, S. B., & Burns, K. (2000). Development and preliminary validation of a brief broad-spectrum measure of trauma exposure: The Traumatic Life Events Questionnaire. *Psychological Assessment, 12,* 210-224.

Lamb, S. (2002). Women, abuse, and forgiveness: A special case. In S. Lamb & J. G. Murphy (Eds.), *Before forgiving: Cautionary views of forgiveness in psychotherapy* (pp. 155-171). London: Oxford University Press.

Larson, D. B., Pattison, E. M., Blazer, D. G., Omran, A. R., & Kaplan, B. H. (1986). Systematic analysis of research on religious variables in four major psychiatric journals, 1978-1982. *American Journal of Psychiatry, 143*(3), 329-334.

Larson, D. B., Sherrill, K. A., Lyons, J. S., Craigie, F. C., Thielman, S. B., Greenwold, M. A., & Larson, S. S. (1992). Associations between dimensions of religious commitment and mental health reported in the *American Journal of Psychiatry* and *Archives of General Psychiatry: 1978-1989. American Journal of Psychiatry, 149*(4), 557-559.

Larson, D. B., Swyers, J. P., & McCullough, M. E. (1997). *Scientific research on spirituality and health: A consensus report.* Rockville, MD: National Institute for Healthcare Research.

Leibniz, G. W. (1890). *Philosophical works* (G. M. Duncan, Trans.). New Haven: Tuttle, Morehouse, & Taylor.

Leslie, R. (1971). *Sharing groups in the church.* Nashville, TN: Abingdon Press.

Levin, J. S. (1994). Religion and health: Is there an association, is it valid, and is it causal? *Social Science & Medicine, 38*(11), 1475-1482.

Marlatt, G. A., & Kristeller, J. L. (1999). Mindfulness and meditation. In W. R. Miller (Ed.), *Integrating spirituality into treatment: Resources for practitioners* (pp. 67-84). Washington, DC: American Psychological Association.

Maugans, T. A., & Wadland, W. C. (1991). Religion and family medicine: A survey of physicians and patients. *Journal of Family Practice, 32*(2), 210-213.

McIntosh, D. N., Silver, R. C., & Wortman, C. B. (1993). Religion's role in adjustment to a negative life event: Coping with the loss of a child. *Journal of Personality and Social Psychology, 65*(4), 812-821.

Miller, J. J., Fletcher, K., & Kabat-Zinn, J. (1995). Three-year follow-up and clinical implications of a mindfulness meditation-based stress reduction intervention in the treatment of anxiety disorders. *General Hospital Psychiatry, 17*(3), 192-200.

Miller, W. R., & Thoresen, C. E. (1999). Spirituality and health. In W. R. Miller (Ed.), *Integrating spirituality into treatment: Resources for practitioners* (pp. 3-13). Washington, DC: American Psychological Association.

Murphy, S. A., Johnson, L., & Lohan, J. (2003). Finding meaning in a child's violent death: A five-year prospective analysis of parents' personal narratives and empirical data. *Death Studies, 27*(5), 381-404.

Neufeldt, V., & Guralnik, D. B. (Eds.). (1988). *Webster's new world dictionary of American English.* Cleveland, OH: Simon & Schuster, Inc.

Pargament, K. I. (1997). *The psychology of religion and coping: Theory, research, practice.* New York: Guilford Press.

Pargament, K. I., & Brandt, C. R. (1998). Religion and coping. In H. G. Koenig (Ed.), *Handbook of religion and mental health* (pp. 112-128). San Diego, CA: Academic Press.

Pargament, K. I., Ensing, D. S., Falgout, K., Olsen, H., Reilly, B., Van Haitsma, K., & Warren, R. (1990). God help me (I.): Religious coping efforts as predictors of the outcomes to significant negative life events. *American Journal of Community Psychology, 18*(6), 793-824.

Pargament, K. I., Ishler, K., Dobow, E. F., Stanik, P., Rouiller, R., Crowe, P., Cullman, E., Albert, M., & Royster, B.J. (1994). Methods of religious coping with the Gulf War: Cross-sectional and longitudinal analysis. *Journal for the Scientific Study of Religion, 33*(4), 347-361.

Pargament, K. I., Olsen, H., Reilly, B., Falgout, K., & Ensing, D.S. (1992). God help me: (II.) The relationship of religious orientations to religious coping with negative life events. *Journal for the Scientific Study of Religion, 31*(4), 504-513.

Park, C., Cohen, L. H., & Herb, L. (1990). Intrinsic religiousness and religious coping as life stress moderators for Catholics versus Protestants. *Journal of Personality & Social Psychology, 59*(3), 562-574.

Plante, T. G., & Manuel, G. M. (1992). The Persian Gulf War: Civilian war-related stress and the influence of age, religious faith, and war attitudes. *Journal of Clinical Psychology, 48*(2), 178-182.

Powell, L. H., Shahabi, L., & Thoresen, C. E. (2003). Religion and spirituality: Linkages to physical health. *American Psychologist, 58*(1), 36-52.

Pritt, A. F. (1998). Spiritual correlates of reported sexual abuse among Mormon women. *Journal for the Scientific Study of Religion, 37*(2), 273-285.

Ragan, C., Malony, H. N., & Beit-Hallahmi, B. (1980). Psychologists and religion: Professional factors and personal belief. *Review of Religious Research, 21*(2), 208-217.

Reinert, D. F., & Smith, C. E. (1997). Childhood sexual abuse and female spiritual development. *Counseling & Values, 41*(3), 235-245.

Remle, R., & Koenig, H. G. (2001). Religion and health in HIV/AIDS communities. In T. G. Plante & A. C. Sherman (Eds.), *Faith and health: Psychological perspectives* (pp. 195-212). New York: Guilford Press.

Rynearson, E. (1995). Bereavement after homicide: A comparison of treatment seekers and refusers. *British Journal of Psychiatry, 166*(4), 507-510.

Schuster, M. A., Stein, B. D., Jaycox, L., Collins, R. L., Marshall, G. N., Elliott, M. N., Zhou, A. J., Kanouse, D. E., Morrison, J. L., & Berry, S. H. (2001). A national survey of stress reactions after the September 11, 2001, terrorist attacks. *New England Journal of Medicine, 345*(20), 1507-1512.

Shafranske, E. P. (1996). Religious beliefs, affiliations, and practices of clinical psychologists. In E. P. Shafranske (Ed.), *Religion and the clinical practice of psychology* (pp. 149-162). Washington, DC: American Psychological Association.

Sprang, G., & McNeil, J. (1998). Post-homicide reactions: Grief,, mourning and post-traumatic stress disorder following a drunk driving fatality. *Omega - Journal of Death & Dying, 37*(1), 41-58.

Stephen Ministries. (2001). Media Fact Sheet. Available online at <http://www.stephenministries.org/pdfs/ssmediafactsheet.pdf>.

Tedeschi, R. G., & Calhoun, L. G. (1996). The Posttraumatic Growth Inventory: Measuring the positive legacy of trauma. *Journal of Trauma and Stress, 9*(3), 455-471.

Thoresen, C. E., Harris, A. H., & Luskin, F. (2000). Forgiveness and health: An unanswered question. In M. E. McCullough, K. I. Pargament, & C. E. Thorsen

(Eds.), *Forgiveness: Theory, research, and practice* (pp. 254-280). New York: The Guilford Press.

Tix, A. P., & Frazier, P. A. (1998). The use of religious coping during stressful life events: Main effects, moderation, and mediation. *Journal of Consulting & Clinical Psychology, 66*(2), 411-422.

Weaver, A. J., Kline, A. E., Samford, J. A., Lucas, L. A., Larson, D. B., & Gorsuch, R. L. (1998). Is religion taboo in psychology? A systematic analysis of research on religion in seven major American Psychological Association journals: 1991-1994. *Journal of Psychology & Christianity, 17*(3), 220-232.

Weaver, A. J., Samford, J. A., Kline, A. E., Lucas, L. A., Larson, D. B., & Koenig, H. G. (1997). What do psychologists know about working with the clergy? An analysis of eight APA journals: 1991-1994. *Professional Psychology - Research & Practice, 28*(5), 471-474.

Witvliet, C. V. O., Phillips, K. A., Feldman, M. E., & Beckham, J. C. (2004). Posttraumatic mental and physical health correlates of forgiveness and religious coping in military veterans. *Journal of Traumatic Stress, 17*(3), 269-273.

Woods, T. E., Antoni, M. H., Ironson, G. H., & Kling, D. W. (1999). Religiosity is associated with affective and immune status in symptomatic HIV-infected gay men. *Journal of Psychosomatic Research, 46*(2), 165-176.

Worthington, E. L., Sandage, S. J., & Berry, J. W. (2000). Group interventions to promote forgiveness: What researchers and clinicians ought to know. In M. E. McCullough, K. I. Pargament, & C. E. Thorsen (Eds.), *Forgiveness: Theory, research, and practice* (pp. 228-253). New York: The Guilford Press.

PART III:
GROUP MODELS

Chapter 12

Integrating Small-Group Process Principles into Trauma-Focused Group Psychotherapy: What Should a Group Trauma Therapist Know?

D. Rob Davies
Gary M. Burlingame
Christopher M. Layne

INTRODUCTION

Watching hockey has always been a popular pastime, especially when admired players are on the ice. Having a unique and well-developed skill, such as speed or a hard slap shot, makes players solid contributors to the team's level of performance; however, when necessary, the team can still succeed without them. In contrast, truly great players are adept in virtually all areas of play and are able to combine multiple skills in ways that almost always result in one or more goals per game. These players have honed their skills to the point that they have mastered a number of roles-and use this versatility to "step up" as the situation demands.

Front-line trauma therapists can also benefit from the versatility that accompanies becoming proficient in more than one area of treatment. Trauma treatment is complex and often requires interdisciplinary teamwork to address its multifaceted nature. Few circumstances are more complex than the aftermath of large-scale catastrophic events that result in hundreds, if not thousands, of individuals in need of care. In such cases, a variety of trauma treatment models may be employed, some of which target victims' initial reactions; others focus on the intermediate and long-term sequelae. In addition to contrasting trauma treatment models along the spectrum of therapeutic focus and time frame for implementation, further distinctions may be made according to the treatment modality used. Specifically, some trauma treatment models rely exclusively upon an individual format (e.g., Foa & Rothbaum, 1998), others utilize small groups (e.g., Layne, Pynoos, et al., 2001; Saltzman, Pynoos, Layne, Steinberg, & Aisenberg, 2001; Stubbenbort,

Donnelly, & Cohen, 2001), and still others utilize large community-based groups that can reach scores of individuals at a time (Mitchell & Everly, Chapter 13 in this volume). Knowledge and experience in the underlying theories, treatment objectives, and accompanying therapeutic skills within each domain are required to maximize the coherence, versatility, and therapeutic impact of interventions based within a given modality.

The small-group treatment literature contains a number of components with significant potential for augmenting the quality and effectiveness of trauma-focused group intervention protocols. More specifically, the small-group process and group dynamic literature extends back over seven decades, providing a rich body of knowledge and experience from the disciplines of both group psychology and psychotherapy. Examples of small-group processes include emergent or structured patterns of member interaction (e.g., participatory versus resistant), as well as factors that are thought to contribute to a group's therapeutic effectiveness (e.g., cohesion, interpersonal feedback, members adopting both the helper and helpee roles). The term *group dynamics* typically refers to predictable changes in the group structure (norms and member roles), feeling tone (climate), and/or working stages. Although not without its methodological problems, gaps, and inconsistencies (cf. Burlingame, MacKenzie, & Strauss, 2004) this research has "pockets" of theoretical and clinical knowledge that may increase a trauma-focused group therapist's versatility and capacity to deal with therapeutic challenges. Notably, the roots of trauma literature, although not continuous, extend back to crisis intervention work carried out nearly a century ago (e.g., Stierling, 1909; Salmon, 1919). Because the trauma treatment and small-group process literatures have followed largely independent trajectories, many trauma theorists and clinicians have not been formally trained in their practical integration and use. Indeed, studies regarding how these two literatures might effectively cross-fertilize with respect to theory, empirical findings, principles, and practices—especially regarding undertakings as specific as trauma-focused group treatment for terrorist disasters—have scarcely been mentioned in the literature. Accordingly, this chapter constitutes a preliminary and illustrative effort to bridge these two rich literatures.

In this chapter, we focus on ways in which the small-group process literature may inform and complement the practices of existing group-based trauma treatment protocols. Our basic premise is that, although not yet sufficiently mature to offer *definitive* recommendations, the small-group process and treatment literatures hold considerable promise for generating ideas, suggestions, and other insights that may complement the effectiveness of existing group-based trauma treatment programs. We propose that there are a number of group process-related principles and insights that trauma clinicians are already using or that they may readily integrate into existing treatment protocols with minimal impact on the integrity of the protocol itself. We further propose that, if deliberately incorporated and judiciously used, these principles can increase therapeutic versatility and problem-solving abilities, the comprehensiveness and accuracy of case conceptualization skills, and therapeutic leverage and treatment effectiveness.

In light of such observations, this chapter presents and discusses a selected set of principles with good track records within the small-group literature for increasing the effectiveness of group treatment, independent of the specific therapeutic orientation used. The chapter begins with an overview of a number of advantages associated with adopting a group-based modality in the treatment of psychological and psychiatric disorders in general. We then briefly review recent findings regarding the relative effectiveness of interventions based on formal theories of change compared to placebo groups, and discuss the implications of these findings for identifying and understanding the mechanisms of change underlying the positive treatment effects observed across a variety of trauma treatment modalities. We subsequently focus on two illustrative principles—group development and interpersonal feedback—that show promise for complementing existing trauma treatment protocols. These principles are then concretely applied in the form of a practical case example, illustrating how small-group process principles can be integrated into, and complement, a representative "gold standard" group-based trauma treatment protocol. We conclude with a discussion of the incremental benefits that may accompany the integration of group-based knowledge and therapeutic skills with formal trauma treatment work, and propose avenues for future research.

WHY SHOULD A TRAUMA THERAPIST BE INTERESTED IN SMALL-GROUP PROCESS PRINCIPLES?

Because becoming skilled in the deliberate incorporation of small-group process principles takes time and effort, some trauma therapists may justifiably ask, "What advantages does group-based treatment have over other modalities that I am already familiar with and good at?"

Perhaps the most obvious benefit of group treatment is that it consistently emerges as the most resource-efficient (both in terms of cost and personnel) means of providing clinical services when compared to other treatment modalities. Indeed, some experts have predicted that this trend will continue for the foreseeable future. Among patients using nationwide managed-care systems, group treatment is predicted to constitute nearly 40 percent of all patient visits over the next ten years (Roller, 1997, p. xi). The pragmatism of the group-based format, both in terms of cost savings and other resource allocation, appears especially applicable to disaster settings, where the number of exposed individuals needing immediate care can exceed the number of mental health workers many hundredfold. In such acute situations, the options may be reduced to a group-based format or no substantive treatment at all, except possibly to a select few who can afford the out-of-pocket expense or have unusually good health care (and mental health care) coverage.

Despite the appeal of certain treatments in certain circumstances, neither the cost-efficiency of a given treatment program nor its capacity for broad dissemination, in and of themselves, constitute compelling evidence for its widespread use. This fact is especially valid in the wake of disasters, which so deplete survivors' internal and material resources that many need the most potent treatments available (Monnier & Hobfoll, 2000). Indeed, what is the social value of providing large numbers of people with efficient access to a watered-down treatment, as some theorists, and prospective patients, no doubt believe?

Fortunately, group-based psychotherapy has consistently emerged as an efficacious treatment for a wide variety of disorders and disabilities (Burlingame, MacKenzie, & Strauss, 2003). As reported by Fuhriman and Burlingame, "the general conclusion to be drawn from some 700 studies that span the past two decades is that the group format consistently produced positive effects with diverse disorders and treatment models" (1994a, p. 15). In addition, studies that directly compare individual-based treatment to group-based treatment generally conclude that the two modalities are equivalent in their field effectiveness and efficacy.

Drawing on more than 150 published studies, recent meta-analytic investigations suggest that group and individual psychotherapy formats are statistically equivalent in treatment effectiveness across a wide variety of disorders and that both formats provide improvement that exceeds the gains made by clients in wait-list control conditions (Burlingame, Fuhriman, & Mosier, 2003; McRoberts, Burlingame, & Hoag, 1998; Tschuschke, 1999). Thus, the best available evidence indicates that group-based psychotherapy is not simply cost-effective but also effective in the field and efficacious in carefully controlled studies. However, these results can and should be applied only at the aggregate level; they should not be generalized downward with the same degree of confidence to specific populations such as those defined by age, psychiatric condition, or ethnicity. Indeed, within some populations (e.g., corporate management), group therapy posts inferior results.

This spontaneous emergence may be especially prevalent when working with trauma-focused groups, which tend to elicit strong cognitive and emotional reactions from the members. These emergent processes or member reactions may interfere with the treatment objectives and session agenda if they are not appropriately planned for and managed. As Ettin (1992) wisely observes,

> All too often, group process is actually ignored and/or viewed as an interruption to the didactic presentation. The leader may become anxious, frustrated, or even hostile when group processes emerge and disrupt the more ordered educational format. . . . No group of breathing souls stands still to be educated and "psychologized." To use the group medium effectively and to maximize the learning opportunities, the leader must be attuned to the movement and stirring within the group. (pp. 241-243)

WHAT ARE THE "ACTIVE INGREDIENTS" IN GROUPS THAT PRODUCE THERAPEUTIC CHANGE?

Why are groups effective and efficacious? The short answer is that experts have some good leads, but no one knows for sure. Indeed, the precise mechanisms by which group treatment exerts its therapeutic effects have been the topic of much debate. In an effort to clarify these mechanisms, Burlingame, MacKenzie, & Strauss (2003) have recently proposed a framework for identifying the most significant sources of therapeutic gain in group therapy and the nature of their interrelationships. The model considers five domains that have been linked to outcomes in group treatment: formal change theory, member and leader characteristics, small-group processes, and group structure. These authors observe that the *most studied* influence on therapeutic outcomes in group treatment is that of the *formal theory of change*—that is, the theoretical framework (cognitive-behavioral, psychodynamic, etc.) that undergirds the intervention. This conclusion is not surprising, given the pervasive influence that the formal theory of change exerts throughout all phases and levels of treatment, including assessment and diagnostic procedures, case conceptualization, the selection and use of specific therapeutic interventions, treatment monitoring strategies, and methods for evaluating treatment effectiveness or efficacy. Based on their review of the literature, Burlingame and colleagues conclude that a variety of formal change theories can be used effectively to treat different psychiatric disorders and patient populations.

The assertion that group-based treatment has positive effects across a range of theories of change is consistent with recent reviews within the trauma literature. In their review of the trauma-focused group treatment literature, Foy et al. (2000) concluded that, regardless of the specific formal change theory employed, group psychotherapy is associated with favorable outcomes across a range of symptom domains. This is an intriguing conclusion, especially when one reflects on the pervasive influence that the formal change theory exerts on virtually every aspect of how a therapy group is conducted. As a brief illustration, a cognitive-behavioral group is likely to spend much more group time in didactic instruction than a process group, in which the focus is directed toward here-and-now, member-to-member interactions. Given the striking differences in theoretical assumptions, treatment objectives, and therapeutic interventions, it is remarkable that these dissimilar modalities have produced statistically equivalent outcomes in the extant literature.

How can group psychotherapy be effective, irrespective of the type of undergirding formal change theory or the absence thereof? A major clue has emerged from the aforementioned conclusions of Burlingame, MacKenzie, & Strauss (2003), who observe that

> a large number of studies contrasted formal change theories with control groups that *deliberately* or *inadvertently incorporated group process principles.* In virtually every case, patient improvement in the control group rivaled active treatment! The replication of this intriguing finding across three patient populations (mood; social phobia; elders) and different research teams, argues for its robustness. (p. 661, italics added)

This conclusion not only alludes to methodological problems within the group treatment literature that render it impossible to draw strong inferences relating to between-group differences, but also raises questions about the effects of more "generic" small-group processes that abound across a variety of theory-based groups.

In particular, findings that "placebo-control" conditions (i.e., groups in which members simply interacted with one another) are generally equivalent in therapeutic effectiveness compared to groups based on a formal theory of change suggests that *group processes are, in and of themselves, instrumental in producing therapeutic change.* This conclusion rests on the observation of Burlingame, Mackenzie, & Strauss (2003) that improvement in patients participating in either formal change or group process conditions exceeded the degree of improvement found in patients in wait-list control conditions. It is important to note that these findings should not be taken to mean that a guiding formal theory of change is irrelevant to the realization of positive treatment effects. Rather, what is needed is not the rejection of theories of change as a legitimate and potent means of producing therapeutic benefits in group-based treatment. Instead, there should be integration of such models with theories of group processes that can account for the latter's added (and, it is hoped, additive or interactive) effects. In particular, a theory is needed to explicate the *mechanisms of change* associated with "generic" group processes that produce therapeutic effects of magnitudes that rival effects produced by formal theory-based groups and individual-based interventions irrespective of formal theoretical orientation.

In summary, assuming that the effects of theory-based and process-based interventions are at least partly additive, it is reasonable to propose that the deliberate incorporation of group process principles *into* group interventions based on a formal theory of change may increase their effectiveness compared to either approach alone. This assertion leads to the questions, "Just what are these group process principles?" and "How can we harness these principles to increase our therapeutic leverage within our own group-based trauma treatment protocols?" These questions address the second main focus of the group literature identified by Burlingame, MacKenzie, & Strauss (2003): the nature and operation of small-group processes.

What do we mean by small-group process principles?

We previously differentiated small-group processes from group dynamics by suggesting that the former relates to group-specific mechanisms of change (e.g., therapeutic factors) while the latter relates to structural features (e.g., group development) that predictably change over the life of the group. What we have not yet disclosed is the absence of consensus in the small-group literature regarding how to define and differentiate either of these organizing rubrics. Indeed, many writers use these two terms interchangeably—a practice which unfortunately obfuscates and symbolizes the disarray currently found in the group process literature. Predictably, this adds confusion to discussions of group process principles and has led some reviewers (e.g., Bednar & Kaul, 1994) to openly express embarrassment at the mental health field's apparent inability to define the most fundamental conceptual elements associated with the mechanisms of change in the group format.

The absence of a consensual theory or framework that effectively organizes the conceptual and empirical knowledge regarding group process or dynamic constructs is a major shortcoming in the field. Moreover, definitional confusion continues to exist with regard to some of the most basic and common group processes, such as cohesion (Burlingame, Fuhriman, & Johnson, 2002). Nonetheless, we contend that sufficient definitional clarity and empirical support exists to propose that certain group processes and dynamics are highly relevant to the practicing clinician—principles and illustrations of which we have summarized elsewhere (Fuhriman & Burlingame, 1994b). For instance, although plagued with definitional challenges, sufficient knowledge is available to suggest empirically based guidelines for interventions that may maximize some aspects of group cohesion (Burlingame, Johnson, & MacKenzie, 2002).

Most recently, we have proposed a framework that identifies some of the particular processes and structural dynamics that should constitute elements of a general theory of group-specific mechanisms of action (Burlingame, MacKenzie, & Strauss, 2004). This framework reflects processes (e.g., therapeutic factors, group cohesion, communication patterns) and structural dynamics (group development, subgroups, norms, roles) that *naturally* emerge from the interactive nature of group treatment. This framework also asserts that certain group processes—such as interpersonal feedback and self-disclosure—can be *directly affected* by group-level interventions, such as pregroup training, early structured group exercises, and therapeutic contracts. An important contribution to this framework is its incorporation of Fuhriman and Burlingame's proposal (1990) that certain mechanisms of action, such as disclosure, are common to multiple treatment formats (i.e., individual and group), whereas others appear to be uniquely group specific (such as members acting in both helper and helpee roles).

The interaction between common and format-specific mechanisms of action adds to the complexity of a group leader's task. For example, self-disclosure is generally regarded as a positive mechanism of change in both individual and group treatment. In individual therapy the therapist facilitates client self-disclosure within the guidelines of a formal change theory. However, the group context adds an additional perspective to a leader's response to, and use of, client self-disclosure. For instance, too much self-disclosure within a group can be as countertherapeutic as too little. Specifically, too much disclosure early in the group can lead to patterns in which the overdisclosing client is viewed by other members of the group as socially inappropriate. Such individuals run the risk of being identified as a group deviant (i.e., one who violates group norms). Group members who are consistently identified as group deviants are often subsequently dismissed or discounted by the other group members, increasing the risk of negative treatment outcomes for the ostracized member. Moreover, group members who overdisclose in early sessions are also at risk for feeling overwhelmed when they face group members in subsequent sessions. These members are at higher risk for refraining from further disclosure or for dropping out of group treatment. The empirical literature (e.g., Stockton & Morran, 1982) suggests that client self-disclosures should be both reciprocal and graduated. This literature further suggests that the timing and content of disclosures are decisive elements of group process that a leader must manage effectively in order to maximize the benefits of group treatment and minimize the risk of iatrogenic effects.

Thus, the fundamental argument for the existence of unique dynamics and processes rests on findings that the specific nature of interactions between group members affects the therapeutic change process for good or ill. Although the nomenclature used to label these features varies by writer (e.g., social microcosm, therapeutic community, group dynamic)—thereby adding confusion to the literature—the interactive interpersonal environment is the bedrock upon which the beneficial properties rest. For instance, the social milieu in group-based trauma treatment provides a unique opportunity for members to learn from one another, which often results in awareness that they are not alone in their struggles—a phenomenon known as *universality*. Groups can provide clients with the opportunity to use this interactive social environment as they assume the dual roles of both givers and receivers of help. In turn, this *role flexibility* requires members to focus on "real time," here-and-now interpersonal experiences within the group session.

Trauma therapists who appreciate small-group processes and dynamics view the *group* as an entity larger than the sum of its individual members or the specific protocol used to guide treatment. These therapists acknowledge that the collective, interactive properties of the group have potent effects on members that extend well beyond interventions associated with the formal change theory. In fact, those who embrace the importance of group processes have advocated that formal change theories developed for the individual format may require modification when delivered in a group format (Bloch & Crouch, 1985). Wilfley, Frank, Welch, Spurrell, & Rounsaville (1998) describe the transition issues that arise when one takes an empirically supported treatment developed for individual treatment and modifies it for a group format. Clearly, the number of group-specific factors to consider in such undertakings can be overwhelming to the neophyte group leader. Students in our group therapy classes indeed often complain that there are simply too many elements to attend to and that they are tempted to give up or ignore most of them. Our response is that, like any complex skill, it takes practice to master the intricacies of good group treatment.

Wayne Gretzky, affectionately celebrated as hockey's "Great One," relates that he started playing hockey at age two and subsequently spent all his spare time playing, until hockey became *second nature* to him (Gretzky &

Reilly, 1990). What follows are a few examples of group process principles that merit focus, with the aim of making them "second nature" intervention strategies for group leaders.

GROUP DEVELOPMENT

A large body of theoretical and empirical literature suggests that groups with stable memberships pass through distinct developmental stages (MacKenzie, 1994, 1997; Wheelan, 1994, 1997). Although a number of different frameworks have been proposed, many theorists put forward the existence of predictable stages and attendant processes that may be monitored and harnessed to maximize therapeutic effectiveness. MacKenzie (1994) contends that a group development perspective is critical in designing treatment strategies that take into account the unique environment of each stage. Beck and Lewis (2000) added empirical support to MacKenzie's argument by reporting the existence of predictable changes between different stages of group development and specific therapeutic processes, as measured by several well-known behavioral observation systems.

One group development model that has received considerable empirical study is MacKenzie's (1994, 1997), which consists of a four-stage descriptive framework: *engagement, differentiation, work,* and *termination.* Given its prominence, it is used here to illustrate the principle of group development. The primary group task, unique group processes, and indicators of success for each developmental stage proposed by this model are summarized in Table 12.1.

The *engagement* stage consists of the first few sessions. During this stage, members may be reluctant to share with the group; they typically engage in cautious and conservative interactions, and often avoid intense exchanges between themselves and with the group leader. Given that members who withdraw excessively during this stage are at risk of poor outcomes or dropping out of therapy, a primary task of the group leader is to create a sense of safety and order in the group.

One method for creating a safe environment is to carefully structure the task and activities of the first few sessions. Structure may take the form of providing concrete information regarding the group's goals and how it will operate over time, thereby reducing members' uncertainty. The leader may control the depth and amount of member self-disclosure by engaging in an exercise that elicits specific, reduced-risk information about a trauma. For example, one trauma/grief-focused treatment protocol, designed for groups with six to ten members, invites participants to "briefly describe an experience that you came here to work on" as an early go-around activity designed

TABLE 12.1. Group Development Stages and Process Considerations

Stage of Group Development	Group Task	Process to Expect	Indicators of Success
1. Engagement	Create a sense of membership in the group and its goals	Create sense of safety and order Intermember exchange Supportive therapeutic factors (universality, altruism, acceptance, and hope)	Firm sense of commitment to group All members have participated in some self-disclosure Group norms laid out by leader
2. Differentiation	Develop pattern for conflict resolution Develop tolerance for negative emotional atmosphere	Shift from commonalities to differences between members May be discomfort or disagreement expressed toward the leader	Reworking of group norms so that they become property of group as a collective Deepening sense of group membership and participation
3. Work	Vigorously address individual and group problems	Shift to greater individual work and personal challenge Issues of control and concerns of overinvolvement in work of the group	Increased closeness among group members as they work on more difficult issues
4. Termination	Consolidate gains and transfer group processes to outside support system	Unfinished business Loss and grief over termination of group	All members participate in termination process

to promote a sense of commonality among group members (Layne, Saltzman, Steinberg, & Pynoos, 2003). Such interventions are consistent with the observations of Bednar and his colleagues, who warned that "lack of structure in early sessions . . . feeds clients' distortions, interpersonal fears, and subjective distress, which interferes with group development and contributes to premature client dropouts" (Bednar, Melnick, & Kaul, 1974, p. 31).

The benefits of bringing adequate structure to a group early are well documented, providing confidence for its identification as a fundamental group process principle (Burlingame, Fuhriman, & Johnson, 2002; cf. Kaul & Bednar, 1994). Dies (1994) summarized this literature, noting that groups with early structure are more cohesive and have increased levels of helpful self-disclosure (see also Caple & Cox, 1989; McGuire, Taylor, Broome, Blau, & Abbott, 1986). Nonetheless, too much of a good thing can lead to trouble. For instance, Fuehrer and Keys (1988) conclude that, although too little structure may leave groups floundering, too much structure can both reduce members' feelings of ownership for group accomplishments and dampen group cohesion.

The observation that leader-imposed structure should be calibrated according to group development leads our discussion to the next stage, *differentiation* (see Table 12.1). A few working assumptions underlie *differentiation*. Because group therapy is often the only treatment many clients receive, some evidence shows that members must promptly "own" or become responsible for their (and the group's) progress in achieving treatment goals and objectives rather than being passive observers in the therapy process. If they do not actively pursue their own treatment objectives, they run the risk of becoming a spectator in group activities, as these are orchestrated by the group leader. Several reviewers have concluded that the most effective use of structure is characterized by an increase at the beginning stage of group therapy, followed by a tapering off in the middle and end stages (Bednar et al., 1974; Dies, 1993; Stockton & Morran, 1982). Notably, the presence of some level of structure throughout the entirety of the therapeutic process is associated with beneficial effects, suggesting that evolution into an entirely leaderless format is not a desired end point of group therapy (Kaul & Bednar, 1994).

An important by-product of the *differentiation* stage of group development is member-to-member conflict. Specifically, members' ownership of their individual- and group-level treatment goals in the *differentiation* stage can lead to limited conflict, as group norms transition from leader established to group member owned (see Table 12.1). These emerging intramember differences provide an opportunity for group leaders to introduce and model constructive conflict-management strategies while increasing members' tolerance for negatively valenced emotions. Special consideration should be given to trauma groups wherein members may be especially sensitive to overt conflict, and indeed may experience intense intragroup disagreement as retraumatizing (see Layne et al., Chapter 9 in this volume). Recent research findings (see Burlingame, Johnson, & MacKenzie, 2002) suggest that overt intragroup conflict may not be a necessary or valid indicator of successful group functioning at this stage. Rather, the assumption of *member responsibility* for their own treatment *early on* (e.g., session 3 or 4 in a twelve-session group) appears to be a critical component of effective group treatment protocols. Nevertheless, the empirical literature continues to report manifest conflict as a developmental stage of treatment in structured (e.g., cognitive-behavioral; Castonguay, Pincus, Agras, & Hines, 1998) and more process-oriented groups (Yalom, 1995).

The next two stages of group development, *work* and *termination*, are less remarkable in many respects because their processes are primarily dictated by the formal change theory guiding treatment. As noted above, if the first two stages are successfully negotiated, group members tend to report higher levels of interpersonal closeness and cohesion, both of which are associated with greater member improvement. Nonetheless, three cardinal principles that permeate the relevant literature are (1) the central importance of the *group* as the primary vehicle for change; (2) member-to-member interaction as the most direct *group* mechanism of change; and (3) the leader as an indirect agent of change (cf. Burlingame et al., 2004). Accordingly, focus is upon the group, the group leader, *and* the individual member's role in group development.

An important consideration to group development, then, hinges on the effect of the therapist's structuring during different points of a group's evolution. One way to come to terms with development and therapist structuring is to frame them using a second dimension of group dynamics introduced in Table 12.1: *group norms*. Group norms are the consensual standards that describe which behaviors should and should not be performed in a given context (Forsyth, 1999). These norms are established early in the group's development and persist for long periods of time. Laboratory and clinical research has demonstrated that, once in place, norms persist for the life of the group and are resistant to change (Bond, 1983; Jacobs & Campbell, 1961; Lieberman 1989).

Findings from the group literature suggest that the responsibility for creating healthy norms falls primarily on the therapist. Interestingly, groups appear to establish norms as a natural consequence of their operation. As Yalom (1995) notes, "Wittingly or unwittingly, the leader *always* shapes the norms of the group and must be aware of this function. The leader *cannot not influence norms;* virtually all of his or her early group behavior is influential" (p. 111).

These conclusions both illustrate the interrelationships between group development, therapist structuring, and norm development, and underscore the need to balance competing therapeutic interests and caveats. As applied to group-based trauma treatment, trauma group therapists can employ their knowledge of group norms and structuring activities during the engagement stage to create beneficial group norms. These norms may, in turn, exert an enduring influence on the types, amounts, and timing of subsequent group processes. Group therapists should thus give careful attention to the structuring methods they employ, especially in the engagement phase, as this will by extension shape the group's norms. As a cautionary example, therapists who choose to engage almost exclusively in highly structured interventions (e.g., psychoeducation) at the beginning of a group intervention may be inadvertently establishing norms that dictate heavy reliance on group leaders, member-to-leader interaction, and member passivity. Once these norms are established it will be difficult to persuade members to interact with one another and become responsible for their own and the group's treatment goals. As we summarize this section on group development, we leave readers with a few questions to ponder as they review group-based trauma models:

- Does the model actively encourage the development of an esprit de corps in members regarding the group as a whole?
- If so, are the interventions used by the model congruent with general principles associated with stages of group development?
- How and when does the leader use structuring interventions, and what group norms might develop as a result?
- Is there a point when the leader deliberately uses less structure, thereby encouraging member involvement and ownership of treatment goals?
- Is there evidence for members "owning" the treatment goals of the group and actively applying them to their individual goals, as opposed to taking on a "spectator" role?

INTERPERSONAL FEEDBACK

A second process that consistently emerges as a potent therapeutic factor in the small-group literature is member-to-member feedback. Were it not for uniquely group-based therapeutic processes of this sort, group as a treatment modality would possess efficiency as its sole incremental benefit. For the purposes of this chapter group feedback is defined as *any interaction between two or more group members in which reactions or responses to a particular activity or process are communicated.* Feedback is considered to be particularly beneficial because of its demonstrated potency, its well-researched effects, and its many potential applications in trauma-focused group treatments.

The therapeutic advantages associated with systematically implementing interpersonal feedback into trauma groups may be viewed from two perspectives. The first perspective focuses on the empirical links that have been established between interpersonal feedback, patient improvement, and related therapeutic processes. Although impressive, a distinct limitation to this attempt to "cross-fertilize" the trauma and small-group literatures is that these findings are primarily drawn from nontraumatized patient populations.

Empirically Identified Benefits
of Feedback Exchange

Individual studies addressing the effects of interpersonal feedback on small-group process are too extensive to be reviewed adequately within this chapter. Fortunately, three recent reviews provide complementary and converging conclusions regarding the impact of interpersonal feedback. The first review is Kaul and Bednar's (1994) summary of nearly a decade of research. They conclude that "feedback from other members is commonly accepted as a critical therapeutic factor in group treatment and is so widely held that we would be embarrassed to mention it to an audience of group leaders" (p. 161; for a longitudinal perspective providing further empirical support for feedback, see Bednar & Kaul, 1986; Kaul & Bednar, 1994).

Rex Stockton and Keith Morran's research team has been programmatically studying interpersonal feedback for nearly three decades. They recently summarized this empirical literature (Morran, Stockton, & Teed, 1998), proffering conclusions regarding the effects of interpersonal feedback as well as evidence-based recommendations on therapeutically enhancing feedback. They concluded that feedback *over and above* that which naturally occurs in groups is related to more successful group and individual outcomes. More specifically, the authors state that interpersonal feedback has been empirically linked to heightened motivation for change, greater insight into how one's behavior affects others, increased comfort in taking interpersonal risks, and higher ratings of satisfaction with the group experience.

A final review of the empirical literature suggests *why* interpersonal feedback may be related to better patient outcomes in group treatment. Burlingame, Fuhriman, & Johnson (2002) summarized the cohesion literature from two perspectives: the first established group cohesion as a strong predictor of patient improvement. For instance, member-to-member cohesion is strongly related to improved patient outcomes (Hurley, 1986; Marziali, Munroe-Blum, & McCleary, 1997; Raskin, 1986; Rugel & Barry, 1990; Widra & Amidon, 1987; Wright & Duncan, 1986), increased client insight (Sexton, 1993), and retention of members (MacKenzie & Tschuschke, 1993; MacKenzie, 1987). Group cohesion is also associated with greater self-disclosure, member "ownership" of group functioning, and the capacity of members to tolerate negative affect and stress—capacities that are particularly vital during group developmental stages (especially the *differentiation* stage) that are typified by higher rates of conflict (Castongauy et al., 1998; Fuehrer & Keys, 1988; Tschuschke & Dies, 1994). It should be noted that some studies have shown an apparently harmless, even beneficial, drop in cohesion during the *differentiation* stage, as members start to challenge one anothers' distortions and seek to own the group and its goals (Kivlighan & Lilly, 1997; Kivlighan & Shaughnessy, 2000). This dip in cohesion is perhaps needed to allow group members the freedom to work on difficult issues. Nonetheless, cohesion has been linked in so many studies to positive treatment outcomes, and to so many beneficial group processes, that it is hard to imagine a successful group operating without high levels of cohesion.

Burlingame, Furhiman, & Johnson's (2002) second perspective on the therapeutic effects of cohesion sheds light on potential cause/effect relationships between cohesion and other therapeutic group processes, including feedback. Notwithstanding their identification of methodological weaknesses in the literature, the authors note that at least three important findings have considerable relevance to successful group-based treatment (cf. Tschuschke & Dies, 1994). First, members who report high levels of cohesion are rated by independent observers as making more self-disclosing statements. Second, higher levels of self-disclosure are related to higher levels of interpersonal feedback between members. Third, higher member ratings of closeness and cohesion are positively related to an increased likelihood that members will accept feedback from one another (Greller & Herold, 1975). Taken together, these findings suggest that although group cohesion may not necessarily cause members to engage in higher levels of member-to-member disclosure, member-to-member feedback, and greater receptivity to that feedback, it is at minimum an important marker of these highly beneficial group processes.

Indeed, in an earlier review of the feedback literature, Kivlighan (1985) shows that group cohesiveness, credibility, desirability, and group interaction may be enhanced by a sequence of feedback in which the positive precedes the negative. This notion is elaborated by Burlingame, Furhiman & Johnson (2002), who conclude that group cohesion, defined as high positive emotional relatedness between group members, appears to promote a propensity to disclose important and meaningful material. These disclosures, in turn, lead to more frequent and intense feedback among group members. In summary, despite a lack of clear understanding in the literature regarding the causal processes underlying the observed relationship between group cohesion and member-to-member feedback, it is clear that feedback is integral to a spectrum of beneficial group processes, perhaps especially cohesion and, by extension, to positive treatment outcome. Finally, the therapeutic effects of feedback may be transtheoretical. In his review of therapists' judgments of what constituted a therapeutic and/or essential ingredient for affecting positive treatment outcomes in individual treatment, Goldfried (1980) reported that feedback was identified across a variety of theoretical orientations. This finding suggests that feedback may serve as a "common factor," or an integral element of group treatment, whether the theoretical orientation is cognitive behavioral, psychodynamic, humanistic, or another (cf. Burlingame et al., 2004).

Unique Benefits of Feedback for Trauma-Group Members

The second perspective on the therapeutic advantages of interpersonal feedback reflects the unique value that these exchanges may have for members of trauma groups. Feedback as a therapeutic tool may be an especially effective intervention with trauma-exposed clients for at least two reasons. First, many individuals with histories of significant trauma tend to exhibit signs either of estrangement from others and/or general social withdrawal (e.g., Foy et al., 2000). These behavior patterns may, in turn, remove withdrawn individuals from social contexts wherein corrective and/or supportive feedback may be given or exchanged. Second, trauma-exposed individuals and similar larger social groups may exhibit the so-called trauma membrane, characterized by the passive or active exclusion of people who are perceived not to have experienced the same, or similar, traumatic events (Lindy, 1985). This phenomenon may manifest itself in the exclusion of other group members and therapists who have the potential to help, and even in therapists rejecting other potential help givers based on the rationale that they cannot fully relate to, or effectively assist them with, their trauma-related experiences because these potential allies have not had, or endured, the same events (e.g., Lindy & Wilson, 2001).

Facilitating member-to-member interactions within the group may be an effective tool for intervening with trauma-exposed groups in general, and especially with those who have erected "trauma membranes," precisely because it both recruits and facilitates therapeutic exchanges between individuals who have "been in the same boat" (Gottlieb, 1996). For example, a common goal of group-based interventions among individuals who exhibit signs of isolation, alienation, and diminished emotional responsiveness is to increase the degree and comfort with which they take interpersonal risks, such as when making appropriate disclosures to fellow group members. Another advantage conferred by member-to-member interaction is that it helps to redefine and delimit the role (and, ideally, any perceived intrusiveness) of group leaders to that of *indirect* agents of change as they facilitate *direct member-to-member* exchanges (cf. Fuhriman & Burlingame, 1990, 1994a).

A considerable body of evidence suggests that group feedback is a potent therapeutic tool for increasing members' insight into the manner in which their own behavior influences their interpersonal relationships; as pointed out by Yalom (1995), members who are exposed to group feedback become more aware of important aspects of their interpersonal behavior and its influences. These include members' strengths and limitations; distortions in their interpersonal perceptions, assumptions, beliefs, and attitudes; and ways in which their own behaviors elicits unwanted and unintended responses from others.

Group feedback to individual members has also emerged as a powerful tool for challenging and changing members' distorted or otherwise maladaptive cognitions. This is a particularly relevant goal among trauma-exposed clients, many of whom report or manifest a variety of cognitive distortions. For instance, common cognitive distortions encountered with this population include maintaining the basic beliefs that the world is not a safe place, that one is "jinxed" and should thus avoid others as a means of protecting them, that interpersonal relationships are exploitative and unhelpful, and that one will not live long despite efforts to avoid high-risk activities (Janoff-Bulman, 1992; Foa, Ehlers, Clark, Tolin, & Orsillo, 1999; for examples of treatments utilizing group-based principles to change maladaptive cognitions see Cohen et al., 2001; Foy, Glynn, Ruzek, Riney, & Gusman, 1997; Layne et al., 2003). As noted by Foy and colleagues, group members may identify the faults in others' thinking when they cannot do so with their own. Moreover, by identifying distortions in others' thinking, members build not only their perceptions of efficacy and self-worth by providing valuable feedback but also often develop insight into their own problematic thinking patterns (see Gottlieb, 1996, 2000). In summary, feedback has repeatedly helped clients gain an accurate picture of how they come across to, interact with, and elicit desirable or undesirable reactions from the people around them.

Empirically Grounded Principles
for Interpersonal Feedback

If a trauma therapist found the aforementioned literature convincing, a next consideration would be how to systematically implement interpersonal feedback exchange within an existing group protocol. Our experience in teaching students, and even some veteran group leaders, about effective group leadership is that many lack knowledge regarding how to facilitate feedback exchange in groups beyond that which occurs spontaneously. As with group development, the literature suggests some empirically grounded principles to guide the group leader.

The first and most obvious consideration regarding the active fostering of feedback is the interactive nature of the group environment. In contrast to individual therapy, in which only the therapist provides feedback, group-based modalities provide a forum in which feedback may be given from a diverse array of sources—namely, other members of the group (see Figure 12.1).

This broader range of potential sources of feedback may constitute a fertile arena within which change-inducing experiences can take place. For example, Yalom (1995) found that members most frequently identified the expression of strong affect between group members as a "critical incident" that constituted a positive turning point in their treatment. Moreover, there is limited evidence to suggest that member-to-member feedback is more "meaningful" than that received from therapists (Flowers, 1979). Dies (1994) summarized the apparent power of member-to-member feedback exchanges by observing, "Ironically, when individual clients view group treatments as successful, they typically cite the helpfulness of client-to-client interactions, but when treatments fail, therapists are generally implicated" (p. 138).

More pointed recommendations regarding the management of members' feedback exchanges in small groups come from a series of studies conducted by Morran et al. (1998). These authors experimentally manipulated feedback valence (positive versus negative), member readiness, and methods of delivery. The findings from these studies led to empirically based

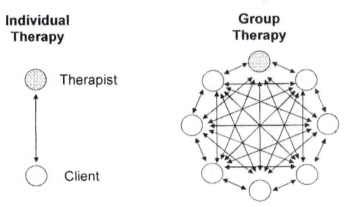

FIGURE 12.1. The Complexity and Potential Richness of Group Feedback As Compared to Feedback in Individual Therapy

recommendations for facilitating the giving, receiving, and therapeutic impact of feedback. Examples include the following.

1. During early sessions, positive feedback should be emphasized. In contrast, middle and later sessions should strive for a balance between positive and corrective feedback.
2. To increase the likelihood that it will be well received, corrective feedback should be preceded by, or sandwiched between, positive feedback messages.
3. Feedback is most effective when focused on specific and observable behaviors.
4. Care should be taken to assess receptiveness before delivering corrective feedback messages to given individuals.
5. Group leaders should model feedback exchange while also facilitating constructive member-to-member feedback.

In summary, the empirical picture of the putative causes, effects, and correlates of interpersonal feedback in small-group treatments is repeatedly associated with factors that directly predict outcome (e.g., cohesion), motivation for change, and other components of success in small-group treatment. Moreover, interpersonal feedback is a well-researched group-specific intervention and has matured to the point that it now yields empirically based guidelines for group leaders. Although our argument for its consideration is based primarily on empirical findings, we end with a clinical example of its application.

In a recent group supervision meeting, two leaders who were running a structured outpatient psychoeducational group made the case that incorporating member-to-member feedback exchanges would detract from the time and focus necessary to present and assimilate their material. After a rather frustrating initial period in which members complained that "these sessions feel like we're being doused with information from a fire hose without getting the time to figure out how to apply it in our lives," the leaders restructured the group to allow time for feedback and other group processes. In subsequent supervision meetings, both leaders remarked that the information they presented was better received and retained by the group members following the restructuring. Similarly, group members reported higher levels of satisfaction with their group experience and increased group cohesion.

INTEGRATING SMALL-GROUP PROCESS PRINCIPLES INTO EXISTING GROUP-BASED TRAUMA TREATMENT PROTOCOLS: A CASE EXAMPLE

How can the information reviewed above be used to benefit trauma group therapists and the help they provide? It should be noted that it is not our intention to look for flaws in the well-conceived and time-tested protocols presented within this and other volumes. Instead, we are suggesting ways in which these and other empirically supported treatment programs may potentially be enhanced by actively harnessing relevant small-group process principles. As stated above, sufficient evidence exists to suggest that some group properties indeed transcend theoretical frameworks—that is, they are inherent properties of the group format, and their effectiveness is not contingent on being paired with any specific theory of change. Indeed, seasoned group clinicians are well aware that if one does not foster the therapeutic properties of groups (e.g., feedback, group development, norms, and cohesion), member attrition is likelier, along with nontherapeutic group processes and iatrogenic effects.

Having outlined general principles regarding group development and feedback, we will now briefly describe a "gold standard" trauma group protocol and then comment on ways in which small-group process principles might be integrated above and beyond those already incorporated into the protocol.

Overview of a "Gold Standard" Group-Based Trauma Treatment Protocol

Trauma-focused group therapy (TFGT) (Foy et al., 1997; Schnurr et al., 2003) is a thirty-session group-based treatment protocol designed specifically for the treatment of war veterans suffering chronic PTSD. The primary objectives of TFGT are to enhance members' control of chronic PTSD symptoms and to improve overall quality of life (Foy, Ruzek, Glynn, Riney, & Gusman, 1998). Groups are comprised of six members and two leaders; treatment is delivered in weekly 1.5- or two-hour sessions for seven months, followed by five monthly "booster sessions" and phone calls.

TFGT Treatment Strategies and Interventions

Psychoeducation about PTSD. TFGT places a strong emphasis on psychoeducation regarding peri- and posttraumatic reactions. This emphasis includes universalizing "fight, flight, or freeze" patterns of peritraumatic

crisis reactions, and normalizing individual differences in ongoing physiological and subjective reactions to trauma-related cues. The therapeutic rationale for engaging in exposure therapy and cognitive restructuring activities is also explained. Efforts are also directed toward educating family
members and significant others regarding the short- and long-term effects
of trauma and the rationale undergirding TFGT.

Coping skills training. TFGT strongly emphasizes the development and
use of coping skills to promote posttraumatic recovery. Early in treatment,
veterans are trained in the use of the subjective unit of distress scale
(SUDS), and they complete a coping resources self-assessment. In subsequent sessions veterans work to develop and refine a variety of coping skills
and resources, including managing traumatic memories, developing problem-solving skills, practicing memory/concentration, and learning emotional self-regulation skills, including anger management and emotional
sensitivity. Veterans also work to develop the size and accessibility of their
social support networks during times of need. Group activities may at times
focus on developing members' spiritual/philosophical coping resources, including religious beliefs and activities, relationships with organized religion, and worldview. Physical health and fitness are emphasized, including
diet, exercise, sleep, and the reduction of high-risk behaviors, especially
smoking and excessive drinking.

Prolonged therapeutic exposure. TFGT utilizes prolonged imaginal therapeutic exposure as a means of desensitizing trauma-related fears and decreasing maladaptive avoidance/escape responses to trauma-related cues. An
important undergirding of TFGT is the assumption that both the member
occupying the "hot seat" and fellow group members profit from the effects
of therapeutic exposure, via direct and vicarious means, respectively. In
essence, a group-level "intervention" takes place as members bear witness,
nonjudgmentally, to each member's recounting of his significant life experiences, followed by supportive feedback (Foy et al., 1998). These exchanges
take place in the form of a group-based scene construction exercise, conducted with each individual member in the form of a three-act play. In Act I,
the member occupying the "hot seat" describes the events immediately preceding the tragedy; Act II consists of a vivid description of the actual
trauma; and Act III consists of describing the aftermath, impact, and meaning of the traumatic event. All scenes are described in the present tense, utilizing all senses, for a period spanning thirty to forty-five minutes to ensure
an adequate dose of exposure. Other group members are encouraged to stay
in the "here and now," to think about similarities between the event being
described and their own traumatic experiences, and to prepare an encouraging comment to deliver during a feedback period following the conclusion
of each narrative. Two consecutive cycles of narrative construction are con-

ducted, allowing each member two spaced trials of prolonged direct therapeutic exposure, in addition to intervening trials of prolonged vicarious exposure to other members' narratives.

Cognitive restructuring. TFGT employs cognitive restructuring exercises on two levels. The first focuses on modifying maladaptive cognitions surrounding the traumatic event featured in the narrative construction exercise. The second focuses on identifying and modifying negative core beliefs "learned" during the traumatic event and their subsequent psychosocial impact. The majority of cognitive restructuring activities are conducted immediately following the second prolonged therapeutic exposure trial. The group leaders begin by leading a review of objective facts in a search for "key points" that led up to the tragic outcome, with particular emphasis given to the degree to which the narrator could have predicted and controlled those critical events. A variety of techniques are then used to identify and modify negative cognitions that are linked to such distressing emotions as helplessness, rage, guilt, shame, and fear. These techniques include direct disputation by the group leaders; positive feedback; counterarguments; persuasion to adopt alternative viewpoints by other group members; and reexamination and self-disputation on the part of the focal group member. The group leaders then guide the member in question to examine the connection between negative beliefs acquired during the traumatic event and subsequent social and occupational functioning. This exercise concludes by aiming to modify one or more negative core beliefs in anticipation of its beneficial impact on work, social, and family functioning. These activities are supplemented by homework exercises in which members construct autobiographies designed to facilitate continuity in members' interpersonal relationships, experiences, and self-image before, during, and after the traumatic event.

Relapse prevention. The last third of TFGT sessions are dedicated to relapse prevention, the primary aim of which is to identify and enhance factors responsible for maintaining control over the long term (Parks, Anderson, & Marlatt, 2001). Emphasis is placed on promoting members' knowledge, perceptions of coping self-efficacy, and coping repertoires to encourage members to identify, predict, and mobilize coping resources when confronted with high-risk situations. Specific activities include revising distortions concerning perceived positive consequences of resuming the problem behavior, while accurately assessing the intermediate and long-term consequences of relapsing. Efforts are also directed toward identifying high-risk situations and the triggers they contain, distinguishing between a "lapse" (temporary setback) and a "relapse" (major setback), and developing anger management skills. Additional foci include identifying, improving, and learning to access important interpersonal relationships during times of need, developing healthy self-care habits, and developing emergency coping plans.

APPLICATION: INCORPORATING SMALL-GROUP PROCESS PRINCIPLES INTO TFGT

Preliminary Comment

Foy et al. (1997) are well aware of many beneficial group process principles, as indicated by their deliberate incorporation into the TFGT protocol. Evident group processes include facilitating member-to-member feedback, promoting group cohesion, and establishing the reactions of fellow group members as a "norm" for evaluating what is expectable. Building on these accomplishments, following are ways in which group development and interpersonal feedback principles may be additionally incorporated that contribute to the quality of the group climate and thereby increase the range of therapeutic options and leverage available to the group leaders. We partition our analysis into the early, middle, and late stages of group development, respectively.

Early Group Development ("Engagement" Stage)

One way in which the group process literature may inform trauma-focused interventions such as TFGT is through establishment of group norms. In particular, a balance should be sought between (1) early structure-promoting interactions such as the provision of psychoeducational information and descriptions of the content and format of the group and (2) establishing group norms that encourage direct member-to-member constructive feedback and active participation. Given the tendency of group norms to persist once established and the strong emphasis that TFGT places on early psychoeducation, care must be taken to avoid the formation of a "teacher/student" norm that precludes other beneficial forms of group interaction. That said, how could one provide early structure, establish a group norm of member interaction, and still adhere to the treatment objectives of the TFGT protocol?

One solution might be to engage in an early structuring intervention designed to promote member-to-member interaction and positive feedback. This could take the form of a short psychoeducational segment focused on teaching members to appropriately give and receive constructive feedback within the group. This presentation could be paired with a brief handout describing the elements of effective feedback and a process-focused discussion designed to increase members' comfort with, and understanding of, providing constructive feedback. Morran et al. (1998) suggest that such discussions are best conducted by first inviting members to discuss their personal feelings about giving and receiving feedback. Such "barrier-focused" discussions carry the secondary benefit of modeling self-disclosure and facilitating member-to-member interactions, both of which are essential elements of positive feedback exchanges. Such interventions can be conducted in as little as twenty minutes, yet generate large, enduring payoffs via the establishment of therapist-directed group norms. These activities also promote structure in an appropriate manner, generate positive group-related expectations, and blend in with a psychoeducational format while not ignoring vital group processes. Importantly, the norms established by this intervention tend to encourage spontaneous member-to-member interaction, as opposed to turn taking, which may promote stagnant interactions and is linked to poor group functioning (Karterud, 1988).

A second way in which the group process literature may complement TFGT is through the incorporation of structured feedback exercises. The empirical literature suggests that by the third or fourth group session members are generally comfortable enough with one another to benefit from this type of intervention (Morran et al., 1998). A representative exercise may consist of encouraging group members to write a positive statement about each of the other group members on a small card and to exchange them. Each member could be encouraged, but not forced, to share some of the feedback he or she received and to discuss its personal impact. This simple exercise can be powerful, to the extent that some group members keep and cherish their feedback cards for years after the group terminates. Therapists can harness this beneficial process by facilitating and reinforcing positive feedback exchanges early in the group. These interventions allow the group leader to model the giving of appropriate positive feedback and to frame feedback exchanges in a positive light. These early interventions can be gradually tailored in later sessions to facilitate the exchange of both spontaneous positive feedback and corrective feedback that is sandwiched between positive messages.

An example of therapist-directed feedback is found in one of our graduate student classes on group therapy. We presented a short didactic lecture on feedback principles and then asked the group to engage in an experiential exercise. The group was asked to write one positive and one corrective feedback statement about the other members on a 3 × 5 card and then to share this with one another. This intervention generated a spontaneous heart-felt discussion about the excitement and apprehension each felt in giving feedback and "taught" our lesson much more effectively than the didactic portion had. More important, a leader-established norm was implemented for constructive and open feedback exchanges, which in turn helped group members to gain valuable information about their skills as group leaders and clinicians over the course of the group.

Notably, the leaders should take group development factors into account when designing feedback exercises. In particular, asking group members to exchange feedback too early in the group may hinder group development. This may be a particularly relevant concern for trauma-focused groups, whose members may be especially sensitive to themes of safety, security, predictability, and control, and who may perceive emotional exchanges as aversive. Conversely, development of feedback exchanges later in a group's developmental history may be impeded by entrenched group norms that discourage direct member-to-member interactions and open disclosure.

A third way in which group process-based interventions may assist during early group formation draws on cohesion-building activities. In keeping with the observation that increasing one beneficial group process will often evoke others, the structured exercises described above that involve the processing of feelings relating to giving and receiving feedback, followed by the actual exchange of feedback in member-to-member interactions, will likely increase group cohesion as well (Stockton, Rohde, & Haughey, 1992). Further, member disclosure is positively associated with group cohesion, as the personal level of detailed disclosure increases, the perceived interpersonal distance between members tends to decrease (Bunch, Lund, & Wiggins, 1983; Kirshner, Dies, & Brown, 1978). Similarly, facilitating the exchange of emotional expressions between group members is also linked to increased levels of group cohesion (Burlingame, Furhiman, & Johnson, 2002). Notably, the TFGT protocol encourages detailed sharing of traumatic experiences in the form of a prolonged therapeutic exposure exercise designed to decrease trauma-related reactivity through habituation. This exercise may concomitantly harness beneficial group processes, especially cohesion, to the extent that members are encouraged to give constructive feedback in the form of "here and now" reactions to members' disclosures. Additional group-level benefits may be realized through the use of member-to-member interactions relating to the identification and disputation of trauma-related cognitive distortions.

This example illustrates the benefits of disclosures accompanied by positive feedback exchanges. An adolescent female group member disclosed to a group of predominantly males (including three boys and two adult male group leaders) that she had a history of multiple trauma exposures relating to male-perpetrated sexual assaults. These experiences ranged from being sexually molested by extended relatives during her childhood, to an attempted rape at a community park, to witnessing a brutal gang rape while at a rave party. She then stated that her life plans actively excluded men, to the extent that she intended to adopt children and raise them on her own. These disclosures were met with sympathetic expressions of sorrow and muted horror by the young men in the group. Later, in an individual session, the young woman stated that receiving the boys' sympathetic reactions to her story helped her to "believe that not all men want only one thing." With the group's ongoing support, she began to accept dating invitations and to modify her life plans to include active involvement with men.

Middle Stage of Group Development (Differentiation and Work Stages)

The establishment of group norms that promote feedback, member-to-member interactions, and cohesion early in the group's development sets the stage for therapeutically potent interactions during the group's working phase. The TFGT protocol specifies that leaders assume a comparatively active role during this phase while assisting members in selecting a trauma narrative and then assisting with trauma narrative scene construction and thought disputation. Nevertheless, group development literature suggests that member-to-member interactions be encouraged to the extent possible during this phase so as to increase client ownership of and responsibility for treatment. These types of interactions are found in the Foy et al. (1997) protocol in a variety of places, such as following the trauma scene construction of a focal member, in the form of members' disclosure of sympathetic emotional expressions, and sharing perceived similarities across members' experiences. The member-to-member interactions in this session may indeed be as therapeutic, although utilizing different curative mechanisms, as the exposure-based portion of the narrative task. Thus, care should be taken to allocate sufficient group time for member-to-member interactions in the form of positive feedback and constructive criticism.

An additional way in which the group process literature may contribute to the middle phases of TFGT falls within the domain of cognitive restructuring activities. TFGT relies on direct disputation by the group leaders; positive feedback, counterarguments, and persuasion by other group members to adopt alternative viewpoints; and reexamination and self-disputation on the part of the focal group member. The group member is then directed toward examining the connection between negative beliefs acquired during the traumatic event and his or her subsequent social and occupational functioning. We suggest that the therapeutic impact of these activities may be maximized to the extent that group members appropriately take the lead. By so doing, members will both actively apply the material and increase the degree of internalization of one another's feedback.

In a second case example, a group member consistently prevented others from emotionally connecting with him by keeping the conversation light and superficial. This reluctance to discuss weighty issues consequently impeded several important group processes. A number of group members seemed annoyed with his style yet reluctant to give feedback or to take responsibility for the direction the group was going. One of the leaders quickly took this responsibility upon himself and repeatedly gave feedback accompanied by psychoeducation, but still with little effect on the member's behavior or on the group process. Eventually, the leader realized that his primary influence had resulted in norms of leader-to-member interaction, coupled with a lack of group ownership of its goals and process. After stepping back and allowing the group to address the problem themselves, members supportively but strongly challenged the reluctant member. The resulting interactions yielded insight as the resistant member disclosed that the rest of the group seemed fragile and thus needed him to protect them from intense interactions. He was surprised to learn that other members did not need or want his "protection," and given the credibility of the source— the group members themselves—he internalized the feedback. In subsequent sessions he modified his behavior and, with the group's help, began to explore his own sense of frailty coupled with his strong wish to protect others from harm.

Another benefit of member-to-member interactions is the realization that they have the power to help others and, by extension, to help themselves. By helping their cohorts, members develop the capacity (and sense of responsibility) to provide effective aid to others as well as to themselves (Burlingame & Davies, 2002). This phenomenon is referred to in the self-help group literature as the "helper-therapy principle" (Riessman, 1965). The chance to help others with a similar problem is often a catalyst for personal change, as suggested by findings that the number of helping statements made by one member to other group members is positively correlated with patient improvement (Roberts et al., 1999).

Final Phases of Group Development

The small-group process literature may also complement TFGT in its final phase with respect to using social support to fortify relapse prevention. Specifically, high levels of cohesion and member-to-member feedback may be positively related to the development of healthy interpersonal relationships, both inside and outside the group (Bloch & Crouch, 1985). This interpersonal learning is developed by group members' actual experiences with one another in the group. It can then be generalized, using repeated practice exercises coupled with feedback and problem-solving activities, to other relationships and domains of members' lives outside the group. Notably, TFGT and other trauma-focused treatments focus on relationship building among group members as an essential tool for learning how to relate in healthy ways with other life figures. This focus on relationship building can lead to much more powerful and life-changing experiences than those furnished by comparatively abstract psychoeducational activities alone.

In a program-effectiveness evaluation of a UNICEF-sponsored postwar program for adolescents conducted in 2000-2001 (Layne, Saltzman, et al., 2001), students participating in focus groups identified the skill of communicating openly, and effectively seeking support from others, as the most beneficial things they gained from the group. Student comments included, "We could speak about everything," and "This group has helped me arrange my home life and to talk with my parents and tell them what is the problem. I tell them my opinions now." A group leader stated, "This program gave [the students] the opportunity to speak about their traumatic experiences for the first time. Many students have never been in a position to speak to anyone, even those they are closest to." Gains in support-seeking/communication skills were also frequently identified; one group leader identified these skills as the *single most helpful* program component.

Facilitating member-to-member feedback is also a potent tool for attaining additional TFGT goals during the final phase. For example, members may help one another identify, predict, and mobilize coping resources when confronted with high-risk situations as they develop and problem solve their coping plans. In addition, members are encouraged to assist one another in disputing and revising distorted expectations concerning the positive consequences of resuming problem behavior and in accurately assessing the real intermediate and long-term consequences of relapsing. These may include identifying "high-risk" situations and the triggers they contain. Group members may also prove instrumental in developing anger-management skills as they model appropriate problem-solving skills and self-statements while disputing erroneous attributions concerning the motives of others. Last, group members may assist one another in developing healthy self-care habits and emergency coping plans. Given that many or all of these activities may involve the use of constructive criticism, care must be taken to precede and/or sandwich criticism between positive feedback and to use other therapeutic interventions (e.g., cutting off or modification of feedback; see Morran et al. (1998) for a more complete description of these techniques) to increase the likelihood of a positive outcome.

OBSERVATIONS AND RECOMMENDATIONS

Overview and Recapitulation

We have observed elsewhere (Burlingame et al., 2004) that extant group protocols can be divided into three clusters within the small-group process literature. First is the protocol that recognizes and actively calibrates interventions against the unique properties of the group format. Foy's TFGT model is a good example of this treatment protocol. A second cluster of protocols adopts a high-level recognition of group processes but does not actively modify interventions or treatment goals with an eye toward fostering format-based considerations. For instance, these models may discuss the importance of cohesion and assess it using a self-report member instrument. However, the primary emphasis is upon mechanisms of change that are guided by the formal change theory rather than format-specific considerations. The third and final cluster makes no mention of group-specific mechanisms of change or modifications of the treatment model that may be required if it is converted to a group format.

The bulk of the extant group models fall into the latter two clusters (Burlingame et al., 2004), leading us to wonder if extant protocols are simply not attending to basic group operations. There are least three possible responses to this unsettling question. The first is that group properties (e.g., climate, norms, development, etc.) may be perceived by practitioners as so integral to group-based intervention that protocol theorists feel no need to mention such "elementary" or basic operations. A second possibility may derive from the observation that the lion's share of empirical evidence associated with group process literature is derived from studies emphasizing a particular theoretical orientation (e.g., process, encounter, and dynamic groups). Perhaps authors of protocols using different formal change theories (e.g., cognitive, behavioral) may see little relevant crossover between these principles and their group protocols. Finally, recent evidence (Fuhriman & Burlingame, 2001) suggests that only a small percentage of mental health training programs provide specific education in the group format; thus, knowledge regarding this literature may simply be insufficient. Our consultation experience with group therapists leads us to believe that all three explanations have merit.

An additional limiting factor is the aforementioned disorder that exists in the current group process literature. Good group process studies are infrequent, scattered across over 160 professional journals (cf. Burlingame, Kircher, & Taylor, 1994), and lacking in a consensually accepted taxonomy, which limits our ability to organize knowledge regarding group-specific mechanisms of change. In short, the acquisition of knowledge and expertise is hampered by our inability to steer interested students and clinicians to focused and consistent areas of the literature. Nevertheless, we have pointed the interested reader to some of the available summaries of salient findings on specific constructs.

An argument could be made that the group processes we have discussed are simply the common factors or nonspecific effects of the group format, and can thus be dismissed on the grounds that they operate equally well across all group treatments. But a potent counterargument comes from Hyun-nie and Wampold (2001), who recently estimated that such common factors might account for up to nine times more patient improvement (as measured by relative effect sizes) than the specific formal change theory used. Moreover, dramatic differences in patient improvement have repeatedly been attributed to reliable differences between therapists (Lambert & Ogles, 2003). A possible explanation for these reliable therapist effects is their use (or lack thereof) of common or nonspecific therapeutic factors (cf. Burlingame et al., 2004). Thus, an empirically based case can be made for the importance of paying careful attention to small-group properties that have consistent empirical track records linking them to patient improvement.

DIRECTIONS IN FUTURE RESEARCH AND TRAINING

We began this chapter with the example of a dedicated hockey star who, through sustained practice, developed versatility and expertise in multiple areas of play. We then drew an analogy to trauma treatment, suggesting that expertise in a variety of trauma interventions, coupled with modality-specific skills, may substantially increase the effectiveness of the trauma therapist. We subsequently illustrated how two group-specific processes—drawn from readily available reviews of the literature—may be profitably incorporated into group-based trauma treatment. For the interested (and at least partially persuaded) reader, questions do arise: "Where would I go to hone my group-specific skills?" and "Where should the field go to empirically test the propositions of this chapter?"

We alluded above to a currently pressing paradox: Although the use of groups in clinical practice is on the rise (Taylor et al., 2001), specialized training in group-based interventions for new mental health professionals (e.g., in the fields of psychology, social work, and psychiatry) is declining (Fuhriman & Burlingame, 2001). Fortunately, training opportunities are available through professional associations such as the American Group Psychotherapy Association (AGPA), the Association for Specialists in Group Work (ASGW), and Division 49—Group Psychology and Group

Psychotherapy of the American Psychological Association (APA). For instance, AGPA offers practicing professionals a course on the principles of group treatment and a certified group psychotherapy (CGP) program to ensure at least minimal training and competence in group work. Although AGPA has historically focused primarily on "clinical" practice alone, it increasingly offers training based on the empirical literature. ASGW has established professional standards (cf. Conyne, Wilson, & Ward, 1997) that articulate the requisite knowledge and skills required to deliver different types of groups (e.g., psychoeducational, psychotherapy, counseling, etc.). Thus, resources are available.

In many respects, the research base pertaining to small-group processes is underdeveloped. A few research teams scattered around the globe are simultaneously exploring group- and model-specific mechanisms of action and their relationship to outcome. For example, our recent research has explored the relationship between patient characteristics and outcome using the Group Selection Questionnaire (Davies, Seaman, Burlingame, & Layne, 2002). Preliminary results are promising and indicate that client expectations for the group experience, potential for group deviancy, and predisposition to actively participate are predictive of positive individual outcomes in a school-based psychoeducational group for war-traumatized children. However, the collective output from these research teams is insufficient to adequately address the burgeoning number of group-based treatment models in the extant literature. These investigators have, along with others (e.g., Bednar & Kaul, 1986), argued that researchers should invest more effort in expanding the theoretical and empirical group process literature.

Given the attention herein to methodological problems extant within the small-group process literature, it seems reasonable to propose one avenue for "disentangling" the contributions of "spontaneous" small-group process principles (versus formal theories of change) from treatment outcome. Specifically, the available treatment-effectiveness literature suffers from methodological problems, making the formation of definitive conclusions regarding the differential effectiveness and relative contributions of theories of change difficult. As a result, the field may benefit from a series of studies using a 2×2 [(formal theory of change versus no formal theory of change) × (small-group processes versus no small-group processes)] design that would generate four experimental cells: (1) A formal theory of change-based groups that facilitates group processes; (2) a formal theory-based group that suppresses group processes; (3) a control group that facilitates group processes; and (4) a second control group that suppresses group processes. A nested design that would accommodate comparisons across formal theories of change is also possible, wherein "type of formal theory of change" would be "nested" within the first factor. This design would facilitate the partitioning and comparison of differential effects relating to the main effects of a formal theory of change, small-group processes, and their interaction. In addition, direct comparisons between the individual cells would be informative in the sense that it would test for formal theory of change × group processes, formal theory of change alone, group processes alone, and group membership alone, respectively. Unfortunately, such designs are not yet found in the extant small-group literature.

CONCLUSION

In this chapter, we have proposed two main reasons why trauma therapists should consider the group format as a treatment of choice and consequently seek to master fundamental small-group process principles. One reason is that most studies have found the group format to be equal in effectiveness/efficacy compared to the individual format, and superior in cost-effectiveness. A second reason is that a sufficient body of evidence exists to support the assertion that the group format has specific therapeutic properties *independent* of those proposed by formal theories of therapeutic change. Whether these group-specific therapeutic properties are additive, interactive, or even attenuated when combined with therapeutic properties associated with a given formal theory of change remains to be investigated. Notwithstanding, some group-specific properties are sufficiently well documented to allow unambiguous recommendations and suggestions for increasing the overall effectiveness of group-based treatment protocols. We have focused on several of these in the concluding section of this chapter, specifically group development and interpersonal feedback, and discussed concrete ways in which they can be applied to group-based trauma treatment protocols. We have proposed that, once mastered, these group principles can be readily integrated into existing treatment protocols with minimal impact on the integrity of the protocol itself. Our hope is that if integrated and used wisely, these principles will increase therapeutic flexibility, leverage, and treatment effectiveness, and decrease the risk of iatrogenic effects. The task remaining is to gather empirical evidence to test these propositions—a process whose results we eagerly await and to which we hope to contribute in coming years.

REFERENCES

Beck, A. P., & Lewis, C. M. (Eds.). (2000). *The process of group psychotherapy: Systems for analyzing change.* Washington, DC: American Psychological Association.

Bednar, R. L., & Kaul, T. J. (1986). Experiential group research: Results, questions, and suggestions. In S. Garfield & A. Bergin (Eds.), *Handbook of psychotherapy and behavior change* (3rd ed., pp. 631-663). New York: John Wiley & Sons.

Bednar, R. L., & Kaul, T. J. (1994). Experiential group research: Can the cannon fire? In A. Bergin & S. Garfield (Eds.), *Handbook of psychotherapy and behavior change* (4th ed., pp. 631-663). New York: Wiley & Sons.

Bednar, R. L., Melnick, J., & Kaul, T. J. (1974). Risk, responsibility, and structure: A conceptual framework for initiating group counseling and psychotherapy. *Journal of Counseling Psychology, 21,* 31-37.

Bloch, S., & Crouch, E. (1985). *Therapeutic factors in group psychotherapy.* London: Oxford University Press.

Bond, G. (1983). Norm regulation in therapy groups. In R. Dies & K. R. MacKenzie (Eds.), *Advances in group psychotherapy: Integrating theory and practice* (pp. 171-189). New York: International University Press.

Bunch, B. J., Lund, N. L., & Wiggins, F. K. (1983). Self-disclosure and perceived closeness in the development of group process. *Journal for Specialists in Group Work, 8,* 59-65.

Burlingame, G. M., & Davies, D. R. (2002). Self-help groups. In M. Heissen (Ed.), *The encyclopedia of psychotherapy* (Vol. 4, pp. 31-35). New York: Academic Press.

Burlingame, G. M., & Fuhriman, A. (1990). Time-limited group therapy. *Counseling Psychologist, 18,* 93-118.

Burlingame, G. M., Fuhriman, A., & Johnson, J. E. (2002). Cohesion in group psychotherapy. In J. Norcross (Ed.), *Psychotherapy relationships that work* (pp. 71-88). New York: Oxford University Press.

Burlingame, G. M., Fuhriman, A., & Mosier, J. (2003). The differential effectiveness of group psychotherapy: A meta-analytic perspective. *Group Dynamics: Theory, Research and Practice, 7*(1), 3-12.

Burlingame, G., Johnson, J., & MacKenzie, K. R. (2002, February). We know it when we see it, but can we measure it? Therapeutic relationship in group. Paper presented at the annual meeting of the American Group Psychotherapy Association, New Orleans, Louisiana.

Burlingame, G., Kircher, J., & Taylor, S. (1994). Methodological considerations in group psychotherapy research: Past, present, and future practices. In A. Fuhriman and G. Burlingame (Eds.), *Handbook of group psychotherapy* (pp. 41-80). New York: John Wiley.

Burlingame, G. M., MacKenzie, K. R., & Strauss, B. (2003). Small group treatment: Evidence for effectiveness and mechanisms of change. In M. Lambert, A. E. Bergin, & S. L. Garfield (Eds.), *Handbook of psychotherapy and behavior change* (5th ed., pp. 647-696). New York: John Wiley & Sons.

Burlingame, G. M., MacKenzie, K. R., & Strauss, B. (2004). *Evidence based group treatment: Matching models with disorder and patients.* Washington, DC: American Psychological Association.

Caple, R. B., & Cox, P. L. (1989). Relationships among group structure, member expectations, attraction to group, and satisfaction with the group experience. *Journal for Specialists in Group Work, 14,* 16-24.

Castongauy, L. G., Pincus, A. L., Agras, W. S., & Hines, C. E. (1998). The role of emotion in group cognitive-behavioral therapy for binge-eating disorder: When things have to feel worse before they get better. *Psychotherapy Research, 8,* 225-238.

Cohen, J. A., Greenberg T., Padlo S., Shipley C., Mannarino A. P., & Deblinger, E. (2001). Cognitive behavioral therapy for childhood traumatic grief. Unpublished Treatment Manual, Drexel University College of Medicine, Pittsburgh, Pennsylvania.

Conyne, R. K., Wilson, F. R., & Ward, D. E. (1997). *Comprehensive group work: What it means & how to teach it.* Alexandria, VA: American Counseling Association.

Davies, D. R., Seaman, S., Burlingame, G. M., & Layne, C. (2002, February). Selecting adolescents for trauma/grief-focused group psychotherapy. Paper presented at the annual meeting for the American Group Psychotherapy Association, New Orleans, Louisiana.

Dies, R. R. (1993). Research on group psychotherapy: Overview and clinical applications. In A. Alonso & H. I. Swiller (Eds.), *Group therapy in clinical practice* (pp. 473-518). Washington, DC: American Psychiatric Press.

Dies, R. R. (1994). Therapist variables in group psychotherapy research. In A. Fuhriman and G. M. Burlingame (Eds.), *Handbook of group psychotherapy: An empirical and clinical synthesis* (pp. 114-154). New York: Wiley.

DiGuiuseppe, R., & Tafrate, R.C. (2003). Anger treatment for adults: A meta-analytic view. *Clinical Psychology: Science and Practice, 10,* 70-84.

Ettin, M. F. (1992). *Foundations and applications of group psychotherapy: A sphere of influence.* Needham Heights, MA: Allyn and Bacon.

Flowers, J. V. (1979). Behavioral analysis of group therapy and a model for behavioral group therapy. In D. Upper & S. M. Ross (Eds.), *Behavioral group therapy, 1979: An annual review* (pp. 5-37). Champaign, IL: Research Press.

Foa, E. B., Ehlers, A., Clark, D. M., Tolin, D. F., & Orsillo, S. M. (1999). The Posttraumatic Cognitions Inventory (PTCI): Development and validation. *Psychological-Assessment, 11,* 303-314.

Foa, E. B., & Rothbaum, B. O. (1998). *Treating the trauma of rape: Cognitive-behavioral therapy for PTSD.* New York: Guilford.

Forsyth, D. R. (1999). *Group dynamics* (3rd ed.). New York: Brooks/Cole.

Foy, D. W., Glynn, S. M., Ruzek, J. I., Riney, S. J., & Gusman, F. D. (1997). Trauma focus group therapy for combat-related PTSD. *In Session: Psychotherapy in Practice, 3,* 59-73.

Foy, D. W., Glynn, S. M., Schnurr, P. P., Jankowski, M. K., Wattenberg, M. S., Weiss, D. S., Marmar, C. R., & Gusman, F. D. (2000). Group therapy. In E. Foa, T. Keane, & M. Friedman (Eds.), *Effective treatments for PTSD: Practice guidelines from the International Society for Traumatic Stress Studies* (pp. 155-175, 336-338). New York: Guilford Press.

Foy, D. W., Ruzek, J. I., Glynn, S. M., Riney, S. J., & Gusman, F. D. (1998, November). Trauma-focused group therapy. In M. J. Friedman & P. P. Schnurr (Chairs),

Two approaches to the group treatment of chronic PTSD. Symposium conducted at the ISTSS Specialty Training Conference, Washington, DC.

Fuehrer, A., & Keys, C. (1988). Group development in self-help groups for college students. *Small Group Behavior, 19*, 325-341.

Fuhriman, A., & Burlingame, G. M. (1990). Consistency of matter: A comparative analysis of individual and group process variables. *The Counseling Psychologist, 18*, 7-63.

Fuhriman, A., & Burlingame, G. M. (1994a). Group psychotherapy: Research and practice. In A. Fuhriman & G. M. Burlingame (Eds.), *Handbook of group psychotherapy: An empirical and clinical synthesis* (pp. 3-40). New York: Wiley.

Fuhriman, A., & Burlingame, G. M. (Eds.). (1994b). *Handbook of group psychotherapy: An empirical and clinical synthesis.* New York: John Wiley & Sons.

Fuhriman, A., & Burlingame, G. M. (2001). Group psychotherapy training and effectiveness. *International Journal of Group Psychotherapy, 51*, 399-416.

Goldfried, M. R. (1980). Toward the delineation of therapeutic change principles. *American Psychologist, 35*, 991-999.

Gottlieb, B. H. (1996). Theories and practices of mobilizing support in stressful circumstances. In C. L. Cooper (Ed.), *Handbook of stress, medicine, and health* (pp. 339-356). Boca Raton, FL: CRC Press.

Gottlieb, B. H. (2000). Selecting and planning support interventions. In S. Cohen and L. G. Underwood (Eds.), *Social support measurement and intervention: A guide for health and social scientists* (pp. 195-220). London: Oxford University Press.

Greller, M. M., & Herold, P. M. (1975). Sources of feedback: A preliminary investigation. *Organizational Behavior and Human Performance, 13*(2), 244-256.

Gretzky, W., & Reilly, R. (1990). *Gretzky: An autobiography.* New York: Harper Collins.

Hurley, J. R. (1986). Leader's behavior and group members' interpersonal gains. *Group, 10*, 161-176.

Hyun-nie, A., & Wampold, B (2001). Where oh where are the specific ingredients? A meta-analysis of component studies in counseling and psychotherapy. *Journal of Counseling Psychology, 48*(3), 251-257.

Jacobs R. C., & Campbell, D. T. (1961). The perpetuation of an arbitrary tradition through several generations of a laboratory microculture. *Journal of Abnormal Social Psychology, 62*, 649-658.

Janoff-Bulman, R. (1992). *Shattered assumptions: Toward a new psychology of trauma.* New York: Free Press.

Karterud, S. (1988). The influence of task definition, leadership, and therapeutic style on inpatient group cultures. *International Journal of Therapeutic Communities, 9*, 231-247.

Kaul, T. J., & Bednar, R. L. (1994). Pretraining and structure: Parallel lines yet to meet. In A. Fuhriman and G. M. Burlingame (Eds.), *Handbook of group psychotherapy: An empirical and clinical synthesis* (pp. 155-190). New York: Wiley.

Kirshner, B. J., Dies, R. R., & Brown, R. A. (1978). Effects of experimental manipulation of self-disclosure on group cohesiveness. *Journal of Consulting and Clinical Psychology, 46*, 1171-1177.

Kivlighan, D. M. (1985). Feedback in group psychotherapy: Review and implications. *Small Group Behavior, 16,* 373-385.

Kivlighan, D. M., & Lilly, R. L. (1997). Developmental changes in group climate as they relate to therapeutic gain. *Group Dynamics, 1,* 208-221.

Kivlighan, D. M., & Shaughnessy, P. (2000). Patterns of working alliance development: A typology of client's working alliance ratings. *Journal of Counseling Psychology, 47,* 362-371.

Lambert, M. J., & Ogles, B. M. (2003). The efficacy and effectiveness of psychotherapy. In M. Lambert (Ed.), *Bergin and Garfield's handbook of psychotherapy and behavior change* (5th ed., pp. 139-193). New York: John Wiley & Sons.

Layne, C. M., Pynoos, R. S., Saltzman, W. R., Arslanagic, B., Black, M., Savjack, N., Popovic, T., Durakovic, E., Music, M., Campara, N., Djapo, N., & Houston, R. (2001). Trauma/grief-focused group psychotherapy: School-based postwar intervention with traumatized Bosnian adolescents. *Group Dynamics, 5,* 277-290.

Layne, C. M., Saltzman, W. R., Burlingame, G. M., Davies, R., Popovic, T., Durakovic, E., Music, M., Campara, N., Djapo, N., Wolfson, L., & Pynoos, R. S. (2001, November). *UNICEF technical report: Effectiveness of the UNICEF School-Based Psychosocial Program for War-Exposed Adolescents.*

Layne, C. M., Saltzman, W. R., Steinberg, A. S., & Pynoos, R. S. (2003). *Trauma/grief-focused group psychotherapy for adolescents.* Los Angeles, CA: National Center for Child Traumatic Stress.

Lieberman, M. A. (1989). Group properties and outcome: A study of group norms in self-help groups for widows and widowers. *International Journal of Group Psychotherapy, 39,* 191-208.

Lindy, J. D. (1985). The trauma membrane and other clinical concepts derived from psychotherapeutic work with survivors of natural disasters. *Psychiatric Annals, 15,* 153-160.

Lindy, J. D., & Wilson, J. P. (2001). Respecting the trauma membrane: Above all, do no harm. In J. P. Wilson & M. J. Friedman (Eds.), *Treating psychological trauma and PTSD* (pp. 432-445). New York: Guilford.

MacKenzie, K. R. (1987). Therapeutic factors in group psychotherapy: A contemporary view. *Group, 11,* 26-34.

MacKenzie, K. R. (1994). The developing structure of the therapy group system. In H. Bernard & R. MacKenzie (Eds.), *Basics of group psychotherapy.* New York: Guilford Press.

MacKenzie, K. R. (1997). Clinical applications of group development ideas. *Group Dynamics: Theory, Research and Practice, 1*(4), 275-287.

MacKenzie, K. R., & Tschuschke, V. (1993). Relatedness, group work, and outcome in long-term inpatient psychotherapy groups. *Journal of Psychotherapy Practice and Research, 2,* 147-156.

Marziali, E., Munroe-Blum, H., & McCleary, L. (1997). The contribution of group cohesion and group alliance to the outcome of group psychotherapy. *International Journal of Group Psychotherapy, 47,* 475-497.

McGuire, J. M., Taylor, D. R., Broome, D. H., Blau, B. I., & Abbott, D. W. (1986). Group structuring techniques and their influence on process involvement in a group counseling training program. *Journal of Counseling Psychology, 33,* 270-275.

McRoberts, C., Burlingame, G. M., & Hoag, M. J. (1998). Comparative efficacy of individual and group psychotherapy: A meta-analytic perspective. *Group Dynamics: Theory, Research, and Practice, 2,* 101-117.

Monnier, J., & Hobfoll, S. E. (2000). Conservation of resources in individual and community reactions to traumatic stress. In A. Y. Shalev and R. Yehuda (Eds.), *International handbook of human response to trauma* (pp. 325-336). Dordrecht, the Netherlands: Kluwer Academic Publishers.

Morran, D. K., Stockton, R., & Teed, C. (1998). Facilitating feedback exchange in groups: Leader interventions. *Journal for Specialists in Group Work, 23,* 257-268.

Parks, G. A., Anderson, B. K., & Marlatt, G. A. (2001). Relapse prevention therapy. In N. Heather and T. J. Peters (Eds.), *International handbook of alcohol dependence and problems* (pp. 575-592). New York: Wiley.

Raskin, N. J. (1986). Client-centered group psychotherapy, part II: Research on client-centered groups. *Person-Centered Review, 1,* 389-408.

Riessman, F. (1965). The "helper" therapy principle. *Social-Work, 10,* 27-32.

Roberts, L. J., Salem, D., Rappaport, J., Toro, P. A., Luke, D. A., & Seidman, E. (1999). Giving and receiving help: Interpersonal transactions in mutual-help meetings and psychosocial adjustment of members. *American Journal of Community Psychology, 27,* 841-868.

Roller, B. (1997). *The promise of group therapy: How to build a vigorous training and organizational base for group therapy in managed behavioral healthcare.* San Francisco, CA: Jossey-Bass.

Rugel, R. P., & Barry, D. (1990). Overcoming denial through the group: A test of acceptance theory. *Small Group Research, 21,* 45-58.

Salmon, T. W. (1919). War neuroses and their lesson. *Journal of Physiology, 66,* 993-994.

Saltzman, W. R., Pynoos, R. S., Layne, C. M., Steinberg, A., & Aisenberg, E. (2001). Trauma/grief-focused intervention for adolescents exposed to community violence: Results of a school-based screening and group treatment protocol. *Group Dynamics: Theory, Research, and Practice, 5,* 291-303.

Schnurr, P. P., Friedman, M. J., Foy, D. W., Shea, T. M., Hsieh, F. Y., Lavori, P. W., Glynn, S. M., Wattenberg, M., & Bernardy, N. C. (2003). Randomized trial of trauma-focused group therapy for posttraumatic stress disorder: Results from a Department of Veterans Affairs cooperative study. *Archives of General Psychiatry, 60,* 481-489.

Sexton, H. (1993). Exploring a psychotherapeutic change sequence: Relating process to intersessional and posttreatment outcome. *Journal of Consulting and Clinical Psychology, 61,* 128-136.

Stierling, E. (1909). *Psycho-neuropathology as a result of a mining disaster March 10, 1906.* Zurich: University of Zurich Press.

Stockton, R., & Morran, D. K. (1982). Review and perspective of critical dimensions in therapeutic small group research. In G. M. Gazda (Ed.), *Basic approaches to group psychotherapy* (pp. 37-85). Springfield, IL: Thomas.

Stockton, R., Rohde, R. I., & Haughey, J. (1992). The effects of structured group exercises on cohesion, engagement, avoidance, and conflict. *Small Group Research, 23,* 155-168.

Stubbenbort, K., Donnelly, G. R., & Cohen, J. A. (2001). Cognitive-behavioral group therapy for bereaved adults and children following an air disaster. *Group Dynamics, 5,* 261-276.

Taylor, N. T., Burlingame, G. M., Kristensen, K. B., Fuhriman, A., Johansen, J., & Dahl, D. (2001). A survey of mental health care providers and managed care organization attitudes toward, familiarity with, and use of group interventions. *International Journal of Group Psychotherapy, 51,* 243-263.

Tschuschke, V. (1999). Gruppentherapie versus einzeltherapie—gleich wirksam (Group versus individual psychotherapy—equally effective)? *Gruppenpsychotherapie und Gruppendynamik, 35,* 257-274.

Tschuschke, V., & Dies, R. R. (1994). Intensive analysis of therapeutic factors and outcome in long-term inpatient groups. *International Journal of Group Psychotherapy, 44,* 185-208.

Wheelan, S. A. (1994). *Group processes: A developmental perspective.* Boston: Allyn & Bacon.

Wheelan, S. A. (1997). Group development and the practice of group psychotherapy. *Group Dynamics, 1,* 288-293.

Widra J. M., & Amidon, E. (1987). Improving self-concept through intimacy group training. *Small Group Behavior, 18,* 269-279.

Wilfley, D. E., Frank, M. A., Welch, R., Spurrell, E. B., & Rounsaville, B. J. (1998). Adapting interpersonal psychotherapy to a group format (IPT-G) for binge eating disorder: Toward a model for adapting empirically supported treatments. *Psychotherapy Research, 8,* 379-391.

Wright, T. L., & Duncan, D. (1986). Attraction to group, group cohesiveness, and individual outcome: A study of training groups. *Small Group Behavior, 17,* 487-492.

Yalom, I. D. (1995). *The theory and practice of group psychotherapy* (4th ed.). New York: Basic Books.

Chapter 13

Critical Incident Stress Management in Terrorist Events and Disasters

Jeffrey T. Mitchell
George S. Everly Jr.

A little help, rationally directed and purposefully focused at a strategic time is more effective than more extensive help given at a period of less emotional accessibility.

Lydia Rapoport,
Crisis Intervention Theorist and Associate Professor,
School of Social Welfare, University of California, 1962

INTRODUCTION

Crisis intervention has been utilized throughout history to assist people in distress. There are many references in the Bible, in ancient secular writings, in historical documents, and in modern popular literature that portray people coming to the aid of others in pain and distress (Lee, 2001; National Bible Society, 1994). However, crisis intervention was not developed as an organized psychosocial approach to offering acutely needed assistance until the beginning of the twentieth century. Edwin Stierlin (1909), for example, discussed the use of a crisis intervention approach for distressed people involved in a 1906 European mining disaster. In the United States, the National Save a Life League was formed in 1906 to reduce the prevalence of suicides (Everly, 1999). Two world wars and a series of horrific disasters introduced further developments in the field of crisis intervention (Artiss, 1963; Kardiner & Spiegel, 1947; Lindemann, 1944; Parad & Parad, 1968; Salmon, 1919; Solomon & Benbenishty, 1986). Of particular note is the pioneering work of Erich Lindemann (1944) and Gerald Caplan (1961, 1964). Other theorists and practitioners followed and made significant contributions. Rapoport (1962) and Parad & Parad (1968) carefully defined basic terms and outlined clearly defined stages of crisis. They also clarified the use of crisis intervention with families. Salby, Lieb, and Tancredi (1975) and Ruben (1976) provided practical management steps for various crises. Frederick (1977) and Farberow and Frederick (1978) were well-known for their crisis-oriented work in both the suicide prevention and disaster management fields.

Crisis intervention from the early 1900s until 1974 was predominantly an individual, "one-on-one" approach. It was utilized to help people deal with acute grief reactions after disasters (Lindemann, 1944) or to assist soldiers in combat situations (Kardiner & Spiegel, 1947; Salmon, 1919).

Except for battlefield applications in World War II (Appel, Beebe, & Hilger, 1946; Holmes, 1985; "Glen Srodes," 1984) and Ruben Hill's 1958 work with families, group crisis intervention techniques were not utilized much until the development of critical incident stress management (CISM) programs in the mid 1970s.

Before describing group interventions in CISM, the reader must understand the context in which these techniques are appropriate. Group crisis interventions are merely components of a complete crisis intervention package. They should be carefully and strategically applied only within the context for which they were designed, and only when a situation warrants their use. They should be utilized by trained providers, who know how, why, when, and under what circumstances they are to be applied.

To avoid misinterpretations of the nature and limitations of crisis intervention one must understand its intended applications and goals, and the circumstances that necessitate group support. The application of psychological intervention within hours or a few days of exposure to a traumatic event is not without controversy. Sometimes referred to as the "debriefing controversy," this debate appears to be based as much upon confusion of terminology as tactical considerations. We will, therefore, define some basic terms used in the CISM field. This background material is intended to increase understanding and proper utilization of group crisis interventions.

Crisis Intervention

Crisis intervention is an active, temporary supportive process to assist people in acute emotional distress. As medical first aid is not a replacement for definitive medical care, crisis intervention is neither psychotherapy nor a substitute for psychotherapy; it is support, not a cure (Neil, Oney, DiFonso, Thacker, & Reichart, 1974).

Also known as "emotional first aid" (American Psychiatric Association, 1964), crisis intervention is not used only by mental health professionals.

The support services of crisis intervention may be provided by members of the clergy, police officers, firefighters, emergency medical personnel, nurses, soldiers, and many others. It is not true, on the other hand, to say that just anyone can provide "emotional first aid." Organized and structured crisis services require specially trained people with reasonable communications skills who are "other centered," and outgoing. Gerald Caplan, who wrote much of the foundational literature on crisis intervention, often suggested the use of paraprofessionals as well as professionals in the application of crisis management techniques (Caplan, 1961, 1964).

Since a person in crisis is usually susceptible to positive or negative changes, crisis intervention consists of techniques to aid people to make progress instead of regressing in the face of distress (Wolman, 1973). Crisis intervention is a systematic approach to return a distressed person or a group to adaptive life functions.

Critical Incidents

If human beings were not subjected to awful events that disrupt their cognitive and affective balance and overwhelm their abilities to cope, there would be no need for the field of crisis intervention or its subset, CISM. Crisis intervention was born of adversity.

Critical incidents are traumatic events that cause powerful emotional reactions in people who are exposed to them. Disasters and terrorist attacks are among the most distressing for communities, but they are not the only critical incidents. Each profession can list its own worst-case scenarios. Emergency services organizations (fire services, emergency medical services, and law enforcement agencies), for example, usually list the "Terrible Ten." They are

1. Line-of-duty deaths
2. Suicide of a colleague
3. Serious work-related injury
4. Multicasualty/disaster/terrorism incidents
5. Events with a high degree of threat to personnel
6. Significant events involving children
7. Events in which the victim is known to the personnel
8. Events with excessive media coverage
9. Events that are prolonged and have a negative outcome
10. Any unusually powerful, overwhelming, or distressing event

Critical incidents endemic to other organizations may vary substantially from those presented here. The airline industry, for instance, might place extreme turbulence, passenger violence, "close calls," and hijackings on its list.

Crisis

A crisis is an acute emotional *reaction* to a powerful stimulus or demand. It is a state of emotional turmoil. As Rapoport, an early theorist, pointed out, "A state of crisis is not an illness" (Rapoport, 1962, p. 23).

Two main types of crises can occur. The first is *maturational*. Maturational crises are those that come about as a result of development, maturation, experience, growth, or aging as a human being. In other words, maturational crises come about as a natural consequence of passing through the stages of life. "Particular stages of development are crisis periods during which the individual is susceptible to change. During these times the person may progress or regress" (Wolman, 1973, p. 384). Retirement, to cite one example, may be a crisis for some people because they must shift from the intense activity of daily work to a whole new set of interests and priorities.

The second type of crisis is *situational*. These crisis reactions are the result of events such as disasters, death, accidents, fires, violence, terrorist attacks, and damage to property. Events such as these are critical incidents for those involved. Crisis reactions in these cases are a direct response to a critical situation.

Three main characteristics are evident in any crisis.

1. The relative balance between a person's thinking and emotions is disrupted.
2. One's usual coping methods fail to work in the face of the critical incident.
3. Evidence of mild to severe impairment occurs in the individual or group involved in the crisis.

Several factors may make a crisis more or less difficult to manage. The suddenness of the incident as well as its intensity and duration may affect how difficult it is to manage. Other factors include the age of the person involved and the availability of resources for handling the situation (Caplan, 1969; Slaikeu, 1984).

Stress

Stress is defined as a state of cognitive, emotional, and physical arousal (Everly 1989). Arousal is caused by exposure to some actual or perceived demand or stimulus in the environment, and accompanies the crisis reaction. Once the intellect, emotions, and body are aroused, changes occur in a person's behavior. Intense fear, for example, activates running-and-hiding behaviors. Reasonable levels of stress are essential for survival and a healthy life. When stress gets out of control, however, it can be a powerfully destructive force.

One may think in terms of four major stress categories:

1. General stress
2. Cumulative stress
3. Critical incident stress
4. Post-traumatic stress disorder (PTSD)

Both *general* and *critical incident* reactions are normal responses to stress. General stress occurs as a result of the demands of everyday living. It may be a positive, driving force for a person or a group; it may also be a distressing force that causes people to change or take actions to protect or care for themselves. People usually deal with general stress, recover from it, and move on in life. As long as the stress is not excessive or prolonged, the individual can stay healthy and productive (Selye, 1956).

Critical incident stress is also normal. It is an expected reaction in healthy people to an unusually traumatic event. It is not a pleasant reaction, despite its normality. A normal reaction does not mean there is an absence of pain; the pain alerts the individual to the fact that the situation demands attention, and is part of a characteristic human drive toward survival. Critical incident stress is simply a heightened state of arousal that results from exposure to some powerful traumatic event (Everly & Mitchell, 2001).

The other two types of stress, cumulative and PTSD, are not normal pathways of stress. Both may produce significant life disruptions. If they continue without treatment, the stage is set for deterioration in health and performance.

Cumulative stress is pathological. If an individual experiences an excessive accumulation of unresolved general stress, he or she is more prone to develop physical illness and/or mental deterioration. Cumulative stress may escalate until a person develops persistent physical and emotional problems which require professional medical and mental health interventions (Everly, 1989; Jones, 1981).

PTSD is the most destructive form of stress. It comes about as a direct result of unresolved critical incident stress and typically requires mental health intervention. Six criteria must be met for a diagnosis of PTSD and are paraphrased as follows (American Psychiatric Association, 1994).

1. *Exposure to a horrible, threatening, or disgusting event.* The same events that initiate the critical incident stress reaction may also bring about PTSD. Of course, PTSD starts only when the critical incident is unresolved.
2. *Intrusion symptoms.* A person sees, hears, smells, tastes, or feels some aspects of the event repeatedly. The person may also have distressing dreams, or nightmares, or trouble controlling obsessive thoughts.
3. *Avoidance.* Aversion to reminders of the event, including attempts to avoid places, people, conversations, or circumstances that are reminiscent of the trauma.
4. *Excessive arousal.* Trouble sleeping, resting, or relaxing; or the frequent anticipation of further harm.
5. *Considerable disruption of normal life.* Patients have trouble both at home and at work. They feel stuck, insecure, and unable to participate in life as they had before the traumatic event.
6. *Symptoms must last at least 30 days* for a diagnosis of PTSD to be made.

The discussion in this chapter focuses on the critical incident and PTSD categories and how group interventions are applied to those circumstances.

CRISIS INTERVENTION

Goals

Crisis intervention has five primary goals:

1. To stabilize and control the situation.
2. To mitigate the impact of the traumatic event.
3. To mobilize the resources needed to manage the experience.
4. To *normalize* (depathologize) the experience.
5. To restore the person to an acceptable level of adaptive function.

Crisis intervention is helpful in stabilizing both the situation and the subsequent intense emotional reactions. Crisis workers aim at containment of the situation and they attempt to manage whatever elements of the situation can be managed. An immediate effort is made to neutralize or mitigate the psychological impact of the crisis. This is often accomplished simply by providing practical information. Once well-informed, individuals in crisis are typically better able to control situations and make appropriate decisions. (Slaikeu, 1984).

Crisis intervention aims to mobilize a range of helpful resources for those in distress, and also normalizes, or "demedicalizes," the negative or deeply disturbing reactions to the crisis situation. It does *not* aim at eliminating all symptoms of distress that may be encountered by those exposed to a traumatic event. Instead, crisis intervention assures traumatized individuals of the normal nature of their symptoms, and emphasizes that most people recover. Crisis intervention provides practical mechanisms by which people may resume normal daily functions while they work their way through a crisis.

Principles

Salmon (1919) established several core principles of crisis intervention. The first is *proximity*. The management of a crisis should occur close to the person's operational area or within an area familiar to them. Soldiers, for instance, should receive intervention services near the front lines on which they serve. However, it should go without saying that any emergency support must be provided in a safe area to avoid disruptive distractions or further trauma.

The second core principle is *immediacy*. An individual in a crisis state cannot wait very long for help. The sooner help is delivered, the more positive the outcome is likely to be. Immediacy must be tempered by issues of safety, timing, and location as well as other factors. A reasonable degree of immediacy ensures a greater potential for recovery from the traumatic experience. Crisis services provided to soldiers during combat immediately and near the front lines enabled 65 percent of them to return to combat, often within three days of experiencing a severe stress reaction. If the same services were withheld until "shell-shocked" soldiers were brought to a rear hospital (approximately twenty-four hours later), only 40 percent of those soldiers returned to combat, and only a small percentage returned to the front lines within three or four days (Brown & Williams, 1918; Holmes, 1985; Salmon, 1919).

The third core principle is *expectancy*. This principle means that expectancy of a positive outcome should be established early in the intervention. For example, the crisis intervener might say, "I know that this destructive fire is very difficult for you to manage right now, but our experience indicates that most people do recover from fires like this and are able to return to work after a period of time."

Other authors have added to the core crisis intervention principles. Over time *brevity, simplicity, innovation,* and *pragmatism* have augmented the list, bringing the current number of core principles to seven (Appel, 1966; Artiss, 1963; Kardiner & Spiegel, 1947; Noy, 1991; Solomon & Benbenishty, 1986).

Brevity refers to the short-term application needed during crisis intervention. From one to five brief contacts (between fifteen and sixty minutes each) may be necessary to calm the person or group and guide them toward recovery, and most contacts are delivered within one to three weeks. Crisis intervention rarely proceeds beyond eight brief contacts without recommending that distressed individuals seek more in-depth professional care.

People cannot usually tolerate complexity during a crisis. For this reason, it is important to keep intervention procedures as *simple* as possible. Limit the number of people who have contact with the crisis victims and use concise, easily understood directions that are repeated if a person appears confused. Reduce the number of stimuli in the environment to the lowest possible level. Present a distressed person with only one issue at a time that requires a decision, especially concerning anything of a serious consequence. Assure security and privacy, but allow contact with family members and friends who are familiar to the victim.

Despite the fact that there are reasonably clear guidelines for crisis intervention, some circumstances may require *innovative* approaches. Innovations in crisis intervention approaches during terrorist or disaster operations, for example, are not unusual. Under extraordinary circumstances, ordinary rules often do not apply. Crisis interventionists must be able to assess the situation and respond appropriately to it even if some alterations must be made in routine crisis intervention procedures. A substantial rationale should be in place, however, whenever these alterations are made. In one case, an open-sided tent had to be erected near a disaster site and tables and chairs brought in because family members vigorously refused to go to a more comfortable location and were standing in the rain waiting for some word on the fate of their loved ones. The tent provided a relatively private rest area for the family with a clear view of the scene.

Finally, it is important to be *pragmatic*. If the person or group in crisis cannot carry out the suggestions presented by a crisis worker in their time of need, it is unlikely that those suggestions will be useful. In fact, they may be perceived as counterproductive or it may be assumed that the person who made the suggestions is either ill-informed or insensitive.

Crisis interveners should follow several additional principles.

1. One should intervene only within his or her training levels.
2. Do not attempt to discuss issues that cannot be managed within the allotted time.
3. Avoid delving into excessive details about the situation.
4. After crisis intervention, recovery is apparent or a referral is considered.
5. Keep in mind the goals of crisis intervention:
   a. Mitigation of impact of a traumatic event.
   b. Acceleration of recovery processes in people with normal reactions.
   c. Identification of individuals who may need additional support (Everly & Mitchell, 1998).

Crisis Intervention Is Not Psychotherapy

Although therapeutic elements are evident in crisis intervention, crisis intervention is not psychotherapy. Crisis intervention is "psychological first aid" and thus is focused on support, not cure. The main differences between crisis intervention and psychotherapy are outlined in Exhibit 13.1.

Not All Crisis Interventionists Are Mental Health Professionals

Paraprofessionals have had a role in crisis intervention services from its beginning. They have been used in educational institutions, disaster relief services, and in emergency medical services, fire/rescue, and law enforcement organizations (Bard, 1970; Bard & Ellison, 1974; Blau, 1994; Frederick, 1977; Samuels & Samuels, 1975; Sawyer, 1989).

Another factor that makes the CISM program unique is its team approach. The teams may be partnerships between mental health practitioners, peer support personnel, and clergy members. Each team has a mental health professional as its clinical director. Usually, several other mental health professionals serve on each team.

The next group of CISM team members is called *peer support personnel* and are drawn from various organizations. Firefighters, paramedics, police officers, communications personnel, nurses, military personnel, and business, industrial, school, or community members are among those on a typical list of team members. Many clergy members also serve on CISM teams (Mitchell & Everly, 2001a).

Psychological Debriefings

A wide range of supportive crisis interventions as well as psychotherapeutic interventions have evolved since the mid-1970s. They are de-

EXHIBIT 13.1. Differences Between Psychotherapy and Crisis Intervention

Psychotherapy	Crisis Intervention
Context	
Reparation, reconstruction, growth	Prevention, acute mitigation Restoration to adaptive function
Strategic foci	
Conscious and unconscious sources of pathology	Conscious processes and environmental stressors/factors
Location	
Safe, secure environment	Close proximity to stressor Anywhere needed
Purpose	
Personal growth and development	Emotional "first aid" to reduce distress and assist the person in crisis to return to a state of adaptive functioning
Temporal focus	
Present and past	Here and now
Providers	
Mental health professionals	A trained, outgoing person who cares for people and has a desire to help those in a state of crisis Paraprofessionals Mental health professionals
Provider role	
Guiding, collaborative, consultative	Active, directive
Timing	
Typically within weeks to months or years after the development of a problem that interferes with normal life pursuits; delayed, distant from stressor	During a critical incident and in the immediate aftermath of an exposure to the event; immediate, close temporal relationship to stressor
Duration	
Eight to twelve sessions for short term; months to years of sessions for as long as needed for long term	Three to five contacts some of which are only minutes in length; maximum contacts, usually eight
Goals	
Symptom reduction; reduction of impairment; correction of pathological states; personal growth; personal reconstruction	Stabilization; reduce impairment; return to function or move to next level of care

scribed in the literature in various ways. Among the many descriptive terms utilized are early interventions, group meetings, support sessions, debriefings, single session debriefings, psychological debriefings (PD), and critical incident stress debriefings (CISD). The terms are not always describing the same processes. Some are individually based; some are group-oriented. Some would be more correctly classified as counseling or psychotherapy rather than crisis intervention. This is especially so when the targets of these interventions are primary sufferers such as victims of sexual assault, severe burns, or serious auto accidents who have highly unique needs and issues defined in part by their catastrophic nature.

These terms have not always been used consistently. They have been mixed with and/or substituted for one another in a sometimes haphazard manner. Some of the inconsistency is due to the fact that academicians, researchers, and practitioners have been at odds over these terms and their meanings for at least a dozen years. Moreover, practitioners may be doing one thing but calling it something else.

Some interventions are not only individual (as opposed to group), they are single-session and completely ad hoc, not connected with other forms of intervention or even referrals for further care (Kenardy et al., 1996; van Emmerik, Kamphius, Hulsbosch, & Emmelkamp, 2002).

In contrast, CISM interventions are part of a systematic approach to managing traumatic stress. CISM utilizes a "package" of interventions. For example, a quick assessment may be followed by one-on-one contacts. These, in turn, may be followed by a small group defusing on the first day and a standardized group critical incident stress debriefing (CISD) within a few days. Significant other support services may be offered in some circumstances. Follow-up individual sessions may then be provided, and, if necessary, referrals for additional care can be made (Dyregrov, 2003; Mitchell & Everly, 2001a).

OVERVIEW OF THE FOUR MAIN
CISM GROUP INTERVENTIONS

Many types of crisis-oriented groups are used. This chapter does not endeavor to provide in-depth coverage of each of these group processes. That would be impossible. For the sake of simplicity and to help reduce the confusion in terminology described in the previous section, this chapter will concentrate on four specific group interventions within the CISM model. Two of them are large-group interventions and two are small-group interventions. Like tools in a toolbox they are used for different purposes.

The two large-group interventions are called demobilization and crisis management briefing and are used to provide information and guidance. The two small-group interventions, called defusing and CISD are useful in assisting a small group to discuss or process a shared traumatic experience. These four terms will be defined briefly, and described in further detail in subsequent sections.

Demobilization

Demobilization is a very brief, large-group informational session specifically used for operations personnel (e.g., firefighters, police officers, and other first responders) who are being released from their first work-related exposure to a disaster. The main purpose is to provide useful information (i.e., common symptoms, suggestions for coping, and directions for obtaining additional help), normalize reactions, and guide personnel toward recovery from the critical incident. Personnel may also receive food and rest.

Crisis Management Briefing

Crisis management briefing is a large-group informational session provided to twenty or more people who have been exposed to a distressing traumatic event. It is usually provided to employees of corporations or uninjured community members in an area where a tragedy has occurred. Accurate, current, and practical information is the primary goal of this particular service. People need to know what happened in their community and what the fire and law enforcement or relief services are doing about it. Community leaders or representatives of the police, fire service, or other organizations join together with mental health professionals to present information and to guide people. Suggestions are provided regarding safety and health issues and handouts are distributed.

Defusing

Defusing is a small-group process to be used within hours after a homogeneous group has endured the same traumatic event. Because it is so close in time to the crisis event, the participants are usually noticeably distressed. A shortened version of a CISD, some refer to it as a guided, brief (thirty- to forty-five minute), storytelling time. Its primary objective is to allow a brief discussion of the event and to supply the group with practical information that may move them toward recovery. Group members are informed that their reactions are normal and that most will recover fairly quickly. The team (often made up of peer-support personnel) offers suggestions regarding sleeping, eating, activity levels, and contact with loved ones. A defusing is useful in assessing the group members to determine who needs more assistance.

Critical Incident Stress Debriefing

The critical incident stress debriefing (CISD) is a specific, seven-phase group crisis intervention process provided by a specially trained team. CISD is designed for a homogeneous group, to mitigate the impact of a traumatic event on the group members. It is typically provided several days after the crisis and lasts between two and three hours. The extended time allows a more detailed discussion of the event than the defusing. The team leaders provide information on the typical physical and psychological impact of the event and the many techniques that can be used to reduce stress reactions. Efforts are made to normalize those reactions and specific suggestions are offered to enhance individuals' stress management capabilities. It is most useful as a screening tool to determine if any group members need additional support or perhaps a referral for therapy.

All four of the group crisis interventions just defined are simply components of a comprehensive package of crisis intervention services. Each has a specific purpose.

CRITICAL INCIDENT STRESS MANAGEMENT

Critical incident stress management (CISM) is an integrated, comprehensive, multicomponent crisis intervention program. CISM services cover a broad spectrum, from precrisis preparation (i.e., education for emergency operations personnel and communities, policy and protocol development, and planning), and on-scene support, through postcrisis intervention and referral for mental health assessment and treatment. The comprehensive nature of CISM gives it the flexibility to be applied within many settings and with many demographic groups. It represents several steps beyond previous crisis intervention applications (Everly & Mitchell, 1999; Flannery, 1998).

Group crisis interventions have always been an integral component of the CISM program. Even CISD, the best known of the group interventions, is not a single, stand-alone intervention. It is but one of a collection of coordinated and linked techniques that are intended to mitigate the impact of a traumatic event and to accelerate normal recovery processes (Everly & Mitchell, 1999).

Components of CISM

CISM providers must choose the right techniques and build them into an overall strategy. No single technique will be effective for or applicable to all people under all circumstances and at all times (Everly, 1999).

Comprehensive CISM services should be available before, during, and after a crisis and they must also cover services for individuals, groups, organizations, and families. The following represents a sample list of CISM techniques and applications.

- Individual interventions
  — education
  — on-scene support
  — one-on-one crisis intervention
  — phone contacts
  — work-site visits
  — contacts at home
  — referrals
- Group interventions
  — education
  — demobilization for large groups (usually emergency workers) immediately after the first exposure to a disaster
  — Crisis management briefings (CMB)
  — defusing on the same day as the event
  — CISD, usually several days to a few weeks after an event
  — predeployment briefings
- Environmental interventions
  — family education
  — family crisis support
  — family death or injury notification
  — organizational support
  — advice to command/supervisors
  — support good management practices
  — enhancement of unit cohesion and function

The following list contains the ten most important CISM techniques.

1. Precrisis preparation and education
2. On-scene support services
3. Large group "demobilization" for operations personnel in major incidents

4. Large group "crisis management briefing"
5. Defusing for small groups immediately after a traumatic event
6. CISD a few days to a few weeks later
7. One-on-one crisis interventions
8. Family support services, administrative/organizational advisement and support
9. Follow-up services
10. Referrals for those who may need more help

Group Processes in CISM

Crisis intervention is, by definition, often applied under far-from-ideal circumstances after people have been exposed to extraordinary events. Group processes may be utilized to manage group crisis needs. Following is a list of those group interventions.

1. Demobilization (large group)
2. Crisis management briefing (large group)
3. Defusing (small group)
4. Critical incident stress debriefing (small group)
5. Follow-up meetings (either large or small group)
6. "Debrief the Debriefer" groups (Mitchell & Everly, 2001b)

LARGE-GROUP PROCESSES

Disasters and terrorist events generate a huge response on the part of emergency services organizations. The bombing of the Alfred P. Murrah Federal Building in Oklahoma City on April 19, 1995, and the attacks on America on September 11, 2001, for example, caused thousands of operations personnel to be deployed into circumstances connected to those devastating and dangerous events. Although not all workers were subjected to intense stressors sufficient to produce powerful reactions, substantial numbers of them suffered a serious enough impact to warrant a range of CISM services, including large-group interventions.

Individually based crisis interventions and small-group services are too time-consuming to be utilized extensively in a major incident such as a large-scale terrorist event. Realistic alternatives must be offered. The most useful intervention for large groups of operations personnel is demobilization immediately after the first exposure to the event, and crisis management briefings (CMB) on subsequent exposures, which may include repeated work assignments at the disaster site or "Ground Zero" as the World Trade Center site became known after the September 11 attacks.

In a crisis, the number of victims, survivors, community members, and witnesses, can easily reach into the tens of thousands of frightened, confused, distraught people. Individual crisis intervention services for each person are not practical. The task would be monumental and there are insufficient numbers of properly trained crisis interventionists to manage it. Moreover, the cost would be enormous. One of the large-group CISM interventions, CMB, is a viable alternative to individually based services for the bulk of the populations needing to be served in the acute phases of a disaster.

Demobilization

Rationale and Objectives

Disasters produce surprises, an atmosphere of confusion, and substantial disruption to normal personal routines or procedures. Emergency personnel are not exempt from such disturbances and they need information and guidance to assist them in dealing with the unusual circumstances they encounter in disaster work. The primary objective of a group demobilization session is to impart essential information and practical guidelines for disaster workers after being shocked by the disaster site.

Demobilization objectives include the following.

- Lower tension and anxiety.
- Normalize possible stress reactions.
- Provide practical information for managing potential future stress reactions.
- Allow a transition from disaster work to normal work activities.
- Provide information to help reduce immediate stress reactions.
- Allow CISM team an opportunity to assess group needs for additional support.
- Establish links to additional support after the incident ends.
- Encourage positive expectations about the outcome; e.g., "Although this is difficult and painful for you now, the experience of others who have worked at disasters indicates that most emergency workers recover and return to normal home and work functions."

Demobilization tasks include the following.

- Gather groups of homogeneous personnel (e.g., firefighters in one group, emergency department personnel in another, and trauma nurses in another, etc.) in a safe and secure environment reasonably near the scene of a disaster after they have been released from their work at the site.

- Provide a ten-minute information session.
- Provide a twenty-minute rest and food opportunity.
- Supervisors provide information about release to go home or reassignment to other, nondisaster-related duties.

Demobilization Group Member Selection Criteria

Demobilization groups are formed from groups that performed the same type of work at the disaster scene for about the same amount of time. Engine or truck companies, rescue squads, platoons, perimeter control teams, security teams, search teams, body recovery teams, fire suppression teams, and penetration teams are all examples of cohesive types of groups that the demobilization staff typically targets.

Demobilization Composition and Preparation
for Entry into the Group

- The sessions are facilitated by one CISM team member. Group size varies from twelve to fifty depending on work assignments and how closely they worked together. Unusually large groups may be accommodated from time to time. The larger the number of workers at a disaster scene during the first few hours, the greater the number of CISM team members needed to provide demobilizations. Usually, four to eight providers take turns presenting the sessions.
- Each CISM member should have the same informational outline to be presented.
- Additional staff may be necessary to keep the food stocked during the sessions.
- One CISM team member should serve as the demobilization service manager.
- Two large rooms are needed—one room for information, another for food. School buildings, hotels, and churches can often accommodate demobilization programs.
- Numerous circles of chairs should be placed around the room.
- Ample supplies of food, water, and juices should be available in the food room.
- There must be sufficient parking to accommodate numerous emergency vehicles.

- A "check-in" person should keep track of the operational units that arrive at the demobilization center. Note: no efforts are made to identify or track individuals. The only information that is necessary is the unit number or identifier and assurance from a group leader that the personnel from that unit are accounted for.
- CISM staff must deny media personnel access to the facility.
- Handout material should be available for all participants
- Obtain permission from the highest ranking officers for the demobilizations. CISM teams should not do anything that interferes with staffing or operations.
- Homogeneous work crews being released from the scene should know that they are going to the demobilization center for information and food.

Demobilization Group:

Structure:

- Two separate large rooms are required for (1) information, and (2) food and rest.
- Demobilizations last thirty minutes. Ten minutes for information; twenty for food.
- The primary purpose of demobilization is to present useful information. The presenter should be a benign authority figure who stands before the group.
- Groups will arrive or leave as they complete their work or the demobilization.
- Handout material is distributed at the end of the informational section.

Phases of the Model:

- Check the group in and make sure that all members are present.
- A CISM team member presents a ten-minute information session on potential stress reactions that might be expected by group members. The group is told that only ten minutes will be necessary. They are also told that the information is important because it may help them to manage the stress produced by their work at the scene. Emphasis is placed on the normal nature of stress symptoms, should they occur. It is also emphasized that it is okay if people do not have symptoms.

For example, the presenter may say, "Hello, my name is John and I am a member of the local CISM team. Thank you for all your hard work. Your organization has asked you to come together for a brief time so that you can receive some useful information that might help you recover from this event. Many of you will not have any of the normal symptoms I am going to describe and that is okay. Even when the signs of distress show up, most people recover quickly. These stress signs do not suggest weakness or mental disturbance."

- Cognitive, physical, emotional, and behavioral symptoms are discussed.
- People are assured that the stress symptoms are expected and normal.
- The presenter lists a number of suggestions that can lower tension and anxiety.
- Practical information is presented on stress management techniques:
  — Limit caffeine intake.
  — Eat nutritious meals.
  — Avoid alcohol.
  — Talk to family members, friends, and colleagues.
  — You do not have to speak to the media.
  — Remain active and occupied when you are not on duty or resting.
- Additional help is encouraged if symptoms are intense or prolonged.
- Attendance at the CISD as a follow-up is encouraged.
- The presenter asks for any comments or questions. Note: It is rare for tired operations personnel to ask questions or to make comments at this point. When discussions come about in a demobilization session, they are usually brief.
- If there are no comments or questions, the presenter gives out a stress handout.
- Disaster workers are tired and they are not likely to express their emotions.
- Anyone needing more support may be approached after the session.
- Summary comments are made by the presenter and handouts are distributed.
- The personnel are then directed to the food service room for a brief rest.
- Supervisors release the group to home or nondisaster duties.

Additional Information on the Group Demobilization Process

The demobilization process is a primary stress prevention technique that is applied only once—immediately after operations personnel complete their first work shift at a large-scale disaster. Demobilization is most useful in the early, chaotic, acute phase of an event when workers are least prepared for the graphic sensory overload they encounter.

Demobilizations are designed for those who respond first to a scene of great destruction, especially when there are human casualties. Demobilization is followed up with individual crisis intervention services or a CMB upon subsequent exposures to the disaster site.

The demobilization process is a limited, time-sensitive, large-group crisis intervention tool. Immediacy is essential. This timing component has been confirmed by many emergency operations personnel who state that immediately after their first shift is the best time to demobilize (Mitchell & Bray, 1990).

After demobilization, the operations personnel are either sent home or assigned to alternative duties away from the disaster site. A minimum of six hours is required before personnel return to the disaster site. This six-hour time frame has been established by asking personnel what they need most after their initial work at a disaster. They state that information, guidance, food, and rest are their greatest needs. Research on the effects of fatigue also bears this out (Holding, 1983; Mitchell & Bray, 1990).

If disaster situations are prolonged beyond the first two or three shifts, workers are usually recycled to the scene multiple times. Repeated use of the demobilization process is inappropriate in prolonged incidents and should be abandoned in favor of other types of support, if needed. In fact, it may be dangerous and disruptive for the operations personnel to have further demobilization. Rescue workers state that they do not like repeating the same process. They have suggested that the demobilization process may unnecessarily raise the sensitivity of some workers to their own stress reactions when it is provided more than once, at which point it simply becomes a case of "overhelping." Additional demobilization processes make rescue workers work harder to suppress their emotions (Mitchell & Bray, 1990).

The demobilization technique was designed for natural or accidental disasters, terrorist events, major wild fires, and large search operations. It is not intended for small events involving only a few people. Other CISM interventions such as defusing would be more appropriate under those circumstances.

The demobilization process is a limited CISM tool. It does require follow-up services. Individual contacts, CMBs, CISDs, family support, post-incident education, and other services are typical after disaster work. Several research reports testify to the significant benefits of combined crisis intervention services over single services (Bordow & Porritt, 1979; Flannery, 2001; Richards, 2001).

Crisis Management Briefing

Rationale and Objectives

Estimates show that disasters produce at least ten noninjured victims for every actual deceased or injured victim; in large metropolitan areas, the numbers may be far higher (North & Pfefferbaum, 2002; Schuster et al., 2001). Management of the psychological needs of a huge number of people in disasters and terrorist events is an overwhelming task. One-on-one services and small-group sessions are out of the question. Providers would be quickly inundated and rendered ineffective. Crisis management briefing (CMB) offers a viable alternative to those services.

CMB Objectives

- Provide information to large numbers of people involved in the same event.
- Provide safety directives, and psychological guidance.
- Engender increased cohesion and morale in groups.
- Lower tension, anxiety, and a sense of chaos.
- Control rumors.
- Provide suggestions for coping and a return to adaptive functions.
- Assess the group and determine group needs.
- Identify subgroups or individuals in need of additional assistance.
- Begin the process of connecting people to various resources.

CMB Tasks

- Gather large homogeneous groups together in a safe place.
- Community leaders present information on the current situation.
- Community leaders describe management actions that officials are taking.
- Mental health professionals or members of a CISM team present practical stress management information.
- Answer questions and ensure that people know what will be helpful as they attempt to handle the situation. For example, it may be suggested that people move into a temporary shelter during a major gas leak.

CMB Group Member Selection Criteria

In large community groups heterogeneity is likely to be present, but a CMB team should strive for as much homogeneity as possible. Attendees should be drawn from the same work group, the same organization, or same neighborhood, so that some unifying factors are present.

Composition and Preparation for Entry into the CMB Group

- Determine the target groups in need of information and guidance.
- Meet with community, organizational, or emergency services leaders.
- Anticipate potential problems.
- Plan a delivery strategy in advance, and decide on specific details.
- Ensure that the presentation team members are closely aligned.
- Develop a list of specific information and guidelines for presentation.
- Generate a list of potential referral resources for follow-up.
- Find an appropriate presentation site.
- Inform participants about the location and time of the meeting.
- Arrive early so that presenters can discuss last-minute details.
- Decide if the group is homogeneous enough to allow a brief discussion.
- Determine if the group is too heterogeneous and if the discussion is unwise.

CMB Group

Structural considerations:

- CMBs usually last approximately forty-five minutes to an hour. More questions or a detailed discussion may lengthen some CMBs slightly beyond an hour.
- CMBs may be provided before people see the destruction to their homes, or communities (e.g., evacuees returning to their homes after a flood).
- CMBs may be given more than once. This is especially true if circumstances change for some weeks following a major disaster. Provide a CMB if sufficient new information becomes available.
- CMBs may be provided as long as people are not in danger.
- A team of at least two people, one a leader from the community or organization and the other a mental health professional, presents the information and the guidelines for recovery.

- The presentation from the leader of an organization or community is first, then the mental health professional presents.
- The brief discussion or question-and-answer period follows.
- Summaries, additional guidelines, and directives conclude the session.

Phases of the model:

1. Gather the group together
   - Be sure the presentation team is on site, briefed, and ready to work according to an established plan.
   - Assemble the group members.
   - Let the participants know why they have been brought together and what is going to occur during the session, e.g., "Good evening, ladies and gentlemen. We have brought you together this evening to discuss the awful experience you have encountered today. We have a team of people with some important information for you and some specific instructions which should help you to get through this experience. I am with the local crisis response team, and I will be filling you in on information about managing the stress aspects of this situation. We will be here about an hour. We hope the session is informative and helpful."
2. Present facts and information
   - The officials (i.e., firefighters, law enforcement, administrator, school official) present accurate, current, and practical information regarding the situation. If it is a medical situation, a physician may address the group. In any case, the reliable organizational representative should have technical credibility in the eyes of the group receiving the CMB.
   - Care is taken not to breach boundaries of confidentiality or to disseminate information that might have a negative impact on emergency operations.
   - Rumors should be dealt with in addition to the known facts. The presenter confirms what is known at the time of the CMB. When specific information is unknown, the group should be told that.
   - Anticipate group needs and issues. Address these concerns immediately.
   - Specific guidelines and instructions are offered. For example, a physician might give practical guidelines for preparing food or cautioning that people not to drink contaminated water. Law en-

forcement officials might advise the group about security measures. All information should be accurate, current, timely, truthful, verifiable, relevant, and practical.

- When questions are asked, the presenter should do his or her best to discourage irrelevant or hostile questions, or counterproductive statements. Do everything possible to prevent the session from deteriorating into name-calling, accusations, or leadership bashing.
- If answers to questions are unknown, let people know that all relevant information may not yet be available, e.g., "We do not have sufficient information to respond to it at this time." If someone is already working on the issue, then say so. In any case, assure the questioner and say, "We will get back to you as quickly as we can on this matter."
- In homogeneous groups, longer question-and-answer periods may be appropriate. The more heterogeneity, the less controlled the questions tend to be. Caution is advised when managing a discussion or questions.
— If a group is already hostile either because of previous political issues in the community or because of some circumstances associated with the trauma, a very limited question-and-answer period should be allowed, or perhaps none at all. A haphazard ventilation of anger and hostility in a group setting, at this early stage in a traumatic event, is more likely to be counterproductive.
— Avoid a question-and-answer period if the group members are too shocked or stressed to remain under emotional control. The team is attempting to set up an informational program—not to establish a therapeutic relationship.

3. Discuss reactions to stress
- The CISM team representative (usually a mental health professional) focuses on common stress reactions. Major signs and symptoms of distress are anticipated and presented (a handout is also distributed that contains such information). Normalize and *demedicalize* such reactions. When appropriate, suggest physical and emotional responses to help counteract stress reactions. If a participant asks, "Am I a coward? I was shaking terribly," the CISM presenter might answer, "Shaking and 'freezing-up' are uncomfortable stress reactions. But they do not indicate cowardice. They are caused by high levels of natural stress chemicals that enter a person's bloodstream during a period of extreme distress."
- Adjust remarks to suit the needs of each group.
- Emphasize the normal nature of stress reactions. Explain that some will not have any noticeable reactions, which is also just fine.

4. Discuss coping mechanisms and resources
   - Questions and answers about stress reactions are now handled.
   - The crisis interventionist suggests practical guidelines and instructions.
   - Handout material related to stress and coping is distributed. The handouts list common reactions and suggestions for the management of stress and movement toward recovery. Lists of individual and community resources should also be made available.
   - Summary remarks are made and the session concludes.
   - In a prolonged incident, the CMB process may be presented regularly. For example, in one multiple-day disaster the community residents were offered a CMB each morning and each evening to keep the group abreast of new developments. If it is presented periodically, variations in content, team members, and handouts must be incorporated into CMB presentations.
   - Follow-up is essential. Individuals identified as needing additional assistance should be given one-on-one crisis intervention services or a referral to mental health professionals after the session. Some groups may need an appropriately timed critical incident stress debriefing (described as follows), family support, or other forms of assistance.

Additional Information on the Crisis Management Briefing

The CMB is a versatile, large-group intervention that can be used to inform and guide people in the aftermath of a traumatic event. It has been applied in communities and school situations as well as in businesses, emergency services units, and the military (Newman, 2000). The CMB is but one component of a comprehensive system, and this should not be utilized as a "stand-alone" technique from other CISM interventions. Follow-up interventions should be in place.

The CMB may be thought of as a form of a "town meeting." It is provided for the sole purpose of crisis intervention with a large group of people. It may be utilized as often as necessary while an event is evolving (Mitchell & Everly, 2001b).

SMALL-GROUP PROCESSES

Small-group processes present greater challenges to CISM teams. Unlike the large-group interventions, which focus on providing information to a generally passive audience, small-group crisis interventions are active processes. More interaction takes place between the participants and the team. Emotional material may surface and must be dealt with appropriately. For example, a CISM leader might say, "This event has had a very personal impact on you. I would be happy to meet with you after the session. I have some thoughts on your situation that I would like to share with you individually." Information provided to the group must be in accord with the specific needs of the group. The intervention team must be vigilant to identify individuals in the group who need additional support or a referral. The small-group processes of defusing and CISD require teams of at least two CISM-trained personnel (Mitchell & Everly, 2001a).

Defusing

Rationale and Objectives

The word "defusing" means to render something harmless before it can do damage, as in defusing a bomb. When applied by a CISM team, defusing is a guided group discussion intended to render the crisis as harmless as possible.

Defusing is a small-group process provided by a CISM team to a homogeneous group within eight to twelve hours of a critical incident when people are most vulnerable to further hurt and intensified distress. After a relatively brief window of opportunity natural defenses get reorganized and the influence of external help is lessened. After about twenty-four hours they may be ready for assistance, but by then defusing would be the wrong procedure. As soon as one day later, groups need more sophisticated interventions such as CISD to help them recover from the traumatic experience (Campfield & Hills, 2001; Dyregrov, 2003; Mitchell & Everly, 2001a; Robinson & Murdoch, 2003).

Defusing objectives:

- Mitigation of the impact of the traumatic event on the group.
- Acceleration of the recovery process and a return to adaptive function.
- Assessment of the need to provide CISD or other crisis services.
- Establishing a link to professional referrals, if necessary.

Defusing tasks:

- Target small, homogeneous groups with the same level of traumatic exposure.

- Create multiple, different defusing groups to maintain homogeneity.
- Provide defusing groups as soon as possible after a trauma event. Typically, there is an eight- to twelve-hour window, but always provide on the day of the event.
- Obtain supervisor approval when staffing or operations is impacted.
- Quickly bring a minimum two-person team to the location of the defusing.
- Obtain essential information about the event and its impact on the personnel.
- Learn the role played by personnel in rescue and related efforts.
- Gather the group together.
- Introduce the CISM team and discuss what you will be doing, and why.
- Lay out the ground rules of the defusing (e.g., confidentiality, the right *not* to participate, and instructions to speak for *oneself,* not another person).
- Go through defusing (about forty-five minutes) and conclude the session.
- Schedule follow-up one-on-one sessions with participants who you think need additional support.

Criteria for Selection of Defusing Group Members

Target groups for defusing should be small and homogeneous, and have been exposed to the same significant distressing experience, which has caused shock, intense emotional reactions, and disturbances in performance (e.g., "freezing up"). Several small defusing groups are better than having one large heterogeneous group. The following criteria should be met.

- The group must have completed its work at the trauma scene. If not, they will find the defusing process a distraction and will resent the time it takes.
- Group members should have had roughly the same exposure to the event. For instance, those who worked on patient care or body recovery have had a vastly different experience than those who provided perimeter control at the same scene.

Composition and Preparation for Entry into Defusing Group

- Response to a request for defusing is rapid.
- Intervention teams have very little briefing about the incident. They enter a "hot" picture while they are in a "cold" condition. They func-

tion with only the sketchiest of information. Sometimes the event may still be unfolding.
- Often the scene personnel are still hyperaroused since they have just come away from their exposure to the incident.
- A "neutral" conference room or classroom, free of distractions, should be obtained, in which to conduct the defusing. Defusing processes should be held away from the scene.
- The CISM team should have sufficient copies of handouts to distribute.

Defusing Group

Structural considerations:

- The defusing process is managed by a minimum of two support personnel.
- In many organizations paraprofessionals handle the defusing process without professional mental health personnel. They obtain supervision later.
- The defusing process may last between thirty and forty-five minutes.
- Only those who were actually involved in the event should be present.

Phases of the model:

1. Introduction
   - Introduction of the CISM team.
   - Definition and description of the defusing process and its goals.
   - Motivation of group.
   - Description of defusing ground rules.
   - Reassurance that the process is not investigative.
   - Encouragement that participants finish the defusing session.
2. Exploration
   - Ask participants to briefly describe what happened to them in the event.
   - Ask pertinent follow-up questions.
   - Engage the participants in a conversation about the experience.
   - Assess, by observing and listening, the need for more assistance.
3. Information
   - Reassure the participants that stress reactions they may experience are normal.
   - Accept and summarize the exploration of the material discussed by the group.

- Teach a variety of stress survival skills.
- Emphasize proper nutrition and the avoidance of alcohol, caffeine, nicotine, fatty foods, sugar, and salt in excess.
- Also emphasize exercise, recreation, rest, and having someone who is trusted to talk to about the event.
- Provide additional information that fulfills group needs.

Additional Information on the Defusing Group Process

Because it is applied so soon after the traumatic event, defusing may be all that is necessary to restore some groups to adaptive functions. In other words, it might be just the right help at the right time. A well-run defusing process sometimes eliminates the need for a CISD. However, just as often, a CISD will be necessary. If so, the CISD is enhanced by having had the defusing first. Defusing calms the reactions of the personnel and allows time to properly set up the CISD.

On some occasions, such as disasters, defusing would be an inappropriate CISM tactic because it is too time-consuming. In a disaster, there are too many people who need assistance, and defusing cannot serve all of them. If demobilizations are being provided during a disaster, they are an acceptable substitute for defusing.

Experience indicates that defusing would not be the right intervention in the case of a line-of-duty death of an emergency worker, for instance. It is not powerful enough to manage the emotional impact of such a loss. A modified CISD, which has five phases instead of the usual seven (eliminating the "thought" and "symptoms" phases), is much more suitable for dealing with the initial impact of a tragedy (Mitchell & Everly, 2001a).

Follow-up services are always necessary after a defusing to ensure that the personnel are managing and recovering from their stress reactions. If the personnel in the defusing have "unfinished business" as well as intense reactions to the experience or no reactions whatsoever, a full seven-phase CISD is indicated (Mitchell & Everly, 2001a).

Critical Incident Stress Debriefing

Rationale, Objectives, and Further Definitions

CISD is a small-group, psychoeducational, crisis intervention process that may be utilized up to several weeks after a critical incident, depending on the circumstances. In disaster situations, it may be applied much later, as was the case in New York City after the September 11, 2001, attacks. Although there were some CISD sessions held for businesses very shortly after the attacks, the majority of emergency services personnel were not given CISD for four to six months after the tragedy because they were still involved in active operations at the scene.

CISD is a structured and guided group technique within the CISM strategy. It is not a stand-alone technique; it is used within the context of CISM's comprehensive, systematic, multi-faceted approach to crisis intervention, and it is most effective when used within that context (Everly, Flannery, Eyler, & Mitchell, 2001; Flannery, 2001; Richards, 2001).

CISD is group crisis intervention. It is *not psychotherapy,* nor is it a substitute for psychotherapy. CISD is a *support and assessment* service which has more to do with mitigating the impact of a traumatic event than it does with curing anything. Any suggestion that the CISD process could erase all symptoms of traumatic stress in the people who have been exposed to a critical incident is wishful thinking.

CISD is the most complex of the CISM interventions. A well-trained team, including a mental health professional and several peer support personnel, provides the service. The meeting may last from one and three hours depending on the number of attendees and the intensity of the event.

Active participation in a CISD should be voluntary. On certain occasions, however, a supervisor may require attendance. The organization makes the decision to mandate the CISD not the CISM team. Supervisors recognize that a CISD is, like all other CISM services, about enhancing unit cohesion and performance. Some supervisors are concerned that a few of their personnel might "slip through the cracks" and suffer more without the CISD. They may order attendance in hopes that their personnel will benefit from the educational aspects of the process, as well as the support of a group.

Participants in a CISD should be motivated by the team to assist fellow group members. Sometimes a single comment one person makes in the group can make a world of difference for others. Group members should be advised that their participation in a CISD has less to do with themselves and more to do with how their fellow workers may need to hear their input about the critical incident. People who might not participate under other circumstances will often avidly participate for the good of their colleagues.

Ultimately, active participation is up to the individuals attending the CISD. If they strongly object to attending, they should be excused without punitive action.

Group members are offered two opportunities to contribute to the discussion. The first is the "fact phase," and the other is the "thought phase." If anyone decides to attend the group, but chooses to remain silent throughout any or all of the phases, that choice is fully respected. No one is pressured to speak during a CISD.

CISD Objectives

- Mitigation of the impact of a traumatic event.
- Acceleration of normal recovery processes in normal people who are having normal reactions to a totally abnormal event.
- Assessment of the group with the objective of identifying those who may need additional assistance, and facilitating referrals to the appropriate resources.

CISD Tasks

- Assess the need for CISD.
- Develop a minimum two-person team. In organizations that use peer support personnel, care must be taken to carefully choose CISM peers who match the background of the majority of the group (e.g., profession, gender, experience, rank, etc.). That is, if you are working with firefighters then you should have CISM-trained fire service peers to assist the mental health professional.
- Obtain an appropriate room for the CISD.
- Gain supervisor approval.
- Bring the homogeneous group together.
- Carefully proceed through all seven phases of the CISD.
- Begin follow-up services immediately after the session.

Criteria for Selection of the CISD Group Members

- CISD groups should be homogeneous. Sharing the same school class or work group, same organization, and same profession is an important aspect of successful CISD.
- The situation must be completed or entering an entirely new phase. The acute phase of the incident must be complete if the CISD is to be effective.
- Group members should have had the same level of exposure to the trauma.
- The group should be psychologically ready to participate in a CISD.

Composition and Preparation for Entry into CISD Group

- Target homogeneous groups.
- The CISD team reviews any information on the event including pictures, incident reports, newspaper accounts, and verbal descriptions by organization leaders.
- A secure room should be established.
- Administrative and supervisor endorsement should be obtained.
- Refreshments should be supplied after the session is concluded.
- Follow-up services must be started immediately after the session.

CISD Group

Structural considerations:

- The typical CISD group is between four and twenty people. Unusual circumstances may justify a somewhat larger group, but caution is advised when such decisions are made. Smaller groups (less than twenty) are preferable.
- A CISD may last between one and three hours.
- Length of time for a CISD depends on number of attendee and intensity of event.
- The general rule is one debriefer for each five participants in attendance.

Phases of the model. CISD follows a specific seven-phase format. Two structures are at work in a CISD. The first is that of psychological domains, the second consists of procedural steps.

The phases are structured in specific procedural steps so that they may facilitate the transition of the group from the psychological domain of cognition to the psychological domain of affect. Although some of the phases may be predominantly in the cognitive domain, emotional content can come up. The group leaders must always be aware of what domain the participants are in regardless of the phase of the CISD.

The seven phases of the CISD process are

1. *Introduction*
   - CISM team members introduce themselves.
   - Intervention workers describe the CISD process.
   - Participants are encouraged to speak about their experiences.
   - Participants are motivated to help one another.

- Participants are reminded that their active involvement is voluntary and that even if they were ordered to attend, they have a right not to speak.
- The CISD team discusses ground rules

2. *Fact Phase*

The title of this phase may cause some readers to think that many specific details about the event are sought by the team. This is not the case. Rather, detailed renditions about specific critical facts are not required. Indeed, detailed descriptions of the event are discouraged. The CISM team only needs a very brief, thumbnail sketch of the event. A few sentences from the participants are usually sufficient. An analogy may be useful here: Consider the fact phase as a diving board at a swimming pool. Its primary function is to get people into the pool. It is not the essence of swimming. Likewise, the fact phase merely gets the participants talking, but is not the essence of the CISD. The question that is typically asked in the fact phase goes something like this: "Please tell our team members who you are; what your role was during the incident; and a very brief description of what happened from your point of view. In other words, paint us a very broad picture of the experience as you saw it. Lots of details are not needed."

3. *Thought Phase*

The thought phase is one in which the participants may begin to take some "ownership" for the experience. The fact phase is really a collection of items, observations, or events that are external to the person. But, when one is asked, "What was your first thought or your most prominent thought once you realized you were thinking?" people must then "own" some piece of the experience. Facts are external to the person, thoughts are internal.

The thought phase is a transition from the cognitive domain to the affective domain. It allows a gradual movement into the somewhat uncomfortable realm of discussing emotionally laden content.

4. *Reaction Phase*

The reaction phase is the heart of CISD. Frequently, it is also the longest. Although tears are sometimes shed in a debriefing, they are not necessary to suggest that the CISD has been successful. This phase has a great deal of emotional content whether or not tears are shed. Anger, frustration, sadness, loss, confusion, and a number of other emotions may emerge at this point.

In the reaction phase the question is asked, "What is the very worst thing about this event for you personally?" Another way to ask the question is, "If there was one thing that you wish you could magically erase with the outcome remaining about the same, what would be the one part you would most want to be erased?"

5. *Symptoms*

Most CISM teams experience the symptoms phase as the shortest. By this time in the process, the participants are growing tired, and therefore tend to offer only one or two symptoms that they are experiencing as a result of the traumatic event. Some of the participants remain quiet during this phase. The question that initiates this phase is, "How has this tragic experience shown up in your life?" or "What cognitive, physical, emotional, or behavioral symptoms have you been dealing with since this event?" One way to move through this phase is to ask the participants to raise their hands if they are experiencing sleeplessness, irritability, frustration, anger, or changes in eating or other daily behaviors. The team members should focus on common symptoms associated with exposure to traumatic events. The CISM team should listen carefully during this phase since they will be using the signs and symptoms of distress presented by the participants as a kicking-off point for the "teaching phase."

6. *Teaching*

Important issues to be addressed by the CISD team in this phase include:

- Normalizing of the symptoms brought up by participants.
- Addressing symptoms that group members did not address in the symptoms phase.
- Providing information on practical stress management tactics.
- Explaining group members' reactions.
- Engendering optimism toward the future hope of recovery in group members.
- Discussing issues brought up by the participants such as the grief process, suicidal ideation, length of time usually required for recovery.
- Discussing steps people may take to recover from the stressful experience.
- Addressing any other topic that appears pertinent to the group's needs.

7. *Reentry*

The reentry phase is the time for participants to ask any additional questions they may have or to make any final statements about the traumatic experience. At the same time, the CISM team is doing whatever it can to bring the CISD process to a conclusion. They will be summarizing what has happened in each of the phases of the process. Final explanations, information, action directives, guidance, and thoughts must be given to the group. At this point, handouts are also distributed. When it is over, participants should feel that the CISD process has come to a conclusion, not simply to an arbitrary end. The loose ends need to be tied up and participants should have a sense of what they can do that will be helpful for them. The group members should not feel that there remains substantial unfinished business (Mitchell & Everly, 2001a).

Each of the seven phases has a specific purpose and each is linked to the other phases. They are utilized in the order given. The overall strategy in CISD is to move the group from the cognitive domain through the affective domain and back again to the cognitive. Skipping steps disrupts the process and may cause it to fail. Likewise, rushing through the steps is an unwise use of the CISD process and may cause harm.

To be successful in managing CISD, several conditions must be present:

1. The group must be homogeneous.
2. The critical incident must be concluded.
3. The group going through the CISD must have approximately the same exposure to the traumatic event.
4. The CISM team that provides CISD must be properly trained.
5. The CISM team should be experienced with the CISD process.

Table 13.1 distinguishes between the four main CISM group interventions and summarizes some key application points. Typical trigger mechanisms, target populations, and time frames have been chosen. The table should be viewed only as a set of general guidelines, actual applications may vary.

Special Situations Requiring Alterations in CISD

Several complicated situations may occur such as dealing with six- to twelve-year-old children, line-of-duty deaths, suicide of a colleague, multiple-incident CISDs, and significantly delayed interventions in which the CISD model stays the same, but the prompts in the various segments of the debriefing are altered to accommodate the needs of the groups. These specialty situations are beyond the scope of this chapter. They also require specialized training that cannot be adequately accomplished within a book chapter.

TABLE 13.1. CISM Group Intervention Summary

Intervention	Trigger	Activity	Target	When	Duration
Demobilization	Disaster or other large-scale incident	Passive (information only)	Large group (staff or disaster workers)	After first shift or after their first exposure to event	Ten-minute lecture plus twenty minutes rest and food
CMB	Any distressing event impacting large group	Semiactive (information plus short Q and A)	Large group (useful with any large group)	Anytime (before exposure, during and after event. Can be repeated)	Forty-five minutes to one hour
Defusing	Event impacting small homogeneous group	Active (loosely guided discussion)	Small group	Within eight to twelve hours after the event	Forty-five minutes
CISD	Event impacting small homogeneous group	Very active (structured team guided, seven-phase discussion)	Small group	twenty-four hours to one week	One to three hours

FOLLOW-UP GROUPS

Rationale and Objectives

It would be unreasonable and impractical to suggest that a single-session group process is going to manage all the issues and problems individuals experience after a traumatic event. Follow-up ensures that the group process has maximal effect.

Follow-up in the CISM field is most often accomplished through individual contacts such as one-on-one conversations, phone calls, or brief visits to a person's work site. Follow-up group meetings, however, are sometimes used when the group itself appears to be in need of additional support.

Follow-Up Group Objectives

- Assess progress in the time (usually one week) since the CISD.
- Assess whether any members of the group are suffering intense or unexpected reactions.

- Answer additional questions that may have arisen since the CISD.
- Provide any additional information needed on issues of recovery.
- List any potential resources from which group members might benefit.
- Be available to individuals who appear to need additional support.

Follow-Up Group Tasks

- Reassemble the group that went through the CISD.
- Briefly review the last meeting (e.g., "When we met with you last week at the debriefing, one of the major issues was. . . . How is that issue now?").
- Determine if new issues have arisen.
- Assess what has worsened or improved since the CISD.
- Answer questions.
- Suggest possible resources, if appropriate.

Criteria for Selection of Follow-Up Group Members

The only criteria for the selection of group members for the follow-up group is that the personnel were present for the CISD.

Composition and Preparation for Entry into Follow-Up Group

Group follow-up sessions are typically held approximately once a week after the CISD. They do not follow any particular format nor do they require any special preparations. They tend to be conversational and very supportive of the group members. They can easily be provided by a CISM team of two people.

The follow-up group meeting starts with a very brief introduction of the CISM team conducting the session. Then the team leader may say something like, "You were here for the CISD last week. We want to see how things are going since then. We will provide you with additional information, which may be helpful as you move onward from the tragedy. In the CISD, we covered what happened, your thoughts on the event, the worst parts, and the signs and symptoms of distress you were experiencing. Hopefully, some of the things you learned in the CISD session were helpful. Today, we are just going to review what may have occurred since our last contact with you."

Follow-Up Group

Structural Considerations

- Follow-up meetings generally run approximately one hour.
- A two-person team is required for the intervention.
- Sessions may be followed up by one-on-one interventions or referrals for therapy.

Phases of the Model

1. The team introduces itself.
2. Team members lead the group through a range of questions such as:
   - Has anything changed substantially since last week?
   - Was the CISD helpful to you?
   - In what ways did it help?
   - What could have been done better to help you?
   - Anything you would suggest (lessons learned) for yourself or for your organization for future events?
   - Has anything changed for the worse?
   - If so, what things have made conditions worse?
   - Have any new questions come up that you would like us to answer?
   - Have any new signs or symptoms of distress come up for you?
   - Is there anything in particular that you need help with?
   - What has your organization done specifically related to the tragedy since the CISD?
   - Has anything the organization done been helpful?
   - Has anything the organization done been unhelpful?
   - Has anything been better since the CISD?
   - What do you think needs to happen now to put this event behind you?
   - Anything you might say to another person in your group that you think would be helpful?
   - Anything that you would like to bring up that has not been adequately covered yet?
   - Anything else that we need to discuss in this meeting?

As questions are presented and responses are made, certain issues will emerge. The CISM team members must be prepared to engage in ongoing conversations that explore the issues. Every effort must be made to find something that helps resolve the most distressing issues for the group. Individual contacts may occur after the meeting.

TAKING CARE OF THE TEAM MEMBERS

The previous section addressed the follow-up needs of members of a group that had been through a CISD. This section will address an equally important issue: supporting the CISM team members. Failure to care for the CISM team sets them up for vicarious traumatization. This condition can result in suffering for the team members as well as team member dysfunction and premature withdrawal from the CISM team.

There are many approaches to providing support for team members. One-on-one support by other team members is certainly among the most common. Easy access to the clinical director of the team or its coordinator may also be helpful. Some people need to request time off from the team when they are going through personal difficulties or when they are having reactions to the material they have encountered while providing CISM services. Sometimes they just need to be off the pager or call-down list for a while.

In-service education programs or advanced training in CISM may be helpful in re-motivating team members who have grown weary listening to the pain of others. Education and training helps to build confidence and the skills required to assist others.

Acknowledging, thanking, and appreciating CISM team members goes a long way in helping to keep CISM personnel on the teams and functioning well. CISM team members who are well trained, acknowledged, and valued by their leaders and by the organizations they serve, generally last longer and provide greater support to others than those who are feeling abused and unappreciated.

A group process called "Debrief the Debriefers" is widely used by CISM teams to take care of their own team members who have worked hard and may have been exposed vicariously to pain associated with CISM work. The "Debrief the Debriefers" session is usually held immediately after the CISM team has completed a CISD or other intense CISM work. In situations involving prolonged operational periods, the debriefing of the debriefers may be delayed a few days to a week to allow CISM team members some recovery time before the discussion. The purpose of the meeting is to make sure that team members are feeling well before they go home or return to normal duties after the trauma in which they have been involved.

"Debrief the Debriefers" sessions have three goals:

1. To prevent negative reactions such as vicarious traumatization, cumulative stress, and the effects of negative self-judgment.
2. To teach and reinforce skills demonstrated during the CISD.
3. To "practice what we preach" regarding supporting people by making sure we support our own team members.

Each session has three phases. The first phase is the "Review Phase." Questions in this phase usually include items such as:

- How did the CISM services you performed go in general?
- Did you feel confident?
- How do you feel the CISM team did?
- How do you feel you did?
- Do you feel you made any mistakes that could be corrected next time?
- How did the group participate?
- Any important themes that came out of the interventions?
- Anyone or anything you are worried about?

The leader of the session may discuss what made the sessions go well and what can be done to improve interventions in the future. Suggestions on other ways to handle things may also be made by the members of the team.

The second phase is the "Response Phase." This phase elicits comments on the self-perception of team members and any concerns they may have about their performance. Typical questions may include:

- What did you say that you wished you had not?
- What did you not say that you wish you had?
- Is anything having an emotional impact on you during the interventions or now?
- Any particular parts that really cause you some pain?
- How are you doing with that now?
- Is there anything you feel particularly proud of in retrospect?

The session leaders guide a discussion of the self-impressions held by the CISM team members about their performance and provide assurance that no serious errors were made during the interventions. Alternatives for interventions in future events are provided. This is a good time for the team leaders to teach new techniques or to reinforce things that were done very well.

The third phase is titled the "Remind Phase." This phase concentrates on after actions related to the CISM interventions. For example, the following need to be addressed:

- Who have you met in the CISM interventions that need follow-up services?
- Have arrangements been made to provide those services?
- What are your personal plans to take care of yourself in the next twenty-four hours?
- What will it take for you to be able to "let go" of this CISD and put it behind you?

"Debriefing of the Debriefers" is also an opportunity to teach CISM team members new skills. It maximizes the chances for the team to return home without excessive self-doubt and it assures members that they are an asset to the CISM team. Involvement in a CISM team ought to be a rewarding experience, not a punitive one. Hence the debriefing should help reassure team members that the services they provided to others were indeed helpful and that they do not cause undue harm to the CISM team members (Potter & LaBerteaux, 2001).

THE RESEARCH

As with psychotherapy and employee assistance programs, crisis intervention is a practice ahead of its science. In this section we will examine some of the key issues relevant to the "science" of early intervention.

The successful application of CISM early intervention techniques, especially in groups, requires many elements to be in place. Group interventions are more complex than they appear on the surface. One reason for the complexity is that effective group interventions combine many elements. Many of these interrelated elements have not, unfortunately, been considered in a good deal of the research to date. Future research must address these elements within the context of the group or group interventions will be threatened by misunderstanding and applications that are inappropriate.

The following is a list of the most common factors or elements that influence the success or failure of group interventions. They are drawn from a broad range of group and crisis intervention literature.

- Used for intended purposes (Hiley-Young & Gerrity, 1994).
- Carefully timed (Burns & Harm, 1993).
- Recipients must be "psychologically ready" for help.

- Level of involvement in a crisis is important (Hiley-Young & Gerrity, 1994; Raphael & Wilson, 1993).
- Level of psychological distress (Raphael & Wilson, 1993).
- Level of preexisting distress (Raphael & Wilson, 1993).
- Homogeneity of group (Mitchell & Everly, 2001a).
- Mission completion (Mitchell & Everly, 2001a).
- Environment (Dyregrov, 1997, 2003).
- Leadership training (Dyregrov, 1997, 2003).
- Leadership skills (Dyregrov, 1997, 2003).
- Nature of group (Mitchell & Everly, 2001a).
- Existence of hidden agendas in a group (Dyregrov, 1997).
- Administrative support of the group (Mitchell, 1986).
- Group altruism (Dyregrov, 2003).
- Group cohesion (Dyregrov, 1997; Mitchell & Everly, 2001a).
- Group insight (Dyregrov, 2003).
- Ability to express catharsis (Dyregrov, 1997, 2003; Mitchell & Everly, 2001a; Wollman, 1993; Yalom, 1970).

All of the crisis intervention tactics of CISM are skill based. One would therefore expect outcomes to be correlated with the training, skill, and experience of the providers. This is an important matter. Arendt and Elklit (2001) reviewed seventy studies on "debriefing" and "psychological debriefing" and state

> Comparisons of studies adhering to and diverging from the original description of PD revealed that the effect obtained seems to depend on deviations from the traditionally defined features. The fact that deviations influence the results obtained is significant and a strong argument for a return to the use of PD in accordance with the original defining features . . . since a preventive effect has been found only when used in this way. (pp. 433-434)

Studies Not Supportive of Early Intervention

Concern over the effectiveness of early intervention in general and psychological debriefings arose in the relevant literature with the publication of two studies. McFarlane (1988) reported on the longitudinal course of posttraumatic morbidity in the wake of bush fires. The study found that acute posttraumatic stress was predicted by avoidance of thinking about problems, property loss, and not attending undefined forms of psychological debriefings. Chronic variations of post-traumatic stress disorder were, however, best predicted by premorbid, nonevent-related factors, such as a family history of psychiatric disorders, concurrent avoidance and neuroticism, and a tendency not to confront conflicts. The delayed-onset posttraumatic stress group not only had higher premorbid neuroticism scores, and greater property loss, but also attended the undefined debriefings. These factors were causally and inextricably intertwined.

Kenardy and colleagues' (1996) investigation, the second of the studies, is often cited as evidence for the ineffectiveness of debriefings. It purported to assess the effectiveness of stress debriefings for sixty-two "debriefed helpers" compared with 133 who were apparently not debriefed after an earthquake. The "debriefed" group had more severe traumatic stress scores at thirteen months, yet the authors state, "We were not able to influence the availability or nature of the debriefing . . ." (p. 39). They continue, "It was assumed that all subjects in this study who reported having been debriefed did in fact receive posttrauma debriefing. However, there was no standardization of debriefing services . . ." (Kenardy et al., 1996, p. 47). In fact, neither Mitchell's CISD, nor Dyregrov's PD, had been taught to rescuers or to the mental health professionals who served them at the time of either of these studies (R. Robinson, 2002, personal communication).

Beyond the initial studies, the primary scientific foundation for recent criticisms of early intervention, especially "debriefing," is found in the Cochrane Library's Cochrane Reviews. Citing the results of the Cochrane Library Review of randomized controlled trials (RCTs) (Rose, Bisson, & Wessely, 2002; Wessely, Rose, & Bisson, 1998) and selected derivative reviews (Litz, Gray, Bryant, & Adler, 2002; van Emmerik et al., 2002), some researchers have reached the conclusion that early psychological intervention (especially debriefing) is ineffectual, and may even cause harm. A few individuals have gone so far as to suggest that early intervention after disasters and mass violence should be discontinued.

Recently, the validity of the Cochrane Reviews (the primary source of scientific criticism for early intervention) has been called into question. Olsen et al. (2001) analyzed fifty-three Cochrane Reviews, and found that in 17 percent of the cases the evidence did not fully support the conclusions; in 23 percent of the cases the conduct or presentation of the review was unsatisfactory. The authors conclude that readers of the Cochrane Reviews "should interpret the reviews cautiously" (Olsen et al., 2001, p. 830). In concert with this declaration, the British Department of Health (2001) practice guidelines has "acknowledged concerns over the validity of the Cochrane report," (p. 24), specifically with regard to the Cochrane Review of early intervention in that many of the studies of early intervention including debriefing have not ensured the quality or the operational fidelity of the interventions they chose to study.

Yule (2001) argues that published reviews bear little resemblance to the interventions as they are actually practiced. The most problematic aspect of the Cochrane Reviews has been the failure to adhere to the prescribed operational protocols by using interventions "loosely based" (Rose, Brewin, Andrews, & Kirk, 1999, p. 796) on outdated (Mitchell, 1983) intervention protocols wherein the "integrity of the intervention was not . . . assessed" (Rose et al., 1999, p. 796), "there was no standardization of the debriefing services," and where researchers "were not able to determine . . . the quality of the debriefing," or even if the intervention actually took place (Kenardy et al., 1996, p. 47). In fact, the Cochrane Review's constituent studies all violated the prescribed intervention protocols by attempting to apply a group crisis intervention protocol to individuals one at a time (Rose et al., 2002). The argument that there is no appreciable consequence for applying group intervention protocols in individual intervention format defies logic, clinical acumen, and a corpus of literature (Dyregrov, 1997, 1998, 2003; Geiger, 2001; Krupnick, 2001; Yalom, 1970). A closer look at several of the Cochrane Review Constituent studies may be useful.

Bisson, Jenkins, Alexander, & Bannister (1997) randomly assigned 110 patients with severe burns to either a debriefing group or a control group. The clinical standard group debriefing was abandoned for an individual adaptation. The goal of the randomization was not met in that the debriefed individuals had more severe burns and greater financial problems than the nondebriefed individuals, thus direct comparison was inappropriate. The debriefed group had more severe traumatic stress scores at thirteen months postintervention. The authors concede the differences between groups at pretest were "associated more strongly with poorer outcome as measured by the IES at 13 months than were [debriefing] status" (p.79).

Hobbs, Mayou, Harrison, & Warlock (1996) performed a randomized trial of debriefings for 106 (fifty-four debriefed; fifty-two control) auto accident victims. Randomization failed to achieve equivalent groups for comparison. The individuals who were debriefed had more severe injuries and spent more days in the hospital. Both factors predicted poorer psychological outcome. The clinical standard group process was abandoned to employing individual debriefings. The individuals receiving the debriefings had higher traumatic stress scores at follow-up. These data have been used to argue that debriefing may be injurious, yet the actual traumatic stress scores were not in a clinical range (indicative of significant distress) at any time, and the overall change went from 15.13 to 15.97 (clinical ranges begin around 26). Such a change has no clinical significance whatsoever. Mayou, Ehlers, & Hobbs (2000) suffers from the same methodological flaws.

Lee, Slade, & Lygo (1996) used a single-session individual debriefing approach to women who had suffered a miscarriage. The authors concluded that the debriefing was ineffective as a treatment for the symptoms of depression. The descriptions of the debriefing used in this study do not equate to debriefing in the CISD operations manual (Mitchell & Everly, 2001a).

Abbreviated Review of Supportive Studies

In a naturalistic quasi-experimental investigation of the CISD intervention, Bohl (1991) assessed the use of CISD with police officers after a critical incident. Police officers who received a CISD within twenty-four hours of an incident ($N = 40$) were compared to officers without CISD (thirty-one). Those with CISD were found to be less depressed, less angry, and had less stress symptoms at three months than their nondebriefed colleagues.

In a follow-up, Bohl (1995) studied the effectiveness of CISD with thirty firefighters who received CISD compared with thirty-five who did not. At three months, anxiety symptoms were lower in the CISD group than in the non-CISD group.

In an early study of the CISD process it was found that in a sample of 288 emergency, welfare, and hospital workers, 96 percent of emergency personnel and 77 percent of welfare and hospital employees stated that they had experienced symptom reduction which was attributed partly to attendance at a CISD (Robinson & Mitchell, 1993).

After a mass shooting in Texas in 1994 in which twenty-three people were killed and thirty-two were wounded, emergency medical personnel were offered CISD within twenty-four hours. A total of thirty-six respondents were involved in this longitudinal assessment of the effectiveness of CISD interventions. Recovery from the trauma appeared to be most strongly associated with participation in the CISD process. In repeated measures at eight days and again at one month anxiety, depression, and traumatic stress symptoms were significantly lower for those who participated in CISD than for those who did not (Jenkins, 1996).

After Hurricane Iniki in Hawaii, Chemtob, Tomas, Law, & Cremniter (1997) did pre- and posttest comparisons of forty-one crisis workers in a controlled, time-lagged design. The pretreatment assessment of the second group was concurrent with the posttreatment assessment of the first group. The intervention was a CISD and a stress management education session. The intervention reduced posttraumatic stress symptoms in both groups.

In another naturalistic quasi-experimental study emergency personnel working the riots in Los Angeles in 1992 that followed the "Rodney King trial" were either given CISD or not depending on the choice of command staff. They had worked at the same events. Those who received CISD scored significantly lower on the Frederick Reaction Index at three months after intervention compared to those who did not receive it (Wee, Mills, & Koehler, 1999).

In 1994 over 900 people drowned in the sinking of the ferry, *Estonia.* Nurmi (1999) contrasted three groups of emergency personnel who received CISD with one group of emergency nurses who did not receive CISD. Symptoms of post-traumatic stress disorder were lower in each of the CISD groups than the non-CISD group.

Everly & Boyle (1999) performed a meta-analysis of five studies on CISD in an attempt to resolve equivocal data on the genre of psychological debriefings that had appeared in recent literature. They concluded that the group CISD model of debriefing was capable of exerting a mitigating effect upon symptoms of stress and trauma.

Two quasi-experimental studies have attempted to investigate the cost-effectiveness of early intervention. The first was a study performed on bank employees after a violent robbery. In the study (Leeman-Conley, 1990), a year with no assistance for employees was compared with a year in which a CISM program was used. Employees fared better with the CISM program. Sick leave in the year in which the CISM program was utilized was 60 percent lower. In addition, workers' compensation was reduced by 68 percent.

Western Management Consultants (1996) did a cost-benefit analysis on a CISM program for nurses in rural areas of northern Canada. The study involved 236 nurses (41 percent of the workforce). The CISM program was instituted as a means of controlling stress reactions. Sick-time utilization, turnover, and disability claims dropped dramatically after the program was put in place. The cost-benefit analysis showed $7.09 (700 percent benefit) was saved for every dollar spent on building the CISM program.

Many other such studies exist. A full presentation of the literature is outside the scope of this chapter. The reader is therefore directed to the reviews already performed on CISM (Hiley-Young & Gerrity, 1994; Dyregrov, 1997, 1998; Flannery, 2001; Everly et al., 2001). The following paragraphs summarize the research issues in the CISM field.

With the exception of randomized controlled studies by Deahl et al. (2000) and Campfield and Hills (2001), studies supportive of CISM and the small group CISD are all quasi-experimental designs. Randomized controlled trials are certainly encouraged, however, the opportunity to conduct them under disaster field conditions may be extremely difficult or impossible (Jones & Wessely, 2003).

1. CISM represents an integrated multicomponent emergency mental health system. Emergency mental health services characterized by early intervention have been utilized since World War I in the military, and most recently by organizations such as the American Red Cross, National Organi-

zation for Victims Assistance, Florida State University's Green Cross Project, and the International Critical Incident Stress Foundation. These organizations and many others including the Salvation Army, the Federal Bureau of Investigation, the United States Marshals Service, all branches of the United States military forces, and the Bureau of Alcohol, Tabacco and Firearms and Explosives, have been successfully providing actual crisis and disaster mental health services in the field, not a laboratory, for several decades.

2. Critics of early psychological intervention most commonly cite the Cochrane Reviews to support their position. The authors of the most recent Cochrane Review of psychological debriefing (Rose et al., 2002), have concluded, "We are unable to comment on the use of group debriefing, nor the use of debriefing after mass traumas" (p. 10). In fact, most of the constituencies of Cochrane investigations represent "single session counseling with medical patients." As Arendt and Elklit (2001) point out, "No evidence has been found for the effectiveness of PD as an individual treatment of direct victims" (p. 434). Studies of debriefings of individuals fail to serve as a valid comparison from which generalizations may be made to other forms of debriefing.

3. The only randomized controlled trial (RCT) of CISD versus a non-CISD control condition, yielded compelling evidence of the effectiveness of CISD when combined with preoperations training conducted with British soldiers (Deahl et al., 2000).

4. Flannery's Assaulted Staff Action Program (ASAP) (a CISM program) was chosen as one of the ten best programs in 1996 by the American Psychiatric Association. A ten-year review of ASAP practice revealed ASAP CISM to be clinically effective (Flannery, 2001). Seven years of peer-reviewed, published ASAP literature has been largely ignored by those opposed to early intervention because of this quasi-experimental approach.

5. One study found both CISD and CISM effective in reducing distress among bank robbery victims. (CISM was found to be more effective than CISD.) (Richards, 2001).

6. A RCT study comparing the effectiveness of early CISD intervention to delayed CISD intervention found early intervention CISD to be more clinically effective (Campfield & Hills, 2001). This finding is consistent with the observations of Salmon (1919), Artiss (1963), and Solomon & Benbenishty (1986) in the military, as well as the pioneering theoretical formulations of Lindy (1985) in his descriptions of the "trauma membrane."

7. After considering relevant research, the American Psychiatric Association's 1989 Task Force on the Treatment of Psychiatric Disorders (Swanson & Carbon, 1989) concluded over a decade ago that crisis intervention

offers a "sound approach" and is an "effective model" for assisting people in emotional distress.

8. The Consensus Workshop on Mass Violence and Early Intervention (NIMH, 2002) recommended the use of a wide variety of interventions depending upon which was best suited to the needs of the specific disaster.

9. Despite its apparent effectiveness as determined by case studies' empiricism, as well as controlled investigations, early psychological intervention has the potential to exacerbate psychological distress and should be applied cautiously and by well-trained personnel.

10. Early psychological intervention is not a substitute for psychotherapy, but can serve as a useful platform for identifying those who may benefit from treatment.

11. The majority of individuals exposed to a traumatic event will rebound by their own natural recovery mechanisms. Early psychological intervention may serve as a useful "safety net" for those who have greater difficulty in the recovery process.

12. Finally, Dyregrov (1998) offered a most interesting perspective into the "debriefing debate," suggesting there may be more issues than merely science at work: specifically, crisis intervention may threaten the foundations of traditional practice in mental health.

MISCONCEPTIONS REGARDING CRISIS INTERVENTION AND CISM

Several misconceptions exist about crisis intervention and CISM. The eight main misconceptions are as follows.

1. *Crisis intervention and its subset CISM are the same as psychotherapy, or the procedures may be used as a substitute for psychotherapy.* Crisis intervention, instead, is psychological first aid and its primary goal is preventing the exacerbation of posttrauma symptoms (Caplan, 1964).

2. *Crisis intervention and CISM may be done by the untrained.* As in any specialty in psychology, one would expect formal training and supervision (Dyregrov, 2003; Robinson & Murdoch, 2003).

3. *CISM (a broad strategy of crisis intervention techniques) is the same as a single group crisis intervention technique known as critical incident stress debriefing (CISD).* Confusing a single approach for a wide-ranging strategic program is counterproductive. From the inception of the CISM field, CISD was only one intervention among many.

CISM is a combination of interrelated techniques (Mitchell, 1983; Mitchell & Everly, 2001a).

4. *All "debriefings" are the same as a CISD.* There are many types of debriefings. They do not necessarily equate to one another. The American Red Cross utilizes one format (Armstrong, O'Callahan, & Marmar, 1991). Many types of debriefings are in use today (Raphael & Wilson, 2000; Dyregrov, 2003)

5. *CISD is a stand-alone, single-session intervention.* CISD was never conceptualized as a stand alone (Mitchell, 1983). Research strongly indicates that combined interventions are far more powerful than any single technique (Chemtob et al., 1997; Deahl et al., 2000; Flannery, 2001; Richards, 2001).

6. *CISD and other CISM techniques require a detailed review of the traumatic experience.* This misconception is directly opposed to the standards of practice that have been developed within CISM (Mitchell & Everly, 2001a).

7. *Crisis intervention and CISM should be able to erase all symptoms of traumatic stress.* This is an unrealistic expectation. The best of therapies would be hard-pressed to achieve that level of success. Studies cited previously show a reduction in symptoms after CISD, and more so with CISM. Total elimination of all symptoms has never been claimed and is not to be expected.

8. *Crisis intervention and CISM are a cure for post-traumatic stress disorder (PTSD).* At least one study has shown that group crisis intervention services including CISD may decrease the potential for PTSD (Deahl et al., 2000). It is unlikely to have curative value.

A Note on Training

The National Institutes of Mental Health (2002) recommended that those who provide crisis intervention services should be adequately trained to do so. The International Critical Incident Stress Foundation, a nonprofit educational organization that coordinates a worldwide network of crisis intervention teams, recommends a total of about 100 hours of initial training, followed by direct supervision (Mitchell & Everly 2001a; ICISF, 2002).

CISM teams must develop skills in six major areas.

1. Assessment of the overall incident and severity of impact.
2. Strategic planning.
3. Skills for assisting individuals.
4. Skills for managing large groups in crisis.

5. Skills for managing small groups in crisis.
6. Follow-up and referral skills.

CISM teams must assess a situation and determine if the stress symptoms displayed by those involved are manageable by peer support personnel or if they need professional care. Then they develop a strategic crisis intervention plan after which they may apply the appropriate skills from the previous list, all of which require training.

CONCLUSION

CISM services including large- and small-group interventions have become widespread. Emergency services organizations, the military, community groups, businesses, industries, school systems, and a wide range of other organizations are utilizing the skills of CISM teams. Numerous pertinent studies have shown the crisis intervention approaches of the comprehensive, systematic, and multifaceted CISM system to be clinically effective.

No one would disagree with the admonitions to use caution in developing and implementing a psychological response program to critical incidents such as disasters or terrorist events. The paramount challenge is to *intervene only where and when needed* and to do so in a manner that does not disrupt natural recovery mechanisms. As with psychotherapy, early psychological intervention should be based upon need, rather than merely the occurrence of some untoward event.

Since it is unlikely that the majority of people exposed to a disaster or a terrorist event will need much beyond useful operational and self-care information services, the large-group processes of demobilization and crisis management briefing are the most useful interventions, and they have been discussed in detail in this chapter. Similar to emergency medical services, early psychological intervention, including group processes, may serve as a useful platform for providing simple reassuring information when indicated, or for the identification of those who may require subsequent and more intensive psychological assessment and/ or intervention.

The more intense small-group processes of defusing and CISD, when used according to the current standards of practice and within the context of a package of crisis intervention procedures, are potent tools in mitigating stress reactions. They have a proven track record of accelerating recovery from trauma. Finally, they are useful screening opportunities for those who need more.

REFERENCES

American Psychiatric Association (1964). *First Aid for Psychological Reactions in Disasters.* Washington, DC: Author.

American Psychiatric Association (1994). *Diagnostic and Statistical Manual of Mental Disorders,* Fourth Edition. Washington, DC: Author.

Appel, J.W. (1966). Preventive psychiatry. In A.J. Glass & R.J. Bernucci (Eds.), *Neuropsychiatry in World War II* (pp. 373-415).Washington, DC: US Government Printing Office.

Appel, J.W., Beebe, G.W., & Hilger, D.W. (1946). Comparative incidence of neuropsychiatric casualties in World War I and World War II. *American Journal of Psychiatry, 102,* 196-199.

Arendt, M. & Elkit, A. (2001). Effectiveness of psychological debriefing. *Acta Psychiatrica Scandanavia, 104,* 423-437.

Armstrong, K., O'Callahan, W., & Marmar, C. (1991). Debriefing Red Cross disaster personnel: The multiple stressor debriefing model. *Journal of Traumatic Stress, 4,* 581-593.

Artiss, K. (1963). Human behavior under stress: From combat to social psychiatry. *Military Medicine, 128,* 1011-1015.

Bard, M. (1970). *Training Police As Specialists in Family Crisis Intervention.* Washington, DC: Law Enforcement Assistance Administration, National Institute for Law Enforcement and Criminal Justice.

Bard, M. & Ellison, K. (1974, May). Crisis intervention and investigation of forcible rape. *The Police Chief, 41,* 68-73.

Bisson, J.I., Jenkins, P., Alexander, J., & Bannister, C. (1997). Randomized controlled trial of psychological debriefings for victims of acute burn trauma. *British Journal of Psychiatry, 171,* 78-81.

Blau, T.H. (1994). *Psychological Services for Law Enforcement.* New York: John Wiley & Sons, Inc.

Bohl, N. (1991). The effectiveness of brief psychological interventions in police officers after critical incidents. In J.T. Reese, J. Horn, & C. Dunning (Eds.), *Critical Incidents in Policing, Revised* (pp.31-38). Washington, DC: Department of Justice.

Bohl, N. (1995). Measuring the effectiveness of CISD. *Fire Engineering, 148,* (8), 125-126.

Bordow, S. & Porritt, D. (1979). An experimental evaluation of crisis intervention. *Social Science and Medicine, 13,* 251-256.

British Department of Health (2001). *Treatment Choice in Psychological Therapies and Counseling.* London: Crown.

Brown, M.W. & Williams (1918). *Neuropsychiatry and the War: A bibliography with Abstracts.* New York: National Committee for Mental Hygiene.

Burns, C. & Harm, I. (1993). Emergency nurses perceptions of critical incidents and stress debriefing. *Journal of Emergency Nursing, 19*(5), 431-436.

Campfield, K. & Hills, A. (2001). Effect of timing of critical incident stress debriefing (CISD) on posttraumatic symptoms. *Journal of Traumatic Stress, 14,* 327-340.

Caplan, G. (1961). *An Approach to Community Mental Health.* New York: Grune and Stratton.

Caplan, G. (1964). *Principles of Preventive Psychiatry.* New York: Basic Books.

Caplan, G. (1969). Opportunities for school psychologists in the primary prevention of mental health disorders in children. In A. Bindman & A. Spiegel (Eds.), *Perspectives in Community Mental Health* (pp. 420-436). Chicago: Aldine.

Chemtob, C., Tomas, S., Law, W., & Cremniter, D. (1997). Post disaster psychosocial intervention. *American Journal of Psychiatry, 134,* 415-417.

Deahl, M., Srinivasan, M., Jones, N., Thomas, J., Neblett, C., & Jolly, A. (2000). Preventing psychological trauma in soldiers. The role of operational stress training and psychological debriefing. *British Journal of Medical Psychology, 73,* 7-85.

Dyregrov, A. (1997). The process in critical incident stress debriefings. *Journal of Traumatic Stress, 10,* 589-605.

Dyregrov, A. (1998). Psychological debriefing: An effective method? *TRAUMATOLOGYe, 4*(2), Article 1.

Dyregrov, A. (2003). *Psychological Debriefing: A Leader's Guide for Small Group Crisis Intervention.* Ellicott City, MD: Chevron Publishing Corporation.

Everly, G.S. Jr. (1989). *A Clinical Guide to the Treatment of the Human Stress Response.* New York: Plenum.

Everly, G.S. Jr. (1999). Emergency mental health: An overview. *International Journal of Emergency Mental Health, 1,* 3-7.

Everly, G.S. Jr. & Boyle, S. (1999). Critical incident stress debriefing (CISD): A meta-analysis. *International Journal of Emergency Mental Health, 1,* 165-168.

Everly, G.S. Jr., Flannery, R.B., Jr., Eyler, V., & Mitchell, J.T. (2001). Sufficiency analysis of an integrated multi-component approach to crisis intervention: Critical incident stress management. *Advances in Mind-Body Medicine, 17,* 174-183.

Everly, G.S. Jr. & Mitchell, J.T. (1998). *Critical Incident Stress Management: Assisting Individuals in Crisis: A Workbook.* Ellicott City, MD: International Critical Incident Stress Foundation.

Everly, G.S. Jr. & Mitchell, J.T. (1999). *Critical Incident Stress Management: A New Era and Standard of Care in Crisis Intervention.* Ellicott City, MD: Chevron Publishing Corp.

Everly, G.S. Jr. & Mitchell, J.T. (2001). *Critical Incident Stress Management: Advanced Group Crisis Interventions, a Workbook,* Second Edition. Ellicott City, MD: International Critical Incident Stress Foundation.

Farberow, N.L. & Frederick, C.J. (1978). Disaster relief: Worker's burnout syndrome. In National Institute of Mental Health (Ed.), *Field Manual for Human Service Workers in Major Disasters.* Washington, DC: US Government Printing Office.

Flannery, R.B. (1998). *The Assaulted Staff Action Program: Coping with the Psychological Aftermath of Violence.* Ellicott City, MD: Chevron Publishing.

Flannery, R.B. Jr. (2001). Assaulted staff action program (ASAP): Ten years of empirical support for critical incident stress management (CISM). *International Journal of Emergency Mental Health, 3,* 5-10.

Frederick, C.J. (1977). Crisis intervention and emergency mental health. In W.R. Johnson (Ed.), *Health in Action.* New York: Holt, Rinehart and Winston.

Geiger, B. (2001). Counseling poor, abused, and neglected children in a fair society. *Directions in Clinical Psychology, 13,* 39-51.

Glen Srodes, 79 dies, Chief of Staff of Hospital. (1984, July). *Pittsburgh Post Gazette.*

Hiley-Young, B. & Gerrity, E.T. (1994). Critical incident stress debriefing (CISD): Value and limitations in disaster response. *NCP Clinical Quarterly, 4,* 17-19.

Hill, R. (1958) Generic features of families under stress. *Social Casework,* Vol. 39, Nos. 2 and 3, 139-151.

Hobbs, M., Mayou, R., Harrison, B., & Worlock, P. (1996). A randomized controlled trial of psychological debriefings of road traffic accidents. *British Medical Journal, 313,* 1438-1439.

Holding, D.H. (1983). Fatigue. In G.R.J. Hockey (Ed.), *Stress and Fatigue in Human Performance* (pp. 145-164). New York: John Wiley & Sons, Ltd.

Holmes, R. (1985). *Acts of War: The Behavior of Men in Battle.* New York: Free Press.

ICISF (2002). *CISM course descriptions.* Ellicott City, MD: International Critical Incident Stress Foundation. Available online at: www.icisf.org.

Jenkins, S.R. (1996). Social support and debriefing efficacy among emergency medical workers after a mass shooting incident. *Journal of Social Behavior and Personality, 11,* 447-492.

Jones, E. & Wessely, S. (2003). Forward psychiatry in the military. *Journal of Traumatic Stress, 16,* 411-419.

Jones, J. (1981). (Ed.). *The Burnout Syndrome: Current Research, Theory, Interventions.* Park Ridge, IL: London House Press.

Kardiner A. & Spiegel, H. (1947). *War, Stress, and Neurotic Illness.* New York: Hoeber.

Kenardy, J.A., Webster, R.A., Lewin, T.J., Carr, V.J., Hazell, P.L., & Carter, G.L. (1996). Stress debriefing and patterns of recovery following a natural disaster. *Journal of Traumatic Stress, 9,* 37-49.

Krupnick, J.L. (2001). Interpersonal psychotherapy for PTSD following interpersonal trauma. *Directions in Clinical Psychology, 13,* 75-90.

Lee, C., Slade, P., & Lygo, V. (1996). The influence of psychological debriefing on emotional adaptation in women following early miscarriage. *British Journal of Medical Psychology, 69,* 47-58.

Lee, S.C. (2001). *Backup on The Beat: An Inspiring Collection of Stories, Essays & Thoughts for America's Peace Officers.* Rocklin, CA: Pascoe Publishing.

Leeman-Conley, M.M. (1990). After a violent robbery. *Criminology Australia,* April/May, 4-6.

Lindemann, E. (1944). Symptomatology and management of acute grief. *American Journal of Psychiatry, 101,* 141-148.

Lindy, J. (1985). The trauma membrane and other clinical concepts derived from psychotherapeutic work with survivors of natural disasters. *Psychiatric Annals, 15,* 153-160.

Litz, B., Gray, M., Bryant, R., & Adler, A. (2002). Early intervention for trauma: Current status and future directions. *Clinical Psychology Science and Practice, 9,* 112-134.

Mayou, R.A., Ehlers, A., & Hobbs, M. (2000). Psychological debriefing for road traffic accident victims: Three-year follow up of a randomized controlled trial. *British Journal of Psychiatry, 176,* 589-593.

McFarlane, A.C. (1988).The aetiology of post-traumatic stress disorders following a natural disaster. *British Journal of Psychiatry, 152,* 116-121.

Mitchell, J.T. (1983). When disaster strikes: The critical incident stress debriefing process. *Journal of Emergency Medical Services, 13*(11), 49–52.

Mitchell, J.T. (1986). Teaming up against critical incident stress. *Chief Fire Executive, 1*(1), 24; 36; 84.

Mitchell, J.T. & Bray, G. (1990). *Emergency Services Stress: Guidelines for Preserving the Health and Careers of Emergency Service Personnel.* Englewood Cliffs, NJ: Prentice-Hall.

Mitchell, J.T. & Everly, G.S. Jr. (2001a). *Critical Incident Stress Debriefing: An Operations Manual for CISD, Defusing and Other Group Crisis Intervention Services,* Third Edition. Ellicott City, MD: Chevron.

Mitchell, J.T. & Everly, G.S. Jr. (2001b). *Critical Incident Stress Management: Basic Group Crisis Interventions.* Ellicott City, MD: International Critical Incident Stress Foundation.

National Bible Society (1994). *God's Word for Peace Officers.* Grand Rapids, MI: World Publishing.

National Institute of Mental Health (2002). Mental health and mass violence: Evidence-based early psychological intervention for victims/survivors of mass violence. A workshop to reach consensus on best practices. Washington, DC: NIMH.

Neil, T., Oney, J., DiFonso, L., Thacker, B., & Reichart, W. (1974). *Emotional First Aid.* Louisville: Kemper-Behavioral Science Associates.

Newman, E.C. (2000). Group crisis intervention in a school following an attempted suicide. *International Journal of Emergency Mental Health, 2,* 97-100.

North, C.S. & Pfefferbaum, B. (2002). Research on the mental health effects of terrorism. *JAMA, 288*(5), 633-636.

Noy, S. (1991). Combat stress reactions. In R. Gal & A.D. Mangelsdorff (Eds.), *Handbook of Military Psychology.* Chichester, UK: John Wiley & Sons.

Nurmi, L. (1999). The sinking of the Estonia: The effects of Critical Incident Stress Debriefing on rescuers. *International Journal of Emergency Mental Health, 1,* 23-32.

Olsen, O., Middleton, P., Ezzo, J., Gotzsche, P., Hadhazy, V., Herxheimer, A., Klwijen, J., & McIntosh, H. (2001). Quality of Cochrane Reviews. Assessment of sample from 1998. *British Medical Journal, 323,* 829-832.

Parad, L. & Parad, H. (1968). A study of crisis oriented planned short-term treatment: Part II. *Social Casework, 49,* 418-426.

Potter, D. & LaBerteaux, P. (2001). Debriefing the debriefer: Operational guidelines. In G.S. Everly & J.T. Mitchell, *Critical Incident Stress Management: Advanced Group Crisis Interventions, a Workbook,* Second Edition. Ellicott City, MD: International Critical Incident Stress Foundation.

Raphael, B. & Wilson, J.P. (1993). Theoretical and intervention considerations in working with victims of disaster. In J.P. Wilson & B. Raphael (Eds.), *Interna-*

tional Handbook of Traumatic Stress Studies (pp. 105-117). New York: Plenum Press.

Raphael, B. & Wilson, J. (Eds). (2000). *Psychological Debriefing: Theory, Practice and Evidence.* Cambridge, UK: Cambridge University Press.

Rapoport, L. (1962). The state of crisis: Some theoretical consideration. *Social Service Review,* vol. 36, No. 2, 211-217.

Richards, D. (2001). A field study of critical incident stress debriefing versus critical incident stress management. *Journal of Mental Health, 10,* 351-362.

Robinson, R.C. & Mitchell, J.T. (1993) Evaluation of psychological debriefings. *Journal of Traumatic Stress, 6*(3), 367-382.

Robinson, R.C. & Murdoch, P. (2003). *Establishing and Maintaining Peer Support Program in the Workplace.* Ellicott City, MD: Chevron Publishing Corporation.

Rose, S., Brewin, C., Andrews, B., & Kirk, M. (1999). A randomized controlled trial of individual psychological debriefing for victims of violent crime. *Psychological Medicine, 29,* 793-799.

Rose, S., Bisson, J., & Wessely, S. (2002). Psychological debriefing for preventing post-traumatic stress disorder (PTSD). *The Cochrane Library,* Issue 1. Oxford, UK: Update Software.

Ruben, H.L. (1976). *CI: Crisis Intervention.* New York: Popular Library.

Salby, A.E., Lieb, J., & Tancredi, L.R. (1975). *Handbook of Psychiatric Emergencies.* New York: Medical Examination Publishing Company, Inc.

Salmon, T.S. (1919). War neuroses and their lessons. *New York Medical Journal, 108,* 993-994.

Samuels, M. & Samuels, D. (1975). *The Complete Handbook of Peer Counseling: An Authoritative Guide for the Organization, Training, Implementation and Evaluation of a Peer Counseling Program.* Miami, FL: Fiesta Publishing Corporation.

Sawyer, S. (1989). The aftermath of line-of-duty death. *FBI Law Enforcement Bulletin, 58*(5), 13-16.

Schuster, M., Stein, B., Jaycox, L., Collins, R., Marshall, G., Elliott, M., Zhou, A., Kanouse, D., Morrison, J., & Berry, S. (2001). A national survey of stress reactions after the September 11, 2001, terrorist attacks. *New England Journal of Medicine, 345*(20), 1507-1512.

Selye, H. (1956). *The Stress of Life.* New York: Free Press.

Shneidman, E. & Farberow, N. (1957). *Clues to Suicide.* New York: McGraw-Hill.

Slaikeu, K.A. (1984). *Crisis Intervention: A Handbook for Practice and Research.* Boston, MA: Allyn & Bacon, Inc.

Solomon, Z., & Benbenishty, R. (1986). The role of proximity, immediacy, and expectancy in frontline treatment of combat stress reaction among Israelis in the Lebanon War. *American Journal of Psychiatry, 143,* 613-617.

Stierlin, E. (1909). *Psycho-neuropathology As a Result of a Mining Disaster March 10, 1906.* Zurich: University of Zurich.

Swanson, W.C. & Carbon, J.B. (1989). Crisis intervention: Theory and technique. In *Task Force Report of the American Psychiatric Association. Treatments of Psychiatric Disorders.* Washington, DC: APA Press.

van Emmerik, A.A.P., Kamphuis, J.H., Hulsbosch, A.M., & Emmelkamp, P.M.G. (2002). Single session debriefing after psychological trauma: A meta-analysis. *Lancet, 360,* 766-771.

Wee, D.F., Mills, D.M., & Koehler, G. (1999). The effects of critical incident stress debriefing on emergency medical services personnel following the Los Angeles civil disturbance. *International Journal of Emergency Mental Health, 1,* 33-38.

Wessely, S., Rose, S., & Bisson, J. (1998). A systematic review of brief psychological interventions (debriefing) for the treatment of immediate trauma related symptoms and the prevention of post traumatic stress disorder (Cochrane Review). *Cochrane Library,* Issue 3, Oxford, UK: Update Software.

Western Management Consultants. (1996). *The Medical Services Branch CISM Evaluation Report.* Vancouver, BC: Author.

Wollman, D. (1993). Critical incident stress debriefing and crisis groups: A review of the literature. *Group, 17,* 70-83.

Wolman, B.B. (Ed.). (1973). *Dictionary of Behavioral Science.* New York: Van Nostrand Reinhold Company.

Yalom, I. (1970). *The Theory and Practice of Group Psychotherapy.* New York: Basic Books.

Yule, W. (2001). When disaster strikes—the need to be wise before the event: Early intervention strategies with traumatised children, adolescents and families. *Advances in Mind-Body Medicine, 17,* 191-196.

Chapter 14

Group Intervention
for the Prevention and Treatment
of Acute Initial Stress Reactions in Civilians

Beverley Raphael
Sally Wooding

INTRODUCTION

This chapter deals with the potential for intervention with civilian populations in the acute reaction phase after a terrorist attack or other mass violence. Several factors must be considered in a discussion of interventions following such events. First, the chaos and disruption in the immediate aftermath may make systematic intervention difficult. The situation is further exacerbated by the limited evidence base from which to inform intervention strategies. Second, the sometimes unique social and other phenomena that may occur in the aftermath of such events should be considered: the heroism, altruism, and outpouring of support; the high number of deaths that may occur; the testing of community response by the very nature of such an extreme disaster. There may also be a convergence of well-meaning helpers, including "trauma counselors" of varying persuasions, from whom affected populations may need to be protected. The public nature of such incidents is also significant, as is the influence of the media; the initial "honeymoon" phase of goodwill between the community and politicians, which may be followed by a "disillusionment" phase wherein well-meaning promises of support may be impossible to keep, and the obstacles to rebuilding become fully evident. These factors may complicate community recovery and the vital importance of leadership. Despite these complexities, and although many individuals will be affected by what has happened, the majority will recover, but might do so in different and individual ways, and with very different timelines.

Group mental health interventions have been popularized as part of the response to such incidents for several reasons: to deal with the large numbers involved; to mitigate feelings of distress and isolation; and to prevent or manage adverse mental health outcomes. Psychological debriefing in one form or another is the most popular form of such intervention and has been widely used for emergency mental health response around the globe. Although it was initially developed as part of stress management for emergency service workers and known as critical incident stress debriefing (CISD) (Mitchell, 1983), it has evolved into a range of similar and related techniques, often used under the same name, to respond to any community disaster or emergency. Despite the hopefulness that has driven the rapid expansion of the use of this technique, it has not been justified by follow-up studies measuring its effectiveness for disaster-affected populations.

Several studies meeting the "gold standard" of randomized controlled trials have led to debate about whether this group intervention is appropriate for disaster-affected populations, with many well-designed research studies suggesting that debriefing may worsen outcomes. Reasons for this, and the nature of these studies, will be described as follows. Those who strongly support debriefing suggest that it should still be used because it is frequently *perceived* as helpful. Nevertheless, in the face of the growing number of studies suggesting negative outcomes, the guiding principle should be "first, do no harm." Those providing help in the acute phase may do better to focus on less prescriptive and more practical interventions. Many groups form spontaneously in the aftermath of disasters, and the options for potentially helpful interventions in these settings will also be discussed.

Significant research and clinical interest in the role of mental health interventions in circumstances of terrorism and mass violence has increased, especially since the terrorist attacks of September 11, 2001. The opportunity for early intervention exists and is now a focus, as discussed and concluded in the recent consensus conference on mass violence (U.S. Consensus Workshop on Mass Violence and Early Intervention, 2001). This chapter will consider some of the options for early intervention with civilian populations, focusing on the first month, but taking into account the special constraints and requirements of the immediate days or weeks postincident as well as the acute reaction phase of the first month or so.

Humanitarian aid, prevention of postincident psychiatric morbidity, and treatment of those who decompensate are among the many aims of intervention. Group focus is a way of providing for the large numbers involved, and because people so often come together spontaneously at such a time. Nevertheless, the efficacy of formal group interventions in the prevention and treatment of initial acute reactions among civilian populations has yet to be established. The use of group intervention for response to such initial acute reactions will be considered, taking into account the potential for interventions to assist both formal and informal groups, the rationale for possible effect, and what is known about outcomes.

RATIONALE AND OBJECTIVES OF THE GROUP

In the chaos following a major terrorist attack, the first requirement is to ensure survival. This includes protection from further threat and harm wherever possible, and the management of acute injury that may be life threatening. Tasks for first responders will include retrieval, resuscitation, first aid, and other life-saving interventions. Thus, in the early acute reactive phase, ensuring safety and dealing with any ongoing life threat, including that which may result from injury or other effects of the disaster are imperative. Triage to determine priorities for intervention with mass casualties is also a critical component of ensuring maximum survival and optimal use of available resources.

The concepts of triage, safety, security, and survival also apply to those who have experienced mass violence and are acutely affected by it psychologically. Many different acute reactions have been described in the literature and these must be taken into account in any consideration of intervention for individuals or for groups. Reactive processes described include intense psychophysiological arousal; behaviors that may threaten the survival of the self and others; cognitive distortions and impairment including dissociation; numbness; and intense overt distress, particularly for loved ones from whom one is separated and whose fate is uncertain. Often there is an ongoing sense of fear that the threat will return, with unpredictability, and uncertainty about what has happened and why.

Thus, any group process in this immediate reactive phase must have an intervention rationale that promotes safety, security, and survival. This may be provided to groups that spontaneously gather in the aftermath, usually to seek the support of others, information about what has happened, the whereabouts of loved ones, or how to deal with what has happened. Such spontaneous gatherings may occur in temporary shelters, or anywhere that people find a safe place far enough from the epicenter of the incident. Groups brought together in this immediate aftermath period are likely to have as their rationale the provision of information, and mobilizing of actions to assist others or ensure ongoing safety.

The second rationale for each of these groups is that there is the need for a spontaneous sharing of experience and knowledge about the event and the planning of actions, i.e., what to do next. These spontaneous processes should be supported, as they represent natural strengths and resilience that can facilitate the recovery processes.

Triage of individuals in the immediate, acute reactive phase should take into account factors that will affect the capacity to function both immediately and later, and may provide a basis for identifying those individuals at heightened risk so that subsequent follow-up and targeted interventions may be provided for them. This triage or preliminary screening may also take place through the group. However, significant change occurs over the days and weeks following an acute single incident. The "ABC" of psychological first aid involves ensuring safety and security, offering comfort and support, and also involves (A) identifying and dealing with very high levels of *arousal;* (B) monitoring and protecting persons from *behaviors* that may place them or others at risk; and (C) assessing and dealing with *cognitive distortions* or dysfunctions. People whose high levels of arousal, behavioral disturbance, or cognitive problems interfere with their functioning are unlikely to benefit from forced group processes.

The third objective of group intervention applied to disaster-affected civilian populations in the very acute phase is group psychological or stress debriefing. The most widely practiced form of this is critical incident stress debriefing (CISD), mentioned earlier and developed by Mitchell (1983). This structured review process was developed for emergency service personnel and aimed at lessening the risk of adverse mental health outcomes related to the stressors of their work experience, including disaster response. This model of structured group review to populations other than emergency workers has had widespread application. Although many perceive it to be helpful, particularly emergency workers, there is not yet evidence of its efficacy among emergency responders (Raphael & Wilson, 2000), although it may play a role as a component in broader stress management process in contexts where stressors are not as high.

The aim of group psychological debriefing in the various existing models that have evolved is to systematically review the experience of the disaster—in this case, a terrorist attack, with the goal of mitigating its traumatic impact. This model also has educational aims, such as promoting adaptive coping styles and advising about traumatic stress effects. The mutual support among groups of survivors is seen as beneficial as well, and is another reason for this intervention.

Another group of individuals described in the literature are bereaved survivors waiting for confirmation of the death or possible survival of their loved ones. Weisaeth (1998, personal communication) has described such supportive group processes, put in place while the bereaved await news while housed near the site of a plane crash. The group members may be provided with information, taken to the site, and protected from media intrusions. This type of group, in a protected setting, is likely to facilitate bonding about the experience and may lead to longer-term patterns of mutual support and advocacy, which may facilitate recovery.

As noted, any of these groups, the majority of which arise spontaneously, can be perceived as helpful and may assist those affected to deal with the immediate reactive phase following the impact of a terrorist attack or other disaster. As discussed, this will be most prevalent after mass violence in which survivors are likely to gather together spontaneously in the immediate aftermath. If a terrorist attack involves chemical, biological, or nuclear weapons, different uses of individual or group intervention would be required and would need to focus on concurrent management of any physiological effects, and perhaps separation or quarantining of individuals to avoid spreading disease, should there be the use of infective agents.

ACUTE INITIAL REACTION PHASE

Once survival, safety, and shelter are secured, the tasks of any group or individual interventions for those affected by terrorism or disaster are those related to the particular stressors experienced. Thus, assessment is the first task and provides possibilities for gathering information to delineate those at heightened risk, and to ensure follow-up and targeted interventions where needed.

Major differences exist between conceptualizing and providing intervention in the postterrorism context or other circumstances of mass violence when compared to groups in circumstances that are less acute. The needs of many individuals at this stage may be very concrete. First, a number of practical tasks will need to be addressed, often alongside psychological assessment. These may include registration of victims, providing financial aid, assisting in finding shelter, making funeral and burial arrangements, gathering information, or collecting objects for evidence and forensics. Practical assistance may also provide a vehicle for establishing trust, thereby making psychological interventions more acceptable.

Whether for individuals or groups, such practical postincident necessities make this acute phase very different from traditional group psychotherapy interventions. This is exemplified by our experience meeting traumatized survivors and bereaved families and individuals who had helped search for their loved ones after the Bali nightclub bombing in October 2002. Each person had to be interviewed by police for purposes of collecting evidence regarding the crisis. Some needed practical financial assistance. All sought information regarding loved ones and friends, even as information was being gathered from *them* by authorities. Some waited in

Bali for body identification, to be able to "come home" with their lost family member. On the returning flights, many were distressed and shared this with others. Natural groups of those who had shared the experience evolved. Pre-existing groups had also been formed among the bereaved: for instance, a number of football teams celebrating the season's end with a Bali holiday had lost several members in the bombing. Relatives and affected persons met at Sydney and other airports in some informal groups and some family settings, and they then dispersed.

Assessment can first be made at a population level when estimating the services that will need to be put in place. For instance the number of deaths, destruction of homes and communities, and the shocking and horrific nature of the event, would all be indicators of increased risk for psychiatric morbidity. At the individual level there are a number of frameworks for systematic assessment that can be undertaken with interviews, screening questionnaires, and systematic questions. (Many of these systematic measures are documented elsewhere in this book and will not be dealt with in detail here.) Nevertheless, screening is likely to be relevant in the clinical sense in these settings, alongside other clinical processes (U.S. Consensus Workshop on Mass Violence and Early Intervention, 2001).

Tasks to be Accomplished by Group or Other Intervention with Civilians

In situations following terrorism or mass violence, the tasks of assessment include:

- Understanding the affected person's or group's degree of exposure in the incident, and the particular stressors experienced.
- Assessing the reactive processes as evidenced in history and mental state that have occurred and are occurring over time, then making a judgment as to how these trajectories may represent progressive recovery and adaptation, risk and resilience, or developing pathology.
- Determining social factors in terms of relationships, social support, or isolation. These may be most prominent and appreciable in families or groups. Also relevant are the broader social responses to, and cognitive construction of, the incident and its course and projected outcomes.
- Assessing strengths and vulnerabilities, including adaptive coping styles, past effective mastery of severe incidents, and personal mental health history.

- Evaluating personal hopefulness and assumptions about the future: these are among the variables that correlate most strongly to outcome (Lewin, Carr, & Webster, 1998).
- Examining personal, social, and occupational functioning.

Although this is an outline of the ideal parameters, the chaos and disruption at personal and community levels in the immediate aftermath of a terrorist attack or other disaster may make it very difficult to collect and systematically document even the most basic information. Nevertheless, such systematic information is vital to follow up, to improve understanding, and to evaluate the effectiveness of any individual and group interventions.

The first set of tasks is identifying stressors and adaptations that will then be the focus of any group or individual interventions. The second set of tasks relates to defining interventions that are likely to be effective for those individuals that have been determined to need them. These interventions will need to address the different stressor components. Those of particular relevance are the following:

1. *Death encounter.* The impacts of a life-threatening event or of witnessing gruesome deaths or mutilation of others such as loved ones are criterion A stressors for PTSD (APA, 2000) and are likely to be associated with reactive processes in domains of reexperiencing, avoidance, and arousal with heightened risk for the development of posttraumatic stress disorder or other morbidity.
2. *Loss.* The loss of a family member or loved one is the major stressor, although other losses may also be very significant. The reactive processes are those of grief, yearning for, and preoccupation with the lost person, angry protest, sad review, and mourning. The morbidity is likely to be more in the depressive spectrum (Shore, Tatum, & Volmer, 1986).
3. *Dislocation and/or separation from family, home, or community.* Disruption of familiar settings may also increase anxiety, irritability, and sense of insecurity. One's physical roots and literal "foundation" may be lost, mirroring and emphasizing the psychological upheaval caused by the disaster.
4. *Human malevolence.* Coming to terms with the intent of others to kill and harm is a particularly difficult stressor. That someone would attack and kill, discriminately or indiscriminately, leads to fear, anger, disillusionment, and a search for meaning and justice. Moreover, terrorism is, by definition, unpredictable and violent. The aim of terrorism is to create fear and uncertainty and to disrupt the security of fam-

ily, familiar places, and behaviors. The reactive processes over time are not well studied or understood, but these factors alone may make resolution of the traumatic event itself very difficult. Cycles of revenge and violence (whether fantasized or real) are possible adverse consequences.

Any one or all of these stressors may complicate the reactions of those exposed to terrorism, for instance, with traumatic bereavements. The design of the interventions provided will depend on the profile of risk and need, as identified above, and those interventions demonstrated to be most effective.

CRITERIA FOR SELECTION OF MEMBERS FOR GROUP INTERVENTION OR INDIVIDUAL PROGRAMS

Some criteria for group intervention will be described. It should be reiterated, however, that many groups come together spontaneously, and may seek advice, support, help, or psychological intervention, and it may be optimally provided in this naturally occurring group context.

First, group members may be selected because they belong to the same community, neighborhood, or workplace; i.e., they share common backgrounds. Such groups would aim to reinforce community strengths and cohesion, and the role of community members in recovery. There are no documented studies of such groups or interventions, but they may form the basis of a strategy for adaptation. They have been described as powerful components of community social recovery and provide familiar networks of social support that may help individuals deal with their adversity. However, when such groups share experiences of a disaster there may be very different exposures. Anecdotal reports of debriefing groups in such circumstances describe this heterogeneity of exposures. Hearing of other peoples' experiences may compound psychological trauma, particularly if some group members have not, themselves, been significantly exposed.

Second, group members may be selected because of shared stressor exposures. For instance, some groups may be psychologically shocked with the death encounter, with risk of PTSD, perhaps associated with early acute stress disorder (Bryant, Sackville, Dang, Moulds, & Guthrie, 1999). Decisions for group as opposed to individual intervention will be complex because of differential exposures, distinct personal meanings, and different coping styles. Current effective programs involve cognitive behavior therapy (CBT) and graded exposure in discussion of the stressor experience. This may be difficult if the group has members who are very vulnerable, or who have different time lines of adaptation. Group intervention may be utilized for supportive strategies focused on action and problem solving, strengths, self-help, and coping, while individual interventions might focus on relevant cognitive interventions.

Groups for bereaved persons are likely to involve emotional support and sharing of experience, with guidance on managing specific aspects through funeral, coronial, and related processes, and may include responding to children's needs. Because bereavements in these circumstances are traumatic, and in the case of terrorist attacks, represent homicide, many of the features of response will be similar to those reported for "ordinary" homicide, (although very public in the case of terrorism and mass violence). Some bereaved may have been directly affected; they may have survived an incident in which their loved one died, and thus have the stress of that personal encounter with death as well as the grief of their loss. They may be experiencing "survivor guilt," struggling with the issue that survival may have come at the cost of others' lives (Lifton, 1967; Raphael, 1986).

Others may have been distant from the incident and have come to its site. In the acute phase the bereaved are likely to be unable, or not ready to process the psychological work of grief and mourning their loved one. Rynearson (1996, 2002) and Rynearson, Favell, & Saindon, (2002) have developed a program of group psychotherapeutic intervention in twelve sessions, which also encompasses practical guidance for homicide victims, but is usually implemented months after the deaths. This type of program contributed to the care of the bereaved during the September 11, 2001, terrorist attacks in New York City. Other stressors are not clearly supported by specific group interventions except with the broad themes of mutual support, information, and strategies for action.

This provides an outline of some of the groups that may be utilized in the aftermath of a terrorist attack or disaster, and the sorts of people who may be selected to join them. However, there are several key factors that must be taken into account for inclusion and exclusion of people for group interventions in the acute phase of initial reaction, over the early weeks, first month, or even later:

1. Some affected persons may be so acutely distressed that they are unable to tolerate or benefit from group interventions and should not be expected to participate in this way at this early stage. They may benefit later from group programs, such as that of Rynearson (1996). Initially, however, they are likely to require individually based supportive interventions.

2. Some persons may have profound and multiple problems of a practical nature and will not be able to engage psychologically because of

their focus on the tasks related to meeting these needs. Later group interventions may or may not be relevant.

3. Some people affected by such incidents rely strongly on denial in the initial phase, and their coping styles make them both unwilling to accept either individual or group interventions, or to benefit from them. It may be inappropriate for these coping strategies to be challenged.

4. People may be excluded on the basis of maturity level, language, cognitive function, physical capacity (if severely injured), or other reasons that make it either inappropriate or impossible for them to function in groups at this time.

5. Children should not be separated from their parents to participate in groups unless these groups are part of their school or other familiar social structure. Groups for children in the school setting may be helpful and have been shown to provide opportunities to deal with disaster experiences. Family groups may be helpful but with emphasis on mutual support and practical tasks in the very acute phase.

COMPOSITION AND PREPARATION
FOR GROUP INTERVENTIONS

As noted earlier in this chapter, composition of groups may vary and the preparation will be linked to this and any proposed purposes. Groups of civilians may be familiar with one another, such as people who normally come together in a neighborhood or work together in a social or other context. This may include groups of people who have shared similar experiences in the incident or were exposed to similar stressors. They may be coincidental groups who have been caught together in the experience and are in some ways joined by it.

Outreach to those surviving tragic events has been welcomed by some, rejected by others. For example, invitations for group discussions may be viewed by survivors as helpful, but only if easy to access, or offering a specific purpose with which they can readily identify. A therapeutic purpose may not be seen as relevant by people who do not see themselves as ill, whereas prevention may be better understood. Whatever is decided, the purpose of any group intervention should be clearly defined and identifiable.

Some examples of specific purposes are the following.

1. The group is to provide information about . . .
2. The purpose is to share experience about what has happened; to find out more.

3. The group's purpose is to provide and set up ongoing support for one another.
4. The purpose is to learn about what will help those involved to cope with what has happened and its aftermath.
5. The group is here to help people manage the stress of the experience and to prevent what occurred from leading to longer-term problems.
6. The group is about determining future actions and priorities, so as to:
   - help recovery,
   - get over what has happened, and
   - know what to do and how to prepare; for instance, where to go in response to terrorism, disaster, or a future attack.

In this acute situation and with civilian populations of diverse backgrounds, either familiar, naturally occurring, or similarly exposed groups may all potentially benefit from these general statements of purpose. Specific and individual needs may also be met. Such group processes are frequently perceived as helpful and productive, however their effectiveness in terms of preventing adverse outcomes needs to be further evaluated.

Another important aspect of preparation will be to define the frequency and duration of group meetings with a strong emphasis on briefer interventions and short-term or staged goals. With civilian populations a smaller number of group sessions in the acute phase is likely to be adequate for the purposes discussed, with any longer-term needs and related purposes evolving from this starting point.

GROUP INTERVENTION APPROACHES
FOR ACUTE REACTIONS IN CIVILIANS:
THE IMMEDIATE PHASE AND BEYOND

As noted, groups, at the time of reactive process, have been spontaneous, familiar, shared-experience, or work groups. A range of interventions has been applied for both groups and individuals.

The immediate phase may range from the time of the incident up to two weeks after. The most universal application of groups in this very early phase has been debriefing, in one form or another. The Mitchell model of CISD is the most widely used and involves a series of structured phases. Primarily used with emergency service workers, for whom it was developed, it is a structured response for a structured and somewhat militaristic emergency service provider system. It has been reported as helpful by many in diminishing symptoms of stress on self-report (Mitchell & Everly, 2000). The authors of this program have expanded it into an organizational stress management system called critical incident stress management (CISM). Some recent reports suggest CISM may be helpful to workers with respect to alcohol use, and sometimes arousal, but there have been no randomized controlled trials demonstrating prevention of postincident psychiatric morbidity. Indeed, Avery and Orner report that this system did not prevent emergency workers' problems, and that it was only perceived as helpful by those who needed it least (Avery & Orner, 1998a, b).

The question then arises as to the application of this type of group debriefing technique to the civilian survivors of disasters. Many factors make it difficult to predict whether there would be transferability of this model to civilian populations in the circumstances following an episode of mass violence, terrorism, or disaster. First, these populations are not an established "team" with established group processes, knowledge of one another, shared experiences of mastering past crises, or skills, knowledge, or training to assist effective response. Furthermore, they may experience very diverse stressors, including shock, uncertainly, psychological trauma, loss, grief, dislocation, and reactions to human malevolence. Psychological debriefing, particularly, in its more structured forms, relies on the concept of an established group, and has risks when applied to individuals brought together in other ways and in different circumstances.

These risks include:

- Retraumatization, or exposure to the trauma of others' stories in addition to the issues already faced by the individual, as, for instance with a woman debriefed after a disaster who could not understand why she was having ongoing traumatic symptoms after listening to the horrific experiences of others more directly exposed.
- Inappropriate timing in terms of the trajectory of a reactive process of response to the incident. The group intervention may interfere with natural recovery processes, as, for instance with a man who had dealt with other incidents by initial denial and subsequent careful discussion when he was "ready."
- Active learning of the concept of being traumatized and expectations of developing a disorder when the debriefing involves a focus on trauma symptoms, as with persons who subsequently have a ruminating focus on this trauma list, with expectation of problems and a failure to use other resources for recovery.
- Leading people to mistakenly believe that this process of review will be all that is required to deal with their experience, as with the woman "debriefed" after a disaster who could not understand why she continued having ongoing problems after she had "been debriefed."

- Application for acutely bereaved persons when it is totally inappropriate for both the reactive processes of grief and the timing, as with a family who became fixated on its trauma and could not grieve.
- Application to populations where physical or other survival is a greater priority, as when (difficult as it may be to picture) after an episode of mass shooting, debriefers tried to apply their debriefing concept while the perpetrator was still shooting at the site.
- Creation of expectations of pathological outcomes.

In addition, the research, which has in any way considered the broad use of debriefing for civilian populations after disaster or incidents of mass violence, has shown no benefit and possibly adverse outcomes. This is exemplified in the Kenardy & Carr study (2000), which reported on the victims of the December 28, 1989, Newcastle (Australia) earthquake. It appeared that those debriefed, if anything, showed a greater risk of problems in the longer term. Civilian victims of a bus crash also showed little benefit (Watts, 2000). Ursano, Fullerton, Vance & Wang (2000) report that "talking through" appeared to have been perceived positively but showed little benefit. These researchers also report that there may have been a reinforcing cycle in which debriefers received positive feedback, and thus felt a reason to continue doing it, though evidence of benefit was lacking. This mutual positive effect and engagement may be part of the "honeymoon phase" of affiliative behaviors in which social barriers are diminished and people demonstrate warm engagement with others, even strangers, who have been through the experience with them, or who are trying to help. Although this is positive and part of the social response after such incidents, it does not necessarily mean that group interventions such as debriefing will be effective, although it does attest to the positive human support that is mobilized.

Other group debriefing processes have been used in military or emergency services settings and evaluated to a degree. One that is described by Weisaeth (2000) is called group stress debriefing (GSD). Weisaeth emphasizes the usefulness of the team leader. The team leader facilitates the group's sharing of their experience, as with operational review. Weisaeth suggests that the team leader be trained to facilitate such sharing. With severe incidents there may be expert mental health backup. Such established groups, or teams, have "comparative advantages for handling and processing the effects of severe stress exposures" (Weisaeth, p. 45). Moreover, Weisaeth suggests, the GSD may be "part of an ongoing internal group process" (p. 45). He highlights the different role of such groups and group interventions with structured professional teams, and even, to a lesser degree, as Weisaeth indicates, "reserve teams"—that is, those who are emergency service volunteers or military reservists who work in teams or groups.

This is very different from "natural" or random groups, which are also likely to be at higher risk of traumatization, particularly if they have no specific roles and tasks, nor training or preparation. Weisaeth's conceptualization of this field comes from experience with a wide range of populations affected by disaster and other severe incidents and agrees with the limited findings available for civilian populations. His findings are also consistent with Alexander's work with police and others following an oil rig disaster where debriefing was integrated with the work of professional teams in an occupational health and safety framework (Alexander, 2000).

Ursano et al. (2000) have worked in a number of disaster situations and examined models of debriefing and their possible benefits as interventions in these acute contexts. These researchers looked at two questions: (1) Who attends debriefing groups when they are offered? (2) Does "natural" debriefing (i.e., spontaneously talking about what has happened with friends, family, spouse, significant others, and co-workers) help? With respect to the first question, Ursano et al. (2000) found that those with high exposure, previous disaster experience, and women (as opposed to men) were more likely to attend. The benefits were not clear but it appeared that those who had resources such as past experience and supportive networks were more likely to attend and, even with high exposure, may not have "needed" to be debriefed.

With respect to "natural" debriefing the Ursano et al. study reported that talking about the disaster to friends and intimates took place frequently at two months postdisaster, but had decreased by seven months afterward. Higher talking scores were associated with higher exposure and higher PTSD levels at two months, suggesting that this talking, if equated to debriefing, may not, among other factors, contribute to decreasing the risk of PTSD, although this is by no means the only possible explanation. What such findings mean for debriefing groups in the wider community is difficult to establish. However, it is clear that from the studies in which all the subjects were either emergency workers or military personnel, where debriefing groups may have been expected to have benefit, it was not demonstrated.

These findings are similar to those of Kenardy & Carr (2000). Their study with civilian populations was not found to be of benefit. They suggest that individual personality traits such as sensitivity and neuroticism may increase trauma reactivity following debriefing. Of course, as discussed earlier, it is frequently argued that debriefing should still be provided because these findings do not come from definitive, randomized controlled trials. Nevertheless, "first, do no harm" is a primary aim, and those providing help in the acute phase may be better focused on less prescriptive and more practical interventions.

The effectiveness of debriefing has also been tested in individuals who have suffered severe burns and motor vehicle accidents. In studies following motor vehicle accidents, Hobbs & Mayou (2000) found that at follow-up, those who received the debriefing intervention had a significantly *worse* outcome than controls. The authors concluded there was no evidence that debriefing was an effective early intervention with this group. In a series of studies with survivors of massive trauma following three separate tour bus crashes, Watts (2000) could also find no evidence of debriefing's effectiveness for survivors or the bereaved. He also reported that there were "risks of compounding distress and interfering with recovery" (p. 142).

In a different context, Bisson, Jenkins, Bannister, & Alexander (1997) showed in a controlled trial of debriefing with acute burn victims that the debriefing group was *worse at follow-up* than controls. These findings, similar to those of Hobbs & Mayou (2000), are of importance because both groups represent the types of injuries that may follow a terrorist attack. The injured may be vulnerable to psychiatric disorders in association with the psychological as well as physical injuries. These studies are also of interest as they deal with nonmilitary and nonemergency service persons, i.e., civilian populations.

Research has shown that an additional stressor is associated with the handling of body parts and disaster victim identification. Although these are usually tasks carried out by experts, at a time of catastrophe, the general public or untrained volunteers may be required to assist with the handling of the dead, the mutilated, and possibly, body parts. These stressors may be associated with heightened risk of PTSD and other morbidity. Although debriefing has been applied for this stressor circumstance (e.g., Deahl, Gillham, Thomas, Searle, & Srinivasan, 1994; Deahl, 2000), it has not been shown to prevent morbidity either with soldiers or other workers, unless perhaps when integrated with occupational stress management programs, a buddy system, and general psychiatric support (Alexander, 2000). Generalizing from emergency response personnel to civilian populations provides little guidance as to what would be likely to ameliorate or prevent potential morbidity associated with this stressor.

In studies of violent assault, another potential stressor experience in a terrorist attack, debriefing does not mitigate potential for morbidity (Rose & Bisson, 1998; Rose, Brewin, Andrews, & Kirk, 1999). Of more significance is the work of Bryant's and colleagues (Bryant et al., 1999), which indicates that cognitive behavioral interventions implemented for those at high risk after the very acute period and in the early weeks after the episode may offer the best approach to decreasing risk of developing PTSD

(McNally, Bryant, & Ehlers, 2003). These issues and their importance for any group-based intervention will be discussed later.

The Cochrane Collaboration (Wessely, Rose, & Bisson, 1998) concluded that there was no evidence to support the benefits of single-session debriefing. Litz, Gray, Bryant, & Adler (2002), in a comprehensive review, report no demonstrated benefit from the early intervention debriefing model in the acute reaction phase after trauma, and that it cannot be recommended. Moreover, there is the risk for poorer outcomes in association with it. The recent consensus conference on early intervention after mass violence also concluded that single-session debriefing is not recommended (U.S. Consensus Workshop on Mass Violence and Early Intervention, 2001).

These issues then raise the question of what interventions may be of benefit, for individuals or groups, in the acute reactive phase during the first month after a terrorist attack or other mass violence. In considering these options it is useful to examine, first, the opportunities for any group interventions and what rationale would be used to support them; and, second, what individual interventions may be effective, and for whom, among the civilian population. Some evidence supports individual interventions, but there is a need for systematic research to determine what group programs would be effective in dealing with the larger populations likely to be affected in the immediate aftermath of mass violence or other terrorism. Central to all these issues is the importance of recognizing human resilience. Interventions should not interfere with this, nor *create the expectation* of pathological outcomes.

The concept and importance of human resilience has been emphasized by a number of researchers. Recently, Bonanno (2004) reiterated that not all people exposed to loss and trauma will suffer pathological outcomes. He further explores the concept of resilience; that it is not uncommon, that it is different from the process of recovery, and that there may be a number of pathways to resilience. He challenges views that it is essential to express grief, and highlights the significance of multiple and sometimes unexpected pathways to resilience after violent and life-threatening events, including hardiness, self-enhancement, repressive coping, positive emotion and laughter, and encompasses these in a "broader conceptualisation of stress responding" (Bonanno, 2004, p. 26).

AFTER THE EARLY DAYS AND DURING THE FIRST MONTH

Many of the same themes apply to the first month after the event, but after the initial very acute phase there is the need to provide follow-up outreach, assessment, and, where indicated, intervention. Even taking into account the broader framework, there is a dearth of studies that could inform interventions during the first month or months apart from those which are guided by a small number of studies, such as Bryant, Harvey, Dang, & Basten (1998) and Bryant et al. (1999). There is an urgent need for much further research in this field (Katz, Pellegrino, Pandya, Ng, & DeLisi, 2002). Indeed, as Dembert and Simmer (2000) conclude on the basis of an extensive review and their own clinical experience, the only studies of group interventions postdisaster, apart from debriefing, are descriptive and do not provide guidance for evidence-based practice. The most likely appropriate interventions based on current knowledge will be described as follows, using the stressor component as the principal rationale, but recognizing that multiple concurrent stressor experiences are the norm, particularly following an incident such as a terrorist attack.

- *Death encounter.* Individual intervention could be recommended (Bryant et al., 1999) for those determined to be at high risk of PTSD due to elevated levels of acute stress disorder symptomatology. Other cognitive behavioral and exposure interventions may be of value to assist with the consequences of an assault, injury, and so forth. No studies have investigated these interventions postdisaster, nor their application in group settings. Nevertheless, individual interventions, particularly for those where symptomatology is continuing at high levels after the early week or weeks, should be considered.

- *Loss is frequent and a number of studies highlight the opportunities for bereavement counseling, potentially alongside practical assistance.* Lindemann (1944) first described bereavement-focused interventions after the disaster of the Cocoanut Grove nightclub fire in November 1942, and wrote of the intervention "facilitating normal grief" rather than ongoing pathological patterns, through "grief work." Lindy, Green, Grace, and Tichener (1983) described psychotherapeutic interventions to survivors following a nightclub fire. They observed that in traumatic bereavements such as these, there was often a need to deal with the psychological trauma first, before the person could grieve. This outreach was provided by skilled psychotherapists in the months after the disaster, rather than in the earliest hours, days, or even weeks (i.e., the acute reaction phase). Nevertheless, both these descriptive reports deal with civilian populations and their potential needs after disasters and are not unlike the Australian experience following the Bali nightclub bombing in 2002.

Bereavement interventions were delivered to the survivors of the Granville train disaster of January 18, 1977, in Australia in the acute phase and subsequently (Raphael, 1977, 1979-80). Independent follow-up (not a controlled trial) suggested that outcomes were influenced by whether or not the bereaved had been able to see the body of their deceased loved ones, the degree of perceived support, and whether the bereaved was the parent, especially the mother, of an adult child who died. Interventions appeared to assist to a limited degree (Singh & Raphael, 1981).

A recent review by Schut, Stroebe, van den Bout, & Terheggen (2001) highlights the limited number of controlled trials of bereavement interventions, for individuals or groups. Benefits are most likely when those identified to be at heightened risk of adverse outcomes are provided with interventions in the early stages, and some benefits have been shown for children, widows, and for parents following the death of a child. Nevertheless, the methodologies were limited in many studies. The benefits of groups were less clear, and in some instances the intervention groups tended to worsen outcomes. This may be because early interventions may interfere with resolution or focus the person on ruminating about his or her trauma or grief experience (Solomon, Neria, & Witztum, 2000).

Traumatic bereavements are highly likely postincident, and not easily dealt with by focusing on trauma alone (Raphael & Wooding, 2004). Some reasons why the bereaved are at higher risk are that the deaths are unexpected, often violent, and usually untimely. Normal social support systems may be disrupted and thus not available; the bereaved may be overwhelmed by the other stressor experiences during or as a consequence of the incident, including economic survival and other practical matters. Of particular importance is trauma concerning the bodies of the deceased, which may not be found, or may be so badly disfigured that identification is by forensics only, all of which occurred on September 11, 2001. The need to say good-bye, to have a ritual of funeral and memorial, and a place where the loved one "rests" are all part of grieving processes and extremely difficult in such circumstances. Bodies may also be contaminated or infected, or religious practices may be impossible to fulfill. All such factors make the grief more likely to be complicated and may lead to greater risk of unfavorable outcomes, even when there is an outpouring of community recognition and support.

In the bereavements following the Bali nightclub bombing, these issues have been of particular significance, moreso when more than one family member was killed. Individual support has included outreach to returning families, counseling to facilitate the experience of receiving the deceased's body and the necessary coronial processes, and follow-up individual counseling for those affected and their families. Years later, there is significant ongoing need, but the intervention programs and support (both practical and psychological) have not been readily available. Group interventions have not been provided except to families, or when several members of a preexisting group—as in the case of the football teams—could benefit from group support.

POTENTIAL PROBLEMS

As noted earlier, groups in this context are quite different from the usual psychotherapeutic group. Relationships may be intense, but some participants may be numb and withdrawn. Different exposures may mean distress is too great to handle and may also interfere with the capacity to respond to others' needs. Groups may also split or scapegoat some members. Some members may dominate with a forceful drive to prove their story and that their way of coping is the most powerful. There may even be competition for the "greatest victim status." Groups may not be sensitive to different needs and different time lines of members. Pressure to disclose may cut across what is for one person or another adaptive coping. Structure, tasks, and information may be helpful, as may clear definition of purpose and what is to be achieved. Nevertheless, groups may be neither appropriate nor helpful for therapeutic purposes in the postincident period, but rather provide mutual support, information sharing, and planning actions to address needs.

The person providing intervention to the group and to the individual may become overwhelmed. Alternatively, he or she may overidentify, attempt to undo, or develop distancing, burnout, or compassion fatigue. Those working with such groups should have support and systems of their own to assist them to deal with the many personal issues that are likely to arise with such work.

Group leaders should have a supportive role in the group, and should be skilled and knowledgeable about trauma, grief, and postdisaster phenomena, indicators of risk and resilience, and what might be appropriate interventions. The trauma of the incident may open past wounds and deprivation, and those providing mental health interventions need to be prepared to deal with this in limiting ways. Interventions that encourage regression are inappropriate as they may reawaken the helplessness of the disaster experience, as well as early trauma and deprivation. The key issue is still the psychodynamic and social parameters or response, reinforcing strengths, having clear goals, and ensuring any intervention does not harm. Longer-term problems require attention and a negotiated framework for subsequent treatment.

COMBINED INDIVIDUAL AND GROUP TREATMENT

During the work of early intervention in the acute reaction phase there are often considerable indications for individual intervention. At the same time, there is value of both naturally formed and familiar groups that come together for mutual support or specific tasks. Thus, a combination of the two approaches may be used. Initial interventions in the acute reaction phase will need to fit with other demands in affected persons' lives, and should not separate them from these realities. Interventions should be short-term, with specific focus and a small number of sessions, and they should be assessed for their effectiveness. Sessions may be up to two hours in duration, and consist of four to six sessions over the first four to six weeks postdisaster. Weekly sessions or, if the need is acute, twice weekly sessions, are suggested, but weekly sessions of longer duration carry expectations of containment and support. Specific planning for longer-term interventions should be done when these are indicated.

CONCLUSION

Groups will preexist or occur naturally in the acute phase, postterrorist attack or other disaster. Groups can be used, but their limitations should be recognized. The direct application of group psychotherapy models in this acute phase is inappropriate although the skills and knowledge of group and interpersonal dynamics are vital. Nevertheless, groups may provide their members mutual support, supply information, and develop strategies for recovery. Targeted interventions for those at high risk of trauma, grief, and other stressors need to be delivered on the basis of what is known about the specific stressors and needs of individual group members, as well as what might be effective in dealing with the impact (both short- and longer-term). Research is urgently needed to evaluate the current interventions and the evidence and rationale for their use, and to extend our knowledge of what is most efficacious in this very important and emerging field.

REFERENCES

Alexander, D. (2000). Debriefing and body recovery: Police in a civilian disaster. In B. Raphael & J. Wilson (Eds.), *Psychological Debriefing: Theory, Practice and Evidence* (pp. 118-130). London: Cambridge University Press.

American Psychiatric Association. (2000). *Diagnostic and Statistical Manual of Mental Disorders* (Fourth Edition, Text Revision). Washington, DC: Author.

Avery, A. & Orner, R. (1988a). First report of psychological debriefing abandoned. The end of an era? *Traumatic Stress Points. International Society for Traumatic Stress Studies, 12,* 3, Summer.

Avery, A. & Orner, R. (1988b). More on debriefing: Report of psychological debriefing abandoned. The end of an era? *Australian Traumatic Stress Points,* Newsletter of the Australian Society for Traumatic Stress Studies, July, 4-6.

Bisson, J., Jenkins, P.L., Bannister, C., & Alexander, J. (1997). Randomised controlled trial of psychological debriefing for victims of acute burn trauma. *British Journal of Psychiatry, 171,* 78-81.

Bonanno, G. (2004). Loss, trauma, and human resilience: Have we underestimated the human capacity to thrive after extremely aversive events? *American Psychologist, 59,* 20-28.

Bryant, R.A., Harvey, A.G., Dang, S.T., & Basten, C. (1998). Treatment of acute stress disorder: A comparison of cognitive-behavioral therapy and supportive counselling. *Journal of Consulting and Clinical Psychology, 66,* 862-866.

Bryant, R.A., Sackville, T., Dang, S.T., Moulds, M., & Guthrie, R. (1999). Treating acute stress disorder: An evaluation of cognitive-behavior therapy and supportive counselling techniques. *American Journal of Psychiatry, 156,* 1780-1786.

Deahl, M.P. (2000). Debriefing and body recovery: War grave soldiers. In B. Raphael & J. Wilson (Eds.), *Psychological Debriefing: Theory, Practice and Evidence* (pp. 108-117). London: Cambridge University Press.

Deahl, M.P., Gillham, A.B., Thomas, J., Searle, M.M., & Srinivasan, M. (1994). Psychological sequelae following the Gulf War: Factors associated with subsequent morbidity and the effectiveness of psychological debriefing. *British Journal of Psychiatry, 165,* 60-65.

Dembert, M.L. & Simmer, E.D. (2000). When trauma affects a community. In R.H. Klein & V.L. Schermer (Eds.), *Group Psychotherapy for Psychological Trauma* (pp. 239-264). New York: The Guilford Press.

Hobbs, M. & Mayou, R. (2000). Debriefing and motor vehicle accidents: Interventions and outcomes. In. B. Raphael & J. Wilson (Eds.), *Psychological Debriefing: Theory, Practice and Evidence* (pp. 145-160). London: Cambridge University Press.

Katz, C.L., Pellegrino, L., Pandya, A., Ng, A., & DeLisi, L.E. (2002). Research on psychiatric outcomes and interventions subsequent to disasters: A review of the literature. *Psychiatry Research, 110,* 201-217.

Kenardy, J.A. & Carr, V. (2000). Debriefing post disaster: Follow-up after a major earthquake. In B. Raphael & J.P. Wilson (Eds.), *Psychological Debriefing: Theory, Practice and Evidence* (pp. 174-181). Cambridge: Cambridge University Press.

Lewin, T.J., Carr, V.J., & Webster, R.A. (1998). Recovery from post-earthquake psychological morbidity: Who suffers and who recovers? *Australian & New Zealand Journal of Psychiatry, 32*(1), 15-20.

Lifton, R.J. (1967). *Death in Life: Survivors of Hiroshima.* New York: Random House.

Lindemann, E. (1944). Symptomatology and management of acute grief. *American Journal of Psychiatry, 101,* 141-148.

Lindy, J.D., Green, B.L., Grace, M., & Tichener, J. (1983). Psychotherapy with survivors of the Beverly Hills Supper Club fire. *American Journal of Psychotherapy, 37,* 593-610.

Litz, B.T., Gray, M.J., Bryant, R.A., & Adler, A.B. (2002). Early intervention for trauma: Current status and future directions. *Clinical Psychology: Science and Practice, 9*(2), 112-134.

McNally R.J., Bryant R.A., & Ehlers, A. (2003). Does early intervention promote recovery from posttraumatic stress? *Psychological Science in the Public Interest, 4,* 2, 45-79.

Mitchell, J.T. (1983). When disaster strikes. . . The critical incident stress debriefing process. *Journal of Emergency Medical Services, 8*(1), 36-39.

Mitchell, J.T. & Everly, G.S. (2000). Critical incident stress management and critical incident stress debriefings: Evolutions, effects and outcomes. In B. Raphael & J. Wilson (Eds.), *Psychological Debriefing: Theory, Practice and Evidence* (pp. 71-90). London: Cambridge University Press.

Raphael, B. (1977). The Granville train disaster: Psychological needs and their management. *Medical Journal Australia, 1*(9), 303-305.

Raphael, B. (1979-80). A primary prevention action program: Psychiatric involvement following a major rail disaster. *Omega, 10*(3), 211-226.

Raphael, B. (1986). *When Disaster Strikes: How Individuals and Communities Cope with Catastrophe.* New York: Basic Books.

Raphael, B. & Wilson, J.P. (Eds.) (2000). *Psychological Debriefing: Theory, Practice and Evidence.* Cambridge: Cambridge University Press.

Raphael B. & Wooding, S. (2004). Early mental health interventions for traumatic loss in adults. In B. Litz (Ed.), *Early Intervention for Trauma and Traumatic Loss* (pp. 147-178). New York: The Guilford Press.

Rose, S. & Bisson, J. (1998). Brief early psychological intervention following trauma: A systematic review of the literature. *Journal of Traumatic Stress, 11,* 697-710.

Rose, S. Brewin, C.R., Andrews, B., & Kirk, M. (1999). A randomised controlled trial of individual psychological debriefing for victims of violent crime. *Psychological Medicine, 29,* 793-799.

Rynearson, E.K. (1996). Psychotherapy of bereavement after homicide: Be offensive. *Psychotherapy in Practice, 2,* 47-57.

Rynearson, E.K. (2002). *Retelling Violent Death.* New York: Brunner-Routledge.

Rynearson, E.K., Favell, J., & Saindon, C. (2002). Group intervention for bereavement after violent death. *Psychiatric Services, 53,* 1340.

Schut, H., Stroebe, M.S., van den Bout, J., & Terheggen, M. (2001). The efficacy of bereavement interventions: Determining who benefits. In M.S. Stroebe, R.O. Hansson, W. Stroebe, & H. Schut (Eds.), *Handbook of Bereavement Research: Consequences, Coping, and Care* (pp. 705-737). Washington, DC: American Psychological Association.

Shore, J.H., Tatum, E.L., & Volmer, W.M. (1986). Psychiatric reactions to disaster: The Mount St. Helens experience. *American Journal of Psychiatry, 143,* 590-595.

Singh, B. & Raphael, B. (1981, April). Post disaster morbidity of the bereaved: A possible role for preventive psychiatry. *Journal Nervous & Mental Disease, 169*(4), 203-212.

Solomon, Z., Neria, Y., & Witztum, E. (2000). Debriefing with service personnel in war and peace roles: Experience and outcomes. In B. Raphael & J. Wilson (Eds.), *Psychological Debriefing: Theory, Practice and Evidence* (pp. 161-173). London: Cambridge University Press.

Ursano, R.J., Fullerton, C.S., Vance, K., & Wang, L. (2000). Debriefing: Its role in the spectrum of prevention and acute management of psychological trauma. In B. Raphael & J. Wilson (Eds.), *Psychological Debriefing: Theory, Practice and Evidence* (pp. 32-42). London: Cambridge University Press.

U.S. Consensus Workshop on Mass Violence and Early Intervention. (2001). Mental Health and Mass Violence: Evidence-based early psychological intervention for victims/survivors of mass violence. Available online at http://www.nimh .nih.gov/publicist/massviolence.pdf.

Watts, R. (2000). Debriefing after massive road trauma: Perceptions and outcomes. In B. Raphael & J. Wilson (Eds.), *Psychological Debriefing: Theory, Practice and Evidence* (pp. 131-144). London: Cambridge University Press.

Weisaeth, L. (2000). Briefing and debriefing: Group psychological interventions in acute stressor situations. In B. Raphael & J. Wilson (Eds.), *Psychological Debriefing: Theory, Practice and Evidence* (pp. 43-57). London: Cambridge University Press.

Wessely, S., Rose, S., & Bisson, J. (1998). A systematic review of brief psychological interventions ("debriefing") for the treatment of immediate trauma-related symptoms and the prevention of posttraumatic stress disorder (Cochrane Review). *The Cochrane Library,* 3, Oxford: Update Software.

Chapter 15

Present-Centered Supportive Group Therapy for Trauma Survivors

Melissa S. Wattenberg
William S. Unger
David W. Foy
Shirley M. Glynn

INTRODUCTION

In the wake of terrorism, survivors must balance two competing tasks: to make sense of the trauma, and to reestablish a secure foundation in the face of intrusions that overwhelm the current time frame. Present-centered group therapy (PCGT) emphasizes attention to current emotional responses, needs, situations, and wants, supporting recalibration of trauma-based symptoms, attitudes, and behaviors. An alternative to exposure-based, dynamic "uncovering," and formal skills-building treatments, this supportive approach provides a context that gently reorients members toward current coping through the group environment, incorporating a modicum of structure and trauma-specific interventions.

Literature

A review of literature examining five studies (including a total of over 800 participants) on the efficacy of present-centered or supportive group therapy models for trauma treatment indicates improvements in symptoms such as depression, anxiety, and self-esteem. The groups in these studies met weekly, with duration of treatment ranging from six to twenty weeks. Three studies used control groups; of these, one used random assignment. However, specific posttrauma sequelae, such as post-traumatic stress disorder (PTSD), were not measured in several of these studies (Foy et al., 2000). Two additional studies also support the efficacy of PCGT. A study by Classen, Koopman, Nevill-Manning, & Spiegel (2001) randomly assigned fifty-three adult women reporting childhood sexual abuse to either trauma-focus group therapy (TFGT), PCGT, or waitlist groups. Results indicated that both TFGT and PCGT showed significant improvement of trauma-related symptoms, while the waitlist controls did not. In the most recent study making direct comparisons of TFGT and PCGT (Schnurr et al., 2003), 360 male veterans with chronic combat-related PTSD were randomly assigned to TFGT or PCGT for a series of twenty-five weekly therapy sessions. Post-treatment assessments of PTSD severity and other measures showed significant improvement from baseline for both TFGT and PCGT. However, rigorous *intention-to-treat* analyses found no overall differences between the two types of group therapy on any measure of outcome. Although average improvement was modest in both treatments, roughly 40 percent of subjects showed clinically significant change. Although analyses of data from a subset of TFGT participants receiving an *adequate dose* of exposure therapy suggested greater reduction in PTSD symptoms for these participants, as contrasted with the aggregate of participants in the present-centered condition, this finding is balanced by the fact that treatment dropout was considerably higher for participants randomized to TFGT than for those randomized to PCGT. These results indicate that PCGT is a relatively "user-friendly" modality, readily implemented and delivered, and reasonably accessible for trauma survivors at a variety of treatment stages. A number of reasons lead one to suspect that the therapeutic factors of supportive groups may be particularly applicable to PTSD. Although present-centered therapy was initially developed as a PTSD-adapted active treatment control, the trauma-specific psychoeducation and interventions within this adaptation can be seen to magnify the typical therapeutic factors, as applied to a trauma-survivor population.

Rationale

Present-centered group therapy bridges the realities of life before and after a terrorist event, legitimizing members' return to nonemergency functioning. The therapy environment afforded by PCGT, through its emphasis on consistent access and attention to the current context, allows an opportunity for the trauma-influenced worldview to gradually readjust, resulting in functioning that more accurately reflects current information. PCGT is historically connected to humanistic and experiential psychotherapies, which emphasize *authentic experience in the present*. Humanistic psychotherapy, in particular, suggests that contact with an interpersonal environment of acceptance and emotional safety allows a "natural," integrative healing process to take place. This supportive attention enhances personal strengths, makes room for human frailty, and increases awareness of emotional experience *in the moment*. The result is a more fully functioning person who has access to a broader, more flexible repertoire of coping behaviors and more complete information about the environment. Yalom (1985) describes how the stable yet fluid complex interpersonal environment of group therapy provides compelling context cues and feedback in the present. PCGT enlists these therapeutic elements in treating the sequelae of trauma. Powerful cues from the group context compete with intrusions from the trauma (e.g., recurring traumatic memories or flashbacks), priming members toward greater focus on their current lives and more balanced expectations for the future. Relying on the intrinsic therapeutic factors of group psychotherapy, and incorporating trauma-specific present-orienting interventions, PCGT mobilizes the strengths and competence of group members toward reduction of interference from symptoms and trauma-based attitudes.

Schema Theory and PCGT

The attention to the present inherent in this model can be understood in terms of schema theory (McCann & Pearlman, 1990; Neisser, 1976), which describes the process of

- actively and automatically engaging a complex environment;
- selectively searching out new information, guided by past experience/concepts;
- modifying previously learned information by accommodating and integrating new information;
- recalibrating expectations based on the integration of new information with former knowledge and concepts;
- reorienting to the environment based on these integrative changes;
- redirecting selective search for information based on this reorientation; and
- accurately and flexibly responding to selected cues in the environment, based on ongoing integration of new information and repeated reorienting.

For individuals with PTSD, trauma-based intrusions, affects, and attitudes interfere with this automatic process of digesting and assessing new information. Past events are revisited, often with a sense that they are, to at least some extent, recurring in the present. These recurrences tend to distort, or even replace, aspects of the current environment. Cues in the present are then overlooked, as the intrusions appear more pressing. Trauma-based attitudes are reinforced, and trauma-based behaviors tend to be reinitiated even once the trauma has subsided. In these ways, intrusive symptomatology disrupts the normal feedback loop (Rosinski, 1977) that allows adjustment of behavior to the demands of the environment. Trauma survivors tend, at least partially, to lose their ability to observe themselves in relation to their current physical and interpersonal surroundings. It becomes difficult for them to gauge their own impact, or accurately assess the actions and motivations of others.

Trauma-based concerns that override everyday experience can be roughly divided into four major themes:

1. Sense of danger or threat.
2. Sense of powerlessness.
3. Loss of connection with the community of other human beings, accompanied by
   a. *Shame. Shame-based interactions* are interpersonal dynamics dominated by a sense of being "less than," inadequate, intrinsically lacking. Helen Block Lewis (1971) distinguished shame from guilt, as a more diffuse, whole-person loss of worth that contaminates the sense of self, as opposed to regret over an act of commission or omission, "The experience of shame is directly about the *self*. . . . In guilt, the self is not the central object of negative evaluation, but rather the *thing* done or undone" (p. 30); "Shame is about the whole self" (p. 40, italics in original). Manifestations of shame include embarrassment, humiliation, chagrin, *shame-rage* (an angry reaction spurred by shame, that usually backfires and results in greater shame), and mortification, but also, scorn and disdain, as a projection of shame onto others.
   b. *Alexithymia. Alexithymia* refers to two symptoms of psychic numbing: (1) inability to put feelings into words; and (2) lack of awareness of emotional experience. Although these definitions differ, treatment for these conditions is largely the same.
4. Loss of sense of purpose and meaning in life.

These themes inform the facilitators' understanding of the group dynamics, and may be identified and discussed in the group without reference to specific details of the trauma. As these themes and related emotions are explored, members learn to distinguish trauma-based intrusions from information retrieved from the current environment. As the informational "feedback loop" is restored, the trauma-based worldview can be put in perspective, and behavior more suited to nonemergency functioning can gradually emerge.

Identification of these themes as they emerge in the group, and recognition of corresponding trauma-based emotional experiences such as fear, helplessness, alienation, and betrayal, help members distinguish their trauma-related intrusions from realistic, situation-specific reactions to the current environment. Members then are encouraged informally, through continued focus on affect, to reconnect the feedback loop in the current context, increasing awareness of information that contrasts with trauma-based assumptions about themselves, others, and the world.

PCGT, then, offers a "corrective" context in which to examine and modify the effects of trauma. As survivors engage in this context, they gradually see themselves more clearly in connection to their environment, perceive the actions and motives of others more accurately, become more aware of their own feelings and behavior, and are better able to put the trauma into a historical time frame rather than persistently reliving it. In terms of the four trauma-based themes, group members begin to feel safer, more empowered, more connected, more effective at expressing themselves, and better able to establish a sense of meaning and purpose in their lives.

PURPOSES OF PRESENT-CENTERED GROUP THERAPY

PCGT targets the psychological sequelae of trauma as they interfere with orientation to the present, current functioning, and quality of life. These sequelae fall into two categories, as shown in Exhibits 15.1 and 15.2.

An advantage of PCGT is that it is flexible enough to be applied to a variety of settings and purposes. In programs for treatment of psychological

EXHIBIT 15.1. Symptoms That Diminish Attention to Everyday Life

1. The three symptom clusters of post-traumatic stress disorder
   - Reexperiencing
   - Avoidance/numbing (including alexithymia)
   - Hyperarousal
2. Associated depression and dysthymic disorder
3. Associated panic disorder, often with agoraphobia
4. Loss of skills, including
   - Interpersonal skills
   - Emotional awareness/expression
   - Communication skills

EXHIBIT 15.2. Trauma-Based Alterations in Beliefs, Attitudes, Habits, and Behaviors

1. Cognitive and emotional schemas affected by trauma, including:
   - Safety versus risk/threat
   - Empowerment versus helplessness
   - Connection versus isolation/alienation
   - Meaning versus loss of meaning, aimlessness
2. Shame resulting from violation of expectations, both during the trauma and in its aftermath
3. Guilt about the traumatic event, e.g., survivor guilt, guilt over not being able to help others, self-blame at victimization
4. Secondary trauma (i.e., negative reception from community and/or primary group soon after the traumatic experience)

trauma, PCGT may serve as the primary therapy modality, or as adjunctive to other therapies, e.g.,

- as an introduction and preparation for further therapy;
- as a support for compliance with concurrent treatment (e.g., individual or group trauma work, formal skills-building, work therapy); and
- as posttrauma-focus support for application of gains to current life.

Objectives

The main objective of PCGT is to support movement from a trauma-based worldview toward a broader perspective that incorporates information from the current environment, and applies it to improving current functioning and quality of life. To this end, PCGT

1. Validates the life-altering impact of the trauma, without revisiting specific trauma memories.
2. Provides a safe, consistent environment in which to begin to feel effective again.
3. Allows a sense of common experience and community to develop, counteracting the isolation, alienation, numbing, and disconnection that trauma survivors often experience.
4. Addresses intrusions, as well as trauma-based attitudes and habits.
5. Encourages expression of affect, decreasing alexithymia.

6. Diffuses the sense of shame that often accompanies traumatic experience.
7. Encourages interpersonal perspective taking.
8. Inoculates against potential "secondary trauma" from other less-supportive contexts in the members' lives, following the initial trauma, and offers a receptive environment for those who initially had negative experiences posttrauma.

GROUP CHARACTERISTICS

PCGT utilizes an active, facilitative leadership style, generally using cotherapy (usually two facilitators). Emphasis is on client strengths, process-encouraging interventions, and a combination of pragmatic and "here-and-now" focus with a modicum of structure. The process of change is regarded as gradual and incremental. Level of confrontation is generally low to moderate, and a sense of interpersonal comfort is supported. Unlike exposure-based and other uncovering therapies, PCGT directs focus away from the details of the precipitating event, while acknowledging and validating the impact of the trauma. Interventions aim at exploring "midrange" affects (e.g., frustration, uneasiness, sadness, happiness, hurt) that pertain to everyday life, and diffusing more extreme affects (e.g., rage, terror) related to hyperarousal.

Group Dynamics

Development of a Working Alliance

PCGT facilitators are active and proactive in relation to group members. Their roles are well-defined but flexible from which to model adaptability and provide a consistent source of support. Facilitators remain accessible—neither uncommunicative or distant (which may elicit confusion, a sense of abandonment, or even paranoia), nor overinvolved, intrusive, or controlling (which may elicit overdependence and passivity). Transferences are addressed as they arise, but diffused so that they rarely become the main focus of the group. Facilitators strive for authenticity within the boundaries of their clinical role, maximizing a sense of connection, and offering clear, reliable interpersonal cues. The aim is to reduce ambiguity for clients whose worldview is already compromised by trauma.

Psychoeducation in early sessions serves a variety of important functions, not the least of which is demystifying the sequelae of trauma, thereby fostering hope, connection, and a sense of empowerment. Trauma survivors tend to experience a sense of a "fall from grace," a feeling of being help-lessly "chosen" for terrible misfortune, which in turn engenders a sense of shame, loss of control, and a profound sense of alienation at being "abandoned" to the trauma. Development of symptoms posttrauma further separates survivors from nontraumatized peers and family members, intensifying the sense of shame and alienation. By normalizing the response to trauma through explaining common trauma-based symptoms and attitudes, facilitators create a sense of connection to both other members and the larger human community, countering this characteristic shame and alienation.

As facilitators share their knowledge base, they also establish their own credibility, demonstrating to the group that they are both knowledgeable about and interested in the sequelae of trauma. Rather than lecturing, they foster discussion. They allow members to develop a sense of effectiveness through active participation in the group, and investment in the group's growing social context. These measures help facilitators negotiate the therapeutic alliance with clients who tend to have difficulty with trust and connection. Communication from facilitators may be more formally polite in early sessions, in order to clearly establish an authoritative, but nonauthoritarian, stance; however, as the group progresses, facilitators may at times become more informal, playful, and direct. At the same time, facilitators gradually step back and become less active, in order to permit group members to take more responsibility within the group process.

Role of Facilitators. Cotherapy is strongly recommended for the sake of continuity and group stability, and as a buffer against vicarious traumatization. The cotherapy team affords opportunities for the facilitators to support each other both in the group and in preparing for it. As they plan together, facilitators may discuss their own reactions to the group, using these reactions to inform their choice of intervention. Cotherapists may also take different positions within the group, representing conflicting views that need to be integrated as trauma-based beliefs and attitudes are processed and reframed. They may model "agreeing to disagree," representing opposite sides of ambivalence expressed in group, and may take different roles in setting limits within the group. The cotherapy team is typically facilitative of group process rather than explicitly directive. Both therapists need to be relatively flexible in role, exchanging task leadership and socioemotional leadership. They model communication, assertion, mutual respect, and healthy capacity for emotion. Because this modality has more "free space" to unfold spontaneously, it requires more nimble intervention on the part of therapists than might be necessary in more structured groups. For instance, they intervene actively when reenactments threaten retraumatization. They maintain ultimate responsibility for violation of group rules and consequences. Unless there is another primary provider, they take responsibility for handling emergencies, risk issues, and other treatment issues. They also serve as a bridge to other providers within the setting, as well as to providers outside the setting (with required releases).

The facilitators' work extends to self-care, and modeling self-care for group members. Through working with survivors, facilitators encounter the issues with which their clients are grappling (in addition to any related personal issues)—loss of meaning, sense of helplessness, fear, anger, and despair—and must process and integrate their own reactions in order to protect their personal well-being and remain available to help members address these issues. As hyperarousal can be contagious among group members, facilitators also are susceptible to this same heightening of arousal. The cotherapy team offers mutual support for managing these affects, for validating the value of the work, and for encouraging a balance between work life and meaningful personal activity.

Members' relationships. A supportive relationship among members is fostered within the group, characterized by exchange of feedback, development of affective connection and expression among members, empathetic listening, and reciprocal perspective taking. Although high levels of confrontation are not encouraged, assertive communication is modeled and promoted. Polarized suballiances within the group are discouraged, as are overgeneralized identification (e.g., "We're all the same"), and "us-them" mentality (see Case Example 8, later in chapter). Members may have relationships with one another outside the group, but the impact of these relationships on the group is "grist for the mill."

Fundamental interventions in PCGT. Broadly construed, present-centered interventions work toward restoring the trauma-compromised "feedback loop," to allow members more accurate and ready contact with their everyday lives. To this end, the focus of these interventions is twofold:

1. Encouragement of greater awareness and expression of midrange affects.
2. Direction of attention to members' presentation within the social environment, including
   a. verbal and nonverbal expression of symptoms and trauma themes; and
   b. changes toward greater flexibility and spontaneity, as trauma-based reactions recede.

This combined emphasis on the outer social environment and the inner, emotional environment supports a return to normal social exchange. While this emphasis may be standard fare in supportive groups, PCGT incorporates a number of formal yet easily integrated reflective techniques from experiential and person-centered therapies that strengthen this process for trauma survivors. Prouty (1994) developed a method for working with individuals suffering profound loss of contact with the social environment (e.g., psychotic clients); his techniques can be readily adapted to more functional clients whose trauma issues interfere with accurate perception of cues. These simple, direct interventions are used judiciously in the group process, along with more complex interventions. They include

- verbal body reflection (noting and describing body posture, gesture, or voice quality in nonjudgmental terms);
- nonverbal body reflection (physically mirroring a posture or movement to communicate effort to understand the emotion or stance behind it), often accompanied by verbal body reflection;
- repetition of a statement or sound made by the client (whether a sigh or a sentence—not as a reflection, but word-for-word, e.g., CLIENT: "Why me?" FACILITATOR, with emphasis: "Why me?")
- Word-for-word reflection a la Rogers, e.g., CLIENT: "I feel so angry!" FACILITATOR: "You feel so angry!"

Use of traditional, client-centered reflective paraphrasing to clarify and elaborate the thoughts and feelings that may be underlying simple statements (e.g., CLIENT: "Why me?" FACILITATOR: "You ask yourself, why did this happen? Is it something about me? It seems so unfair, something that just should not have happened") is also useful in PCGT. Use of these interventions with higher functioning individuals cuts through intellectualizing and self-criticism, accentuating affect and immediate experience. Integrated into the therapy genuinely rather than mechanically, these reflections deepen the group process, and reassure members that the facilitators are with them in these subjectively perilous, affect-laden moments. This basic mirroring involved in these techniques also provides clear evidence of connection for individuals who feel more uncertain about themselves and the world than they may let on. Most importantly for trauma survivors, these interventions allow group members to develop perspectives on what they are communicating to others, completing the compromised feedback loop, and providing evidence that their actions and statements are taken seriously and have interpersonal significance.

Another useful approach to enhancing connection, from an intrapersonal perspective, is Gendlin's focusing technique (1981, 1996). This experiential technique identifies a *felt sense* in the midsection of the body, an inner experience that is sensed rather than immediately verbalized. The felt sense is distinct from physical tension or physiological arousal; it is an affective sensation roughly akin to a "gut feeling." Attention is directed to the area of the body between the throat and just above the navel, often after a period of relaxation and "clearing a space" to remove distracting thoughts and feelings from attention. Authentic emotional experience evolves from sitting with this inner sense and allowing words or images to spontaneously emerge. The gradual labeling of inner experience encourages tolerance of emotion and increases awareness of affective nuances. It is especially powerful for addressing alexithymia. Continued examination of the felt sense may lead to a "felt shift" or change in the placement and quality of the inner experience, with corresponding insight and emotional relief. Focusing can be introduced either formally or informally (see Case Example 2). Although the group therapy format does not lend itself to lengthy focusing sessions, brief focusing exercises can be integrated into group discussion as a model for examining affective experience. For example:

CLIENT: "I feel so angry!"

FACILITATOR: "You feel really angry . . . and where in your body do you experience that anger? Check in the center of your body, between your throat and just above your navel. . ."

CLIENT: "In my chest—here—it feels like it wants to explode!"

FACILITATOR: "So, right here, in your chest (indicates with gesture) . . . like it wants to explode . . . does that seem to just fit the way it feels? Can you check back with that feeling in your body, and see if it fits perfectly, or if there is anything else there now?"

In any group therapy for trauma survivors, symptoms and trauma-based themes may manifest at the level of a single individual, or at the level of the whole group. Correspondingly, interventions can be made at either of these levels. Although the examples address individuals within a group, group-level reflections or focusing can also be very useful, e.g.,

- *Reflective paraphrasing:* "From the tone of the group today, it seems there's a really angry feeling in the group—is that right?"
- *Focusing:* "Everyone seems kind of quiet today. . . . Let's take a moment, and see what that's all about—what's the feeling here, in the room, right now? What do you notice? . . . Where do you feel it, in your body? . . . Okay, so here, in your chest—and in your chest, too . . . and you—chest, and right here, the solar plexus? . . . Does any word come up that just seems to fit what you're feeling, what we have here today, in the room?"

Conceptual interpretive comments along trauma themes are also made at a group (as well as individual) level, e.g., FACILITATOR: "So, today, we've been talking about how disappointed many of you are in the way your families are responding to your grief. It seems as though the group is dealing with that sense of disconnection, or alienation. Does it feel like that to you?"

Skill development and optional additional structure. PCGT is flexible enough to incorporate structured skills-building material; however, the use of this material is adapted to a discussion format, and is used to further group process, or to enhance the comfort level of the group. Demand on members is typically limited, with no testing for mastery of material. Homework is generally absent or minimal, although members may be encouraged to practice skills outside the group and report back. Some skill areas such as interpersonal relating, self-expression, increased tolerance of the higher end of the midrange emotions, goal setting, or problem solving, can be integrated into group discussion relatively flexibly and informally. Other skill areas such as assertiveness/communication training, anger management, affect management, focusing, grounding, and thought stopping, may initially require more introduction and/or training, and so may be designated as the main topic of a particular session. Additional psychoeducation about PTSD (e.g., psychophysiology of PTSD) may be included as well.

CHALLENGES IN PCGT

Trauma-Based Group Dynamics

Although trauma-related symptoms and attitudes may respond well to group interventions, they may also generate specific patterns of group behavior that can interfere with or even derail the group process. Members may be motivated to avoid therapeutic change because it means giving up responses initiated in the interest of survival, under conditions of threat and danger. Members may also be motivated to avoid acknowledging vulnerability, disability (or any degree of impairment in functioning), dependence on therapy, or a subjective sense of being "crazy." Groups of trauma survivors with PTSD may tend to develop a premature, nontherapeutic cohesion (Parson, 1985, p. 13) that can lead to "therapeutic impasses, intensification of symptoms, acting-out, and even the demise of the group." This premature cohesion represents a form of pseudomutuality that prevents differentiation and avoids true intimacy and sharing of feelings (and so, termed here, *pseudocohesion**). "Feelings of narcissistic endangerment" can set off "paranoid defenses in the group's social structure," leading to "general fragmentation-prone functioning," accompanying "loss of self-cohesion" (p. 16). This pattern is self-enclosed and self-reinforcing, as the pseudocohesive group environment generates increased reexperiencing, which in turn generates avoidance, numbing, and arousal that contribute to greater pseudocohesion. Members affirm one anothers' trauma-based emotional and cognitive schemas (e.g., loss of trust, authority issues), creating an alliance that ignores, demotes, or co-opts the facilitators. The group develops an "us-them" mentality that supports the status quo against change and discourages connection to the world outside the group. The result is loss of trust, immobilization of therapy within the group, and diminished sense of responsibility as group members merge.

Interventions to diffuse pseudocohesion include a cautious paradoxical approach, expressing the negative side of the ambivalence; respectful confrontation; and overt limit setting, with review of group purpose and guidelines (some of which are demonstrated in Case Example 8). Use of additional structure, whether psychoeducation or skills building, can be useful in reestablishing the facilitators' legitimate authority, as well as concretely setting a therapeutic course for the group.

Trauma Triggers in Group

If a trauma trigger occurs in the group or the treatment setting, the group may become preoccupied with reexperiencing, and overwhelmed with trauma-based affects. If such events persist or go unaddressed, members may become depressed and apathetic. They may attend irregularly, or drop out. Grounding techniques are helpful in responding immediately when a member or the entire group is triggered. In addressing a persistent pattern of triggers arising in group, facilitators offer psychoeducation about trauma reminders, and set limits regarding overdisclosure of trauma details. Occasional inadvertent shifts into discussion of the trauma can be processed in a present-centered fashion, focusing on current affective responses to the disclosure (see Case Example 1).

**Premature pseudocohesion* is a term that has been coined to describe the phenomenon of unhealthy, pseudomutual alliance among members that works against therapeutic change. It was derived from Parson's (1985) description of counterproductive, "premature cohesion" in therapy groups for Vietnam Veterans. The prefix "pseudo" has been added to indicate that this phenomenon does not represent true cohesion, but a more primitive merging that does not acknowledge individuality, and therefore does not allow reciprocity or mutuality in relationship (hence also use of the term, *pseudomutual*).

Present-Centered Focus

Maintaining present-centered focus can be a challenge in this group model. Members who are triggered may spontaneously start to disclose; facilitators, not wanting to invalidate the members' trauma experience, may feel uncomfortable redirecting them. It takes skill and practice to validate while redirecting. However, as the group focus is clear from the outset, consistent, gentle but firm redirection actually reassures members, and will result in greater trust in the group and facilitators.

Case Example 1: Maintaining Present-Centered Focus and Dealing with Triggers in Group

Goals include

- Identifying triggers.
- Refocusing on symptoms.
- Verbally reflecting body posture.
- Grounding (breathing, eye contact, contact with chair, olfactory stimulus).
- Interrupting.
- Redirecting to symptoms.
- Setting limits.
- Offering brief psychoeducation.
- Enlisting group.
- Turning group back to members.
- Focusing on coping with symptoms.

MEMBER 1: I've been having a hard time going to the parts of the building where it all went down . . .

MEMBER 2: Well, I tried to go back to the cafeteria to eat, because I've been eating out ever since the "incident"—you know, the cafeteria is where I had to barricade myself and my co-workers when the terrorists were taking over the building. We were listening to the shots, and then there was a huge rush of sound, and then moaning and crying, I assume, the wounded . . .[members appear uncomfortable, move around in their seats, Member 2 is talking loudly and intensely, appearing agitated].

FACILITATOR 1: [identifies "trigger" and begins to refocus on symptoms] Sam—you're telling us you were triggered after you went back to the cafeteria—

MEMBER 2: Yes—I went to the cafeteria, and the food was no good as usual, and I lost my appetite, and left, and I kept thinking about what happened that day, and—you know, I can still hear the screams, and the shots—

FACILITATOR 1: [directly intervening, using body reflection to engage member's attention] Sam—you seem to be getting into the trauma, and you look agitated—and the group members, do, too—[some members nod]. [Introducing grounding] See if you can ground yourself, and come back to the group—

FACILITATOR 2: Okay, let's see if anyone wants to do the grounding—[Facilitator 2 offers olfactory stimulus for grounding, members pass it around]

MEMBER 2: Sorry—I don't really want to go into it, it was just in my head . . .

MEMBER 1: It's okay with me if he wants to talk about it.

MEMBER 2: Well, it's just that it's been coming into my head ever since I tried to eat there—I was even looking around for terrorists. . .

FACILITATOR 1: [Interrupts directly, after several attempts to address the central issues of the discussion] Sam—I am going to interrupt again—can you take a deep breath [makes deliberate eye contact with him] and come back to the present, with us? You can feel the sides of the chair . . . look around the room, make eye contact [Member 2 does so]. [Redirecting to symptoms] Sam, how much have you been having thoughts about this? Is it coming back to you a lot?

MEMBER 2: Yes—just about every day . . . . It's hard to concentrate . . .

FACILITATOR 2: Does anyone else in the group have any suggestions for Sam on how to cope with intrusive memories and flashbacks?

[Members share coping techniques]

FACILITATOR 2: [Setting limits, facilitator returns to address group regarding the suggestion to "talk about it" for Member 2 to "talk about it"—offers brief psychoeducation and explanation] I want to get back to what Jane said, when she said it was okay with her if Sam talks about it—I appreciate your willingness, Jane, to listen to Sam. If it seemed as though we were cutting him off, we were—and it's not because we don't want to hear about it—it's because it seemed to us that Sam was getting triggered further by discussing it, and the group was, too [addresses members, enlisting group]. How was that for the rest of you, that we redirected Sam? [members express variety of opinions].

MEMBER 1: Well, I was getting kind of triggered, too, but I thought maybe he needed to get it off his chest.

FACILITATOR 2: The things that have happened to members during the attack are very important, and we don't mean to take away from it. But in this group, we have contracted to stick with the present, and if we diverge too much from our purpose, then it will be hard for you to know what to expect, and hard to trust us to keep things emotionally safe here.

FACILITATOR 1: But, if we thought Sam, or anyone else here, really needed to talk about it, and that that would be the most therapeutic thing for him, we would do something to make that happen—like encourage him to talk with his individual therapist about it, or arrange another group with a different focus. The fact is, in my opinion, Sam was talking about the trauma because he felt triggered, and was having symptoms, and he was looking pretty upset and distressed, and needed help getting back to the present.

MEMBER 2: You have a point—it's not that I love discussing it, believe me . . . [he reaches for the "grounding" scent]; I hadn't been thinking about it as much, but going to eat in the cafeteria brought it right back. Well, I think I might steer clear of the cafeteria for now, because I don't like how it affected me. I don't think I'll be going back there for a while.

FACILITATOR 1: Okay—you don't need to go beyond your comfort level— we can work here on how to deal with being triggered, so you can gradually work into it, if you want to [returns group to members, re: coping]. How are other members dealing with having to go into, or avoid, areas of the building because of the attack?

SELECTION CRITERIA

As in any trauma group, members are selected based on significant trauma history, ability to tolerate a group format, and cognitive ability to understand the purpose and requirements of the treatment. However, for PCGT, it is *not* necessary for prospective members to produce concrete and "workable" trauma memories.

General Inclusion Criteria

Members must be willing to work within the present-centered format, to receive feedback from other members and from group leaders, to discuss symptoms and affective experience, and to accept redirection away from the details of trauma. Symptoms may be acute or chronic. Members must also be able to *contract for safety*—affirm to facilitators or themselves that they will not act on suicidal or aggressive urges. They must be able to observe the group guidelines (e.g., attending group sober, maintaining respect for other members and group leaders, including no violence, no threats or name-calling, etc., and refraining from business transactions with other members). Members must also agree to maintain confidentiality of other group members, and to accept limits to confidentiality. Exhibits 15.3 and 15.4 indicate circumstances under which PCGT may be preferred over TFGT, and criteria for exclusion from PCGT.

Inclusion and Exclusion of Comorbid Disorders

This modality easily accommodates individuals with accompanying mood disorders, with the exception of acute mania. Episodes of substance abuse may be more recent than in other trauma-related therapies, but a period of sobriety of three months is preferred, and more advanced groups may require at least six months sobriety. This criterion may be modified if the group is intended for dual diagnosis (e.g., trauma and substance abuse), or if substance abuse treatment is available concurrently. Inclusion of individuals with comorbid psychotic disorders is not recommended, unless the entire group is adapted for individuals with psychosis.

EXHIBIT 15.3. Factors Favoring PCGT Inclusion

1. Member is new to therapy, or has had negative prior therapy experiences
2. Stability is an issue, e.g.,
   * Current unstable mood
   * Acute exacerbation of PTSD
   * History of bipolar disorder or significant depression
   * Difficulty managing affect
3. Member has completed trauma-focus work and needs subsequent support, or needs additional support during trauma-focus individual or group treatment
4. History of multiple prior traumatic experiences
5. Strong preference for current-day focus
6. Member has current or recent stressful circumstances, e.g.,
   * Acute or severe medical problem in self or significant others
   * Concrete resource issues such as housing issues, financial problems, legal stressors
   * Recent or expected loss, e.g., death or illness in family, divorce
   * Especially high risk or demanding situation at home or work
   * Member belongs to a high-trauma-risk profession (e.g., paramedic, police officer, firefighter)
   * Member is disabled by symptoms to the point of being unable to work

---

EXHIBIT 15.4. Exclusion Criteria for PCGT

1. Active substance abuse interfering with ability to attend to treatment (impairing concentration, diverting attention, preventing engagement in group)
2. Psychotic symptoms interfering with attention to group/interpersonal functioning
   - Acute paranoia or delusions
   - Pervasive hallucinatory experience
   - Mania
3. Factors interfering with benefit from group therapy
   - Prominent sociopathy
   - Cognitive impairment (serious head injury) preventing engagement in group environment
4. Inability to tolerate the views of others
5. Inability to refrain from aggressive gesturing or threats
6. Inability to accept limits, e.g.,
   - Inability to commit to "safety" (regarding homicidal or suicidal impulses, for outpatient groups)
   - Unwillingness to accept redirection to present-day issues

The Clinical Interview

The PCGT interview is a modified standard psychosocial examination that screens for inclusion/exclusion criteria, and also provides information on factors that might affect group role and anticipated adjustment. History-taking includes more recent experience of trauma (since the terrorist event), secondary trauma (distressing events occurring immediately after the trauma), and prior history of trauma and/or neglect. The extent of inquiry into the details of the trauma may be adjusted to the comfort level of the client. It is also important to assess for current or past mood swings, as lability may influence both attendance and performance. The interview covers risk issues that can affect treatment, such as danger to self or others, substance abuse, and history of violence and victimization. In addition, interviewers should ask about other issues that have implications for treatment, such as psychiatric and treatment history, structure of current family and family of origin, and parenting issues. The interview also covers work and educational history, and financial issues. Current and upcoming stressors should be identified, and goals and level of motivation should be assessed informally. Past therapy and nontherapy group experiences (e.g., work, family, school) that might affect adjustment to the current group are examined as well.

Assessment instruments include:

1. Impact of Events Scale—Revised (IESR; Weiss & Marmar, 1996), a paper-and-pencil measure that can be quickly administered and scored.
2. The Clinician Administered PTSD Scale (CAPS) (Blake et al., 1995). Used when a more extensive PTSD interview is required, with specific scoring rules for level of symptomatology to meet criteria, establishing:
   - exposure to a criterion A traumatic event and the requisite diagnostically related responses (e.g., terror, helplessness, horror);
   - intensity and frequency of the seventeen PTSD symptoms outlined in the DSM-IV; and
   - associated beliefs and emotions typically found in trauma survivors (e.g., survivor guilt).

Because interview-based and self-report instruments for PTSD usually have a high rate of concordance, many clinicians prefer to utilize a brief self-report tool as part of the documentation of the presence of PTSD and as a foundation for ongoing symptom assessment:

- The Los Angeles Symptom Checklist (LASC; King, King, Leskin, & Foy, 1995), offering severity scores for PTSD and depression
- The Trauma Symptom Inventory (TSI; Briere, Elliott, Harris, & Cotman, 1995)
- The Posttraumatic Stress Diagnostic Scale (PDS; Foa, Cashman, Jaycox, & Perry, 1997)

In addition, it can be useful to assess level of depression, using the Beck Depression Inventory (BDI; Beck, 1978).

COMPOSITION OF AND PREPARATION
FOR ENTRY INTO GROUP

Homogeneous and Heterogeneous Characteristics

Homogeneity is desirable for characteristics affecting ability to participate comparably in the group process. Members should be within close range regarding ability to interact verbally, respond cognitively and emotionally, and draw on personal resources. They should also be within range regarding level of emotional and interpersonal functioning (which may be related to severity/recency of the targeted trauma, previous trauma history, personality disorder or other significant comorbid disorder, or cognitive factors affecting ability to regulate and express emotion) and degree of symptomatology from the trauma. Extreme disparities in level of functioning can result in unhealthy comparisons among members, exacerbation of a sense of shame among both higher- and lower-functioning members, and stalemate in the group process. However, modest differences among members allow distinctions in personal identity to develop within the group, so that members may more clearly understand the impact of trauma, and distinguish symptoms from defining personal traits. Distinctness of members also allows greater acknowledgment of the healthy aspects of individuals in the group, as opposed to development of a trauma-based identity among members that can defy differentiation (e.g., "We are all alike. . . . Our trauma defines us").

Heterogeneity regarding personality variables, age, gender, sexual orientation, ethnic and religious background, vocational and educational histories, political beliefs, socioeconomic status, level of intelligence (from low average to very superior) is not only permissible but usually desirable. Members may also vary considerably regarding prior treatment experience, unless the group is provided within a "stage of treatment" model (e.g., the Judith Herman/Victims of Violence model, which suggests a progression from introductory, preparatory work before addressing trauma, and supportive work following trauma work (Herman, 1992). Motivation levels can be heterogeneous—more motivated members will model for others; however, a predominance of motivated members is preferable. Even higher levels of divergence are desirable on characteristics that allow members to benefit from diverse perspectives in the group, such as personal style, ethnicity, and religion. Such differences encourage flexibility and range in learning, coping, and problem solving.

For dual diagnosis groups, members should have comparable levels of the particular comorbid disorder, whether for substance abuse, major mental illness, or other disorder. For members who are court-referred, level of legal issue should be roughly equated. Generally, it is inadvisable to include court-referred and voluntary members within the same group. If groups are provided within a particular organization, members should be of comparable hierarchical status, and supervisor/supervisee pairs should not be in the same group. Placement of employees who work closely together may compromise the boundaries of both group treatment and work environments. If circumstances require that they be placed together, strategies for avoiding boundary violations should be discussed at the outset.

Preparation for Entry

Preparation begins during the intake interview, and is completed in an individual pregroup meeting with facilitators in advance of the member formally joining the group. The interview process involves a modicum of introductory psychoeducation about the impact of trauma, normalizing the experience of trauma survivors, and reducing the sense of shame surrounding symptoms and need for treatment. The interviewer (who may or may not be the group facilitator) presents the present-centered focus and rationale as the interview begins, and assesses the individual's receptiveness to this modality. If results of the interview and assessment indicate that PCGT is the treatment of choice for the client, the client is then referred for an individual pregroup meeting with one or both facilitators.

The individual pregroup meeting establishes the final decision for group entry, and orients members to the group. Facilitators explain the benefits and expectations of group therapy, review group guidelines, and discuss the roles and responsibilities of group members and facilitators. Confidentiality and limits to confidentiality are also discussed in this meeting (as well as in the first group session), and members are expected to make a commitment to maintaining the confidentiality of other group members. Policies concerning attendance and excused absences are established. Facilitators check in with members about current stressors that may affect participation in group. If members are interested in participating, and accept the requirements of the group, they are formally admitted into the group.

Characteristics of the Group

PCGT is flexible regarding many aspects of its format and application. Weekly meetings are standard; however, in an intensive program, groups may meet several times per week or even daily. As aftercare to more intensive treatment, groups may meet every two weeks or monthly. If the group's purpose is to provide introduction to treatment or initial stabilization, a short-term model of six to sixteen sessions may be offered. As the major treatment modality, or when utilized for posttrauma work support, PCGT may run from six months to more than five years. This kind of group may operate on a cohort model, or may open admissions periodically (in which case, psychoeducation may be repeated). Session time is uniform each week, but may be scheduled from sixty to ninety minutes. The example that follows uses sixteen sessions followed by four tapering sessions (two sessions every other week, then two monthly sessions). Each session runs ninety minutes.

The format of each session generally proceeds as follows, depending on duration of the group:

1. Opening the group (five to ten minutes): includes welcoming members, announcements, and option of "grounding" exercise.
2. Brief check-in (five to ten minutes): deals with reactions to prior group and any issues interfering with group participation (note: check-in is *not* a report on the week from each group member, which would slow the group process).
3. Agenda setting (five to ten minutes) may be done informally, or on a board or easel.
4. Group discussion/processing (fifty to sixty-five minutes) informally addresses issues raised by members and, sometimes, by facilitators (including offer of psychoeducation on an "as-needed" basis); discussion is interactive, with members giving and receiving feedback.
5. Review of group (ten minutes) includes summary and feedback regarding session, plan for self-care, preparation for making the transition from group to life outside group, and any plans for upcoming sessions.

Phases of Group

In PCGT, content varies both according to stage of group and to the specific issues presented by group members. Psychoeducation is systematically presented only in the first few sessions (although it may be added in other sessions as needed, per facilitators' determination of need based on problems arising in the group, and/or group member feedback). The group process is consistent with the normal stages of group development; what is specific to this approach is that orientation to the present develops more fully as the group evolves. Phases proceed as follows:

Phase 1: *Group Forming and Norms:* Introducing the present
Phase 2: *Cohesion and Mutuality:* Connecting to the present in group relationships
Phase 3: *Differentiation and Authenticity:* "Regrouping" to incorporate new information about self, others, and the world
Phase 4: *Consolidation and Termination:* Time frame and timelessness in the group's perspective

Group development through these phases can be tracked by noting progress along the four major trauma-related themes listed in Exhibit 15.1. Juxtaposed against a trauma-based worldview that tends toward isolation, shame, threat, helplessness, and loss of meaning, the present-centered orientation encourages authentic connection to others, a realistic sense of safety, and a sense of empowerment. The perspective of this group also affirms life posttrauma in ways that create new meanings and values.

These themes build on one another; a sense of safety allows members to explore connection, while connection inspires a sense of empowerment. In turn, increased connection and empowerment help restore meaning to members' lives. Facilitators may either explicitly or implicitly reference these themes in helping group members develop more freedom from trauma-based symptoms, behaviors, and attitudes. These themes provide facilitators a meta-level framework for understanding both group dynamics and members' individual issues, independent of whether members specifically identify them in their goals. Tables 15.1 and 15.2 show the likely relationships of these themes to symptoms.

- Reexperiencing compromises a sense of safety, generates a sense of helplessness, and challenges accepted meaning of life experiences.
- Numbing and avoidance limit the sense of connection and self-expression.
- Hyperarousal interacts with a sense of safety by activating a sense of danger or risk.
- Working on goals such as a member looking for a new job despite trauma-based fears tends to activate issues of safety, empowerment, connection, and sense of purpose and meaning.
- Working to improve family relationships activates issues of connection and meaning.
- Work toward establishing concrete resources such as housing activates safety and empowerment issues.

Gradual Change in Behavior and Attitude

In terms of readiness for change, members will vary along the dimensions identified by Prochaska and DiClemente (1984): precontemplation,

TABLE 15.1. Relationship of Trauma-Based Themes to PTSD Symptoms

Symptom/Theme	Safety	Empowerment	Connection/Self-Expression	Meaning
Reexperiencing	X	X		X
Numbing/Avoidance			X	
Hyperarousal	X			

TABLE 15.2. Relationship of Trauma-Based Themes to Goal Areas

Goal Area/Theme	Safety	Empowerment	Connection/Self-Expression	Meaning
Vocational Issues	X	X	X	X
Relational/Family Issues			X	X
Concrete Resources	X	X		

contemplation, planning/action, and maintenance. A group member may gain simply by acknowledging a symptom he or she has been keeping to himself or herself, while another may take concrete action, such as overcoming a fear of returning to work by sending out resumes. Members tend to cycle through a variety of readiness stages across phases, although Phase I tends to be associated with contemplation-level readiness for change, and members are most likely to make more active changes in the third and fourth phases of group. Level of readiness for change may also vary by issue (and its related trauma theme), even within one individual. For example, a member may work on becoming able to leave the house after dark again (theme of safety; planning/action stage), yet be unable to acknowledge a sense of numbing (theme of connection; precontemplation stage). Facilitators should start with members' current levels of readiness when it comes to goals, supporting successive approximations toward the next level of readiness. Group discussion will tend to "cross-pollinate" themes and stages of change, as members model movement in areas that other members had not yet considered. Table 15.3 shows the trauma themes as they tend to interact with levels of readiness for change:

PHASES OF PCGT

Phase I: Group Forming and Norms: Introducing the Present

Objectives

- Introduce members.
- Establish group norms and expectations.
- Begin to explore affects.
- Encourage recognition and acceptance of differences among members.
- Identify current and upcoming stressors in members' lives.

- Encourage movement from *Readiness for Change Stage* of:
  —precontemplation to contemplation, through psychoeducation,
  —contemplation to planning/action, through goal setting.
- Through psychoeducation, create common language and shared information to
  —demystify trauma,
  —develop group bond through shared understanding,
  —identify characteristics of trauma-based worldview.

TABLE 15.3. Trauma-Based Themes As They Interact with Stages of Readiness for Change

Stage of Change/ Theme	Safety	Empowerment	Connection/Self-Expression	Meaning
Precon-templa-tion	Over- or underprotective of self; overprotective re: others; defensive in group	Immobilized; has diminished awareness of own impact on others and in tasks	Disconnected from others, self; little self-expression; alexithymic	Losing hope, lacking sense of meaning; remaining at a superficial level
Contem-plation	Increased sense of safety in group; "downshift" in over/underprotective stance, defensiveness	Increased sense of effectiveness in group; some awareness of own impact on others and in tasks	Increased empathy, tentative self-expression in group; improved perspective taking, reduced alexithymia	Increased hopefulness without specific focus; able to find purpose/ meaning in group interaction
Planning/ Action	Able to identify "safety zones" in life outside group; taking appropriate risks; invested in good self-care	Sense of empowerment established; taking action in goal areas	Expressive and involved in and outside group; improved interpersonal relationships, further decreased alexithymia	Able to identify sense of purpose in activity in and outside of group; invested in spiritual and/or community activity or meaningful work
Mainte-nance	Sustaining good self-care, expanding "safety zones"	Sustaining/expanding goal-related activity	Sustaining/expanding relationships, self-expression, diminishing alexithymia	New values and beliefs established; sustaining/expanding community and/or spiritual activity or meaningful work

Themes

- *Safety:* Introductions are conducted so as to increase trust and a sense of safety in the group. All members share something about themselves, and all acknowledge (either verbally or through their presence in group) that they have been subjected to trauma. Facilitators must be active in directing the group process, establishing a trusting relationship with group members through direct communication. They help the group establish a *trauma barometer,* noting trauma-based reactions both in the group and in members' stories about their lives, and help members engage in problem solving about how to create a greater sense of safety.
- *Connection, empowerment, and meaning:* Initially, psychoeducation helps members join with one another over symptoms and trauma themes, enhancing members' sense of connection. It also provides a framework from which they may begin to make sense of their inner experiences, encouraging both empowerment and reconstruction of meaning. Formulation of coping plans at the end of early sessions also helps members develop an empowering expectation that symptoms can become manageable. As the group continues, facilitators should regularly elicit members' feeling responses, supporting development of a shared language around affective experience that helps counter alexithymia. When members start to communicate more effectively, they are able to further establish connection to one another. Facilitators encourage a healthy mutuality, i.e., a relationship among members that acknowledges, respects, and appreciates differences as well as similarities, thereby countering any tendency toward premature pseudocohesion. Any tendency for members to overidentify with one anothers' unhealthy behaviors is diffused. Facilitators note any shame-based interactions among members, and must intervene quickly in any hostile or shaming interchanges that could compromise connection and increase the likelihood of early termination. These efforts support a healthy cohesion and foster the therapeutic alliance. As members begin to affirm the therapeutic stance that trauma-based feelings and behaviors can be addressed and transformed, the sense of empowerment in the group develops further.

Sessions 1 Through 5

Session 1: Members introduce themselves, and may be asked to provide some information about their current lives and what they hope to gain from the group. Rationale for the group is given, focusing on the need to stay in the present. As members have already been interviewed and have accepted the present-centered focus, members are encouraged to help one another refocus if the details of the trauma are raised in the group. Facilitators may ask members to briefly describe the nature of their trauma, without specific details; or reference to the trauma may be omitted entirely, if members already understand that they have a specific type of shared trauma or have shared the same traumatic event. They review and discuss group guidelines, usually distributing a handout for ongoing reference, and outline the group format. They also offer brief psychoeducation about trauma and PCGT, using this information to foster group discussion. The purpose of this session is to establish the therapeutic tone of the group, and to help members begin to bond as a group and as individuals who have endured shared or related traumatic experiences. During check-out, members are asked to identify some simple plans for coping during the week, as well as ways to transition from group to life "outside" the group as they leave.

Structured Content and Materials

Introductory exercise. A well-worn and simple group-forming exercise can help foster a sense of connection and support the present-centered focus as members introduce themselves. Members may be asked to say something about a "'favorite" in their current life, such as a favorite food, movie, or activity. This exercise serves to remind members that they are not their trauma, and that life after trauma is possible.

Handouts of group guidelines and confidentiality standards. These materials set expectations for maintaining the group's therapeutic stance, and include establishment of physical and emotional safety, a commitment to confidentiality, an understanding of limits to confidentiality, and a shared sense of responsibility for the group environment.

Grounding exercises. Grounding involves focused contact with the sensory world in a way that "changes the channel" from trauma memories to present-day stimuli, as a means of maintaining emotional safety. For example, use of olfactory stimuli as grounding activates the most immediate and least verbally mediated sensory modality, competing with the sensory memories represented in reexperiencing. Facilitators may offer one or more bottled scents (usually oils). Use a scent that is not in any way reminiscent of the trauma. Members may also be encouraged to make eye contact, feel the sides of the chair, plant their feet firmly on the ground, or focus on an object in the room and describe it. All these measures offer antidotes for reexperiencing symptoms in the moment, and may be introduced when members show signs of intrusive memories, flashbacks, or dissociation (as signaled by a member looking off, appearing agitated, suddenly talking about the trauma, becoming unusually withdrawn, or moving suddenly without apparent purpose). Facilitators may use grounding as needed in sessions, or may incorporate grounding rituals at the start and/or end of sessions, beginning in session 1. As a starting ritual, grounding can help reduce the triggering potential of the group itself, given the trauma-related purpose of the group and the experience of being with other survivors.

Optional structure; coping lists. A simple and useful ritual for group check-out asks members to report what they are going to do immediately after they leave group, as a signal to transition into the rest of the day. This exercise sets the expectation that members will develop their own rituals for reentering the rest of life after the group session is over. It helps to contain the intense emotion that may be generated in group, and encourages members to attune to their own lives once they leave the group.

Sessions 2 and 3: Facilitators provide more extensive psychoeducation about trauma, symptoms of PTSD, common comorbid disorders such as panic, depression, or substance abuse, trauma-based attitudes and beliefs, and cognitive and emotional "mindset" adopted in response to trauma. Psychoeducation varies between formal and informal, with emphasis on discussion. The content in these sessions validates members' experiences with trauma and the resulting symptoms, and provides a sense of common experience among the members, to facilitate trust and reduce shame. These sessions also model discussion of inner experience, thoughts, and feelings, to begin addressing the alexithymia associated with PTSD.

Materials and Structure to Support Discussion

- *Session 2:* Handouts of the DSM-IV criteria for PTSD and associated disorders (See Appendix)
- *Session 3:*
  —Handouts of common issues that survivors face (Exhibit 15.5)
  —Trauma themes and cognitive and emotional schemas affected by trauma (McCann & Pearlman, 1990; see Exhibit 15.6) are used to stimulate their own responses on these dimensions, adding to Session 3 materials. They can also be used in Session 4, as an aid to goal setting.
  —Phenomenology of trauma and recovery (effects of a shattering experience on image of self and world), useful for Session 3, or as additional psychoeducation, as needed, in later sessions (see Figure 15.1)

**EXHIBIT 15.5. For Session 3: Common Issues
That Survivors Face**

- Sense of isolation and alienation
- Feelings of guilt—e.g., survivor guilt, self-blame over trauma—inconsistent with actual facts of responsibility for the event
- Guilt or shame about changes in functioning, e.g., in families, at work, socially, at home
- Emotional numbing; varying between lack of emotion and extreme emotions
- Sudden or persistent fear in the absence of immediate threat
- Increased anger, and increased risk of acting on anger, out of proportion to immediate situation
- Intensified concern about physical safety of self and loved ones
- Interference with capacity for close interpersonal relationships
- Difficulty with sexual intimacy
- Feelings of lack of empowerment and lack of control
- Feelings of worthlessness, low self-esteem
- Loss of spirituality, sense of meaning in life
- Sense of loss—not only regarding people, or concrete things you depended on—but regarding your own dreams, innocence, goals for future, and sense of the world as you knew it

(Adapted from Shea, Wattenberg, & Londa-Jacobs, 1997)

Session 3 or 4: Optional; information on the psychophysiology of trauma may be culled from texts on trauma as a useful supplement, such as fight-or-flight response, stress response, or biochemistry of trauma.

Sessions 4 and 5: Facilitators complete psychoeducation and assist members with setting realistic goals. It is understood that goals set at this early phase may evolve or change as the group continues, and will need to be revisited periodically. For some members, goal achievement may occur in a linear, step-by-step fashion. However, for others, progress may occur in a nonlinear, apparently haphazard manner, e.g., making no progress at all toward the chosen goal of finding a job, but progressing unexpectedly in relationships with friends and family. The purpose of these sessions is to establish a working atmosphere in which the group supports the progress of each member. Goal setting also addresses the sense of helplessness associated with trauma, asserting that change is possible, and that members can be instrumental in their own lives.

EXHBIT 15.6. Trauma Themes and Cognitive Schemas

Trauma Themes, Session 3

Threat versus safety (including trust issues)
Disconnection and shame versus connection and self-expres-
sion (including alexithymia versus ability to express feelings)
Helplessness versus empowerment
Loss of meaning versus meaning making

Effects of Disruption of Cognitive Schemas in PTSD (adapted
from McCann & Pearlman, 1990)

Loss of meaning
Loss of community (isolation)
Sense of helplessness
Loss of trust
Authority issues
Control issues/loss of control
Loss of self-esteem
Fear of intimacy
Fear of vulnerability
Sense of danger (safety issues)

Structured Content

Goal Setting: Members are assisted in setting small, attainable goals that
they can review and continue to evaluate and refine throughout the group. This
task may be approached informally, or it may be structured through recording
specific steps toward goal achievement, projecting a time frame for these goal-
related steps, and reviewing progress at predetermined intervals. Early on,
goals may reflect the contemplation stage; later, members may move toward
greater action on plans for change. This process offers continuity to the group,
and also provides a record of how goals may evolve over time. Goal review is
recommended at sessions 10 and 11, and sessions 15 and 16.

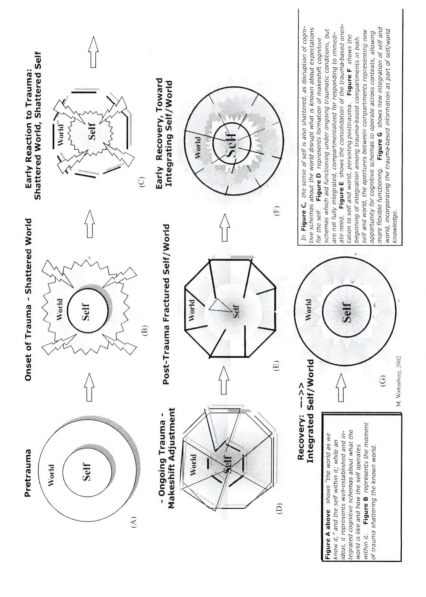

Pretrauma

(A)

Onset of Trauma – Shattered World

(B)

Early Reaction to Trauma: Shattered World, Shattered Self

(C)

– Ongoing Trauma – Makeshift Adjustment

(D)

Post-Trauma Fractured Self/World

(E)

Early Recovery, Toward Integrating Self/World

(F)

Recovery: ––>> Integrated Self/World

(G)

M. Wattenberg, 2002

Figure A above shows "the world as we know it," and the self within it; while an ideal, it represents well-established and integrated cognitive schemas about what the world is like and how the self operates within it. **Figure B** represents the moment of trauma shattering the known world.

In **Figure C,** the sense of self is also shattered, as disruption of cognitive schemas about the world disrupt what is known about expectations for the self. **Figure D** represents formation of makeshift cognitive schemas which aid functioning under ongoing traumatic conditions, but are not fully integrated, compartmentalized for responding to immediate need. **Figure E** shows the consolidation of the trauma-based orientation to self and world, persisting posttrauma. **Figure F** shows the beginning of integration among trauma-based compartments in both self and world, the apertures between compartments representing new opportunity for cognitive schemas to operate across contexts, allowing more flexible functioning. **Figure G** shows new integration of self and world, incorporating the trauma-based information as part of self/world knowledge.

FIGURE 15.1. Self-World Diagram for Trauma and Recovery

535

Phase II: Cohesion and Mutuality: Connecting to the Present in Group Relationships

Objectives

- Encourage healthy cohesion, including differentiation among members.
- Encourage movement toward contemplation of significant issues previously discussed in precontemplation stage.
- Encourage movement toward planning/action for issues previously discussed in contemplation stage.
- Support changes in behavior, attitude, and coping within the group process.
- Support shifts in trauma-based worldview.
- Address new stress events as they occur in members' lives.

Themes

- *Connection and Meaning:* As Phase I comes to an end and the second phase begins, the shared problem solving and supportive differentiation in the group process allow true cohesion to occur. The group evolves a shared set of values, including a distinctive shared sense of humor, and a shared language and style for addressing inner experiences and feelings. These developments help members manage and transcend the initial shame over having symptoms and having been traumatized. Reduced interference from shame allows greater spontaneity and authenticity. Members establish a broader and deeper self-perception, as distinguished from a predominantly trauma-based concept of self, and are freed up to work more actively toward therapeutic change. Connecting interpersonally and integrating new perspectives on the trauma's effect on their lives, members establish a foundation for reconstructing meaning in their lives.
- *Empowerment:* As the sense of shame diminishes, effective group functioning leads to a sense of empowerment. In the process, members witness and validate one anothers' strengths. Facilitators shift toward a less active stance, although they comment actively when trauma-related patterns emerge that could intrude in the group process. Members work more independently toward their goals, refining them as they go along. They take responsibility in the group process, engaging one another actively, with little prompting. They make adjustments in their behavior within the group in a way that is consistent

with their goals, and are increasingly able to confront one another in a respectful, supportive manner. Often, new stressors emerge, and members have the opportunity to address them in new ways that result in an increased sense of empowerment; at the same time, it is important to identify possible triggers in these new stressors, and to address potential trauma-based reactions such as fear and helplessness. Group feedback concerning these stressors begins to establish the group itself as a source of empowerment for members.

- *Safety:* Emotional safety progresses in this phase, as well. However, as members allow themselves to be more vulnerable with one another, there is potential for a new level of conflict, which can in turn trigger trauma-related distress. Facilitators need to intervene strongly if the group shows signs of destabilizing in response to such interpersonal demands, in order to prevent the group from becoming polarized, and to help members refrain from reenactment of the trauma (which may result in secondary trauma and/or premature termination).

Sessions 6 Through 9

Session 6: As the less formally structured work of the group begins, content typically centers on current coping and problem solving around hard-to-manage situations, affects, and trauma-related symptoms.

- Facilitators review the session format.
- Issues remaining from the previous group are reviewed.
- Facilitators ask for brief summary statements ("headlines") to identify issues to put on the agenda for the session's group discussion.
- Facilitators ask the group to prioritize the topics for discussion.
- Discussion is opened.
- Discussion typically lasts forty-five minutes to an hour. The facilitators reserve approximately ten minutes prior to the end of the session for "check-out" (group review), during which members address any strong reactions engendered by the group, acknowledge and consolidate what they have gained from that group, and generate coping plans.

Sessions 7 through 9: Facilitators actively support the shift to the less-structured discussion format by encouraging feedback among members. They offer additional psychoeducation or informal skills building as needed. Discussion often leads to informal problem solving around life issues and management of symptoms. Group support is key in "brainstorming" a variety of options for members whose trauma-based reactions may imbue already difficult situations with an intensified sense of helplessness. When members feel stymied, anxious, and stuck, group brainstorming reminds them that they are not as limited in their options as they were during the trauma. Facilitators guide the group problem solving to help members separate current problems from the past trauma. Although the discussion and interactive group process become the focus of the group, facilitators may also offer formal skills building, as needed, in these or later groups, for example,

- Stress management/relaxation.
- Thought-stopping, through use of an image or external stimulus to interrupt intrusions or negative thoughts.
- Focusing.
- Problem solving using a more structured form of brainstorming, followed by evaluation of pros and cons; assigning weights or rank order to options; selecting and applying a strategy, and reporting back to the group on its effectiveness.

Phase III: Differentiation and Authenticity: "Re-Grouping"—New Information About Self, Others, World

Objectives

- Review and refine goals.
- Encourage generalization of behavior change to life outside the group.
- Focus on maintenance of gains made, contemplation of new goals, and moving into planning/action stage regarding goals previously contemplated.
- Support more integrated worldview, incorporating new information.
- Encourage continued tolerance of differences among members.
- Review recent stress events and current coping.
- *Optional:* Additional skills building (as previously described).

Themes

- *Empowerment and meaning:* As the group advances, it begins to function as a fluid organism, and becomes a source of empowerment and meaning in itself. Change may occur more rapidly during this phase, as more forward-moving members demonstrate that they are revising their trauma-based worldviews, checking out possibilities and options, incorporating new information, and discarding outmoded notions. Members work to accept the range of their experiences, and establish paths of recovery that take this range into account. They work more effectively with current stressors, with group support. This process increases a sense of empowerment for the group as a whole, as well as for individual members.

- *Safety:* Members who are progressing more slowly may feel some pressure at this point. New challenges emerge, as members may become more competitive and confrontive. Facilitators need to support assertive rather than aggressive expression of disagreement, and model agreeing to disagree, to help members acknowledge and accept their differences.
- *Connection:* The group shifts and "regroups" based on a new commonality—the integration of new, nontrauma-related information into a worldview that also incorporates the shared trauma. At this stage of group, members tend to report affective experience more readily, and with greater articulation. In this phase, facilitators will frequently be able to defer to the activity of group members, yet must intervene if members become polarized around more subtle representations of shame and rigidity.

Sessions 10 Through 13

Sessions 10 and 11: By this time, members have most likely had to field difficult situations that challenge their developing skills, and have made adjustments in their coping strategies. Discussion tends to focus on feedback among members about effective ways to make such adjustments.

Structured Content—Goals Review

During these sessions, members formally review the goals they set in sessions 4 and 5, refining and setting new goals as informed by the challenges the members have encountered. Those who have made active changes in their behavior within the group are encouraged to carry those changes into their personal lives. Members who have moved more slowly are encouraged to try out new behaviors in group.

Sessions 12 and 13: In these sessions, members begin to put the trauma into perspective, as commitment to current life is reaffirmed. They develop a greater awareness of trauma-related reactions, and respond differently to triggers; for example, they may plan to deliberately honor trauma "anniversaries," instead of becoming immobilized by them. Group members, as they become more secure in their sense of individuality as well as in their bond with one another, are well-positioned to accept one another more fully and to appreciate their differences. If resentments over differences linger for some members, facilitators should work to restore perspective taking, identifying and validating the divergent points of view expressed by members, and setting the expectation that members can "agree to disagree." Formal skills building may be utilized to enhance communication and to slow down any "rush to judgment" that interferes with this process of healthy differentiation in the group (see as follows). In this phase, members' increasing recognition and acceptance of their own symptoms, thoughts, and feelings will tend to enhance their relationships with family and friends, and may lead to improved relationships at work as well. Facilitators should help members acknowledge gains, identify stress events occurring since group began, address relapses, encourage reinvestment in coping strategies, and support learning of further coping strategies.

Options for more formal skills building include assertiveness training and anger management.

Phase IV: Consolidation and Termination: Time Perspective and Timelessness in the Group's Perspective

Objectives

- Terminate effectively.
  - —Address both current and trauma-based affects raised by termination
  - —Develop coherent narrative of group's history
  - —Acknowledge attachments and relationships in the group
  - —Conduct exchange of feedback among members, and between members and facilitators
  - —Acknowledge successful coping with stressors since group started
  - —Establish termination ritual that acknowledges the significance of the group experience
- Review goals, consider further goals.
- Continue putting planned changes into action, and maintain changes already made.
- Continue work toward generalizing behavior change to life outside group.

- Review status of stress events that have occurred since the group started, and plans for continued coping with these or new stressors.
- Plan further treatment as needed.
- Consolidate changes in worldview based on more integrated incorporation of new information.

Themes

- *Meaning and empowerment:* As the group moves toward termination, members reflect on the meaning of termination, the group experience itself, and their accomplishments within it. The strengths of the group, achievements of individual members, and members' weathering of recent stress events, create a sense of empowerment that can generalize to life after group, as long as the group can process this shared history sufficiently. If the meaning of the group experience and the significance of members' participation in it are not sufficiently addressed, members may feel aimless, helpless, and inconsequential, as they are likely to have felt during the trauma.

- *Meaning and safety:* If termination is not processed adequately, another risk is that the loss of the group will leave members feeling abandoned and insecure—at the whim of a capricious world—as many have felt during trauma. Therefore, it is essential that facilitators help members consolidate and honor the meaningful aspects of the group experience, and the sense of security that they can carry into the future as a result of the sense of safety and trust derived from the group experience.

- *Meaning and connection:* Trauma-based loss of meaning tends to imbue endings with a sense of aimless disconnection. To establish the group as meaningful beyond its ending, facilitators should encourage processing of the relationships among members, as well as members' contributions to the unique character of the group itself. Development of a narrative about these relationships establishes the group as a time-limited yet timeless experience, one that has closure in the personal history of the members, yet lives on through its unique meaning in members' lives. The work of termination offers potential for a corrective emotional experience through its contrast with the traumatic experience, offering a sense of completion and transcendence that were lacking in the circumstances surrounding the trauma and its aftermath. An ending that is meaningful and connected offers hope to trauma survivors that their whole lives need not be dominated by trauma.

Sessions 14 Through Tapering

Session 14: Members begin to reflect on the group's history, and the experience of forming a group together. Facilitators should congratulate members for both concrete progress on goals, and general changes in expressiveness, openness, and spontaneity. They should encourage members to exchange feedback about changes they have seen in one another, and about the experiences they have had together in the group, including events in the group's development (e.g., "What do you think you will remember most when you look back on this experience?"). Members review progress and areas for continued work, putting the group experience into perspective as part of their own life (and treatment) history. They discuss continued application of gains to life outside the group. Facilitators encourage discussion of feelings about the group coming to a close, the sense of loss about the ending, feelings about relationships established within the group, goals fulfilled and unfulfilled, and upcoming plans. If members avoid the subject of termination, facilitators may need to introduce it into the session's agenda, modeling continued engagement and appropriate expression of feelings about it, and helping members identify trauma themes that may emerge in this context, such as

- feeling inconsequential (as they may have during the trauma);
- feeling helpless;
- experiencing intensified sense of loss due to losses during the trauma;
- feeling betrayed or abandoned;
- "numbing" to avoid emotionally painful experience; and
- needing to "take preemptive action," for example, terminating before the end of the group.

As facilitators help members remain engaged in the process of termination despite trauma-related reactions, members have an opportunity to experience the ending of the group in a healthy way that contrasts with endings that occurred due to the trauma. Facilitators may need to underscore this contrast between "then and now," pointing out that the ending is not happening chaotically or in crisis, and that members have an opportunity to be fully present as they say good-bye to one another, rather than shutting down as they are likely to have done during the trauma. Although interpreting group members' trauma-related reactions is important in this session, it is equally important for facilitators to offset any sense of urgency or abandonment by actively helping members plan for life after the group, and assisting with any need for follow-up treatment.

Sessions 15 and 16: Facilitators continue to foster discussion of the group's ending and the transition to tapering, as members refine future goals and solidify any plans for future treatment. They also reinforce coping strategies used between sessions, as members may feel more tenuous about their coping as they anticipate the loss of group support. As members further explore the meaning of their good-byes, facilitators credit members for their ability to tolerate and express strong emotion. Encouraging members to observe their own engagement in the termination process creates another level of distinction between losses sustained under overwhelming traumatic conditions and this new ending. It also allows the group to recognize the progress of individual members and the group as a whole. As they acknowledge the loss of the group during these sessions, members may also indirectly process the losses sustained during the trauma. Whether the loss is to death of others, physical injury, or personal functioning, facilitators support this grief work while maintaining the group's orientation, emphasizing the emotional impact of the losses, the struggle to honor them in a meaningful way, and the search for closure.

Structured Content—Goals Review

Members formally review progress on goals to-date, and formulate how to carry their goals into the next phase of their lives. They are likely to be at the action or maintenance stage of behavior change at this point. Further goals work can be anchored in plans for further treatment, and concrete plans for maintenance of recent gains.

Formal Good-Bye Ritual (optional)

During theses sessions, and during tapering, facilitators may formalize part of the termination by going around the room asking members to state what they have gained from the group, what they had wished for but didn't get, what they wish for others, and/or what they continue to want for themselves in the future. One or two of these questions may be chosen in one session. One variation of this exercise involves passing around an object (anything small, pleasant, and not associated with the trauma) and speaking while holding the object. This variation emphasizes the bond among members through the act of passing one to another, and enhances meaning through giving ceremonial "weight" to members' feedback. This ritual can be incorporated during the discussion section of the group, or may be used as an extended check-out, and can run from ten to thirty minutes, as desired.

Tapering Sessions 1 through 4: The first two of these sessions occur at two-week intervals following the end of the weekly group. The third and fourth meet at one-month intervals. Tapering sessions tend to have the quality of a reunion. Members have a chance to recognize that they miss one another, and bring one another up-to-date on their life events. The group narrative continues to develop during these sessions. Goals are supported in informal discussion, and members continue to receive support for their plans, including problem solving about potential obstacles and the need to manage any periods of exacerbation in symptoms. Small exacerbations are normalized as part of the transition to ending, and members are encouraged to rely on the coping strategies they have developed, as well as to develop additional strategies. Feelings about the termination are processed further, and final good-byes are said.

Structured discussion: As during sessions 15 and 16, members may be formally asked to share

- changes they have seen in themselves and in one another;
- feedback for the group as a whole, for other members, and for the facilitators; and
- goals and hopes for the future.

Structured termination rituals: Groups may honor the termination within the final tapering session, with any or a combination of

- a formal celebration, e.g., potluck party in the group;
- certificates of completion in the final session, as a closure ritual; and
- structured good-bye ritual, as described under sessions 15 and 16.

CLINICAL INTERVENTIONS

Clinical Management of Trauma-Based Affects and States

Trauma can have a powerful impact on subsequent perception of emotion. Survivors' inner experiences tend toward high-level arousal states such as terror, panic, and rage, and disconcerting absence of affect, as represented by numbing, dissociation, or *freezing*. Reexperiencing brings up strong affective responses that are more related to the memory of traumatic events than to present-day experience. In reexperience, midrange affects in response to everyday life, such as hurt, anger, sadness, happiness, uneasiness, and even milder levels of fear, are diminished or simply dismissed. The result is disconnection from information about the current environment that is normally afforded by affective nuances, and the domination of experience by affects associated with trauma. Alexithymia often develops, most likely secondary to the shutting down of higher cognitive processes due to hyperarousal. Trauma survivors then continue to live in a world dominated by trauma, despite absence of immediate threat.

Responding Therapeutically to Alexithymia

Although PCGT validates the impact of trauma in the current lives of survivors, it focuses away from the details of the terrorism experience in order to diffuse extreme affects related to hyperarousal. Its aim is to help members experience and explore midrange affects that tend to shut down in the aftermath of trauma. PCGT interventions counter trauma-based reactions, encouraging identification of and attention to nuances in emotion. This repeated attention promotes *successive approximations* toward broader emotional awareness, which serve to reduce alexithymia and enhance reconnection to the current environment. Members are encouraged to perceive the feelings in others, as well, beginning with a sense of identification with one another. Members may also be enlisted to help others identify affects that are hard to express (see Case Example 2).

Case Example 2: Interventions for Alexithymia

Goals include

- Making reflective statements to encourage expression of feelings.
- "Shaping" expression of emotion through successive approximations to feeling statements.
- Asking questions to elicit emotion accompanying a thought or behavior.
- Reflecting near-feeling statements.
- Returning to discussion of feelings when group deflects feelings.
- Using Gendlin's focusing technique.
- Using verbal and nonverbal body reflecting of identified "felt sense" in the body, to support the focusing process.
- Redirecting feelings outside the focusing area of the body to the focusing area between the throat and the base of the diaphragm.
- Verbally reflecting body posture.
- Physically mirroring body posture/movement (usually with accompanying verbal reflection of posture).
- Returning issues of individual member to group to process.
- Sharing of affect elicited from group members (using focusing).

MEMBER 6: Sometimes I don't want to come to group. . . . I don't know why. And when I get here, it's like I just want to get it over with.

FACILITATOR 1: [reflecting] So sometimes you really don't want to come in—and was today one of those days?

MEMBER 6: Actually, it was. I almost didn't get out of work in time to get here.

FACILITATOR 1: And right now you just want to get this over with.

MEMBER 6: Yes—I think so—something like that, anyway.

FACILITATOR: [Going for an approximation to a feeling] Okay—so let's pay attention to that. What's that "I want to get it over with" feeling? What's it like?

MEMBER 6: It's like . . . I don't know. It's not a good feeling. Can't put it into words, though. . .

FACILITATOR 1: Can anyone help Irene out?

MEMBER 2: Yes, I feel that way a lot of the time. Like I'm looking at the clock, wondering why it doesn't seem to move.

MEMBER 6: Yes, that's what it's like. It's like that a lot, at work, too, and at home. Ever since the attack—everything is different. I doubt it ever will be the way it was. Sometimes I am just going through the motions.

FACILITATOR 1: [asking for accompanying feeling] Everything is different—is there a feeling that goes with that?

MEMBER 5: I've been feeling that way lately—like I am just going through the motions.

MEMBER 2: Yep. That's it.

FACILITATOR 1: [reflecting a possible feeling] Going through the motions. What's that like?

MEMBER 6: It's just like—just EXISTING.

MEMBER 5: Yeah, it is like not really feeling anything. Blah.

FACILITATOR 1: [reflects] Okay, so a "blah" feeling, like not really feeling much of anything [asking for focusing of attention to locus of feeling in the body]. So let's try something—if we all look in this part of our bodies [indicates throat to above navel]—this is a technique called focusing, for examining feelings a little more—for looking at what that feels like in your body. If you look in here—from around your throat, to just above your navel . . . and just take a few breaths—where do you feel it? Just take your time, and—

MEMBER 5: [interrupting] Well, I think it's like, when I was at work last week, and someone asked me if I minded if he could take lunch early—and I said, 'Whatever!'—and that's all I felt like saying. . .

FACILITATOR 1: [interrupting in order to bring it back to feelings and focusing] Okay, Jane—that's another good example. Now, I would like all of you to try this, to see if it helps shed some light on what 'whatever!' feeling is all about. So I would like us to do some focusing right here—can you bear with me?

MEMBER 5: Okay, no problem.

FACILITATOR 1: So, right here—just take a moment, and don't force any words or feelings on it . . . and see what comes up when you focus on that feeling—'whatever'. . . like "just going through the motions" [pauses for about five to ten seconds]. Does anyone get a feeling, in your body?

MEMBER 2: Yes—I do—it's right here—right under my ribs.

FACILITATOR 1: [reflects] Under your ribs—okay. Anyone else?

MEMBER 5: I tried "whatever"—and it feels like a weight, in my chest.

FACILITATOR 1: A weight in your chest—okay. Now—can you take a moment, see what it feels like? [turns to Member 2; Member 2 nods]. Can you describe it?

MEMBER 2: It's like . . . a dull feeling. A dull empty feeling.

FACILITATOR 1: Okay—good—a dull empty feeling, right here [body reflection—Facilitator 1 indicates her own body just below her ribs]. Now take a moment, and can you just stay with that sensation below your ribs, and check back [indicates place in body] and see if that fits exactly—"a dull empty feeling."

MEMBER 2: [pauses] . . . Yes—dull and empty . . . and like an ache—but now—it's in my head, too—

FACILITATOR 1: Okay, it's moving to your head?

MEMBER 2: Yes—it feels like cloth or something, inside my head—

FACILITATOR 1: A feeling like cloth inside your head [brings it down into the "focusing" area of the body] can you bring that down into the center of your body?

MEMBER 2: I'm trying, but it's still in my head.

FACILITATOR 1: Okay. [Member 2 could not move the sensation from his head into the center of his body. Facilitator 1 asks for accompanying feelings in "focusing" area of body]. See if there is a feeling in the center of your body that goes with that feeling like cloth in your head.

MEMBER 2: [pauses] Yep . . . the feeling under my ribs is gone, and there's a feeling in my chest . . . like a heavy aching feeling right here—a heartache.

FACILITATOR 1: Heartache—right here [indicates her own heart area—nonverbal body reflection]—can you check back to see if that fits?

MEMBER 2: [nods] That's it—it's heartache—right here.

(Facilitator 1 moves on to other members [turning from the individual issue, to the group]; the focusing exercise assists members with alexithymia, and allows feelings of sadness and grief to be expressed, rather than absence of feeling.)

Therapeutic Responses to Reexperiencing and Hyperarousal

Members may react to either internal stimuli (such as hunger, tiredness, physical pain) or external stimuli (such as comments made in group, weather conditions, loud sounds, certain smells) with reexperiencing symptoms. Intrusive memories are likely to occur in the group in response to such triggers, leading to hyperarousal and its concomitant emotional distress and distractibility. Signs of reexperiencing may be relatively subtle, including discontinuation of eye contact, or intensification of the "thousand-yard gaze"; or they may be dramatic, accompanied by aggressive or fearful posturing, anxious foot tapping, loud talking, sudden movement, or panic attack. When arousal is so high that members cannot focus, or when a member is distracted by reexperiencing, facilitators must help to diffuse these experiences. As discussed earlier, body reflection (Prouty, 1994) can be useful in making contact with members who are registering hyperarousal or reexperiencing. Facilitators also model more relaxed or less aggressive body postures when intervening. These interventions (see details in Case Examples 3 through 5) help to reduce arousal and interrupt reexperiencing in affected members, as well as diffusing likely contagion of fight-or-flight response among members. Once identified, the intrusive memory and/or exacerbated arousal is linked to possible trauma triggers. Techniques for affect management may be discussed, including practice of relaxation techniques to help reduce arousal levels, generally. Subsequently, when members experience high-level arousal, the group assists in locating more subtle, antecedent affective states that may have led to the arousal, such as an uneasy feeling, a feeling of discomfort, feeling hurt, or feeling confused.

Case Example 3: Interventions for Hyperarousal

Goals include

- Using nonverbal cues such as gestures and body posture to reinforce therapy message.
- Verbally reflecting body posture.
- Physically mirroring body posture.

- Reflecting group nonverbal response to behavior.
- Associating members' nonverbal behavior with affective meaning.
- Interrupting potentially hostile exchanges.
- Modeling stepping back from strong trauma-based affects, using body posture.
- Reinforcing of interpersonal boundaries.
- Checking back with group for affective responses to hyperaroused member.
- Using relaxation exercise.

In an early-phase session, Member 1 begins to express frustration about his job, and escalates:

MEMBER 1: That's all I want to say about my job, because I don't plan to keep it for long.

MEMBER 2: Why, Ozzie? Your department runs well, and you just said you are up for a promotion—don't let all this get in your way.

MEMBER 1: What do you know about it? Runs well? Nothing runs well any more! I can't take that place! [sits forward in his seat]. To hell with our little jobs, I just want out! [slams his fist against the arm of his chair]. That's it! [Speaks loudly] I don't want to talk about it any more!! CASE CLOSED!!

FACILITATOR 2: [makes a "hold-it" gesture with hands] Ozzie—[verbal reflection of body language] you're raising your voice, sitting forward in your chair [Facilitator 2 moves forward in his chair in body reflection] and slamming your hand down [gestures "slamming" in slow-motion—nonverbal reflection of body language]. [Reflecting group nonverbal response to behavior, and associating it with affective meaning] and everyone in the group jumped; it felt intimidating—I don't know if you realized it.

MEMBER 1: [hesitates, but defensive] Hey, I'm not trying to intimidate anyone [adjusts back in chair], I just can't deal with this B.S. . . . [beginning to escalate again].

MEMBER 2: Hey, man, fine with me, I won't say anything to you again. . . .

FACILITATOR 2: [intervenes in potential hostile exchange] Whoa! [models stepping back to gain perspective, physically sitting back]. Let's sit back and notice what's going on now [reinforcing interpersonal boundaries]. We are all new to the group, so no one knows anyone well. Ozzie, Sam was giving you feedback, which is useful in group. It may seem like B.S. to you, because—and I'm making a guess here—after what you've been through, a lot of things may seem like B.S., really trivial and meaning-less. I think Sam was trying to support you, even though you aren't ready to hear it. Am I right, Sam? [Member 2 shrugs, then nods].

MEMBER 1: Sorry—it's not personal to you [appears deflated, fidgets]. I have been giving my friends a hard time, too. I am lucky I have any friends left—and I have a lot of great friends, they really are sticking around.

FACILITATOR 2: That's great that you have supportive friends. You know, you still seem anxious, and the group appears kind of jumpy, too [checks out affect with group, since other members look distressed; Members agree they feel agitated, and Facilitator 2 offers a brief relaxation exer-cise].

FACILITATOR 2: It's been an intense group so far. Let's try to bring ourselves back down. Can we do three deep breaths, breathing out all the way, and breathing in to the count of four, out to the count of eight? [Leads brief relaxation exercise].

Case Example 4: Interventions for Reexperiencing

Goals include

- Verbally reflecting body posture.
- Using nonverbal (physical) reflection of body posture/movement.
- Offering clarifying questions about symptom.
- Enlisting group.
- Verbally encouraging shift in body posture.
- Using grounding techniques.
- Offering breathing techniques/relaxation.
- Shifting out of arousal-related body postures.
- Using olfactory interference (scent) as grounding (see Case Example 5).

During group discussion, Member 3 starts to look away:

FACILITATOR 1: Amy, you're looking off, there [verbal body reflection—looks off to the side herself, gestures to side with hand—nonverbal body reflection]—where are you now?

MEMBER 3: (caught off guard) Oh—no—well, yes—I was thinking about what happened at our office, you know, that day—the faces. . .

FACILITATOR 1: [clarifying degree of reexperiencing] You're thinking about it—or were you kind of 'back there'?

MEMBER 3: Well—when Ozzie was talking loudly, I thought about the terrorists yelling—Sorry, Ozzie—it's not your fault, I've been really jumpy, too—and then I was looking out there [gestures toward window], and I thought I saw someone . . . you know—from THEN. . . . I see faces sometimes. . . . I know I'm weird.

FACILITATOR 1: [enlisting group] Does anyone else find this happening? Does it seem weird to you? Seeing something suddenly from the attack?

MEMBER 4: It happens when I don't want to think about it. Then I get irritable when my wife asks me what's wrong—sort of like the way Ozzie acted today—that's how I feel inside, and that's maybe what my wife sees. So I can identify with Ozzie. AND Amy.

MEMBER 6: And you're not weird, Amy. We've all been there.

MEMBER 5: I sometimes I see a flash of something, like someone rushing past, when I'm on the subway; I don't like crowds anymore. It reminds me of what I saw that day.

[Facilitators move to "grounding" techniques, breathing techniques, and direction to shift out of arousal-related postures]

Case Example 5: Grounding Exercises

Goals include

- Shifting awareness to feet and their connection to the floor.
- Intensifying contact with floor: (asking members to move their feet back and forth against the floor; modeling movement physically).
- Redirecting attention to physical contact with the chair; intensifying that contact.
- Using exhalation (releasing breath, if held).
- Providing psychoeducation.
- Using grounding technique through focused attention on an object.
- Making deliberate eye contact to increase connection in the present (modeled by facilitator, verbally encouraged).
- Interrupting traumatic memories/intrusions through use of olfactory interference with scent.
- Checking in with members regarding reexperiencing symptoms to determine efficacy of the grounding exercise.

FACILITATOR 1: [after members have shown signs of being triggered] Let's do a grounding exercise to deal with reexperiencing: everyone—feel your feet on the ground; you might want to move your feet back and forth, heel to toe inside your shoes, to feel the ground underneath you, and then, wiggle your toes; press your feet against the ground. Great. Make sure you keep breathing, especially, breathing OUT. We all tend to hold our breath when we're anxious. And that takes us out of the present, out of our bodies. So breathe OUT! Now, feel the chair supporting you, and feel the arms of the chair, press your arms a little into the arms of the chair, so you can really feel it. No, we keep our eyes open, so we don't 'space out'—you can look at that vase on the table, or any spot in front of you.

FACILITATOR 2: [makes deliberate eye contact with members to encourage connection in the present] I'm going to pass around these two scents [smells them himself] as we did at the start of the group. This is vanilla [passes it] and this is orange.

[Members engage in olfactory grounding; affect shifts as members discuss the scents and the nontrauma associations they evoke].

MEMBER 4: [making eye contact, looking around at other members] I like the orange smell—it reminds me of orange groves when I was in California last summer. It was a relaxing time, not much like all this. Thanks for reeling me back in.

[Facilitators check with members for changes in level of reexperiencing in the group].

Responding Therapeutically to Numbing

When a member experiences an absence of affect, or *psychic numbing,* facilitators must first determine whether the numbing is largely a reactive state, or a consistent trait. If the former, the group may explore possible "triggers" leading to the numbing, and any other symptoms such as reexperiencing, or a high level of arousal, that could have contributed. Feelings that may have occurred just prior to the numbing are examined, and grounding techniques may be used. If the numbing has been a consistent symptom over a period of months or years, it may be necessary to *shape* feelings through successive approximations. Interventions may include: asking whether any feeling is associated with feeling numb, such as a "blah," depressed feeling, or an empty or cold sensation; asking what feeling seems to be missing, since noticing the absence of feeling suggests a belief that some feeling should be present; asking what the member feels *about* being numb, for example, when it is hard to enjoy happy occasions or to cry at a funeral; asking members to consider whether there is any advantage to the numb state; and asking other group members to generate possible feeling responses for the numb member, based on what the feeling "in the room" might be, or what the members might feel if they were in the numb member's shoes. Interventions like these interject the possibility of emotion, and potentiate any subtle emotional experience of which the member had been unaware (see Case Example 6).

Case Example 6: Intervention for Numbing

Goals include

- "Shaping" through successive approximations to emotions; reinforcing near-feeling statements.
- Pulling for greater affective content.
- Enlisting feedback from the group.
- Contrasting the numb state with former emotion, giving permission for authentic emotion and expression.
- Asking member to retrieve memory of former feelings/actions.
- Using body reflection—gestures, body posture, sighing.
- Inquiring about feelings about the numb state (injecting potential for feelings).
- Reflective listening.
- Amplifying of emotion expressed, if it appears understated.
- Modeling expression of feelings.
- Validating feelings.
- Pulling for greater specificity in expression of feelings.
- Validating shift in emotion (including use of nonverbal and verbal reflection, verbalization/amplification of member's nonverbal response).

MEMBER 6: So I used to love my job, but now it's just 'nine-to-five.' I can't get excited about anything.

MEMBER 5: What about your kids—you were showing me their pictures—you're still close to them, right?

MEMBER 6: Oh, sure, I love the kids . . . but I really don't think they need to be hanging out with me, especially the way I am these days. I don't even hug them anymore, really. They're more independent—they don't even really need my help with homework. And I'm thinking their father should take them for the summer. We've already done everything we can do in New York City about fifty times, anyway.

FACILITATOR 2: [successive-approximations approach to feelings; going with the most feeling-related statement, and pulling for emotion] When you said, 'They don't need to be hanging out with me'—what's your feeling there?

MEMBER 6: I don't really know; just nothing, really. I just figure I'm not much fun to be around. Either I'm yelling, or I'm looking all around, getting overprotective. They can't stand it. I don't blame them, either.

FACILITATOR 1: [enlisting group] Does anyone else find this happening? [several members nod] What's that like?

MEMBER 4: It's not like a 'nine-to-five' thing—it's like I am alert all the time; and my family can't relate to me—I'm sure I'm not much fun for them, either.

FACILITATOR 2: [going for approximation of feelings when client can't produce feelings] What's that like for you, to feel your not much fun to be with anymore?

MEMBER 6: Well, that's just the way it is—I'm not!

MEMBER 4: I guess I'm not, either.

FACILITATOR 2: [going to the point of change, for contrast between current numb state and past (presumably) more accessible state] Since the terrorist attack on your job, right? Before that, were you more fun?

MEMBER 6: Oh, sure, I was a lot of fun—for my friends, and for the kids, too—I used to practically MAKE them have fun. I used to tell them, 'take time to smell the roses.' Like I said, we went EVERYWHERE.

FACILITATOR 2: [asking Member 6 to recall formerly loving feelings] And you hugged them?

MEMBER 6: [more animated] Sure I did, all the time!

FACILITATOR 2: So it's a big difference in you, since then.

MEMBER 6: Sure, it is.

FACILITATOR 2: What's it like for you, to be so different?

MEMBER 6: [frowns] Not good. I don't like it.

FACILITATOR 2: [reflects her most feeling-oriented statement] You don't like it. [Member 6 shakes her head; Facilitator 2 then shakes his head (body reflection)]. Can you say more about what you don't like about it?

MEMBER 6: Well, there's NOTHING I like it about it, actually. It's like a one-hundred-eighty-degree turn around, overnight. I hardly know who I am anymore.

FACILITATOR 2: [injecting the potential for a feeling about the numb state] It sounds really hard.

MEMBER 6: It IS really hard [shows more affect in her face].

FACILITATOR 2: Do you miss that person, the one you remember before the terrorist attack?

MEMBER 6: [struggling with some emotion] Sure, yes, I do, definitely [considers]. Yes, I miss that person, from before.

FACILITATOR 2: [pulls for most affective content] What do you miss the most?

MEMBER 6: [sighs] I don't know. . .

FACILITATOR 2: [Body reflecting, sighing with Member 6] So, what's that—that was a sigh there—what were you thinking of—something that you miss the most?

MEMBER 6: [nods] Hugging the kids [tears up].

FACILITATOR 2: [nods—body reflection; reflects her words and amplifies what she has said or suggested so far] Hugging the kids. It's SO hard to not be able to hug the kids now.

MEMBER 6: [nods, cries a little] I really MISS that.

FACILITATOR 2: [nods] You REALLY miss hugging the kids. . . . That's what's bringing the tears [Member 6 nods]. It's such a SAD feeling. . . ? [modeling labeling of feelings].

MEMBER 6: Sad . . . yes . . . very sad . . . but not only sad. . . . Seems like I'm MAD . . . so mad—that all that was taken AWAY from me! . . . Like I feel . . . CHEATED! Yes, that's it [cries some more]—I feel CHEATED!

FACILITATOR 2: [validating her feeling, reflecting, amplifying slightly, emphasizing specific expressive terms about her feelings, over more generic terms] So that's it—cheated! Sad, but also, SO angry at being CHEATED.

MEMBER 6: [nods] That's it! Whew! I don't like feeling this way, but I'm glad I got it off my chest.

FACILITATOR 2: [validates shift in emotion—verbal and nonverbal reflection] Whew! Seems like a relief!

Managing Dissociation

Dissociation may lead to loss of interpersonal contact, even without specific trauma-based thoughts or imagery. Signs of dissociation can include confusion, withdrawal, vagueness, "automatic" quality, or distracted, aimless behavior. Grounding techniques, as described for reexperiencing, are also useful with dissociation. The affected member is encouraged to discuss his or her inner experience, and return to contact with the group.

Clinical Management of Trauma-Based Worldview

Trauma-Based Transferences

Transference toward facilitators in PCGT is addressed as it arises, but diffused so as to minimize it as a focus of the group. Transferences commonly revolve around issues of trust, authority, control, connection, meaning, and shame; members may question whether the facilitators are trustworthy, challenge their authority, engage them in control struggles, or deflect shame issues by attributing inadequacy to facilitators. A nondefensive but firm stance on the part of the facilitators, including a willingness to acknowledge their own imperfections, goes a long way toward diffusing trauma-related transference (see Case Example 7).

Case Example 7: Interventions for Trauma-Based Transference and Trauma Themes of Safety and Connection

Goals include

- Connecting trauma-based issues to specific trauma themes.
- Providing psychoeducation about themes.
- Refocusing away from details of trauma.
- Replacing details of trauma with a more general trauma theme.
- Validating trauma-related emotion.
- Identifying triggers.
- Refocusing on the present.
- Diffusing trauma-based thinking and emotion.
- Differentiating past and present.
- Diffusing trust issues by inviting openness and assumption of responsibility.
- Predicting trust issues.
- Reframing the need to develop trust gradually.
- Acknowledging intrusions.
- Identifying issues as trauma-based.
- Reframing trauma-based issues as an obstacle to be overcome rather than endorsed.
- Predicting that trauma-based issues may interfere.

FACILITATOR 2: [during agenda setting] Why don't we check back in on goals today? Goals can change while we're in the group, and it's important to get a sense of what everyone wants to work on. And it will also help us get to know one another better.

MEMBER 4: [irritable] This is the fifth time since we started group that you mentioned we don't know one another well. I don't know why you keep saying that. We've been in group together for a while—weeks—I feel I know everyone well, and I trust everyone here. I don't really know what you're getting at, but it seems as though you don't want us to trust one another!

FACILITATOR 2: [acknowledging themes of connection/disconnection, and trust as a representation of emotional safety] It seems what I said really bothered you. I know you are starting to trust one another, although I don't think trust is automatic. Also, I want to let the group know that getting to know one another takes time.

MEMBER 4: That's what I mean. You sound very negative. I just don't agree at all.

MEMBER 3: Ezra, I think it's okay—we are getting to know one another just fine. Trust is something that is harder for me than it was before. But I am giving everyone here the benefit of the doubt.

MEMBER 4: [not reassured] I don't know . . . I don't know . . . I don't know why it bothers me so much. I just wish you would stop saying that we don't know one another.

MEMBER 5: It makes me feel a little nervous, too, actually. I don't know why, because we haven't been together all that long.

[Facilitators identify trauma themes that may play out in group.]

FACILITATOR 1: [supporting Facilitator 2, her cotherapist referencing an aspect of the trauma that is being triggered by an otherwise neutral phrase] Ezra—it's not that we don't know one another at all. We do. It seems as though that phrase makes you feel unsafe—like the 'enemy from within.'

MEMBER 1: Yeah, because there WAS an 'enemy within' in our midst.

FACILITATOR 1: So is that it? Ezra, could that be what's getting to you about that phrase?

MEMBER 4: Is it? You tell me! Okay, okay—sure, it could be—it really could be. Because, as I told you in the interview—I did his [the terrorist] orientation—it was with a pretty large group—and the terrorist, the guy on the 'inside,' Joseph, or whatever his real name was—he was there. I didn't hire him, but I oriented him to the job. Just seems like I should have been able to tell . . .

MEMBER 5: Well, I knew one of the terrorists slightly, too, and it gives me the creeps. I guess that's why that statement also makes me nervous. And for a while—sorry, Ozzie—after Ozzie raised his voice in group, I wasn't too sure of him. But now I feel fine about everyone. It's just . . .

MEMBER 1: I'm not offended—I didn't mean to trigger anyone. But I feel betrayed about the guy being in my department—and I don't blame anyone here. But someone IS to blame . . . and that person just isn't in this room.

[Refocusing away from details of trauma, redirecting to trauma themes.]

FACILITATOR 1: [validates feelings, notes trauma theme in a general way, moving away from specifics; returns to the current context of the group, identifying and diffusing the trauma themes] So, Ozzie, and Jane, you have some strong feelings about this, too. 'The enemy within' could easily be a theme for this group, given your experience [members nod]. [Reassures group by taking legitimate authority, diffusing potential for trauma-based conflict among group members] So we, as facilitators, will try to be aware of these themes and bring you back to what's really happening in the group, here and now, when that theme comes up.

[Separating past from present.]

FACILITATOR 1: [continuing, differentiating past and present] No one in this group needs to be an enemy to anyone else in this group. And Kevin and I did screen the members of this group carefully. So if you do have any problems with the membership of this group, 'the buck stops here!' [diffuses trust issues by inviting openness and taking responsibility]. Just let us know if you're feeling uncomfortable with us, or having trouble trusting us [predicting that trust issues may arise]. Because you are likely to have other doubts about Kevin and me at some point, and it is really important to bring them up, instead of just distancing from the group.

MEMBER 6: I can't imagine really distrusting either of you. You're here to help us.

FACILITATOR 2: [supporting Facilitator 1 on the issue of trust] We are here to help, but we are also human beings, and we can make mistakes. So please let us know when something bothers you—just as Ezra did. We can take it!

MEMBER 4: You know, I guess I know now what made me uncomfortable, Kevin, when you kept saying we don't know one another well. I kept hearing, 'you can't trust everyone here'—which is not what you were really getting at. And also, it made me feel that you, the facilitators, didn't feel confident about knowing us, either.

FACILITATOR 2: [reframes the getting-to-know context realistically, acknowledging the intrusion of trauma issues] It's okay to be getting to know one another, you can't know everyone right away, even though it would feel safer [identifying the issue as trauma-based, and an obstacle to be overcome rather than endorsed; predicting that the trauma-based issue may interfere]. That means you may have more anxiety than you would otherwise have, getting to know one another; but that's because of the trauma, not because of the members here. So let's just bring it up as it arises, okay? [members agree].

Self-Care for Facilitators Working with Trauma-Based Affects and Transferences

Therapists must have an outlet for their own personal issues (including trauma issues), and for their own reactions to working with trauma survivors. Although PCGT does not involve direct discussion of trauma, members' trauma-based affects (e.g., anger, terror) can be powerfully communicated, and can be experienced as contagious; facilitators are not immune from fight-or-flight arousal responses, which tend to be perceived within any human social group as signaling danger or threat. In addition, members' trauma-based attitudes, derived from overwhelmingly disillusioning and disheartening experiences, will present a challenge to facilitators' worldview, and will require personal work to incorporate the actuality of trauma in the world. Therapists who are trauma survivors themselves may find that if they can gain sufficient perspective, their own experiences can be used to inform their work. At the same time, personal experience with trauma can pose a number of challenges, including

- assuming more commonality of experience than actually exists (leading to missing the nuances of group members' unique reactions);
- impatience with members whose style of coping departs from the facilitator's;
- activation of sense of helplessness, leading to a need to "rescue" members;
- activation of avoidance responses, leading facilitators to tune out painful affects;
- counter-avoidant responding, leading facilitator to allow or encourage members' trauma disclosure; and
- triggering of facilitator's trauma memories in unanticipated ways, resulting in acute distress for the facilitator.

Even if facilitators are not trauma survivors, they are susceptible to vicarious traumatization (Figley, 1995). Facilitators need to support each other in dealing with trauma-based affects. Use of formal clinical consultation is extremely valuable for this kind of work. Facilitators should have at least informal consultation with other clinicians; a sense of connection to a clinical community serves as a buffer against *secondary trauma*. Personal therapy is, of course, invaluable in maintaining perspective while working with trauma survivors. Other forms of self-care are also important, including adequate time off, regular recreational and creative outlets, healthy practices that facilitate relaxation (to counter the agitation involved in trauma work), regular social contact, and investment in a primary social group (i.e., family or personal life).

Managing Shame and Losses to Self

In addition to trauma-based symptoms, members may experience extremely powerful affects stemming from the helplessness and horror experienced during the trauma. Sources of shame include

- being unable to help others (or help "enough") during the trauma;
- feeling confused during the trauma;
- feeling used in the context of terrorism;
- feeling dehumanized, expendable, and devalued during the trauma;
- feeling compromised in terms of basic human needs that were challenged during the trauma;
- being symptomatic after the trauma; and
- being poorly received after the trauma.

The impact of trauma on the ordinary needs for safety, control, and predictability creates a sense of a "fall from grace," as basic expectations are violated, and things happen that "shouldn't happen to anyone." When trauma violates the basic needs for predictability and control, it also violates the common subjective assumption that an effective and "good person in a just world" will not be victimized (Janoff-Bulman, 1992). Left with a profound sense of helplessness, group members may also be wracked with unreasonable self-blame. They may feel "jinxed," as though they are "magnets for trouble." This loss in sense of self, combined with actual changes due to PTSD symptoms, may be seen as further evidence of contamination. In addition, to have survived when others died or were seriously injured underscores both the fragility of life, and the randomness of life-threatening events, engendering loss of meaning. If survivors witnessed death or serious injury during the trauma, or were injured themselves, loss of sense of body integrity may contribute to shame and a sense of contamination. These psychological assaults compromise members' capacity to value their own lives, to connect meaningfully with other people, and to invest in significant activities. Facilitators are likely to see multiple expressions of these affects underlying obstacles to development of a working alliance. Initial psychoeducation about these trauma themes can be used to predict and normalize these issues, restoring a sense of emotional safety. Facilitators' awareness and use of these themes to gently interpret interpersonal conflicts and avoidance will help members access group support rather than retreating further into alienation. Facilitators can help members recover from shame issues through rehumanizing outlets such as creative expression, humor, and a sense of belonging and acceptance within the group, as well as through community contacts (see Case Example 8).

Group Transference in Premature Pseudocohesion

Members may be motivated to avoid therapeutic change because it means giving up responses initiated in the interest of survival, under conditions of threat and danger. In any group with significant common experiences, a period of overidentification within the group, and alienation from others outside the group, may be natural. However, inflexibility and crystallization of these features may signal the start of premature pseudocohesion, warranting early intervention. Without intervention, premature pseudocohesion will likely precipitate reenactments of the trauma, with entrenched trauma-based group roles and limitation of healthier coping (see Case Example 8 for interventions).

Reenactment of the trauma can threaten the group with overwhelming affects, a sense of being unsafe in the group, and potential for secondary trauma. For example, a member who reacted during the trauma by taking action tells a member who was immobilized during the trauma that he "didn't do enough." Facilitators should step in quickly to reframe and normalize the range of reactions during a trauma: the sense of helplessness in the trauma situation, which motivates the "active" member's comment; the "immobilized" member's discomfort with his own reaction; and the sense of respective scorn and shame pertaining to the judgment the "active" member has pronounced upon the "immobilized" member. If reframing does not occur, the "immobilized" member may feel alienated from the group, and there is a risk of losing this member from the group. Although other members may stay quiet during such an interaction, there is a loss of emotional safety for all members, who are reminded of their own vulnerability during the trauma, and who may experience other group members, facilitators, or both, as undependable.

When reenactments do occur, the facilitators may return the group to therapeutic functioning by reframing the interaction in terms of the trauma, identifying the reenactment and its components. This reframing allows for: understanding, a chance for members to disengage from the reenactment, and reassignment of appropriate responsibility for action rather than over-attributing responsibility or developing an exaggerated sense of helpless-ness. For example, members may be held accountable for their behavior in group, but not for freezing or activating during trauma.

Entrenched Roles

Entrenched roles, rigid characteristic responses to managing the emo-tional sequelae of trauma, tend to lend themselves to reenactments. Facilita-tors help members recognize these patterns and enlarge their range of roles and accompanying emotions:

Identifying and diffusing trauma-based interactive styles and roles. PCGT discourages assumption of entrenched, rigid, trauma-based roles, such as Victim, Aggressor, Protector/Caretaker/Rescuer, Challenger/Rebel, Cynic/Critic (adapted from Davies & Frawley, 1992). Members may report that aspects of their personal lives revolve around these roles; for example, one member may characterize himself as a victim in a personal relationship, eliciting other members to respond as protector. These roles can be re-framed as stopgap protective responses to the trauma, which now interfere with recovery. Often, it is effective to express appreciation for the function of these roles, while encouraging greater role flexibility and authentic ex-pression of emotion. Facilitators also intervene by identifying the trauma themes beneath these roles, e.g., the sense of helplessness, alienation, fear, or betrayal, allowing more direct access to emotion without exploring de-tails of the trauma. As these trauma-based yet authentic emotions are shared more directly, rigid roles become less necessary, and it becomes more ac-ceptable to express a variety of emotions, including nontrauma-based ones.

Encouraging healthy interactive styles and roles. Members are encour-aged to take responsibility for their participation, and for responding with interest in and respect for one another's individuality. The emphasis on af-fect in this model supports attention to the roles members assume in group, and to what extent rigidity in these roles, often intensified by the experience of trauma, limits emotional and interpersonal experience. Facilitators gently challenge entrenched roles that prevent members from allowing themselves to be vulnerable in the context of the group. Members who tend to take a passive stance (e.g., help seeker, listener, follower) are supported in becom-ing more active, while members who tend to dominate the group (e.g., helper, leader, monopolizer) are encouraged to become more receptive. As members become more flexible in their behavior, they can be more spontaneous and genuine in their connection within the group. This progression toward more authentic interaction helps prepare members to reinvest in the roles available to them outside of group, such as family member, friend, co-worker, supervisor, supervisee, with greater mutuality and flexibility.

Case Example 8: Interventions for Shame-Based Premature Cohesion, Entrenched Roles, and Reenactments

Goals include

- Addressing overidentification.
- Making inclusive statements to counter scapegoating.
- Validating and normalizing emergence of trauma-based issues.
- Taking the trauma-based side of the ambivalence.
- Reframing.
- Predicting trust issues.
- Verbal reflecting.
- Using humor to diffuse shame-based affects.
- Encouraging specifics rather than trauma-based generalizations.
- Supporting cotherapist (when challenged by member).
- Encouraging use of names, rather than generic categorization.
- Interpreting trauma themes at a group level.
- Separating traumatic past from present.
- Reframing entrenched trauma-based behaviors by labeling them as such.
- Empathic reframing of feelings or vulnerability behind entrenched protective stance.
- Pulling for more differentiated cohesion.
- Countering potential for scapegoating.
- Giving potentially scapegoated members a voice.
- Setting limits.
- Encouraging acceptance of feedback.
- Encouraging respect among members.
- Providing facilitator feedback.
- Reinforcing acceptance of feedback.

FACILITATOR 1: As you get to know one another better . . .

MEMBER 5: We already know one another—we're all alike, we've had the same experience . . .

MEMBER 2: Yeah—after what we've been through, we don't have to say anything . . . we just know! [slaps member next to him on the shoulder].

MEMBER 4: Yeah, I can tell just by looking in your eyes . . . some of us have been there!

FACILITATOR 2: [addressing overidentification, and including all group members, to counter possible move toward scapegoating] You do all have the trauma in common; but, at the same time, I think each of you is a distinct individual, with his or her own way of seeing the world.

MEMBER 4: Sure, we're unique, but—none of us trusts anyone!

MEMBER 2: I don't trust anyone outside of here—but two or three people here I would trust with my life! And that's saying a lot!

MEMBER 5: Exactly! And, to be completely truthful [to Facilitator 1] YOU I trust . . . but frankly [to Facilitator 2] you're going to have to do a lot to earn MY trust . . . and that's just the way it is, I have to be honest with you . . .

FACILITATOR 2: [validates trust issue, normalizing it, empathizing with the nontrust side of the ambivalence] I do think it takes a long time to trust someone, and so I am not asking or expecting you to trust me all at once. [Reframing the process of developing trust.] At the same time, I don't know if I can prove to you that I am trustworthy—over time we will get to know one another better, and we may, I hope, be able to develop some trust [predicting likely intrusion of trust issues at a later date]. But I am sure I will disappoint you at times, as well.

MEMBER 3: Why do you say that, Sam? Why don't you trust Claire?

MEMBER 5: It's really nothing personal. . . . It's just that I think that, well, Kevin—if there were any problem here, I know he would respond. But Claire—well, no offense, she's a nice person—but she looks like she's had 'cookies and milk' all her life.

FACILITATOR 2: [reflects, avoids getting defensive] It sounds as though you think of Kevin as someone who can be there for you in an emergency— but you're not so sure about me! [uses humor to diffuse] 'Cookies and milk,' huh? [sets the stage for direct, specific feedback in place of generalizations] But, Sam, I hope you let me know, specifically, if there are times you feel I'm not getting it.

MEMBER 5: Sure, I'll let you know.

FACILITATOR 1: [supports cotherapist, and using humor] Fortunately— since Claire is my cotherapist—I know I can depend on her to pull me out of trouble if I get off base here! It's great to be trusted, Sam, but I don't have any illusions—I am sure there are some days when you might have your doubts about me, and I hope you can let me know that, too.

MEMBER 5: Well, like I said, it's not personal. Actually, with me, I guess I'm old-fashioned—I just don't think women—especially young women—can back me up the way a man can. So—it's hard to trust, if you know what I mean.

MEMBER 3: So you have a thing with women, huh? Does that apply to me, too?

MEMBER 5: No, not you, Irene—you're one of the guys, as far as I'm concerned—you seem like you can handle yourself. I'm talking about the young women—no offense, ladies!

FACILITATOR 1: [inserting use of names, in place of generic grouping by shared characteristic] You mean Jane, and Amy?

MEMBER 5: Sorry, I'm not so good with names . . . but yeah, and no offense intended!

FACILITATOR 1: [interpreting trauma-based responding, at group level] It seems as though all of a sudden, today, we're talking about the group as though we are an emergency response team [makes supportive statement inclusive of all members]. I personally would feel comfortable with everyone here, if there were an emergency [works to separate trauma from present]. But, to my mind, this group is about something quite different, it's about recovery, and trying to 'live well'—the best revenge, you know!? So I wonder if it's so useful to be measuring your fellow group members according to what you would expect from them in an emergency.

MEMBER 2: I don't have a problem with Jane and Amy. But I do think we need to stay on guard—you never know what's going to happen!

FACILITATOR 2: [interprets trauma theme] Ezra, it's perfectly natural you would feel that way—it's in keeping with that theme we talked about, how your sense of safety was violated, and now it's hard to let your guard down, even here.

MEMBER 4: I will NEVER let MY guard down! I know what can happen if I DO!

FACILITATOR 2: [reframes member's statement as trauma-based example] Exactly, Ozzie—you feel that way too. In fact, it seems as though the whole group feels that way, right now—it's kind of an extension of hypervigilance.

MEMBER 5: Yeah, well, I am never going to let myself be caught off guard.

FACILITATOR 1: [reframes using empathic statement] That sounds so TIRING, Sam.

MEMBER 5: It is.

FACILITATOR 2: [pulling for cohesion at a more differentiated level] How many of you here feel tired out by having to stay hyperaroused?

[Members respond affirmatively; group joins around how painful it is to be hypervigilant, have their sleep disturbed, feel anxious all the time, etc., upset their families with their anxiety. . . .]

FACILITATOR 1: [returning to the issue that could potentially evolve into scapegoating] Jane and Amy have been very quiet about their reaction to Sam's comments. Sam, you did kind of single them out—

MEMBER 5: I'm sure they know I didn't mean anything by it—don't you?

FACILITATOR 1: [giving permission for the targeted members to express discomfort; setting limits and engaging feedback function of the group] And it's usually uncomfortable to be singled out that way—even the way you called them 'ladies'—you may not think it was rude, but it sounded like you were putting them down. But you can ask them—and I think it's important for you to be able to really hear the answer, and not get defensive. It's important feedback for you.

MEMBER 5: Bring it on, I'm not worried.

FACILITATOR 1: [persisting in expectation of respectful exchange of feedback] Sam, take a moment to step back a bit—I would really like you to hear what Jane and Amy have to say.

MEMBER 5: Okay, okay, I'm all ears. I guess I was a jerk, eh?

MEMBER 1: Sam, I did feel singled out. I knew it wasn't personal, because you said it was about us being women—but I felt like a second-class citizen in the group, and I didn't feel like saying anything—like I didn't have the right to say anything, almost . . . in a way, I felt the way I did back then . . . alone, and like there was nothing I could say that would make a difference.

MEMBER 6: I was really angry, and I was afraid I would go overboard, so I just didn't say anything—not that I didn't have the right, but I know, these days, when I start to get mad . . . well, it's too easy to lose control.

MEMBER 5: I hear you! I'm like that, too. I guess we're not so far apart after all. If it's any consolation, my daughters have been telling me the same thing—that I'm not respecting them or something. So you're not off-base, not at all. I kinda thought because, well, excuse me for saying it, but because you were ladies . . . young women, sorry—I thought it was different for you, like you wouldn't feel as hardcore as I do. [To Member 1] Sorry, I didn't mean to make you feel like a second-class citizen—or like you did back then. I've just been out of sorts lately.

MEMBER 1: Well, thanks, I appreciate it. But, one thing—I think you owe Claire an apology, too—for that 'cookies and milk' comment. As far as

I'm concerned, she and Kevin have been just about the same in the way they treat us, so there was no reason to single her out, either.

MEMBER 5: Oh—I didn't think she was mad! She didn't seem offended! Did I offend you? I'm sorry if I did.

FACILITATOR 2: [offers feedback] Thanks, Sam. Was I offended? Well, I was aware you were reacting to me in a way that didn't seem to have much to do with who I am and what I am actually like in this group. I was glad you felt you could bring up your feelings; and I especially want you to know you can always let me know if I do something that concerns or upsets you. I did feel stereotyped when you made that 'cookies and milk' statement, and that wasn't a good feeling—although it was kind of funny, the way you said it!

MEMBER 5: Yeah, well, I do have a way with words! And I have been jumping to conclusions lately—not thinking things through.

FACILITATOR 2: [positively reinforcing acceptance of feedback] Well, thanks, Sam—Kevin gave you a challenge, to hear all this feedback, and you really rose to the occasion. And you did it in a way that showed you were really listening, and taking things in—and not jumping to conclusions, this time [encouraging development of healthy cohesion, enlisting group]. A few members have mentioned that today—that it's easy to be reactive. Does anyone else notice that, in your life? [Members respond with examples, group focuses around more authentic discussion.]

Scapegoating represents one form of reenactment. It occurs when distinctions among members appear to threaten the common, trauma-based identity of the group. Members take on specific roles from the trauma itself and play them out, resulting in a sense of the "enemy" entering the group. It is more likely to emerge when there is a sense of loss regarding one of the facilitators, e.g., extended or repeated absence, serious illness, or pending departure that has not been addressed adequately, leading to a sense of shame at being abandoned, or intensifying a sense of contamination ("I must be driving people away from me"). In these latter instances, the group directs anger toward one of its members rather than toward the facilitator in question.

When scapegoating is extreme, it can intimidate the out-of-favor member or members to the point of driving them from the group, or at best, silencing them within the group. A group that threatens potential extrusion of a member signals an unhealthy group process, warranting quick and direct intervention. This situation also requires reexamination of the cotherapy relationship, and of alliances between members and facilitators. Facilitators may intervene by encouraging the group to direct resentments and fears toward the facilitators, acknowledging any inconsistencies in their own availability. Attributing difficulties among members to identifiable process factors allows members to rejoin one another in a meaningful way. Facilitators identify divisions within the group in terms of trauma-related issues such as trust, control, and safety. They intervene in any reenactments, set limits on victimizing behavior, and identify trauma-based interactions such as victim-aggressor-rescuer configurations, encouraging reestablishment of healthy, flexible roles. Trauma-based themes and affects such as "the enemy within," shame issues, and sense of contamination from the trauma, are explored. Facilitators support the targeted member, and redirect the attacking members toward their own feelings and behavior, reframing the process in terms of the group as a whole rather than the personalities of individual members. Facilitators also model tolerance and ability to consider a variety of viewpoints, demonstrating the valuing of each individual, and explaining that each member has strengths and frailties that could become a target of sustained negative group attention in an unhealthy process. Once the process has been redirected and members have been reassured, the attacking members can be encouraged to take responsibility for hostile statements or actions, while the targeted member can be supported in expressing a reaction to the formerly skewed process. The aim is for a more authentic process that allows honest expression among members without shaming any one person or subgroup (as demonstrated in Case Example 8).

Sense of Vigil As Transference

Following trauma, a sense of "keeping vigil" may interfere with a return to "life as usual." This vigil typically begins as a way to maintain a sense of control in the face of helplessness, to preserve a sense of meaning when normal meaning was lost, to maintain connection through continued acknowledgement of those who have suffered or died, and, paradoxically, to maintain a sense of safety by never again being caught off guard. In the aftermath of trauma, survivors may be unwilling to allow themselves to relax again, hesitating to embrace what seems like a false sense of security. Going on with life may also appear to betray experiences of horror and loss during the trauma, invoking survivor guilt. Group members may maintain "witness" for others who died as a result of terrorism. They may also hold a vigil for personal losses of innocence and faith. Members may show a variety of "stuck" patterns, e.g., a "thanks-but-no-thanks" attitude regarding suggestions, or repeatedly posing questions that "stump the therapist." These patterns serve to underscore the enormity of trauma (and the futility of therapeutic efforts). Some members may generate a series of crises that distract from loss of meaning. A crisis also presents a new, urgent situation to which facilitators and the group can respond meaningfully, as a test balloon for the larger trauma. Yet another meaning-related pattern involves assuming a "bigger-than-life" stance, in order to transcend annihilation by trauma. Members with this stance may deny any sense of vulnerability. They may reject help, insisting that other people need help more than they do. Or they may belittle members who do express vulnerability. They may find ways to discount the facilitators, e.g., treating them as a "buddies," or sexualizing them.

Four major needs that emerge in the aftermath of trauma tend to come into conflict initially with therapy, based on this sense of vigil:

1. *The need to protest injustice* creates secondary gain for behaviors such as venting, returning to details of the trauma without attempting resolution, and challenging the authority of facilitators.
2. *The need to resist loss of meaning* can kindle existential issues, as survivors represent the trauma, its importance, and its senselessness. Group members may struggle with despair, and concomitant depression.
3. *The need to protect others from contact with trauma* can inject the group with irrelevant or even boring material. Despite the present-centered nature of the group, other trauma survivors can be a trigger. Members may monopolize or "filibuster" to avoid having to interact at a deeper level with other survivors.
4. *The need to remember and represent the dead and wounded (survivor guilt)*. The survivor can have a sense of being undeserving of therapy, feeling that even being alive and physically well when others have died or been seriously wounded is more than he or she has a right to. In addition, traumatic events typically carry with them a sense of betrayal, suggesting the absence of any protecting spiritual presence. The burden then falls to the survivor to take on responsibility for the remembrance of these individuals and not falter until things have been put right. Group members may be halfhearted about therapy of any kind, and may even feel a need to continue suffering in order to honor those lost to the trauma.

Interventions for Sense of Vigil

Identifying the sense of vigil, and positively reframing the themes embedded within it, allows facilitators to join with members' concerns without colluding with them. They help members make realistic assessment of likelihood of current risk, and help them express the trauma-based themes and needs behind the vigil (see Case Example 9).

Case Example 9: Interventions for Sense of Vigil

Goals include

- Reframing member's self-attributions in terms of vigil.
- Identifying vigil-related theme.
- Expanding on the vigil theme, highlighting and reflecting vigil-related material.
- Amplifying.
- Directly identifying vigil.
- As members identify vigil, encouraging scaling down the level of sacrifice.
- Encouraging exploration of the "negative" side of ambivalence.
- Interpreting vigil in terms of trauma theme.
- Separating past and present.
- Reiterating encouragement of smalls steps toward present orientation and progress, in terms of trauma themes.
- Instilling hope of progress.
- Encouraging specific steps to secure change process.

MEMBER 1: I've been having more nightmares, and not sleeping much, besides that . . . anxious a lot, can't seem to concentrate—any ideas what I can do about it?

FACILITATOR 1: A couple weeks ago we were talking about this, Ezra, and you were going to try to build relaxation exercises into your schedule. Have you been able to do that?

MEMBER 1: Well, I tried it, but I really didn't get anywhere.

FACILITATOR 2: Ezra, that's pretty much how it goes at the start—it's like learning an instrument or a sport—it's the practice that makes it work.

MEMBER 1: I'm sure that's true, but I don't know that it's for me—I feel kind of like a wimp when I try to do it—like it's not masculine, or something—I guess I'm more of an old-fashioned guy than I realized!

MEMBER 2: Really, Ezra? So I guess it's not all roses being a man, after all! For myself, I have been enjoying doing yoga, and it has helped me feel calmer overall, when I do it regularly. Maybe you would like yoga?

MEMBER 1: I don't know—it seems like it's a waste of time—I don't really need it—there are so many other things that take up my time and attention—especially with my family. . .

FACILITATOR 1: Well, you're right, of course, Ezra—you are the only one who can set your priorities, and we don't want to tell you what to do. It's just that you have been experiencing a lot of agitation over the past couple of months, and I think you actually said it was affecting your home life to an extent.

MEMBER 1: Well, my wife has been getting worried that I'm not sleeping. And I feel too uptight sometimes to talk with my kids the way I used to.

FACILITATOR 1: [reframes member's self-attributions in terms of vigil] Right—so I wonder what this is about—I know you said you're old-fashioned, but really, you seem like a pretty enlightened guy, and you seem secure with yourself [identifies vigil-related theme]. Is it, like, you feel somehow that you don't deserve to feel more relaxed?

MEMBER 3: Yeah, Ezra—that's a good point—it seems as though you take care of everyone but yourself.

MEMBER 1: I don't know, I don't know—it's not exactly the idea of being undeserving, really . . . but I do feel, in the overall scheme of things, and considering what happened . . . I feel like I'm indulging myself if I try to do an exercise like that.

FACILITATOR 2: [expanding on the center of member's statement, highlighting and reflecting vigil-related material] Okay, so considering what happened, you feel self-indulgent if you do relaxation exercises?

MEMBER 1: I guess that's right—I mean, I'm still alive, I can be with my family—not everyone can say that. So what do I have to complain about? I feel as though I'm just whining about my insignificant problems.

MEMBER 4: You know, Ezra, you have a point—sometimes I find it hard to do the things I need to do for myself—I figure, 'what's the point?' I mean, since the attack, everything seems so trivial . . . and so senseless! I mean, if people's lives are threatened for some cause that doesn't even matter, how does that make sense?

MEMBER 3: Who says things are supposed to make sense?!

MEMBER 4: Well, sure, you're right, but that doesn't work for me. I guess I'm the ultimate rationalist—it's hard for me to do anything, if I don't know what the point of it is.

FACILITATOR 2: [amplifying] So you're holding out, until you can set things right in the universe?

MEMBER 4: I don't know about the whole universe, but—yeah, I guess I want things to be set right!

MEMBER 5: It's kind of like some of us are on strike.

MEMBER 2: Actually, that's a good point—when I do yoga, I do kind of feel as though I'm doing something effective—not just for me, but more than that. It's like, in one of the meditations, we send positive energy out into the universe, and receive it back again—and I really like that. It does help me feel that at least there is some positive energy in the universe.

MEMBER 4: Well, that works great for you—but that wouldn't do it for me.

FACILITATOR 1: [directly identifies vigil] It sounds like members of this group are holding a vigil—keeping that light burning for what's good and right and makes sense, in the face of all this chaos.

MEMBER 1: Now that you say that, I think you're right! I do feel I have to hold out, to stand for something, in light of what happened in the attack. It makes it hard to think of myself, unless I'm very upset and can't sleep—and then it really bothers me.

FACILITATOR 2: [encouraging adjustment in the "vigil" strategy, scaling down the level of sacrifice] I wonder if it's possible to carve out a smaller piece of the universe for yourself—and not take responsibility for so much—just tend your own garden?

MEMBER 1: It sounds good—but it's easier said than done.

FACILITATOR 1: [encouraging exploration of the "negative" side of ambivalence] Okay, so what's the 'down' side of releasing yourself from the vigil?

MEMBER 1: I suppose there really isn't a 'down' side.

FACILITATOR 1: [continuing to redirect to 'negative' side of ambivalence] Well, I wouldn't jump to that conclusion too quickly. Anything you commit energy to, like the vigil, must be doing something important—even though it gets in the way in the long run.

MEMBER 2: Well, we all are finding ways to feel that we're doing something—we may not be accomplishing anything, but at least we're not forgetting.

FACILITATOR 2: [interprets vigil in terms of trauma theme] It sounds as though the vigil helps to deal with feeling helpless—if you hold out, you don't have to feel helpless. So if you release yourself from the vigil and allow yourself some self-care . . .

MEMBER 1: I really do hate feeling helpless . . . I don't like thinking of myself as a victim.

MEMBER 4: Aren't we victims, though?

MEMBER 2: I don't feel like a victim—but that's why I like doing spiritual things, like meditation—it helps me feel more powerful.

FACILITATOR 1: [separates past and present] Exactly—good point. Most everyone feels helpless during trauma; but you don't need to feel powerless now [suggests a small step toward present orientation and progress in terms of trauma themes]. And one thing you can do is to take on those small things that take care of you, and help you connect and be present with those close to you.

MEMBER 3: So we should do what we have to do, even if it doesn't make sense anymore?

MEMBER 4: Well, we can't wait until it all makes sense—we could be in suspended animation for a very long time.

FACILITATOR 2: True—we usually recommend that you start small—figure out what does make sense to you, and build on that.

FACILITATOR 1: [instillation of hope] But keep an open mind—maybe you will encounter things that make sense—or maybe you be able to find meaning in what you do.

MEMBER 4: Sounds doubtful. But I will try to do SOMETHING.

MEMBER 5: Yeah, because if we don't DO something, we may never find out if things make sense.

FACILITATOR 1: [encouragement of specific steps to secure change process] So, what's a good first step—for Ezra, and for yourself? [members explore small steps they can take].

USES AND STAGING OF PCGT

PCGT is a flexible model that can be employed in a number of settings and circumstances. In intensive outpatient, partial hospitalization, or inpatient programs for PTSD, this modality often provides the sense of cohesion that supports other, more demanding therapies. PCGT may also be used to address special circumstances, such as crisis intervention, adjustment to life transitions (e.g., retirement, marriage, graduation) affected by trauma, stabilization (e.g., following acute crisis, or after an inpatient hospitalization for PTSD), and sobriety maintenance (e.g., during or following substance abuse rehabilitation).

By virtue of random assignment, studies comparing PCGT and TFGT could not offer information about for whom and under what circumstances each model is most effective. However, the clinical literature (Herman, 1992) suggests that trauma-focus work is most usefully conducted as a mid-stage therapy, after stabilization and a secure foundation for therapy have been established. Of the three stages identified by Herman (1992) (Safety, Remembrance and Mourning, and Reconnection with Ordinary Life), present-centered therapy is consistent with the initial and final stages. The initial stage involves identification of trauma symptoms and issues, reestablishing control in members' lives, and establishing a safe environment. The later stage, provided after trauma-focus work has been completed (or after other treatment has concluded), focuses on interpersonal effectiveness and assertiveness, development of a more integrated sense of self, reconnection with others, and engagement in productive activity that helps to heal the disempowerment experienced during the trauma. PCGT has been developed to address these issues, and can be geared to selectively emphasize stage-appropriate recovery themes. If not assigned for a particular stage of treatment, this model will tend to support the functions of both stages—especially in long-term groups—encouraging stability, and gearing members toward life after trauma.

For terrorism survivors, especially when the trauma occurred in the normal course of everyday life, reestablishing a sense of security in return to the ordinary can be especially challenging. The early stages of therapy help survivors restore a sense of safety in everyday life. The later stages encourage reinvestment in current life, despite the likelihood of continuing in the same or similar, subjectively suspect environment in which the terrorism occurred.

PCGT can also be used adjunctively, during trauma-focus work, for individuals who need additional support during intensive work. Although research may further define issues of staging, PCGT offers a reasonably safe, effective, well-tolerated model for supporting survivors of trauma.

CONCLUSION

The inclusion of present-centered supportive groups among the treatment options for psychological trauma represents an expansion in the understanding of the needs of trauma survivors. The early treatment literature focused on addressing the traumatic events themselves in order to facilitate recovery. Although the present-centered model does not eschew trauma-focus work, it shifts emphasis to recovery through reconnection with the non-traumatic, everyday environment. This broadening of the understanding of trauma treatment can be in part attributed to application of schema theory to psychological trauma. Schema theory allows a more complex and dynamic conceptualization of the relationship between overwhelming stress events and their likely impact on worldview and sense of self. The traumatic "shattering" of self and world requires more than examination of the trauma; worldview and self-image become the focus of treatment. This broader context suggests that PCGT is advantageous as a modality that shores up current functioning and supports development of a secure foundation in the present. The eventual goal is integration of experience, including development of new images of self and world that can incorporate aspects of life prior to the trauma, the fact that the trauma did occur, and investment in life post-trauma. Yet the development of these new, integrated schemas can be a daunting task for survivors, requiring a venturing into a new unknown after the terrifying uncertainty of the traumatic experience. This process is especially overwhelming when trauma-related symptoms have undermined social supports and compromised current functioning. Survivors will often cling to "lessons" learned during the trauma, in part because a trauma-based world is at least a defined world. Although trauma-focus work (especially when cognitive restructuring is incorporated) often moves survivors forward from this entrenched position, the review of traumatic events also reminds them of the reasons for their hard-won trauma-based views. Present-centered group therapy offers a normalizing influence, loosening the hold of trauma-based schemas, and encouraging a gradual broadening of perspective based on attention to the current environment.

APPENDIX

For Session 2: Psychoeducation on PTSD and related disorders (APA, 2000)

*Criteria for PTSD**

1. *The person has been exposed to a traumatic event in which both of the following were present:*
   a. the person experienced, witnessed, or was confronted with an event or events that involved actual or threatened death or serious injury, or a threat to the physical integrity to self and others
   b. the person's response involved intense fear, helplessness, or horror
2. *Intrusive symptoms:*
   a. recurring and intrusive distressing recollections of the event, including images, thoughts, or perceptions

---

*Reprinted with permission from the *Diagnostic and Statistical Manual of Mental Disorders,* Fourth Edition (Copyright 1996). American Psychiatric Association.

   b. recurrent distressing dreams of the event
   c. acting or feeling as if the traumatic event were recurring (includes a sense of reliving the experience, illusions, hallucinations, and dissociative flashback episodes, including those that occur on awakening or when intoxicated)
   d. intense psychological distress at exposure to internal or external cues that symbolize or resemble an aspect of the traumatic event
3. *Avoidant symptoms:*
   a. efforts to avoid thought, feelings, or conversations associated with the trauma
   b. efforts to avoid activities, places, or people that arouse recollections of the trauma
   c. inability to recall an important aspect of the trauma
   d. markedly diminished interest or participation in significant activities
   e. feelings of detachment or estrangement from others
   f. restricted range of affect (e.g., unable to have loving feelings)
   g. sense of a foreshortened future (e.g., does not expect to have a career, marriage, children or a normal life span)
4. *Hyperarousal symptoms:*
   a. difficulty falling or staying asleep
   b. irritability or outbursts of anger
   c. difficulty concentrating
   d. hypervigilance
   e. exaggerated startle response

Associated Disorders

1. *Alcohol and/or drug abuse:* Sometimes people with PTSD may try to rid themselves of or forget painful emotions or disturbing thoughts, numb themselves, or try to help themselves sleep by abusing alcohol or other drugs.
2. *Panic attacks:* Sometimes people with PTSD may have sudden attacks of panic that seem to come out of the blue and that cause the person to feel very afraid, and have a number of physical symptoms, such as
   a. heart racing, irregular heartbeats
   b. sweating
   c. trembling or shaking
   d. shortness of breath or hyperventilation or choking feelings
   e. chest pain or discomfort

    f. nausea or abdominal distress
    g. feeling dizzy, lightheaded or faint
    h. feelings of unreality or being unattached to oneself
    i. fear of losing control or going crazy, fear of dying
    j. numbness or tingling
    k. chills or hot flashes
    l. thoughts like, "I am going crazy," "I am having a heart attack," or "I am going to die."

Sometimes people have these symptoms when they try to leave their homes, or in crowds, wide open spaces, closed places or while traveling in buses, planes or trains, or even while going over bridges or through tunnels.

  3. *Depression:* Very commonly, people who are suffering from PTSD also have depression. Symptoms of depression include
    a. feeling down or blue
    b. lack of interest or pleasure in all or most activities
    c. significant weight loss not due to dieting, or weight gain
    d. sleep problems (either not enough or too much)
    e. feeling keyed up and restless or slowed down and heavy
    f. fatigue and lack of energy
    g. feelings of worthlessness, or extreme guilt
    h. difficulty concentrating or with memory, or indecisiveness
    i. recurrent thoughts of death or suicide or suicide attempts

REFERENCES

American Psychiatric Association (2000). *Diagnostic and Statistical Manual of Mental Disorders* (Fourth Edition, Text revision) Washington, DC: Author.

Beck, A. T. (1978). *Depression Inventory.* Philadelphia: Center for Cognitive Therapy.

Blake, D. D., Weathers, F. W., Nagy, L. M., Kaloupek, D. G., Gusman, F. D., Charney, D. S., & Keane T. M. (1995). The development of a Clinician-Administered PTSD Scale. *Journal of Traumatic Stress, 8,* 75-90.

Briere, J., Elliott, D. M., Harris, K., & Cotman, A. (1995). Trauma Symptom Inventory: Psychometrics and association with childhood and adult victimization in clinical samples. *Journal of Interpersonal Violence, 10,* 387-401.

Classen, C. C., Koopman, C., Nevill-Manning, K., & Spiegel, D. (2001). A preliminary report comparing trauma-focused and present-focused group therapy against a wait-listed condition among childhood sexual abuse survivors with PTSD. *Journal of Aggression, Maltreatment and Trauma, 4,* 265-288.

Davies, J. M. & Frawley, M. G. (1992). *Dissociative Processes and Transference-Countertransference in the Psychoanalytically Oriented Treatment of Adult Survivors of Childhood Sexual Abuse.* Hillsdale, NJ: The Analytic Press.

Figley, C. R. (Ed.). (1995). *Compassion Fatigue: Coping with Secondary Traumatic Stress Disorder in Those Who Treat the Traumatized.* Philadelphia: Brunner/Mazel.

Foa, E. B., Cashman, L., Jaycox, L., & Perry, K. (1997). The validation of a self-report measure of posttraumatic stress disorder: The Posttraumatic Diagnostic Scale. *Psychological Assessment, 9,* 445-451.

Foy, D. W., Glynn, S. M., Schnurr, P. P., Jankowski, M. K., Wattenberg, M. S., Weiss, D. S., Marmar, C. R., & Gusman, F. D. (2000). Group therapy. In E. B. Foa, T. M. Keane, & M. J. Friedman (Eds.), *Effective Treatments for PTSD: Practice Guidelines from the International Society for Traumatic Stress Studies* (pp.155-175). New York: The Guilford Press.

Gendlin, E. T. (1981). *Focusing.* New York: Bantam Books.

Gendlin, E. T. (1996). *Focusing Oriented Psychotherapy: A Manual of the Experiential Method.* New York: Guilford Press.

Herman, J. L. (1992). *Trauma and Recovery.* New York: Basic Books.

Janoff-Bulman, R. (1992). *Shattered Assumptions: Toward a New Psychology of Trauma.* New York: Free Press

King, L. A., King, D. W., Leskin, G., & Foy, D. W. (1995). The Los Angeles Symptom Checklist: A self-report measure of posttraumatic stress disorder. *Assessment, 2,* 1-17.

Lewis, H. B. (1971). *Shame and Guilt in Neurosis.* New York: International University Press.

McCann, I. L. & Pearlman, L. A. (1990). *Psychological Trauma and the Adult Survivor: Theory, Therapy, and Transformation.* New York: Brunner/Mazel.

Neisser, U. (1976). *Cognition and Reality: Principles and Implications of Cognitive Psychology.* New York: W. H. Freeman and Company.

Parson, E. R. (1985). Post-traumatic accelerated cohesion: Its recognition and management in group treatment of Vietnam veterans. *Group, 9*(4), 10-23.

Prochaska, J. O. & DiClemente, C. C. (1984). *The Transtheoretical Approach: Crossing Traditional Boundaries of Therapy.* Homewood, IL: Dow Jones-Irwin.

Prouty, G. (1994). *Theoretical Evolutions in Person-Centered/Experiential Therapy: Applications to Schizophrenic and Retarded Psychoses.* New York: Praeger.

Rosinski, R. R. (1977). *The Development of Visual Perception.* Santa Monica, CA: Goodyear Publishing Company.

Schnurr, P. P., Friedman, M. J., Foy, D. W., Shea, M. T., Hsieh, F. Y., Lavori, P. W., Glynn, S. M., Wattenberg, M. S., & Bernardy, N. C. (2003). A randomized trial of trauma focus group therapy for posttraumatic stress disorder: Results from a Department of Veterans Affairs cooperative study. *General Archives of Psychiatry, 60,* 481-489.

Shea, M. T., Wattenberg, M. S., & Londa-Jacobs, J. (1997). Present Centered Group Therapy Manual, Training and instruction manual for VA Cooperative Study #420 on Group Treatment of PTSD.

Weiss, D. S. & Marmar, C. R. (1996). The Impact of Events Scale-Revised. In J. P. Wilson & T. M. Keane (Eds.), *Assessing Psychological Trauma and PTSD: A Practitioner's Handbook.* New York: Guilford.

Yalom I. D. (1985). *The Theory and Practice of Group Psychotherapy* (Third Edition). New York: Basic Books.

Chapter 16

Cognitive-Behavioral Groups for Traumatically Bereaved Children and Their Parents

Karen Stubenbort
Judith A. Cohen

INTRODUCTION

The terrorist attacks of September 11, 2001, changed the nature of life in the United States. Never before had U.S. citizens been confronted with this level of mass death and destruction on U.S. soil or the accompanying sense of widespread fear and vulnerability due to an act of terrorism. For most, changes in functioning that followed in the days and weeks after the attack, such as anxiety and sleep disruption, were eventually followed by the resumption of day-to-day routine and schedule. For others, symptoms of distress and anxiety persisted. Thus, mental health professionals have been challenged with the task of identifying those persons evidencing symptoms signaling psychosocial disturbance and the need for appropriate interventions.

Thousands of children were directly affected by the terrorist attacks. Some witnessed the planes crashing into the Twin Towers or some other portion of the event, including the cataclysmic collapse of one or both towers. Others were evacuated from their schools in a shower of dust and debris and witnessed thousands of adult New Yorkers and employees running for their lives. Furthermore, thousands of children lost parents or other loved ones in the attack, and are bereaved in the midst of trauma. Many of these children may be experiencing a level of psychiatric disturbance that will require professional help.

Portions of this chapter are derived from: Cohen, J.A., Stubenbort, K., Greenberg, T., Padlo, S., Shipley, C., Mannarino, A.P., & Deblinger, E. (2001). *Cognitive and behavioral therapy for traumatic grief in children: Group treatment manual.* Center for Traumatic Stress in Children and Adolescents, Allegheny General Hospital, Drexel University College of Medicine, Department of Psychiatry, Pittsburgh, PA. Unpublished manuscript.

Traumatic bereavement (TB) typically refers to a condition in which grief and trauma issues are present simultaneously. TB happens when people lose a loved one to sudden and catastrophic death. TB may arise in children who are present when a parent dies, in children who are exposed to graphic details of the death, or if a child is the person who discovers the body. In some instances, children will develop TB in the absence of objective "trauma"; for these children, the loss is great enough to result in trauma-related symptoms. In all instances, symptoms of intrusive posttraumatic reminders and avoidance interfere with the normal process of bereavement, preventing resolution of both the trauma and grief (Layne, Pynoos, Saltzman, Arslanagic, & Black, 2001; Layne, Saltzman, Savjak, & Pynoos, 1999; Nader, 1997; Pynoos, 1992; Rando, 1996). With appropriate treatment, victims who experience trauma-complicated bereavement have reported a significant decrease in levels of psychological distress experienced (Black, 1977; Pynoos, Steinberg, & Wraith, 1995).

Treating children evidencing traumatic bereavement requires skill in both the treatment of trauma and the treatment of bereavement. Treatment with successful outcomes most often also include supportive intervention with the surviving parent(s). The treatment model being presented, trauma-focused cognitive-behavioral therapy (TF-CBT) for children and their parents, is a manualized treatment approach that has been developed from years of empirically based work with trauma victims. Cognitive-behavioral techniques with child abuse survivors (Cohen & Mannarino, 1996a, 1998; Cohen, Mannarino, Berliner, & Deblinger, 2000) and survivors of victims killed in air disasters (Stubenbort, Donnelly, & Cohen, 2001) have been used successfully to treat children and parents evidencing trauma-related symptomatology.

EVIDENCED-BASED FRAMEWORK

The amount of stress a person can manage before normal functioning is interrupted is limited. Posttraumatic reactions, including post-traumatic stress disorder (PTSD) and acute stress disorder (ASD), are characterized by symptoms of increased arousal, avoidance of thoughts, feelings, and/or other external stimuli in order to minimize distressing memories, and reexperiencing phenomena. These reactions occur in some children in response to exposure to overwhelming or life-threatening events (American Psychiatric Association, 2000). This cluster of symptoms has obvious incompatibilities with the mental processes of grieving that accompany bereavement. Specifically, PTSD symptoms may lead to avoidance of thoughts about the deceased person and the way that person died, whereas resolving grief involves addressing the emotional loss and remembering the deceased in a manner fitting to the nature of the relationship. Thus, it is important that a person experiencing traumatic loss work through the shock of the trauma so as to facilitate adaptive grieving. With this in mind, the posttraumatic processes of avoidance and denial must be managed in a way that allows loved ones to anticipate and deal with daily reminders of the deceased (Klein & Schermer, 2000).

The limited capacity for abstract thinking and the cognitive immaturity of children may further complicate the grieving process in those who experience TB. The death of a loved one may result in misguided fears, unrealistic expectations, and inaccurate attributions. Furthermore, the painful affects that accompany loss may be more difficult for children to manage. Finally, when a child loses a parent or a sibling, it typically means that surviving family members and parents are also grieving and, consequently, are less emotionally available, leaving the grieving child that much more vulnerable.

Cognitive therapies are based on the premise that emotional disturbance is a consequence of negative thinking that develops as individuals attempt to apply meaning to events that are emotionally laden (Beck, 1976). Cognitive interventions attempt to address these negative-thinking processes and have been found useful in treating trauma-related symptoms. Given the nature of traumatic memories, one can surmise that they are emotionally laden and difficult to ignore. Furthermore, traumatic events are outside the realm of ordinary experience and therefore, are typically incompatible with existing cognitive constructs. Horowitz (1976, 1979) and Marmar & Horowitz (1988) hypothesize that traumatic memories remain active in memory unless they are successfully assimilated and integrated into existing schematic structures. As long as these memories remain active, they continue to burden individuals with the trauma-related symptoms of increased arousal, intrusive thoughts, and efforts to avoid reminders.

Gradual exposure (GE), an adaption of adult exposure techniques (Deblinger & Heflin, 1996) is used to unpair thoughts, reminders, or discussions of the traumatic death of the loved one, from overwhelming negative emotions such as terror, horror, extreme helplessness, or rage. GE techniques encourage children to talk about their trauma in a supportive environment so that they can gradually experience less emotional arousal. Over the course of several sessions, children are slowly and carefully encouraged to describe their experience, or their understanding of the events surrounding their loved one's death, as well as any thoughts and feelings occurring in regard to their loved one's death. Over the course of several repetitions, children are expected to experience progressively less extreme emotional reactions and physiological reactivity.

GE has been utilized in the treatment of sexually abused children (Cohen & Mannarino, 1993, 1998; Deblinger & Heflin, 1996), children exposed to community violence (Pynoos & Nader, 1988), disasters (Goenjian et al., 1997), and single-episode traumatic events (March, Amaya-Jackson, Murray, & Schulte, 1998). In a review of empirical studies using various methods of exposure, Rothbaum and colleagues (Rothbaum, Meadows, Resick, & Foy, 2000) report that exposure resulted in a significant decrease in PTSD symptoms. Many researchers describe the use of some form of trauma-focused discussion to decrease the emotional valence of traumatic memories and thus, the trauma-related symptoms (e.g., Berliner, 1997; Cohen & Mannarino, 1996a,b; Deblinger & Heflin, 1996; Pynoos & Nader, 1988; Terr, 1990).

The cognitive-behavioral techniques of stress-inoculation training and cognitive-processing therapy have also been found effective in reducing physiologic manifestations of trauma-related symptoms. These manifestations include increased heart rate, increased startle response, hypervigilance, agitation, sleep disturbance, restlessness and irritability, and anger/rage reactions, and may be especially problematic when children experience traumatic reminders (Cohen et al., 2001). Stress-inoculation therapy (SIT) refers to a variety of interventions that protect children from the negative effects of stress. SIT encourages the use of optimal coping skills, including relaxation techniques (deep breathing, positive imagery, and progressive muscle relaxation), which are helpful in reducing the physiologic manifestation of stress and PTSD.

Thought-stopping and thought-replacement skills can also enhance one's sense of control over troublesome negative emotions. Although it is sometimes argued that the active suppression of emotionally laden cognitions may increase avoidance behaviors, and thus increase intrusive thoughts and emotional arousal (Walser & Hayes, 1998), the simple realization that one has control over his or her thinking patterns can be helpful. When children understand that they have the power to control their thinking, they are better able to tolerate troublesome thoughts, and are therefore less likely to resort to avoidance techniques (Cohen et al., 2000).

The cognitive immaturity of children may complicate recovery from traumatic loss. Cognitive processing techniques, including identification of feelings, identification and correction of cognitive misconceptions, and enhancement of children's sense of safety can help address this immaturity. Cognitive distortions, frequent occurrences following traumatic loss, are particularly likely to occur with children, whose abstract thinking abilities are limited (Cohen et al., 2000; Hendricks, Black, & Kaplan, 1993). In particular, children are likely to resort to self-blame, shattered assumptions about themselves and the world, and other distorted cognitions (Janoff-

Bulman, 1985) that may complicate recovery from trauma, or interfere with the bereavement process. Identifying children's distorted thoughts and challenging them with factual information, reframing, and replacement, can reduce attributional errors and the negative emotions they generate (Foa, Hearst-Ikeda, & Perry, 1995; Resnick & Schnicke, 1992).

Finally, the initial six months following exposure to trauma are generally recognized as the most critical for emotional recovery (McFarlane, 1987; Rando, 1993; Sugar, 1988; Terr, 1987). During this time, psychoeducation can be most helpful and is crucial for some children. Providing information regarding the facts about the trauma, the logistic and emotional impact of trauma on one's life, and resources for coping, are critical. Providing this information in a group format can further assist children and their families in normalizing their experiences. The group format provides a forum for normalization and connection as members recognize the shared nature of their trauma, as well as their emotional response to it (Johnson & Lubin, 2000).

OBJECTIVE AND FOCUS OF GROUP

The primary focus of the proposed therapeutic interventions is the relief of trauma-related symptoms so that children in the group may proceed with the normal process of bereavement. Through the group process, children are expected to normalize and validate posttraumatic reactions, increase coping skills, identify and correct inaccurate attributions, and become desensitized to trauma reminders through the use of GE. After the trauma-specific symptomatology has been addressed, the children can attend to the process of bereavement. Once again, the group process is expected to normalize and validate grief reactions and increase coping abilities. Furthermore, a therapeutic focus of the group is the enhancement of support-seeking skills. As the children begin to address the process of moving on in the absence of the deceased, the prosocial process of the group is anticipated to facilitate these efforts.

A suggested time line and sequence for presenting treatment interventions is offered here. This sequence has been used in several controlled trials for children with PTSD and other trauma-related symptoms (Cohen & Mannarino, 1996a, 1998; Deblinger, Lippman, & Steer, 1996). Treatment is divided into two components: (1) trauma-focused, and (2) bereavement-focused. Within these sections are five phases of treatment, each targeting specific PTSD or bereavement-related symptoms. Aside from the fact that trauma interventions typically precede bereavement interventions, therapist judgment should be used to determine if sequence or timing should be adjusted. Treatment is comprised of separate parent and child groups and typically last twelve weeks.

Trauma-Focused Interventions

Phase 1: Orientation to the model, building coping skills—lasting approximately two to three sessions

Phase 2: Cognitive processing—typically one session

Phase 3: Gradual exposure/creating a trauma narrative—typically lasting four sessions. (Although, some children may need more time, especially if they are strongly avoiding thoughts and feelings. When this occurs with a particular child, an individual session may be warranted.)

Bereavement-Focused Interventions

Phase 4: Mourning the loss—lasting approximately two sessions

Phase 5: Redefining the relationship and moving on—typically two sessions

CRITERIA AND SELECTION OF MEMBERS

Evaluation and Assessment

Following a comprehensive psychological evaluation, including psychiatric, social, and developmental history, review of symptom presentation, and DSM diagnosis, children are administered a series of assessments to determine their degree of symptomatology. Suggested assessments for children experiencing traumatic bereavement include the following.

- *The Mood and Feelings Questionnaire (MFQ)* (Angold et al., 1996): The MFQ is a thirty-three-item scale that is used to address depression. This self-report instrument has excellent internal consistency and reliability, correlates highly with other self-report instruments of child and adolescent depression, as well as with diagnostic interviews, and shows sensitivity to treatment effects (Angold et al., 1996; Brooks & Kutcher, 2001; Wood, Kroll, Moore, & Harrington, 1995).
- *The Child PTSD Symptom Scale (CPSS)* (Foa & Tolin, 2001): The CPSS is a children's version of the PTSD Symptom Scale (PSS), a well-validated measure for assessing PTSD severity and diagnosis in adult victims of a variety of traumas. The CPSS has been found to be

a reliable instrument for assessing PTSD following single-incident trauma. In addition, the CPSS includes seven items that assess impairment in functioning, adding to the utility of the measure. This assessment tool demonstrates very good to excellent internal consistency and moderate to excellent test-retest reliability.

- *The Expanded Grief Inventory (EGI)* (Layne, Savjak, Saltzman, & Pynoos, 2001): The EGI is a twenty-eight-item self-report measure of the frequency with which normal and traumatic grief reactions have been experienced during the past thirty days. This inventory is a revised version of the UCLA Grief Screening Inventory (Pynoos, Nader, Frederick, Gonda, & Stuber, 1987), which has been used in a number of published studies of traumatically bereaved children and adolescents. The EGI contains three factor-analytically derived subscales theorized to measure both adaptive and potentially maladaptive bereavement reactions. A preliminary version of the GSI, used with a sample of war-exposed Bosnian adolescents, showed acceptable to very good internal consistency reliability (full scale Chronbach's Alpha = 93, subscale Alphas between .70 and .90). The measure also showed good evidence of convergent validity in its correlations with measures of theoretically related constructs, posttraumatic stress symptoms, loss reminders, trauma reminders, depressive symptoms, and somatic symptoms (r's range from .3 and .6).

- *The Screening for Anxiety-Related Emotional Disorders (SCARED)* (Birmaher et al., 1997): This measure is a forty-one-item self-report instrument that corresponds to DSM-IV anxiety disorders. It has been shown to have excellent internal consistency, test-retest reliability, and convergent discriminant reliability (Birmaher, et al., 1997; Khetarpal-Monga et al., 2000).

Assessments given to the parent or primary caregiver include

- *The Parent Emotional Reaction Questionnaire (PERQ)* (Cohen & Mannarino, 1996b): This instrument was designed to measure parents' emotional reactions to their child's distress, examining such reactions as fear, guilt, anger, embarrassment, and feelings of upset. The PERQ consists of fifteen items that describe specific emotional reactions. Parents are asked to rate each item on a five-point scale regarding how well it describes their emotional response to their child's distress. Internal consistency for the PERQ has been calculated to be .87. Two-week test-retest reliability was .90.

- *Child Behavior Checklist (CBCL)—Parent Form* (Achenbach & Edel-brock, 1983): This measure was developed as a descriptive rating instrument to assess adaptive competencies and behavior problems, and is completed by parents or parent surrogates. Parents provide information on twenty competence items and rate their child on 118 problem items using a 0-1-2 scale (0 = not true, 1 = somewhat true, 2 = very true or often true) for behaviors over the past six months. There is separate scoring for gender and for age groups (ages four to eleven and twelve to eighteen). Test-retest reliability for the CBCL over a one-week period was .88. There are norms for both the social competence scales and the behavior problems scales.
- *The PTSD Symptom Scale (PSS)* (Foa & Tolin, 2001): The PSS is a seventeen-item brief interview instrument that measures PTSD in adults. Although developed relatively recently, it has excellent reliability and validity properties and has been shown to correlate highly with the Clinician Administered PTSD Scale (CAPS), a lengthy, semi-structured interview for assessing PTSD.
- *The Beck Depression Inventory (BDI)* (Beck, Ward, Mendelson, Mock, & Erbaugh, 1961): The BDI consists of twenty-one items considered to be representative of attitudes and symptoms specific to depression in adults. For each of the twenty-one items, subjects must choose which one of four statements most represents how they have been feeling during the past week and current day. The items are presented in increasing order of severity and each individual item is scored 0 to 3, with a total test score ranging from 0 to 63. The retest reliability scores with adolescents and young adults ranged from .67 to .87. Internally consistent scores are excellent, with ranges of .87 to .91 (Brooks & Kutcher, 2001).

The assessment tools lend insight into the nature of the TB-related symptoms; all have been shown to be sensitive to pre- and post-treatment change in traumatized children and adults. The information provided by these assessment instruments, together with that gathered during the psychological evaluation, inform the selection of members for the group.

Inclusion Criteria

The treatment model presented here is designed for children ages eight to thirteen who have experienced the traumatic death of a loved one, which in cases involving young children frequently means the loss of a parent or parent figure. The model also assumes that each child has the presence of a supportive parent or caretaker* who is willing to be involved in treatment.

In the face of traumatic loss, many parents experience difficulty following through with ordinary parenting practices. Some parents have difficulties because they themselves are consumed by trauma-related symptoms. Others become overly permissive because they are worried that they might further harm their children if they are harsh with them. Whatever the reason, a lapse in parenting practices at this time is unfortunate because, in the face of stress, routine and consistency help children to maintain optimal functioning. In fact, regular, consistent parenting practices are particularly important since some children's PTSD symptoms take the form of aggression, angry outbursts, and other negative behaviors (American Academy of Child and Adolescent Psychiatry, 1998). For parents who did not have optimal parenting skills prior to the trauma, developing these skills now may be crucial.

Around eight years of age, most children enter the realm of concrete operations. They are interested in facts and information. Talking and information gathering are extremely important to children of this age group, which makes them excellent candidates for group treatment. As their ability to understand concepts (such as the permanence of death) increases, they are more able to experience the feelings associated with those thoughts. Unfortunately, they are not adept at appropriate expression of such emotion and can easily be overwhelmed. Even though their understanding of death and mourning has increased, they may remain confused about what to do with what they know and how they feel. Thus, they are likely to need assistance in creating concrete ways of mourning or memorializing those they have lost. Group activities that increase affective awareness, encourage information sharing, challenge inaccurate cognitions, and memorialize loved ones may be very supportive.

Most parents of children experiencing TB likely will be dealing with their own trauma and bereavement issues, as they will have known or been related to the deceased. Understanding that parental emotional reaction as well as parental emotional support are critical factors in children's resolution of trauma-related symptoms (Cohen & Mannarino, 1993, 1996a; Deblinger & Heflin, 1996), the proposed model suggests that a parents' group be run parallel to the children's group. Participation of a caring adult communicates commitment to the child. Moreover, the increased understanding and skill that the parent receives through the group is likely to re-

---

*For simplicity's sake, throughout this chapter, adult participants in the children's treatment are referred to as parents, but it should be understood that other caretakers may similarly participate in this treatment.

sult in improved communication and emotional support for the child, thus maximizing therapeutic influence. This parent intervention must be *child focused*. Nevertheless, the parents are expected to benefit from the affiliation, emotional support, and validation that occur in the context of the group.

Exclusion Criteria

Exclusionary criteria include severe psychopathology, developmental delay, or the presence of mental retardation. Any of these criteria would be expected to interfere with the individual's ability to benefit from the cognitive skill development inherent in the model and would likely interfere with the group process.

Children with preexisting psychiatric or medical conditions may experience exacerbations of these difficulties following a traumatic death. To prevent secondary adversities related to these conditions (such as school failure, aggressive behaviors, etc.), therapists should be experienced in diagnosing the full spectrum of child psychiatric disorders, and knowledgeable of appropriate treatments and/or referral resources; and they should address comorbid conditions in a timely manner. In some cases, children may not be candidates for group therapy, as the presentation of related symptoms may compromise the focus of treatment and the group process.

Finally, it is important to recognize the special needs of children who have lost both parents, or a single parent who was the child's sole caretaker. These children have experienced not only trauma and loss, but also are likely to have been displaced from their home, school, peers, and community. Since these children are deprived of the parental support and stability, which normally assists children in adapting to such changes, they have greater challenges to overcome. Unless several of these children can be brought together as a group, it is likely that these children will benefit most from individual treatment.

GROUP COMPOSITION AND PREPARATION

The intervention model being presented is designed for school-aged children (ages eight to thirteen) who have lost a loved one due to terrorist disaster and are manifesting trauma-related symptoms that interfere with the normal process of bereavement. Specifically, the posttraumatic symptoms are interfering with the process of remembering the deceased and mourning the loss. The clinician is advised to consider, individually, each child being referred and make a decision about his or her inclusion in a group based upon the social skills and cognitive abilities of the other children involved. For example, one clinician may choose to keep sibling pairs together, regardless of age, while another may choose to keep younger children's groups (eight- and nine-year-olds) separate from groups comprised of older children (ten- to thirteen-year-olds). The most important homogeneous characteristic lies in the nature of symptomatology being presented. All of the children involved in the group, as well as their parents, will likely meet full or partial criteria for post-traumatic stress disorder, which is interfering with their ability to negotiate the normal bereavement process.

This is not to say that all group members will be experiencing the same degree of avoidance, arousal, intrusive thoughts, etc. On the contrary, the different degrees of symptom presentation is just what makes the group process so successful. For instance, members who are less avoidant will model their ability to talk about issues without significant distress for others. Similarly, members whose cognitions remain accurate and beneficial are able to encourage and support those who struggle with realizing more helpful attributions.

Preparation of Members

As previously noted, each potential member should have a comprehensive psychological evaluation. Furthermore, a therapist should meet with each child and his or her parent(s) to discuss the nature of the group and to elicit and answer questions. During this meeting, the therapist should offer general information about the group, provide a rationale for the group format, and make clear the cognitive-behavioral framework. Furthermore, the therapist should begin establishing a therapeutic relationship during this initial meeting. Along these lines, the therapist should facilitate any member's emotional expression while providing empathy and support. The therapist should also encourage the child or parent to express any concern they might have about joining the group.

ROLE OF THE THERAPISTS

Each group should possibly be run by two therapists. This team approach assists in the task management of the group as well as the facilitation of empathic expression and discussion. Furthermore, tasks in the children's component of the model require the children to split into smaller groups. Sometimes, therapists meet individually with the child/parent dyads. Sharing the duties of managing the small groups and scheduling the individual meetings with the dyads is most helpful.

Therapists should probably have knowledge of the treatment model and prepare for weekly sessions. With this in mind, the therapists must review and familiarize themselves with the weekly exercises. These interventions are designed to address the posttraumatic and grief symptoms in a manner that normalizes and validates the traumatic bereavement experience, enhances coping skills, builds tolerance for the emotional reactions, and increases positive psychosocial adjustment.

THE GROUP

The activities described here are intended to encourage self-disclosure, group alliance, and trust. Furthermore, alliance and trust, in combination with the planned activities, are expected to increase tolerance for trauma-related materials and reminders, normalize and validate thoughts and feelings regarding trauma and grief reactions, and facilitate adaptation. Nevertheless, group dynamics and roles will need to be managed.

As with any instance when groups of people come together, personalities are involved. For this reason, members are likely to retain their individual communication styles. For example, some children may dominate the group, while others fall quietly into the background. The therapists must be aware of group roles that arise and ensure that each group member stays involved.

Children in the age group proposed for this intervention (eight- to thirteen-year-olds) strongly believe in the importance of and adherence to rules. Thus, stating rules, or inviting the children to help make the group rules, assists in the management of the group. For example, having a rule that only one child will talk at a time can be helpful in managing a child who has a tendency to interrupt. It is the therapists' responsibility to ensure that group members adhere to the rules. Each child should be assured an opportunity to participate fully in the group process.

Occasionally, one child in the group will tend to be bullied or scapegoated. Scapegoating may be a method by which the group discharges increased arousal secondary to trauma-related material or the more general group process (Yalom, 1985). The therapists must recognize when a child is being scapegoated and protect that child. Another role that may appear in the group is the caretaker, or the child who tries to protect others from the emotions associated with the TB. The therapist must redirect the caretaking child while modeling empathy and acceptance of these feelings. Furthermore, the gradual exposure process is meant to facilitate discussion of and promote tolerance for trauma and grief reactions. It is the therapist's role to attend to these and other forces that are likely to interfere with group integrity.

Other dynamics in the group are likely reflective of the posttraumatic symptomatology and should be addressed through the group process. For instance, children who are avoidant will be exposed to peers who are more willing to discuss the TB experience. Along with having to listen to others, gradual exposure exercises address avoidance directly. In the process of addressing TB-related material, some children may experience increased arousal, another common posttraumatic symptom. Stress inoculation exercises are intended to protect children from the negative effects of stress, and encourage the use of optimal coping skills.

THE COGNITIVE-BEHAVIORAL GROUP MODEL

As mentioned earlier, in the presence of TB, it is often essential to address and at least partially resolve the trauma issues before the bereavement issues can be successfully processed (Layne et al., 1999; Nader, 1997; Rando, 1996). This may be particularly true when a child is fixated on the most horrifying aspects of the traumatic event or does not have accurate information on how the person died, and thus imagines terrifying scenarios. Often in such children, positive memories of the deceased (a critical aspect of negotiating the bereavement process) may segue into traumatic reminders. Avoidant children may be so detached from their feelings that they are unable to experience their grief. Thus, trauma-focused interventions are typically utilized in the beginning phase of CBT-TB, with bereavement issues addressed later in treatment.

As noted, the group child- and parent-treatment components are provided in parallel. Goals and objectives for each session are coordinated to maximize therapeutic leverage within the participating families. For clarity, the child and parent interventions are described as follows in separate sections. Generally, there are twelve sessions and each group session typically lasts ninety to 120 minutes. The therapists should meet, at least briefly, following each session as a means of providing support and sharing any important information.

CHILDREN'S TRAUMA-FOCUSED INTERVENTIONS

Trauma-focused CBT interventions include components of Stress Inoculation Therapy (SIT), Cognitive Processing (CP), and Gradual Exposure (GE). Specific elements of each component are described as they typically apply in treatment. The order of intervention is somewhat flexible and may be adapted to each group's specific needs, so long as the children's group therapists and parent's group therapists are in agreement. Keep in mind, however, that it is important to address stress inoculation prior to beginning the gradual exposure exercises.

Phase 1: Orientation to the Model, Building Coping Skills

Treatment objectives include

1. Developing a working alliance.
2. Normalizing survivors' distress reactions and establishing fellow members as a comparison group for gauging one's own emotional reaction.
3. Beginning to increase tolerance for trauma-related material by means of gentle exposure exercises and beginning stress-inoculation training.

The therapists should spend time in the initial treatment session orienting the children to the CBT-TB group model. This consists of explaining the reason for the group and what treatment may consist of. The following factors are presented and discussed.

- Someone very important to each of them has died.
- The nature of the death was traumatic.
- The nature of the trauma was intentional (e.g., a group of terrorists planned and purposely carried out acts meant to hurt or kill many innocent people).
- When such a terrible thing happens, people usually have strong feelings and a natural tendency is to not want to talk about it.
- By working with many children who have had such experiences we have learned that talking about these feelings is the best thing to do.
- Sometimes it is especially helpful to talk with a group of others who have had the same kinds of experiences. Everyone in this group has lost someone in similar circumstances.

By using this explanation, group members gently learn the nature of the group. Specifically, children learn right away that their group is comprised of others who have experienced traumatic loss. They learn that the process of the group includes the expectation that each group member will discuss his or her thoughts and feelings about the trauma and loss.

Following the group orientation, the therapists, along with the children, participate in an introductory cohesion-building activity. For example, the children may be asked to make name tags that include something unique about themselves to share with the others (Stubenbort et al., 2001). (It may be helpful if the therapists have stickers, pictures, and/or precut papers of various shapes and colors). Once the children make their name tags, they should be invited to introduce themselves and speak about the special thing they have chosen. An alternative introductory exercise, especially for older children, is to assign each child a partner and have them interview each other for about five minutes. Following these interviews, each child introduces his or her partner to the group.

After everyone has been introduced, each child should be asked to identify the person he or she loved who died in the disaster. The children may also be given the option of saying something additional about the person who died. Following this brief exposure exercise, the therapists should introduce the children to SIT.

Stress Inoculation Therapy

SIT refers to a variety of interventions that protect children from the negative effects of stress, and encourage the use of optimal coping skills. The SIT techniques used in CBT-TB include feelings identification, relaxation, thought stopping, cognitive coping, and enhancement of a sense of safety. As mentioned, many children benefit from mastering these skills prior to embarking on the gradual exposure and cognitive processing components of the model.

Feelings Identification

Feelings identification involves helping the group members to accurately identify and label their feelings, which promotes coping under stressful circumstances. For child survivors of terrorist attacks, the fact that someone intentionally set out to injure or kill their loved ones may lead to intense feelings of anger, fear, hatred, or revenge. The intensity of their feelings may frighten them. Therapists should help the children understand that all of their feelings are acceptable. Feelings are not actions and therefore are not, by themselves, harmful.

Understanding the nature of feelings involves practicing the skill of feelings identification and labeling. To teach these skills, therapists begin by asking the children to name any feeling they can think of. Using a large piece of paper, or a flip chart, a therapist (or one of the children) writes each feeling down. This activity may continue for three to five minutes. Once a list is compiled, therapists instruct the children to practice labeling feelings that occur in diverse situations.

An enjoyable group activity that develops identification and verbalization of feelings is "Feelings Bingo." During this game, each child receives a card of twenty-five squares labeled with "feelings words." One of the therapists randomly selects a card that describes a situation (e.g., receiving an unexpected gift). If a child has a square labeled with a feeling that appropriately describes how someone might feel in this situation (e.g., surprised, excited) they may cover the square. Like ordinary bingo, the first child to cover five squares in a row wins.

Similar to "Feelings Bingo" is a feelings version of the "Go Fish" game. Each child receives up to ten "feelings cards." The "fish pond" has descriptive situations written on cards. The children take turns pulling a card from the fish pond. If they have a feeling word that matches the descriptive card, they may take that pair and set it aside. The first player to make pairs out of all their cards is the winner.

Relaxation

Another set of skills learned and practiced in SIT is relaxation exercises. Relaxation exercises are helpful in reducing the physiologic manifestations of stress and PTSD, which may become especially prominent when children experience traumatic reminders, and may occur during the gradual exposure (GE) exercises described as follows. Deep breathing, or "belly breathing," is very easy to teach children and very easy for them to understand. If there is room, the therapist may instruct the group to lie on the floor. Otherwise, the children may be asked to lean back in their chairs with their legs extended. They are then asked to close their eyes, and breathe in deeply, so that the lower abdomen protrudes during inhalation, and recedes during exhalation. Younger children may be helped to understand this by having them keep their eyes open and setting an object such as a small book or stuffed animal on their lower abdomen. When they can watch the object rise and fall while inhaling and exhaling, they will see that they are doing the belly breathing correctly. To slow the breathing process and increase the effectiveness of the exercise, the therapists may instruct the children to count to five each time they inhale or exhale. Along with slowing the breathing, counting to five directs attention to the task of breathing and away from other thoughts.

Therapists may also want to further direct attention by instructing the children in a guided imagery exercise during the deep breathing practice, to further divert them from distressing thoughts. Guided imagery may be accomplished by first guiding the children to relax and then asking them to choose a place (preferably one that is calm and relaxing, such as a beach or a winter cabin) that they would like to go to. The children are then instructed to close their eyes and imagine being in that place. The children should be slowly instructed to consider smells, sounds, colors, and other sensations that add detail to their images.

Another option is to engage the children in a group imagery exercise. During this activity, the group agrees upon a "place" that they would like go to. The children may simply engage in the imagery exercise, or they may want to draw individual pictures or a group picture of what this place may look like. In either case, the therapists should ask questions that prompt the children to add extensive detail to the pictures. The greater the detail, the more effective the imagery in directing attention away from distressing thoughts and inducing a calm state. Once the children have successfully practiced guided imagery, they are told that they can go to their special "place" whenever they are feeling particularly stressed. Guided imagery is especially helpful when children are having sleep difficulties.

Progressive muscle relaxation, another SIT technique, can also be particularly helpful to children who are experiencing somatic symptoms or having difficulty falling asleep. Some children may understand the concept of muscle relaxation with a simple analogy such as, "stand like a tin soldier versus a Raggedy Ann doll." However, others may need specific instructions on how to progressively relax different muscle groups. During this exercise, children are told to tense and then relax, one set of muscles at a time, starting with the toes, then foot, then ankle, etc., all the way up to the head, until every body part has been progressively relaxed. Once again, progressive relaxation may be helpful to children with TB, because the directed attention precludes focusing on distressing thoughts.

Thought Stopping

Directing attention away from distressing thoughts is the goal of thought stopping. Although it may seem contradictory to use *both* thought-stopping *and* gradual exposure in the same treatment model, children come to understand the important difference between the two strategies. Specifically, thought stopping is useful when the children need to be focused on things going on around them, such as at school or during interactions with friends. Trauma-focused attention (the GE process) can be reserved for therapy sessions. Thought stopping teaches children first and foremost that they have control over their thoughts. For children initially overwhelmed by intrusive reminders or distortions related to the trauma and loss, simply learning that they are in charge of their thoughts can be enormously helpful. Thought stopping is accomplished by simply interrupting an unwanted thought. Children can be taught to give themselves a cue word ("stop") or action (finger snapping) to assist thought stopping.

Some children prepare for thought stopping by having a positive image ready; this is similar but less complex than guided imagery. Various activities facilitate this. For example, the children may want to make a group mural depicting objects that prompt them to recall happy thoughts or events. Younger children may enjoy a bead-stringing activity (Stubenbort et al., 2001). For the latter, therapists will need to purchase a large bag of craft beads, preferably in a variety of colors and interesting shapes. The children are instructed to choose and string beads that remind them of happy thoughts. They are then encouraged to keep their beads under their pillows or in their pockets so that they can touch them and be mindful of happy thoughts when distressing ones are interfering with sleep or daily functioning.

Positive Self-Talk

Positive self-talk is another skill designed to entice children to focus on positive and helpful thoughts as opposed to distressing thoughts. This consists of practicing the skill of focusing on the positive instead of the negative aspects of any given situation. One could easily argue that there is nothing positive to be found following a terrorist disaster. Furthermore, it may take considerable time for surviving family members to come to the point where they find themselves stronger, more compassionate, or even appreciative of the outpouring of sympathy and assistance they experience. However, children who are coping well may benefit from focusing attention on the fact that, despite great adversity, they *are* coping. Positive self-talk interventions consist of helping children recognize and verbalize the manner by which they are coping, particularly when they are feeling discouraged. As a group activity, children may be encouraged to generate a list of positive coping statements, such as "I can get through this," or, "I still have a loving family." This list can then be used to make "encouragement banners" to hang around the group room. Another activity would be to instruct each child to say something positive about the person he or she is sitting next to.

Enhancement of a Sense of Safety

Finally, in the wake of an act of terrorism, children are likely to be experiencing a general loss of trust and safety. Explore safety concerns and examine social supports. Specifically, encouraging discussion around whom the children count on in school, at home, in the community, etc., enhances a practical awareness of support and safety in the here and now.

Phase 2: Cognitive Processing

Treatment objectives include

1. Continuing to strengthen group cohesion.
2. Educating the children about the distinction between thoughts and feelings.
3. Educating the children about the relationship between thoughts, feelings, and behaviors.
4. Increasing the children's awareness of the nature of automatic thoughts.

Many people do not realize that by having control over their thoughts, they are able to change their feelings and behaviors. Educating children about the connections between thoughts, feelings, and behaviors is an essential element of cognitive processing (CP). The first step in this process is to review and practice feelings identification. The next step is to help children recognize the difference between thoughts and feelings. For example, a child is asked, "How would you feel if the person you thought was your best friend didn't invite you to his or her birthday party?" In this example, the child answers, "I would think he or she hated me." The therapists would need to point out that the child had just named a *thought* rather than a *feeling*. The therapists may need to suggest feelings words in order to help the child understand the distinction (e.g., "sad," "disappointed," "rejected," "unloved" are feelings that may have been evoked by the absent birthday party invitation.). The therapists may need to use several examples before the children can make the distinction for themselves.

Once the children seem to understand the distinction between thoughts and feelings, the next step is helping them become more aware of the nature of "automatic thoughts." A therapist may say the following:

> There are some thoughts that we have so often, we aren't even aware
> we are having them. These are called "automatic thoughts" because

they just pop into our heads and we think everyone else probably has the same thoughts at similar times. But some automatic thoughts are not helpful, and they leave us with feelings that hurt us rather than help us.

At this point, therapists present children with several examples for which they must identify a thought and a feeling that would arise from that thought. One scenario could be, "Your mother just blamed you for something that your little sister did." Encourage the children to take turns coming up with helpful and unhelpful thoughts for each scenario. For example, an unhelpful thought might be, "She always takes her side." Feelings that may arise from this thought are "hurt" or "mad." An alternative and more helpful thought could be, "She will understand once I explain what really happened." The accompanying feeling would be "hopeful."

The final step in explaining cognitive processing is to help children recognize the relationship between thoughts, feelings, and behaviors, as well as the relationship between our behaviors and how other people act in response to us. It is sometimes very helpful to use a diagram that is similar to a math problem:

EVENT + THOUGHTS about the event → FEELINGS and,
EVENT + THOUGHTS about the event + FEELINGS → BEHAVIORS

Since children of this age are in school, the diagram gives them a format that they are familiar with and thus, is easy to remember. Using, once again, the example of a big sister being blamed for something that her little sister did,

Thought: "Mom always takes her side."
Feeling: Hurt, mad.
Behavior: Stomping a foot, crying, and saying "I hate you!"
Mother's response: Child gets sent to her room.

and, alternatively (preferably)

Thought: "She'll understand once I explain what really happened."
Feeling: Hopeful.
Behavior: Explaining to Mom what happened.
Mother's response: Listening and apologizing.

Whenever possible, the scenarios should be identified by the children. In this manner, the exercise is more likely to mimic real-life experiences. Furthermore, it increases group cohesion as the children identify with one another's life experiences. However, unless one of the children spontaneously gives an example of his or her thoughts and feelings related to the trauma/death, processing of these cognitions should be done in conjunction with GE techniques in the next phase of treatment.

Phase 3: Gradual Exposure: Creating a Trauma Narrative

Treatment objectives include

1. Strengthening group cohesion.
2. Normalizing members' perception of the trauma.
3. Normalizing members' distress reaction to the trauma and establishing fellow members as a comparison group for gauging one's own reaction.
4. Strengthening members' perception that the group is a safe place to disclose and explore painful reactions to the traumatic loss.
5. Increasing members' tolerance for trauma-related information.

GE has been used in the treatment of sexually abused children (Deblinger & Heflin, 1996; Cohen & Mannarino, 1993), children exposed to community violence (Pynoos & Nader, 1988), disasters (Goenjian et al., 1997), and those exposed to single-episode traumatic events (March et al., 1998). The goal of this intervention is to gradually unpair thoughts, reminders, or discussions of the traumatic event from overwhelming negative emotions that accompany them. Over the course of several sessions, children are encouraged to describe the traumatic events they experienced in increasing detail, along with any thoughts and feelings they may have experienced during these times, and thoughts and feelings they may be having while they discuss it.

Introducing Gradual Exposure

Prior to starting GE sessions, the children (and the parents, as will be discussed) must understand the basis of this intervention. Children and parents are likely to have concerns about directly discussing the events surrounding the death of their loved one. PTSD-based avoidance may play a factor in this reticence, or it may simply be due to the discomfort that is commonly experienced in discussing sad and painful events. The following rationale can be used to explain the intervention:

It is very hard to talk about painful things. In fact, many people wonder if it is a good thing to bring back memories of sad things. We know, though, that people don't just forget about upsetting memories. If they did, they would not be having any problems, and they would not come to therapy in the first place. It's like when you fall off a bicycle and skin your knee on the sidewalk, and dirt and germs get into the wound. You can act like it never happened and leave it alone and hope that it gets better all by itself. Sometimes that works fine. But, usually the wound ends up getting infected. Infections don't usually get better by themselves. They get worse and worse.

Your other choice is to wash the wound out very carefully, getting all of the dirt and germs out of there. It hurts at first, but once the wound is taken care of the pain goes away. It doesn't get infected, and can begin to heal.

GE is like cleaning out the wound. It might be a little painful at first, but we go carefully. We try to go at just the right pace, so that it never hurts more than a little bit and you can let us know at any point if we are going too fast for you, and we will slow down. (Deblinger & Heflin, 1996)

Group Activity: Descriptive Narrative

Children are typically able to understand the rationale for GE once they hear the wound analogy. Thus, the therapists should encourage the children to begin the GE process. As a *group* activity for children who experienced the repercussions of the same traumatic event, the children create a group book that tells about the disaster. One of the therapists takes charge of writing the group book on a poster board or flip chart as the children contribute to telling the story. Some children are likely to be hesitant, while others may try to rush through the process in order to get it over with. The goal is to create a narrative of the events on the day of the terrorist attack. Through use of empathic statements ("This is very difficult to talk about") and encouraging questions ("What happened next?"), the therapists prompt the children to contribute to the narrative while seeking increasingly greater detail. Although it is essential that each child examine his or her own thoughts and feelings experienced during the disaster, the first step is to create the trauma narrative. Interrupting the flow may make it harder for some children to focus on the events and may even encourage avoidance of the details of what occurred.

Writing the group descriptive narrative of what happened is likely to take a whole treatment session. It depends upon how willing the children are to describe the details, whether there are disagreements that must be resolved, and how long a time period is covered in their description. Individual children often have specific memories that may disagree with some of the more general aspects of the trauma narrative. Therefore, they should be told ahead of time that each of them will have the opportunity to put specific details into their own narratives. Introducing the group activity of creating the trauma narrative may go something like this

> Okay, today we are going to write a book about what happened on the day your loved ones died. Now, we know that each of you has your very own story to tell, and you will have a chance to do this in your own way, when you make your very own book. But there are parts of this story that are the same for all of you, and that is where we will begin. I'm going to write everything on this flip chart. I will start at the very beginning of the very first sentence, but you know the story best and I want you to add what happens next. "It was a beautiful, sunny, day. . .'"

At this point the children are encouraged to complete the first sentence and to continue adding to the narrative until the facts surrounding the event have been addressed. At various intervals, the therapist reads aloud what has been written thus far. This is helpful in both desensitizing the children to verbalizing details of the trauma and deaths, and also in keeping them focused on the details of the description. Another option is to have the children take turns reading the narrative. Either way, they are experiencing exposure through the retelling of the trauma and deaths with the goal being to progressively lessen emotional reaction and physiological reactivity. If any of the children appear to experience a high degree of reactivity, the therapists will assist the group in using deep breathing or relaxation techniques. Furthermore, if any of the children experience a high degree of anxiety, and do not respond to SIT interventions, the therapists may want to schedule an individual session or refer them to individual rather than group therapy.

Once the children have completed the group description of what happened on the day of the traumatic event and they have read it in its entirety, they will form smaller groups and begin the process of creating their individual books. To assist the children in creating their individual books, the therapists will provide each child with a copy of the group narrative. This version of the group narrative will have been copied onto ordinary paper, preferably having one or two sentences per page, thus leaving room for the addition of each child's personal thoughts, feelings, and drawings. The therapists may also want to prepare a series of questions to help the children add details to their own books. For example, questions such as, "Where were you when you first heard about the event?" and, "Who told you about the dying?" may make the children more aware of their own personal details. It may also be helpful if the therapists were to display the feelings word list that the children generated in the earlier sessions.

In the smaller group settings, therapists will assist the children in adding feelings, thoughts, and other details to their individual books. The small-group experience of sharing assists in normalizing the emotional experience and builds alliances. At some point in the writing of their books, the children are encouraged to describe their worst moment or worst memory and include this in the book. Each child is encouraged to describe this in as much detail as possible, including drawing a picture of this memory. While doing this, some children may experience fear, revulsion, sadness, or anger. Therapists should encourage the children to write these feelings, and to describe any physical sensations that accompany these feelings. Children are typically able to tolerate this part of the GE task, so long as they have gradually become desensitized to other aspects of the trauma. Nevertheless, it is helpful for the therapists to remind the children that these are feelings that are related to something that has happened in the past, not something that is occurring in the present. The children may again be encouraged to use relaxation techniques as they complete these tasks. Also, thought stopping may be used to remind the children they have control over their thoughts, and they may benefit from a brief distracting thought (such as the day's events at school). Once again, if any child appears significantly distressed, an individual session, or a referral to individual therapy may be warranted.

Given the intentionality of terrorist attacks, it may be helpful to reconvene as a group in order to explore some of the children's thoughts and feelings about this prior to completing the individual GE books. Therapists can point out to the children that thoughts and fantasies about revenge and rescue are normal, and are indicative of how much the children wish that the events had not happened at all. It may be emphasized at the same time that they cannot change the past, but they do have the ability to change some things in the present and the future by their own actions. As an ending to each child's individual GE book, the therapists prompt the children to include a page titled "My Wishes for the Future," or "My Happy Ending." Therapists then encourage the children to think about specific ways to achieve symbolic corrective action in the present and future.

Once the individual GE books are completed, therapists assist each child in selecting a section that he or she would like to read aloud to the group. Using individual narratives, feelings, and thoughts, the group can then use this information to add to the group GE book. Finally, the children may choose to write a corrective ending for the larger GE book. For example, the final page of the group book may be titled "Our Hopes and Dreams for the Future."

Trauma-Focused Cognitive Processing

Treatment objectives include

1. Increasing coping skills surrounding the trauma-related material.
2. Examining and correcting of attributions (assignments of fault, blame, or cause) and other inaccurate and unhelpful cognitions surrounding the trauma-related material.
3. Strengthening group cohesion and normalizing reactions to traumatic loss as members share their thoughts and feelings.

During this phase of cognitive processing, the children are encouraged to explore their thoughts about the trauma/death, and to challenge and correct cognitions that are either inaccurate or unhelpful. Inaccurate cognitions are thoughts that are either absolutely false (e.g., "It's my fault my father got killed"), or impossibly unrealistic (e.g., "I should have warned my father that this terrorist attack was going to happen"). Inaccurate cognitions sometimes include rescue or hero fantasies, which represent children's attempts to gain control in the wake of an uncontrollable situation. This is a common response to posttrauma fears of a world out of control. However, attempting to gain control over unpredictable events at the cost of blaming oneself is rarely helpful in promoting optimal adjustment. Unhelpful cognitions may be inaccurate, as in the previous example, or they may be technically accurate (e.g., dying in a fire is terribly painful), but likely to contribute to ongoing distress. Replacing such cognitions with equally accurate but more helpful thoughts, may aid children's ability to tolerate trauma or loss reminders.

The children have already been familiarized with the concepts of cognitive processing (changing thoughts to revise feelings and behaviors). After reviewing these cognitive processing skills, therapists begin to directly explore and correct trauma-specific cognitive errors. Both inaccurate and unhelpful cognitions about the traumatic death will have been identified as the children created their GE books. At this point, therapists ask the children to read portions of their books, with a focus on the thoughts being expressed and on the feelings that the thoughts evoke. The group then explores whether particular thoughts are accurate and helpful. For example, a child may write, "I should have told my mother not to go to work that day. Then, she would not have died." The following dialogue illustrates how progressive logical questioning is used to address this inaccurate cognition:

THERAPIST: Hmm. I'm wondering about this thought. I'm wondering how it might make you feel.

CHILD #1: It makes me feel sad. It makes me feel like what happened to my mom is my fault.

THERAPIST: I can certainly understand why you are feeling sad. But, I'm wondering if your thought about being able to warn your mother is helpful for you.

CHILD #2: If we would have warned them, none of us would be here right now!

THERAPIST: That's true. But did *anyone* know that the terrorist attack was going to happen? I mean, if people would have known—like if the president would have known, don't you think he would have warned everybody?

CHILD #1: Someone just should have known. I should have known—she's my mother! She always seemed to know. She always warned me!

CHILD #3: Yeah, that's just the way it is with Moms.

THERAPIST: Help me understand this. Are you guys saying there were probably signs, or warnings, or something obvious that you or someone else should have seen to indicate that there was going to be a terrorist attack? Is that what you mean?

CHILDREN: Yes!
I don't know.
Maybe.
We should have known.

THERAPIST: So, you should have known. You should have been able to predict it—like predict the future?

CHILD #1: Well, not predict the future. But I wish I knew.

THERAPIST: I understand that. *Wishing* you would have known and *thinking* that you should have known are very different things. And these thoughts create very different feelings. I think many of us wish we would have known what was going to happen. I, too, wish we had known those terrorists were coming. But we didn't know. We were surprised. And that was part of their plan . . . to surprise us. So, how do you feel when you say you *wish* you would have known?

CHILDREN: Sad, very sad.

THERAPIST: How do you feel when you say you should have known?

CHILDREN: Guilty, bad, horrible.

THERAPIST: Which thought is more accurate, more helpful?

CHILDREN: Wishing.

In addition to self-blame for the traumatic event, some children may be struggling with intrusive, horrifying thoughts about the agony and suffering experienced by their loved ones prior to death. These accurate but unhelpful cognitions are perceived by some as necessary to truly deal with the situation. In truth, focusing on horrifying realities or possible realities of the trauma/death may impair a person's ability to optimally cope with the trauma and loss. It is important to help the children realize that they need not perseverate on these distressing and unhelpful thoughts. Therapists can suggest alternative thoughts that may be more helpful. For example, it may be more helpful for a child to think that his or her parent's final thoughts were thoughts of loving the child as opposed to terror-filled or painful thoughts. The use of progressive logical questioning can help children realize that they can choose to think about the same situation in different ways, and some ways are more helpful than others.

Joint Parent–Child Session

Treatment objectives include

1. Increasing the child's perception of the parent as an optimal source of emotional support.
2. Increasing the parent's awareness of the child's perception of the trauma.
3. Facilitating communication and emotional exchange between parent and child regarding trauma-related material.

This model includes at least one joint session in which the child and parent meet together, with a therapist, to read the child's GE book. This session typically occurs following the trauma-focused phase and is scheduled apart from the regularly scheduled group session (which will still be meeting). Usually, one joint session is sufficient, although in some cases, two sessions may be optimal to address a trauma-focused GE book. The parent and therapist should make this decision together.

Prior to the joint sessions, the children will have had the experience of reading portions of their books and getting feedback within the group. Furthermore, the parents will have heard about the book during their group sessions. Preparing the parents for the joint session is critical to increasing their ability to emotionally tolerate the actual reading of the book. This is not to say that the parents will not experience any emotional distress. But, prior exposure to the book increases the likelihood that the parents will not engage in uncontrollable sobbing or extreme avoidant coping mechanisms. Thus, each parent will be able to offer emotional and verbal support to the child during the joint session

Each child-parent session should include only one therapist. The session typically lasts about one hour. The therapist will first meet with the child for about fifteen minutes. During this time, the therapist prompts the child to read the GE book out loud. Reading the book aloud to the therapist helps prepare the child for the task of reading the book to the parent. The child should also be told that he or she will be asked to read the book aloud to the parent when the parent joins the session. The therapist may then encourage the child to generate a list of questions that he or she would like to ask the parent regarding the trauma, the book, or treatment thus far. The goal behind this is to begin to open the lines of communication regarding trauma-related material. Following the meeting with the child, the therapist meets with the parent for about fifteen minutes. During this time the therapist reviews the child's book with the parent and answer any questions the parent may have. This gives the parent an opportunity to respond emotionally to the book and to formulate supportive answers to their child's questions. The parent may also have questions for the child, and the therapist may help the parent prepare these. Finally, the therapist meets with the child and parent together for the remainder of the session.

The joint session begins with the child reading his or her book to the parent. Some children are reticent to do this. Occasionally, a child may be so distressed over the idea of reading the book aloud to the parent that the therapist will find it best to do the reading. In any case, the parent and the therapist should praise the child for the courage it took to write the book and share it with the parent. The general rule for this session is to allow the parent and child to communicate about the book. Other activities that take place during the joint session include practicing safety skills and sharing questions about the trauma. The therapist's role is to facilitate communication, especially if unhelpful cognitions arise.

During their respective group sessions during this week, the parents and children will have the opportunity to discuss the experiences they had in the joint sessions, their feelings about the sessions, how the child or parent coped with exposure, and anything that is relevant.

CHILDREN'S BEREAVEMENT-FOCUSED INTERVENTIONS

Phase 4: Mourning the Loss

Treatment objectives include

1. Increasing group cohesion through collaborative exercises.
2. Normalizing distress reactions related to the loss.
3. Increasing understanding of and challenging misconceptions about death.
4. Identifying and discussing unfinished business or ambivalent feelings about the deceased.

The bereavement process typically involves specific phases: mourning the loss, recognizing and resolving ambivalent feelings about the deceased, preserving the deceased in memory, and redefining the relationship and moving on (Rando, 1996; Worden, 1996). However, children must have some understanding of death before they are able to begin to mourn.

Even after addressing the traumatic nature of the event, the children may continue to have difficulty understanding and talking about death. This may be due to cognitive and affective immaturity. It may also be the result of adult modeling. The loss of one parent typically means that the surviving parent is grieving and, consequently, may be less emotionally available for the child. Furthermore, adults often harbor the misconception that children are resilient and will forget about trauma if left alone. Thus, they may not talk with their children about the death, thereby modeling avoidance.

Finally, it is not known what happens when someone dies, and there are various belief systems that must be respected. For this reason, this phase of treatment begins with a psychoeducation activity. For example, the therapist may ask the children to draw pictures about what they believe happens when someone dies (Stubenbort et al., 2001). When the drawings are complete, each child should be asked to share his or her picture with the group. The therapists can use this activity to educate the children about death and to correct any misconceptions that arise, with respect for individual cultural and spiritual beliefs.

Once the children have had the opportunity to talk about their pictures and address their beliefs about death, they are directed to generate a list of the feelings that people have when someone they love dies. The next step is to encourage each child to talk about his or her loss. One way to do this is to ask the children to draw a picture, or talk about how they learned of the death of their loved one. Therapists encourage the children to refer to the list of feelings words. This may help them express themselves, while increasing tolerance for bereavement-related material. Furthermore, it increases group cohesion, as it is likely that the children will find a shared emotional experience in this activity.

Once the children have explored their feelings and beliefs about death, it is necessary to move on to mourning the loss of the relationship that they had with the deceased. This requires remembering, identifying, and naming the many things that they did with their loved ones. This may include basic caregiving tasks along with special events, such as birthday parties, holidays, and the like. Remembering also includes addressing "unfinished business." For example, one child may have had an unsettled disagreement with the deceased just prior to the death, another may have wanted to say, "I love you" just one more time. Finally, it is also important to address the loss of things that would likely have occurred in the future, such as graduations, weddings, and other meaningful events that occur in the course of growing up.

One method of addressing memories is to have the children write letters to their deceased loved one. Therapists can assist the children in this task in several ways. For example, the therapists may have a letter "template" from which the children can fill in the blanks. This method is especially helpful for younger children, who may have difficulty with writing in general. A second approach might be to assign each child a partner with whom he or she will discuss his or her loss; express feelings, talk about memories, and share other thoughts. Once the children are acquainted with their partners, they are instructed to write letters to the deceased on each other's behalf (Stubenbort et al., 2001). This method is especially helpful for children who are having considerable difficulty talking about their own feelings pertaining to the loss. The task of communicating on behalf of another is likely to increase their ability to communicate more personally. Finally, the children may be encouraged to write their letters on their own, with the simple direction of expressing their feelings, thoughts, memories, and expectations for the future. No matter the method of writing the letters, the children should be encouraged to say all of the things that they wished they'd had an opportunity to say to their loved ones. When the letters are complete, the children should share their letters with the group, while therapists listen for, and prepare to address unhelpful cognitions, which may disrupt the process of bereavement.

Another aspect of mourning is the ability to focus on positive aspects of the relationship shared with the deceased. This can be both a very painful, but positive experience. For this exercise, therapists request that the children bring a photo or some sort of memorabilia to group. Each child is then encouraged to share his or her particular item with the group and describe why that item was chosen. Following this sharing activity, the children are given the opportunity to begin preserving memories, beginning with the item that they brought to group. Some children may choose to make a special frame for their photograph (Stubenbort et al., 2001) or a memory box, which consists of photos, keepsakes, poems, or other writings that assist in remembering loved ones (Worden, 1996). Younger children may need the help of older siblings or a surviving parent to assist them with memory tasks.

Phase 5: Redefining the Relationship and Moving On

Treatment objectives include

1. Increasing group cohesion through collaborative exercises.
2. Beginning the process of placing the deceased in memory.
3. Beginning the process of moving on with life.
4. Making meaning of the loss.
5. Anticipating and planning for reminders of the trauma and loss.

Once the children have addressed mourning the loss and preserving memories, it is time to begin to redefine the relationship and initiate moving on. This requires the children to begin accepting that the relationship with the loved one is not an interactive one in the present, but rather a relationship of memory (Wolfelt, 1991). One activity to assist in this task is a balloon activity (Stubenbort et al., 2001) during which the children are given two drawings of balloons. In one drawing, the balloon is floating away in the air, and in the other the balloon is anchored on the ground. The floating balloon represents things the child has lost, while the anchored balloon symbolizes many of the things that the child still has, including memories. The children are asked to fill each balloon with words or drawings that describe what they have lost, and what they still have. Once the balloons have been filled in, the group comes together to share their balloon pictures. The activity and the resulting discussion allow the children to concretely recognize that, although memories remain in the present, the interactive relationship is gone.

To support the ability to move on, it is important to help the group anticipate and prepare for trauma and loss reminders. One activity to facilitate this is the creation of a time line or a "Circle of Grief" calendar. For this activity, the children are invited to draw a large circle on a piece of paper. Beginning with January 1 at the top of the page, and going clockwise around the circle, the children are encouraged to record the significant days of the year, such as birthdays, holidays, and other events they suspect may lead to reminders of the trauma and/or the loss. The children should be told that they may continue to add other important dates to the circle as time goes on. The Circle of Grief empowers children to better understand emotional shifts and manage events that may lead to a change in feelings over the months and years following the loss (personal communication with Robert Zucker, MA, Director of Caring Communities Respond, Western Massachusetts, November 15, 2002). Remind them to think about things such as the weather, seasonal changes, and other significant events (such as the beginning of the school year, as in the events of 9/11) that surround the day of the trauma or memorial service, as often these things trigger memories.

The final phase of the bereavement process is making meaning of the loss. Although it may be difficult to imagine that one could find meaning in the wake of a terrorist disaster, doing so enables the children to integrate the trauma into their existing identity and future worldview. For example, some children may be able to identify ways in which they have become stronger since the trauma. Others may find that their family has grown closer. Various activities can assist children in making meaning of the loss. Younger children enjoy the task of "Making Three Wishes" or creating a "Happy Ending" story. Older children appreciate the idea of helping other children. Thus, asking them questions such as, "If you met other children whose parents died like yours did, what would you tell them?" or, "What do you think was most helpful about therapy?" Older children are also likely to benefit from a *corrective activity* such as writing a column for their school paper or writing a letter to a local news editor.

Finally, it is important to let the children know that they have done an incredible job and have begun working through a terrible situation. They need to remember that there will be times when they will experience feelings of loss and sadness. The therapists can assist the children in remembering this by teaching them about the "Three Ps". This includes the following:

1. *Predict* to the children that they will have times of sadness and grief throughout various points in life. These times may be triggered by loss reminders, trauma reminders, or events, such as birthdays or anniversaries that they have identified as prompting feelings and memories of the deceased.
2. *Plan* for how to optimally cope with these times. This plan may include talking to a parent or other significant person, using a specific relaxation technique, visiting a memorial site, looking at the bereavement book, or any other activity that will bring comfort.

3. Give yourself *permission* to experience these feelings at any point in life, and to express these feelings without construing them as a sign of pathology. (Personal communication with Sue Padlo, LCSW, Senior Psychiatric Clinician, Center for Traumatic Stress, Pittsburgh, PA, November 8, 2002.)

Parents also need to learn and practice the Three Ps and to reinforce them in their children. They, too, should plan to seek support at those times.

PARENTS' TRAUMA-FOCUSED INTERVENTIONS

Frequently, when children are experiencing traumatic bereavement, their parents are also suffering because they too have known or are related to the deceased. Prior to the onset of the parents' group, therapists need to emphasize that this model is child-focused. Thus, any parent who is experiencing severe PTSD or other psychiatric symptoms may need to be referred to individual treatment. This is especially important if a parent's symptoms are significantly compromising his or her emotional availability or judgment to the point that the therapists believe symptoms are interfering with adequate parenting practices. For the most part, severe symptomatology would have been recognized at the time of evaluation and appropriate referrals would have been made. If this is not the case, the therapists can address concerns with any parent individually in a supportive and nonjudgmental manner.

As mentioned, the parent interventions parallel those of the children. In this manner, the parents are aware of the content covered in the children's sessions, and are optimally prepared to reinforce/discuss this material with the children between sessions. Various studies demonstrate the benefit of actively involving parents in the treatment process (Cohen & Mannarino, 1998, 2000; Deblinger et al., 1996). Thus, adding a parent component enhances the effectiveness of TB-focused CBT interventions. Start each session by "checking in," that is, asking how the parents have been doing since the last treatment session.

Phase 1: Orientation to the Model, Parenting, and Building Coping Skills

Treatment objectives include

1. Developing a working alliance.
2. Normalizing distress reactions and establishing fellow members as a comparison group for gauging one's own emotional reaction.

3. Facilitating optimal caretaking practices.
4. Establishing the parents as the ultimate source of support for their children.

Just as in the first session with the children, the therapists should spend time in the initial session orienting the parents to the CBT-TB model. This consists of explaining the philosophy of this approach, and should include the following factors:

- Someone very important to each of them and their children has died.
- The nature of death was traumatic.
- The nature of the trauma was intentional (e.g., a group of terrorist-planned and purposely carried out acts meant to hurt or kill many innocent people).
- Their children are having significant PTSD symptoms related to this experience.
- These symptoms need to be addressed in order to proceed to the bereavement issues.
- One characteristic of PTSD is a tendency not to want to talk about it.
- By working with many children who have had such experiences experts have learned that talking about these feelings is the best approach.
- The intervention process will encourage the parents and their children to talk about their thoughts and feelings in a gradual, supportive manner so that they will be able to tolerate the discomfort associated with such discussion.
- Sometimes it is especially helpful to talk with others who have had similar experiences. In this case, everyone in the group has lost someone in similar circumstances.

The therapists for the adults should assure the parents that they are working in collaboration with their children's therapists. With this in mind, any feedback the parents hear from their children is welcomed. Furthermore, parents should be encouraged to share any concerns that arise as their children proceed through therapy.

Building Parenting Skills

The parenting skills included in this model are basic and easy to learn, but nevertheless have been found to have great impact on parenting abilities and behavior problems in parents of children experiencing stressors such as sexual abuse (Cohen & Mannarino, 1996a; Deblinger et al., 1996). These skills include the use of praise, active ignoring, effective time-out procedures, and contingency reinforcement schedules (behavior charts).

Attention-seeking behavior. Because children seek their parents' attention, the first skill reviewed with the parents is the use of praise or positive attention. Most parents believe that they praise their children frequently and consistently, but in fact many parents do not attend to their children when they are behaving *well*. More often, parents pay attention to their children in the form of hollering or criticizing only when they engage in negative behaviors.

Therapists should begin this phase of treatment by having parents generate a list of things their children do well, or reasons they are proud of their children. Once this list has been generated, the therapists should ask whether parents recall recently praising their children for any of these behaviors. Many parents have no trouble recalling praising their children for specific things, such as getting an "A" on a test. However, most parents admit that their children's more general positive behaviors are taken for granted. In fact, some parents will say that they do not believe they should praise their children for behaviors that are expected of them, such as picking up after themselves. Other parents may have difficulty identifying any positive behaviors in their children. Therapists should encourage these parents to find something praiseworthy. This may be something as simple as sitting quietly.

The next step is to instruct the parents in the use of effective praise. This includes:

- Praise specific behaviors so that their children are able to identify what it is that pleases them.
- Provide praise as soon as possible after the behavior has occurred.
- Praise desired behaviors consistently.
- Don't turn praise into criticism (e.g., "Thank you for asking politely, instead of just taking things like you usually do").
- Praise with the same level of intensity that is used for criticism.

Once they understand the specifics of praise, the group can practice by role-playing various situations. Finally, therapists instruct the parents to focus on actively praising their children for any positive behaviors that they notice in the coming week, and note the effect of this praise on their children's mood and subsequent behavior. If a child responds negatively to an instance of praise, the parent should take this opportunity to practice active ignoring.

As mentioned, children seek attention from their parents—even if the attention they receive is negative, such as yelling. Thus, parents often reinforce negative behaviors in their children by giving them the attention their children were hoping to elicit. Active ignoring is the act of consciously withdrawing attention from undesired behaviors. In order to increase the instance of positive behaviors, parents should learn to ignore negative behaviors and praise positive behaviors. Of course, parents should never ignore behaviors that may be dangerous, but in most cases, children's behaviors are simply unpleasant and are meant to get their parents' attention.

Parents should be warned that once they begin ignoring their children's negative attention-seeking behaviors, their children might try harder to get their attention by increasing these behaviors. The parents might find it easier to tolerate and ignore any increase in attention-seeking behavior if they keep this thought in mind. As long as the behaviors are not harmful, the parents should continue to ignore them. On the contrary, if the negative attention-seeking behaviors become hostile or dangerous in any way, parents may need to resort to the use of time-out.

Time-out. The purpose of the time-out is twofold. Time-outs deprive children of the attention they are seeking. Time-outs also interrupt children's negative behaviors and allows them time to regain emotional and behavioral control. The proper use of a time-out includes the following: The child is calmly instructed to discontinue a *specific* behavior (e.g., "Please stop kicking"). If the behavior continues, the child is warned, only once, that if the behavior continues he or she will need to go to time-out. If the behavior continues, the parent should calmly and quietly escort the child to the time-out area. Time-out should occur in a quiet area, free of stimulation, such as a corner or a hallway (sending a child to his or her room often does not work, because children have toys, televisions, and other distractions). The child should be placed in time-out for about one minute for every year of age. A timer is helpful because it further removes the parent from interacting with the child. The timing begins when the child has quietly settled. When time-out is over, the parent should not lecture the child about the undesired behavior. Rather, the parent should offer praise as soon as the child engages in some positive behavior. Thus, the child learns that good behavior leads to positive parental attention whereas bad behavior leads to time-out (no attention).

Parents need to be instructed to explain time-out to their children before using it for the first time. Specifically, parents should tell their children they must obey requests to stop particular behaviors, or they will be placed in time-out. Children should be shown where the time-out area is. They should be told that when they are in time-out they must sit quietly for a designated period of time (e.g., four minutes). Children should also be told that they will know when the time is up because the timer will ring (or they will be told).

When used properly, time-out is an effective behavior management technique, especially for younger children. Parents often see rapid improvements in their children's minor behavior problems. Thus, they feel more competent about their own parenting skills. More serious behaviors may be addressed through the use of contingency reinforcement programs, otherwise known as behavior charts.

Behavior charts. Behavior charts are useful for decreasing undesirable behaviors and making children more aware of desired behaviors. Briefly, points are awarded in the form of stars, stickers, check marks, and the like, when a child engages in specified behaviors; and the child is rewarded a chosen privilege when he or she earns a specified number of points. The details of this intervention are provided elsewhere (Bloomquist, 1996). Therapists are urged to refer to this resource for specific instructions.

Finally, if a child exhibits severe behavioral problems, or if the therapists find that a particular child's behaviors are becoming a main focus of the group, referral for ancillary treatment to address those problems may be advisable.

Parents are instructed in the use of the same SIT interventions as their children. This includes deep breathing, progressive muscle relaxation, thought stopping, cognitive coping, and enhancing the parents' sense of safety. With this knowledge, parents can practice and reinforce these skills in their children. They are also likely to benefit from using these techniques for managing their own stress and anxiety. All of the SIT techniques may be taught to the parents in a manner similar to that used with their children. In addition, parents who are burdened by intrusive distressful thoughts may benefit from meditation (the practice of focusing one's attention on a single predetermined thing) and/or paradoxical intention (choosing to think upsetting thoughts for a predetermined amount of time, after which they stop such thinking). Parents may choose to use meditation to reduce stress and increase their ability to control their thoughts. The use of paradoxical intention gives parents permission to attend to worrisome thoughts in a controlled manner and thus makes it easier to stop the intrusive nature of these thoughts.

Phase 2: Cognitive Processing

Treatment objectives include

1. Continuing to strengthen group cohesion.
2. Educating parents on the distinction between thoughts and feelings.

3. Increasing parents' awareness of the nature of automatic thoughts.
4. Promoting the understanding of the relationship between thoughts, feelings, and behaviors.

Educating the parents about cognitive processing may be done in the same manner that was used for the children, using examples the parents can identify with and eliciting alternative thoughts and outcomes. As with the children, the parents are shown how to distinguish between thoughts and feelings. Parents are encouraged to identify their negative thoughts and replace them with more positive thoughts that may lead to less upsetting feelings and more productive behaviors. As in the children's session, therapists do not attempt to elicit trauma-specific thoughts at this session. If thoughts and feelings specific to the trauma/death do arise, the therapists should utilize the appropriate CP techniques to explore and correct cognitive distortions.

Phase 3: Gradual Exposure/Hearing the Children's Trauma Narrative

Treatment objectives include

1. Strengthening group cohesion.
2. Normalizing members' perceptions of the trauma.
3. Normalizing members' distress reactions to the trauma and establishing fellow members as a comparison group for gauging one's own reaction.
4. Strengthening members' perceptions of the group as a safe place to disclose and explore painful reactions to the traumatic loss.
5. Increasing members' tolerance for trauma-related information.

To begin the GE component, therapists encourage the parents to talk about their specific bereavement customs. This topic gently encourages the parents to discuss aspects of the trauma and death. Following this discussion therapists explain the GE procedure using the same "wound" analogy that it is used to explain the procedure to the children (see Phase 3 of the children's sessions). Parents may have concerns that the procedure will be stressful for themselves and their children. It may be helpful to predict that it will indeed be a difficult aspect of therapy. In fact, it may help to let the parents know that some children may resist coming to group and may even transiently show more symptoms during this phase of treatment. The parents are instructed to let therapists know if this occurs so they can share this with the children's therapists. It is also important to let the parents know that most children tolerate GE very well if it is correctly calibrated and they are given appropriate support from therapists and parents. The parents are encouraged to praise their children for tolerating the sessions and talking about these painful experiences. The therapists should also praise the parents for attendance and support of their children.

Finally, along with resolving the children's PTSD symptoms, a critical goal of GE is to facilitate open communication between the parents and their children regarding the trauma and death. The parents should understand that their ability to provide emotional support to their children, along with their ability to tolerate discussion of the trauma will encourage their children to talk with them about any problems that arise in the future. Most parents are eager to accomplish this goal, and support the GE procedures when explained in this manner.

Once the children begin creating the group GE book, the therapists share portions of the book in the parallel parent sessions. The children's book is shared with the parents so that they are able to tolerate hearing their children's upset thoughts and feelings in preparation for hearing and seeing their children's *individual* GE books during the joint sessions. Some parents may have an urge to "correct" inaccuracies in the children's books. The therapists must explain that the point of the book(s) is not factual accuracy, but to desensitize the children to upsetting thoughts and images. As further preparation for the joint sessions, the therapists should encourage parents to talk about their own experience of the trauma. The therapist may encourage the parents to talk about a specific aspect of the event in order to focus the discussion and give each adult an opportunity to talk. Finally, parents should have adequate time to gain their composure at the end of each GE session, prior to rejoining with their children.

Trauma-focused cognitive processing

Treatment objectives include

1. Increasing coping skills surrounding the trauma-related material.
2. Examining and correcting inaccurate attributions and unhelpful cognitions surrounding trauma-related material.
3. Strengthening group cohesion and normalizing reactions to traumatic loss as members share their thoughts and feelings.
4. Strengthening the parents' ability to respond to their children's cognitions about the trauma and death.

During the GE sessions with parents, therapists will recognize various cognitive errors and/or distortions regarding the trauma or the children's well-being. Similar to the interventions with the children, the parents should be encouraged to examine these thoughts. They should then be asked to identify thoughts about the traumatic event, the death of the loved one, or their children's well-being and to use the cognitive interventions to understand the impact of those thoughts on their feelings and behaviors. The therapists should model progressive logical questioning to address these unhelpful thoughts. Finally, the therapists should ask the parents if they are aware of any cognitive distortions that their children may have and encourage the group to role-play effective challenging of their children's cognitions.

PARENTS' BEREAVEMENT-FOCUSED INTERVENTIONS

Phase 4: Mourning the Loss

Treatment objectives include:

1. Increasing group cohesion through normalizing distress reactions related to the loss.
2. Increasing understanding, and challenging misconceptions about traumatic bereavement.
3. Identifying and discussing unfinished business or ambivalent feelings about the deceased.

Therapists must understand and respect parents' individual religious and cultural beliefs, and provide a nonjudgmental, accepting setting for discussing issues that may arise in this regard. Furthermore, therapists must take time to help the parents understand developmentally appropriate childhood perceptions of death. Often, parents need considerable help in understanding children's reactions to death. They are likely to be upset or confused if their children show very little emotion about the death of the loved one. Therapists can address these concerns from a developmental perspective, as well as symptomatic perspective (e.g., avoidance). The parents must also be aware that it is possible a child is attempting to shield his or her parent from knowing how upset he or she truly is.

Although the CBT-TB treatment model is child-focused, parents do, to some extent, address their own trauma and bereavement issues in treatment. Therapists should encourage the parental expression of feelings related to the death of the loved one, and assist them in addressing and, to some extent, resolving personal bereavement issues. As parents become more comfortable discussing these feelings (including ambivalent feelings about the deceased if applicable), the therapists should emphasize the importance of parents expressing their own feelings to their children in an age-appropriate manner. This models to the child that it is okay to talk about death, the deceased loved one, and even to express negative feelings about loved ones, alive or dead.

Many of the bereaved parents will have fond memories of their relationships with the deceased, while others will have had less than perfect relationships. Thus, many in the group may be struggling with various thoughts and feelings related to their losses. To encourage discussion of their relationships, the parents are asked to bring a photo or a piece of memorabilia that represents something about their relationship with the deceased. This object is then used to promote discussion of their losses in the context of day-to-day activities, feelings, and memories. The parents are encouraged to reflect on their entire relationships with the deceased and not just the good times, the difficult times, or the manner in which they died. This discussion is designed to assist in preserving positive memories of the deceased, normalizing the ambivalent feelings that frequently accompany loss, and prompting empathic and supportive conversation among group members.

Several parents may be struggling with significant guilt or regret over things they did or did not do with or for the deceased. These feelings may be exaggerated by the traumatic or sudden manner of death. There are usually legitimate reasons for past feelings and behaviors, and it is important that they not lose sight of this in the height of emotions following the traumatic loss. When members are unable to come to terms with these issues in the course of group discussion, a "best friends role-play" may be helpful (Deblinger & Heflin, 1996).

A "best friends role-play" encourages parents to be kinder in their self-assessments. During this activity, they may break into dyads and imagine that they are one anothers' best friends. In role-reversal style, parent Number 1 tells parent Number 2 the disparaging, guilt-ridden things that parent Number 2 has been saying about himself or herself. For example, parent Number 1 may say to parent Number 2, "I was a terrible wife. I was so involved in my career. There were so many days when I wished that I had never gotten married or had kids. . . . I never really showed him that I appreciated him. And now I'll never have the chance." Parent Number 2, the "best friend," then responds in a manner meant to correct cognitive distortions and incorrect attributions. For example, "Does enjoying your career make you a terrible wife? Didn't your career and income contribute to the family's well-being? I'm guessing your husband appreciated the fact that he wasn't solely responsible for the family's well-being. . . ." Through this role-play, the parent often is more supportive of the "best friend" than he or she has been of himself or herself. The parents are encouraged to be their own best friends as they continue to negotiate their own and their children's bereavement issues.

In addition to grieving the loss of their loved ones, parents are frequently grieving the losses that their children must bear, such as being raised in a single-parent home. The parents' group discussion of these issues is geared to normalizing feelings and encouraging problem solving in anticipation of difficult times in the future. Some parents may experience feelings of resentment and anger when they consider these losses. At the same time they may feel guilty for having these feelings. Therapists should encourage discussion of these feelings through normalizing and cohesion-building activity.

Parents should understand that their children did not have the same relationship with the deceased as they did. Thus, their children may have very different feelings about the deceased. Therapists can make the parents aware of some of the ambivalence that their children have expressed, and help them recognize that some children may need to resolve their own "unfinished business" with the deceased. Even in the presence of ambivalence, parents should encourage their children in recalling and preserving positive memories of the deceased. Therapists can explain the concept of "benevolent intent," specifically, the idea that the deceased meant well and wanted good for the child, even in the presence of negative characteristics. This concept does not take away the negative characteristics, but allows the child to more accurately appraise the relationship. Finally, the parents are encouraged to accept their children's views of the deceased as being valid for their particular child, and to focus on ways to help their children resolve any "unfinished business" with the deceased, regardless of whether they share an identical view of the person.

Phase 5: Redefining the Relationship and Moving On

Treatment objectives include

1. Increasing group cohesion through collaborative exercises.
2. Beginning the process of placing the deceased in memory.
3. Beginning the process of moving on in life.
4. Making meaning of the loss.
5. Anticipating and planning for reminders of the trauma and loss.

With the understanding that ambivalent feelings and unfinished business are common following loss of a loved one, especially when that loss is unexpected, therapists ask the parents to consider communicating some of these feelings to their loved ones in the form of a letter. The parents should be told that they will be invited (but not required) to share their letters with the rest of the group in a future session. They should also consider sharing their thoughts and feelings with their children. In this manner, they are modeling communication as a positive way of coping.

Similar to their children, the parents need to prepare for reminders of the trauma and loss. Parents are encouraged to identify dates, events, and sensations that are likely to activate trauma and/or loss reminders. Like their children, the parents may want to make calendars or prepare some concrete way to identify potential triggers. Along with tangible triggers, parents must be made aware of other means by which trauma reminders may occur. For example, if a spouse typically wore a certain cologne, that scent may prompt intense feelings of sadness or loss.

Parents are encouraged to find meaning in the loss and committing to present relationships. Some parents may choose to turn their most difficult experience into something positive by engaging in activities that may help others to cope when faced with similar tragedies. An example of this is the USAir Flight 427 Air Disaster Support League, established by survivors of the victims of that airline crash, which occurred on September 8, 1994. The League has traveled around the United States, providing assistance and support to family members of other airline crash victims, and has had a major impact on how airlines address the needs of these survivors (Stubenbort et al., 2001).

Therapists also remind parents that they have experienced times of sadness and grief throughout various points in life. These times may be triggered by trauma or loss reminders, which they can prepare for, to a certain degree (such as birthdays, anniversaries, etc.). They should be reminded of the "Three Ps." Specifically, they should be *predicting* and *planning* for times of recurring sadness and loss, and giving themselves and their children *permission* to express these feelings without worries that they are pathological.

Closure

Finally, as a closing activity, parents and children may want to hold a memorial service. Many children with TB struggle with cognitive distortions about the meaning and consequences of death resulting in body disfigurement, dismemberment, or fragmentation of body parts. In the absence of a body, as may be the case following a disaster, memorial services may make the death more real to the child and thus facilitate the mourning process. The children's memorial service gives them the opportunity to orchestrate their own special tribute to their loved ones. Each child in the group chooses or is assigned a special task (e.g., bringing a snack, writing an "announcement," preparing invitations for the parents). The nature of the memorial service itself will be up to the children involved. Some may choose to write a group poem, or each child may wish to write a poem. Another group of children may choose to make a memorial of photos and give each child the opportunity speak. Therapists should ensure that each child participates in the service and has an opportunity to memorialize his or her loved one.

Some children may still be hesitant to share their activities with their parents for fear of creating emotional distress in their parents. They should understand that if their parents cry during the memorial service, they are crying because they are sad about the loss, not because the children made them cry. Furthermore, it is recommended that the therapists for the adults apprise the parents of the nature of the memorial service beforehand. Thus, if any parent has a strong emotional response, they will experience the support of their group and not overwhelm the children.

CONCLUSION

Uncomplicated bereavement in childhood, even when it involves the loss of a parent, does not appear to place children at increased risk for ongoing mental illness, provided they are able to resolve their loss and experience adequate parenting following the death (Harrington & Harrison, 1999). However, if the attachment to the deceased is not changed, specifically, if the emotional investment in the deceased cannot be transferred to others, it may interfere with normal development (Rando, 1988). In instances of traumatic bereavement, the traumatic aspects of the loss are likely to interfere with the normal process of bereavement. Thus, children with TB need treatment that addresses both the trauma symptoms and the bereavement issues that they are experiencing.

The treatment model presented was developed to address these children's needs. We chose to adapt trauma-focused CBT for use in this population, because of all the interventions that have been scientifically evaluated for treating childhood PTSD, CBT has the most support (Cohen et al., 2000). The inclusion of parents in the children's treatment is believed to be critical to the well-being of the children. Studies of sexually abused children have shown that enhancing parental support of the child (Cohen & Mannarino, 1998, 2000) and helping the parents resolve their own emotional distress (Cohen & Mannarino, 1996a) predicts better outcomes for children. The group experience is predicted to further enhance both emotional child and adult support.

Dyregrov (1997) describes the importance of the group experience following mass trauma. Specifically, the group has the potential to normalize and validate cognitive and affective reactions, provide a supportive peer environment, and serve as a resource for its members. Furthermore, it has been found that group intervention facilitates relationships that may be maintained following the group experience, thus preserving benefit following intervention. Group intervention may provide the most efficient and effective form of therapy following a terrorist disaster. The bonds that are formed may carry the survivors through the long process of bereavement and ongoing difficulties related to the nature of the disaster.

REFERENCES

Achenbach, T.M. & Edelbrock, C.S. (1983). *Manual for the Child Behavior Checklist and Revised Child Behavior Profile.* Burlington, VT: University of Vermont, Department of Psychiatry.

American Academy of Child and Adolescent Psychiatry (AACAP) (1998). Practice parameters for the diagnosis and treatment of posttraumatic stress disorder in children and adolescents. Judith A. Cohen, principal author. *Journal of the American Academy of Child and Adolescent Psychiatry, 37*(Suppl. 10), 4S-26S.

American Psychiatric Association (2000). *Diagnostic and Statistical Manual of Mental Disorders* (Fourth Edition, Text Revision). Washington, DC: Author.

Angold A., Costello E.J., Messer, S.C., Pickles, A., Winder F., & Silver D. (1996). Development of a short questionnaire for use in epidemiological studies of depression in children and adolescents. *International Journal of Methods in Psychiatric Research, 5,* 237-249.

Beck, A.T. (1976). *Cognitive Therapy and the Emotional Disorders.* New York: International Universities Press.

Beck, A.T., Ward, C. H., Mendelson, M., Mock, J., & Erbaugh, J. (1961). An inventory for measuring depression. *Archives of General Psychiatry, 4,* 561-571.

Berliner, L. (1997). Intervention with children who experience trauma. In D. Cicchetti & S. Toth (Eds.), *The Effects of Trauma and the Developmental Process* (pp. 491-514). New York: Wiley.

Birmaher, B., Khetarpal, S., Brent, D., Cully, M., Balach, L., Kaufman, J., & Neer, S.M. (1997). The Screen for Anxiety Related Emotional Disorders (SCARED): Scale construction and psychometric characteristics. *Journal of the American Academy of Child and Adolescent Psychiatry, 36,* 545-553.

Black, D. (1977). Children and adolescents: Treatment of children and families. In D. Black, M. Newman, J.H. Hendricks, & G. Mezey (Eds.), *Psychological Trauma: A Developmental Approach* (pp. 281-287). London: Gaskell.

Bloomquist, M.L. (1996). *Skills Training for Children with Behavior Disorders: A Parent and Therapist Guidebook.* New York: Guilford Press.

Brooks, S.J. & Kutcher, S. (2001). Diagnosis and measurement of adolescent depression: A review of commonly utilized instruments. *Journal of Child and Adolescent Psychopharmacology, 4*(11), 341-376.

Cohen, J.A. & Mannarino, A.P. (1993). A treatment model for sexually abused preschoolers. *Journal of Interpersonal Violence, 8,* 115-131.

Cohen, J.A. & Mannarino, A.P. (1996a). A treatment outcome study for sexually abused preschooler children: Initial findings. *Journal of the American Academy of Child and Adolescent Psychiatry, 35,* 42-50.

Cohen J.A. & Mannarino, A.P. (1996b). Factors that mediate treatment outcome of sexually abused preschool children. *Journal of the American Academy of Child and Adolescent Psychiatry, 35,* 1402-1410.

Cohen, J.A. & Mannarino, A.P. (1998). Interventions for sexually abused children: Initial treatment findings. *Child Maltreatment, 3,* 17-26.

Cohen, J.A. & Mannarino, A.P. (2000). Predictors of treatment outcome in sexually abused children. *Child Abuse and Neglect, 24*(7), 983-994.

Cohen, J.A., Mannarino, A.P., Berliner, L., & Deblinger, E. (2000). Trauma-focused cognitive-behavioral therapy: An empirical update. *Journal of Interpersonal Violence, 15,* 1202-1223.

Cohen, J.A., Stubenbort, K., Greenberg, T., Padlo, S., Shipley, C., Mannarino, A.P., & Deblinger, E. (2001). *Cognitive and behavioral therapy for traumatic bereavement in children: Group treatment manual.* Center for Traumatic Stress in Children and Adolescents, Department of Psychiatry, Allegheny General Hospital, Pittsburgh, PA. Unpublished manuscript.

Deblinger, E. & Heflin, A.H. (1996). *Treating Sexually Abused Children and Their Non-Offending Parents: A Cognitive Behavioral Approach.* Thousand Oaks, CA: Sage.

Deblinger, E., Lippman, J., & Steer, R. (1996). Sexually abused children suffering posttraumatic stress symptoms: Initial treatment outcome findings. *Child Maltreatment, 1*(4), 310-321.

Dyregrov, A. (1997). The process in psychological debriefing. *Journal of Traumatic Stress, 10,* 589-605.

Foa, E.B., Hearst-Ikeda, D., & Perry, K.J. (1995). Evaluation of a brief cognitive-behavioral program for the prevention of chronic PTSD in recent assault victims. *Journal of Consulting and Clinical Psychology, 63,* 948-955.

Foa, E.B. & Tolin, D.F. (2001). Comparison of The PTSD Symptom Scale-Interview Version and the Clinician Administered PTSD Scale. *Journal of Traumatic Stress, 13*(2), 181-192.

Goenjian, A.K., Karaya, I., Pynoos, R.S., Minassian, D., Najarian, L.M., Steinberg, A.M., & Fairbanks, L.A. (1997). Outcome of psychotherapy among early adolescents after trauma. *American Journal of Psychiatry, 154,* 536-542.

Harrington, R. & Harrison, L. (1999). Unproven assumptions about the impact of bereavement on children. *Journal of the Royal Society of Medicine, 92,* 230-233.

Hendricks, J.H., Black, D., & Kaplan, T. (1993). *When Father Kills Mother: Guiding Children Through Trauma and Grief.* London: Routledge.

Horowitz, M. (1976). *Stress Response Syndromes* (Second Edition) New York: Jason Aronson.

Horowitz, M. (1979). Psychological response to serious life events. In V. Hamilton & D.M. Warburton (Eds.), *Human Stress and Cognition* (pp. 235-263). New York: Wiley.

Janoff-Bulman, R. (1985). The aftermath of victimization: Rebuilding shattered assumptions. In C.R. Figley (Ed.), *Trauma and Its Wake* (pp. 15-35). New York: Brunner/Mazel.

Johnson, D.R. & Lubin, H. (2000). Group psychotherapy for the symptoms of posttraumatic stress disorder. In R.H. Klein & V.L. Schermer (Eds.), *Group Psychotherapy of Psychological Trauma* (pp. 141-169). New York: Guilford Press.

Khetarpal-Monga, S., Birmaher, B., Chiappetta, L., Brent, D., Kaufman, J., Bridge, J., & Cully, M. (2000). The screen for child anxiety related emotional disorders (SCARED) convergent and divergent validity. *Depression and Anxiety, 12,* 85-91.

Klein, R.H. & Schermer, V.L. (2000). Introduction and overview: Creating a healing matrix. In R.H. Klein & V.L. Schermer (Eds.), *Group Psychotherapy for Psychological Trauma* (pp. 3-46). New York: Guilford Press.

Layne, C.M., Pynoos, R.S., Saltzman, W.S., Arslanagic, B., & Black, M. (2001). Trauma/grief-focused group psychotherapy: School based post-war intervention with traumatized Bosnian adolescents. *Group Dynamics: Theory, Research & Practice, 5*(4).

Layne, C.M., Saltzman, W.S., Savjak, N., & Pynoos, R.S. (1999). *Trauma/grief-focused Group Psychotherapy Manual.* Sarajevo, Bosnia: UNICEF Bosnia & Herzegovina.

Layne, C.M., Savjak, N., Saltzman, W.R., & Pynoos, R.S. (2001). UCLA/BYU Expanded Grief Inventory. Unpublished instrument, Bringham Young University, Provo, Utah.

March, J.S., Amaya-Jackson, L., Murray, M.C., Schulte, A. (1998). Cognitive behavioral psychotherapy for children and adolescents with PTSD after a single-incident stressor. *Journal of the American Academy of Child & Adolescent Psychiatry, 37*(6), 585-593.

Marmar, C. & Horowitz, M. (1988). Diagnosis and phase oriented treatment of post-traumatic stress disorder. In J. Wilson, Z. Harel, & B. Kahana (Eds.), *Human Adaptation to Extreme Stress: From the Holocaust to Vietnam* (pp. 81-102). New York: Plenum Press.

McFarlane, A.C. (1987). Posttraumatic phenomena in a longitudinal study of children following natural disaster. *Journal of the American Academy of Child and Adolescent Psychiatry, 26,* 764-769.

Nader, K.O. (1997). Childhood traumatic loss: The interaction of trauma and grief. In C.R. Figley, B.E. Bride, & N. Mazza (Eds.), *Death and Trauma: The Traumatology of Grieving* (pp. 17-41). New York: Hamilton Printing Company.

Pynoos, R.S. (1992). Grief and trauma in children and adolescents. *Bereavement Care, 11,* 2-10.

Pynoos, R. & Nader, K. (1988). Psychological first aid and treatment approach to children exposed to community violence: Research implications. *Journal of Traumatic Stress, 1,* 445-473.

Pynoos, R.S., Nader, K., Frederick, C., Gonda, L., & Stuber, M. (1987). Grief reactions in school age children following a sniper attack at a school. *Israel Journal of Psychiatry and Related Sciences, 24,* 53-63.

Pynoos, R.S., Steinberg, A.M., & Wraith, R. (1995). A developmental model of childhood traumatic stress. In D. Cicchetti & D. Cohen (Eds.), *Manual of Developmental Psychopathology Vol. 2: Risk, Disorder and Adaptation* (pp. 72-95). New York: Wiley.

Rando, T.A. (1988). *How to go on Living when Someone You Love Dies.* New York: Lexington Books.

Rando, T.A. (1993). *Treatment of Complicated Mourning.* Champaign, IL: Research Press.

Rando, T.A. (1996). Complications of mourning traumatic death. In K.J. Doka (Ed.), *Living with Grief After Sudden Loss* (pp. 139-160). Washington, DC: Hospice Foundation of America.

Resnick, P.A. & Schnicke, M.K. (1992). Cognitive processing therapy for sexual assault victims. *Journal of Consulting and Clinical Psychology, 60,* 748-756.

Rothbaum, B.O., Meadows, E.A., Resick, P., & Foy, D.W. (2000). Cognitive-behavioral therapy. In E.B. Foa, T.M. Keane, & M.J. Friedman (Eds.), *Effective Treatments for PTSD: Practice Guidelines from the International Society for Traumatic Stress Studies* (pp. 320-325). New York: Guilford Press.

Stubenbort, K., Donnelly, G.R., & Cohen, J.A. (2001). Cognitive-behavioral group therapy for bereaved children following an air disaster. *Group Dynamics: Theory, Research and Practice, 5,* 261-276.

Sugar, M. (1988). A preschooler in disaster. *American Journal of Psychotherapy, 42,* 619-629.

Terr, L.C. (1987). Childhood psychic trauma. In J.D. Noshpitz (Ed.), *Basic Handbook of Child Psychiatry* (Vol. 20, pp. 262-270). New York: Basic Books.

Terr, L.C. (1990). *Too Scared to Cry.* New York: Harper & Row.

Walser, R.D. & Hayes, S.C. (1998). Acceptance and trauma survivors: Applied issues and problems. In V.N. Follette, J.I. Rusek, & F.R. Abueg (Eds.), *Cognitive-Behavioral Therapies for Trauma* (pp. 256-277). New York: Guilford Press.

Wolfelt, A. (1991). Children. *Bereavement Magazine, 5*(1) 38-39.

Wood, A., Kroll, L., Moore, A., & Harrington, R. (1995). Properties of the Mood and Feelings Questionnaire in adolescent psychiatric outpatients: A research note. *Journal of Child Psychology and Psychiatry, 36,* 327-334.

Worden, J.W. (1996). *Children and Grief: When a Parent Dies.* New York: Guilford Press.

Yalom, I.D. (1985). *The Theory and Practice of Group Psychotherapy* (Third Edition). New York: Basic Books.

Chapter 17

Combining Cognitive Processing Therapy with Panic Exposure and Management Techniques

Sherry A. Falsetti
Heidi S. Resnick
Steven R. Lawyer

INTRODUCTION

In this chapter we describe multiple channel exposure therapy (M-CET), a group treatment developed by Sherry Falsetti and Heidi Resnick, and its application to post-traumatic stress disorder (PTSD) and comorbid panic attacks. This treatment was developed to meet the needs of clients who had difficulty doing trauma-focused therapy because high levels of emotional arousal would often trigger panic attacks, which were very fearful to these clients. The treatment approach is designed to include exposure in all three major response channels: cognitive, behavioral, and physiological (Falsetti, Resick, Davis, & Gallagher, 2001). M-CET integrates components of cognitive processing therapy (CPT; Resick, Nishith, Weaver, Astin, & Feuer, 2002; Resick & Schnicke, 1992) that address changes in cognitive schema following traumatic events. These include cognitive restructuring and writing about the memory of the traumatic event to reduce symptoms of PTSD and altered belief systems that result from the aftermath of traumatic events. In addition, M-CET includes adapted components of Barlow and Craske's Mastery of Your Anxiety and Panic (MAP) treatment package, a highly effective treatment for panic disorder (Barlow & Craske, 1988; Barlow & Craske, 2000; Barlow, Gorman, Shear, & Woods, 2000). The MAP treatment includes in-depth psychoeducation about the physiology of panic, cognitive restructuring related to overestimation, and catastrophizing panic

Research for this chapter was supported by Grant Number R21MH053381-03 from the National Institute of Mental Health.

attacks; and provides exercises that allow for exposure to the physical sensations of panic. As adapted within M-CET, the theoretical rationale posits that panic attacks may have been initially experienced *during* the traumatic event but are currently experienced with or without identified event-related cues which elicit fear (Falsetti, Resnick, & Davis, 2005). Finally, M-CET incorporates in vivo exposure exercises to promote habituation to PTSD- and panic-related situational cues. This component of M-CET is similar to the in vivo exposure component developed by Veronen and Kilpatrick (1983) in their stress inoculation treatment to reduce anxiety and other negative mental health outcomes following rape.

The M-CET treatment, implemented in a twelve-session group format, is associated with significant reduction of PTSD and panic attack diagnoses compared to the prevalence of these diagnoses in individuals participating in a minimal attention control group (MA). All participants met criteria for PTSD and panic attacks following assault incidents such as rape, aggravated assault, and homicide of a loved one (Falsetti et al., 2001; Falsetti et al., 2005). This treatment approach may be highly relevant for persistent PTSD and panic attacks that may be experienced by those exposed to terrorist incidents such as the September 11, 2001, attacks on the World Trade Center and the Pentagon or other similar events.

RATIONALE AND OBJECTIVE OF THE GROUP

M-CET is based on a theoretical model of PTSD that is similar to the information processing model proposed by Foa and Kozak (1986) and also based on Lang (1968; 1977), as well as earlier models based on classical and operant conditioning following exposure to a traumatic event (Keane, Zimering, & Caddell, 1985; Kilpatrick, Veronen, & Resick, 1979). This model proposes that a *fear network* comprised of information about stimuli associated with event exposure and cognitive, behavioral, and physiological responses and perceptions of danger associated with stimulus-response relationships can develop following exposure to a traumatic event. Any element of the fear network can then serve as a signal to escape or avoid danger. Prolonged exposure treatment for PTSD using both imaginal exposure to trauma-related memories and in vivo exposure to situational cues theoretically allows for accessing of the fear network and thus for corrective learning (Foa & Rothbaum, 1998). Prolonged exposure treatment has been associated with significant reductions in chronic PTSD compared to a wait-list control group: 60 percent of treatment participants no longer met criteria for PTSD after treatment, whereas all control participants still had

PTSD. In addition, treatment gains were maintained over the course of a twelve-month follow-up (Foa et al., 1999).

M-CET also incorporates components of CPT based on the rationale that cognitive schema about major beliefs (e.g., views of safety, trust, power/competence, esteem, and intimacy) are likely to be altered following exposure to a traumatic event, just as prior beliefs may be radically changed or the view of the event itself distorted (Resick & Schnicke, 1992). Data from a randomized controlled treatment outcome study comparing prolonged exposure (PE), CPT, and MA indicated that 53 percent of those in either CPT or PE no longer met criteria for PTSD posttreatment. These treatments were also associated with significant reductions in major depression in the treatment groups. In contrast, only 2 percent of the MA group no longer met full diagnostic criteria for PTSD. These gains were maintained in the treatment groups throughout a nine-month follow-up period.

CPT treatment provides education about how beliefs may be affected and how these changes can negatively influence mood and behavior. A major goal of this approach is to help the individual integrate or accept the event as it occurred, without distortion, within a more moderated or functional system of beliefs about oneself, others, and the world. As noted, components of this treatment, adapted and integrated into the M-CET manual, include an explanation of how beliefs may shift following trauma and how these shifts may lead to development of PTSD. In addition, exercises are adapted from CPT to help clients identify the connections between events, thoughts, and feelings. Exercises are also included that guide clients to challenge or question current thinking patterns possibly associated with symptoms of PTSD or panic (or other emotional reactions and behaviors), as well as thoughts directly related to the traumatic event. Clients are directed to challenge whether thoughts surrounding a current situation are actually a response to the traumatic event versus an independent appraisal of a new situation. M-CET also includes assignments that require clients write about their memories of the traumatic event, adapted from CPT, to allow for habituation of anxiety or other distress associated with the event. The writing assignments are done at least twice as homework (over the course of two sessions). Clients are instructed to incorporate memories of sensory experiences across all modalities in their writing along with thoughts and emotional and physical reactions that occurred during the incident. They are further instructed to read the written accounts daily and to rate their distress upon reading what they wrote so that the therapist can monitor whether anxiety decreases across repeated exposures. Such habituation or reduction with repeated exposure is based on the rationale that writing about the event is a form of imaginal exposure with similar effects to those of prolonged exposure. Writing may also allow for processing of the event as it occurred, without distortion. Finally, an adaptation unique to M-CET includes having participants read their written account aloud to the group. Sharing the accounts allows for therapeutic exposure related to fears of others' negative perceptions.

In addition to the focus on treatment of PTSD, M-CET also emphasizes the etiology and treatment of panic attacks that are theorized as developing or occurring during initial exposure to a potentially traumatic event (Falsetti, Resnick, Dansky, Lydiard, & Kilpatrick, 1995). As noted in the current *Diagnostic and Statistical Manual of Mental Disorders* (Fourth Edition, Text Revision) (American Psychiatric Association, 2000), panic attacks are defined as sudden and brief episodes of intense fear or discomfort and can occur in conjunction with any of the anxiety disorders. At least four or more symptoms of panic are required to meet criteria for a panic attack. These panic symptoms can include rapid heart rate, shortness of breath, dizziness, trembling or shaking, sweating, choking, nausea or abdominal distress, depersonalization or derealization, numbness or tingling sensations, hot flushes or chills, chest pain or discomfort, fear of dying, and fear of going crazy or losing control. Panic attacks may be relatively common among those exposed to traumatic events. For example, Falsetti and Resnick (1997a, b) found that 69 percent of those seeking treatment for mental health problems related to crime or other traumatic events met criteria for panic attacks. During or in the immediate aftermath of a traumatic event, such reactions could potentially be very functional in terms of survival. However, it was hypothesized that such reactions could persist and lead to fear of the physical reactions themselves (Falsetti et al., 1995), as proposed in the Foa and Kozak (1986) model. In this model, the physiological reactions would signal danger as part of the fear network formed after exposure to a traumatic event. In addition, Falsetti et al. (1995) proposed that physiological arousal might cue thoughts that occurred during the event (e.g., fear of dying) that might further influence the perception of such reactions as dangerous. Perceptions that something might be wrong with one's body might also arise due to such associations. Chronic arousal related to PTSD could lower the threshold at which panic attacks might occur. Thus, individuals who may or may not have experienced acute panic attacks during the event itself might be more prone to develop panic attacks later, due to greater vulnerability to panic in response to developing lowered thresholds to external stressors.

Data from several studies indicate that panic reactions and panic attacks are prevalent during or shortly following exposure to a traumatic event. Resnick, Falsetti, Kilpatrick, and Foy (1994) found that 90 percent of rape victims assessed within seventy-two hours postassault met full criteria for panic attacks. Similarly, Bryant and Panasetis (2001) found that 53 percent of civilian trauma survivors reportedly met criteria for panic attacks during the traumatic event. Reported panic symptoms occurring during the event were significantly positively correlated with measures of acute stress disorder and panic assessed between two and twenty-eight days postevent. Specifically, the correlation coefficients between the Physical Reactions Scale (PRS; Falsetti & Resnick, 1992) and the Acute Stress Disorder Scale (Bryant, Moulds, & Guthrie, 2000) was .71, $P < .01$; the correlation between the PRS and the Impact of Events Scale (Horowitz, Wilner, & Alvarez, 1979) Intrusion subscale was .59 and the avoidance subscale .63. Data from a study of residents of Manhattan shortly after the September 11, 2001, attacks indicate that a large percentage of those near the disaster reportedly experienced panic attacks during the events or soon after learning of them (Galea et al., 2002). Among a random sample of 1,008 Manhattan residents, 12 percent reported panic attacks acutely following the event. Importantly, report of a peri-event panic attack was a significant predictor of PTSD diagnosis five to eight weeks following the attacks. Estimated prevalence of PTSD in that study was 7.5 percent, while 9.7 percent of the sample reported symptoms consistent with a diagnosis of major depression. Peri-event panic attack was associated with an odds ratio of 7.6 for prediction of PTSD, after controlling for race, prior stressors in the previous twelve months, loss of possessions due to the attacks, and residence below Canal Street (within close proximity to the World Trade Center). Report of a peri-event panic attack was also a significant predictor of depression, after controlling for other predictors (Galea et al., 2002). Similarly, in terms of PTSD findings, Schlenger et al. (2002) reported an estimated prevalence of PTSD among those in the New York City metropolitan area of 11.2 percent, one to two months postevent.

Data reviewed across other previous studies of natural, technological, and mass violence disasters indicate that PTSD, major depression, and anxiety are the most common disorders identified following exposure (Norris, 2002). Although Norris reports that panic disorder rarely occurred following such disaster, it was unclear how often panic disorder was actually assessed across studies. Similarly, there was no report of panic attack prevalence following exposure within that review of studies. Based on the findings from Galea et al. (2002), such panic reactions are likely to be prevalent in the acute aftermath of exposure and to be predictive of longer-term symptomatology.

Thus, M-CET may be a useful treatment approach for those exposed to terrorist attacks. Although M-CET was not specifically developed for treatment of postterrorism reactions, it is directly adaptable to such situations because it was designed for use and evaluated in a group format with clients who have been exposed to diverse traumatic events and who currently experience PTSD and panic attacks. Psychoeducation during treatment focuses on elements of traumatic events that may be common risk factors for PTSD across a variety of precipitating scenarios. Based on the results of Kilpatrick et al. (1989), such elements include fear of death or injury during an event, receipt of injury, and injury or death of a significant other. These characteristics are common to many types of traumatic events including physical or sexual assault, combat, or other civilian trauma incidents including terrorist attacks.

Preliminary results from a controlled treatment outcome study comparing M-CET to a wait-list control group (Falsetti et al., 2001) indicated that this may be an effective treatment for PTSD and panic attacks. Future research will have to be conducted to evaluate efficacy relative to other established treatments for PTSD, including prolonged exposure (Foa et al., 1999). In the initial study (Falsetti et al., 2001), participants were randomly assigned to either twelve weeks of once-weekly M-CET group therapy ($n =$ 12) or an MA group ($n = 15$) that received bimonthly supportive phone counseling. During treatment, participants reported a range of multiple traumatic events and the treatment groups were not restricted to those who had experienced one type. Participants focused on the event that they felt was most relevant to their current symptoms. All participants were women; all met criteria for current PTSD and panic attacks at least three months posttrauma. All subjects who completed the control condition were offered free treatment after the completion of their participation. Eleven participants who served in the control group elected to participate in treatment following the initial phase of the study. These individuals thus served as the control group, then as treatment participants.

At posttreatment, only 8.3 percent of subjects in the M-CET treatment condition met criteria for PTSD according to the Clinician Administered PTSD Scale (CAPS), compared to 66.7 percent of subjects in the MA group, indicating a significant difference at posttreatment between the groups. Analyses also revealed that panic attacks and related symptoms decreased significantly. At the posttreatment evaluation, 93.3 percent of the MA group subjects reported experiencing at least one panic attack in the past month, compared to only 50 percent of the treatment group (X^2 (1, 25) $= 6.51$, $P < .01$). Data also indicate that those in the treatment group reported significantly less frequent panic attacks compared to the control group over time as well as less fear of panic attacks and less interference with activities due to panic symptoms. Both groups improved significantly over time in terms of symptoms of depression.

In summary, M-CET can be delivered easily in a group format. It is designed for use and has been evaluated with clients who are experiencing chronic symptoms of PTSD and panic attacks. M-CET includes components of CPT (Resick & Schnicke, 1992) to address a range of beliefs about safety, trust, control, and esteem that may be negatively affected in the aftermath of a traumatic event. In addition, M-CET includes repeated writing and reading assignments about the event, adapted from CPT, allowing for prolonged exposure to trauma-related memories. M-CET differs from CPT and prolonged imaginal exposure (as traditionally implemented for treatment of PTSD) in that it also includes comprehensive psychoeducation about panic reactions that may occur during exposure to a traumatic event which may persist as part of a cognitive fear network. In addition, M-CET includes interoceptive exposure (i.e., exposure to exercises that bring on physical sensations similar to those experienced during panic attacks). Interoceptive exposure, adapted from the MAP program developed by Barlow and Craske (1988), is conducted prior to imaginal exposure, which allows clients to gain mastery related to reduced fear of panic sensations. Finally, similar to the in vivo exposure component of stress inoculation treatment (SIT; Veronen & Kilpatrick, 1983) and the in vivo exposure component of prolonged exposure (PE; Foa et al., 1999) clients are guided through in vivo exposure to feared trauma-related situational cues.

DESCRIPTION AND TASKS TO BE ACCOMPLISHED

Based upon the findings that panic attacks may occur acutely and may persist among a significant subset of those exposed to traumatic events, it was hypothesized that a graduated approach to exposure treatment might be optimal for certain clients. These clients can potentially experience excessive arousal or panic during exposure to traumatic memories (Falsetti et al., 2005; Resnick & Newton, 1992). Such excessive arousal might interfere with successful processing of traumatic memories, as well as integration of corrective information about anxiety responses during exposure treatment. This hypothesis would be consistent with Foa and Kozak's (1986) review indicating that a certain, carefully calibrated level of arousal is necessary for therapeutic exposure; and that excessive arousal might prevent attention to and integration of corrective information about the event or associated anxiety responses.

M-CET is thought to be a more graduated approach because it includes in-depth education about the learning theory and cognitive network models (Foa & Kozak, 1986) of how both PTSD and panic attacks may develop and persist following exposure to a traumatic event. In addition, M-CET adapts components of Barlow and Craske's (1988, 2000) MAP treatment for panic, and implements these components prior to conducting imaginal exposure to the traumatic memory. Thus, the tasks to be accomplished within M-CET include provision of the treatment rationale and psychoeducation; instruction on exercises that help clients learn to challenge distorted cognitions; exposure to physical sensations of panic (interoceptive); imaginal exposure to the traumatic memory via the use of repeated writing and reading assignments; and exposure to trauma-related situational cues (in vivo). The specific tasks to be accomplished are described as follows for each session of the treatment protocol. Some clients may require more than one session to master the content covered in the individual sessions. In those cases treatment can be conducted over a longer period of time.

Treatment begins with an introductory session focused on the nature of the treatment approach and psychoeducation about PTSD and panic attacks. For homework, clients are asked to monitor panic attacks and PTSD symptoms and to write about cues associated with traumatic events that currently bring about fear reactions. They are also asked to write about the impact the event has had on their thinking. The second, third, and fourth sessions are devoted to continued psychoeducation and skills development and mastery. Homework throughout sessions always includes monitoring of PTSD and panic attacks, along with exercises to practice skills learned in the individual sessions. Goals and tasks for the second session include education about the physiology of breathing and the effects of breathing through the diaphragm and at a slower pace for reducing the general level of anxiety. Breathing exercises are modeled and taught in session. The third session includes education about overestimating the probability of dangerous or aversive events occurring following exposure to a traumatic event. Clients are also educated about and instructed to complete exercises examining the relationships between events, thoughts, and emotional reactions and behaviors.

The fourth session includes psychoeducation about the cognitive distortion of catastrophic thinking. Take-home worksheets are also introduced at this session. These include not just connections between events, thoughts, and feelings/behaviors, but also questions to evaluate the validity of the thoughts within the sequence. For example, if a woman is at a party and thinks "everyone knows that I am a rape victim," she may feel afraid, angry, or ashamed and she might not speak to others. Challenging this thought involves evaluating the evidence for and against the objective reality of the thought. The technique includes questioning whether the thought represents distortions, including going to extremes in thinking and overgeneralizing from a previous situation to the current situation. These worksheets are incorporated as homework along with panic and PTSD ratings throughout the remainder of treatment, plus new exercises related to specific sessions.

The first modality in which exposure occurs is the physical channel of responding. Interoceptive exposure is introduced in the fifth session of M-CET. Following the MAP program procedures, clients are asked to engage in a series of exercises including hyperventilation for one minute, spinning in a chair, and stair stepping. These exercises are designed to elicit physical reactions (e.g., heart palpitations) similar to those that occur during panic attacks and which may have occurred during exposure to the traumatic event. Particular exercises are identified that are similar to the sensations the client experiences during panic attacks and that are associated with anxiety. These exercises are then practiced repeatedly as homework across remaining treatment sessions until anxiety is decreased in response to the relevant sensations. After progression toward mastery of this exposure to the physiological symptoms occurs, clients move on to engaging in imaginal exposure (the cognitive channel of responding) to the traumatic memory. This is done by repeated writing and reading about the event.

In the sixth treatment session, the first written account of the traumatic event (and the sensory details) is assigned. Clients are asked to write about the event as soon as possible following this session and to read their account at least once each day. In the seventh session, clients read aloud the written account. Difficult parts of the memory are discussed, and cognitive distortions are identified and discussed. The writing assignment is then completed a second time, with instructions to read it as homework on a daily basis. In the eighth session, the writing assignment is again discussed and clients are introduced to areas of thinking that may have been affected by exposure to the traumatic event. The first content area addresses how beliefs about safety may have been affected, and challenging thoughts related to safety beliefs are assigned. In subsequent sessions, various other handouts are assigned and discussed along with cognitive restructuring homework related to the impact of trauma on beliefs about trust, power/competence, and esteem.

Sessions 9 through 12 focus on in vivo exposure, which is the final stage of treatment. This involves exposure to real-life (in vivo) situations and cues that are currently avoided but that are not in themselves realistically dangerous. Three such situations are identified and exposure hierarchies are developed. Homework includes exposure to the situational cues on the first hierarchy prior to moving to the more challenging second and third hierarchies. This allows for exposure in the behavioral channel of responding.

This prior exposure to physical sensations allows clients to have a more manageable level of arousal for processing the traumatic memory and situational cues. A similar rationale is involved in conducting interoceptive exposure prior to in vivo exposure for those who have current panic attacks and fear of those sensations. It remains a question whether patients who report both PTSD and panic attacks benefit more from treatment using interoceptive exposure prior to imaginal and in vivo exposure or whether they would do as well with imaginal and in vivo exposure alone. However, preliminary results suggest that M-CET may provide a promising treatment program for a subset of PTSD patients who experience panic attacks. In session 11, clients are asked to write again about the impact of the traumatic event on their thinking related to safety, trust, power/competence, and esteem. In session 12, the final session, clients compare their initial writing assignment to their current version. Changes in thinking are discussed by the individual and other group members. In addition, any problems encountered in treatment are discussed. Problem areas are identified and planning is done for continued goals that the individual can work on after conclusion of the group. Education about the impact of possible future stressors on symptoms is discussed as well as strategies for coping with such events.

CRITERIA FOR SELECTION OF MEMBERS

Evaluation and Assessment

Structured interview assessment of history of exposure to traumatic events, PTSD, and panic attacks should be conducted before and after treatment and at longitudinal follow-up points. For assessment of history of exposure, the Clinician Administered Trauma Assessment for Adults Interview can be used (Resnick, 1996). An abbreviated version of this measure has been demonstrated to have good test-retest reliability, with Kappas ranging from .82 to 1.00 for specific events assessed (Vielhauer, Findler, Schnurr, Garcia, & Spiro, 1998). The Trauma Assessment for Adults interview includes sensitive screening questions that behaviorally define incidents of sexual and physical assault and other potentially traumatic events that may occur over the lifespan. Follow-up questions are included to assess age of first and most recent incidences of given events, as well as perceived life threat or injury in relation to specific events.

Useful structured interviews for the assessment of PTSD include the Clinician Administered PTSD Scale (CAPS: Blake et al., 1990) and the Structured Clinical Interview for DSM-IV (SCID-CV; First, Spitzer, Gibbon, & Williams, 1997). CAPS includes separate versions to assess lifetime and current PTSD. Each version includes measures of frequency and intensity of PTSD symptoms. In addition, CAPS allows for assessment of the impact of symptoms on social and occupational functioning. Cutoff scores for CAPS have been developed and there is strong agreement between this measure and SCID (.89) in diagnosing PTSD (Blake et al., 1995). SCID has been used in many studies with victims of crime and those exposed to other types of traumatic events.

The panic disorder section of the Anxiety Disorders Interview Schedule-Revised (ADIS-R: Di Nardo & Barlow, 1988) assesses detailed information regarding panic symptoms. In addition to symptom ratings, this scale includes an assessment of the history of panic attacks, how panic attacks, are handled, and distress experienced as a result of panic. In addition to the structured interview assessment of PTSD and panic attacks, assessment of basic demographic characteristics should be performed, as well as assessment of psychiatric and medical history, including use of psychotropic medications.

Self-report measures to assess frequency and intensity of a range of symptoms including PTSD, depression, and panic attacks should be used as well. The measures described as follows were included in the initial M-CET treatment study (Falsetti et al., 2001). The Modified PTSD Symptom Scale Self-Report (MPSS-SR: Falsetti, Resnick, Resick, & Kilpatrick, 1993) is a modification of the PTSD Symptom Scale developed by Foa, Riggs, Dancu, and Rothbaum (1993). This version was modified to assess frequency and severity of symptoms. The scale is comprised of seventeen items that correspond to PTSD symptom criteria in the DSM-IV. Falsetti et al. (1993) investigated the validity and reliability of this scale and reported high internal consistency (.96), as well as good concurrent validity with the SCID. The Physical Reactions Scale (PRS; Falsetti & Resnick, 1992) is a twenty-item self-assessment of the frequency and severity of current (past twenty weeks) panic symptoms. In addition, it also assesses panic attacks in relation to a traumatic event. Fears about what these symptoms indicate, such as a heart attack, or going crazy, are also assessed. To assess symptoms of depression, the Beck Depression Inventory (BDI; Beck, Ward, Mendelsohn, Mock, & Erbaugh, 1961) is included. The BDI is a twenty-one-item self-report questionnaire to assess depression. This scale has been used in numerous research projects on depression (Beck, Steer, & Garbin, 1988), and has also been used to assess depression in victims of crime (Atkeson, Calhoun, Resick, & Ellis, 1982; Foa, Rothbaum, Riggs, & Murdock, 1991; Resick & Schnicke, 1992). Beck et al. (1961) reported a split-half reliability of .93.

Inclusion Criteria

Major inclusion criteria for the initial efficacy study of M-CET (Falsetti et al., 2001) were the presence of PTSD as determined by the CAPS (Blake et al., 1990) and SCID (First, Spitzer, Gibbon, & Williams, 1997) interviews, and report of panic attacks within the previous month on the ADIS-R

(Di Nardo & Barlow, 1988). Clients taking psychotropic medications were included if they were stabilized on their medications, with no changes over the previous three months. Individuals were included if they had experienced a traumatic event at least three months prior to assessment, as research suggests that spontaneous recovery is most likely to occur within the first three months (Kilpatrick & Calhoun, 1988; Foa, Rothbaum, & Steketee, 1993). The groups were not homogeneous with regard to trauma type; the type of traumatic event to which a client had been exposed did not in itself determine inclusion or exclusion from the group. Thus far, however, the groups have been restricted to female participants, based on the idea that some types of traumatic events such as sexual assault might be more difficult to address in mixed-gender groups. Specifically, women are less likely to attend and to disclose issues about sexual assault with men present; there is nothing inherent in the treatment content that would restrict its use to women only. M-CET has been used successfully with men in single case studies (e.g., Falsetti & Resnick, 1995) but has yet to be evaluated in a group format.

Exclusion Criteria

In the past, treatment outcome research was often criticized for excluding complex cases and people with comorbid psychiatric disorders. Therapists in practice wondered about the usefulness of such research, as most of their cases were complex and contained comorbid conditions. In an effort to determine how well M-CET worked in the "real world" and not just on very specific, clean cases, patients with complex histories or personality disorders were not excluded. Exclusionary criteria did include active psychosis, mental retardation, suicidal or parasuicidal behavior, and drug or alcohol dependency. Someone who is actively psychotic cannot participate appropriately in this type of a group setting. The homework involved in the treatment requires a reading level of at least sixth grade, and a fair amount of writing that a patient with mental retardation may not be able to successfully master. Patients who were suicidal or parasuicidal were also excluded as these issues would need to be addressed and require more attention than a group setting could provide. Finally, individuals with drug and alcohol dependencies were also excluded because these disorders are associated with poor attendance and do not allow for appropriate exposure therapy. Patients who were using (but not abusing) alcohol or substances were not excluded, but were encouraged them to abstain during the course of treatment.

COMPOSITION AND PREPARATION
FOR ENTRY INTO GROUP

Several factors must be considered in the composition of a group. Here, homogenous factors such as symptom presentation, functional impairment, type of trauma and past trauma history, and hetergeneous factors such as diversity of race and ethnicity are explored.

Homogenous Characteristics

Symptom Presentation

The primary presenting problem for all participants of an M-CET group should be PTSD with comorbid panic attacks. A high level of motivation is required for M-CET to be most effective. Individuals without these principle complaints may find that this treatment does not target their specific problems. Terrorism victims who have PTSD and other comorbid problems (e.g., major depressive disorder, substance abuse, obsessive-compulsive disorder) should be referred elsewhere to receive psychological and/or pharmacological interventions that are appropriate for their presenting problems. Ensuring a focus on PTSD and panic symptoms provides a common ground on which to provide the M-CET treatment and maintains focus on symptoms that are amenable to modification in a short-term group. Even with a relatively acute focus on panic and PTSD symptoms, there will likely be significant variability in symptom presentation among the group.

Functional Impairment

As with other emotional-behavioral problems (e.g., depression), the level of functional impairment that victims of terrorist activity suffer may vary significantly. Functional impairment following traumatic events may be associated with numerous elements of the trauma and/or the victim. An individual's functioning should be considered in the context of the group before M-CET treatment is recommended. Individuals with functional impairment that is too great may not benefit because they have difficulty attending to the group process (e.g., severe cognitive impairment and/or distractibility). Moreover, such individuals may reduce group effectiveness for other members by taking up significant portions of group time or creating other distractions. If functional impairment is marked, individual psychotherapy should be considered. A group that contains individuals with a "reasonable" degree of functional impairment may be useful; higher-functioning members can serve as behavioral models for lower-functioning ones.

Type of Trauma

Very little is known about the effectiveness of combining victims of different types of trauma in the same group. However, it is known that the nature of the precipitating trauma can promote qualitatively different symptom profiles among individuals with the same psychiatric diagnosis. For example, rape victims often suffer sexual dysfunction and concerns that would not likely be relevant to individuals with PTSD due to an armed robbery. Vietnam veterans with PTSD often present with feelings of anger, betrayal, and hostility that are qualitatively different from the feelings of physical assault victims with PTSD. Moreover, rape victims and combat veterans are often reticent to talk about their traumas to those who haven't "been there" and who they believe might not understand or empathize with their trauma. For these reasons, some therapists find it helpful to restrict membership to individuals with certain kinds of trauma to promote group cohesion and therapeutic disclosure.

At first glance, one may view victims of the same terrorist event (e.g., the September 11, 2001, World Trade Center attack) as having the same experience. However, many different "types" of trauma may emerge, as victims of the same event may have greatly different trauma experiences. Some victims will have been physically injured, others will have witnessed the event, and still others will have lost friends or loved ones, among the many other possibilities in so enormous a disaster. Therapists must be aware of and sensitive to these many different experiences and outcomes, all resulting from the same event. Also, for many victims, this may not be the only traumatic event ever experienced.

In the efficacy study of M-CET on panic and PTSD symptoms, Falsetti et al. (2001) recruited women with diverse (e.g., sexual assault, motor vehicle accidents) trauma histories and reported no difficulties discussing their disparate kinds of trauma. This may be because the focus of M-CET treatment is on symptoms of PTSD and panic rather than objective characteristics of the trauma per se. Therefore, differences in the kinds of traumas that terror victims have suffered may not have prognostic importance in the context of M-CET treatment, as long as therapists are sensitive to the different experiences and do not assume because all members were victims of the same terrorist event, that their experiences were identical.

Trauma History

Given that approximately 50 percent of men and women experience one or more traumatic events over their lifetime (Kessler, Sonnega, Bromet, Hughes, & Nelson, 1995), there is a strong likelihood that many potential M-CET group members will present with multiple traumas. The extent of previous trauma may be important for understanding an individual's symptoms because such previous traumatic experiences may play a role in the development of PTSD (Acierno, Resnick, Kilpatrick, Saunders, & Best, 1999; King, King, Foy, Keane, & Fairbank, 1999). M-CET, like most other cognitive-behavioral treatments, is quite flexible and capable of integrating previous trauma experiences into terrorism-related treatment. Special consideration should be given to survivors of childhood sexual abuse (CSA) in this context. As noted by Cloitre (1998), CSA survivors pose important challenges because issues related to interpersonal trust and betrayal may be central to their trauma experience. Patients for whom CSA might come to the forefront may benefit from supplementary treatment after M-CET has been completed. In our treatment outcome study, however, many of the women reported a history of CSA and responded well to the twelve-week treatment, so supplementary treatment may not be necessary for all CSA survivors.

Heterogeneous Characteristics

In general, cognitive-behavioral therapy (CBT) groups are most effective when the membership contains a balanced mix of homogenous and heterogeneous characteristics that maximize group cohesion while simultaneously establishing the conditions for members confronting irrational fears and challenging dysfunctional thought patterns. Especially in the case of terrorist activity involving mass destruction or violence, therapists may be presented with a demographically diverse group of people seeking treatment for related psychological problems, which may have implications for the process of treatment. For example, there may be intense animosity toward individuals from certain ethnic groups or religious backgrounds viewed to be responsible for terrorist activity within the group as well within society at large.

Perspectives on this issue have typically been expressed in the context of individual psychotherapy in which therapist characteristics (e.g., gender, race, physical characteristics) bring about trauma-related responses in individual psychotherapy patients. In many cases this is referred to as therapist-client mismatching and specifically refers to placing a therapist who shares a characteristic, such as race, with the perpetrator, such that the client may feel mismatched. Root (1996) cautioned that therapist-client mismatching can damage rapport between the therapist and client. Lyons, Cho, & Brown (2001) examined patient responses to mismatched therapists in a sexual assault group (i.e., a male cotherapist) and among Korean/Vietnam veterans (an Asian cotherapist). Some patients reacted negatively at first, refusing to participate, but many who accepted the arrangement responded well to the therapists. The authors reported that the use of mismatched therapists decreased patients' avoidance of such individuals in the natural environment. This was not a controlled investigation. These reports do not speak to the effects of mismatched group members. Group members may be more tolerant of mismatched individual group members, especially if there is good rapport among the group as a whole.

In theory, a diverse group membership provides important and unique opportunities for therapeutic change in this context. From a behavioral perspective, group members that resemble the alleged perpetrators of the terrorist act may provide other group members with therapeutic exposure necessary to change trauma-related emotions (e.g., fear, anger) and behavior (e.g., avoidance). From a cognitive perspective, a mix of group members provides the conditions for alteration of trauma-related dysfunctional schemas (Resick & Schnicke, 1993) in which individuals from certain backgrounds are viewed uniformly as "dangerous." Group members may also have an exaggerated sense of vulnerability if they believe they were targeted because of their own ethnicity, religious background, or political views. A demographically mixed treatment group may facilitate emotional processing of irrational fears of future danger, based on group members' characteristics.

STRUCTURAL CONSIDERATIONS OF THE GROUP

In this section we will consider the impact of short-term versus long-term groups, closed versus open membership, and the development of a working alliance. We will discuss each of these factors in relation to M-CET.

Short-Term versus Long-Term

No consensually endorsed criteria exist for determining what is considered "short-term" or "long-term" therapy. Therefore, a better way to characterize groups in this context may be to describe them as "session-limited" or not. Most session-limited treatments use approximately twelve to twenty group sessions. MacKenzie (1993) recommends that at least six sessions be used for group treatments so that the benefits of group cohesion may be realized. Cognitive-behavioral treatment packages have commonly used a session-limited framework. Importantly, many of these treatments have benefited from empirical validation as efficacious treatments for PTSD, anxiety disorders, and other behavioral problems. Exposure-based treatments such as M-CET have fared especially well when applied to PTSD and other anxiety disorders (see Chambless & Ollendick, 2001). Of course, the absolute number of M-CET sessions is amenable to change (e.g., if extra focus on a particular component is deemed important).

Closed versus Open Membership

In closed groups, membership stays generally the same from the beginning to the end of treatment. Group members may choose to leave the group (for any number of reasons), but new group members are not added once the group has been formed. In "open" groups, membership may change from session to session, as group members may attend all, some, or a few sessions, depending on their needs. Though both kinds of therapy have advantages, a closed-group format is used for M-CET groups. Maintaining continuity in the group membership is essential because skills learned during M-CET treatment build from session to session forming a comprehensive package of coping techniques that ensure maximal treatment effects. For this reason, group members are encouraged to attend all sessions. Although research has not been conducted that isolates the relative importance of M-CET treatment components, missing one or more treatment sessions may attenuate treatment effectiveness.

Development of a Working Alliance

A working (or therapeutic) alliance refers to the extent to which group members feel a positive collaborative relationship with the therapist and group members. Bordin (1979) suggests that the working alliance is made up of three components: (1) the emotional connection between therapist and client, (2) mutual agreement on the goals for the intervention, and (3) mutual agreement about the tasks that are used to achieve those goals. The zeitgeist of group therapy holds that a working alliance among group members and clinician is fundamental to client change (e.g., Yalom, 1975) and a strong working alliance is a good predictor of treatment outcome across psychological problems (e.g., Orlinsky, Grawe, & Parks, 1994). However, it is not clear whether the working alliance predicts treatment effects or if positive effects improve the working alliance (e.g., DeRubeis & Feeley, 1990).

Historically, cognitive-behavioral therapists have placed less importance on the therapeutic relationship than have therapists from psychodynamic-interpersonal perspectives, though CBT clinicians do form strong therapeutic alliances (Raue, Castonguay, & Goldfried, 1993). The relative impact of the working alliance in M-CET has not been evaluated, though it is assumed that a good working alliance is beneficial to therapy. The explicit work done early on in treatment regarding the foci, goals, and tasks involved in M-CET should ensure a strong working alliance, assuming that treatment goals and client goals are relatively parallel. In fact, maximizing patients' understanding of the treatment rationale may be useful in getting the patient "on board" and may help prevent early withdrawal from the group.

COMPONENTS OF TREATMENT

As previously stated, M-CET has been conducted in both group and individual format. However, in our research, this treatment was conducted exclusively in groups and was found to be very effective. Throughout the twelve-week treatment, clients are asked to monitor reexperiencing, avoidance, and arousal symptoms, and to record the number of panic attacks each day. These are graphed so that clients can monitor their own progress over time. As can be seen from the Patient Manual (Falsetti & Resnick, 1997a, b) excerpts, this treatment was not designed specifically for survivors of terrorism disasters. Rather, it is symptom focused and is designed for the treatment of PTSD and panic attacks, regardless of etiology. Research indicates that the greater number of traumatic events experienced, the higher the risk for PTSD. Thus it is quite likely than many people in a group have experienced multiple traumatic events, even if the presenting problem was a terrorist disaster. Following is a session-by-session description of the M-CET treatment.

Session 1

The first session of treatment is designed to provide education about traumatic events and reactions to trauma, including education about PTSD, panic attacks, depression, and increased drug and alcohol use. Patients are also provided with a treatment rationale using learning and information processing theories. The remainder of session 1 is used to discuss what the client can expect from treatment, the importance of attending sessions and completing homework, and the importance of monitoring symptoms. In the first session the assigned homework is to complete the monitoring forms for PTSD symptoms and panic attacks (these assignments are given throughout treatment), to write about the meaning of the traumatic event, and to complete a worksheet that assists in identifying conditioned cues.

Following is an excerpt from the Patient Manual that describes reactions to trauma (Falsetti & Resnick, 1997a, b):

> Mental health reactions that include depression, panic attacks, post-traumatic stress disorder, and coping by using drugs or alcohol occur at high rates among those who have had rape or physical assault crimes and who have had a loved one who has been murdered or violently killed. At least 1 in 3 women who have been raped or physically assaulted and 1 in 4 women who have had any crime experience will develop posttraumatic stress disorder, described below. Those who have had ongoing mental health problems for at least 3 months or more may need mental health treatment to help them recover from the effects of crime. Women who have had more than one trauma may be at even greater risk of mental health problems. Again, you are not alone. Many people suffer from these reactions due to traumatic experiences.

> *What is Posttraumatic Stress Disorder?*

> Posttraumatic stress disorder (PTSD) is an anxiety disorder that can develop after experiencing a traumatic event. This event could be rape, physical assault, a serious car accident, the murder of a family member or close friend, a natural disaster, military combat, a terrorist disaster, or some other sort of traumatic event in which you experienced, witnessed, or were confronted with an event that involved actual or threatened death or serious injury to you or someone else and which caused you to react with strong feelings of fear, helplessness, or horror.

> There are three main sets of symptoms to the diagnosis of PTSD: reexperiencing, numbing and avoidance, and arousal symptoms. *Reexperiencing symptoms* include having thoughts about the event even when you weren't trying to think about it, distressing dreams about the event, acting or feeling as though the traumatic event was occurring again, feeling very distressed when something reminds you of the traumatic event, and experiencing physical symptoms that you experienced at the time of the traumatic event (e.g., heart racing, sweating) when something reminds you of the traumatic event.

> *Avoidance and numbing symptoms* include trying to avoid thoughts, feelings or conversations that remind you of the traumatic event, avoidance of activities, places, or people that remind you of the

traumatic event, not being able to remember important parts of the traumatic event, and not feeling very interested in participating in activities that used to interest you. Another avoidance and numbing symptom that you may be experiencing includes feeling like you don't have much of a future, such as thinking you will never get married or have children or a career, or not expecting to have a normal lifespan. Finally, you may also notice that you are unable to feel all of your emotions since the traumatic event. An example of this would be feeling as though you are unable to have loving feelings.

The final set of PTSD symptoms are *arousal symptoms*. These include difficulty falling or staying asleep, irritability or outbursts of anger, difficulty concentrating, feeling on guard even when there is no reason to be, and being more easily startled by unexpected or sudden noises. You do not have to have all of these symptoms to have PTSD. In order to be diagnosed with this disorder you would only have to experience one of the reexperiencing symptoms, three of the avoidance and numbing symptoms, and two of the arousal symptoms for a least one month after the trauma.

What Are Panic Attacks?

Panic attacks are characterized by a feeling of intense fear or discomfort in which four or more of the following symptoms are experienced:
- palpitations, pounding heart, or increased heart rate
- sweating
- trembling or shaking
- shortness of breath or smothering feelings
- choking feelings
- chest pain or discomfort
- nausea or stomach distress
- feeling dizzy, unsteady, lightheaded or faint
- feelings of unreality or being detached from oneself
- fear of losing control or going crazy
- fear of dying
- numbness or tingling sensations
- chills or hot flushes

Panic attacks develop very suddenly and usually reach a peak within about 10 minutes. You may have panic attacks when reminded of the trauma in some way. Some examples of this would be if you read something in the newspaper that reminded you of the traumatic event you experienced or if you saw someone who reminded you of the person who attacked you and then you had a panic attack. You may also have some panic attacks that seem to be out of the blue, that are not related to the trauma in any way. For example, you may have a panic attack when shopping at a mall, or when sitting home relaxing, or you may even experience panic attacks while sleeping.

What About Other Symptoms?

The treatment that you will be participating in is specifically designed to treat the symptoms of PTSD and panic attacks. Of course that is not to say that these are the only symptoms or problems that you may be experiencing as a result of the traumatic event(s). You may also have symptoms of depression, such as feeling down or sad, weight loss or gain, fatigue or loss of energy, feelings of worthlessness, or recurrent thoughts of death. These symptoms are common in individuals who have experienced traumatic events, but should decrease as your other symptoms decrease. If these symptoms continue to be a significant problem when you have completed this treatment, your therapist may recommend additional treatment specific to depression.

Another common problem associated with PTSD is an increase in alcohol and/or drug use following a traumatic event. This may be a form of self-medication to avoid painful thoughts and feelings about the event. If you or your therapist think you may be physically dependent on drugs or alcohol, it is important to get treatment for this prior to undergoing M-CET. Your therapist will probably already have carefully assessed your use of drugs and alcohol to determine if it will interfere with participating in this treatment. If your drug or alcohol use has increased, but your therapist has determined that this treatment is still right for you, try to refrain from use during the 12 weeks of treatment and always attend sessions sober.

Session 2

Session 2 begins with reviewing the homework that was assigned in session 1, emphasizing the importance of homework. Causes and characteristics of PTSD and panic symptoms are reviewed and any questions the client has are addressed. This emphasizes the importance of the readings and assists in refreshing the client's memory about how symptoms may be trauma related. Also in session 2, education about the connection of breathing, panic attacks, and the traumatic event is provided. This lays the groundwork for breathing retraining. The main points presented here are how hyperventilation may lead to panic attacks and how the fight-or-flight reaction during a traumatic event can lead to conditioning of panic symptoms. In addition, clients are asked to complete a brief questionnaire about overbreathing in the Patient Manual. This helps the therapist to determine if overbreathing is playing a role in the client's panic attacks and if the overbreathing may be trauma related. Following is an excerpt from the Patient Manual (Falsetti & Resnick, 1997a, b).

Breathing, Panic Attacks, and the Traumatic Event

Barlow and Craske highlight in their treatment for panic disorder (*Mastery of Your Anxiety and Panic,* 1998) the importance of breathing. We will be discussing the main points from their treatment here and teaching you their method for breathing retraining. They reported that 50 to 60 percent of people who panic show some signs of overbreathing or hyperventilation. Overbreathing can either be one of the first symptoms of a panic attack or it can develop later into the panic attack. For people who have experienced traumatic events, overbreathing may also be a conditioned cue. If you think back to the traumatic event, do you remember what your breathing was like during that time? If you were very frightened, chances are that your breathing was affected. You may have been holding your breathing, or taking in big gulps or air, or breathing quickly and shallowly. How do you breathe when you feel frightened or stressed out in other situations? Because this type of breathing may be paired with the traumatic event you experienced, it may cause feelings of fear and danger similar to what you experienced at the time of the trauma.

In addition to the possible connection between overbreathing and the traumatic event, overbreathing causes a whole range of other physiological symptoms. These symptoms are not in and of themselves dangerous, but they may be frightening on their own, or again they may remind you of other physical sensations that you experienced at the time of the traumatic event. To help us to determine if overbreathing is an important symptom for you, let's go through some questions that will help us sort this out.

Are You Over-Breathing?

1. At the time of the traumatic event do you remember holding your breath, taking in big gulps of air, or breathing quickly and/or shallowly?

2. Was your breathing restricted in any way during the event (for example, did a perpetrator have his hands on your throat, or for rape victims, were you forced to perform oral sex)?
3. Did you feel like you were suffocating during the event?
4. In general, do you often feel short of breath or have the feeling that you are not getting enough air?
5. Do you ever experience a feeling of suffocation?
6. Do you sometimes experience chest pain, tingling, prickling, and/or numbing feelings?
7. Do you find that you often take in big gulps of air or yawn or sigh a lot?

If you answered yes to any of questions 1-3, then your over-breathing symptoms of panic may be associated with the event that you experienced, which can cause these symptoms to remind you of the trauma. If you answered yes to any of only questions 4-7 then your overbreathing symptoms may not be paired directly with the trauma, but the effects of overbreathing may bring on other symptoms that are associated with the trauma, or the physical sensations brought on by overbreathing may be in and of themselves frightening, just as they are for someone who has panic disorder, but has not experienced a traumatic event.

Finally, clients are taught how to breathe from the diaphragm. This gives each client a useful skill that may reduce physiological arousal and gives a sense of control over the body. Homework for this session includes monitoring of PTSD and panic symptoms, and practicing breathing retraining exercises.

Session 3

Session 3, like all sessions, begins with a review of the homework assigned in the previous session. In addition, clients are asked to work further on breathing retraining skills by slowing down the rate of breathing. In this session the connection of events, thoughts, feelings, and behaviors is introduced. For homework, clients are asked to continue monitoring their PTSD and panic symptoms, to continue practicing breathing retraining and slowing down breathing, and to complete their worksheets for trauma- and panic-related thoughts.

An excerpt from the Patient Manual, and an example of a completed worksheet used to assist identifying distorted cognitions, follows.

The Connection of Events, Thoughts, Feelings, and Behavior

Psychiatrists and psychologists have known for some time that it is not just what happens to us that affects how we feel, but it is also how we *interpret* events that affects how we feel. Here's an exercise to complete which will help to demonstrate this point. Pretend you were walking down the street and passed an acquaintance. You said hello, but this person said nothing back. What would you think, or in other words, how would you interpret the event? How would you feel— happy, sad, mad, or scared? Depending upon your interpretation you may have very different emotions. If you thought this person was mad at you, you might feel scared. If you thought the person was being a snob, you might feel angry. If you thought the person didn't see you, you might not experience much emotion at all about it. As you can see, several different emotions can result from the same event, depending upon your interpretation of the event.

The experience of a traumatic event can lead to some errors in thinking about current or future events. You may have a tendency to overgeneralize or to overestimate the probability of future trauma. Sometimes people who have been victimized walk around every day certain they are going to be attacked again each and every day. When they look at the *probability* though they see that the chances of that happening each and every day is often times very small. For example, a rape victim who was raped once when she was 20 also had 365 days a year for 19 years before she got raped that she was not raped and it had been 10 years since her rape, so she had another 3650 days after the rape that she was not raped. So, it didn't make sense for her to tell herself, "I can't go to the grocery store, because I just know I'll see the man that raped me," when the chances of that happening were very small.

Other examples of distorted thinking include using extremes and overgeneralizations such as "never, always, forever, must, every time, need should, and can't." Examples of extremes would be, "I can never trust anyone again," or "There aren't any men that I can ever feel safe with again." This is your mind's way of trying to protect you from future danger. However, the long term results of such thinking can leave you feeling very lonely and isolated. Look back at your first writing assignment. What overgeneralizations or extremes in your thinking can you identify? Have you predicted negative events for the future at rates that are higher than the facts suggest?

In addition to errors in thinking related to the traumatic event itself, you may also have errors in thinking about your panic attacks. Many people overestimate the probability of something bad happening as a result of panic attacks. Examples of this include thinking they will die or go crazy from their panic attacks. Sometimes people think they are having a heart attack, or that they will faint as a result of their panic attacks. It is important to look at all of the facts, not just how you are feeling, because your body may send false danger alarms. Panic attacks may feel terrible, but the facts are that they will not cause a heart attack, or cause you to go crazy. Have you ever had a heart attack from a panic attack? Have you ever gone crazy as a result of a panic attack? It is important to look at all the factual evidence. In future weeks you will be learning how to evaluate your thinking to determine if you are overestimating the probability or if you are overgeneralizing as a result of the traumatic event. For this week pay attention to how events, thoughts, and feelings are connected. To help you to do this, complete an "Event/Thoughts/Feelings Worksheet #1" each day on either an event that occurs that day, a panic attack, or the traumatic event you experienced.

Here are some examples of completed worksheets to get you started.

Example 1: Events/Thoughts/Feelings Worksheet #1

A. Event (something happens):
   The World Trade Center was attacked by terrorists and my sister was working there and was killed.
B. Thoughts/Beliefs (what you say to yourself about what happened):
   These people (the terrorists) are horrible and will stop at nothing. It will keep happening again. What if I lose someone else or there is a horrible nuclear war. I just couldn't take it. I can't take it that I lost my sister.
C. Feelings/Behaviors (how you feel sad, mad, glad, or scared, and what you do when you say the above (B) to yourself?
   I feel scared and angry. I am glued to CNN to see what is happening next.

Example 2: Events/Thoughts/Feelings Worksheet #1

A. Event (something happens):
   A friend calls to ask me to go out to dinner.

B. Thoughts/Beliefs (what you say to yourself about what happened):

I say to myself, "What if I have a panic attack from being around so many people in a crowded restaurant? Maybe I shouldn't go, I'll just embarrass myself and then my friend won't ask me to do anything again."

C. Feelings/Behaviors (How you feel sad, mad, glad, or scared, and what you do when you say the above (B) to yourself?

I feel scared and sad. I call her up and make up an excuse why I can't go.

These worksheets are intended to assist clients in making the connections between events, their thoughts about those events, and how they feel. In the next session, clients are taught how to challenge their own thinking.

Session 4

At session 4, after reviewing homework, the clients and therapist discuss distortions in thinking associated with trauma. A series of questions are introduced in worksheet format to assist clients in challenging distorted thinking. This builds on the worksheets of the previous session. Each client can usually relate to a particular type of cognitive distortion, for instance a tendency to overestimate the probability of something bad happening, or to disregard important aspects of a situation. For example, one client who was abducted and raped by a serial rapist blamed herself that other women were raped after her, and disregarded the fact that she had been the only victim able to identify the perpetrator and therefore contribute to his conviction. The therapist should assist each client in identifying a particular way of distorting that seems to be habitual, so the client may pay particular attention to this question on the worksheet as he or she completes homework for the week. For homework, clients are asked to complete worksheets on catastrophic thoughts and to continue to practice breathing retraining. An example of the worksheet completed by a client who lost her sister in the 9/11 attack on the World Trade Center is reproduced in Exhibit 17.1.

Session 5

Session 5 is devoted primarily to interoceptive exposure exercises aimed at desensitizing patients to physiological panic sensations. Exercises from Barlow and Craske's (1998) PCT are practiced in the session and ratings regarding the intensity of the sensations, the similarity to panic, and anxiety at

EXHIBIT 17.1.
Events/Thoughts/Feelings Worksheet #2

A. Event (something happens): *The World Trade Center was attacked by terrorists and my sister was working there and was killed.*

B. Thoughts/beliefs (What you say to yourself about what happened): *These people (the terrorists) are horrible and will stop at nothing. It will keep happening again. What if I lose someone else or there is a horrible nuclear war. I just couldn't take it. I can't take it that I lost my sister.*

C. Feelings/behaviors (What you feel—sad, mad, glad, or scared, and what you do when you say the above (B) to yourself). *I feel scared and angry. I am glued to CNN to see what is happening next.*

Complete the following about your thoughts in B.

1. What is the evidence for this thought?
    There have been other terrorist attacks. There is talk of a possible war with Iraq and they have nuclear, biological, and chemical weapons.
2. What is the evidence against this thought?
    The terrorists believe what they are doing is for a reason. As horrible as all of this is I am somehow getting through each day. Our country has now been alerted and is taking action.
3. Rate the probability that this thought is true on a scale from 0 (not at all true) to 100 (completely true).

    These people (the terrorists) are horrible and will stop at nothing. Probability: 90
    It will keep happening again. Probability: 90
    What if I lose someone else or there is a horrible nuclear war. I just couldn't take it. I can't take it that I lost my sister. Probability: 40

4. Am I using words that indicate extremes in my thinking, for example: always, never, should, forever, need, must, can't, every time?
    Saying I can't take it—even though there are days that I feel like I won't make it through I always do and I have to for my children and my sister's children.
5. Are my thoughts based on feelings rather than facts?
    Both. Based on my scared feelings about the possibility of something like 9/11 happening again and based on the fact that there have been other terrorist acts.
6. Am I overgeneralizing from a single or past incident?
    No, there have been several terrorists acts and the threats are real.

EXHIBIT 17.1. *(continued)*

7. Am I disregarding important aspects of the situation?
    Yes, I'm disregarding that the United States is the most powerful nation in the world and has many allies, even if it does have enemies. We are also now better prepared to deal with any future attacks and there have not been any on U.S. soil in several years now.
8. What else could I say to myself instead of initial thought or belief in B?
    Although I feel devastated at the loss of my sister I may be overestimating the risk of me and my family being in imminent danger on a daily basis from acts of terrorism. It might be good for me to watch less CNN and see other things happening around me with my children that are good.
9. If I were able to say this how would I feel?
    Less angry and scared, more able to focus on some of the positives in life.

the sensations are elicited to develop a hierarchy for homework exercises. Interoceptive exercises include hyperventilating, breath holding, tensing muscles, spinning in a chair, breathing through a straw, and stair stepping. Cognitive restructuring work to help clients identify overestimations, catastrophic thinking, and other cognitive distortions is also completed. Lastly, clients are instructed to begin applying breathing retraining in anxiety-provoking situations. Homework includes continuing to complete worksheets for cognitive restructuring, applying breathing when feeling anxious, and practicing interoceptive exposure exercises.

Session 6

At session 6, work is continued with interoceptive exposure exercises and cognitive exposure is introduced. We have found that patients with PTSD and comorbid panic attacks are generally successful in quickly reducing their fear to the physiological symptoms of panic attacks. If the client is successful in decreasing fear in the initial two interoceptive exercises on the hierarchy, then we move on to the next two. Practice these in session in case the client forgets how to perform the exercises. To begin the cognitive exposure component, patients are asked to write about the traumatic event. The rationale for exposure to the traumatic memory is very important in motivating the client to complete this assignment. The first writing of the traumatic event is often the most difficult therapy assignment. Clients should be encouraged to begin this assignment as soon as possible, to read the account at least once a day, and record their anxiety after the writing and each reading of the account.

Following is an excerpt from the Patient Manual explaining the writing exposure:

Facing the Trauma

As we talked about when we reviewed the symptoms of PTSD, avoidance of thoughts and feelings about traumatic events is quite common. It makes logical sense to avoid reminders of the trauma because these thoughts and feelings can be very painful. In the short term, this may lessen the pain for a brief time, but in the long run you never get to process these thoughts and feelings, so the pain attached to the memories can be just as strong as after the event occurred.

During this session we will be working on helping you become less fearful of the traumatic memory and the strong emotions that are attached to it. It is important to remind yourself that it is a memory that you are dealing with now—the event is over and is not happening again. Sometimes the feelings attached to the event are so strong that it can feel as though the event is reoccurring, but you must remind yourself that it is a memory you are dealing with now.

This week you will begin to process the traumatic memory through writing about the event you experienced. This will allow you to face the memory and the strong feelings attached to it will begin to decrease as you process the memory. Think of this process of facing the memory as similar to watching a scary movie on television. If you were to turn on the TV and catch a glimpse of a scary movie, feel frightened and immediately change the channel, you would remain frightened of that movie. At the moment when you changed the channel, you would feel less frightened because you were no longer watching it, but you would still be frightened by the movie itself if you were to change the channel back to the movie. Now let's say that instead of changing the channel, you forced yourself to sit through the movie. You might feel really frightened while you were watching it. If you kept watching the same movie over and over, however, how do you think you would feel? Let's say you watched that same scary movie 10 nights in a row. Do you think you would continue to be scared of the movie? You might in fact even become a little bored with the movie— something called *habituation* would happen. This means that your body would not keep responding with fright because it would learn that nothing bad is really going to happen to you—that it is just a movie.

Let's apply this reasoning to your traumatic memory. If you remember part of what happened to you in a memory, but you immediately push it away, that is similar to changing the channel when the scary movie comes on. It provides some immediate relief from the fear, but in the long run you remain fearful of the memory and your body never gets a chance to learn that just because the memory is there does not mean you are in danger again. In other words, your body doesn't habituate to the fact that it is a memory you are dealing with now and that you are in no immediate danger. However, if you allow yourself to fully face that memory, like watching the scary movie, your fear will decrease over time.

This week you will be asked to write about the trauma you experienced as a way of replaying that memory. Use all of your senses. Write about any sounds you remember hearing, any smells that you can recall, and any physical sensations that you remember. Write in as much detail about what you saw, what the perpetrator looked like and was wearing, and what kind of day it was. Also write about your surroundings and yourself in as much detail as possible. As you write allow yourself to feel your feelings. Cry if you feeling like crying, or hit a pillow if you feel angry. Do this assignment when you know you can have some privacy. Write this as soon as you can, then read it to yourself at least once a day. Use the monitoring form to record your anxiety level. Just as with watching a scary movie over and over, your anxiety should decrease as you practice reading this assignment.

Session 7

Session 7 focuses on the traumatic event writing homework. If treatment is conducted in a group format, clients read their assignments to one another. If the client is being treated individually, he or she is instructed to read the assignment aloud to the therapist. In both group and individual treatment, it is important to discuss the writing homework and identify which parts of the event were most painful to recall. It is also important to identify any cognitive distortions that may interfere with processing the memory accurately (e.g., "I must have done something to deserve being raped. I should never have gone on a date with him.") The therapist then assists the client in using cognitive restructuring skills to examine the evidence, and to challenge and correct distortions. This session is also used to review any remaining interoceptive exposure exercises that need to be completed. If the client has completed these exercises, the therapist should provide reinforcement for completing this phase of treatment and note any decreases in panic attacks and avoidance behaviors that are associated with this. For homework, patients are asked to write about the traumatic event a second time. This time clients are also asked to identify any emotions they notice as they write. It is not unusual for clients to remember more details in the second account and to report reliving emotions such as fear and anger. Therapists should explain that writing about the event a second time will not be as difficult as writing about the event the first time, although it will not be easy; and to note any decreases in anxiety that took place the first week as a result of reading the account. Following is a modified account (to protect identity) from a client who assisted at the World Trade Center and was looking for his best friend.

Exposure Worksheet

I got the call from my best friend's wife on the evening of September 11, 2001. Ben, her husband and my best friend, was missing. I had been in their wedding just last year and knew they were expecting their first child. My heart raced, as she told me that Ben had gone into work early that morning to get caught up on some things—he worked in the first tower. I felt numb, not Ben, this couldn't really be true. Maybe he was just there helping and hadn't had a chance to call home yet. I told Jan I would fly up as soon as flights resumed. I got off the phone and called the airline—Delta, I think—could I make a reservation? All flights were canceled and they would have to rebook people who'd been delayed first. Should I drive? What should I do? Two days passed. I don't remember much from them. Then I think it was September 13 I flew to NYC. The flight crew had heard I was an ex-cop and one of the attendants cried with relief to have me on the flight. I volunteered to help with the clean up and because of my background they let me. It was horrific. Now I've worked in big cities before and have seen a lot—gruesome murder scenes. Nothing, nothing could prepare anyone for this. It's how I would imagine something to look after a nuclear holocaust. The sky was dark from all of the debris, the air—the stench is suffocating and all the stuff in the air. I can't breathe. We look for bodies, but don't really find bodies—we find pieces, things—a ring, a shoe, a hand. Nothing seems real and at the same time everything is so painfully real. I'm telling you, it was like the end of the world. I go to Jan's house. For a week we have no news of Ben. We hold each other and cry. It is such a strange thing to hold your best friend's wife and there isn't a damn thing you can do. I'm not sleeping, not really eating much, yet I feel compelled to go back day after day. Day 10 someone tells me I have to take a break. My hands are bloody, even with wearing gloves, from digging. We go a little outside the site and it looks like life is still going on, but everyone is a shell, zombielike if you look up close, going through the motions. We never found Ben. He is among the missing. I can't stop seeing the images, the pieces, the smoke in the air, the choking feeling. They haunt me at night, I can't breathe during the day. Ben has a son now. Where the hell is he!?!

Session 8

At session 8 the second exposure writing assignments are read aloud and discussed. Clients and therapist further identify distortions and difficulties about the event, such as self-blame that they were unable to prevent the event, or generalizations from this event to all situations or groups of people (e.g., "All Islamic men are dangerous"). The clients and therapist should also discuss interoceptive exposure exercises if any were assigned the previous week. At this point in treatment, the client should be less fearful of physical sensations, and as a result may also be experiencing a decrease in panic attacks. If the client has successfully completed all interoceptive exposure exercises and has also experienced a decrease in panic attacks, the therapist should verbally reinforce this progress. New material in this session includes how trauma may have affected the client's sense of safety. Pertinent to safety issues is discussion of conditioned cues, changes in thinking, and changes in the ability to recognize dangerous situations. Facts about crime and basic safety tips, if appropriate, are also discussed. For homework, clients are asked to continue monitoring symptoms and to record panic attacks, apply breathing, read a handout on safety and a handout of crime facts and strategies for prevention and coping.

Session 9

Session 9 introduces in vivo exposure to trauma- and panic-related cues. Clients develop fear hierarchies for at least three in vivo target fears and work through these hierarchies as their homework in the remainder of the sessions. Cognitive distortions also continue to be addressed. This session focuses on trust issues and includes a handout that is modified from Resick and Schnicke (1993) to broaden the applicability from rape victims to victims of other traumatic events. For homework, clients are asked to complete the Events/Thoughts/Feelings #2 worksheet, to address distorted trust cognitions, and to begin in vivo exposure to the first target fear. Following is an excerpt from the Patient Manual discussing in vivo exposure to conditioned cues.

Exposure to Panic- and Trauma-Related Cues

This week you will also begin to expose yourself to situations that are not dangerous, but which make you feel anxious or fearful because they are associated with either the traumatic event you experienced or with panic attacks. In session your therapist will have asked you to write down any places, events, or situations that you still avoid. If you have not yet done this, take the time now to write these down. Your therapist should have also helped you to pick the top three target fears that you will be working on over the next couple of weeks. On a scale of 1 to 100 (0 = no anxiety and 100 = extremely high anxiety) rate how much fear/anxiety each of these produce. This week we will begin with the least anxiety provoking event, place, or situation that you avoid.

Target Fears

1. _____
2. _____
3. _____

Your therapist will also help you to develop what is called a *fear hierarchy*. This hierarchy is a step-by-step program to help you to expose yourself to the feared situation and reduce your anxiety and avoidance. Step 1 should be the step that causes the least anxiety. You will practice this step until your fear reduces to at least a 10, then you will move on to Step 2, and so on, working your way to the actual feared event, situation, or place. An example of this for a fear of going out to a restaurant (where you went on a date prior to being raped) with a large group because you are afraid you will panic might be to first go just for coffee with one friend at a time when the restaurant is not crowded. Your second step might be to go with two friends for coffee and dessert during a time when the restaurant is a little more crowded. Your third step might be to go with two friends for dinner, and the final step would be to face your fear and go with a large group. This is very similar to how you reduced your fear of the panic sensations when you first started with those exercises that were the least anxiety provoking and worked your way up to the exercises that caused the most anxiety. You will also be rating your anxiety as you do these exercises.

Session 10

This session includes review of the worksheets to address distorted trust cognitions and to discuss progress on in vivo exposure. If exposure to Target Fear 1 was successful, then the therapist can discuss a plan for in vivo exposure to Target Fear 2. Cognitive work focuses on power/competence issues and clients are asked to discuss the potential effects of trauma on beliefs about power/competence. Often clients describe one of two extremes: feeling very helpless or thinking they have to be completely in control of everything. Homework includes completing in vivo exposure exercises to Target Fear 2 and completing worksheets to address cognitive distortions about power/competence.

Session 11

Session 11 begins with a review of the homework. Any difficulties in doing in vivo exposure or cognitive restructuring may be discussed at this time. A plan for in vivo exposure to Target Fear 3 may also be discussed. New material covered in session 11 includes identifying cognitive distortions in the area of self- and other-esteem. The homework assignment is to complete in vivo exposure exercises to Target Fear 3. A handout on esteem is also assigned and clients are asked to apply cognitive restructuring skills for esteem problems that are trauma related.

Session 12

Session 12, the final treatment session, is devoted to reviewing treatment, discussing problems encountered in treatment, discussing relapse prevention, and making plans for what clients may want to continue working on independently. Following is an excerpt from the Patient Manual discussing maintenance of gain and relapse prevention.

Maintenance Planning

If you have made progress during your treatment, you will of course want to maintain these gains. People sometimes worry that once they are no longer in treatment that they will feel bad again. This is generally not true, but there are things that you can do to maintain your progress. First, it is important to remind yourself that anxiety is normal and that the goal of your treatment was not to eliminate all anxiety, but rather to decrease your panic attacks and symptoms of posttraumatic stress disorder. You have learned many skills to reduce these symptoms including breathing retraining and challenging your thinking. It is important to continue to use these skills as needed to maintain your progress. Another key to maintaining your progress is to continue to expose yourself to situations that you previously avoided for fear of panicking or because it reminded you of the trauma. It is very important not to fall back into previous avoidance patterns. When you experience anxiety practice the skills you learned in treatment.

High Risk Times

There will be times in your life when you are more vulnerable to either panic attacks or experiencing a recurrence of some of your symptoms of PTSD. This is most likely to happen at times of high stress or during times when some of your core beliefs which were shattered by the trauma are challenged by life situations. Stress can increase your physiological arousal level. This increased arousal level can make it more likely that you will experience physical sensations that are similar to what you experienced when you had panic attacks. It is important to recognize this increased arousal and work to decrease it through proper breathing and correcting any catastrophic thinking. Even if you do have a panic attack again, it is important not to catastrophize about it. In fact, you are likely to have a panic attack again at some time in your life. This does not mean that all of your hard work has been wasted, or that you will continue to have panic attacks. Instead, it may be a sign that you need to focus on taking care of yourself and practicing some stress reduction techniques and review your work from treatment to keep you on track.

PTSD symptoms, such as nightmares or intrusive thoughts are most likely to occur when events happen in your life that touch on beliefs that were affected by the traumatic event you experienced. This could be anything from beginning to try to trust someone again, to your daughter hitting adolescence and feeling scared for her. If someone you know is traumatized in some way this could also trigger symptoms for you. If this occurs it does not mean you are "going downhill" or that you will continue to feel worse. It may be helpful to review your workbook from treatment, though, and to practice the skills you learned in treatment. If you really find yourself struggling with some things, it might be helpful to call your therapist and set up an appointment for a booster session to get you back on track. Most times the recurrence of symptoms is temporary and you may be able to get through it in only one or two sessions at times of stress.

You have worked hard in treatment and have chosen to take control of your life! Congratulations on finishing this program! There may be areas that you think would be helpful for you to continue to work on with a therapist, or you may be at a point where you are ready to go it on your own. Either way congratulate yourself for having the courage to deal with some very frightening and difficult memories and be proud of the work you have accomplished.

As explained earlier, this treatment was not developed specifically for terrorist-related disasters, thus the excerpts showed examples of many types of trauma. However, through some of the example homework assignments, use of this treatment with survivors of terrorism who are suffering from PTSD and panic attacks is demonstrated.

FUTURE TREATMENT DIRECTIONS

M-CET is a group cognitive behavioral treatment that was developed for PTSD and panic attacks. Preliminary data have indicated that it is an effective treatment. We believe this treatment may be applied effectively to survivors of terrorist disasters who are suffering from PTSD. However, there are specific issues to be considered, and of course M-CET still needs to be tested with this specific population.

REFERENCES

Acierno, R., Resnick, H., Kilpatrick, D.G., Saunders, B., & Best, C.L. (1999). Risk factors for rape, physical assault, and posttraumatic stress disorder in women: Examination of differential multivariate relationships. *Journal of Anxiety Disorders, 13*(6), 541-563.

American Psychiatric Association. (2000). *Diagnostic and Statistical Manual of Mental Disorders* (Fourth Edition, Text Revision). Washington, DC: Author.

Atkeson, B.M., Calhoun, K.S, Resick, P.A., & Ellis, E.M. (1982).Victims of rape: Repeated assessment of depressive symptoms. *Journal of Consulting and Clinical Psychology, 50,* 96-102.

Barlow, D.H. & Craske, M.G. (1998). *Mastery of Your Anxiety and Panic.* San Antonio, TX: Psychological Corporation.

Barlow, D.H. & Craske, M.G. (2000). *Mastery of Your Anxiety and Panic (MAP-3).* San Antonio, TX: Psychological Corporation.

Barlow, D.H., Gorman, J.M., Shear, M.K., & Woods, S.W. (2000). Cognitive-behavioral therapy, imipramine, or their combination for panic disorder. *Journal of the American Medical Association, 283,* 2529-2536.

Beck, A.T., Steer, R.A., & Garbin, M.G. (1988). Psychometric properties of the Beck Depression Inventory: Twenty five years of evaluation. *Clinical Psychology Review, 8,* 77-100.

Beck, A.T., Ward, C.H., Mendelsohn, M., Mock, J., & Erbaugh, J. (1961). An inventory for measuring depression. *Archives of General Psychiatry, 4,* 561-571.

Blake, D.D., Weathers, F.W., Nagy, L.M., Kaloupek, D.G., Gusman, F.D., Charney, D.S., & Keane, T.M. (1995). The development of a Clinician-Administered PTSD Scale. *Journal of Traumatic Stress, 8,* 75-90.

Blake, D., Weathers, F., Nagy, L., Kaloupek, D., Klauminzer, G., Charney, D., & Keane, T. (1990). *Clinician Administered PTSD Scale (CAPS).* Boston: National Center for PTSD, Behavioral Sciences Division.

Bordin, E.S. (1979). The generalizability of the psychoanalytic concept of the working alliance. *Psychotherapy: Theory, Research & Practice, 16*(3), 252-260.

Bryant, R.A., Moulds, M., & Guthrie, R. (2000). Acute stress disorder scale: A self report measure of acute stress disorder. *Psychological Assessment, 12,* 61-68.

Bryant, R.A. & Panasetis, P. (2001). Panic symptoms during trauma and acute stress disorder. *Behaviour Research and Therapy, 39,* 961-966.

Chambless, D.L., & Ollendick, T.H. (2001). Empirically supported psychological interventions: Controversies and evidence. *Annual Review of Psychology, 52,* 685-716.

Cloitre, M. (1998). Sexual revictimization: Risk factors and prevention. In V.M. Follette, J.I. Ruzek, & F.R. Abueg (Eds.), *Cognitive-Behavioral Therapies of Trauma* (pp. 278-304). New York: Guilford Press.

DeRubeis, R.J., & Feeley, M. (1990). Determinants of change in cognitive therapy for depression. *Cognitive Therapy & Research, 4*(5), 469-482.

Di Nardo, P.A. & Barlow, D.H. (1988). *Anxiety Disorders Interview Schedule–Revised (ADIS-R).* New York: Phobia and Anxiety Disorders Clinic, State University of New York at Albany.

Falsetti, S.A. & Resnick, H.S. (1992). *Physical Reactions Scale.* Charleston, SC: Crime Victims Research and Treatment Center, Medical University of South Carolina.

Falsetti, S.A. & Resnick, H.S. (1995). Helping the victims of violent crime. In J.R. Freedy and S.E. Hobfoll (Eds.), *Traumatic Stress: From Theory to Practice* (pp. 263-285). New York: Plenum.

Falsetti, S.A. & Resnick, H.S. (1997a). Frequency and severity of panic attack symptoms in a treatment seeking sample of trauma victims. *Journal of Traumatic Stress, 10,* 683-689.

Falsetti, S.A. & Resnick, H.S. (1997b). Multiple channel exposure therapy: Patient manual. Medical University of South Carolina.

Falsetti, S.A., Resnick, H.S., Dansky, B.S., Lydiard, R.B., & Kilpatrick, D.G. (1995). The relationship of stress to panic disorder: Cause or effect? In C.M. Mazuer (Ed.), *Does Stress Cause Psychiatric Illness?* (pp. 111-147). Washington, DC: American Psychiatric Press, Inc.

Falsetti, S.A., Resnick, H.S., & Davis, J.L. (2005). Multiple Channel Exposure Therapy: Combining cognitive behavioral therapies for the treatment of posttraumatic stress disorder with panic attacks. *Behavior Modification, 29,* 70-94.

Falsetti, S.A., Resnick, H.S., Davis, J.L., & Gallagher, N.G. (2001). Treatment of posttraumatic stress disorder with comorbid panic attacks: Combining cognitive processing therapy with panic control treatment techniques. *Group Dynamics: Theory, Research, and Practice, 5,* 252-260.

Falsetti, S.A., Resnick, H.S., Resick, P.A., and Kilpatrick, D.G. (1993). The modified PTSD symptom scale: A brief self-report measure of posttraumatic stress disorder. *The Behavior Therapist, 16*(6), 161-162.

First, M.B., Spitzer, R.L., Gibbon, M., & Williams, J.B.W. (1997). *Structured Clinical Interview for DSM-IV Axis I Disorders-Clinician Version (SCID-CV)*. Washington, DC: American Psychiatric Press.

Foa, E.B., Dancu, C.V., Hembree, E.A., Jaycox, L.H., Meadows, E.A., & Street, G.P. (1999). A comparison of exposure therapy, stress inoculation training, and their combination for reducing posttraumatic stress disorder in female assault victims. *Journal of Consulting and Clinical Psychology, 67,* 194-200.

Foa, E.B. & Kozak, M.J. (1986). Emotional processing of fear: Exposure to corrective information. *Psychological Bulletin, 99,* 20-35.

Foa, E.B., Riggs, D.S., Dancu, C.V., & Rothbaum, B.O. (1993). Reliability and validity of a brief instrument for assessing post-traumatic stress disorder. *Journal of Traumatic Stress, 6,* 459-473.

Foa, E.B. & Rothbaum, B.O. (1998). *Treating the Trauma of Rape: Cognitive-Behavioral Treatment of PTSD*. New York: Guilford Press.

Foa, E.B., Rothbaum, B.O., Riggs, D.S., & Murdock, T.B. (1991). Treatment of posttraumatic stress disorder in rape victims: A comparison between cognitive-behavioral procedures and counseling. *Journal of Consulting and Clinical Psychology, 59,* 715-723.

Foa, E.B., Rothbaum, B.O., & Steketee, G.S. (1993). Treatment of rape victims. *Journal of Interpersonal Violence, 8,* 256-276.

Galea, S., Ahern, J., Resnick, H., Kilpatrick, D., Bucuvalas, M., Gold, J., & Vlahov, D. (2002). Psychological sequelae of the September 11 terrorist attacks in New York City. *New England Journal of Medicine, 346,* 982-987.

Horowitz, M., Wilner, N., & Alvarez, W. (1979). Impact of event scale: A measure of subjective distress. *Psychosomatic Medicine, 41,* 209-218.

Keane, T.M., Zimering, R.T., & Caddell, J.M. (1985). A behavioral formulation of posttraumatic stress disorder in Vietnam veterans. *The Behavior Therapist, 8,* 9-12.

Kessler, R.C., Sonnega, A., Bromet, E., Hughes, M., & Nelson, C.B. (1995). Posttraumatic stress disorder in the National Comorbidity Survey. *Archives of General Psychiatry, 52,* 1048-1060.

Kilpatrick, D.G. & Calhoun, K.S. (1988). Early behavioral treatment for rape trauma: Efficacy or artifact? *Behavior Therapy, 19,* 421-427.

Kilpatrick, D.G., Saunders, B.E., Amick-McMullan, A., Best, C.L., Veronen, L.J., & Resnick, H.S. (1989). Victim and crime factors associated with the development of crime related posttraumatic stress disorder. *Behavior Therapy, 20,* 199-214.

Kilpatrick, D.G., Veronen, L.J., & Resick, P.A. (1979). Assessment of the aftermath of rape: Changing patterns of fear. *Journal of Behavioral Assessment, 1,* 133-148.

King, D.W., King, L.A., Foy, D.W., Keane, T.M., Fairbank, J.A. (1999). Posttraumatic stress disorder in a national sample of female and male Vietnam veterans: Risk factors, war-zone stressors, and resilience-recovery variables. *Journal of Abnormal Psychology, 108*(1), 164-170.

Lang, P.J. (1968). Fear reduction and fear behavior: Problems in treating a construct. In J.M. Schlien (Ed.), *Research in Psychotherapy* (Vol. 3) (pp. 90-102). Washington, DC: American Psychological Press.

Lang, P.J. (1977). Imagery in therapy: An information processing analysis of fear. *Behavior Therapy, 8,* 862-886.

Lyons, J.L., Cho, L., & Brown, S.A. (2001, November). *Issues in Race/Gender Matching Between Therapist and Trauma Group.* Poster session presented at the annual meeting of the International Society for Traumatic Stress Studies, New Orleans, Louisiana. December 6-9.

MacKenzie, K.R. (1993). Time-limited group therapy and technique. In A. Alonso & H.I. Swiller (Eds.), *Group Therapy in Clinical Practice* (pp. 423-447). American Psychiatric Press: Washington, DC.

Norris, F.H. (March 2002). 50,000 disaster victims speak: An empirical review of the empirical literature, 1981-2001. National Center for PTSD. Available online at: http://www.istss.org/terrorism/victims_speak.htm.

Orlinsky, D.E., Grawe, K., & Parks, B.K. (1994). Process and outcome in psychotherapy: Noch einmal. In A.E. Bergin & S.L. Garfield (Eds), *Handbook of Psychotherapy and Behavior Change* (Fourth Edition) (pp. 270-376). New York: John Wiley & Sons.

Raue, P.J., Castonguay, L.G., & Goldfried, M.R. (1993). The working alliance: A comparison of two therapies. *Psychotherapy Research, 3*(3), 197-207.

Resick, P.A., Nishith, P., Weaver, T.L., Astin, M.C., & Feuer, C. (2002). A comparison of cognitive-processing therapy with prolonged exposure and a waiting condition for the treatment of chronic posttraumatic stress disorder in female rape victims. *Journal of Consulting and Clinical Psychology, 70,* 867-879.

Resick, P.A. & Schnicke, M.K. (1992). Cognitive processing therapy for sexual assault victims. *Journal of Consulting and Clinical Psychology, 60,* 748-756.

Resick, P.A. & Schnicke, M.K. (1993). *Cognitive Processing Therapy for Rape Victims: A Treatment Manual.* Newbury Park, CA: Sage Publications.

Resnick, H.S. (1996). Psychometric review of trauma assessment for adults (TAA). In B.H. Stamm (Ed.), *Measurement of Stress, Trauma, and Adaptation* (pp. 362-365). Lutherville, MD: Sidran.

Resnick, H.S., Falsetti, S.A., Kilpatrick, D.G., & Foy, D.W. (1994). *Associations between panic attacks during rape assaults and follow-up PTSD and panic attack outcomes.* Paper presented at the 10th Annual Meeting of the International Society for Traumatic Stress Studies, Chicago, Illinois. November 4.

Resnick, H.S. & Newton, T. (1992). Assessment and treatment of post-traumatic stress disorder in adult survivors of sexual assault. In D. Foy (Ed.), *Treating PTSD* (pp. 127-164). New York: Guilford Press.

Root, M.P. (1996). Women of color and traumatic stress in "domestic captivity": Gender and race as disempowering statuses. In A.J. Marsella, M.J. Friedman, E.T. Gerrity, & R.M. Scurfield (Eds.), *Ethnocultural Aspects of Posttraumatic*

Stress Disorder: Issues, Research, and Clinical Applications (pp. 363-387). Washington, DC: American Psychological Association.

Schlenger, W.E., Caddell, J.M., Ebert, L., Jordan, B.K., Rourke, K.M., Wilson, D., Thaliji, L., Dennis, J.M., Fairbank, J.A., & Kulka, R.A. (2002). Psychological reactions to terrorist attacks. Findings from the National Study of Americans' Reactions to September 11. *Journal of the American Medical Association, 288,* 581-588.

Veronen, L.J. & Kilpatrick, D.G. (1983). Stress management for rape victims. In D. Meichenbaum & M.E. Jarenko (Eds.), *Stress Reduction and Prevention* (pp. 341-374). New York: Plenum.

Vielhauer, M., Findler, M., Schnurr, P.P., Garcia, G., & Spiro, A. (1998). *Preliminary findings on the Brief Trauma Interview.* Poster session presented at the annual meeting of the International Society for Traumatic Stress Studies, Washington, DC, November 20-23.

Yalom, I.D. (1975). *The Theory and Practice of Group Psychotherapy.* New York: Basic Books.

Chapter 18

Trauma/Grief-Focused Group Psychotherapy with Adolescents

William R. Saltzman
Christopher M. Layne
Alan M. Steinberg
Robert S. Pynoos

RATIONALE AND OBJECTIVE OF THE GROUP

Over the past decade, the UCLA Trauma Psychiatry Program has developed specialized group and individual therapy intervention programs for children and adolescents who were exposed to trauma and/or traumatic loss, and who suffer from chronic distress and impaired functioning in school, peer, and family settings. Our programs have been used in postwar Bosnia, at Columbine High School in Littleton, Colorado, after the tragic shootings there, in New York City following the 9/11 attacks, and in school districts in communities affected by high levels of community violence.

Recent studies on the prevalence and impact of traumatic exposure among youth (Kilpatrick, Saunders, Resnick, & Smith, 1995; Lorion & Saltzman, 1993; Saltzman, Pynoos, Layne, Steinberg, & Aisenberg, 2001b) show that almost one-quarter of adolescents living in the United States have been directly exposed to or have witnessed significant violence, and that as many as 20 percent of these youths experience chronic posttraumatic symptoms. Our clinical experience surveying large numbers of adolescents suggests that the majority of these youths are neither identified nor provided with appropriate psychosocial support services (Saltzman et al., 2001b). Despite the profusion and variety of short-term crisis intervention services

Support for this chapter was provided by UNICEF Bosnia and Hercegovina, the Family Studies Center and Kennedy International Studies Center of Brigham Young University, and the UCLA Trauma Psychiatry Bing Fund.

The authors gratefully acknowledge the assistance of Josh Britton with portions of the literature review.

that crop up in the aftermath of a disaster or large-scale traumatic event such as a terrorist bombing, there are often far fewer intermediate and long-term services for youths who show signs of severe persistent posttraumatic distress reactions and developmental disturbance.

Recent findings underscore the importance of providing appropriate therapeutic services to trauma-exposed youths who show signs of significant post-traumatic distress and developmental difficulties three months or more posttrauma. In particular, evidence suggests that a significant proportion of youths who experience severe and persisting posttraumatic stress reactions are at risk for experiencing chronically severe disorders such as post-traumatic stress disorder (PTSD) or depression, if they are not given specialized intervention (Pynoos, Steinberg, & Piacentini, 1999). Of particular concern, severe and persistent posttraumatic stress symptoms are markers of risk for serious developmental disruption and a host of adverse outcomes, including reduced academic achievement, aggressive, delinquent, or high-risk sexual behaviors, and substance abuse (March & Amaya-Jackson, 1994; Saigh, Yasik, Sack, & Koplewicz, 1999; Kilpatrick, Aciermo, Saunders, Resnick, & Best, 2000). These empirically based observations are also consistent with findings from clinical work with many youths whose persistent hyperarousal symptoms place them at risk for academic difficulty and failure. In particular, sleep difficulties (and resulting tiredness) and distractibility at school are linked to teachers' and students' self-reports of dozing off in class, difficulty paying attention, and inability to recall academic material when required to do so in class or during exams. The following vignette, and case studies throughout this chapter, are drawn from our school-based programs in postwar Bosnia and in high-crime communities within California. Treatment was provided to the young man described as follows at one of the participating schools in southern California. This case illustrates the insidious way that hyperarousal symptoms can disrupt a young life.

A young man was referred to the school counseling office for academic difficulties one year after a near-fatal incident in which he was mistaken for a gang member and shot in the chest while walking to school. When asked to describe what was the hardest part about returning to school, he described feeling "scared" and "very watchful" every day as he walked to and from school: "I look out for cars I don't recognize and am always looking for places I can hide—you know, behind bushes and walls, and shops that I can duck into if a car pulls up. I don't think about the future. I just try to survive for another day. It takes me a long time to calm down once I get to school." When asked to describe why he often felt sleepy at school, he stated, "I try not to be scared of violence and death, so after my family goes to bed, I stay up and watch violent movies, like the *Faces of Death* videotapes. But that makes me feel so 'wired' that I can't get to sleep for hours, like until three or four in the morning, so I'm tired a lot at school."

Other correlates of trauma exposure suggest that traumatized youths are at increased risk for experiencing a variety of interpersonal difficulties that may serve as significant and persistent sources of stress in the aftermath of violent events. For instance, many traumatized youths show symptoms or signs of social withdrawal, depression, or anxiety. Of particular developmental concern, traumatized youths may engage in generalized patterns of traumatic avoidance and/or withdrawal from important interpersonal relationships, which, in turn, may lead to a marked restriction of developmentally salient activities, tasks, and experiences, leading to alteration and stunting in the developmental trajectory (Pynoos, Steinberg, & Wraith, 1995). The traumatic avoidance of friends or family members may create significant disruptions in interpersonal relationships in which others assume the role of trauma or loss reminders. In addition, intrusive questions by peers and family members concerning the event may also constitute significant social distress by serving as potent trauma or loss reminders, thereby complicating efforts to recover.

A sixteen-year-old female student was referred to a school counseling center two days after witnessing her best friend, with whom she was walking home after school, being shot and seriously injured in gang-related cross fire. For several weeks following the incident the student often presented once or twice a day at the school counseling center where she tearfully described being barraged with questions by fellow students who "keep on asking me to tell them about what happened." During the course of her participation in a trauma-focused therapy group for students exposed to community violence, the student described, with an apparent mixture of anger and sadness, the breakup of her friendship with the young woman who was shot: "We used to spend lots of time together and would talk about everything. But after it happened, she told me that my being around makes her think about when she got shot, and she's mean to me now and says she doesn't want to be around me. I told her that I was sorry about what happened but that it wasn't my fault. But she said it didn't matter. So I told her, 'Whatever!' So she's not my friend anymore. We just ignore each other now."

In contrast, other youths may become increasingly irritable and aggressive following trauma or traumatic loss and may engage in impulsive and self-destructive behaviors.

A fourteen-year-old male student was referred for school-based trauma/ grief services following the death of his mother. The presenting problem listed in the referral was recurrent suspensions for aggressive and defiant behavior, including starting fights with peers, yelling at teachers, and storming out of classrooms. The student's behavior during the course of treatment was volatile; he often came to group sessions with bloody and torn knuckles from fighting or from hitting walls. In the pregroup interview, he described deliberately walking into areas controlled by rival gangs as a means of provoking fights. Later in the group, the student disclosed that his mother had been an intravenous drug user who had died after receiving what he believed was a lethal overdose of heroin from her boyfriend.

Traumatized youths—particularly those with histories of multiple trauma exposures—may develop enduring negative expectations with regard to themselves ("I must have deserved this," "I'm a helpless loser," "It's all my fault," "I'm jinxed"), others ("You should never trust people," "People will shoot you for a pair of shoes"), social institutions and authorities ("Nobody can protect me," "The police are out to get me"), and their future ("I'll be dead before I'm twenty, so to hell with it"). Such *traumatic expectations* can profoundly disrupt normal development by corroding self-confidence, undermining close interpersonal relationships, reducing initiative and pro-social engagement, and hindering preparations for the future in school, home, social and work settings. Such cognitions may also lead to marked passivity, pessimism, and/or maladaptive coping in the face of daily challenges and trauma-related adversities (Layne, Pynoos, & Cardenas, 2000).

A high school student was referred to a school-based counseling center for serious academic problems. She presented as severely depressed in an initial clinical interview. When asked about stressful experiences she had been through in the past year, she described an incident in which she had been forced to watch helplessly as her girlfriend was assaulted by a violent gang of girls: "They grabbed her by the hair and punched her and pushed her down on the sidewalk. Then a big girl sat on her chest and grabbed her by the hair and started smacking her head against the cement over and over again. I was just standing there and crying and wanting to help her, but I knew that if I tried to, they would just beat me up, too." Subsequent interventions were focused on therapeutically addressing her pervasive feelings of helplessness and guilt over her perceived disloyalty to her friend at a time when that friend desperately needed protection.

The UCLA Trauma/Grief-Focused Intervention Program was designed to identify and provide specialized psychological services to youths whose histories of violence exposure place them at significant risk for severe, persisting posttraumatic distress reactions and developmental disruptions. The program was developed to address trauma-related distress and disturbances associated both with exposure to a single focal event and with multiple traumatic exposures over time. Early versions of the program were implemented in post-earthquake Armenia (Goenjian, et al., 1997), in poverty- and violence-ridden regions in the Los Angeles area (Layne et al., 2000; Saltzman et al., 2001b), and in postwar Bosnia and Hercegovina (Layne et al., 2001). The protocol has since been implemented in the New York City area post 9/11, Long Beach, California, and other sites (Saltzman, Layne, Steinberg, Arslanagic, & Pynoos, 2003). Program-effectiveness data from Bosnia indicate that participation is associated with significant reductions in posttraumatic stress and depressive symptoms in addition to complicated grief reactions. Furthermore, reduction in posttraumatic stress and depression symptoms is linked to a range of positive outcomes, including improved compliance with authority in the classroom, increased interest in school, better peer relationships, and reduced school anxiety/withdrawal (Layne et al., 2001). Qualitative analyses indicate that participation in the program is linked to a number of broad positive outcomes within the school environment as well as within group members' peer and family relationships (Burlingame, Fuhriman, & Johnson, 2000).

The UCLA program includes a system for screening children and adolescents in school or clinical settings, a manualized twelve- to twenty-four-week (variable content and duration) trauma/grief-focused group psychotherapy protocol, guidelines for adjunctive individual and family therapy, and methods of evaluating treatment outcomes. The Leader's Guide for the treatment protocol is titled: The UCLA Trauma/Grief Program for Adolescents (Layne, Saltzman, Steinberg, & Pynoos, 2003) and is designed for use either in group or individual treatment modalities. This 300-page manual contains detailed guidelines for each session along with background materials and directions for conducting parent sessions. The program is comprised of four modules that can be combined, according to the unique needs of the given group or individual. The overarching therapeutic aims are

1. to provide useful psychoeducation regarding reactions to trauma and loss;
2. reduce the frequency, intensity, and degree of interference associated with posttraumatic distress reactions, including and especially posttraumatic stress symptoms, depressive reactions, and grief reactions;
3. enhance effective coping and positive adaptation in relation to distressing reminders, secondary adversities, and other stressful life events and circumstances;
4. enhance the quality, frequency, duration, and/or timing of supportive transactions with appropriate life figures; and

5. reduce trauma-related developmental derailment and encourage movement toward normal developmental progression.

In this chapter, we describe the content and organization of the UCLA program, and provide recommendations for its successful implementation within a variety of school and community-based settings. We first describe the composition and therapeutic objectives of the program, giving emphasis to five treatment foci developed by Pynoos and his colleagues (1995). Next, we recommend criteria for selecting group members and describe how to prepare youths for successful participation in group treatment. A detailed description of the group sessions is then provided with concluding comments on the combination of individual, family, and group treatment modalities.

THEORETICAL MODEL

The UCLA trauma/grief-focused treatment protocol is based primarily on a developmental psychopathology model developed by Pynoos et al. (1995). This model is consistent with other ecologically and developmentally based formulations of the determinants of posttraumatic adjustment in youths (e.g., Garbarino,1992; Garbarino & Kostelny, 1996). Collectively, these models posit that the course of posttraumatic adjustment in children and adolescents is influenced by numerous psychological and socioenvironmental risk and protective factors embedded within the pretrauma, peritrauma, and posttrauma ecologies. They further propose that these factors must be systematically targeted in intervention efforts. Drawing on this literature, Pynoos et al. (1995) propose that the treatment of trauma-exposed children and adolescents should address five therapeutic foci:

1. Traumatic experiences
2. Trauma and loss reminders
3. Loss, bereavement, and the interplay between trauma and grief
4. Secondary adversities
5. Developmental impact

Traumatic Experiences

A comprehensive intervention for traumatized adolescents should systematically address both objective and subjective features of traumatic experiences. This focus often begins with psychoeducation regarding age-appropriate reactions to trauma and loss. These activities help teenagers identify trauma- and loss-related distress reactions and difficulties, and re-duce maladaptive perceptions that their reactions are bizarre or indicative of personal shortcomings. Treatment should also utilize principles of pro-longed therapeutic exposure involving the repeated retelling of the trauma experience in a safe and supportive environment to promote habituation and tolerance for traumatic memories while decreasing avoidance of trauma- and loss-related cues. These retellings create repeated opportunities to re-visit the traumatic experience via narrative exposure exercises, in which group members are guided to weave together objective and subjective fea-tures of the traumatic experience to render a coherent, temporally ordered narrative. Guided exploration of the worst moments of the experience helps to clarify and restructure cognitive distortions and maladaptive beliefs, in-cluding misunderstandings and misattributions linked with excessive guilt and shame, and to identify distorted or unhelpful trauma-related expecta-tions regarding self, others, social agencies and institutions, and the future.

Trauma and Loss Reminders

Trauma and loss reminders are thought to serve as two of the pathways through which traumatic events and losses exert persisting effects on long-term psychosocial adjustment (Pynoos, Nader, Frederick, & Gonda, 1987). *Trauma reminders* are defined as stimuli, either internal or external to survi-vors of trauma exposure, that resemble or symbolize aspects of traumatic events and thereby elicit distressing reactivity (including emotional, cogni-tive, physiological, and/or behavioral reactions) associated with past trau-matic experiences.

In trauma/grief-focused group treatment, an adolescent boy disclosed that he was confronted with a distressing trauma reminder each morning when he dressed in front of the bathroom mirror. The reminder consisted of a prominent scar running up his neck, the result of emergency surgery he received following a gang-related stab wound. The student described the distressing intrusive image he experienced each morning as he saw his scar in the bathroom mirror: "First I see this hand coming over my shoulder with a knife, and then I try to grab the knife and stop it, but I can't. Then I see it going into me—and I start feeling really sick to my stomach."

In contrast, *loss reminders* are defined as cues, either internal or external to the individual, that evoke grief-related reactions in individuals with a his-tory of significant loss. They are presumed to do so by directing attention either to the continued absence of the lost object or to life changes conse-quent to the loss (see Schein et al., 2005).

A young woman disclosed within a treatment group, "I am going through a rough time this month. My mother's birthday and the anniversary of her death in a car accident eight years ago are both coming up. They are within ten days of each other. I miss her a lot and feel sad this time of year."

Targeting trauma and loss reminders directly in treatment is thought to assist in interrupting one of the mechanisms by which traumatic events and losses exert their adverse influences over time. Treatment objectives relating to trauma and loss reminders include (1) facilitating the identification of current and future trauma/loss reminders; and (2) clarifying the links between distressing reminders, the traumatic experiences and/or losses they evoke, and distress reactions, including coping behavior. Efforts to increase adaptive coping include facilitating cognitive discrimination between the present and past, increasing tolerance for expectable reactivity, reducing unnecessary exposure to disturbing reminders, and developing appropriate support-seeking and anxiety management skills for the periods before, during, and after unavoidable exposure to distressing reminders.

Loss, Bereavement, and the Interplay Between Trauma and Grief

The traumatic death of a family member or close friend often generates a complex interplay between the processes of posttraumatic adjustment and bereavement. In particular, if the death involves violence, mutilation or disfiguring injury, or other extremely tragic elements, engagement in "normal" and adaptive grieving processes may be compromised. Adaptive grieving processes vulnerable to traumatic intrusions and avoidance include

- positive reminiscing about and remembering the deceased,
- processing painful emotions associated with accepting and adapting to the loss,
- making meaning of the death,
- participating in grief rituals,
- learning more about the deceased,
- renegotiating the relationship with the deceased, and
- other processes that promote accommodation to the death. (Jacobs, 1999; Rando, 1993; Worden, 1996)

These healing processes may be disrupted by intrusive distressing memories and emotions linked to the violent, tragic circumstances of the death; avoidance of cues linked to the death or to the deceased; and/or numbing of emotional responsiveness (Pynoos, 1992). The result is a form of complicated grief that may interfere with adaptation to the permanent physical absence of the loved one, increase the risk for severe and persisting distress and developmental disturbance, as well as increase the risk for maladaptive coping (Webb, 2002). Many adolescents from the Bosnian program presented with these difficulties (Layne, Saltzman, & Pynoos, 2002).

We learned that my uncle had been killed by the paramilitaries. My grandfather was sitting up all night and crying—nobody could console him until morning, when he went with some others in search of him. They found him, lying dead out in the forest with other men from his village. His belly was riddled with bullet holes. He had torn his shirt to bandage himself. He was all stiff, still holding handfuls of leaves tight in his hands. This is a memory that I will not forget for a long time.

As a first objective, treatment of complicated bereavement must therapeutically process the traumatic circumstances of the death in order to reduce distressing trauma-related intrusions, reduce maladaptive avoidance, and free up psychological resources needed for bereavement. Specific therapeutic tasks in this phase include providing psychoeducation about grief reactions and the course of bereavement and reframing grief reactions as beneficial processes that help us accommodate the physical absence of loved ones. In cases in which group members report an inability to access positive, nontraumatic images of the deceased, a group activity is conducted wherein members construct a nontraumatic mental representation of the deceased to facilitate remembering and reminiscing. Other therapeutic tasks include increasing tolerance for current and future loss reminders, making healthy lifestyle changes, broadening social support to further accommodate to the loss, and addressing conflicts over past interactions that evoke feelings of regret, guilt, shame, or anger. This therapeutic work aims to promote acceptance of traumatic losses, mobilize adaptive coping resources, and facilitate the bereavement process.

Secondary Adversities

Effective treatment for trauma-exposed individuals must also address the adversities and vulnerabilities that traumatic events can generate, as well as the preexisting adversities that may interfere with posttraumatic recovery. Clinical research studies indicate that a series of adverse life changes often follow in the wake of a traumatic event or death of a loved one, typically consisting of such events and circumstances as financial hardships, family estrangement or dissolution, relocation, and adolescents' assumption of adult responsibilities (Layne et al., 2001). Such secondary adversities may exacerbate the impact of trauma and traumatic loss on current levels of adaptive functioning and thereby prolong posttraumatic distress. Accordingly, intervention efforts targeting this therapeutic focus are directed toward identifying current difficulties, developing pragmatic problem-solving strategies to cope with adversities, and enhancing social skills needed to contend with the aftermath of trauma and loss. These social skills include identifying specific types of needed support, locating those appropriate sources, and *asking* for what is wanted.

Developmental Impact

A comprehensive trauma/grief-focused treatment program for children and adolescents must therapeutically address not only the sequelae of trauma exposure, but also the impact these factors have upon ongoing developmental processes and tasks (Saltzman et al., 2003). Of particular therapeutic concern, traumatization may *delay, interrupt,* or in some cases *prematurely accelerate* the initiation of developmentally appropriate tasks. This developmental disturbance may manifest in a variety of forms. These include uneven development of the self-concept; distortions or maladaptive changes in beliefs about the self, others, the world, and the future; disruptions in primary relationships; decreased academic performance; and alterations in future ambitions, planning, and preparation.

Posttraumatic stress reactions pose a particular developmental challenge to adolescents. As they look forward to becoming adults, adolescents are especially prone to interpret many of their distress reactions as regressive or "childlike." Others interpret their reactions as signs of "going crazy," of being weak, defective, or different from their peers. The group leaders' task with respect to these developmental challenges is to move the adolescents toward an understanding that their reactions are expected, adaptive in the face of danger, "adultlike," and therapeutically addressable. Guided by an understanding of normative developmental competencies, tasks, transitions, and expectations, treatment should identify missed developmental opportunities, support the resumption of compromised developmental activities, facilitate an active future orientation, and challenge and change maladaptive belief systems.

RATIONALE FOR GROUP TREATMENT

The rationale for embedding adolescent trauma work within a group setting is based on a recognition of the potentially synergistic nexus among three beneficial factors: (1) the cost efficiency and effectiveness of group treatment, (2) the capacity of trauma-focused group work to promote such beneficial processes as normalization, validation, member-to-member feedback, vicarious exposure to other members' experiences, and (3) the potent influence of the peer group among adolescents (Davies, Burlingame, & Layne, 2005). Skillful group work has the potential to facilitate mutual understanding between members, bridge interpersonal estrangement, and promote the normalization and validation of emotions that pose a special challenge to adolescents' emerging self-concepts, including shame, guilt, and the desire for revenge. Group work also enriches members' capacities to give and receive support and to speak authentically and genuinely about their experiences. They develop and practice social skills to help them select with whom, when, and how much of their experience to disclose. These skills help them to be self-protective and maintain closeness to selected others so that they may call upon support when confronting distressing reminders and difficulties.

CRITERIA FOR SELECTION OF MEMBERS AND PREPARATION FOR ENTRY INTO THE GROUP

Evaluation and Assessment

Selection of appropriate candidates for group treatment utilizes a three-stage process (Saltzman et al., 2003). The first stage consists of administering a set of screening measures to determine the extent of exposure to a range of traumatic events and associated distress reactions. These measures cover topic domains including community violence and trauma exposure (e.g., shootings, beatings, kidnappings, car accidents, illnesses, medical procedures, disasters, and the death of friends or family members; Saltzman Steinberg, Layne, & Pynoos, 1999), posttraumatic stress reactions (Pynoos, Steinberg, et al., 1999), depressive symptoms (Birleson, 1981), and complicated grief reactions (Layne, Saltzman, & Pynoos, 2003). The intent is to identify students with chronic and debilitating forms of posttraumatic stress, comorbid depression, and functional impairment related to a trauma or loss.

Given the emphasis of the program to prevent or remediate persistent distress and developmental disturbance, risk screening for large-scale trauma typically commences at least two to three months posttrauma and may be administered to individuals, circumscribed groups, or entire student populations. Children or adolescents who endorse items indicating significant trauma or loss exposure, and/or significant levels of current distress as indicated by elevated scores on one or more of the distress measures are then invited to participate in an individual screening interview. This interview is designed to verify the survey results, to explore functional impairments at school, home, and with peers, and to assist in determining appropriateness for the group psychotherapy program. This determination is based, in large part, on the inclusion and exclusion criteria described as follows. A semi-structured guide for the screening interview is used to ensure consistency in group assignments and referrals (Layne, Wood, Saltzman, & Pynoos, 1999; Layne, Wood, Steinberg, & Pynoos, 1999; Layne, Wood, Saltzman, Steinberg, & Pynoos, 1999).

A pregroup clinical interview is the critical third component to the screening process and is an important vehicle for preparing members for entry into the group. Whenever possible, this interview is conducted by one of the group leaders to provide an opportunity for the clinician and prospective group member to form a working alliance, clarify expectations of what the group will be like, build positive expectations, address client concerns, and develop personal treatment goals. This interview also provides the group leader with in-depth information concerning the objective and subjective features of the individual's trauma or loss experience, predominant negative emotions (e.g., guilt, shame, rage, revenge), an inventory of trauma or loss reminders, and a description of developmental disturbances and current psychosocial adversities. Because traumatized youths have frequently experienced more than one traumatic event (Layne et al., 2000), the clinical interview is also used to develop a hierarchy of traumatic experiences and losses according to their degree of associated psychosocial impairment, and their appropriateness for group-based work. Finally, the interview provides a forum for the transmission of psychoeducation about traumatic stress and complicated bereavement; it also initiates construction of a shared vocabulary about the trauma and/or loss that forms the basis for a common conceptual framework regarding the meaning and impact of traumatic experiences in group members' lives (Layne et al., 2001).

Inclusion Criteria

A portion of the inclusion criteria is unique to trauma treatment. Other criteria are drawn from the small-group treatment literature (Burlingame et al., 2000).

Exposure and Current Distress

The three essential criteria for admission into group treatment are (1) current or past exposure to a stressor event of sufficient magnitude to evoke a posttraumatic distress response (specifically, an event that meets DSM-IV

PTSD Criterion A1: "The person experienced, witnessed, or was confronted with an event or events that involved actual or threatened death or serious injury, or a threat to the physical integrity of self or others";* and (2) evidence of clinically significant current distress. This distress may assume many forms, including and especially posttraumatic stress symptoms, depressive symptoms, and intense grief reactions. A third criterion is evidence of significant functional impairment and/or disruption in developmentally sensitive domains or the completion of developmental tasks. These developmental disruptions include, among others, and depending on age and context, uneven development in aspects of the self; changes/distortions in beliefs about the self, others, the world, and the future; disruptions in primary relationships; disruptions in romantic relationships and preparation for family life; decreased academic performance; and alterations in future ambitions, planning, and preparation (Pynoos et al., 1995).

Information regarding trauma exposure and associated distress is typically gathered using screening surveys and follow-up screening interviews (Saltzman et al., 2003). Depending on the population and context, measures of trauma exposure range from screening tests of community violence exposure (Saltzman, Steinberg, Layne, & Pynoos 1999), war trauma exposure (Layne, Stuvland, Saltzman, Steinberg, Pynoos, 1999), and sexual abuse (Briere, 1995). Data regarding current distress are often obtained using the UCLA PTSD Reaction Index (Pynoos, Rodriguez, Steinberg, Stuber, & Fredericks, 1999), a standardized measure of depression such as the Depression Self-Rating Scale (Birelson, 1981), and where appropriate, a measure of grief reactions (Layne, Saltzman, Steinberg, & Pynoos, 2003). In a follow-up screening interview, items from these distress measures are systematically probed to confirm the presence of current distress and to identify areas of functional impairment and/or developmental disturbance. In U.S. samples, confirmed scores of 35 or higher on the UCLA Reaction Index and 30 or higher on the Depression Scale are considered evidence of clinically significant distress, although lower cutoff scores (down to 25 on the Reaction Index) have been used to identify youths with moderate distress and impairment.

Postacute Difficulties with Adaptation

In the immediate aftermath of a traumatic event or death of a loved one, children and families experience acute destabilization that requires extensive adaptation across a variety of psychological and functional domains,

---

*Reprinted with permission from the *Diagnostic and Statistical Manual of Mental Disorders,* Fourth Edition (Copyright 1996). American Psychiatric Association.

including emotional adjustment and behavioral accommodation (Pynoos & Nader, 1988; Worden, 1996). Based on this observation, we recommend *against* engaging in intensive uncovering psychotherapy during this period, especially within a group setting. Instead, acute intervention services should focus on providing support, promoting daily care routines, and buttressing parental resilience through such interventions as providing psychoeducation, training in coping skills, and mobilizing resources within the family, peer group, community, and in mentored relationships (Saltzman et al., 2003). Given these recommendations, if a group candidate reports a recent trauma or loss, an assessment of the level of individual and family restabilization following the trauma should be made. A general guideline to follow is that at least three months should pass after a trauma or loss before a youth is considered for inclusion in trauma/grief-focused group psychotherapy. During the interim, individual and family forms of intervention and support are the treatments of choice.

Current Psychological Discomfort and Motivation to Change

The presence of current psychological distress and functional impairment increases the likelihood of positive treatment response. The presence of these factors, coupled with the hope of obtaining relief and improving their daily lives, provides the most reliable source of motivation for undertaking the often arduous work of acquiring new coping skills, reprocessing traumatic events, and problem solving current life adversities.

Ability to Tolerate Anxiety and Stress

Working on trauma and loss experiences in the company of fellow youths who share similar histories can provoke strong reactions in group members who often cope by avoiding situations or people that elicit painful memories and emotions. Trauma narrative construction exercises are essentially prolonged exposure trials designed to elicit tolerable increments of distressing traumatic memories over repeated trials in order to reduce reactivity. Because listening to fellow group members' personal narratives can produce moderate to high levels of anxiety in group members, it is important to select individuals who can tolerate these emotionally laden group exercises so that reexposure does not lead to retraumatization (Layne et al., 2005).

Capacity to Give and Receive Assistance

Facilitating trauma/grief-focused groups with adolescents in particular is made more challenging by age-specific imperatives to avoid showing vulnerability, dependency, or apparently regressive behaviors. To counteract these often powerful forces, the group must become a cohesive and supportive holding environment—an outcome made possible only if individual group members have the basic capacity for relationship and the ability to give and receive assistance. Although this capacity varies greatly in children and adolescents; youths admitted to the group should possess a basic level of this capacity and should be able to benefit from modeling and shaping of interpersonal awareness and supportive interpersonal exchanges.

Parental Support

State and federal laws mandate that written parental consent must be obtained prior to a minor's participation in any of the program activities. Parents or guardians are also asked to attend at least two family sessions, held at the beginning and midway through the group therapy program. These parent-focused sessions are designed to help parents learn about trauma and loss and the ways these experiences may influence their child's behavior and adjustment. Therapists work with parents to help them learn to become more sensitive to their children's needs, and provide effective support during and after the program.

Exclusion Criteria

To ensure that prospective group members are not ill suited to group-based work because of their fragility, risk for disruptive behavior, or a lack of empathic and supportive capabilities, a number of exclusionary criteria are included in the selection protocol. These include the following.

Recency of Trauma Exposure or Loss

As described in the inclusion criteria for postacute adaptation, candidates whose trauma or loss experiences are too recent, or who are still struggling with acute stabilization issues, are generally excluded from group participation. Instead, they are offered support-oriented services, usually in the form of individual sessions. Once stabilized in individual treatment, they are reevaluated for group participation during the next program cycle.

Sexual Abuse Victims and Trauma/Abuse Perpetrators

The standard of care for childhood and adolescent sexual abuse requires a specialized course of therapy with clinicians trained in this particular field. The need for confidentiality and sensitive handling of this particular form of trauma, especially for children and adolescents, also indicates the need for specialized individual and group treatment. The UCLA Trauma/Grief Group Psychotherapy Program is not designed for sexually abused individuals, though physically abused individuals may be admitted to groups whose roster includes other similarly abused individuals. Similarly, trauma or abuse perpetrators require types of specialized services not provided within this protocol.

Severe Comorbid Psychological Conditions

Prospective clients with severe debilitating psychological conditions including psychoses, severe affective disorders, and significant homicidal or suicidal tendencies are generally excluded from the group therapy program. These conditions tend to interfere with group work, place the individual at heightened risk, and indicate the need for more intensive services.

Substance Abuse or Substance Dependence

Prospective clients who indicate in the screen or interview that they abuse or are dependent on substances are generally excluded from the group therapy program and referred to local specialized services.

Refusal or Extreme Discomfort

In the latter part of the interview, the clinician summarizes the prospective client's current symptoms and expressed difficulties, describes the group program, and, if appropriate, invites the child or adolescent to participate. If the prospective client refuses or expresses extreme discomfort with participating in a group, then he or she is offered alternative services. Likewise, if a parent who has been informed about the program and his or her child's needs refuses to give consent for the child to participate in the group program, alternative services are offered.

Inability to Participate in Group Tasks

During the screening interview, the clinician decides whether the prospective client possesses sufficient impulse and behavioral control to participate in group tasks. If at a school, the clinician may draw on information provided by teachers, counselors, and school archival data relating to disruptive behavioral problems. Conversely, if working in a clinic, the therapist usually has access to intake materials and statements from parents relating to behavioral and interpersonal abilities and problems.

GROUP COMPOSITION

Groups are typically comprised of six to eight members and two leaders. Complementary roles for "Leader 1" and "Leader 2" are described in the treatment manual with specific activities and duties prescribed for each. Leader 1 is the primary group leader, whose role is to facilitate the group's basic therapeutic processes and personal exploratory exercises. In contrast, the role of Leader 2 is more supportive, consisting essentially of leading didactic and psychoeducational activities, and monitoring group members. Moreover, while Leader 1 is directing the trauma narrative reconstruction exercises or facilitating group discussions, the main task of Leader 2 is to attend to the level of participation and emotional reactivity of group members and to intervene as necessary. The role of Leader 2 may involve directing the discussion to include a withdrawn member, speaking individually to a group member who appears extremely upset, or following a group member out of the room who has left during an evocative narration. These approaches are consistent with recent recommendations from the small-group process literature (Burlingame et al., 2000), which describes the interaction of group members as the most *direct mechanism* of change, the group as the *primary vehicle* of change, and the group leaders as the *indirect agents* of change. Skilled group leaders in both roles thus work through and with the group by facilitating group structural characteristics and interactive processes that are maximally therapeutic.

Homogeneity in Trauma versus Loss

The overall objective in comprising treatment groups is to maximize group homogeneity. Greater similarity among group members in terms of developmental level and type/severity of traumatic exposure leads to more rapid group cohesion and supportive exchanges, and allows more in-depth and open exploration of trauma and loss experiences (Saltzman, Pynoos,

Layne, Steinberg, & Aisenberg, 2001a). In early implementations of the UCLA Trauma/Grief Group Psychotherapy Program, members who had experienced a traumatic death and members who had experienced trauma alone, without the death of a friend or family member, were combined in the same groups. The needs of these two populations of youths were sufficiently divergent to necessitate offering separate groups for each. Currently, the "primary trauma groups" are comprised of members who have experienced trauma without loss, and the "primary grief groups" are comprised of members whose current difficulties stem from loss-related issues. Separating groups into these two tracks can sometimes be a complex clinical decision in instances where a group member has experienced both primary trauma and death, or a case in which a death occurred in traumatic or violent circumstances. In these cases, a placement decision must be made that weighs the degree to which each event is contributing to psychosocial distress, functional impairment, and associated developmental disturbance.

Homogeneity in Severity of Trauma Exposure

Another issue relating to group homogeneity involves composition of groups based on the severity of trauma or losses. Early in the development of this model, groups combined youths across levels of severity (e.g., including those who had been the direct victims of shootings and beatings, those who had witnessed such acts, and others who had lost a parent or close friend). In these groups, members with less severe exposure tended to reduce their involvement while articulating thoughts such as "I shouldn't take up group time, my problem is almost nothing compared to his." Now, groups are assembled with exposures in a similar range to avoid this kind of self-marginalization.

Homogeneity in Cognitive and Developmental Levels

Trauma/grief groups are both psychoeducational and process oriented. As such, it is helpful to select group members who are somewhat cognitively and developmentally matched. Broad divergence along these dimensions can lead to some group members becoming bored or disenchanted as less advanced members require more behavioral management, have difficulty maintaining focus during extended trauma narrative activities, require significantly longer periods of time to understand or apply learned skills, or have more difficulty expressing affect and providing support to other group members.

Considerations Regarding Group Heterogeneity

If a program is designed to provide treatment to adolescents in the after-math of a focal traumatic event such as a school shooting, then group mem-bers will naturally share at least one common traumatic experience. In the absence of a focal event, however, a single group may be comprised of indi-viduals reporting a broad variety of trauma or loss experiences. For exam-ple, one recent school trauma group was comprised of members reporting the following exposures: witnessing the shooting death of a peer, witness-ing an uncle being badly beaten, being forced to take cover during a drive-by shooting, and being present during an armed robbery in a convenience store. Though the experiences varied greatly, the general degree of severity of the events was judged to be within a therapeutically workable range.

We have also found that putting boys and girls from different cultural backgrounds into the same group tends to promote better group process as does coeducational grouping in itself. In fact, evidence suggests that same-gender groups post *lower* levels of cohesion (Burlingame et al., 2000). In school and clinical settings, generally more boys than girls are referred for nonsexual forms of violence and trauma exposure. As a result, occasionally, groups consist of all boys. For a variety of reasons, these groups have tended to be less cohesive and supportive of in-depth trauma work. Thus, although having girls present often initially evokes boys' competitive and aggressive posturing, group leaders report that the coeducational groups generally afford greater emotional openness and support among group members. Therapists also report that groups with culturally diverse memberships help to broaden members' perspectives of their own experiences and difficulties, and promote a more accepting ethic, especially in regard to supporting the expression of emotions and reframing help-seeking behavior as a sign of strength.

STRUCTURAL CONSIDERATIONS FOR GROUP TREATMENT

The modular design of the UCLA Trauma/Grief Group Psychotherapy Program permits a flexible approach to the number of sessions in a course of treatment as well as the relative amount of time given to a particular skill or area of focus. This accommodates the varied needs of groups. These in-clude groups distinguished by single acute traumas versus those with multi-ple and chronic trauma, younger adolescent groups who need much more initial skill development versus groups comprised of more mature adoles-cents, and groups that focus primarily on trauma versus those focusing pri-marily on grief or some combination of trauma and grief. These varied "trajectories" through the treatment protocol can result in groups that meet for only twelve to fifteen weeks, as was the case for a recent school-based group comprised of youths dealing with intermediate levels of exposure to a single traumatic event. This flexible approach can also accommodate groups convened for youths with complex histories of trauma and loss that meet for thirty sessions. Depending on the clinical judgment of the group leaders, specific modules can be highlighted or deleted, and specific content areas designated for a single session may be expanded to multiple sessions. A notable exception to this latitude occurs in the context of research protocols in which the length and content of participating groups are standardized.

The length and content of a group may also vary because of the setting in which the treatment is conducted. School-based groups generally may meet for a single class period (forty to fifty minutes), and students must be sufficiently emotionally reconstituted at the end of each group session to enable them to rejoin the ongoing school day. The time required for transitioning activities at the end of the session may further diminish the available working time. Because most of the sessions are designed to take about sixty minutes to complete all of the prescribed activities, a single session may require multiple meetings to complete in a school setting.

A key characteristic of the trauma and grief groups described in this chapter is that they have *closed membership*. Once constituted, additional members should not be added, and drop outs should be vigorously discouraged through careful group member selection, the identification of valued personal goals (e.g., "what I want to get out of my group experience") coupled with a clear statement of expectations during the pregroup interview, and negotiation of a group contract as an early group activity. The creation of a consistent and delimited group membership also assists in building the level of trust and group cohesion needed to undertake the trauma narrative work in Module II. For this reason, at least three, and up to eight sessions, are prescribed for Module I preparatory to Module II work, both detailed later in this chapter.

In recognition of the importance of ritual and to promote a sense of consistency and security, the treatment protocol provides a common format for all group sessions. Each session consists of a check-in exercise, a review of the previous session's activities, one or more group activities, an assigned practice exercise, a check-out exercise and, if necessary, a transitional activity. This standardized format is designed to foster a sense of predictability and security, to identify pressing current problems that may compromise members' ability to fully participate in the session, and to ensure that members can successfully regulate their emotional and behavioral responses within the allotted time period. Periodic expanded check-in procedures, consisting of reviews of how the group is doing, provide individual members with an advocating voice in monitoring the group and influencing its course. The check-in exercises are usually simple group rounds in which each member is asked to respond to a question such as, "What is something good or fun, and something not so good or fun, that happened to you during the past week?" or, "Share with the group where you are right now on the 'Anxiety Thermometer,'" (ranging from 0 to 100 degrees) or, "Tell us if there is anything going on for you that makes it difficult to be 100 percent present today." The central group activities—the topics and weekly practice exercises—vary from session to session. The check-out exercises usually involve a final group round in which members are asked to share their current level on the "Anxiety Thermometer" and to share something that was meaningful or something that was learned in that session. A lighthearted and fun transitional activity may then be shared for five to ten minutes to help group members "shift gears" and prepare to return to their school day. Appropriate activities include group Pick-Up Sticks, the block-building game "Jenga," inviting each member to share a joke or a funny experience, or collectively telling a story or drawing a picture on a poster or chalkboard.

The treatment protocol is also supported by media materials, including student handouts and poster-board illustrations of the session agenda and of significant concepts and skills. The sessions are semistructured and consist of a choreographed dialogue for Leaders 1 and 2, coupled with group psychoeducational and therapeutic processing activities. These include psychoeducational discussions, skills demonstrations and practice exercises, trauma-processing work, problem-solving activities focusing on current life events, and processing spontaneously introduced material (e.g., stressful exams, anniversaries of a loved one's death).

OVERVIEW OF GROUP THERAPY MODULES

Module I

As shown in Table 18.1, the treatment manual is divided into four modules, each of which focuses on a specific content domain. As noted, the modules can be linked in various combinations according to the specific needs of each group. Module I, comprised of three to six sessions, is designed both to reduce acute distress and to provide a foundation for upcoming trauma- and grief-focused therapeutic work. This foundation consists of psychoeducation and the acquisition of coping skills, including cognitive restructuring (thought/emotional regulation), support seeking, and development of an adaptive coping plan for anticipated stressful events. Module I

TABLE 18.1. Treatment Manual

	Module I (3-6 sessions)	Module II (4-8 sessions)	Module III (4-8 sessions)	Module IV (4 sessions)
Group Phase	Opening	Working Through		Termination
Module Title	Group Cohesion, Psychoeducation, and Basic Coping Skills	Constructing the Trauma Narrative	Coping with Traumatic Loss and Grief	Refocusing on the Present and Looking to the Future
Therapeutic Tasks	1. Welcome and introduction (program overview, barriers, group contract, posttraumatic stress, depression, and grief reactions) 2. Learning about trauma and loss reminders (how I react to, and cope with, reminders) 3. Learning coping skills 4. The event-thought-feeling link 5. Identifying and challenging distressing thoughts ("Three Steps to Taking Charge of Your Emotions") 6. Support seeking ("Five Steps to Getting Support")	**First:** Preparing for trauma narrative work (constructing the group narrative, constructing my personal trauma timeline) **Middle:** Constructing the trauma narrative (prolonged therapeutic exposure; develop a vocabulary for communicating about the trauma) **Final:** Exploring the worst moments (prolonged therapeutic exposure; using trauma reminders to understand the nature and personal meaning of traumatic experiences; cognitive restructuring of cognitions associated with guilt and shame; exploring intervention fantasies)	1. Learning about grief (grief reactions, loss reminders, and grief processes/tasks) 2. Understanding grief reactions: Focus on anger 3. Understanding grief reactions: Focus on guilt 4. Remembering and reminiscing 5. Guided imagery: Retrieving a nontraumatic image of the deceased 6. Adjusting to a world in which the deceased is absent 7. Planning for difficult days (relapse prevention) 8. Saying good-bye in a good way	1. Resuming developmental progression 2. Problem-solving current life ("Three Steps to Solving a Problem") 3. Dealing with problems that are not my job to fix 4. Saying good-bye in a good way

690

also includes activities specifically designed to promote positive group structures and processes, including the establishment of positive group norms (e.g., direct member-to-member interactions, reciprocity, group ownership, and confidentiality), developing group cohesion, and increasing personal commitment to participate by identifying and sharing personal treatment goals.

Module II

Module II is comprised of four to eight sessions and constitutes the "working through" phase of the protocol. This module is dedicated to re-processing traumatic experiences using trauma narrative construction and cognitive restructuring as primary therapeutic tools. The narrative construction exercises are designed to promote habituation to traumatic memories and to trauma reminders in the form of prolonged therapeutic exposure to members' narratives of selected traumatic experiences. Each group member is given the opportunity to recount one or more traumatic experiences on at least two separate occasions: the first round of exposure is devoted to narrative construction, in which a chronological sequence of events is developed. The second round of exposure is devoted to more in-depth reprocessing of the "worst moments" of a traumatic experience and to identifying and re-structuring maladaptive cognitions associated with the event. As deemed appropriate, a range of other therapeutic interventions may be used to augment or facilitate the work, including in vivo exposure to trauma reminders; audiotaping and listening to the narratives as prescribed homework exercises; and narrative storytelling through artwork, writing, drama, or music. The number of sessions allocated to this module is flexible, based on an understanding that group members may vary in their levels of need to retell their narrative, depending on the severity, complexity, and duration of the referent trauma.

Module III

Module III, consisting of four to eight sessions, was specifically developed for groups comprised of members who have experienced the traumatic death of a close friend or family member. The sessions are designed to provide information regarding the interplay between trauma and grief reactions and to facilitate grief processes that promote positive adaptation to traumatic loss (Rando, 1993; Worden, 1996). These processes include identifying personal loss reminders, expressing a range of grief-related emotions, and accommodating to life changes consequent to the loss through adaptive coping.

Module IV

Last, Module IV is comprised of four sessions and constitutes the "closing" phase of the group's work. Its primary therapeutic objectives are to promote movement toward normal developmental progression and to facilitate appropriate leave taking. This is accomplished through a variety of developmentally oriented activities, including applying problem-solving skills to contend with current life adversities, developing skills for enriching relationships, identifying and restructuring maladaptive core beliefs, developing a relapse-prevention plan, and forming specific goals and plans for the future. In addition, special emphasis is given to termination in recognition that parting may in itself constitute a trauma and/or loss reminder. Part of this therapeutic emphasis involves discriminating between the type of sudden "traumatic good-bye" that members likely had experienced with the death of their loved one and the "planned good-bye" afforded in the group.

As noted, the program is designed to allow multiple treatment "pathways" depending on group members' unique trauma/loss histories, levels of current distress, and developmental maturity. Groups comprised of members with primary trauma without a history of multiple or chronic exposure may, for example, conduct only the first three sessions of Module I, then conduct eight sessions of Module II, skip Module III, and conclude with all sessions of Module IV. In contrast, groups comprised of members with multiple trauma exposures not involving the death of loved ones may benefit from all six sessions of Module I, followed by Modules II and IV. In this scenario, the more intensive work in Module I will likely promote a higher degree of mastery of anxiety management skills and provide additional "germination" time for group cohesion and positive group norms to develop. Last, primary bereavement groups may choose to conduct all sessions of Modules I, III, and IV.

MODULE I: GROUP COHESION, PSYCHOEDUCATION, AND BASIC COPING SKILLS

The (three to six) sessions of Module I are directed toward several important therapeutic objectives. The first is to generate group cohesion; the second is to establish positive norms including those of mutual support, respectful understanding, group ownership, direct member-to-member interactions, and ability to relate to the group as a specialized forum for "doing the work I need to do to get on with my life." These group norms are particularly important given that group work requires a degree of self-revelation capable of eliciting intense self-consciousness related to how the adolescent is perceived and/or accepted by others. The emotional work inherent in group activities also places demands on members' maturing self-regulatory capacities. This work is carried out against the "developmental backdrop" of members' ongoing age-related developmental transitions, which can regularly distract the group from trauma-focused work. The art of conducting the group thus entails monitoring each adolescent's current concerns and developing them, as appropriate, into opportunities to practice coping skills, while keeping the group primarily focused on the therapeutic tasks of each session. If external issues prove too distracting for group work, then additional one-on-one time may be scheduled with individual group members outside of the formal group meeting time.

The following sections will describe each of the sessions in Module I, focusing only on the primary activities and not on the regular features of the weekly sessions (which include the check-in, review of homework, checkout, and transitional activity.)

Session 1: Welcome and Introduction

The initial session is directed toward engaging group members in constructing positive group norms to ensure the group is a safe and supportive place that will facilitate the work that needs to be done. An overview of the program is first provided, which is designed to directly challenge the often-held adolescent belief that one cannot get better. This attempt to bolster members' confidence is accompanied by therapist "straight talk" about the ups and downs that can be expected over the course of therapy and beyond. The exercises are also constructed to take advantage of adolescents' facility to share interests, experiences, and their day-to-day lives with one another. The therapist underscores the concept that individual group members are the "experts" in knowing what going through these experiences is like, while accentuating the commonalities among members' experiences.

Treatment Objectives

Content objectives for this session include describing the program to members and its relevance to their circumstance and needs, mutually establishing a group contract with clear expectations regarding leader and member roles, describing the format of each session, providing psychoeducation on common posttraumatic and grief reactions, and assigning homework on self-monitoring changes in feeling states.

The session begins with ice-breaking activities and an opportunity for each group member to briefly share his or her trauma or loss experience. The goal is to build group cohesion by highlighting shared experiences and current difficulties, and to help group members acknowledge one another as true peers. Group members sometimes feel compelled to talk about their trauma or loss experiences in great depth and detail during the initial sessions. However, sharing potentially raw and evocative narratives at this early stage can be harmful to individual group members and to the unfolding group process (Saltzman et al., 2001a). Consequently, group leaders should gently redirect such premature retellings with the assurance that there will be ample opportunities later on to share these important experiences.

Therapists should praise members' current coping efforts during the initial session to challenge commonly held perceptions that members are deficient, weak, or defective. For example, the therapist can say:

> You've all been through very difficult experiences, and yet your presence here tells me that, in many ways, you are still holding it together. That is a very impressive accomplishment, given the severity of your experiences and your losses. We would like to spend some time now hearing about some of the excellent coping skills that you have essentially figured out on your own. Ask yourself, "How do I deal with the anxiety, the sadness, and the fears that come up? How do I deal with the painful memories and the many things that remind me in upsetting ways of what happened?"

As an initial step toward helping group members assemble a personal coping plan, members are encouraged to take note of what they are currently doing that they find helpful. In a later session, members will examine the short- and long-term consequences of various coping responses to more accurately discriminate between helpful and unhelpful coping strategies.

Following an overview of the program and a discussion of group rules, group members participate in a specialized card game that is designed to help them identify their own posttraumatic and grief reactions, and to recognize the ways in which these reactions interfere with their lives. The session wrap-up includes remarks designed to summarize the session and the overarching goals of the program.

> This psychotherapy group is designed to help you become more aware of your posttraumatic and grief reactions. As we share more as

a group about the difficulties we experience, we give the gift of telling our fellow group members that they are not alone. It takes courage to share this kind of personal information, and I am already impressed with how you are willing to support one another in this way.

This group is not a place where you will be "analyzed" or expected to dig up things from the distant past that don't affect your current lives. It is a focused and practical group in which you will learn specific skills to help you manage trauma- and stress-related anxiety, deal with painful memories and feelings connected to your trauma and losses, and gain more control over your lives. We will help one another heal by learning new skills, sharing our stories, and finding alternatives to behaviors that aren't very helpful or may even be destructive.

Practice Assignment

The homework for this first session builds on members' increased awareness of personal posttraumatic and grief reactions. It consists of a form in which members (1) describe situations during the week in which they experience these reactions, (2) describe what is going on "outside" of them such as trauma or loss reminders, and (3) what is going on "inside" them by identifying and rating the intensity of salient emotions.

Excerpts from Session 2: Learning About Trauma and Loss Reminders

Treatment objectives for this session include learning about ways in which trauma and loss reminders can evoke posttraumatic and grief reactions; identifying and sharing personal reminders; and linking distressing reminders, distress reactions, and coping responses.

Following the check-in, group leaders lead a brief presentation on trauma and loss reminders and facilitate a group discussion. This discussion is based on the week's homework exercise, in which members share their most upsetting reminders, describe the situations in which they occur, and express the thoughts and feelings they evoke. The next activity, titled "How I React to and Cope with Distressing Reminders," extends the discussion to the identification of specific coping reactions used by group members to deal with trauma or loss reminders. Group leaders then reframe coping behaviors to distressing or painful circumstances as learned responses that may become automatic or habitual responses used outside of conscious awareness. They underscore the distinction that some are helpful, such as seeking appropriate support, engaging in distracting activities, or using positive self-talk; whereas others are maladaptive, such as drinking, using drugs, or engaging in high-risk behaviors.

Practice Assignment

Worksheets are distributed to help structure the session's discussion and are used for the week's homework. The assignment asks members to identify specific occasions during the week in which they encounter an upsetting reminder, and then to use the queries on the sheet to describe what was going on "outside of me" and "inside of me." Members then describe the specific coping response(s) they used.

Excerpts from Session 3: Learning Coping Skills

Treatment objectives for this session include evaluating current coping strategies with respect to their short- and long-term consequences, developing adaptive personal coping strategies, and preparing members to use these strategies by practicing them in the session. An initial activity involves a discussion of the pros and cons of group members' current coping strategies.

A young man who was invited to participate in trauma-focused group treatment after being robbed and badly beaten by gang members reported that he often compulsively watched television or played computer games for many hours a day. He disclosed that doing so kept his mind off his terrible experience and gave him something to do as he increasingly avoided activities outside his home. With the group's help, the young man began to see that this coping strategy provided distraction from his boredom and loneliness in the short-term, but exacted a high long-term price as his life became more isolated, monotonous, and unfruitful.

To transition to the next activity, termed "Developing a Personal Coping Strategy," leaders express therapeutic surprise at how well group members have done given what they have been through, calling attention to the many effective coping responses already in use among the membership. The leaders then suggest that each member work on developing a personal coping tool kit, with specific strategies for coping before, during, and after a stressful event. Group leaders point out that the selection of coping strategies is a very personal matter, given that some strategies will suit some individuals and some circumstances, but not others. Selected coping techniques are then practiced in the group, including such relaxation techniques as abdominal breathing and progressive muscle relaxation, and cognitive techniques, such as calming self-talk, differentiating "then" from "now," and distraction.

Practice Assignment

The homework for this week is for group members to use one or more coping strategies and to report back on the success of their efforts using a specialized worksheet.

As noted earlier, groups may choose to use only the first three sessions in Module I and then proceed to Module II. This is most appropriate for groups whose membership has experienced a single, relatively recent traumatic event or loss. For groups whose membership has experienced multiple or chronic forms of trauma or loss, an additional three sessions are provided to allow more time for the group to cohere before the more intensive narrative work, and to equip group members with a larger repertoire of coping and anxiety management skills. These sessions teach a three-step cognitive model for regulating distressing thoughts and feelings, and a skill for eliciting support from appropriate members of their social networks.

Excerpts from Session 4: The Evemt → Thought → Feeling → Behavior Connection

The primary treatment objective in this session is to help members gain insight into ways in which their thoughts influence how they feel. This relationship is especially prevalent in trauma- or loss-related thoughts and feelings, given that certain "hurtful" or "unhelpful" ways of thinking about experiences can set into motion intense negative feelings and self-defeating behaviors. Group activities begin by focusing on developing members' emotional awareness and augmenting their vocabularies for describing emotional states. Some groups require intensive work in this area, particularly youths with chronic experiences of trauma or loss who have learned to constrict their emotional reactions and awareness to avoid distressing thoughts or sensations. Additional illustrations and activities focus on understanding how "automatic" thoughts operate, and attempt to explain how thoughts, feelings, and behaviors are connected. Group activities then focus on differentiating "helpful" from "hurtful" thoughts, and identifying members' personal hurtful thoughts.

The basic skills described provide the foundation for the three-step cognitive model that is the focus of the latter half of Module I. The model is presented as a technique for taking charge of one's emotions, and entails asking oneself the following questions: "WHAT am I feeling?"; "WHY am I feeling this way?"; "HOW can I feel better?"

Step 1 ("WHAT am I feeling?") involves learning to identify specific feeling states. Special attention is given to monitoring downward changes in one's mood, as such shifts often signal that a hurtful thought has crossed one's mind. Step 2 ("WHY am I feeling this way?") entails looking "outside" oneself for stressful situations, such as trauma or loss reminders, and inside oneself for hurtful thoughts, to understand the source of the downward shift in mood. Due to its greater complexity, Step 3 ("HOW can I feel better?") is addressed in Session 5.

To help demonstrate the connection between thoughts and feelings, a series of discussions are conducted in the groups with the aid of illustrations. For example, group members are asked how the young man in Figure 18.1 might feel differently depending on what he was thinking. In Figure 18.2, group members are asked to identify possible thoughts that may result in the listed feelings.

In the next series of exercises, the group is presented with the illustration of a pyramid with three building blocks labeled Situation, Thoughts, and Feelings. In one case Figure 18.3 (a) the pyramid is unbalanced, with the "Feelings Block" teetering on the ill-fitting "Situation and Thoughts Blocks." This illustration underscores the premise that when one's thoughts (what's going on inside), are out of step or a distortion of the situation (what's going on outside), such as entertaining exaggerated thoughts of self-blame or unreasonable perceptions of danger, then unbalanced and distressing emotions can result, such as excessive fear, anger, or helplessness.

This technique is then integrated with the three-step model, which invites members to notice when they experience downward shifts in their mood or feelings and to write the name of the dominant feeling in the top block ("WHAT am I feeling?"). In the next step ("WHY am I feeling this way?"), members look outside of themselves and describe key elements of their circumstances—especially distressing reminders—in the Situation Block of the pyramid, and then look inside of themselves for hurtful thoughts. If found, these thoughts are written in the Thoughts Block of the pyramid. The goal in Step 3 ("HOW can I feel better?") is to more realistically align the Thoughts Block with the Situation Block (i.e., to clarify and challenge distorted or otherwise hurtful and unhelpful thoughts) and to create a more balanced and positive feeling (Figure 18.3b). Individual practice using actual scenarios is provided by the group leader to enhance facility with the three-step model.

(a)

(b)

FIGURE 18.1. Illustration Used in Group Therapy—Identifying Feelings

Thought	Emotion
_____	ANGRY
_____	SAD
_____	OK

FIGURE 18.2. Illustration Used in Group Therapy: Identifying Thoughts

Practice Assignment

During this week, group members are given worksheets with pictures of balanced and unbalanced pyramids, and directed to fill in the blocks based on real-life situations in which they experience a strong distressing emotion (at least a 70 or 80 on the Anxiety Thermometer). This entails completing Step 1 ("WHAT am I feeling?") and Step 2 ("WHY am I feeling this way?") of the three-step skill.

Excerpts from Session 5: Identifying and Challenging Distressing Thoughts

The primary objectives of this session are to refine skills identifying trauma-related distressing thoughts, and to learn skills for challenging these

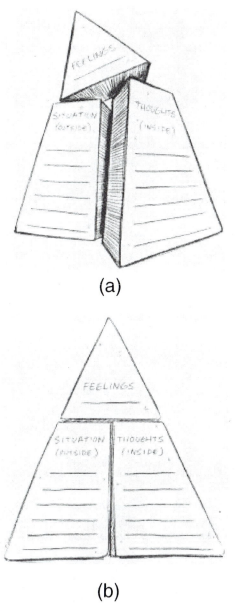

(a)

(b)

FIGURE 18.3. Illustration Used in Group Therapy—Pyramid

"hurtful" thoughts and replace them with more realistic or constructive, "helpful" thoughts. These skills comprise Step 3 ("HOW Can I Feel Better?") of the three-step model. Following the check-in, group leaders focus the discussion on the distressing situations members identified as part of their homework. Special attention is directed toward identifying trauma or loss reminders and participants attendant helpful or hurtful thoughts and emotional reactions. This interactive, often playful activity helps members to identify and express habitual hurtful thoughts, and to connect the links between situations, thoughts, feelings, and behaviors.

The group protocol is designed to provide flexibility in the degree of sophistication of skills learned, depending on the group's general developmental level. Materials include both a basic and a more advanced version of the cognitive restructuring task that comprises Step 3. The cognitive approach to "taking charge of your emotions" presented in Sessions 4 and 5 is the most challenging skill-based aspect of the program. To make this material more accessible, especially to younger group members (ages eleven to twelve) and older adolescents with developmental delays, a series of playful activities, illustrations, and graphic aids are included in the student workbooks.

A follow-up activity familiarizes members with common trauma-related hurtful thoughts and assists members in identifying their own distressing thoughts and beliefs. Group members are given copies of a "Hurtful Thoughts Checklist" that has clusters of common self-defeating thoughts, including: "No one understands me" (feeling unloved and unwanted); "It will always be like this" (hopelessness); "I can't protect myself" (helplessness); "I can never allow myself to feel safe or to relax" (preoccupation with danger); "You can't trust anybody" (distrust); "I'm a loser" (self-criticism); and "I don't deserve to be happy" (self-blame). Group members are invited to check off or write down their most common hurtful thoughts. Once completed, group members are then invited to share their "favorite" hurtful thoughts. The discussion that ensues may be humorous, while increasing group cohesion and self-awareness. By the end of this activity, each group member should have a short list of personal hurtful thoughts and beliefs in anticipation of the next skill-building exercise.

The next series of activities introduces a basic skill, displayed in a worksheet, for challenging and replacing distressing thoughts. This approach involves *listing* the hurtful thoughts elicited by stressful situations, challenging the hurtful thoughts, and developing more realistic and/or helpful thoughts with which to replace them. Skills used to challenge hurtful thoughts include evaluating the evidence for and against the thought, looking for an alternative explanation or perspective, examining the advantages versus disadvantages of holding the thought, and identifying cognitive distortions within the thought. Following is a demonstration of the use of this worksheet with several examples drawn from the group.

A sixteen-year-old group member was walking home from school with his older sister when several gang members cornered them, demanding their money and jewelry. When he hesitated, a gang member grabbed him around the throat and smashed his face into the wall. Bleeding and crying, the group member cowered on the ground and surrendered his wallet and watch while the gang members taunted and threatened him. His sister finally offered to give the gang leader her phone number if they stopped. Afterward, the group member exhibited pronounced bouts of angry and aggressive behavior at home and at school. He withdrew from many of his friends and treated his sister in a hostile manner.

In the group, the member disclosed that he felt "nervous and angry when I'm around gangbangers." His thoughts included, "I can't protect myself" and, "Everybody knows about what happened and thinks I'm a coward." With the group's help, these thoughts were challenged by discussing the evidence for and against the hurtful thoughts. Fellow members helped him to devise more realistic and helpful thoughts, including, "I did the smartest thing I could do under the circumstances," and "There are things I can do to keep myself safe."

Practice Assignment

Copies of the "Changing Hurtful Thoughts" worksheet are given to group members with the understanding that they will fill out at least one during the week. This involves focusing on a distressing situation they encounter during the week, working through the steps of identifying possible hurtful thoughts or beliefs, challenging these thoughts, and replacing them with more realistic and helpful thoughts. This skill requires ongoing practice and is carried forward, refined, and reviewed for the remainder of the group.

Excerpts from Session 6: Recruiting Effective Support

The primary objective of Session 6 is to train group members to elicit specific kinds of support as needed from family, friends, and adult mentors. Key skills focus on the use of a five-step model for getting support by identifying what type of support is needed, deciding which relationship to turn to, asking at the right time, using an *I-message*, and sincerely thanking the support giver. We often start this session with an interactive discussion regarding the various kinds of support (e.g., emotional support, companionship, advice, obtaining a favor), while inviting members to map out their current support network. Group members then contrast their *current* social support map with their desired support map, noting areas where desired support is deficient.

Practice Assignment

Group members are then given a handout defining the support-seeking technique and are asked to use the model at least once during the week. Each group member then identifies a likely situation in which he or she will need support, and a target person outside the group on whom to try the communication skill. The remainder of the session is dedicated to practicing the skill using the anticipated problem situation.

MODULE II: CONSTRUCTING THE TRAUMA NARRATIVE

Therapeutic Rationale

Module II builds on the concepts of trauma and loss reminders by integrating them into trauma narrative work. Trauma reminders are often linked to the "worst moments" and other central features of the traumatic experience, thus serving as links to appraisals of danger in teenagers' current lives, the fluctuating course of posttraumatic distress, and avoidant or protective behaviors. Moreover, without a proper understanding of the manner in which traumatic experiences influence their own ongoing adjustment, adolescents may continue to react to trauma reminders with: excessive appraisals of immediate danger; insufficient cognitive discrimination between past versus present and between dangerous versus innocuous circumstances; difficulty modulating distress reactions; and, inappropriate protective or intervention behaviors. Of particular concern, traumatized adolescents may over- or underestimate the magnitude of danger and threat in the course of their daily lives. These appraisals are often accompanied by intense emotional reactions such as fear, shame, guilt, anger, or rage.

In accordance with their traumatic expectations, traumatized youths may respond to trauma and loss reminders, and to potentially dangerous situations, in ways that place them at risk for severe and persisting distress reactions and developmental disturbance. In particular, traumatized adolescents may avoid distressing reminders associated with trauma or losses, although such avoidance may impose a high developmental cost. Conversely, these youths may overreact to trauma-related cues and circumstances with excessive aggression, recklessness, or other risk-promoting behaviors (sometimes termed *traumatic reenactment*). Both trauma-specific avoidance and exaggerated response behaviors may constrict the adolescents' *life space*. Often unnoticed, these constrictions may result in the loss of important developmental opportunities, such as forming healthy peer relationships and preparing for future professional life via participation in extracurricular activities. In essence, trauma-related preoccupation with, avoidance of, or preparation for protective action detracts from the capacity of adolescents to adaptively cope with both ongoing developmental *and* trauma-generated challenges and adverse life changes.

Module II will thus use posttraumatic distress reactions and traumatic experiences as tools for understanding one another. In particular, the content of selected posttraumatic stress symptoms serve as markers of salient moments of the adolescents' traumatic experiences that are in need of in-depth therapeutic processing:

> In a clinical interview, a fatherless Bosnian youth reported that the mere sight of an old family friend and war buddy of his father continued to evoke highly distressing images of the night the friend came, drenched in blood, to his family's doorstep to tell them of their father's death. The youth then sorrowfully recounted that this friend had, an hour before, witnessed his father being blown apart by a mortar shell several meters away while both were walking home from their army post, and that he immediately knew that his father was dead upon seeing his buddy, bloodied and alone, standing at the door.

Conversely, exploring details of the traumatic experience helps shed additional light on the meaning of posttraumatic symptoms:

> In a pregroup interview, a Bosnian girl reported, "I'm not doing too well. I always feel bad on rainy days like today." Later in the group, while narrating her account of being expelled from her village to become a war refugee, she commented, "I hadn't remembered until now that it was raining when we fled from our home. Now I understand why I always get upset on rainy days." She was subsequently able to use this insight to improve her mood, stating, "A rainy day for me now is just a rainy day. I don't have to make more out of it than that."

The trauma narrative work is designed to progressively build adolescents' abilities to deal constructively with trauma reminders, avoidance behavior, and reenactment behavior in current and future life situations. Nevertheless, a significant barrier to engaging in this work is found in traumatic avoidance of the narrative construction process itself. Indeed, the first time adolescents discuss the personal details of a traumatic experience, they will likely reexperience a range of distressing emotional reactions as they confront the grim reality of the threat, severe injury, destruction, and death to which they were exposed and the accompanying intense distress through which they passed.

A question commonly posed by youths who are reluctant to retell their stories is, "Why should we talk about something that happened a long time ago and that makes us feel bad?" The response is twofold: First, prolonged repeated therapeutic exposure is a "best practice" tool for reducing trauma-related distress and associated avoidance (Cohen, Berliner, & March, 2000). Second, the roots of avoidance, persistent distress reactions, and associated developmental impact are embedded in the personal details of the traumatic experience. Indeed, adolescents are often not aware of the full extent to which their ongoing distress, maladaptive behavior, and developmental problems are linked to aspects of their traumatic experience.

Thus, although this module will prove emotionally challenging for the adolescents and therapists, it constitutes a critical tool for reducing distress and promoting adaptive coping in the aftermath of trauma and loss. As a consequence, group leaders must be prepared to hear everything, however horrifying or painful. In parallel fashion, group members must, over time, become prepared to speak in direct and genuine terms about the traumatic details and their personal thoughts, feelings, and attributions of meaning— therapeutic work that will draw heavily on the adolescents' sense of courage. Because most adolescents did not feel courageous during the traumatic experience itself, accomplishing this demanding therapeutic work may indeed help to restore this critical dimension to the adolescent self-concept.

Why does this module refer to *constructing* a trauma narrative? A traumatic experience, however short in duration, is complex, with many different moments of changing external events and circumstances and accompanying internal events. These internal events include potentially moment-to-moment changes in appraisals of danger; emotional valence and intensity; physical sensation and reactivity; and, efforts at protective intervention. We have observed that adolescents' traumatic memories often feature these themes in some form, but often not in an integrated, coherent manner. In our experience, many adolescents' initial narratives tend to be brief, somewhat journalistic, fact-oriented accounts that lack both emotional depth and give little mention to therapeutically salient themes and moments of the experience. Others involve emotional exclamations (e.g., "It was so awful, it was so awful!"), without accompanying descriptions of the traumatic details that give such emotions relevance and meaning.

A therapeutic construction of the trauma narrative should aim toward genuine communication that intertwines the external (objective) and internal (subjective) aspects of the event as experienced by the adolescent. The unfolding narrative should include: perceptions of the presence of danger within the external environment; thoughts concerning the meaning of the event and associated efforts to prevent, protect against, or repair the harmful effects; emotional reactions, including terror and horror; helplessness and numbing; and, behavioral actions, whether voluntary or involuntary. Various interventions are used to enhance memory recall, promote narrative coherence, and facilitate in-depth processing of significant traumatic moments. Therapeutic skill is most needed as the narrative construction moves toward an exploration of the "worst moment(s)," cognitive distortions, and their associated developmental impact.

The two primary objectives of the narrative construction exercise are (1) to reduce reactivity to distressing trauma-related reminders/cues and memories, and (2) to increase tolerance for distressing reminders, memories, and emotional reactions. All four modules of the treatment protocol are designed to create the foundation for, or to actively carry out, trauma narrative work, providing both the requisite skills and a graduated set of exposures to trauma-related material within the sessions and via homework practice.

Given that this module involves the most intensive trauma narrative exposure, group members often report that the exercises are challenging; at first, some youths may engage in traumatic avoidance or reenactment behavior. Therapists should help members understand that they will revisit this material within a safe, supportive environment. As with the previous work on trauma reminders, this module has the seemingly paradoxical goal of increasing members' awareness of their traumatic experiences in order to reduce the influence these experiences exert on daily mental activity and functioning.

Trauma-narrative construction is performed by carrying out four therapeutic tasks:

1. Establishing a temporal sequence of the traumatic experience. This chronology should begin by establishing the background context in the adolescent's life, shift to a focus on events just before, during, and immediately after the experience, and conclude with a focus on posttraumatic and/or current trauma-related adversities.
2. Elaborating on the moment-to-moment aspects of the experiences by weaving together objective features relating to the external environment (i.e., "what happened outside of me" comprised of the who, what, when, where, and how of the experience), with the subjective features of the experience (i.e., "what happened inside of me"). These subjective features consist of *thoughts* (e.g., ongoing appraisals of danger; attributions of meaning or consequence; evaluating what to

do to prevent, protect, or repair the effects of harm; shifts in attention and concern from oneself to others), *feelings* (e.g., changes in bodily sensations and physiological reactions, efforts to control one's emotions and bodily reactions), and *behaviors* (e.g., voluntary and involuntary actions, including running, fighting, freezing, crying, and shaking).

3. Identifying "worst moments" within the narrative for more in-depth therapeutic processing. These moments are typically characterized by extreme physical helplessness, recognition of dire irreversible consequence, extreme distress, and agonizing dilemmas involving forced and irreversible choices between dire alternatives.

4. Identifying and challenging maladaptive cognitions, especially those associated with severe persisting distress, including guilt, shame, excessive fear, and helplessness.

The group work in Module II proceeds through three phases. In the initial session(s), group members select the traumatic experience that will serve as the focus of their narrative. The middle sessions are focused on building a coherent narrative and on increasing tolerance for unpleasant or painful images, thoughts, feelings, and physiological reactions. This is accomplished by "weaving" the narrative fabric, using the threads of "What happened outside of me" and "What happened inside of me" to form a coherent and chronologically sequenced narrative. The concluding sessions are devoted to in-depth reprocessing of the worst moments of members' experiences, including a focus on challenging maladaptive cognitions associated with persisting distress and maladaptation.

Conducting the Trauma Narrative

Overview

The trauma narrative exercise is intended to allow survivors to expose themselves to and develop tolerance for their painful memories within a safe and supportive setting. The trauma narrative is, itself, a textual fabric that is weaved out of fragments of what happened *outside* and *inside* of the narrator to create a coherent story. This exercise allows group members to gain a greater sense of meaning and self-mastery through exploring and conceptualizing their experiences in a different way. The narrative procedure also helps the narrator and the larger group to challenge "hurtful" thoughts with more realistic "helpful" thoughts.

In conducting the narrative, it is important to remember that the purpose of the exercise is not to superficially review what happened. Rather, it is to carry out a *prolonged therapeutic exposure* exercise through which the narrator is exposed, in a systematic and gradual fashion, to his or her memories and feelings. If the exposure is of sufficient duration, and if it takes place within the "working range of anxiety," he or she will develop an increased capacity to tolerate these memories and feelings and, in turn, his or her perceived frequency and/or intensity may decrease.

The sessions in this module therefore require special attention to time to ensure that the event is adequately processed with sufficient time remaining to provide closure to the session. Because trauma narrative work elicits traumatic anxiety and other strong emotions, stopping a narrative in the middle can significantly increase the risk of future avoidance. The narrator must have adequate time to move *through* the worst portions of the experience to an exploration of the subsequent consequences (e.g., describing the funeral of a deceased family member; talking about the adversities that the family is facing as a result of a parent's death). Some very long and/or complex narratives may take more than one entire session to construct; in this case, we "unpack" and process a portion of the experience, then move to a discussion of posttraumatic or present adversities. Time is also needed for member-to-member constructive feedback, assigning practice exercises, and conducting the "check-out" procedure. Generally, a forty- to fifty-minute school hour will accommodate one full narrative, including the check-in, homework review, and check-out exercises. Ideally, these sessions should extend to sixty to eighty minutes to allow either a second narrative or to provide time for additional practice in coping skills or transitional activities.

Respective Roles

Group leaders' roles are well-defined during the narrative construction work. The task of Leader 1 is to facilitate the work of the "narrating" group member; the task of Leader 2 is to focus on the other group members to ensure that all are in the "working range" of involvement, being neither too involved nor too uninvolved. *Both* facilitators should model sympathetic understanding, support, and validating respect for the profound meaning and impact of the experience in the life of the narrator.

During the trauma narrative exercise, group members sit in a circle. The narrating group member and Leader 1 should sit facing each other so that they can develop the narrative as a mutual dialogue without being distracted by fellow group members. At the outset of the narrative, Leader 1 reminds the group to hold their comments and supportive reactions until the narrative is concluded. Leader 1 and the selected group member then work as a team to create the narrative, adjusting the pace of the narrative as needed to ensure that the traumatic material is processed in tolerable doses. It is best to work deliberately and systematically, slowing down at the worst moments.

The Narrative Construction Procedure

The procedure by which the trauma narrative is constructed uses a four-point "traumatic square" formed by the angles: *situations, thoughts, feelings,* and *behaviors* (Figure 18.4). This square underscores the basic premise that feelings are the products of thoughts, while emphasizing the importance and connections between the objective (external) and internal (subjective) dimensions of the traumatic experience as these unfold and interact throughout the traumatic event. The traumatic square is also well adapted for therapeutically addressing the "chain reaction" by which distressing reminders and secondary adversities tend to evoke distressing thoughts, emotional reactions, and maladaptive behavior.

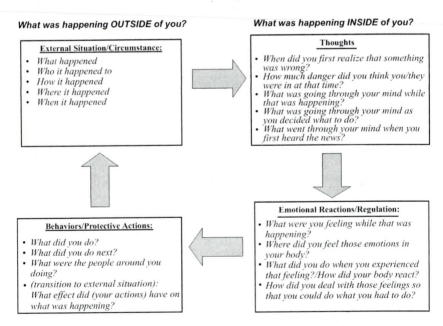

FIGURE 18.4. Trauma Narrative Construction Procedure

A brief list of nine steps for conducting the narrative (the task of the middle sessions) is shown in the following.

1. Start a blank audiotape that will serve as an exposure-based homework exercise (optional).
2. Get a baseline "thermometer" rating: "What number are you at on the 'Anxiety Thermometer?'"
3. Start the narrative construction, commencing with "external" events, to set the stage: "How did that day start? When did you first notice something was wrong?"
4. Continue the narrative construction, using a combination of "objective" and "subjective" questions to construct a story that weaves together what happened *outside* of me with what happened *inside* of me: "When you heard the first shot, what went through your mind? What did you say to yourself? What were you feeling? How did your body feel?"
5. As appropriate, intermittently check on how the narrator is doing, to ensure that he or she is in the "working range of anxiety." As needed, intervene to help the individual regulate his or her thoughts and emotions.
6. After working through the immediate traumatic experience, bring closure to the narrative by exploring the immediate aftermath (e.g., "Who was at the funeral?" "What kind of medical treatment did you receive?"), and to promote reflection upon the adversities they currently face (e.g., "How is your daily life different now that your father is gone?").
7. Conclude the narrative and help the group member psychologically reconstitute and recover.
8. Involve the group in giving positive feedback: "Who would like to share some of their thoughts and reactions with _____? Remember, we listen to personal stories like these with our hearts and try to find things to say that will help our friend who has courageously chosen to share this story with us."
9. If time permits, return to Step 1 and repeat the process with another group member.

The same procedure is followed during the final sessions, with the exception that the narrative work involves revisiting the worst moments in combination with identifying and challenging maladaptive cognitions. This latter task involves an adaptation of the skill "Three Steps to Taking Charge of Your Emotions," which was learned in Module I. This procedure includes

1. Identifying parts of what happened that the narrator continues to feel bad about (e.g., guilty or ashamed, angry at self or others, or confused over how you acted).
2. Using the three-step model, "Three Steps to Taking Charge of Your Emotions" to challenge maladaptive cognitions and replace them with more adaptive beliefs.

    a. Step 1: "What am I feeling?" Ask members, "Can you identify the bad feeling or feelings that you experience about what happened? Can you give (them/it) a name?"

    b. Step 2: "Why am I feeling this way?"

    - Tell members, "Look at what was happening outside of you. What parts of what happened do you have the strongest bad feelings about?"
    - "Next, let's look inside of you—at the way you are thinking about what happened. Can you put your beliefs about what happened into words?" Obtain a baseline rating of how strongly they believe the thought on a scale from 1 to 10 where 1 represents "don't believe it at all" and 10 represents "totally believe it."

    c. Step 3: "How can I feel better?"

    - Instruct members, "Look for parts that don't make complete sense in that they contain distortions (e.g., exaggerated self-blame or responsibility), or are simply not helpful. Also look for other ways of looking at the situation that make more sense to you, or that help you to feel better about what happened."
    - Invite the remaining group members to make additional observations and suggestions.
    - As appropriate, evaluate the "evidence for" and "evidence against" the thought. Invite the group to offer insights in the roles of caring and concerned friends.
    - Evaluate the belief again: "How strongly do you believe the statement now on that same 1-to-10 scale?"
    - Work together to help them select a more helpful way of looking at what happened. Use the group, as appropriate, to make suggestions and to provide feedback. As appropriate, write down the new belief and give it to the member to work on.
    - Evaluate the strength of belief in the new perspective: "How much do you believe this new perspective on that same scale? Does it seem useful to you?"

MODULE III: COPING WITH TRAUMATIC LOSS AND GRIEF

The sessions in Module III are designed specifically for therapeutic work with adolescents who have experienced the death of a loved one. However, additional prompts are built in to allow adaptation for youths who have suffered other significant losses, including the loss of contact with loved ones, waiting for closure concerning a loved one classified as missing, living with a family member suffering from a serious physical or mental disability (e.g., "I miss the man my father used to be"), or the loss of one's homeland. Due to space limitations and the trauma-focused emphasis of this book, we will restrict our description of this module to a general session-by-session overview.

Therapeutic Rationale

Module III is designed to assist the adolescent with his or her responses to the traumatic death of a family member or close friend. The overarching goals of this module are to (1) reduce the intensity and severity of loss-related distress reactions, particularly, complicated bereavement reactions; and, (2) facilitate beneficial mourning processes in an individually, developmentally, and culturally appropriate manner.

Sessions 1 through 4 of the eight-session module are designed to reduce traumatic bereavement symptoms, specifically, distressing and intrusive preoccupation with the violent circumstances of the death, and promote positive reminiscing and remembering. Sessions 5 through 8 are designed to promote adaptive coping and accommodation to the loss. The long-term aim of the module is not to eliminate grief reactions and mourning, as these may indeed continue long into the future. Rather, it is to help adolescents to make adaptive accommodations to the loss in their current lives, and to form a realistic yet constructive outlook and investment in future interpersonal life and activities. Notably, consistent with the tendency of Module II work to evoke a temporary upsurge in posttraumatic distress reactions, Module III work may elicit temporary upsurges in grief reactions.

Psychoeducation regarding the variability of grief reactions is the first cornerstone of the module. Core concepts include (1) grief and bereavement are natural and generally beneficial processes that promote accommodation to the ongoing absence of a loved one in our lives; (2) a wide range of grief responses and courses of bereavement exist, even among group members; and (3) many developmental and cultural factors influence the experience and expression of loss. In particular, grief and mourning are comprised of both private and public dimensions, and thus are interactive processes that involve different responses across varying contexts and across time. By exploring the wide range of responses, group members develop additional insight into the nature of their personal grief reactions, as well as those of other important life figures.

This type of psychoeducation is especially important to adolescents. In our experience, peers and family members, including parents, often do not appreciate how long children's and adolescents' grief reactions may endure. Particularly important clinically, adolescents who entertain such unrealistic expectations about the nature and course of bereavement may develop an altered sense of self, as based on the belief that something is seriously wrong with them.

On the one-year anniversary of his best friend's death, a teenage boy pretended to faint in class rather than allow the class to see his tears as he missed his friend. He later confided to a therapist that he thought he should be "over it by now," because none of his other friends appeared to be experiencing similar grief reactions.

Explore ways in which the traumatic circumstances of a death or disappearance can make bereavement more difficult. In particular, differences in exposure to the death may influence the course of bereavement among family members and friends. Such differences in exposure and associated bereavement processes may indeed create dissonance within family and peer networks that impose additional social adversities on the bereaved. Differentiating between trauma and grief responses provides a framework to understand the interaction of these experiences and how they respectively influence the course of bereavement for the adolescent, family members, and peers.

The first challenge in treating traumatic bereavement, and an ongoing focus in the module, is to *recognize the loss*. Grief over a death requires recognition of the physical reality of the death, even within a larger spiritual or religious context. Traumatic deaths complicate this process across two dimensions. First, the manner of death may be so horrific that it is difficult to concretize and thus to promote acceptance of the death. Alternatively, a traumatic death may result in an absence of physical remains or witnesses, which leaves the reality of death inconclusive in the adolescent's mind. Due to adolescents' preoccupation with their own physical image and bodily transformation, they can be both "grossed out" by, and extremely inquisitive of, the manner and circumstances of the death.

The second treatment focus in this module is *facilitating adaptive reactions to the loss*. This work is carried out in several ways. First, emphasis is given to addressing "suicidal thoughts" (independent of suicidal behavior) that often constitute wishes for reunion with the deceased. Furthermore, and as appropriate, the role that reckless behavior plays as a maladaptive coping tool for regulating grief pangs is explored and challenged.

The grief-related tasks of remembering and reminiscing also present special challenges as adolescents seek to adapt to a loss. These essential grief activities may be disrupted and, in extreme cases, eclipsed by a need for the avoidance of remembering, reflecting on, and talking about the good and bad aspects of the lost person and the lost relationship. This avoidance may interrupt the construction of the realistic "composite image," necessary to serve as the representation with which the mourner interacts in the new memory-based relationship with the deceased (Rando, 1993). A realistic "composite" is essential to healthy grieving, as it symbolizes the nature and personal meaning of the lost relationship.

Young people may also avoid or suppress experiencing the feelings (positive, negative, or ambivalent) that bind them to the deceased. In the absence of processing these feelings—experiencing and expressing them—they cannot be discharged and released, and, subsequently, diminished in their intensity. In turn, without the dissipation of these emotional connections, mourners continue to remain intensely emotionally attached to the deceased, relating to him or her as if he or she were still living and intimately involved in their day-to-day lives. This "as if" pattern of relating to the deceased is both unrealistic and ultimately maladaptive, because it absorbs great psychological resources without the possibility of reciprocation, prohibits adjustment to the loss, and precludes investment in rewarding (i.e., reciprocated and gratifying) pursuits such as positive life activities and interpersonal relationships (Rando, 1993).

The work of remembering and reminiscing about a loved one in the wake of traumatic loss may also interact with the emerging personal identity and developing sense of self over time and across contexts. Traumatic deaths often introduce a sense of historical discontinuity in a person's life, characterized by a sense of no longer being the same person one was before the death occurred. This phenomenon is especially challenging to adolescents who are experiencing the emergence of their personal identities while constructing a more elaborate framework of past, present, and future by which to plan their lives. Bridging this sense of discontinuity through exercises that enhance active reminiscing and connectedness with one's former life and life figures is critical to resuming this important developmental task. In adolescents, the process of remembering the deceased may be marked by ambivalence linked to difficult last interactions with family or friends, or to unpleasant characteristics of their personal histories with the deceased such as abuse, acrimony, or neglect. In some adolescents, behaviors and attitudes that evoked negative responses by the deceased may be maintained as a reunion fantasy.

As part of their ongoing maturation, adolescents are engaging in ongoing reappraisals of parents, siblings, mentors, and friends, and their interpersonal relationships with them. This developmentally driven process can be disrupted by sudden death, thereby creating a dissonant intersection between maturational and grieving processes. Specifically, the grieving process often engenders idealization via selective remembering or selective learning about the deceased at a developmental stage in which an adolescent is "de-idealizing" a parent, mentor, or friend. In addition, adolescence involves a continual renegotiation of the parent-adolescent relationship as the adolescent makes the transition to young adulthood. The loss removes the interpersonal contexts wherein the adolescent engages in a myriad of positive, neutral, and negative interactions with adult figures that serve to promote autonomy and a more integrated and mature perspective of those figures. Consequently, grief-focused sessions are designed to facilitate members' continuation of this process of relationship renegotiation in the absence of their lost loved one. A potential barrier to this work is the assumption that actively moving on with life is tantamount to letting go of the deceased in uncaring, ungrateful, and even "disloyal" ways. Developmental achievements and mileposts, such as graduations, entering college, weddings, and childbirth, may provoke intense grief reactions as they signify that one is moving ahead in the absence of "indispensable" life figures.

Overview of Sessions 1 through 8

Session 1 provides an interactive approach to learning about grief to normalize, validate, and explore members' grief reactions in personally relevant terms. An important exercise in this session uses a playful mode to address and challenge common misconceptions about the grief process.

- How long does it take before I'll start to feel better?
- Will not talking about the deceased help me to get on with my life?
- Should grief reactions steadily decrease day by day?
- Does healthy grieving require that we forget or stop thinking about our loved one who died?
- Will everyone in my family grieve in the same way, and will they "finish" grieving at about the same time?

Session 1 also dedicates time to providing psychoeducation regarding how loss reminders evoke feelings of sadness and longing, with the aim of helping members to better understand, anticipate, and thereby control their grief reactions.

Session 2 focuses on processing feelings of anger that members experience relating to the loss. The core of the session centers on a visual-aid-guided exercise designed to help members understand that grief reactions change over time, that young people may experience many different types of anger in the aftermath of a death (e.g., anger about the way they died, at the perpetrator, at the world, at God, and often at the deceased), and that "hard" feelings of anger can cover up "softer" feelings of sadness, pain, and longing.

Session 3 is dedicated to processing feelings of guilt relating to the death. Core exercises include normalizing guilty feelings and identifying and challenging "unhelpful" thoughts that lead to feelings of guilt or shame. Cognitive restructuring skills are used to challenge grief-related "hurtful thoughts" and are the same skills used in Module I for challenging trauma-related hurtful thoughts. This session also emphasizes the use of coping skills, especially the five-step model for eliciting support from others. In addition, group leaders use plain language (e.g., "after your father died") rather than common metaphorical phrases ("after your father passed away") to concretize the reality of the death.

Sessions 4, 5, and 6 focus on the work of remembering and reminiscing. Sessions 4 and 5 are devoted to facilitating the development of the capacity to reminisce constructively. More specifically, the fundamental goals are (1) to explore the positive and negative consequences associated with avoiding or suppressing memories and feelings relating to the deceased; (2) to practice reminiscing, which includes thinking, visualizing, communicating, and experiencing feelings (both positive and negative) about the deceased, without becoming overwhelmed by intrusive thoughts, images, and negative emotional reactions; and (3) to identify and challenge barriers to reminiscing (in the form of "hurtful thoughts" and "helpful thoughts") so as to allow members to both "hold on" to their memory-based relationships with the deceased, while "moving on" with their lives. The death of a loved one, especially a nuclear family member, may precipitate many changes in the lives of young people. Relationships within and outside of the family may be altered in their size or composition, often with a net loss of important forms of social support. Additional changes may be found in youths' living circumstances, roles, responsibilities, family rituals, and leisure activities.

Session 6 is designed to help group members to identify changes in their relationships and in key life domains since the death. Session exercises are directed toward recognizing life changes brought about by the loss, including a reduction in needed social support and changes in day-to-day life, and identifying ways in which to further accommodate to the loss. For example, members may discuss the need for additional interpersonal relationships and develop problem-solving techniques they can use to deepen their existing relationships or develop new ones.

Session 7 uses relapse prevention as a model for anticipating and developing a coping plan for predictable "setback" events and settings. This session focuses on developing a list of members' personal trauma reminders, loss reminders, and life adversities that elicit distress and associated maladaptive coping responses. The remainder of the session focuses on decreasing perceptions of helplessness, unpredictability, and hopelessness through the development of a positive coping plan. This involves a group discussion emphasizing the idea that, although painful memories and feelings often persist long after a loss, they tend to be predictable due to their links to trauma reminders, loss reminders, and secondary adversities. Following this, the group leaders facilitate a "personal coping plan" activity in which each group member identifies and anticipates high-risk situations, identifies maladaptive coping they will be tempted to engage in, and then develops and practices a more adaptive set of personal coping responses with which to effectively contend with the situation.

Session 8 is a potential ending point for members who do not exhibit signs of developmental disturbance. Treatment objectives for this session include reviewing and reinforcing treatment gains, processing what participating in the group has meant to each member, comparing and contrasting this "voluntary" separation with previous "traumatic" separations, and helping group members to plan and look forward to their lives after the group.

MODULE IV: REFOCUSING ON THE PRESENT AND LOOKING TO THE FUTURE

This final module is designed to assist adolescents in maintaining and building upon their newly developed understanding and skills to improve their present lives and to prepare for the future. This strong emphasis on remediating developmental disruptions is based on recognition that the success with which traumatized and bereaved adolescents are therapeutically treated should not be measured exclusively in terms of the resolution of posttraumatic and complicated grief reactions. Rather, because recently acquired competencies are often disrupted, and, in turn, can interfere with further developmental progression, focus needs to be on the development of skills that will enable group members to remediate these disruptions. Moreover, posttraumatic stress responses can also lead to "traumatic expectations" that may undermine interpersonal relationships, adaptive functioning, and successful future planning. These expectations may also impose trauma-related restrictions on developmental opportunities which, in turn, may interfere with critical adolescent developmental achievements.

In addition, adolescents are often stressed by an array of secondary life changes or adversities created by their traumatic experiences and losses. Traumatized adolescents are often keenly aware of how their adolescence is different than they had imagined it would be. Teens participating in our treatment groups are typically much more invested in talking about the developmental impacts of these secondary adversities than in discussing their distress symptoms per se.

The group format is an especially powerful tool for supporting adolescents as they seek to repair these wounds to development. The group provides a secure and supportive environment in which they can develop and exploit their abilities to problem solve current and anticipated future adversities, and to practice and apply their coping skills. In addition, group processes, including the establishment of a new reference group against whom members may normalize their personal traumatic experiences, distress reactions, and subsequent life adversities, assists members in correcting traumatic expectations about themselves, others, and social institutions with reference to themes of danger, prevention, and protection.

One of the developmental tasks most vulnerable to trauma-induced disruptions is the integration of past, present, and future into an emerging personal identity. The powerful sense of historical discontinuity that trauma and loss can create "freezes" the adolescent into a "traumatized" adolescent personality, as evidenced by such statements as "I miss the person I was before this all happened," and "I don't know who I am anymore." The opening sessions of Module IV are designed to bridge this sense of discontinuity by promoting the resumption of trauma-interrupted activities and by encouraging developmental progression via the initiation of "age-appropriate" activities. Additional emphasis is given to an explicit reflection on traumatic expectations as "lessons that our traumatic experiences have taught us," while problem solving more constructive yet realistic basic beliefs and preparations for the future.

Turning away from a "retrospective" exploration of traumatic experiences and their continuing impact, the module then focuses on pragmatic approaches to dealing with current stresses and adversities. Members are encouraged to consider not only their personal adversities, but to discuss group-level activities that they may collectively undertake to improve their school, community, and generation as it approaches adulthood, marriage, and family life.

Overview of Sessions

The central activities of *Session 1* (resuming developmental progression) are designed to help members identify and reframe pessimistic changes in their beliefs about themselves, others, and the world around them, as the legacy of trauma and loss. Some examples of this altered worldview include

- We no longer feel like we can take our safety for granted.
- It is much harder for us to trust now than before.
- We feel much less in control over our lives.
- We lost much of our self-esteem.
- Our future is more bleak now, and it's harder to get motivated to work hard in school.
- We are a lot more afraid than before, and do not see the world as very safe or predictable.
- We do not have as much confidence in laws and justice.

Ensuing exercises and discussions help group members link their personal "traumatic expectations" with their traumatic experiences and to explore ways in which these altered views may lead to undesirable life consequences. For example:

- If you have no faith in the future, you are unlikely to apply yourself at school.
- If you believe that you have no control over what happens in your life, you are not going to try to accomplish anything worthwhile. So you'll be sitting back and watching your life happen—watch things happen to you. This will probably reinforce your perception that you do not have any control over what happens.
- If you have given up on the idea of getting married and having kids, then you probably will not invest in your relationships, or you will not date.
- If you do not believe in justice at the societal level, then you also may give up on your own morality. If you say that laws are meaningless or irrelevant, you also run the risk of concluding that right, wrong, and personal integrity do not matter any more.

The remainder of the session is devoted to exploring the circumstances under which group members would be willing and able to relinquish their "hurtful" beliefs in exchange for more "helpful" and constructive beliefs.

In *Session 2* group members are taught a basic three-step approach to solving problems that builds upon the three-step model for taking charge of your feelings presented in Module I. In this activity, group members comment on ways in which their lives have changed since the trauma or death, and discuss stressful life events or circumstances that they currently face. Common examples include financial hardships due to parental death or job loss, family moves, increases in family members' responsibilities, increases in family conflict, peer difficulties, and the loss of parental attention or time spent with loved ones.

The group leaders then segue into a presentation of the three-step model comprised of the steps "What am I feeling?"; Why am I feeling this way?"; "How can I make things better?" This model contains a number of subskills, including discriminating between "problems that are my job to handle" and "problems that are someone else's job to handle;" as well as considering one's goals, brainstorming options, and choosing the best option. The remainder of the session is dedicated to problem solving and practice exercises dedicated to identifying current problems and developing a personal coping plan to contend with it.

In *Session 3* group members are taught a practical skill for dealing with feelings of excessive personal responsibility. Following a traumatic event or death, teenagers may have distorted beliefs concerning their personal responsibility for the event and their obligations regarding the well-being and recovery of family members and friends. For example, youths may consider it their job to make a bereaved and depressed parent happy or to "fill the shoes" of a lost sibling. They may also be burdened with responsibilities and worries that are beyond their capabilities and which may adversely affect their lives and development. These issues are addressed in Session 3 in the form of group discussions directed toward identifying and exploring the impact of "hurtful thoughts" such as, "No one can do it but me," "I can't allow myself to be happy when people I care about are feeling bad," or "They're lost without me! I've got to save them!" Following this, several simple heuristics are introduced to help group members discern whether a problem is theirs to cope with, or whether it is the responsibility of someone else:

> One good way is to ask yourself, "Is this something that is happening between me and someone else? Or is this problem mainly something that is happening to someone else or between other people?" If the problem is happening between you and someone else, then it is partly

your job to fix—it is up to the two of you to figure out what to do. If, for example, you get in trouble with a parent for not cleaning your room, then the problem is between you and your parent, and it is your job to fix. What if, on the other hand, your dad or mom just lost his or her job, and he or she is feeling bad. Is it your job to find them a new job? (No—the problem is between them and their boss or workplace, and does not involve you directly.)

We are now going to have an activity to practice telling the difference between problems that are a teenager's job and problems that are not a teenager's job to deal with. I'm going to read a list of problems, and the task of the group is to decide whether the problem is a teenager's responsibility to fix or not. In your mind, I want you to visualize whether the "problem arrow" is pointing directly at you (indicating that it is your job to fix) or whether it is pointing at someone else (indicating that it is his or her job to fix).

Group brainstorming activities are then facilitated to help answer such questions as "What happens when I try to fix other people's problems? How do you cope with problems that are not yours to fix?" and "How do you support someone who has a problem?" These problem-solving discussions assist the group members in developing pragmatic individual action plans and in learning how to provide effective and timely support to loved ones in need.

Session 4 attends explicitly to group termination issues, with an emphasis on making this a nontraumatic leave taking. Group activities are designed to help members to learn through experience that, in many relationships, "saying good-bye" is natural, voluntary, benign, and often reversible. Termination issues may indeed surface throughout the module and may be addressed as they appear. For example, members may show signs of concern regarding whether they are ready to continue without the group's weekly support. Some group members may show signs of regression, or a loss of treatment gains, which often appears as a return to the same level of functioning evident at the time the group began. These behaviors may be gently interpreted:

These setbacks sometimes show how we are feeling about the group coming to a close. They may show that we don't feel sure about whether we are really ready to go on without the group being there for us every week.

During these final sessions, group members may also begin to start new friendships, either with other group members or with others outside the group. These investments may represent members' anticipation of the social losses that will accompany the cessation of the group's meetings, and may also be gently interpreted:

> Sometimes when a group is ending, we want to start new friendships because we want to have relationships that will continue to last. This helps us to feel better when we think about how we'll miss being with the other group members.

Leaders should be prepared for expressions of anger that the leaders are "willing to let the group end." This can also be gently interpreted:

> It sounds like being in the group has been an important part of your life and that you will miss it.

The objectives of this final session include

1. reviewing and reinforcing treatment gains;
2. helping group members to plan and look forward to their lives after the group;
3. allowing group members to process termination issues; and
4. giving special emphasis to differentiating between this "voluntary" separation and previous "traumatic" separations.

An example of an introduction of a process-focused discussion follows:

> Let's first reflect on what we've accomplished during this time we have been meeting together, both as individuals and as a group. Do you remember our first individual interviews together, before the group formed? During those interviews, we asked you to describe a chronology of your life—the important things that happened before the (trauma), during the (trauma), and after the (trauma)—the high points and the low points. (Distribute the time lines to the individual members.)
> We'd like you to take a minute and look over your time lines—how you viewed your life before joining the group. Try to remember what you were like back then—how you were thinking and feeling, and especially how you were thinking about your future. Reflect on your life history—who you were before, how the traumatic event changed you, and who you are striving to be now. We'd like to invite each of you to

comment on what you feel you've been able to accomplish with the group's help.

- Share one or two positive changes that you have made in your life, or in your family's life, while you have been a part of this group.
- Before our first session, did you think that you would be able to accomplish what you have accomplished with the group's help?

The group leaders and members then share their reactions to participating in the group, giving emphasis to a review of the positive changes that each group member has made and to expressing respect and admiration for the courage each has shown. Following, a small "graduation" party with treats is organized and held, complete with a "memento" group picture.

COMBINING INDIVIDUAL AND FAMILY INTERVENTIONS WITH GROUP TREATMENT

A number of circumstances occur in which individual and family therapy can be used to preface and/or augment the group therapy program. A primary consideration during the initial assessment and interview process is to determine the recency of the trauma or loss exposure and the child's current degree of fragility or risk. Clients whose trauma or loss has occurred within the past one or two months are generally poor candidates for group therapy because the group work constitutes a much more evocative and less controlled therapeutic setting than individual counseling. Carrying out an in-depth assessment of youths at least two to three months following an identifiable traumatic event enables us to identify those who are at highest risk for severe, persisting posttraumatic stress reactions and developmental disturbances, while providing clients with sufficient time to consolidate and gain support.

Based on these considerations, we suggest a course of *individual therapy* for youths whose trauma exposure is recent or who appear highly reactive or unstable at the time of the interview. Periodic consultation with the youth and his or her therapist will indicate if and when group therapy becomes a viable option.

In some cases, clients may concurrently participate in group and individual therapy. The treatment modalities have sufficiently unique strengths that one may complement the other, especially if there is some degree of collaboration between the group and individual therapists. Participation in an appropriate trauma or grief group can help children and adolescents find the type of peer support and validation that is generally beyond the reach of individual therapy. For example, hearing from a similarly traumatized peer that ". . . you shouldn't blame yourself, there was nothing you could have done," often carries much more weight than even the best cognitive interventions employed by highly skilled clinicians.

On the other hand, individual therapy has important advantages as well, particularly with reference to the individual tailoring that this modality provides regarding the rate and depth at which traumatic material is explored. Individual treatment also affords greater flexibility with respect to focusing on specific treatment foci as dictated by individual needs, and the one-on-one therapeutic relationship may permit more speedy resolution of issues of trust and confidentiality. The trauma/grief-focused group therapy protocol presented here may be used, with only minor alterations needed, in individual therapy as well.

Family therapy may also be offered concurrently or in place of group therapy. In some circumstances, such as in the injury or death of a family member, family interventions may be seen as the treatment of choice. Participation in group therapy with similarly bereaved peers carries the unique advantage of combating an adolescent's sense of isolation and facilitating the resumption of academic striving and social relationships. Conversely, with respect to traumatic events in which multiple family members are exposed or which has a significant direct impact on the family, family therapy carries the advantage of facilitating recovery via mobilization of the primary family unit.

Concurrent family and group treatment can capitalize on the strengths of each modality. One strategy is to synchronize the content of group and family interventions. For example, during the same week, group and family meetings can focus on trauma/loss psychoeducation, identification of individual and family trauma/loss reminders, and the development of individual and family coping strategies. In our group therapy program, we try to incorporate at least some family involvement through an initial family contact and, if possible, one or two meetings at key stages in the group process, preferably at the beginning of Module II (preparatory to commencing the trauma narrative work) and toward the end of Module IV (preparatory to termination).

CONCLUSION

Studies involving traumatized adolescents in high crime communities in Southern California (Saltzman et al., 2001b) and in postwar Bosnia (Layne et al., 2001) have provided evidence of the effectiveness of the UCLA Trauma/Grief Group Psychotherapy Program. Specific outcomes included reductions in PTSD, depression, and symptoms of complicated grief, as well as improvements to a range of measures of psychosocial adaptation. In the domestic school-based application of the program (Saltzman et al., 2001b), participating students improved their academic performance and incurred fewer disciplinary actions.

The manualized UCLA Trauma/Grief Group Psychotherapy Program continues to evolve as a result of its application in diverse settings. It is being used at multiple sites across New York City, in a major California school district, and is slated for implementation at several additional clinic and school settings in various U.S. cities. The program is being used in the aftermath of single large-scale traumatic events and for chronic exposure to multiple events, as a guide for both group and individual interventions, and is being adapted for use in diverse ethnic communities with unique histories and needs.

Plans for the continued development and availability of the program include:

1. Progressive research on the effectiveness of the intervention with efforts made to "unpack" the contributions made by specific components with specific populations. This will entail a randomized, controlled study in selected secondary schools.
2. Consultation with group process experts to increasingly integrate this important knowledge base with our program design.
3. Expansion of the parental and family portions of the intervention to increase family involvement in the intervention.
4. Publication of the manual and supporting materials in the near future.

REFERENCES

Birleson, P. (1981). The validity of depressive disorder in childhood and development of a self-rating scale: A research project. *Journal of Child Psychology and Psychiatry, 22,* 73-88.

Briere, J. (1995). *The trauma symptom checklist for children manual* (TSC-C). Odessa, FL: Psychological Assessment Resources.

Burlingame, G., Fuhriman, A., & Johnson, J. (2000). *The empirical evidence for developing effective therapeutic relationships in group psychotherapy: A guide to psychotherapy relationships that work.* London: Oxford University Press.

Cohen, J. A., Berliner, L., & March, J. S. (2000). Treatment of children and adolescents. In E. B. Foa, T. M. Keane, & M. J. Friedman (Eds.), *Effective treatments for PTSD: Practice guidelines from the International Society for Traumatic Stress Studies* (pp. 106-138). New York: The Guilford Press.

Davies, R., Burlingame, G., & Layne, C. (2005). Integrating small-group process principles into trauma-focused group psychotherapy: What should a group trauma therapist know? In L. A. Schein, H.I. Spitz, G.M. Burlingame, & P.R. Muskin, (Eds.), *Psychological effects of terrorist disasters: Group approaches to treatment* (pp. 385-423). Binghamton, NY: The Haworth Press.

Garbarino, J. (1992). *Children and families in the social environment.* New York: Aldine.

Garbarino, J. & Kostelny, K. (1996). What do we need to know to understand children in war and community violence? In R. J. Apfel & B. Simmon (Eds.), *Minefields in their hearts: The mental health of children in war and communal violence* (pp. 33-51). New Haven, CT: Yale University Press.

Goenjian, A., Karayan, I., Pynoos, R., Minassian, D., Najarian, L., Steinberg, A., & Fairbanks, L. (1997). Outcome of psychotherapy among early adolescents after trauma. *American Journal of Psychiatry, 154*(4), 536-542.

Jacobs, S. (1999). *Traumatic grief: Diagnosis, treatment, and prevention.* Philadelphia: Brunner/Mazel.

Kilpatrick, D., Aciermo, R., Saunders, B., Resnick, H., & Best, C. (2000). Risk factors for adolescent substance abuse and dependence: Data from a national sample. *Journal of Consulting and Clinical Psychology, 65*(1), 1-12.

Kilpatrick, D. G., Saunders, B. E., Resnick, H. S., & Smith, D. W. (1995). *The National survey of adolescents: Preliminary findings of lifetime prevalence of traumatic events and mental health correlates.* Charleston, SC: Medical University of South Carolina, National Crime Victims Research and Treatment Center.

Layne, C. M., Pynoos, R. S., & Cardenas, J. (2000). Wounded adolescence: School-based group psychotherapy for adolescents who have sustained or witnessed violent injury. In M. Shafii & S. Shafii (Eds.), *School violence: Contributing factors, management and prevention* (pp. 184-211). Washington, DC: American Psychiatric Press.

Layne, C., Pynoos, R., Saltzman, W., Arslangic, B., Savjak, N., & Popovic, T. (2001). Trauma/grief-focused group psychotherapy: School-based postwar intervention with traumatized Bosnian adolescents. *Group Dynamics: Theory, Research and Practice, 5,* 277-288.

Layne, C. M., Saltzman, W. R., & Pynoos, R. S. (2002, April). Grief Reactions: A clinician's perspective. *Marriage and Families.* Provo, UT: Brigham Young University. Available online <http://marriageandfamilies.byu.edu/issues/2002/April>.

Layne, C. M., Saltzman, W. R., Steinberg, A. M., & Pynoos, R. S. (2003). Trauma/grief-focused group psychotherapy manual for adolescents. Unpublished treatment manual. University of California, Los Angeles.

Layne, C. M., Warren, J. S., Saltzman, W. S., Fulton, J., Steinberg, A., & Pynoos, R. S. (2005). Contextual influences on posttraumatic adjustment: Retraumatization and the roles of revictimization, posttraumatic adversities, and distressing reminders. In L. A. Schein, H. I. Spitz, G. M. Burlingame, & P. R. Muskin, (Eds.), *Psychological effects of terrorist disasters: Group approaches to treatment* (pp. 235-286). Binghamton, NY: The Haworth Press.

Layne, C. M., Stuvland, R., Saltzman, W., Steinberg, A., & Pynoos, R. S. (1999). *War trauma exposure scale.* Sarajevo: UNICEF Bosnia & Hercegovina.

Layne, C. M., Wood, J., Saltzman, W. R., & Pynoos, R. S. (1999). *School-based psychosocial program for war-exposed adolescents: Screening and pre-group interview protocol.* Sarajevo, Bosnia & Hercegovina: UNICEF.

Layne, C. M., Wood, J., Steinberg, A., & Pynoos, R. S. (1999). *Protocol for scoring the post-war screening survey and for selecting group candidates.* Sarajevo: UNICEF Bosnia & Hercegovina.

Layne, C. M., Wood, J., Saltzman, W. S., Steinberg, A., & Pynoos, R. S. (1999). *School-based psychosocial program for war-exposed adolescents: Screening interview.* Sarajevo: UNICEF Bosnia & Hercegovina.

Lorion, R. & Saltzman, W. R. (1993). Children's exposure to community violence in Maryland: Following a path from concern to research to action. *Psychiatry,* 56.

March J. & Amaya-Jackson, L. (1994). Posttraumatic stress disorder in children and adolescents. *PTSD Research Quarterly, 4,* 1-7.

Pynoos, R. S. (1992). Grief and trauma in children and adolescents. *Bereavement Care, 11,* 2-10.

Pynoos, R. S. & Nader, K. (1988). Psychological first aid and treatment approach for children exposed to community violence: Research implications. *Journal of Traumatic Stress, 1,* 445-473.

Pynoos, R. S., Nader, K., Frederick, C., & Gonda, L., (1987). Grief reactions in school age children following a sniper attack at school. *Israel Journal of Psychiatry and Related Sciences, 24,* 53-63.

Pynoos, R. S., Rodriguez, N., Steinberg, A. M., Stuber, M., & Fredericks, C. (1999). *Reaction-Index-Revised.* Unpublished psychological test. Los Angeles: University of California.

Pynoos, R. S., Steinberg, A. M., & Piacentini, J. C. (1999). Developmental psychopathology of childhood traumatic stress and implications for associated anxiety disorders. *Biological Psychiatry, 46,* 1542-1554.

Pynoos, R. S., Steinberg, A. M., & Wraith, R. (1995). A developmental model of childhood traumatic stress. In D. Cicchetti & D. J. Cohen (Eds.), *Manual of developmental psychopathology* (pp. 72-93). New York: John Wiley & Sons.

Rando, T. A. (1993). *Treatment of complicated mourning.* Champaign, IL: Research Press.

Saigh, P. A., Yasik, A. E., Sack, W. H., & Koplewicz, H. S. (1999). Child-adolescent posttraumatic stress disorder: Prevalence, risk factors, and comorbidity. In P. A. Saigh & J. D. Bremner (Eds.), *Posttraumatic stress disorder: A comprehensive text.* Boston: Allyn & Bacon.

Saltzman, W. R., Layne, C. M., Steinberg, A. M., Arslanagic, B., & Pynoos, R. S. (2003). Developing a culturally and ecologically sound intervention program for youth exposed to war and terrorism. *Child and Adolescent Psychiatric Clinics of North America, 12,* 319-342.

Saltzman, W. R., Pynoos, R. S., Layne, C. M., Steinberg, A. M., & Aisenberg, E. (2001a). Trauma- and grief-focused intervention for adolescents exposed to community violence: Results of a school-based screening and group treatment protocol. *Group Dynamics: Theory, Research and Practice, 1*(5), 291-303.

Saltzman, W. R., Pynoos, R. S., Layne, C. M., Steinberg, A., & Aisenberg, E. (2001b). A developmental approach to trauma/grief focused group psychotherapy for youth exposed to community violence. *Journal of Child and Adolescent Group Therapy, 11*, 2-3.

Saltzman, W. R., Steinberg, A. M., Layne, C. M., & Pynoos, R. S. (1999). *UCLA Adolescent Trauma Exposure Screening Battery*. Los Angeles: UCLA.

Webb, N. B. (Ed.). (2002).*Complicated grief: Dual losses of godfather's death and parents' separation in helping bereaved children*. New York: The Guilford Press.

Worden, J. W. (1996). *Children and grief: When a parent dies*. New York: The Guilford Press.

Chapter 19

Trauma-Focus Group Therapy: An Evidence-Based Group Approach to Trauma with Adults

William S. Unger
Melissa S. Wattenberg
David W. Foy
Shirley M. Glynn

INTRODUCTION

Rationale and Objectives

The rationale for using group therapy with survivors of traumatic events caused by terrorism is based on both the survivors' needs and the advantages of joining with others in therapeutic work to cope with victimization consequences such as isolation, alienation, and emotional numbing. Group therapy is valuable for survivors of traumatic events caused by terrorism who may feel ostracized from the larger society, or even judged or blamed for their experience. Group members experience relief from anxiety and self-blame with the discovery of shared suffering and acceptance by others (Yalom, 1995). Bonding in a supportive environment with others who have had similar experiences can be a critical step toward regaining trust and moving forward with one's life following a traumatic event. Beyond its obvious cost advantage, group therapy makes use of corrective peer feedback. Peer feedback is a particularly valuable intervention for promoting change in those individuals who fail to acknowledge responsibility for life choices and outcomes, since corrective feedback from one's peers is more easily assimilated than that of a therapist. (Klein & Schermer, 2000).

Group therapies for adult trauma survivors are categorized by the approach to traumatic memories as being either *uncovering* or *covering* types, depending upon whether the therapy promotes in-depth trauma processing (uncovering), or discourages references to traumatic memories in favor of focusing upon issues in the present (covering). Although *covering* and *uncovering* group methods differ in their theoretical models of symptom development and therapeutic intervention, they share a set of key features that build a safe and respectful therapeutic environment. These features include

1. group membership determined by shared type of trauma (e.g., terrorism survivors, combat veterans, or adult survivors of child abuse);
2. validation of the traumatic experience;
3. normalization of trauma-related responses;
4. validation of behaviors required for survival during the time of the trauma; and
5. challenge to the idea that the nontraumatized therapist cannot be helpful through the presence of fellow survivors in the group.

For our discussion we present a review of the literature and a detailed description of trauma-focus group therapy (TFGT), an uncovering approach.

Evidence Supporting Group Therapy

Foy et al. (2000) reviewed published reports of clinical trials of group psychotherapy for adult trauma survivors. The findings indicate that group therapy is typically conducted over ten to fifteen weekly sessions (range = six weeks to one year for the duration of the group), and session length is usually set at one-and-a-half or two hours. Most studies were conducted with female survivors of childhood or adulthood sexual abuse; very few published reports have included male participants.

Overall, the current literature provides consistent evidence that group psychotherapy, regardless of the type, is associated with favorable outcomes across a number of symptoms (Foy et al., 2000). PTSD and depression are the most commonly targeted symptoms, but efficacy has been demonstrated for a range other symptoms as well, including global distress, dissociation, low self-esteem, and fear. A number of significant methodological issues including random assignment of participants, assurance of adequate statistical power, and use of standardized treatment manuals currently constrain the causal inferences that can be drawn about the efficacy of group treatment (Foy et al., 2000). These limitations are particularly important for forming future research in this area.

Cognitive-Behavioral Group Treatment

Each of six studies examining the efficacy of cognitive-behavioral group treatment demonstrated improvements in the level of group members' distress at the end of treatment (Frueh, Turner, Beidel, Mirabella, & Jones, 1996; Lubin, Loris, Burt, & Johnson, 1998; Resick, Jordan, Girelli, Hutter, & Marhoefer-Dvorak, 1988; Resick & Schnicke, 1992; Stauffer & Deblinger, 1996; Zlotnick et al., 1997). The groups represented a variety of trauma populations (e.g., sexual assault, adult survivors of abuse, and combat veterans). The variety of cognitive behavioral techniques represented in these groups included the following: cognitive processing therapy, assertiveness training, stress inoculation, and affect management. The groups met, usually weekly, for a range of six to sixteen weeks. All six studies assessed PTSD symptoms directly (Foy et al., 2000).

Trauma-Focus Group Therapy

TFGT emphasizes the use of two cognitive-behavioral methods, systematic prolonged exposure and cognitive restructuring, to process each group member's trauma experience. Each group member has the opportunity to recount his or her story as others listen. Therefore, group members take part in trauma processing both through directly experiencing their own trauma event, as well as vicariously experiencing those of others. Traumatic experiences may often be encoded differently into our memory and in part at a nonverbal level complete with affective components and visual, auditory, tactile, olfactory, and taste components (Levis, 1990). During therapy, memories of these experiences may need to be retrieved with a different therapeutic approach and be reexperienced rather than discussed (Hegeman & Wohl, 2000).

The TFGT group model encourages both the use of personal narrative and group support; members "stand together" and hear the experiences of others without judgment. Providing psychoeducational material regarding awareness of normal reactions to trauma and utilization of key coping skills bolsters the group members' resources for coping with current and future trauma-related reminders and symptoms (Foy, Eriksson, & Trice, 2001).

Overview

The information presented in the following sections will review issues and treatment considerations for using TFGT with individuals victimized by acts of terrorism based upon a treatment protocol developed for use with combat veterans (Foy, Gusman, Glynn, Riney, & Ruzek, 1997). A recent study evaluating the efficacy of TFGT indicated that combat veterans showed significant improvement in PTSD symptoms following completion of the treatment (Schnurr et al., 2003).

Traumatic events have a broad impact on an individual's daily life including medical, psychiatric, interpersonal, and cultural disruptions that may be expressed by negative alterations of physical, psychological, and social well-being (Herman, 1992). Individuals are often told to forget traumatic events or to move on with their lives and not dwell on the past. For many victims of traumatic events, following this simple philosophy is not possible. Treatment via therapeutic exposure asks the individual to focus on the events and all the memories, affect, and stimuli associated with the incident with the understanding that this may be a painful emotional experience. The term *therapeutic exposure* refers to a treatment approach in which the client, with informed consent, focuses on memories and experiences associated with the reported traumatic event. Therapeutic exposure focuses on traumatic events of extreme affective intensity. A *traumatic event* is defined as an incident during which an individual directly experiences a significant threat to his or her personal safety or the safety of someone else. The individual in turn, experiences extreme fear, horror, and/or distress. Following the incident, the individual reports recurrent intrusive and distressing memories of the event, recurrent nightmares of the event, increased physiological arousal with the presentation of stimuli associated with the event, and/or flashbacks, all of which are considered post-traumatic stress disorder (PTSD) reactions. Symptoms associated with the event are viewed as resulting from the individual's attempts to avoid the painful memories. The group facilitator and other group members serve as the guide and support structure to assist the individual in working through the painful memories.

Although posttraumatic stress reactions involve avoidance of trauma-related memories, TFGT directly confronts memories. Taking place in a therapeutic setting and with the support of the facilitators and other group members, TFGT helps individuals to integrate their experiences into their present lives and to move forward from the experiences.

The cognitive restructuring component of TFGT focuses on modifying the thoughts, beliefs, and attitudes related to traumatic events leading to a decrease in feelings of distress. Some of the thoughts and interpretations about the traumatic incident are altered and the client begins to perceive what happened in a healthier way and develop a new perspective about the experience.

GROUP GOALS

Group goals are divided into treatment goals and in-session goals. The first group treatment goal is a gradual reduction of emotional reactivity to painful memories for each member of the group. Because gradual reduction occurs via extinction of the negative affective drive state, the process of extinction is facilitated by the repeated exposure to traumatic memories during both the scheduled treatment sessions and the self-exposure home-based therapy assignments. All of the group members will take turns presenting their trauma scenes during two of the scheduled meetings. The self-exposure home-based therapy is accomplished using audiotaped recordings made during these actual in-session scene presentations, which group members are assigned to listen to at home throughout the course of the TFGT sessions.

The improvement of group members' abilities to manage symptoms is the second group treatment goal. Individuals in the group work toward accomplishing this goal both during sessions and via home-based therapy assignments. Because work done during trauma-focused sessions may initially cause a temporary increase in nightmares, anxiety, and other symptoms, this possible increase should be discussed during sessions and group members should be taught to use techniques to manage this stress.

The third group treatment goal is for members to increase their abilities to tolerate strong negative emotions. Improved tolerance occurs with practice, as members learn that they can manage high levels of distress by using the skills taught during group sessions (Foy et al., 1997). Once again, practice is necessary to ensure the process of extinction and reduction of negative affects.

The final group treatment goal, through the course of TFGT sessions, focuses group members on the thoughts associated with the traumatic incident. Trauma survivors often have strong feelings of guilt and negative self-statements associated with guilt, as well as other aversive emotions (Foy et al., 1997; Levis, 1990). Distorted interpretations of the incident based on these feelings may also be present. During TFGT members are asked to review their trauma-related feelings and thoughts while the group as a whole provides corrective feedback to each member. The group members learn that their emotions, reactions, and thoughts concerning the incident can be changed.

The major in-session goals for the group facilitators during the first two sessions include developing group cohesion and trust among group members since they may be initially nervous and uncertain about the group process. *Cohesion* may be defined as the degree of group members' interest in one another (White & Freeman, 2000). Since establishing a positive therapeutic alliance is paramount to successful treatment outcome, the facilitators should be open and supportive of the group members. The group facilitators must convey a sense of empathy and concern to accomplish therapeutic tasks (White & Freeman, 2000). Providing the group members with effective stress management skills including deep breathing, relaxation, and thought stopping will enhance this process. The group facilitators should also explain the importance for the group to manage symptoms more effectively using home-based therapy assignments.

The second in-session goal for the group facilitators is to guide group members through a reexperiencing of their traumatic events called the *exposure session*. The group facilitators help direct each group member by asking the individual to place himself or herself back into the event as if it were happening. It may be beneficial to explain that, during their exposure sessions, group members are like actors in a movie scene. During two sessions, each member provides an elaborate description of the event they experienced including the behaviors prior to, during, and immediately following the event. One of the facilitators will prompt each member about feelings and thoughts throughout the scene while other group members remain silent, listening to the presentation until the cognitive restructuring portion of the session. Since it is not unusual that a group member begins to digress to other memories during their trauma scene, the presenter may also require assistance from the facilitators in maintaining focus on the trauma scene. This can be accomplished by redirecting the presentation back to the event. If the newly remembered memories are revealed during the trauma scene, it may be necessary to assess these memories at a later time and use this new information during the second round of scene presentations.

The third in-session goal of trauma-focused treatment is cognitive restructuring in which group members change faulty cognitions associated with the event by listening to and recording feedback of other group members. Emphasis is placed on discussing and assessing negative self-statements as a means to correcting the individual's thoughts or beliefs concerning the event described during the trauma-focus scene. At this point, group members will, each in turn, comment on the scene material presented during the exposure session. The group facilitators guide and prompt the members' responses to the key points of the scene, which involve feelings of guilt, shame, and anger associated with the actions taken, or perhaps not taken during the traumatic event. A particular emphasis is placed on questioning the group members about predictability, controllability, and culpability of the tragic outcome of the scene. Because emotions of guilt and anger associated with the member's current perspective help to maintain avoidance symptoms, anxiety, and depression, the goal of cognitive restructuring is to replace these thoughts with more realistic and constructive ones.

The in-session goal of the final treatment meeting is to assist group members with termination. Members are asked about how they feel about the termination of the group: Were their expectations met? Did their feelings and thoughts about their traumatic event change? Are there key issues that have not changed? How do they plan to continue working on these areas? Because group members will most likely need to continue their work in these areas, facilitators should assist them with developing a plan using the skills and techniques learned during the group.

LEADERSHIP ISSUES

Co-Leadership

Trauma-focused groups require coleadership. Successful cotherapy requires open communication between the group facilitators to resolve differences (Yalom, 1995). Both group facilitators discuss and share clinical decisions made throughout treatment. The role of the group facilitators is to support and direct members' work during each session. Group leaders also facilitate group discussion and keep therapy on track in order to cover scheduled material and prevent avoidance behaviors.

The need for two group facilitators is best illustrated during the trauma-focus treatment sessions. While one of the group facilitators is working with the group member presenting his or her traumatic event, the second group facilitator is monitoring the remaining group members who may experience an increase of symptoms, suffer a panic attack, or dissociate during the presentation. Without disrupting the work of the first group facilitator, the second can provide support or accompany someone leaving the group room, should this unlikely event occur.

Leadership Style

The group facilitators are generally required to be directive and didactic in style especially during the first two sessions. During the trauma-focused treatment sessions, the group facilitators should be supportive and assist group members with the therapeutic exposure scene. At certain times, a confrontational style may also be necessary, especially when clinical judgment suggests that the group member is avoiding memories, cognition, or affect associated with the traumatic event being presented.

Management of Extreme Affect

Some or all group members will also have strong emotional reactions during treatment. Self-report via *subjective units of distress* (SUDS) is utilized during the sessions to assess the level of emotional response of each group member throughout group sessions (Goldfried & Davison, 1976). Group members communicate how calm or tense they feel at any given time throughout treatment on a scale of 0 to 10, where 0 represents complete relaxation and 10 represents maximum tension. Group facilitators will assist group members with both staying in the affect and not avoiding the aversive thoughts and feelings. Instruction in the utilization of positive coping skills to more effectively manage symptoms will also be presented. Repetition of this material during exposure sessions will facilitate extinction. It will also benefit the individual if group facilitators challenge the negative cognition and affects. Group facilitators may utilize relaxation skills, deep breathing, and grounding techniques to assist group members with prolonged episodes of extreme distress at the end of each session. Examples of these techniques are illustrated in the case material presented in the following sections.

After a scene presentation, the group facilitators will prompt the group members with questions concerning the predictability of the tragic outcome. Questions such as, "Was the event foreseeable?" or "Given the individual's knowledge of the circumstances, could he or she predict what would happen?" should be asked. The controllability of the event is also a consideration in alleviating the guilt experienced by the member. Prompts in the form of questions such as, "Was the outcome of the traumatic event controllable?" or "Was there anything that anyone could have done at the time to change the outcome?" may be used. These questions are directed at the group members following the scene presentation. Issues concerning self-blame should be challenged by the group.

Following the cognitive restructuring, each member is asked to develop a coping plan for the remainder of the evening and the time between group sessions. Group members should not review the audiotape for a few days following the group meeting, however, they should review the tape prior to the next trauma-focus session. Each group member is asked to write about the incident, and answer the questions on the form that is provided as a home therapy assignment.

Countertransference Pitfalls

Maintaining compassionate objectivity is essential to trauma-focus work (Ziegler & McEvoy, 2000). Group facilitators must manage their responses to the trauma material being presented by group members as well as their reactions to the group as a whole. However, in structured cognitive-behavioral interventions transference and countertransference are not considered to be essential to the progress in therapy, and unless managed appropriately, can reduce positive treatment effects (Johnson & Lubin, 2000). The presence of a cofacilitator is extremely useful concerning the resolution of issues regarding countertransference. Individual clinicians may themselves identify this response. However, as one cofacilitator monitors the group during the session, he or she may be the first to observe the countertransference. Once identified, a clinical resolution will be determined during treatment planning and case review between both cofacilitators. The cofacilitators must work as a team, providing peer supervision and constructive feedback to each other throughout the treatment protocol. A determination for specific in-session interventions should be made prior to the next scheduled group meeting.

Trauma work often exposes clinicians to profoundly disturbing stories of human suffering and aggression. Much has been written about the necessity of clinician self-care and the need to prevent "secondary traumatization" (e.g., Figley, 1995) in view of this ongoing stress. Similar to natural disasters, terrorist attacks raise the likelihood that clinicians have been exposed to a traumatic event at the same time and in a similar way to the client, i.e., both the clinician and the client have undergone recent direct traumatization. Both may still feel at risk.

Although relatively little has been written to date about this phenomenon, the demands of conducting trauma work under these circumstances can probably not be overestimated (Saakvitne, 2002). Clinicians are challenged with the task of managing their responses to trauma and fears about immediate prospective safety, while attending to clients' concerns as well. Complicated issues such as the clinicians' possible desires to avoid reminders of trauma and how to discern what is considered appropriate clinician self-disclosure about the event are often confronted. Clinicians must have a safe place to deal with their responses to the trauma caused by terrorist acts as well. Clinicians must develop plans to maximize personal safety so as not to interfere with the therapeutic work with clients. Contact with a support network and collaborating, consulting, or seeking peer supervision with other colleagues can be vital.

Many clinicians find that the "work" of seeing clients actually helps restore a sense of normalcy and worth in a very disrupted, anxiety-provoking environment. Confronted with high levels of uncertainty, a manualized therapeutic intervention can be especially useful in providing a clear, comprehensible, accessible intervention structure for both the clinician and client.

CRITERIA FOR SELECTION OF MEMBERS

Evaluation and Assessment

Individuals must meet three major sets of prerequisites to take part in the intensive trauma work outlined in this chapter: PTSD symptoms resulting from the terrorist event, stable living circumstances, and tolerance of intense affect. Thus, to determine if it is the appropriate time for an individual to be participating in this intervention, all three of these domains must first be assessed. Typically, assessment can be conducted in a single meeting with a clinician, who provides the rationale for the items being assessed, then conducts a clinical interview, a mental status examination, and asks for the completion of the Impact of Events Scale–Revised (Weiss, 1996). The Impact of Events Scale–Revised is a paper-and-pencil measure and can be quickly administered and scored.

Should more extensive documentation of PTSD be required, the "gold-standard" assessment tool is the Clinician's Administered PTSD Scale (CAPS; Blake et al., 1995). This semistructured interview begins by eliciting exposure to a criterion A traumatic event and the requisite diagnostically related responses (e.g., terror, helplessness, horror). Inquiries are then made about the intensity and frequency of the seventeen PTSD symptoms outlined in the DSM-IV (APA, 1994). The manual includes scoring rules to ensure that the diagnosis of PTSD has been confirmed, as well as optional questions to elicit other thoughts and emotions typically found in trauma survivors (e.g., survivor guilt).

Inclusion Criteria

Clinicians will need to make decisions concerning which individuals to include in the group by assessing whether potential participants promote appropriate group function and trust. Groups may be comprised of same-sex members or a mixture of men and women, however, groups should be limited to either all adults or all children so that member feedback is at the appropriate level for all participants. If multiple members of a family were involved in an incident, it is recommended that they be placed into different groups. Family members may have very different views of the event and may in fact have complicated negative attitudes and feelings about other relatives' behaviors during the incident, which can wreak havoc with group dynamics. Individuals must of course be agreeable to participation in a group. Group members are required to have the capacity to participate in treatment objectives and to interact effectively with other members. Members must have the ability to tolerate high levels of distress and anxiety, be accepting of the rationale for exposure-based therapy, and have a willingness to share information concerning traumatic experiences. Group members should have stable living arrangements and be able to meet the schedule of treatment sessions.

Exclusion Criteria

Exclusion criteria include but are not limited to: acute psychosis, homicidal or suicidal tendencies, insufficient command of the common group language, other language difficulties, and heart disease/severe angina that might preclude intensive trauma work. Group members who have a tendency to avoid stressors through use of alcohol or other substances or who have difficulty with affect regulation to the extent that they become aggressive toward themselves or others are generally not good candidates for intensive group trauma treatment. If included in the TFGT, these individuals may require a referral for additional supportive services during participation in the group. Cross-cultural factors must also be considered as relevant to group participants; the inclusion of a cofacilitator familiar with these differences is recommended. Other exclusion criteria include individuals with pending litigation and those that may be seeking compensation.

Additional Selection Considerations

Multiple traumatizations over a lifetime are common (Turner & Lloyd, 1995); it can never be inferred that any traumatic incident is the first in any client's life. Any individual exposed to a traumatic event may have already been diagnosed with PTSD from a prior event. Ascertaining that an individual has had a prior formal diagnosis of PTSD is typically not important in assessing readiness for therapy in a group setting. However, one of the tasks of a clinician's preliminary interview is to determine whether an individual has been exposed to prior traumas and then to endeavor, as much as possible, to link current PTSD symptoms with the most recent terrorist experience.

An individual's exposure to prior traumatic events can be ascertained through questioning (e.g., "Have there been times in your life when you thought your life might be in danger? That you might be hurt or die? Tell me about what happened"). In addition, a questionnaire (e.g., the posttraumatic diagnostic scale; Foa, Cashman, Jaycox, & Perry, 1997) with a section that addresses lifetime exposure to specific traumatic incidents may be used. If exposure to a prior trauma has been confirmed, the clinician should then inquire whether there has been an increase in symptoms since the most recent terrorist event (e.g., nightmares about the current event, more irritability, etc.). This information can be used to help educate the individual with regard to the nature of his or her symptoms.

COMPOSITION AND PREPARATION
FOR ENTRY INTO GROUP

Homogeneous Characteristics of Group Members

Once a diagnosis of PTSD has been confirmed, the clinician has two more tasks in ascertaining appropriateness of an exposure-based intervention. First, the clinician must be sure that the client's basic living needs are being met and that he or she can make a commitment to regular group attendance. Because exposure therapy can be very emotionally demanding and involves doing out-of-session assignments, group members usually obtain optimal results when their living situations are stable during participation in the treatment. At a most basic level, this means the group member has a safe place to reside that affords him or her some privacy and place for downtime. Thus, intervention is best deferred if someone is living in a shelter or in temporary housing. Similarly, it is best if the group member clears his or her "emotional calendar" when embarking on this work. That is, if the group member anticipates an emotionally charged event that may be highly distressing or affect attendance during the duration of the group, it is usually optimal to delay the start of the trauma work until this issue is resolved. This includes surgery or treatment for a serious medical problem, divorce or child custody proceedings, and serious illness or imminent death of a loved one.

Heterogeneous Characteristics of Group Members

Group members vary in a number of ways including degree of psychological mindedness, adaptational strengths, active-passive response style, capacity to be independent, sexual orientation, and presentation of symptoms. Some members may have had experiences with psychotherapy; others may be new to therapy. Although some group members may be flexible and enthusiastic about engaging in treatment, others may be suspicious and rigid. Some members may be extremely passive and require frequent prompting and repetition of instructions to complete tasks. Group members may also present with different constellations of symptoms. Some may respond with severe reexperiencing symptoms, others may be emotionally numb. Still other individuals may engage in extremely disruptive avoidance behaviors. All these symptoms can be managed in group therapy and may be a focus of the coping-skills interventions.

Specific Member Preparation Techniques

Trauma work can sometimes lead to short-term exacerbation of PTSD and depressive symptoms. A fundamental principle of the treatment states that by allowing oneself to experience prolonged exposure to previously avoided images and thoughts during a supportive therapy session, anxiety will first heighten and then eventually extinguish. Group members are instructed with regard to the focus of the group during a discussion prior to individual assessment. The individual will be informed that they will be asked to describe their trauma experience in detail including the thoughts, feelings, and bodily reactions during the event. The ability of each individual to manage stress and symptoms will vary. Each individual will be informed that they will experience a short-term increase of negative affect and symptoms, that this is a normal response but one participants need to tolerate for treatment to be successful (Foy et al., 1997). Often, this reaction continues after the formal treatment session has concluded for the day. The group member may experience residual feelings and thoughts about the therapeutic work. He or she engages in assigned out-of-session tasks such as listening to the therapy audiotape or going to a previously avoided location. Teaching coping skills and other stress management techniques to group members is needed in preparation for the management of heightened anxiety.

Some group members may require concurrent individual therapy for support concerning ongoing significant life events, as these issues are not directly addressed in the group treatment protocol. Initial work to reduce substance use or develop better anger management strategies can be critical in preparing someone for the more intense trauma work. Here, a referral for a medication evaluation can also be invaluable.

GROUP

Structural Considerations

The trauma-focus group, consisting of six group members and two group facilitators, will meet for sixteen sessions and should be closed after the first session. The duration of the sessions is at least ninety minutes with the trauma-focus sessions lasting two hours each. Each session has five different components: check-in, review of home-based therapy assignments, session-specific content, presentation of new home-based therapy assignments, and checkout.

In addition to group exposure work, an important component of TFGT is the use of out-of-session, home-based therapy assignments. Group members must leave each session and take their treatment home to do additional work. Because repetition of scene material is critical for therapeutic gain, it is essential that group members listen to their audiotapes at home and complete the self-exposure worksheets. Other home-based therapy assignments are given each session and include a weekly journal and a symptom record. These are important to the group facilitators to assist with the evaluation of each group member's progress during the treatment protocol.

Development of Working Alliance

Development of a positive working alliance begins when the group members take part in a screening/assessment interview. The rationale behind the protocol, nature of the treatment, roles of the group facilitators, and tasks of group members are described. Each group member is instructed in the need to participate in the group process and the benefits of treatment. Group facilitators should maintain a positive working alliance throughout the treatment protocol. This alliance is strengthened by the presentation of positive coping and stress management skills during Session 2. Group facilitators should also utilize a supportive style whenever possible during each of the sixteen treatment sessions.

BEGINNING PHASE
(SESSIONS 1 THROUGH 3)

During Session 1, each group member introduces himself or herself. The treatment protocol is described, behavior guidelines used during group sessions are discussed, and symptom education is provided. During Session 2, coping skills, deep breathing, and relaxation skills are demonstrated. An outline of the trauma scene to be used during Sessions 4 through 15 is developed during Session 3.

Session 1

Therapist Notes

During this session group members are asked to introduce themselves and give a brief description of their background including marital status, family situation, and employment. Providing personal information will help reduce tension, as this is the first group meeting. The facilitators should monitor the group members' behaviors and affects. Some individuals may experience anxiety and confusion concerning the start of treatment and require emotional support. A discussion of the treatment rationale and an outline of the protocol will also help reduce group members' distress. In addition, the group is informed that they will receive instruction in a number of techniques to manage stress prior to the start of direct therapeutic exposure. The session closes with a discussion of the importance of each group member's involvement with, and completion of the group sessions and the home-based therapy assignments.

Objectives include the following.

- Discuss the rationale and the treatment protocol for the TFGT.
- Introduce the use of the SUDS/distress level scale.
- Discuss the use of home-based therapy assignments.

Procedures

Check-in process. Each session will begin with a brief check-in wherein group members answer the question, "Are you ready to work today?" It will be important for the group members to put aside other concerns and concentrate on what's going on in the group. The check-in process is designed to help group members learn to cope with their worries and emotions so that they do not interfere with daily life or work in the group.

SUDS/distress level. The subjective units of distress (SUDS) scale is used to monitor anxiety and stress as part of the check-in process and throughout the sessions. Use of SUDS helps the group members improve self-control by learning to monitor and assess emotional distress in the moment.

Check-out. At the end of each session, members are asked how they plan to cope during the week ahead. A SUDS probe is also done at this time to monitor their condition again via the 0-to-10 scale. The facilitators may also ask, "Are you okay to go home today?" The importance of this response is emphasized to each group member at this time and group members are asked to be open and honest in their responses. This becomes increasingly critical during the later stages of treatment.

Home-based therapy. Weekly home-based therapy assignments will include the completion of a journal page by each group member so that they can record their reactions during the week. This helps group facilitators evaluate each individual's status during the course of therapy. It also helps group members to process their own gains as treatment progresses and to continue the session work at home. Additional session-specific assignments are also given each week. These may include practicing the deep breathing, relaxation, and other coping skills taught during the next group treatment session. The following assignments are distributed at the end of Session 1: complete a journal page (Appendix I), complete the Coping Resources Self-Assessment (Appendix II), fill out the Checklist List of Coping Tools (Exhibit 19.1), and fill out the Negative Coping Checklist (Exhibit 19.2).

EXHIBIT 19.1. Checklist of Coping Tools

Make a check next to any of the following coping tools you have used successfully in the past few months.

_____ Support seeking
    Visit a friend.
    Talk with a supportive family member.
_____ Relaxation exercises
    Deep breathing
    Relaxing imagery
_____ Time-out
    Walk away. Take a deep breath and calm down.
_____ Journal
    Write about the situation and your feelings.
_____ Self-talk
    Be positive. Remind yourself of what you've accomplished.
    Tell yourself that you are a good person.
_____ Regular exercise
    Walk, swim, bike, go to the gym, etc.
_____ Consistent daily routines
    Get up at same time and prepare for sleep at same time.
_____ Thought stopping
    Utilize a "rubber-band snap."
    Visualize a STOP sign.
_____ Self-reward
    Pat yourself on the back. Buy yourself something healthy.
    Treat yourself to something positive.
_____ Distraction through positive activities
    Play a sport, go fishing, go to dinner with a friend, etc.
_____ Support group attendance
    Go to a meeting
_____ Other:

(*Source:* Adapted from Foy et al., 1997)

EXHIBIT 19.2. Negative Coping Checklist

Place a check next to the negative coping behaviors you have used in the past. Circle the ones you are still using. Fill in the blanks with ways of coping you have used that are not listed here.

_____ Isolating _____

_____ Using drugs or alcohol _____

_____ Fighting _____

_____ Quitting jobs _____

_____ Taking risks (e.g., driving too fast) _____

_____ Spending excessive time on the computer _____

_____ Working all the time _____

_____ Watching too much television _____

_____ Spending time on busy work _____

_____ Thinking about suicide _____

_____ Thinking about homicide _____

_____ Driving aimlessly _____

_____ Overeating _____

_____ Other _____

(*Source:* Adapted from Foy et al., 1997)

Session 2

Therapist Notes

During this session the group members are asked to describe their symptoms and how their lives have been changed since the traumatic event, with particular emphasis being placed on the symptoms that are most troubling to each group member. This information is useful to the group facilitators as therapy progresses since each individual has his or her own understanding and experience of PTSD. This discussion allows group facilitators to provide accurate information about the disorder and enables facilitators to predict the reactions of group members during the course of therapy.

Objectives

- Review current symptoms with each group member.
- Discuss the nature of PTSD and provide accurate information about the disorder.
- Discuss the use of coping skills (deep breathing, relaxation, thought stopping) to manage symptoms.

Procedures

Check-in. Each group member is asked to rate how they currently feel on the 0-to-10 scale presented during Session 1. The follow-up question, "Are you ready to work today?" is also asked.

Rationale. Group members are instructed that their present inappropriate behaviors are associated with their high level of distress. During the session several coping techniques will be presented to help them manage stress and deal with their symptoms. These new techniques are in fact skills and will require practice.

Coping Resources Form (Appendix II)

Review responses to Coping Resources form. Spark discussion with group members about strengths, items identified as good, and weaknesses, which are the items marked as poor.

Checklist of Coping Tools and Negative Coping Checklist
(Exhibit 19.1 and Exhibit 19.2)

Review the items from the Negative Coping Checklist. Discuss the consequences of these behaviors. Move to a discussion of how to implement positive coping tools. Focus on the importance of social support noting the difficulty with attempting to manage distress and internal pressure in isolation as opposed to venting off some of the tension by sharing concerns with a significant other.

Deep-Breathing Exercise

Explain to group members that improper breathing decreases the flow of oxygen to the body. This makes coping with stress more difficult and may enhance anxiety, panic attacks, and other negative symptoms. Normalizing breathing will facilitate relaxation and reduce stress.

During this exercise, group members remain seated and are asked to close their eyes. This may be difficult for certain group members, but they should be encouraged to close their eyes, as this will enhance the efficacy of the technique. The focus of the practice is inward and closing the eyes will facilitate the relaxation process by reducing visual distractions.

The facilitators instruct members to take long, slow, deep breaths using their diaphragms. Instruct members to sit up in their chairs, feet on the floor facing front, chin up, arms at their sides in a comfortable position, and taking long slow breaths expand their diaphragms. Instructions are given not to hold their breath but exhale slowly and begin the process again.

Relaxation Training

The group is presented with a relaxation technique that can be used along with the deep-breathing procedure. A tape of the in-session practice may be made for the members to review and practice at home. Other forms of relaxation may be discussed including meditation, progressive muscle relaxation, and music tapes.

Case Example: Relaxation Techniques

F: We are going to do a brief relaxation exercise. Before we begin I would like to get a SUDS level from each of you. How are you feeling right now on the zero-to-ten scale?

We would like you all to begin by closing your eyes. Now find a comfortable position with your feet on the floor and your arms resting gently at your sides. Now take a few long deep breaths using the deep-breathing skills we discussed.

GM 1: This feels strange. I don't like closing my eyes.

F: You may feel a little uncomfortable as we begin, but it will help your inward focus if you close your eyes. You want to reduce as many outside distractions as possible.

GM 2: I've heard about this stuff. You ask us to imagine that we're someplace nice. It sounds like you are trying to hypnotize us.

F: This is not hypnosis. Both techniques do use relaxation and visual imagery, but you are in control at all times. I will be giving you instructions on how to relax more effectively. I will not be making suggestions concerning your actions or thoughts.

GM 2: Okay, I'll close my eyes. Let's get started. This is making me feel nervous.

F: Make yourself comfortable and close your eyes. Take a long deep breath without holding it, and exhale. Now take another breath and exhale. Continue with long, slow, deep breaths.

We want you now to focus on your breathing and feel the tension in your chest with each breath as you inhale. Now focus on the sense of relaxation as you release the pressure and exhale, and feel some of the pressure leave your body with each breath as you exhale. You feel more relaxed with each breath; let a little bit more of the tension go as you breathe in and out.

Now bring your attention to the muscles of your body and relax your muscles. Begin with a focus only on the muscles of your fingers and hands. With each breath, release the tension, letting all the tension go. Feel the stress and tension slowly leaving your body. A warm tingling sensation of relaxation begins to grow.

Now focus on the muscles of your arms, the forearms, and upper arms. Again with each breath continue to relax your muscles. You feel calmer and more at ease with each breath. You feel the warm, soothing sensation of relaxation slowly moving from your fingers and hands and moving into you forearms and upper arms. You feel calmer and more relaxed.

Now focus on the muscles of your shoulders and neck. Release the tension and let the muscles relax. The warm, soothing sense of relaxation is even a little stronger now, you feel calmer, tranquil, and relaxed.

Now focus on the muscles of your head and face. You feel some tension in the muscles of the forehead, around the eyes, the mouth, and jaw. Release the tension and relax the muscles. Feel the soothing and warming sensation of relaxation moving up the muscles of your fingers, hands, arms, shoulders, neck, and into the muscles of the head and face. You feel calmer and more relaxed.

Now focus on the muscles of the chest and stomach. You feel some tension; with each breath the stress and pressure leaves your body as you exhale. As you exhale, say to yourself . . . relax. . . . The tension leaves your body. You feel the warm sensation of relaxation growing deeper and stronger. The muscles become limp and gently droop; you are supported by the chair. You feel calmer and more relaxed.

Now focus on the muscles of your legs. You notice some tension in the muscles of the thighs, calves, and feet. Release the tension and relax, letting all the tension drain away. It gently drains away leaving the warm, soothing sense of relaxation even a little bit stronger now. Feel the muscles gently sag and droop. The tension and pressure leave the muscles of your legs, thighs, calves, and feet; you feel calm, tranquil, and more relaxed.

I would like each of you to now give me a SUDS level. How are you feeling right now?

Okay, continue to relax. Focus on all the muscles of your body. You find that some tension remains. Focus on the tension in your muscles and relax it away. Release the tension, let all the tightness go, it just drifts away with each breath you take. You feel calmer and more relaxed. The warm, soothing sense of relaxation becomes deeper, gently warming the muscles; you feel tranquil, calm, peaceful, and more relaxed.

Now, I want you to stay with that feeling for the next minute. Continue to relax your muscles . . .

I am going to count slowly from one to five. As I count from one to five you will gradually become more alert and awake. When I count five you may open your eyes. You will feel calmer and more relaxed.

One, gently rousing yourself. . . . Two, slightly more alert and awake. . . . Three, slowly bringing yourself back. . . . Four, almost ready to open your eyes. Five, you may open your eyes when you are ready.

How do you feel? Please give me a SUDS level, where are you right now?

Thought-Stopping Technique

Members will develop a technique to interrupt recurring traumatic thoughts. To employ this technique, when they identify recurrent negative ideation, they may silently shout, "stop" to themselves and replace the thoughts with positive coping statements or thoughts. The group members may say, "If I just take it moment by moment I can get through this," or "Everyone makes mistakes, I can learn and grow from this." Group members may also use a rubber band placed around a wrist and snap it as they say silently to themselves, "stop." A strong scent such as ammonia, lemon, pine, or wintergreen may also be used as an intervention to interrupt recurrent negative self-statements or ideation.

Positive Coping

The group members are asked to identify other positive coping behaviors. The group and the facilitators may assist group members having difficulty identifying personal activities or items from using list of coping tools provided.

Coping with Symptoms

The "Coping with Symptoms Record" (Appendix III) is presented and explained. Group members are instructed to record their symptoms during the week along with where and when the symptoms occurred. The negative thoughts associated with the symptoms are listed. Group members identify positive actions, thoughts, and coping behaviors for each situation listed on the form.

Checkout

Group members will also be asked how they plan to cope during the week. A SUDS probe is done again at this time to monitor their condition via the 0-to-10 scale. The facilitators may also ask, "Are you okay to go home today?" and "What did you learn today?" The importance of this response is stressed with each group member at this time. Members are asked to be open and honest in their response.

Case Example: Coping Skills

F: Today we would like to share with you some techniques to manage stress. It will be important to have a plan for coping with tough days and this plan should include social support from family and close friends.

GM 1: What do you mean by social support?

F: It will be important not to isolate yourself from people. Isolation is a symptom of PTSD. You will need to identify people that you can go to when the stress begins to build. People that care for you can help but you will need to ask and let them know what they can do for you.

GM 2: When I feel bad I need to be alone. I can't talk to anyone about it. If I stay in my den I will eventually feel better.

GM 3: No one can help when I am feeling bad. No one can understand. No one can understand unless they were there.

GM 1: I don't want to make my family feel bad. If I talk to them about it they just get upset. My wife sat next to me and I told her what I was thinking about and she started to cry. I had to comfort her. There is no way I could do that to her again.

F: Do you think she doesn't know if you are having a tough time?

GM 1: No, of course she does. We have been married for over twenty years. She knows me and can tell if I am upset. She wants to help but there is nothing she can do about it.

F: So she can tell if you are upset. But you don't let her in on what is happening. She wants to help but you don't let her. How do you think she feels knowing that you are hurting but you will not let her help you?

GM 1: But she can't help me with it, I see it in my head all the time. There is nothing she can do.

F: You are right. She cannot change what happened or take that away, but she can still help you. It depends on the type of help you ask of her. If you are having a rough night with nightmares, when you wake up it is not necessary to go into detail about the dream or what happened to you. You could simply explain that you had a bad dream and if she wanted to help she could sit up with you for a while and talk. There are ways she can support you and help you cope with tough days. This will be good for you and she will feel better because she was able to help, and that will be good for you both.

GM 1: It is hard for me and I just want to be alone.

F: Being alone is different than isolating. Taking time alone for yourself can be a positive means of coping. People can use deep breathing, meditation, or relaxation while alone. This is a way of decreasing stress. When someone isolates himself or herself from others—and you can isolate yourself in a room full of people—emotional pressure and stress can build. People have difficulty turning off recurrent negative thoughts and negative self-statements.

GM 2: That's exactly how I feel. I sit there for hours by myself thinking about what happened and what I did that day. I feel terrible—a lot of guilt and shame about my behavior. I feel like I am going to explode. I don't want anyone around to see that.

F: That is one of the problems with isolation. The pressure builds because you try to hold it in and do not let anyone know about it. But as you said, your spouse can tell if you are hurting. You need a safe place to release or vent off some of the stress.

GM 2: My husband looks at me sitting alone in the living room at night and he just walks away. We don't talk about it.

F: Perhaps this example will help. When people don't have a place to vent their feelings, they become like a pressure cooker without a valve. As long as there is stress or a fire underneath the pressure continues to build. What will eventually happen if there is no vent for the pressure?

GM 1: The pot will explode.

F: That's right. The strength of the steel does not matter. The pot will eventually explode. So strength is not a factor. Even steel will crack and break. You don't know when that will happen, but eventually as you said, the pot will explode.

GM 2: That's what I am afraid of, I don't what to explode. That's why I'm here in this group. I need some help. This has been going on too long.

F: These skills can help manage the stress. When the valve is open the pressure cooker steams along. The fire or stress is still there adding pressure, but with the valve open the pressure is controlled. If the valve closes the pressure begins to build. Someone could even look at the pot and not notice that the valve was closed and think that the pot was doing fine. But if the valve stays closed it will explode.

GM1: That's a fancy way of telling me I have to do this stuff.

F: Yes, because with the valve open, despite the constant pressure, the stress can be managed. There is a big difference between the pot steaming along or exploding.

GM 2: It doesn't change what happened. It won't take away the memories of that day.

F: Therapy will not take the memories away. However, the level of stress and pain can be changed. Part of the answer is to have a plan and a set of skills to help you on the difficult days.

GM 1: Every day is difficult right now. I don't want to get out of bed. I don't want to leave the house. I don't want to go to work. That's not me. I was always active and enjoyed going to work. I liked my job. Now nothing seems important. Nothing matters.

F: These changes happen as people attempt to cope with stress on their own. Individuals develop different behavior patterns and isolation is often a key component. Having a group of skills to use to help cope is part of the answer. Seeking social support is part of the answer. Deciding to come to group is part of the answer. Group is a place to share your feelings and thoughts with other people that share your experience.

GM 1: This feels like the only place that is safe for me to open up. Other people know what happened but they don't understand. I can't talk to them.

F: You can begin the process here. Part of the answer is also doing the trauma-focused work at home, by going over your experiences that day.

GM 3: When I try to talk about it, people just look sad and tell me to forget it. People say you made it, you're okay, it happened months ago, be happy you are alive and get on with your life.

F: The important people in your life want to help but they don't know how. You will have to talk to them and let them know what they can do to help. Family members or friends cannot take the memories away. But they can sit with you, keep you company, take a walk with you, go for a drive or whatever else you feel will help you manage the stress. You will need to seek their support rather than bottling it up inside.

Session 3

Therapist Notes

This session is the first to focus on traumatic memories. Each of the group members is required to outline his or her traumatic experience. It will be very important to conduct a SUDS level for group members after they present their outline. One of the group facilitators should lead the outline of the incident being presented. The other group facilitator should observe the remainder of the group and also document the key components of the outline. Documenting the outline will be a useful tool to help guide the group members during the trauma-focused treatment session. The reactions of the group during the outline presentation may be powerful. Members are urged to remain in the session. Some may have heightened anxiety or panic attack symptoms during this session and wish to leave the room.

Each group member is asked successively to present a brief, ten-minute outline of the event. It may be difficult to limit the group members in their outline. Some may give a full-blown version with all the associated affect, details, and thought processes. It may be necessary to explain that the exposure work will begin at the next session.

Group members may have several memories of different events that occurred during the incident. You may need to help them focus on the portions of the incidents that are most disturbing to them. These typically will be the memories that have the strongest negative emotions, the parts of the event that cause the most distressing nightmares, or the memories that they work hardest at avoiding.

Each member may briefly present two or three traumatic events. If the individual was involved in a single event, different aspects of the event may be used to help the group member focus on the most difficult portion of the experience. For some, hearing the experiences of the other group members may reactivate images and memories of their own traumas. Group therapists should be supportive of the group members but also help them to keep the descriptions brief so that the exercise is completed in the allotted time.

In addition, the group members have the opportunity to utilize the newly acquired skills from Session 2 to manage stress and memories triggered by the scene material. At this point, review the rationale with the group members for the trauma-focused work.

Group members are asked to stop avoiding the memories, since this has been one of the mechanisms that has prevented healing. Memories are connected to many aspects of our everyday lives and common stimuli encountered during the day may become painful reminders of the traumatic experience. The most common reaction to and the most popular advice given are to avoid these stimuli. These reactions may turn into a habitual response and continue until individuals feel trapped by the world around them. Some individuals may feel like running away from everything they know including their families, friends, and jobs. However, they cannot avoid the real sources of their pain as they carry them wherever they go; they are part of their memories and experiences.

The trauma-focused group treatment utilizes exposure to these painful memories. Group members reexperience the traumatic events in a safe environment; a place where they can express their feelings and thoughts. Members learn that they can reduce the intensity of their pain so that it will be less distressing in the future. This enables them to reexamine their thoughts and begin to see the events in a different way. Although the memories will always be sad and uncomfortable to think about, the toxicity and intense pain experienced with the memories will be changed. The future will be less fearful and they will no longer feel trapped by the world around them.

Objectives

- Present the rationale behind the session protocol and help group members understand the scene selection process.
- Each group member presents a brief outline of his or her traumatic event.
- Each member must identify a single event on which to work.
- Elicit SUDS ratings before and after the scene outline.
- Identify coping skills to be used to manage distress.

Procedures

Check-in. At the beginning of Session 3, the group facilitators should conduct check-ins with each group member using the 0-to-10 SUDS scale and collect and review home-based therapy assignments.

Rationale. Present the rationale for trauma-scene selection to the group members, emphasizing that they must deal with painful memories in order for healing to take place. Review the role of avoidance of the traumatic memories as a symptom; avoidance prevents healing and time does *not* heal all wounds.

Scene selection. Each member will briefly describe his or her traumatic experiences. Since some members will have more than one memory, group facilitators will assist individuals in the group with scene selection. Although some members may be reluctant to identify a memory, others may need some assistance in keeping the outline of the incident brief or may feel overwhelmed by anxiety. A fear of being viewed negatively due to their actions during the traumatic incident may also be present.

Case Example: Scene Selection

F: Today we are going to select a scene to be used during the next twelve weeks. The scene should be of the traumatic event. It may be that you have more than one incident or memory. We will help guide you through this outline.

GM 1: I don't know where to start. There was so much happening that day and a lot of it is a blur. I didn't want to think about it. It was too painful.

F: All we want today is a brief outline of the event so we can help you stay focused when it comes time to present it during the trauma-focus sessions. For this particular exercise, you should give just the facts. No elaborate details. We will work with each of you for about ten minutes.

GM 2: I can see it already. I don't like this. I'm getting nervous.

GM 3: Let me go first I just want to get it over with and get out of here.

F: We will help each of you with your outline. We may ask you to pull back and keep it brief because we will not be going into great detail about emotions or thoughts during the incident.

GM 2: I can feel my heart beating faster.

F: Today is also about using the skills we discussed to manage your stress. If you are experiencing anxiety, this would be a good time to use deep breathing or relaxation. Use your skills to cope with the anxiety symptoms.

GM 1: Let's do this. That's why I'm here.

F: Okay, give me a SUDS rating. How do you feel right now?

GM 1: I'm at a seven.

F: Briefly outline your scene.

GM 1: I'm standing on line waiting to get in to see the Liberty Bell. There are a lot of people there that day. I'm with my family and there is a couple trying to decide whether to wait on line or go over to Independence Hall. They can't agree and start talking about all the things they want to do, and they get into an argument.

F: Okay, stay with the outline. You are waiting on line with your family. What happens next?

GM 1: I'm on line and it's moving pretty fast so we're all in a good mood. I'm almost in the building and all of a sudden there is an explosion and people are screaming. I don't know what happened. The explosion was to my left at the corner of the building. People are crying, some people are running around, I see guards with rifles coming. I'm all confused; time seems like it's standing still.

F: You hear the explosion. What happens next?

GM 1: I see people dressed in black. They have guns and start running. There is another explosion in the same area and they start shooting at the guards. The guards shoot back. I get on the ground and I hear gunfire and yelling. I don't know what's going on. I'm scared, confused, it seems like forever, things are moving slowly, all distorted.

F: You are on the ground, and you hear gunfire. What happens next?

GM 1: I hear the gunfire moving off. It gets farther away. Everything gets quiet. (Group member pauses, closes his eyes, and puts his head down.)

F: It's okay, stay with it. You are on the ground and it gets quiet. What happens next?

GM 1: I hear people crying. I hear people starting to move around. I look at my wife and my family. I didn't even think about them when this was happening. I look at my family and they're scared and shaking. I'm shaking.

F: Okay, please stop there. Take a few deep breaths. How are you doing right now? Give me a SUDS rating.

GM 1: A nine. I can see it all. There's more. I saw other things, things that happened to other people.

F: I want you to stop there. Again, take a few deep breaths. Put your feet on the floor; sit back with your hands at your sides. Relax. That's right. Just keep breathing, long, slow, deep breaths.

GM 1: I'm all right.

F: You did fine; give me another SUDS rating.

GM 1: Seven.

F: Very good, continue the deep breathing and relaxation. Let me check in on the rest of the group. Where are you on the SUDS scale?

At the end of each brief outline, group facilitators should have each group member give a SUDS rating. The group members may discuss their feelings and thoughts about the content of the session. Before ending the session, review coping skills and a plan for personal care for the remainder of the day and during the week until the next group session. Give home-based therapy assignments.

- Write in the focus group journal
- Practice coping skills
- Use the coping with symptoms record forms

Checkout. At the end of the session, conduct a checkout and have each member rate how he or she feels on the SUDS Scale. Ask each member, "Have you learned anything today?"

MIDDLE PHASE

During these twelve weeks, one session will be dedicated to each group member's individual traumatic experience. The session duration will increase to two hours to allow for prolonged therapeutic exposure and for corrective feedback from the other group members. Each group member will take a turn for the first six weeks and repeat the scene during the next six weeks. During the second scene presentation, the group facilitators may focus the group members on additional or newly revealed memories of the traumatic experience. The group member may report recalling previously forgotten portions of the event with somewhat different emotional and cognitive valence. Newly revealed memories should be carefully evaluated by the group facilitators and incorporated into the themes of the second scene presentation.

At the start of each session, group members should be informed as to whom will be asked to present their scene. The order of presentation may be altered during the second twelve weeks. For example, if Group Member 3 presents during Session 1, rather than presenting again at session 7 Group Member 3 may not present again until Session 13. The two group facilitators determine the order of presentation each week prior to the session. Group members are not told in advance when they will be asked to take their turn thus preventing the possible disruption of the treatment protocol by the absence of a group member on the day he or she is scheduled to present his or her scene.

Sessions 4 Through 15

Therapist Notes

Group facilitators should test tape recorders and place the microphones and other equipment in a way that will allow for a clean and clear recording of the entire session. If a tape is inaudible, group members should use the self-exposure form as their assignment. This may be done until another tape can be made during the second round of the trauma-focus treatment.

Each group member takes one session during the first round of scene presentations to provide a detailed description of his or her event with an emphasis on using the present tense. The use of the present tense enhances the emotional intensity of the scene. It will be natural for the presenter to use the past tense. The group facilitator leading the exercise should prompt the group member to use the present tense while repeating the details the member is providing. The facilitators should give the following instructions:

> Today we would like you to describe in detail the event you presented in your outline. We want you to focus on the memories that are the most painful to you. It will be difficult, but do not avoid talking about the events that embarrass you, make you feel guilty, or make you feel ashamed. We will be here to guide and support you through this process.
>
> Put yourself in the scene as if it were happening again. Tell us what you see, hear, and feel. We want to know your thoughts and emotions. I will be asking you questions to help you stay with your emotions. This is the time to recall your painful feelings to the best of your ability. Are you ready to start? Start at the beginning and tell us what happened. You have most of the session to do this today. Please begin.

The group member should be allowed to use his or her own words to describe the event. The group facilitators should provide prompts using the terms and descriptions as presented by the group member. It may be necessary to assist the presenter in enhancing the description and content of the memory with prompts. Although the facilitators should focus on questions concerning the individual's emotions and thoughts, asking questions concerning the physical details of the individual's surroundings may be necessary. If the individual pauses, the facilitator may repeat a brief review of the scene content to the point at which the group member stopped. A question can then be used to prompt the individual to continue with the memory. Some useful prompts include:

- Look around you, what do you see?
- What are you thinking?
- What are you feeling?
- How is your body reacting?
- What are you remembering?
- What happens next?

Because the group member may briefly describe or skip portions of the event that are critical components of the memory, the facilitators should assess the group member's presentation for avoidance of key aspects of the memory. Facilitators should probe with questions in the following areas:

- Bodily reactions (trembling, nausea, dizziness, etc.)
- Anxiety, fear, or panic
- Anger, sadness, or depression
- Emotions and thoughts
- Guilt and shame
- Morality and betrayal
- Consequences or outcomes
- Perceptions or judgment by others
- Self-image

Case Example: Scene Presentation

F: I would like you to present your scene today.

GM 1: I knew it. I knew that you were going to start with me. I almost stayed home today.

F: I'm glad you decided to come. It was a tough choice but the right one. I would like you to describe the scene again as you did in your outline, only this time in detail, including your thoughts, feelings, actions, and reactions to the event.

We want you to put yourself back in the scene as if it were happening. We want you to tell us what you see, hear, feel, and think. I will be asking you questions to help guide you to focus on the difficult parts of the experience. I will also be asking you how you are feeling and to give me a

SUDS level from time to time. This is your chance to tell your story, to open up and cleanse the wound. We are here to help. We are listening. Are you ready?

GM 1: Yeah, let's get this over with.

F: You started the scene with standing in line waiting to see the Liberty Bell. It was a clear day. You were with your family. Who was there? Tell me what you see.

GM 1: I'm in line. I'm with my wife and my two kids. We live in Philly but never took the time to go in and see the Bell. We drove by lots of times and have been in Independence Hall, but there was always a line for the Bell and we always left. It's hard to find parking, and it always seemed like we were busy and had other things to do.

F: Okay, you're in line with your wife and kids. How do they look, what are they wearing, how old are they? Look at their faces, what is their mood?

GM 1: My son Mark is fifteen and Jennifer is twelve. My wife would get mad if I told you her age but she is a few years younger than me. We are all wearing shorts, T-shirts, and sneakers. We are all smiling and joking because we got lucky and there was a short line and it was moving fast. I say I picked a good day to come down.

F: You are on line with your family. What happens next?

GM 1: The line is moving fast. A couple comes by and argues about whether to wait on line or go into Independence Hall. We are trying to ignore them. I smile at my wife because we had a similar argument about coming today because the lines are always long.

We are just about to go in and we hear an explosion. I fall over and stay down. I was all confused. I didn't know what was happening . . . (pause)

F: Keep yourself in the scene like it is happening again, you hear an explosion, you fall over, you are confused, you don't know what is happening. What happens next?

GM 1: I smell smoke. I'm still lying down. I hear people yelling and screaming. I see someone dressed in black. He has a gun. He shoots at someone. I can't tell who he shot at but I hear people scream. He yells something I can't understand. I see the guards coming and he shoots at the guards. He starts to run. There is another explosion to my left near where the guards are, I feel the vibration, I feel the shock, I smell the smoke, and the air is black. I have a metallic taste in my mouth.

F: What are you thinking?

GM 1: I'm thinking what is going on? Why are they doing this? (pause) I'm gonna die. (pause)

F: You think you are going to die. Focus on that thought and stay with it. What are you feeling?

GM 1: I'm scared.

F: Where are you right now on the zero-to-ten scale?

GM 1: What?

F: Where are you right now on the zero-to-ten scale we have been using at the start of each session?

GM 1: I'm a ten.

F: Okay, you are on the ground, you see a man with a gun and you hear another explosion. You smell the smoke. You have a metallic taste in your mouth. Look at the man with the gun. What do you see?

GM 1: He is dressed in black. He has a gun and is pointing it. He shoots, I hear it go off and I see the flash.

F: How far away is he? What does he look like?

GM 1: He's about forty feet away on my left. He has dark hair, a dark beard, and mustache. He looks about thirty-five. He looks tall. He's yelling something and waving his hand. I hear someone yell back. I see the guards and they yell something. He yells back and points his gun at someone on the ground. He starts to run away. I see another man in black run by. The guards follow. I hear another explosion to my left. It seems like it is across the street.

F: What are you thinking?

GM 1: These guys are terrorists and wanted to destroy the Liberty Bell. Explosions go off and I'm thinking if I move I could die. I hear shooting and see that they were shooting at people. I don't know how many there are. Maybe there's one behind me. I'm afraid to move. I'm afraid to look. I just want it to stop. I just want it to stop.

F: What happens next?

GM 1: I hear some more gunfire. I hear another explosion. It's farther away. It gets quiet and then I hear people crying. I hear people talking. I hear someone say, "Please everyone stay down until we are sure it's safe. Help is coming." I hear sirens, a lot of sirens. People are getting up. I turn my head. I see my wife. I see my family. I forgot all about them. I see my wife looking at me. She has the children with her. She has her arms around them and they are crying. My children were crying and I didn't even hear them. I didn't even think about them or my wife. She is holding them close to her. I look into her eyes and I have to look away. I can't take the way she looks at me.

F: Look at your wife. Look at your children. Look at her face and into her eyes. What do you see, what are you thinking?

GM 1: I see confusion. I see fear. I feel ashamed. I didn't even think about them. I was so afraid I was only thinking about myself. It's my fault. I wanted to come here today. I decided that we had to go. My family didn't want to come. We all could have died. It's all my fault. I see it in my wife's eyes. I'm not a man. I'm supposed to take care of them, to protect them. All I thought about was myself. I'm a coward. That was my little girl's voice calling, "Daddy." I didn't even know. It was like I was in a fog.

F: Focus on your emotions. What do you feel?

GM 1: Shame. I could hear my little girl's voice calling me and I was so afraid I couldn't move. I couldn't turn my head. I thought that they were shooting people and if I moved they would shoot me. He looked right at me. I saw his face. I saw his eyes. His eyes were crazy. He looked crazy. I thought that he would kill me next.

F: That's right. He is looking right at you. He looks crazy. You see his face. You see it in his eyes. He wants to kill you. Stay with that feeling. (Brief pause.)

Give me a SUDS rating. How do you feel on the zero-to-ten scale?

GM 1: A ten. I feel sick to my stomach. I think I'm going to throw up.

F: Stay in the scene. You see his face and eyes. He is looking right at you. He has a gun. He is shooting people and he's looking right at you. You say to yourself, 'I'm next,' stay with that. (pause)

Okay, I want you to go back to the beginning of the scene. You are standing on line with your family. It's a beautiful day. The line is short to see the Bell and is moving along quickly.

What do you see? What happens next?

GM 1: No, not again. I can't do this again.

F: Keep yourself in the scene. It's an image, a memory. You can do this. You are on line with your family. You are waiting to see the Liberty Bell. What do you see?

GM 1: It's a nice day, sunny and warm. The line is moving fast. Some people are arguing about getting on line.

F: That's right, and then you hear an explosion and you fall to the ground. What happens next?

GM 1: I hear people scream. I hear people cry. I smell smoke. I don't know what is happening. I'm all confused. I see a man dressed in black. He has a gun. He's looking around and is shooting. I hear another explosion. He yells something I don't understand, some other language. He shoots again. He is shooting at people moving, trying to run away. I see a man run and he shoots at him and the man screams and falls. He keeps looking. He looks right at me.

F: What are you thinking?

GM 1: I'm next. If I move he will kill me. I see his eyes. He's crazy. He looks at me and he points the gun right at me. He points the gun right at me. He's going to shoot me. I'm dead.

F: Focus on your body. He looks right at you and he points the gun right at you. See his face. See his eyes. He looks crazy. He could do anything. See that happening. He points the gun right at you. What do you feel? What's going on in your body?

GM 1: I'm shaking. I feel sick, panic. I want to hide, to run. I feel helpless; there is nothing I can do. I see him point the gun at my head. He looks crazy, insane. I hear more screaming and gunfire. I feel all wet, my body can't take it, and it lets go. I think I pass out, but I still hear yelling and the guards come.

   The man runs away and I see another man dressed in black run past me. I see the guards run after them. I hear another explosion. I hear a girl's voice calling, "Daddy." It seems far away like it's coming from somewhere else.

F: You see him point the gun at your head. You think, I'm gonna die. Feel the panic, feel your body let go; you feel it's wet. Feel that and stay with that. (pause)

   Give me a SUDS rating from zero to ten. Where are you now?

GM 1: Ten, I'm at a ten. He points the gun at my head, I see his eyes and I hear a gun shot. I soil myself. I think I'm hit. I think I'm hit.

F: How do you feel?

GM 1: Ashamed. I was so frightened. I thought I was going to die.

F: Now you hear a girl's voice calling, "Daddy." It seems like it's far away. But you recognize the voice. Whose voice is it?

GM 1: It's my little girl. I hear her voice calling me. She's crying. She wants to go home. She wants me to take her home. She's begging me, sobbing and crying. I hear her calling, "Daddy."

   I feel myself coming back. I realize that my family is there with me. I wasn't thinking about them. I was only thinking about myself. I was hiding.

I look at my wife. I see her eyes. I see my children. I see their faces and they are crying, scared, and lying next to my wife. She has an arm around each one of them.

She is holding them close to her.

I see tears in her eyes. She looks at me all confused. She's right next to me. She was saying something but I couldn't understand it.

F: What does she say? Focus, you hear her voice. What does she say to you?

GM 1: She asks me if I'm all right. She's worried about me. She thinks I am hurt.

F: What are you thinking?

GM 1: I'm thinking she's worried about me. She tells me that they were talking to me but I wasn't answering. I'm thinking I forgot all about them and she was worried about me.

F: What are you feeling?

GM 1: Shame, fear, and guilt. It was my fault. I wanted to go to see the Bell. I dragged everyone else there. My family didn't want to go. Everyone had other plans. I insisted that we go. It was my fault. We all could have been killed and it would be my fault. I didn't even think about them, I didn't try to protect them. I was so scared that I wet myself. All I cared about was hiding. I was hiding.

F: See that again. You hear the explosion and you are on the ground. The man points the gun at your head. What happened?

GM 1: I fell to the ground. I don't know what is happening. I see a man with a gun, he is pointing it at people and shooting. I am next to a turnstile. I hide behind it. It's made of stone and concrete. I can see what's happening.

The man looks around. I hear another explosion. He turns and looks at me. He looks right into my eyes. He points the gun at me. I think I'm going to die.

F: You are hiding behind the turnstile. The man turns and looks at you. He points the gun at your head. See that. Stay with that image. (pause)

What happens next?

GM 1: I think I am going to die. I panic. It's like electricity through my body. I think I'm going to pass out; I wet myself. My pants are wet. It feels warm.

I feel like it's the end, the end of everything.

F: What happens next?

GM 1: He runs away and another man runs after him. Everyone is running away.

F: Now focus on your daughter's voice. You hear her calling you. She is crying. She was calling you and you didn't even hear her. Your wife was calling you. You didn't hear her. Your son was crying. All your family was there.

What are you thinking?

GM 1: I am hiding behind the turnstile. I am safe. My family is out in the open. My wife protects them. She covers them with her arms. She holds them close. She keeps them with her so they don't run away. If they move they would be shot. I forget them. I don't do anything to help them. I don't protect them. My wife takes care of them. I'm worthless, a coward. I wet myself I am so scared. My wife is more of a man than I am.

F: What are you feeling?

GM 1: Ashamed, I hate myself. I see my wife's eyes. She's afraid I'm hurt. I'm having a panic attack when she needs me and I forget her. I hide. I take care of myself. I don't think about anyone else.

F: What else are you feeling? Look into her eyes. Hear your children call you. See their faces.

GM 1: Guilt, guilt and anger, I feel anger. I feel anger at myself for not protecting them, anger at the man with the gun. I want to kill him. I want to kill all of them. If I had a gun I would shoot them. I would shoot them all for what they did to me and to my family.

F: Once more, I want you to hear the explosion. You fall down and hide behind the turnstile. You see a man dressed in black with a gun. You hear people scream. He shoots at someone and looks around. He looks right at you.

What happens?

GM 1: Not again.

F: Keep yourself in the scene. Once more, what happens?

GM 1: He looks at me. He looks right into my eyes. He looks crazy. He points the gun at my head and holds it there pointing right at me. I hear him yell something. Someone screams. I think he's going to shoot me in the head. I think I am going to die. I'm scared, shaking; everything is moving in slow motion. It seems like forever that he has the gun on me.

F: Stay with that image. See his face and his eyes. He looks crazy. He shoots other people and you are next. He points the gun right at your head and you think you are going to die. Feel that. Feel the panic. Stay with that image for the next few seconds. (Pause for a minute or two.)

Okay, we are going to break from that scene now. The images fade. You are back with the group. You are in the office.

I want you to take a few deep breaths. Give me a SUDS rating from zero to ten. Where are you right now?

GM 1: A ten.

F: Keep breathing. (Conduct deep breathing exercise.)

After a scene presentation group facilitators prompt the group members with questions concerning the predictability of the tragic outcome. Was the event foreseeable? Given the individual's knowledge of the circumstances could he or she predict what would happen? The controllability of the event is also a factor. Prompts in the form of questions such as "Was the outcome of the traumatic event controllable?" or "Was there anything that anyone could have done at the time to change the outcome?" may be used. Issues concerning self-blame should also be challenged by the group.

Many themes may be presented during trauma scene presentation. Some of the most common ones should be considered (Lebehowitz & Newman, 1996).

- Alienation
- Fear
- Guilt
- Helplessness
- Isolation
- Legitimacy
- Loss
- Influence of culture or society
- Influence of family
- Negative beliefs about people
- Negative beliefs about the self
- Negative beliefs about the world
- Rage
- Shame

These themes may be a focus of the individual's presentation. Repetitions of this material during the exposure session will facilitate extinction. It will also benefit the individual if the group facilitators challenge these cognitions and affects. The other group members may also be prompted to focus on these themes during the cognitive restructuring portion of the trauma-focus session.

The following instructions should be given to the group after the scene presentation:

A moment ago we heard a description of a traumatic event. It took courage to do this today. It is a risk to open oneself up and share the experience. These memories have caused a great deal of distress since the traumatic event happened. Your feedback will be an important source of support for the individual presenting today. It will provide a way to see what happened differently. We heard what happened and _____ feels that he made a bad decision, a decision that caused the traumatic outcome. He feels responsible for this and takes all the blame for what happened. What do you think about the event?

In-Home-Exposure Assignments

Repetition of the scene is a core component for progress in therapy. During the in-session presentation of the scene, an audiotaped recording is made of the entire scene and of the cognitive restructuring. Group members are each given the tape from their exposure along with a Home-Based Therapy Self-Exposure worksheet (Appendix IV) to take home for review. This additional exposure through use of the home-based therapy assignments can result in a change in the negative affects associated with the memories. Group support during in-session cognitive restructuring further enhances extinction with emphasis on the individual's thoughts during the event. Because the individual presenting the scene may be distracted or emotionally overwhelmed during the cognitive restructuring segment of the session, he or she may not be able to focus on or benefit from the supportive comments made immediately following the scene presentation. The group member should not review the audiotape for a few days following the group meeting. However, he or she should review the tape prior to the next trauma-focus session. Reviewing the audiotape provides an opportunity to receive this feedback again and further the process of extinction. It is not uncommon for group members to return after several sessions and report that they heard the supportive comments from the group "for the first time."

Following in-session exposure, each member is asked to develop a coping plan for the remainder of the evening and the time between group sessions. The group member is also given the following instructions and asked to write about the incident, answering the questions on the form provided as a home therapy assignment.

This assignment is an important part of your home-based therapy, and is an important part of your recovery. Listening to the tape and completing the form continues the work you did during the session today. Confronting the painful memories and repeatedly exposing yourself to the emotions, images, and thoughts will promote the healing process. You will also learn that you can manage these distressing feelings and thoughts effectively.

You should review the tape and complete the form at least once a week. This can be done any time during the week, when you feel ready. You can do this assignment more often if you wish, but not more than three times each week since you will need to have time for the exposure to gradually alter the feelings and thoughts associated with your traumatic experience. The form will be an aid to guide you through this process. Please bring the completed forms back to the next group session so that the group can assist you in developing skills and techniques to manage your distress. This may be one of the most difficult decisions you ever make for yourself. Your recovery is under your control. You have the responsibility to continue the healing process. Only you can decide to put yourself into the scene. But you are not alone in this. We are all here to help you, support you, and see you through it.

Objectives

- Complete individual trauma scene work.
- Sustain a minimum of thirty minutes of exposure to key components and themes presented during the trauma-focus session for each member.
- Prevent avoidance of scene material.
- Guide the individual presentation and the group in cognitive restructuring.
- Identify key themes and negative cognitions to be addressed during cognitive restructuring.
- Review the entire group's reactions to the scene material.
- Assess current emotional state via SUDS scale.
- Review the group's comprehension and acceptance of the rationale for the exposure treatment.
- Review treatment compliance.

Procedures

- Perform check-in and elicit SUDS rating scale.
- Describe the task to the group members.
- Identify the group member selected to present the trauma-focus scene.
- Begin the trauma-focus scene.

- Upon completion of the trauma-focus scene, assess each member via the SUDS scale.
- Conduct grounding techniques if necessary.
- Begin group feedback and cognitive restructuring.
- Review coping plans to be used between sessions.
- Assign home-based therapy assignments.
- Checkout.

FINAL PHASE (TERMINATION)

Session 16 is the final group meeting. The role of the facilitators will be to assist group members with termination and transition. Plans for effective coping should be reviewed with each group member, and continued use of the skills learned and the materials provided during the treatment should be promoted.

During this session, facilitators should emphasize the need to continue a routine for emotional and mental health. At times of crisis, symptoms of anxiety and depression may reoccur (White, 2000). It is the group member's responsibility to manage his or her continued recovery, which will include continued use of the audiotapes and the self-exposure forms. Because members of the group may feel sad or experience a sense of loss with the termination of the group, redirection to other social supports or additional treatment should be provided if necessary.

Session 16

Objectives

- Discuss feelings about the termination of the group.
- Ask each group member to describe how his or her memory of the traumatic incident has changed.
- Ask each member if his or her symptoms have changed in any way.
- Review each member's plan to utilize home-based therapy and manage symptoms after termination of the group.

Procedures

- Perform check-in.
- Review home-based therapy assignments.
- Perform checkout.

Before ending the final session facilitators complete a final SUDS rating to gauge the emotional stress group members are experiencing when faced with termination. The facilitators may assist group members with deep breathing and coping skills to manage their distress.

SITUATIONS THAT MEMBERS EXPERIENCE AS DIFFICULT

Difficult Members

Sometimes, a group may include a member that is difficult to manage due to behaviors that disrupt the structure and schedule of treatment. Presentation of a set of guidelines for group behavior during the treatment sessions will assist with the management of difficult group members. These guidelines should include but may not be limited to the following:

- Members must attend each session, on time, or call twenty-four hours ahead if an emergency occurs.
- Missing sessions is grounds for being asked to leave the group.
- Members must attend clean and sober.
- Members must not discuss group content outside of group or after meetings.
- Members must show mutual respect, and not minimize others' experiences.
- No violence or threats of violence.
- No touching other group members.
- No therapizing other group members.
- No leaving the room unless a break is specified by group facilitators.
- No discussion of politics or religion.
- No candy, food, or smoking is allowed.

During sessions, issues may come up that can cause concern for group members and be difficult for them to discuss or hear about. Confidentiality, fear of disclosure, legal concerns, and focusing on differences in traumatic experiences can create stress and distrust between group members, resulting in poor group cohesion and an unwillingness to work together with other group members. In response, individuals may engage in disruptive behaviors such as repeatedly interrupting the speaker. Other individuals may repeatedly dissociate, report "no feelings," or scores of "35" on the SUDS scale. Group cohesion and support are extremely important when discussing such issues.

Difficult Situations

Certain situations may be distressing or cause difficulties for group members during TFGT. These situations will require action by the cofacilitators during treatment sessions. Aside from the stress associated with PTSD, group members frequently have present-day stressors affecting their lives as well. The presence of such stressors may interfere with the individuals' ability to focus on the content of the group. These issues can be identified during the check-in, and facilitators can provide a referral for additional services outside the group if necessary. Identification of these potential distracters is critical to the progress during the group sessions because they may facilitate avoidance and hamper work during therapy sessions if not addressed. Providing a strong rationale for the structure of the treatment at the start of the group process may help. Check-in also prevents stressors from monopolizing group sessions and hinders avoidance of session protocol. Each group is highly structured and the completion of required material is a frequent challenge to group facilitators.

Facilitators must also remind group members that as treatment advances and they expose themselves to traumatic memories, exacerbation of their PTSD symptoms is likely. Group members should use the positive coping skills illustrated during group sessions to manage their symptoms and prepare a plan for a symptom flare-up. Review with members the goals of the treatment and the work to be done, so that they have realistic expectations for their progress in therapy. Group members may feel overwhelmed by the tasks presented during this session, or it may not be clear to them how these simple-sounding skills may help them improve their lives. However, a basic rationale for proactive symptom management is essential for group members to safely and effectively engage in the home-based therapy assignments.

A rationale for embracing treatment at home is also critical for adequate compliance. Explain to group members that progress in treatment requires work—perhaps the hardest work that they have ever done. Emphasize that what they get out of therapy is directly related to the amount of work done. Members who come into group and expect to work for ninety minutes or two hours per week and go home and forget about the group will make little progress. The facilitators can explain, "If you go home and put the weekly assignment on the shelf you are also putting your treatment on hold." Although avoidance is a common aspect of symptom maintenance group members need to continue to process, and not avoid the memories of their trauma. Group members will need to work at home on material covered during treatment sessions each week and reduce behaviors that have perpetuated the pain.

If group members have strong reactions to the content being presented during a session they should be prompted to use the coping skills practiced during prior sessions. Group members may utilize deep breathing or relaxation skills to manage their symptoms. This is also an opportunity to assist members with these techniques in the presence of distressing stimuli. Some individuals may also be anxious about taking their turn to present or be concerned about the type of reactions they may receive from the group and the two facilitators. Others may fear judgment and rejection because of their actions or lack of action during the incident. Taking a final SUDS level at checkout is very important to identifying levels of stress. Asking each member for a plan as to how they will take care of themselves the rest of the day can help prepare them for potential reactions.

Selection of a specific group member for the trauma scene presentation during a particular session should be based on the clinical concerns for the group. Because certain individuals may experience extremely high levels of anticipatory anxiety, a significant increase of symptoms may be experienced as the therapeutic exposure sessions begin. One way to decrease anxiety within these individuals is to select such individuals to go first or very early in the rotation.

During the trauma-focused sessions, group members are informed which individual has been selected to present his or her scene on the day of the presentation. Individuals experiencing extremely high levels of distress and severe symptoms may be selected to present first or very early in the trauma-focused rotation. Therapeutic exposure should lead to an overall decrease of symptom intensity within the selected group member after the first session of trauma exposure. The other group members will continue to experience symptoms, but some relief should occur especially with use of home-based therapy assignments during the following week.

Another consideration of order selection may be the response of each group member to the outline and scene description during Session 3. The SUDS scales ratings of the group members during this process and their reactions to the presentations of other group members may provide information helpful to the selection process. Although some group members will present with coherent and well-defined outlines, others may struggle with the scene outline, provide vague descriptions, and be confused about the procedure and requirements of the task despite repeated clarifications by the group facilitators. These individuals may benefit from a later position in the rotation if given the opportunity to observe other group members taking their turns first.

Change will be extremely difficult for some group members. Treatment will also be difficult and members may not be flexible in their thoughts of self-blame or judgments concerning their behavior. One technique used to challenge members' thoughts is to have them estimate the probability that a particular behavior would lead to a specific outcome. This question could also be directed to the group members listening to the scene presentation. The group consensus should be that it is impossible to predict with complete certainty that any one action will cause a specific outcome. This exercise helps the individual continue the focus on emotions and thoughts associated with the event and provides for additional extinction of the negative affect and an opportunity to change perspective concerning the event.

Based upon past experience with TFGT, disclosing who will be asked to present at the beginning of each session rather than giving advanced notice resolves possible disruptions in the treatment protocol and helps decrease anticipatory anxiety. Advanced notice may cause an individual to feel heightened distress and experience an increase of symptoms during the week prior to his or her scheduled presentation, which may also make it more difficult for him or her during the actual in-session trauma work. Also, individuals may avoid attending the session during which they have been scheduled to present their scene. Subsequently, if an individual misses his or her scheduled turn, it can create stress within the group. The facilitators will either have to ask someone else or add sessions to the treatment protocol to accommodate and maintain the treatment schedule. Adding sessions places a burden both on the group facilitators concerning session content during a group member's absence and on the other group members with regard to time, their emotions, and financial concerns.

Premature Terminations

Premature terminations occur for many reasons. Group members may have significant life events such as a death in the family, an unexpected illness, or some other unpredictable circumstance during treatment. A group member may also leave because he or she no longer wishes to continue with the trauma-focused group.

Following a member's departure, the group facilitators should use the check-in time at the start of each session to review the reactions of the remaining group members' concerning the departure. Group facilitators should be supportive of concerns and their decision to remain in treatment.

Sometimes, individuals who have prematurely terminated may wish to return to treatment. The return of a departed member may be very disruptive to the group process and requires careful consideration by the group facilitators and group members. If a group member has missed more than one session, careful consideration for participation in a subsequent group is recommended. For individuals having missed one or less group sessions, the group facilitators should first consider and decide on a recommendation for a return to group, but approval for a return to group must come from the group itself.

Premature termination also changes the flow of the treatment sessions. The entire group of six members is scheduled for sixteen sessions. Since each individual group member is scheduled to present his or her trauma scene during two different group meetings, twelve of the sessions are designated for trauma-focused work. A departure of one member alters the schedule, and only ten sessions will be required for each member to present the scene twice. The residual two sessions may be offered to two of the remaining group members so that each one may participate in a third trauma-focused scene presentation. As an alternative, additional work on coping skills may be scheduled for these meetings. All variations from the original protocol should be reviewed by the cofacilitators and agreed upon by the group.

VARIATIONS OF THE MODEL

Group members should not be involved in other individual or group psychotherapy services *concerning their traumatic experiences* during participation in the trauma-focused treatment group. Participation in the group is often very demanding of each member's time, resources, and energy.

If necessary, group members may attend peer-led services such as Alcoholics Anonymous (AA) or Narcotics Anonymous (NA). Similarly, marital counseling or other services that do not focus on the trauma history may be considered. Group members may also be involved in case management programs for assistance with housing, finances, and other personal or family needs.

The group facilitators should review medication management because symptom presentations may change during the course of the treatment protocol. Consultations with the prescribing physician are recommended.

FINAL THOUGHTS

Much of the chapter has been devoted to the presentation of a treatment manual for trauma-focused group therapy. The chapter is comprehensive in that it provides basic information and instructions for the delivery of the treatment. However, the chapter is not intended to be an alternative to supervised training, sound clinical judgment, or a firm conceptual model of symptom formation and psychopathology. Without these, when faced with a difficult situation during treatment, clinicians are left asking the question, "What do I do now?"

The Need for Theory

The effective delivery of any treatment protocol requires a comprehensive command of the guiding principles and techniques implemented during the process. This issue has become increasingly critical in today's health care environment given the pressure by outside service management to provide brief "cookie-cutter" treatments. For example, during training with interns and other mental health trainees, supervisors often receive requests to teach new techniques or specific skills. Training directors have students search for a set of tools or a manual to bring into therapy. In fact, the goal of a student's search for training is often for a "toolbox" rather than a conceptual approach with which to implement the tools. I have described this to supervisees as "seeking to be trained as a mechanical technician." Although the technician is an expert with the tools and what can be done with them, the technician's knowledge is limited to the use of the tools to change the individual components of the motor. In other words, the technician changes the "defective" component to fix the problem without having a clear understanding of the entire "automobile." This leads to returning for another appointment when the original problem persists requiring yet another change of components. This process may be repeated several times until the underlying issue is resolved. The true mechanic will test drive the vehicle, listen to the motor, and make the appropriate intervention. This is much less painful for the client and actually improves the relationship with the therapist. In brief, the message is that a manual is only as effective as the understanding of the provider using it. An understanding and expertise with a set of tools is not enough. An understanding of the basic aspects of theory as a conceptual frame for psychotherapy is, and remains, step one in the delivery of treatment

Although a manual can be useful for delivery of a technique, it is not helpful when group members do not follow the procedures. Since each human being is an individual with a wondrous supply of idiosyncratic response tendencies, it is impossible to predict how someone will react to therapy. If we limit ourselves to the parameters of a specific procedure, we run the risk of not knowing what to do when the client's response is not "in the book."

Our discussion is not meant to argue the value of a specific conceptualization of prolonged symptomatic behavior above others. Each clinician will bring his or her own understanding to the implementation of the technique. However, it is presented to assist with the development of an understanding of therapist's tasks for treatment to be effective. Exposure must be presented in a manner that allows for an extinction effect. The key to the technique is *direct therapeutic exposure*. The specific task of the trauma-focus group facilitators will be to expose the group members to as many avoided and fearful cues as possible, in an emotionally safe milieu, while at the same time preventing the occurrence of avoidance behaviors. Preventing avoidance behaviors allows for a longer duration of exposure to the conditioned stimulus (CS) complex, the presentation of a greater portion of the entire CS complex, and a greater extinction effect. The cues are presented until a reduction of the conditioned negative emotional response occurs; the actual duration of the trauma-focus scene is a critical aspect for therapeutic change. The duration of the scene often becomes a major stumbling block for the clinician, as the emotional response from the group member may be extremely intense. To work though the difficult situations encountered during group sessions, an understanding of learning theory will assist group facilitators with answering questions concerning the implementation of TFGT. Clinicians may benefit from a review of relevant literature (Boudewyns & Shipley, 1983; Carroll & Foy, 1992; Levis, 1980, 1995; Mowrer, 1960; Skinner, 1969; Stampfl & Levis, 1967; Solomon & Wynne, 1954; Watson & Rayner, 1920).

The work for the group member is difficult but the reward is great—a reduction or elimination of extremely aversive symptoms and a return to normalized lifestyle. Progress in treatment will allow the group member to recall the event as "sad" or "terrible" without the destructive consequences of overwhelming negative affect.

Trauma work can be extremely demanding for the group members as well as for the group facilitators. However, the rewards for the clinician are also great. With the support of the group and the facilitators, group members accomplish work that they would not be able to do alone. The simple fact is that while in session, they are not alone, and this changes the entire experience. No longer are they isolated and overwhelmed with the need to avoid the pain. Instead, they are instructed to witness their experience and share it with others. The memories are still extremely difficult, but with the help of the group the members can learn to cope and move forward with their lives. The group is a powerful vehicle for therapeutic exposure. Corrective feedback from one's peers further enhances the effect thus altering the embrace of shame, guilt, and fear.

Clinicians doing trauma work should provide for their own needs as well. We must realize that our community has expanded and the world has been forever altered. Clinicians are not immune to the terrorist attacks on September 11, 2001. We all shared concern for the victims of the sniper in October, 2002 in the Washington, DC/Virginia area. Some clinicians also have experienced a traumatic event, and the scenes presented by group members may reactivate strong emotions. As Nietzsche said "When you gaze long into the abyss, the abyss also gazes into you." Clinicians working with trauma survivors expend great energy in an effort to reduce the pain of others. The need to work to maintain our own sense of self and state of being is all the more evident. *Pax mentis* is the wish of the trauma survivor, and it is an ongoing effort for the clinician.

APPENDIX I:
JOURNAL HOMEWORK FORM

Name _____ Session _____

What I learned *about myself* in group this week:

_____

_____

_____

_____

_____

_____

_____

_____

_____

_____

_____

_____

_____

(*Source:* Adapted from Foy et al., 1997)

APPENDIX II:
COPING RESOURCES SELF-ASSESSMENT

Following are some important resources that can help people cope with pressures and problems. How able are you to use these coping resources

Cognitive	**Circle best answers**		
1. Managing traumatic memories?	good	adequate	poor
2. Problem solving/decision making?	good	adequate	poor
3. Memory/concentration?	good	adequate	poor
Emotional			
4. Anger management?	good	adequate	poor
5. Able to experience range of emotions?	good	adequate	poor
Social			
6. Adequate number of people in support network?	good	adequate	poor
7. Able to use network in time of need?	good	adequate	poor
Spiritual/Philosophical			
8. Religious beliefs, activities satisfactory?	good	adequate	poor
9. Relationships with organized religion?	good	adequate	poor
10. Worldview distorted?	good	adequate	poor

Physical Health

11. Management of health problems?	good	adequate	poor
12. Fitness/regular exercise?	good	adequate	poor
13. Risk behaviors, e.g., smoking?	good	adequate	poor

(*Source:* Adapted from Foy et al., 1997)

for yourself? After completing the assessment, rank order the top three coping areas marked "poor" or "adequate" in terms of your priority for change.

APPENDIX III:
COPING WITH SYMPTOMS RECORD

During each week, log the symptoms that you found to be distressing, the situations in which they occurred, the negative thoughts that accompanied the symptoms and situations, and the new coping thoughts and actions you took to address the situations.

#	Symptom	Situation	Negative thoughts	Coping thoughts and actions
1				
2				
3				
4				
5				

(*Source:* Adapted from Foy et al., 1997)

APPENDIX IV:
HOME-BASED THERAPY SELF-EXPOSURE

Name _____ Date/Time _____

Step 1. Find a quiet setting in your home where you will not be disturbed.

- Plan a self-care activity following the exercise. (For example, a walk, sitting out in a favorite place, calling a friend, or spending time with a supportive family member.)
- Begin by relaxing yourself! Briefly RELAX yourself. Clear your mind of all other thoughts. Rate SUDS:

SUDS (Distress) Rating:
        Completely Calm 0 1 2 3 4 5 6 7 8 9 10 Most Distressed Ever
Step 2. Review your traumatic incident.

- *Listen* to your trauma-focus tape. Listen to tape from beginning to end—as if it were happing right now! Just listen to the tape; don't question or interpret.

OR

- *Write* your trauma scene. If written, use blank sheets of paper. Write about the scene from beginning to end—as if it were happening right now! Just write the script; don't question or interpret.

Following trauma scene review, rate SUDS:
SUDS (Distress) Rating:
        Completely Calm 0 1 2 3 4 5 6 7 8 9 10 Most Distressed Ever
Step 3. Process the memory.
a. What new information did you recall?

_____

_____

_____

_____

b. What are your feelings and thoughts right now? (Don't censor!)

_____

_____

_____

_____

c. What is the most painful part(s) of the memory or the most difficult for you to remember or accept?

_____

_____

_____

_____

d. What do you wish to change about the incident? What can you do to change it now?

_____

_____

_____

_____

e. Listen to the feedback from the group. Can you allow yourself to feel that?

_____

_____

_____

_____

Step 4. Take a deep breath and bring yourself out of the memory. Rate SUDS:
SUDS (Distress) Rating:
Completely Calm 0 1 2 3 4 5 6 7 8 9 10 Most Distressed Ever

(*Source:* Adapted from Foy et al., 1997)

REFERENCES

American Psychiatric Association. (1994). *Diagnostic and Statistical Manual of Mental Disorders,* Fourth Edition. Washington, DC: Author.

Blake, D. D., Weathers, F. W., Naggy, L. M., Kaloupek, D. G., Gusman, F. D., Charney, D. S., & Keane, T. M. (1995). The development of a Clinician-Administered PTSD Scale. *Journal of Traumatic Stress, 8,* 75-90.

Boudewyns, P. A. & Shipley, R. H. (1983). *Flooding and Implosive Therapy,* New York: Plenum Press.

Carroll, E. D. & Foy, D. (1992). Assessment and treatment of combat-related posttraumatic stress disorder in a medical center setting. In D. Foy (Ed.), *Treating PTSD: Cognitive-Behavioral Strategies* (pp. 39-68). New York: Guilford Press.

Figley, C. R. (1995). *Compassion Fatigue: Coping with Secondary Traumatic Stress Disorder in Those who Treat the Traumatized.* Philadelphia: Brunner/Mazel, Inc.

Foa, E. B., Cashman, L., Jaycox, L., & Perry, K. (1997). The validation of a self-report measure of posttraumatic stress disorder: The Posttraumatic Diagnostic Scale. *Psychological Assessment, 9,* 445-451.

Foy, D. W., Eriksson, C. B., & Trice, G. A. (2001). Introduction to group interventions for trauma survivors. *Group Dynamics: Theory, Research, and Practice, 5,* 246-251.

Foy, D., Glynn, S., Schnurr, P. P., Jankowski, M. K., Wattenberg, M., Marmar, C., & Gusman, F. D. (2000). Group therapy. In E. B. Foa, T. M. Keane, and M. J. Friedman (Eds.), *Effective Treatments for PTSD* (pp. 155-175). New York: Guilford Press.

Foy, D., Ruzek, J., Glynn, S., Riney, S. & Gusman, F. (1997). Trauma-Focus: Group Therapy. Combat related PTSD: In session. *Psychotherapy in Practice, 3,* 59-73.

Frueh, B. C., Turner, S. M., Beidel, D. C., Mirabella, R. F., & Jones, W. J. (1996). Trauma management therapy: A preliminary evaluation of a multicomponent behavioral treatment for chronic combat-related PTSD. *Behavior Research and Therapy, 34,* 533-543.

Goldfried, M. R. & Davison, G. G. (1976). *Clinical Behavior Therapy.* New York: Holt, Rinehart and Winston.

Hegeman, E. & Wohl, A. (2000). Management of trauma-related affect, defenses, and dissociative states. In R. H. Klein & V. L. Schermer (Eds.), *Group Psychotherapy for Psychological Trauma* (pp. 64-88). New York: Guilford Press.

Herman, J. (1992). *Trauma and Recovery.* New York: Basic Books.

Johnson, D. R. & Lubin, H. (2000). Group psychotherapy for the symptoms of posttraumatic stress disorder. In R. H. Klein & V. L. Schermer (Eds.), *Group Psychotherapy for Psychological Trauma* (pp. 141-169). New York: Guilford Press.

Klein, R. H. & Schermer, V. L. (2000). *Group Psychotherapy for Psychological Trauma.* New York: Guilford Press.

Lebowitz, L. & Newman, E. (1996). The role of cognitive affect themes in the assessment and treatment of trauma reactions. *Clinical Psychology and Psychotherapy, 3,* 196-207.

Levis, D. J. (1980). Implementing the technique of implosive therapy. In A. Goldstein & E. B. Foa (Eds.), *Handbook of Behavioral Interventions* (pp. 92-151). New York: John Wiley & Sons.

Levis, D. J. (1990). The recovery of traumatic memories: The etiological source of psychopathology. In R. G. Kunzendorf (Ed.), *Mental Imagery* (pp. 233-240). New York: Plenum Press.

Levis, D. J. (1995). Decoding traumatic memory: Implosive theory of psychopathology. In W. O'Donohue & L. Krasner (Eds.), *Theories of Behavior Therapy* (pp. 180-206). Washington, DC: American Psychological Association.

Lubin, H., Loris, M., Burt, J., & Johnson, D. R. (1998). Efficacy of psycho-educational group therapy in reducing symptoms of posttraumatic stress disorder among multiply traumatized women. *American Journal of Psychiatry, 155,* 1172-1177.

Mowrer, O. H. (1960). *Learning Theory and Behavior.* New York: John Wiley & Sons.

Resick, P. A., Jordan, C. G., Girelli, S. A., Hutter, C. K., & Marhoefer-Dvorak, S. (1988). A comparative outcome study of behavioral group therapy for sexual assault victims. *Behavior Therapy, 19,* 385-401.

Resick, P. A. & Schnicke, M. (1992). Cognitive processing therapy for sexual assault victims. *Journal of Consulting and Clinical Psychology, 60,* 748-756.

Saakvitne, K. W. (2002). Shared trauma: The therapist's increased vulnerability. *Psychoanalytic Dialogues, 12,* 443-449.

Schnurr, P. P., Friedman, M. J., Foy, D. W., Shea, T., Hsieh, F. Y., Lavori, P. W., Glynn, S. M., Wattenberg, M., & Bernardy, N. C. (2003). Randomized trial of trauma-focused group therapy for posttraumatic stress disorder. *Archives of General Psychiatry, 60,* 481-489.

Skinner, B. F. (1969). *Contingencies of Reinforcement.* New York: Appleton.

Solomon, R. L. & Wynne, L. C. (1954). Traumatic avoidance learning: The principle of anxiety conservation and partial irreversibility. *Psychological Review, 61,* 353-385.

Stampfl, T. G. & Levis, D. J. (1967). The essentials of implosive therapy: A learning theory based on psychodynamic behavioral therapy. *Journal of Abnormal Psychology, 72,* 496-503.

Stauffer, L. B. & Deblinger, E. (1996). Cognitive behavioral groups for non-offending mothers and their young sexually abused children: A preliminary treatment outcome study. *Child Maltreatment, 1,* 65-76.

Turner, R. J. & Lloyd, D. A. (1995). Lifetime traumas and mental health: The significance of cumulative adversity. *Journal of Health & Social Behavior, 36,* 360-376.

Watson, J. B. & Rayner, R. (1920). Conditioned emotional reactions. *Journal of Experimental Psychology, 1920, 3,* 1-14.

Weiss, D. (1996). Psychometric review of the Impact of Events Scale-Revised. In B. H. Stamm (Ed.), *Measurement of Stress, Trauma, and Adapation* (pp. 186-188). Lutherville, MD: Sidran Press.

White, J. R. (2000). Depression. In J. R. White & A. S. Freeman (Eds.), *Cognitive-Behavioral Group Therapy for Specific Problems and Populations* (pp. 29-62). Washington, DC: American Psychological Association.

White, J. R. & Freeman, A. S. (Eds.). (2000). *Cognitive-Behavioral Group Therapy for Specific Problems and Populations.* Washington, DC: American Psychological Association.

Yalom, Irvin D. (1995). *The Theory and Practice of Group Psychotherapy* (Fourth Edition). New York: Basic Books, Inc.

Ziegler, M. & McEvoy, M. (2000). Hazardous terrain: Countertransference reactions in trauma groups. In R. H. Klein & V. L. Schermer (Eds.), *Group Psychotherapy for Psychological Trauma* (pp. 116-140). New York: Guilford Press.

Zlotnick, C., Shea, M. T., Rosen, K. H., Simpson, E., Mulrenin, K., Begin, A., & Pearlstein, T. (1997). An affect-management group for women with posttraumatic stress disorder and histories of childhood sexual abuse. *Journal of Traumatic Stress, 10,* 425-436.

Chapter 20

Psychodynamic Group Treatment

Daniel S. Weiss

RATIONALE AND OBJECTIVE OF THE GROUP

Providing Understanding Safely

The primary objective of the psychodynamic group approach to treating survivors of widespread major traumatic events is to assist individuals in coming to terms with the unique and idiosyncratic meaning of the event to which they have been exposed. Treatment is indicated when survivors are unable to assimilate the impact of the event into a changed internal working model of self and other. This fundamental tenet implies that the primary task for each person who has lived through a traumatic event is to come to terms with what it means to him or her (Weiss, 2001). Each person brings a different life history and set of experiences to bear, and though there is a common set of symptoms marking post-traumatic stress disorder (PTSD), and a common set of thematic concerns after exposure to traumatic stress (e.g., grief, guilt, loss of control, helplessness, rage, fear, and anxiety about the event), not everyone will have the same set or subset of concerns after exposure to trauma. Nonetheless, the power of this group approach lies in the fact that members of a therapy group typically share a number of concerns. Consequently, they can provide unique understanding of thoughts, feelings, and conflicts.

The rationale for conducting this treatment in a group setting is multifaceted. First, a group format is simply an economical and efficient use of precious, skilled, mental-health professional time and expertise. When the need is great, as it has been after large-scale terrorist incidents such as the World Trade Center and Pentagon attacks, or the Oklahoma City bombing, this rationale is essential. A group format also provides an extra measure of safety for participants and leaders.

Even though a member's main concerns may shift with the passage of time, after initially strong reactions have abated, or new thoughts about the trauma's meaning have surfaced, other members remain a source of strong support and emotional connection. As the feelings of helplessness turn into anger, or the desperate need for revenge transforms into sadness and a nearly overwhelming sense of loss and isolation, the importance of the presence of those who have similarly been exposed does not lessen. Because intrusive symptoms (e.g., unbidden images or thoughts, nightmares) and those of avoidance (e.g., numbing, shunning locations or dialogues) oscillate and change (Weiss, 1993), the meaning of the exposure to traumatic terrorism can evolve over time.

The extra measure of safety for group members comes from the development of a cohesive bond that forms as experiences are shared and similar internal processes are recognized. As members listen to one another describe what happened to them and how they have reacted, the commonality of experiences becomes a powerful counter to the feeling of social and emotional isolation so widespread among those who have experienced traumatic events. Correcting the erroneous idea that one is completely alone, and that literally no one else can truly understand how profound the suffering is, can be a key ingredient of group treatment. Members have the ability to convey an authentic appreciation of what postdisaster life has been like, since they too have had to endure a comparable fate.

Another benefit of the group format is the titration of the psychologically painful feelings that accompany repeated and exclusive focus on the traumatic event that typifies one-on-one psychotherapy. The inherent characteristics of a group setting provide an opportunity for taking a breather. As well, groups offer the opportunity, from time to time, for members to assume the role of helper, itself often an important element of growth and recovery.

The group format affords an extra measure of safety for therapists as well (Goodman & Weiss, 2000). Co-therapists can assist each other in dissipating the emotional impact of conducting a trauma group by providing support, validation, consultation, and supervision. Cotherapists can literally share the "therapeutic load." Another benefit, as long as cotherapists maintain a standard of mutual respect, is the opportunity for firsthand reinforcement of good therapeutic work from a respected colleague, something that is usually not possible in individual treatment. As long as collaboration and communication exist between cotherapists, and they are open to feedback from each other, they can take comfort that they will proceed in accord with the ancient directive of treatment providers of any sort: *primum non nocere* (first, do no harm). In a more practical vein, cotherapists can also cover for each other when needed as well as during vacations if the group requires ongoing continuity without a break.

Because the other cotherapist is observing the same phenomena from a different perspective with different starting points or schemas, the process of developing each member's psychodynamic formulation can be richer and deeper. Treatment with two therapists also provides greater opportunities for the emergence of members' parental transference. If the therapists are male and female, this opportunity may well be amplified. Hence, interventions stemming from countertransference reactions can be diminished by having the independent view of a second therapist.

Though the group format seeks to maximize safety, it can also have the opposite impact on members. Rather than being reassured that others have had similar experiences and similar reactions, being aware of the similarities may serve to deepen despair and pessimism, stemming from the perspective that the event was so catastrophic that few escaped unscathed. In turn, this may be taken as evidence that the hope for recovery is small, since the impact of the event was so pervasive. This line of thinking is what terrorists attempt to create, so countering this view may be framed as countering the goals of terrorists.

In a similar vein, the group format can, in some situations and in some individuals, serve to amplify the focus on trauma rather than titrate it. In these cases, listening to others is experienced as traumatizing in itself. Research, however, has not found this to be a pervasive problem (Schnurr et al., 2003). Nevertheless, cotherapists still need to be vigilant about these possibilities not only during the group itself, but also during the screening and intake process.

Providing Education About Common Responses to Traumatic Stress

The conceptual model that we advocate for understanding the response to traumatic life events (Weiss, 1993) is based on one presented by Horowitz (1986). It describes a series of states of mind that begins with the event, moves to an outcry, proceeds to an oscillation between intrusion and avoidance, includes physiological arousal and hyperarousal, and recognizes the presence of numbing as well. One of the values of this model is its assumption that there is a normal response to traumatic life events that may or may not eventuate in PTSD. Group members should understand that the nature of their psychological responses and symptoms are understandable reactions to extreme stress, and that the severity of the responses occur on a continuum in much the same way that degrees of burns are on a continuum of severity. The need to assimilate and accommodate to the changed reality brought about by a traumatic life event and its sequelae, be it from terror, disaster, or single traumatic life events is present for everyone who has been exposed.

Event

In many cases, the nature of the traumatic event is clear (e.g., the destruction of the Pentagon) and what makes the event traumatic is also clear (many lives lost and more threatened). One of the core precepts of the psychodynamic approach, however, is that despite the objective clarity of the event and its aftermath, much of the symptomatic distress experienced after a terrorist event lies in the meaning of the event to the individual. This meaning, specific to each individual, usually needs to be discovered, discussed, and thought about for symptomatic resolution and restored functioning. This is especially the case in circumstances in which the traumatic event has a variety of components, and the specific features of the event leading to symptoms could be one or several of countless factors (e.g., lost body parts; an angry fight before work; witnessing the death of a colleague) of the traumatic event. This fundamental assumption is easy to overlook, and in practice, it unfortunately is overlooked frequently.

Thus, one of the initial tasks for each group member is to help the others clarify exactly what it was about the event that remains a source of symptomatic distress. Frequently, the source of symptomatic distress is an aspect of the traumatic event that is difficult to deal with, but the precise nature of the difficulty remains obscure. Resolute and patient probing and questioning regularly reveals that the nature of the difficulty is linked to psychological issues or early working models of the self in relation to others that have either been dormant prior to the event or have been exacerbated by it.

Outcry

Repeated observation of videotapes and live footage of individuals witnessing a traumatic event or being informed of the outcome of a traumatic event reveals a characteristic bodily response and a concomitant cognitive response that are linked by their common agenda—a wish, and a need, to disbelieve. The typical bodily movement (for whatever reason, appearing to be more typical among women), is to rapidly bring one or both hands to cover the mouth, and to say something such as "Oh no," or "I don't believe it!" The shocking recognition of a potentially vastly new and changed reality is first rapidly apprehended, and then as rapidly rejected and bidden to be untrue.

When individuals receive shockingly bad news over the telephone, it is extremely rare if nonexistent for the response to be "Oh yes!" rather than "Oh no!" This apparently trivial observation, however, makes it abundantly clear what the cognitive task of individuals who have received the news is going to be: to assimilate an unwanted and distressing new reality to existing cognitive structures; and to accommodate and modify cognitive structures so that they can integrate the changed reality. Emphasis on the automaticity and near universality of the outcry can help illuminate why the adjustment to traumatic exposure is so painful, especially for individuals who typically take a more rational and cognitively controlled approach to emotional distress and turmoil. Explaining the functional significance of the outcry and its initial rebuff helps those who have been exposed to traumatic stress understand these reactions.

Oscillation Between Intrusion and Avoidance

The longitudinal course of reactions to traumatic stress is an emotional and cognitive oscillation between accepting the changed reality (forced by intrusion) and rejecting, denying, avoiding, or warding off its recognition and the ensuing implications. This oscillation is evident in the occurrence of intrusive phenomena (e.g., dreams or nightmares, unbidden thoughts or images, and, rarely, flashbacks) experienced by survivors. These emotionally painful occurrences provoke avoidant states, wherein the painful process of changing models of reality is avoided or shunted aside.

Often, the process of oscillation may feel like it has a life of its own, and in many ways it does. Typically, people feel that they are vulnerable to intrusions, and these undermine their sense of personal emotional control. Avoidance, therefore, is embraced with even greater fervor than that based on need for respite alone would warrant. The frequently intense need for a sense of control in a reality that feels out of control can lead to an entrenched avoidance that is very difficult for individuals to modulate.

Cotherapists must also explain the difficulty survivors experience modulating between perspectives of denying and accepting the traumatic event. The first perspective is based on the individual's deliberate choice to avoid reminders or stimuli that are known triggers of painful feelings and memories. The second perspective emanates from the less consciously controlled self-protective mechanisms that allow tolerable doses of the new circumstance and reality to be comprehended. This new reality invariably contains loss of some sort, whether substantive, as in the case of a loved one's death, or the loss of a limb, or some bodily function; or the loss may be more conceptual, such as the loss of innocence or invulnerability.

Because acceptance of loss is universally involved in adapting to traumatic events, the acceptance of the changed reality requires the use of mental mechanisms that allow gradual cognitive changes to occur without intolerable and overwhelming feelings of pain. The most common mechanism individuals employ for coping with loss is the oscillation between acknowledging the intrusive phenomena, which communicate that reality is different, and the dampening or shutting down of emotional reactivity. Cotherapists can explain that the processes of mourning the loss of a loved one who has died of natural causes are fundamentally similar to those experienced in traumatic loss. The distressing psychological processes are virtually universal. The mourning conceptualization helps group members counteract the frightening but often secret idea that the frequent loss of emotional control means that somehow they are losing their minds. Indeed, our working assumption is that the psychological journey of coming to terms with a traumatic exposure is at its core the same journey as that of the mourning process. Consequently, the accumulated knowledge and experience of a mourner can be brought to bear in furthering the understanding that, like all individuals who have experienced loss, group members will experience oscillation between painful moments of experiencing the changed reality and the avoidance of these painful recognitions.

Arousal and Hyperarousal

Among biological reactions to traumatic stress are the psychological reactions that result in the "fight-or-flight" response to acute danger or stress. After a situation involving immediate danger has passed, a state of increased arousal often continues. This increased arousal is manifested by sleep disturbance, difficulty concentrating, hypervigilance, exaggerated startle response, and physiological arousal (especially when confronted by reminders of the trauma). Other reactions related to hyperarousal are irritability and angry outbursts. Though related to the fight-or-flight response, they are also likely part of the well-known angry response to the disruption of attachment (Bowlby, 1973).

Physiological arousal may continue months or years after the initial exposure. Prepare those who have been exposed by providing psychoeducation about the natural history of responses to traumatic stress. Often this preparation is sufficient to reassure group members that their reactions are understandable and not atypical. Nonetheless, continuation of the hyperarousal responses is at the heart of most theories of the genesis of PTSD, including fear-conditioning paradigms and alterations of the hypothalamic-pituitary-adrenal (HPA) axis (Yehuda, 2002). Some researchers have proposed that a kind of reexposure occurs when people think about the event and experience the arousal that accompanied the initial exposure. This more cognitive view (Foa, Riggs, Massie, & Yarczower, 1995; Foy et al., 2001) stipulates that effective interventions include repeated exposure to the traumatic event through thought and discussion without physiological arousal.

The view that arousal is a fear-conditioned response presents problems for approaches to treatment that focus on extinguishing the fearful responses. The extinction view assumes that a response can be "unlearned," however, this is contradicted by the same animal studies that validate the biological processes in the fight-or-flight response. Unlearning appears to be the conditioning of a different response that replaces the fear response but does not eliminate it. This view suggests that exposed individuals are at risk for reemergence of the conditioned response long after it appears to have been extinguished. Indeed, some might argue that delayed onset of PTSD represents just such a reemergence of the conditioned response, or that the waxing of PTSD symptoms following a subsequent life event demonstrates the reappearance of the fear response. How this knowledge articulates to the presence of rational thought and cognition in adults is not yet well understood, but is of considerable interest.

Although not often noted, the major aspects of the fear-conditioning or stress hormone (hypothalamic-pituitary-adrenal axis system, Charney, Deutch, Krystal, Southwick, & Davis, 1993) theories of hyperarousal symptoms are unable to explain the presence of intrusion and avoidance and why they oscillate. Our view of the cognitive and emotional responses of intrusion and avoidance following traumatic stress does not directly address hyperarousal. The presence of hyperarousal symptoms is not easily incorporated into our perspective of the need to revise one's working model of the world. Explaining the role of hyperarousal may require further research. Also, there is an empirical question as to whether the working-through process is as effective for reducing hyperarousal symptoms (as opposed to reducing the distress produced by these symptoms). In fact, it could be argued that neither view provides much explanation for the presence of intrusion and avoidance, the assault upon views of self and safety, or the other themes that typically accompany exposure to traumatic stress. Until further research is conducted, it can be said that these two processes are integral to the diagnosis of PTSD, but a coherent explanation of the presence of both has so far proved elusive. An explanation connecting the two phenomena, or the revision of the conceptualization of the response to traumatic stress to include hyperarousal as an evidentiary (but not definitive) aspect of PTSD is a goal for future research (Meehl, 1995).

Numbing

Numbing is a relatively infrequent presenting complaint of those who seek treatment following exposure to a traumatic event. It is characterized by a profound lack of reactivity to thoughts, images, or conversations about the event, and visits to the site of the trauma (e.g., Ground Zero or the Pentagon). Numbing is not merely increased avoidance; it is an unbidden response beyond voluntary control, much in the same way that intrusion is. The central experience of numbing is the individual's awareness that his or her level of reactivity is much less than she or he would expect and/or want.

Descriptions of being numb include "feeling frozen," "being unable to feel anything," and "having no emotional reactions." Short episodes of this phenomenon are to be expected following a traumatic event, but prolonged states of emotional shutdown are more ominous. In some cases, careful assessment and intervention are required. Indeed, a group approach may be insufficient to break through whatever conflict or obstacle is impeding the naturally occurring oscillation between intrusion and avoidance and therefore may be contraindicated or require adjunctive individual treatment.

CORE ELEMENTS OF INTERVENTION

Encourage Telling of the Story

Track onset of change in affect.

The heart of this approach is giving each group member the chance to tell his or her story, a task which Stern (1985) describes as the key developmental milestone in the attainment of selfhood. This should happen multiple times, in shorter or longer episodes, sometimes dovetailing with another group member's memories and recollections. A key observational window will take place the first time that someone tells his or her story accompanied by a perceptible change in affect.

Repeated observation of individuals relating what happened to them, with the hope of being understood and having their listeners be empathically attuned to their emotional reactions, has confirmed that there is typically a key point in the story where each individual's composure changes, the ability to proceed is disrupted (Milbrath et al., 1999; Stinson, Milbrath, & Horowitz, 1995), and, typically, crying occurs. Remarkably, these moments do not typically occur at the moments in the story where the most "objectively" horrific or distressing aspect of the event is told. Rather, such moments typically reflect the difficulty the individual is having in coming to terms with the way the event unfolded. At these moments, the conflict is most clearly presented.

The traumatic event and its sequelae may stir up issues that have been more or less dormant (Zilberg & Horowitz, 1983), or may be concatenated to issues with which the individual had been actively dealing with prior to the traumatic exposure. In either case, this change in affective tone and the disruption of the storytelling almost always point to the central issue upon which therapeutic work must concentrate. Cotherapists leading the group need to be especially vigilant in tracking this process and formulating hypotheses about what the difficulty in continuing in the retelling of the events indicates. Though by no means universal, such moments often indicate problematic aspects of self-image that have been carried by the individual from childhood.

Discern meaning of change in affect.

After recognizing the moment of disruption in the verbal presentation of the story, the task for the cotherapists, ideally with input from other group members, is to discern the meaning of the affect change. The questions that the cotherapists should pose to themselves in conceiving potential formulations include, "What is so painful here?"; "What is the feeling being experienced yet remaining unspoken?"; and similar questions.

Understand the meaning of the story.

Understanding the meaning the story holds for the presenting group member involves identifying the thematic issue provoking the painful emotional experience. These themes typically involve feelings or judgments about the self. Consequently, they comprise responses such as shame, guilt, humiliation, grandiosity, loss of control, or feelings of unloveability. They typically do not involve the defining characteristics of events that the DSM-IV classifies as potentially traumatic: carrying immediate threat and generating anxiety or fear. These affects more commonly play a role in the powerful reluctance shown by an individual when asked to remember and tell his or her story in the first place. This characteristic unwillingness is fueled by two sources. The first is the conscious recognition that telling the story again will recreate the circumstances of the emotional pain and distress that it had caused. The second less-conscious recognition is that the emotional distress and pain are linked to views of the self, and the fear of confronting warded-off ideas about the relationship between self and others may recur during the telling.

This fear is similar to the fear and anxiety generated by the actual exposure. However, it is different because unlike the traumatic experience itself, where the outcome was not known while the fear, anxiety, and distress were operative, in retelling the story, the individual is typically aware of the outcome and knows it will bring painful memories.

Elucidate internal working models of the character of relationships.

The larger context of internal working models of self and others is as important as the recognition of the pivot from an emotional to an unencumbered retelling of the story. This concept draws largely on John Bowlby's (1988) theories of how the developing child, and later the adult, views himself or herself in relation to important attachment objects.

Not surprisingly, people whose early history contains disruptions of attachment (Tennant, 1988) are likely to view a current disruption to psychological homeostasis through the lens of the earlier disruptions. Whatever conclusions were drawn about the self in those earlier events (e.g., "It's my fault"; "I'm unlovable"; "My assertiveness is hurtful"; "I am unwanted") become operative in the processing of an abruptly changed reality. Because the internal working models of childhood are formed in the context of preoperational formal logic, and are therefore relatively immune to change when urged from an adult logical viewpoint, so too will the meaning of the current traumatic event contain irrational and nonlogical linkages, equivalences, and assumptions. The elucidation of these models is a necessary precondition for the next step in the process, which is the explicit linkage of current reactions to the habitually used models derived from the past.

Link historical events to reactivated self-images.

Therapists should persuasively explain to the group that the ways in which they are reacting to the traumatic event repeats or echoes aspects of a previous problematic moment in their development (Hiley-Young, 1992). Therapists must have a fairly detailed knowledge of the developmental history of the group members and the ways in which they view themselves as vulnerable or "not okay." Such imprecise language as "not okay" is usually preferable for patients, as it is customarily the way in which they may think and talk about flaws and weak points.

Such information may not emerge right away and may have to be rather freely inferred and iteratively corroborated. The insights of other group members, especially if they share some of the same vulnerabilities, can provide an additional avenue to generate and validate this information. Often one group member recognizes in another the same struggle as he or she is enduring and is able to clearly articulate the vulnerability or wounded view of self. Such recognition can generate strong and healing empathy.

Provide compassionate response to suffering.

A powerful tool in initiating, bringing about, and sustaining change in internal working models is to be empathically attuned to the emotional experience generated by the historically derived and now maladaptive models of self and other (Goodman & Weiss, 1998). The role of the cotherapists is to provide a compassionate response to the suffering endured by the individual on whom the attention is being focused, and encourage the other group members to do so as well. This means allowing very painful emotional experiences to unfold without attempting to cut them off, and acknowledging the pain and expressing the wish that such suffering did not occur. Being present and comforting but not falsely reassuring is a relatively new and different response to episodes of strong emotion.

Some therapists and many group members may become particularly uncomfortable with the display of the raw emotional pain, usually weeping or sobbing, that accompanies considering the implications of the traumatic event. Most everyone feels the need to try to curtail the crying, a choice that is countertherapeutic. The typical alternative to this is to remain speechless, reinforcing the sense of loneliness and isolation. Cotherapists should model for both the member telling his or her story, as well as the other members, that tolerating the powerful negative affect is both possible and necessary, and by expressing the sincere wish that the situation were different.

Some group members may attempt to shut down the suffering of a compatriot because of a mistaken belief that comfort is what is most needed in the alleviation of psychological pain. It is crucial to modify such an attitude in favor of the view that true comfort is the result of being deeply understood. Thus, when one is suffering, being understood validates the suffering and mitigates it in a way that trying to stop it or turn it off does not.

Such a task may be very difficult because other group members may become overwhelmed with the intensity of the storyteller's pain. Therefore, it is very important to reassure all parties that such pain typically ebbs in a time frame of seconds or minutes, and that tolerating the expression of such pain is one of the main sacrifices and contributions that individuals need to make to benefit from group treatment.

Modify Internal Models

Differentiate adult and childhood expectations.

One of the tasks that behavior change requires is differentiation of child-hood wishes, expectations, patterns, and habits from those that are more appropriate for an adult. This task would be rather straightforward if it involved only cognitive activity, however, such a view overlooks two important functions. The first is the entanglement of cool cognition with warm or hot emotional reactions; individuals have significantly more control over thoughts than feelings. The second is the understanding that cognition itself is developmental, and the attainment of formal operations and the ability to reason and attribute appropriate causality is always preceded by a world-view that is not capable of comprehending certain relationships and mental operations. Thus, the wish for others to know what one is thinking and feeling and needing without the need to articulate it verbally is a very common and overlearned (even if at an earlier age) distortion of adulthood.

Those exposed to trauma often experience estrangement from others and the inability to have loving feelings due to a sense that others cannot understand or know what they feel or have gone through. Indeed, this is one of the symptoms of PTSD. Though the adult, rational, cognitive view of events does coincide with the reality that one not exposed does not know what it is like to have been exposed, the mismatch appears to be experienced at a much more fundamental level. There are literally hundreds of things that an individual experiences that others have not and cannot, yet these everyday mismatches are rarely grounds for feeling profoundly cut off and misunderstood. Thus, it is important to the progress of treatment to clarify what wishes and feelings are from earlier developmental stages and how they may interfere with more adult perspectives, and the ability to feel connected to others.

An especially salient differentiation in this arena is the notion of blame, culpability, or fault versus that of bad luck, misfortune, or tragedy. Children lack the capacity to understand that the negative actions of important figures in their lives (e.g., parents divorcing) could be due to things having nothing to do with the child. Consequently, if something bad happened, or the child did not receive sufficient love and attention, his or her conclusion is that he or she was responsible in some way. In the realm of traumatic stress or the aftermath of terrorist actions, this idea will likely emerge in the form of "should haves" or "shouldn't haves." Examples are, "I shouldn't have let her go to work with such a bad cold"; "I had a premonition and I should have acted on it"; "I shouldn't have let him stay to rescue others";

"I should have known that the Towers would collapse"; "I shouldn't have yelled at him as he left for work."

Resistance in this arena stems from at least two sources. The first is the developmental legacy. The second is an adult appreciation of the fact that exposure to trauma or terrorism is largely a matter of bad luck. If one accepts the idea that having become a victim is something that is out of one's control, then one is forced to acknowledge that becoming a victim is possible again. Thus, rather than accept the idea that indeed the individual was not at fault, that there was nothing that could have been done, and that it was just tragic misfortune to have become a victim, individuals fiercely cling to the idea that if they had only done, thought, not done, not thought something, then their victimization would have been avoided. Overcoming this distortion is one of the most difficult tasks of trauma work, since it means accepting permanent potential vulnerability.

Differentiate rational from irrational meanings of event.

Despite the real characteristics of the traumatic event and its aftermath (e.g., "My husband died in the collapse of the North Tower"), there are also irrational or fantasied meanings of the event (e.g., "I will never again be happy"; "No matter what I do, I always end up alone"; "I ruin anything that is good for me"). These latter irrational meanings derive from two sources. The first comprises ideas and internal working models of the self and others that preceded the trauma, closely tracking many of the child cognition issues, especially issues of control and responsibility.

The second source, and typically the central focus in the psychodynamic group approach, are derivations of childhood ideas, involving specific beliefs about the consequences to self, and others, of having been exposed to traumatic stress. For example, victims of traumatic exposure commonly hold a belief that they really have no future. This irrational belief is so common and debilitating that it is one of the set of symptoms that confer a diagnosis of PTSD. Nevertheless, the belief is irrational and will need to be clarified in the treatment process. Clarifying the actual meaning does not involve modifying the facts producing the irrational belief; it means confronting the conclusion. Undeniably, the individual's life is inevitably altered, sometimes in profound and (if bodily injury occurs) incapacitating ways. Nevertheless, with a successful treatment outcome, his or her life can continue and progress, albeit in the changed and new fashion demanded by the actual circumstances of the traumatic event.

A specific pair of anxieties that group members often articulate is a view of themselves as overly strong or overly weak. Paradoxically, views of the self as both overly strong and weak often occur in the same individual. As to which view predominates, more often than not, a group member's perspective depends on the nature of a relationship to an important person, and the risks to the relationship if the group member acts in accord with his or her own needs or wishes. In trauma contexts, it may be very difficult for a lay person to accept the idea that he or she should be concerned about being unable to recover because that would mean he or she would move on where others either cannot (because they have died) or have not (because they too share a worry, perhaps feeling guilty, about triumphing).

An important corollary of this phenomenon is that these themes may also produce countertransference. Cotherapists need to be aware of any reluctance they may have in helping members explore and pursue their stories, for fear that they will be revictimizing the members either by evoking painful emotions or by instigating a repetition of feeling or being out of control. This may be especially true when a member is expressing considerable guilt or fear about what it would mean to recover and is resisting moving ahead strongly. Alternatively, the roles of victim and victimizer may subtly switch so that the member is reluctant to expose the leader and/or other members to their trauma, believing that the others are insufficiently strong to tolerate hearing explicit horrifying details. The roles of victim and victimizer are perhaps inextricably built into the process of recovering from trauma. Being alert to the themes of strength and weakness likely will not prevent their appearance, but can appreciably mitigate unnecessary distress.

Explore alternative constructions of self and others.

Once a relatively clear formulation of the issues that an individual is struggling with has been achieved, the task of the cotherapists, with input from other group members, is to help the individual explore more functional beliefs about who he or she is and how he or she relates to others. This exploration can occur in any three combinations: view of self, view of others, and finally, the relationship between the self and others.

The second of these, views of others, may prove surprisingly detrimental to treatment. For example, a group member may worry that if he recovers from the impact of having lost his wife in the collapse of the Towers, his success will irreparably injure his mother who never remarried after the death of his father. If the group members or the cotherapists help him reinterpret his mother's never remarrying as a manifestation of her strength to make her own choice after her loss, this revision may allay his worry that not choosing the same path as his mother would mean that she would be hurt and unable to forgive him for making a different choice. Moreover, since many of these ideas are unspoken but not unknown, it is possible for explicit conversations to occur that help refute these worries, both inside and outside the group.

Cotherapists cannot be certain about what the consequences will be of a candid discussion with a family member or friend outside the group. Consequently, because there is risk in such a course, explicit suggestions are generally to be avoided, even when there is considerable hope that such a conversation would be helpful. A compromise to this is for the cotherapists or other members to encourage the individual to explore what such a real conversation would be like, and then appraise the potential consequences.

Another avenue for the exploration of alternative views of self and other is to focus on what the individual's relationships were like prior to the trauma. In doing so, the leaders can often help clarify which previously unresolved but dormant aspects of the self have been reactivated by the trauma and help the individual to appreciate that these were previously surmounted and therefore can again be dealt with.

The third approach for the exploration of alternative views is an appeal to idealized perceptions about the self and others. Here, the cotherapists and group members push an individual to simply articulate a hoped-for outcome and put aside any assessment of whether attaining the outcome is probable or even possible, e.g., "I want to tell my mother-in-law that I think I need to remarry so my kids can have a mom in their lives." Only after stating the goal, is he asked to state what he believes are the impediments to attaining it. Once the goals and obstacles are clear, it then becomes a matter of tracing back from the goal to clarify where such a journey would begin and what it would comprise.

Providing Setting for Initiation of Mourning Process

Clarify that loss always implies mourning.

Though it has been mentioned several times already, it is important to make clear to group members that recovery from loss must include a period of mourning (Weiss & Marmar, 1993). Often, mourning lasts longer than individuals both wish it to or expect that it will, and it is frequently more painful and difficult than had been imagined.

Refute the notion that suffering is permanent.

The passage of time is an underestimated source of healing from traumatic exposure (and many other hardships). While in the throes of suffering, many group members feel that their anguish will never relent; if they come to the group many years after the occurrence of the traumatic event, their personal experience often supports this belief. Nonetheless, the desire for recovery is what brings people to seek treatment for PTSD even after years of compromised functioning. Even though epidemiological studies estimate that in a lifetime 50 percent of people experience exposure to at least one traumatic event, they also find only 5 percent (of men) and 10 percent (of women) develop PTSD (Kessler, Sonnega, Bronnet, Hughes, & Nelson, 1995). This evidence suggests that most people who are exposed to traumatic stress go on to live productive and rewarding lives.

The length and natural history of the disorder, sometimes spanning decades, make it possible to believe that suffering *will* be permanent. Two factors lead to this belief. First, many will not have sought prior treatment for PTSD, thus, the time course may not be the most optimal or representative description of what could or would happen when PTSD is treated in a timely and effective fashion.

Second, chronic PTSD is a waxing and waning disorder, and in some cases may represent a more or less permanent state of affairs. Even when it is permanent, however, the degree of suffering can be ameliorated considerably from its untreated level.

TASKS TO BE ACCOMPLISHED

In this section, we describe a closed group, twenty-four-session treatment model specifically designed to address PTSD. This model will serve to: (1) function as a set of guidelines for clinicians, suggesting a treatment strategy and identifying essential themes and issues; and (2) provide a standardized treatment paradigm to facilitate research on the effectiveness of group psychotherapy. Premorbid developmental issues not involved in the response to trauma and posttrauma issues are touched upon, but are best treated subsequent to completion of this trauma-focused group. The content of group discussions throughout treatment focuses primarily on identifying and working through traumatic experiences and their impact on present-day adjustment, as seen through working models developed through prior experience. Attention is also given to intragroup dynamics to help facilitate the process of change.

Inclusion Criteria

We have used this approach exclusively with homogenous groups in relation to a traumatic event (e.g., combat, rape, individuals with a history of child abuse prior to being in combat (Goodman & Weiss, 1998). In principle, there is nothing in the approach that would prevent its use in a group that did not have a common experience or trigger, but experience may well not accord with principle. In a heterogeneous group, complaints against a common outside antagonist (e.g., the government, the rapist, abusive parents) might be less salient. If so, cohesion will need to form using another target.

Even in circumstances in which all would agree that the group is homogeneous, such as victims of September 11, 2001, war veterans, etc., detailed consideration of the specific events make clear that the concept of a traumatic event itself is very complicated absent the distressing and arousal reactions to it (Ozer, Best, Lipsey, & Weiss, 2003). The peritraumatic responses of fear, horror, and distress are a basic commonality, and this shared experience may be sufficient homogeneity for a single group.

Other basic criteria for membership include ability to attend group and pay for the treatment, and absence of gross cognitive or affective dysfunction, such that the individual could not usefully participate in the group. Current suicidal or homicidal intents are explicit exclusion criteria as well. Additional criteria that must be considered are substance abuse (alcohol, drugs) and preexisting psychotic disorder, both of which raise major cautions against inclusion. Participants with a prior history of substance dependence must be able to tolerate negative affect and intrusive imagery without jeopardizing sobriety. Prospective members should have at least three months of sobriety prior to beginning the group, and all members must commit to attending group sessions substance free.

Ongoing individual psychotherapy or pharmacotherapy are not contraindications, though for the former an explicit discussion about confidentiality and releases and the plan for communication or absence thereof with any other mental health professionals involved must be addressed prior to the start of the group. For a much more detailed discussion of this issue, the reader is referred to Goodman and Weiss, 2000.

Group Structure

The group should include five to seven members—more than seven would make it unlikely that a secure group process and identity could develop. Ideally, there should be some heterogeneity as to the demographics, e.g., specifics of traumatic exposure, time elapsed since the trauma, age, ethnicity, socioeconomic status. Conversely, it will be helpful if members are homogeneous, as far as is possible regarding level of ego functioning, interpersonal skills, and ability to confront defenses and address and integrate warded-off material.

Attempts to determine prospective members' veracity should be tactfully undertaken prior to their inclusion in the group; e.g., cotherapists should question existence of criminal records, medical records, and collateral information, if possible. Unfortunately, from time to time, individuals' psychopathology leads them to want to become members of a specialized group such as described here when they have *not*, in fact, had the kind of experience that would qualify inclusion. Such "as if" or "wannabe" individuals are sometimes difficult to identify, but their inclusion in a group can be extremely disruptive to the necessary sense of safety and trust. In these instances, other members often sense an artificiality that exposes the dishonesty. Consequently, to the degree possible, the integrity of an individual's presentation should be examined carefully.

Group treatment requires two cotherapists; sessions last one-and-a-half hours. Cotherapists should attempt to structure sessions such that emotional intensity builds but has adequate time to deescalate before the end of each session. An optimal group session may take ten to twenty minutes for an informal check-in or brief warm-up period, followed by fifty to sixty minutes of concentrated work, ending with a ten to twenty minute wind-down phase. At this point, cotherapists may wish to summarize group process or content, or point out connections between different members' experiences or emerging patterns over an individual's lifetime. Alternatively, cotherapists may wish to move the group toward a less volatile topic or actively help a member implement dosing of emotional experiences, (e.g., refocusing onto a more positive topic or giving attention to relaxed breathing). Therapists should acknowledge the work accomplished in that group session, and emphasize the likelihood that when they revisit that incident or those feelings in a later session it will likely not be as intense.

Overall length of group is flexible, but the minimum limit is no fewer than six months. Whatever length is selected, all members need to be informed and in agreement prior to start of the group. Though an open-ended contract is possible, it is usually preferable to structure the contract with designated time points (e.g., every three months) to reevaluate progress of the group. The assessment focuses on deciding if the group should continue for an additional block of time, or whether all agree that it is appropriate to begin the process of ending the group.

Overview

Session Descriptions

Phase I sessions (1 to 6 in a twenty-four-session model) are structured by cotherapists and initially include presentations of psychoeducational material, preparation for group participation, and introduction of themes to be addressed in each session. Cotherapists should actively model frank discussion about disagreements or differences in viewpoints about how to proceed, provide appropriate supportive-probing questioning, and explicitly attend to intragroup process (interpreting parallels to trauma-related issues). This structure is gradually reduced as group members learn to take on this role.

The content of Phase II (sessions 7 to 22 in the twenty-four-session model) are dictated primarily by group members, but the main agenda consists of the telling and retelling of each participant's story; this allows repeated examination of the thoughts and feelings that accompany memories of the traumatic event. Cotherapists must actively direct the process and ensure that relevant themes are raised for consideration during this trauma-focused phase. The leaders explicitly help structure the focus of repeated tellings of the event for the individual participants, asking for concentration on affect, details, reactions, or other aspects as indicated by the participant's difficulties and what has occurred during previous tellings of the story. The expectation during this phase is that a different member will retell his or her story every session. Nevertheless, when either the members or the cotherapists judge that issues of the group process itself need to take priority, process discussions take precedence.

Phase III (sessions 23 and 24 in the twenty-four-session model) explicitly focuses on termination issues, with cotherapists increasing structure through somewhat increased activity. Treatment follow-up, referral for additional treatment, and the potential for a reunion at some time in the future may also be addressed

Group Leadership

Cotherapists have a number of responsibilities to the group. They must titrate the degree of structure and activity in the group to simultaneously create an atmosphere of safety and a sense of trust in their competence. One typical challenge to this is the desire of group members to make their own decisions and deal with issues that arise in the process with little or no input from the cotherapists. A balance between leading and following is called for; collaboration usually accomplishes this goal.

Cotherapists also need to allow for transference and working through of issues with authority figures, if these exist. More active leadership is often necessary early in group development, though whenever it seems appropriate, the leaders should solicit input and commentary from the group members themselves about the current theme, topic, story, or issue.

Cotherapists must acknowledge mistakes and limits of their understanding. Becoming defensive or automatically assuming a "one-up" power position is counterproductive. Statements such as, "You're right. I've not been through your experience, and I can only imagine how terrifying that must have been for you. But I want to try to understand as much of the experience as you are able to share right now," are more powerfully therapeutic than statements that imply the cotherapists are experts and the members are not.

Even though there may be a tendency to want to spare members the pain of exploring associations about various aspects of their trauma, the very heart of the treatment is the careful, compassionate, and watchful but consistent questioning of the meaning of various reactions and associations. Countertransference issues regarding being too nice or too brutal can both interfere here. In these circumstances, the cotherapists must be assertive enough to redirect the discussion. In order to model for members that it is possible to be constructively critical within a close relationship, the cotherapist needs to explicitly verbalize such an intervention.

Acknowledge and utilize the expertise of the other group members whenever possible; empowerment is a fundamental treatment goal. Utilizing the group members as much as possible to process issues such as being late, missing sessions, coming to group intoxicated, and having destructive conflicts with other members or the cotherapists will assist in this goal. Group members may often be the first and best detectors of other members whose agendas are different from the articulated agenda. Misalignments include malingering, faking PTSD for a variety of motives, feigning authentic participation, or other issues in the presentation of some group members.

Cotherapists may be hierarchical or egalitarian, but leaders must model basic interpersonal skills with each other as a prelude to expecting similar interactions with members. The cotherapists should have some dialogue with each other, probably in every session, about choice points, decisions, and other matters. Doing so models collaboration through tolerance of differing opinions, negotiation, amicable disagreements, and other examples of connected interpersonal process.

Strategy for Addressing Traumatic Incidents

Effective treatment involves eliciting an accurate recounting of events, to the degree possible, including pre- and posttrauma issues that play an important part in the story. These may include responses by law enforcement, family, the social milieu in which the trauma occurred, or other issues. Appropriate affective involvement, dosed to make an impact without overwhelming group members or precipitating dissociative reactions, including the very rare frank flashback, usually proceeds as follows: initial anxiety prior to recounting the incident, anxiety and/or tears of pain during the telling, and a kind of "calm after the storm" during which some consolidation occurs.

Coping with and understanding the context in which the event took place will vary as a function of both the traumatic event and the type of group convened. If it is a group for women who have been the victim of a sexual assault, for example, issues such as stranger versus known assailant, use of drugs or alcohol, and clear warning signs or signals that were ignored, are all important aspects to elucidating the story and should be given some attention. For a group dealing with the attacks of September 11, 2001, issues of being on or off duty, doing regular routines or unusual activities, witnessing injury or death to strangers versus individuals known to members (e.g., co-worker, newspaper vendor), are just a few contextual circumstances to elucidate.

Modifying the working model of self and others involved in reaction to a traumatic event is a key ingredient to this approach. Modifications include changes in the cognitive appraisal of internal dialogue about the meaning of the event, specifying lessons learned, or revising the meaning attributed to an event. This process involves the exploration of conscious and unconscious self-concepts related to weak and strong self-representations evoked by the trauma, as these are related to current conflicted views of the self and self-representations from early development.

While listening to patients' retelling of their stories, cotherapists must be attuned to subtle but significant omissions or incongruities in patients' narratives. Group members may also be able to perform this function as group progress solidifies and improves. A key aspect of empathic attunement involves noticing and calling attention to the speaker's comments about his or her own narrative as it unfolds. Cotherapists can accomplish this task, without criticism, simply by being curious: "I noticed that halfway through that sentence you stopped yourself. What were you planning to say?"; "Correct me if I'm wrong, but it looked like you were on the verge of tears when you mentioned your husband's name. What was the feeling underneath?" These comments in the moment serve two functions: first, they give the member an immediate feeling of being understood; second, they set the stage to follow up on associations and related meanings.

Actions taken in the moment for immediate survival must be acknowledged and validated. Notwithstanding the necessity for these actions, *living with* having had to take them is a stumbling block for some patients. Although it makes sense to have gotten out of a burning tower, since that is what survival demanded, patients still struggle with such issues, making acceptance of these realities difficult. In helping to clarify the underlying issue (e.g., survivor guilt, shame about weakness, rage that someone else's panic needlessly caused more death), it is important to beware of countertransference reactions. These may be either overly critical or overly accepting. It is also important to ensure that other group members do not minimize these conflicts. Such minimization is a potentially serious threat to cohesion. For example, saying, in effect, "Don't feel bad, you had no other choice," conveys a fundamental misunderstanding of the other members' internal experience.

When appropriate (generally some time after the initial disclosure and telling of the story), the cotherapists should provide feedback on each patient's recurring pattern of dysfunctional or self-defeating behavior that has emerged more clearly since the traumatic event. Examples of these patterns include overgeneralized mistrust of strangers, reluctance to accept responsibility for choices made, and inconsolability after loss. How this has been replayed throughout a patient's life must be linked with how it is now emerging. Relating this to processes within the group as they are unfolding can increase the therapeutic impact.

Cotherapists should attempt to involve group members in processing the incidents of others, but ensure that painful affect is not prematurely shut down, diverted, or avoided just because it is painful. These events may be used as illustrations of the difference between tolerable modulation of difficult emotional experiences and their complete avoidance. Group process is likely to be more intense during these episodes, therefore, discussions that allow recognition of good faith efforts gone bad, miscalculations in judgment, and basic human error promise to have more lasting impact if raised in these more emotionally charged moments. A question to be answered is how to live comfortably in a present that includes having experienced the unchangeable traumatic experiences of the past. One goal of these discussions is to come to a realistic assessment of one's ability to control outcomes of negative events, ability that neither demands complete control nor bows to complete absence.

Attention must always be paid to defensive intragroup dynamics and behavior. These defenses can be tied to avoidance and the need to regulate painful affective experiences in the educational sessions of the group. Such issues will likely come up regarding the trustworthiness of the cotherapists and other members—is the group a safe place to become vulnerable? A number of examples follow.

1. "It seems like in each group you are ready to give other members all the time and support they need, but it's hard for you to accept or ask for help yourself. Is there a part of you intent on helping others and not feeling you deserve help yourself?" This is an example of individual-level resistance.
2. "Has anybody else noticed that after a painful session like last week's, we all tend to focus on relatively safe topics?" This is an example of group-level resistance.
3. "It seems like when we bring up reminders of the need to work on the telling of your stories here in the group, issues of victimization, being forced to experience painful memories, and anger come up. Are there some feelings you have about us that we should be talking about?" This could be an example of transference issues regarding authority. Members are reluctant to confront parental figures directly.

Initially, many group interactions take the form of member to cotherapist, back to member. The development of group cohesion requires that, as much as possible, the group develop its own identity. This may be helped along by cotherapists deflecting questions posed to them back to the group. This will facilitate greater member-to-member communication, which is an important early goal. The therapists may make comments such as: "Have other members had similar feelings or experiences?" One therapist may make a direct link between one member's experiences and another member's similar experiences or encourage members to speak directly to each other. At the end of a group session, cotherapists may wish to acknowledge comments a member made that were helpful to another member.

Cotherapists should be aware of a tendency to "talk about talking about" specific traumatic incidents and events, versus *actually* talking about them, and should be prepared to confront and interpret this resistance by members, especially during the renditions of their stories. This can be discussed further as group resistance and avoidance, and group members may be enlisted to help manage such issues.

It is also necessary to discuss the impact of anniversaries (e.g., September 11 each year), media reports, or ongoing legal procedures (e.g., compensation fund) that take place during the course of the group. This may be discussed broadly in the context of managing triggers of certain memories, feelings or behaviors, although attempts should be made to keep the focus on individual trauma experiences.

Cotherapists should strongly limit general discussion and global anger about such topics as specific nationalities, the legal or health care system, or other ancillary aspects of the trauma that distract from dealing with painful individual reactions. Anger at political leaders or perpetrators is fair game, especially as it relates to issues of poor parenting and discontent with authority. Often it is helpful to quickly counter, no matter how legitimate these feelings might be, and explain that this anger, energy, and time would better be spent in areas where the participants can actually make a difference, i.e., in focusing on their own lives. When the anger is legitimate, whether toward a victimizing terrorist, a purveyor of anthrax, or other appropriate target, it can be used to facilitate the expression of anger by those who are otherwise afraid of giving voice to their own rage or acting on their wrath. A full discussion of the role that anger plays may be crucial to locating the meaning of the traumatic event to specific group members. Thus, careful explication of why a member is angry may be very helpful, even if it seems on the surface self-evident. Such dialogue allows unspoken issues such as trust, betrayal, guilt, or helplessness to be discussed openly.

Sequence of Group Sessions

The next sections describe the sequence of the group sessions; the model is made up of three phases. The first is an introductory phase, the second and longest is the exploration and modification of the meanings of the trauma through repeated recitations of the events, and the third phase includes summarizing work accomplished and terminating the group treatment. The content of some sessions (primarily in Phase I) is presented in great detail, since its components change little from group to group. Paradoxically, even though Phase II is where the bulk of the therapeutic work occurs, its specifics can be presented only in broad and general terms, since these will vary considerably from group to group and from person to person. Because each member's issues are individualized, the principles of the therapeutic work are the heart of the presentation. Phase III can also be specified in more detail, since the activities in summarization and termination do not vary greatly.

PHASE I

Since an explicit requirement of this treatment model is its flexibility, a feature which differentiates it from other group approaches, mandating the particular session in which material is presented is contrary to our objective, (e.g., a cognitive-behavioral trauma-focused method; Schnurr et al., 2003). Because there is a natural logic and progression of topics that need to be presented in the initial meetings of the group, cotherapists may find themselves in a dilemma given that so much material, ideally, should be presented in the first session of the group. The challenge is to find a good balance between the need to present logistics, ground rules, and other procedural issues prior to beginning actual therapeutic work with the need to start building a therapeutic alliance and group cohesion immediately. In the presentation that follows, the agenda for the first session is daunting. In typical circumstances it is a very full session. If a group spends somewhat more time than anticipated on one facet, adjustments to the agenda will need to be made. Such modifications are quite appropriate and anticipated. Though the tasks are presented in relative order of importance, the cotherapists will need to exercise their judgment about how to modify the idealized model to fit the specific circumstances in the same way a surgeon modifies performing a procedure given what is found inside the patient.

Two additional factors influence how detailed and comprehensive the agenda for the first session will be. The first is how much of the information about ground rules has been presented in an informed consent document. Such documents usually make explicit limits on confidentiality inherent in Tarasoff warnings, child and elder abuse, danger to self or others, and raising psychological or psychiatric issues in the legal activities. Many will also include information about fees, absences, and handling of records under the Health Insurance Portability and Accountability Act (HIPAA) regulations. Each of these factors will lead to some limits on confidentiality. A written informed consent document that each member signs is now considered to be the standard of care for psychologists in California, but also makes good clinical sense elsewhere since it embodies trust, mutuality, and agency. It also reduces the possibility of misunderstandings and reduces ambiguity about boundaries. If much of the ground rules information has been reviewed prior to the treatment session, less time may be needed in the actual session.

The second factor that will influence the first session's agenda is how much of the information has already been presented in the process of screening and evaluation. There will likely be considerable variability in the degree of prescreening from agency to agency and group to group. As with the informed consent, the more information presented prior to the first session, the less time will be needed to review it during the session. Regardless of how much has already been presented, all topics still need to be raised in the first session, if only to refresh agreements already made.

Introductions, Brief Exchange of Names, Identifying Information, and Initial Disclosure

During the first session the focus is on current problems: living situations, relationships, employment and treatment histories, and goals for group. Cotherapists should avoid discussion of the event at this point other than mentioning the most basic information such as a few words about what happened ("I was trapped in the rubble for five hours"; "I had to identify my wife's body"; "I saw a man jump out of the tower and watched him plunge to his death") and when it happened (e.g., "on September 11"; "two months after the towers collapsed"). Firmly and explicitly limit discussion of the who, what, when, where, and why until later, despite the possibility of a strong pull to say more. Limit discussion to about five minutes per member, reminding the group that more in-depth introductions will follow in subsequent sessions, as well as more time for discussion of what happened to each member.

Presentation of Group Ground Rules

The first session should include a very brief overview of the goals and philosophy of the group. Right from the very start, cotherapists should stress the importance of establishing an environment that is safe for all members, including the prohibition of violence and threats. Along these lines there should be a clear differentiation of threats from verbal expressions of anger or strong feeling.

Discussion of alcohol and substance abuse must also occur. Cotherapists need to provide the rationale for abstinence during therapy for anyone who has met diagnostic criteria for alcohol or drug abuse or dependence prior to the group, something that would be determined during prescreening. Continued abstinence for these members is essential since there is a documented history of disorder. For group members without problematic substance use, candid discussion of the risk of turning to the use of alcohol or drugs as a method of avoiding painful feelings stirred up by treatment needs to occur. Cotherapists must obtain a commitment for attendance without being intoxicated from everyone and simultaneously encourage an atmosphere in which participants can honestly discuss slips in abstinence if these occur outside the group. Any other compulsive behavior designed to cope with painful feelings such as binge eating or overworking should be addressed in terms of maladaptive forms of coping. Careful tracking of when slips occur may be helpful in clarifying the reasons behind such behavior.

An accurate picture of the safeguards and limits of confidentiality must be presented or reviewed during the first phase even if it has already appeared in a consent to treatment form. Members should be asked to make a commitment to keep group material confidential, and cotherapists need to articulate a philosophy of treatment to protect confidentiality to the fullest extent possible, but within the legal requirements described earlier.

During this discussion it is important for cotherapists to validate the legitimacy and historical reasons for caution and mistrust. Part of the members' therapeutic tasks involve learning who, when, how, and whether to trust. Incremental trust and "tests" may be appropriate and healthy, but the leaders should beware that members may construct tests that cannot be passed, seeking to confirm distorted notions about untrustworthiness.

Contact between members outside of the group is a complex issue best dealt with directly and openly in the group itself. Mutual support and/or contact could be helpful, but the shunting of out-of-group conflicts, "secrets," or one member calling another expressing suicidal feelings (and thus avoiding raising these issues in the group) are some of the dangers in allowing such contacts. Stress that *the group* is the primary place for members to connect with one another.

Regarding interruptions during sessions, the cotherapists must emphasize the importance of remaining in the room to process distressing feelings and discourage leaving the group in order to avoid discomfort. If leaving the group becomes absolutely necessary, one of the leaders should attempt to accompany the member, quickly deal with the issue, and return as soon as possible.

Other ground rules to consider include these: (1) no food, drinks, or other distractions during session out of respect for person talking and (2) no touching during the group in order to guard against using touch as a substitute for verbal expression and clarification. Although opinions about these issues differ, in the absence of research that supports these policies, we offer them as suggestions. The reasons for the policies, however, are more than suggestions. Respect for other members, minimizing distractions, and exhorting the use of words to express feelings are important principles of our approach.

Regular attendance and advanced notice of anticipated absences are crucial in a setting in which loss, unexpected tragedy, lack of knowledge, and loss of control are major portions of the agenda; this fairly stringent policy signals that some losses can be avoided. As well, this policy reiterates that doing the therapeutic work requires effort and sacrifice on the part of each member. Members are expected to make every effort to attend each session and notify therapists as soon as unavoidable absences are anticipated. (Unexplained absences should be commented upon in session by leaders if not raised by group members.)

Cotherapists must clarify their availability (and the limits of such availability) for emergencies between sessions. Information about these limits as well as information for twenty-four-hour emergency room or hotline should be given out prior to the first session.

Check-In and Checkout Procedures

Check-in is a time to assess if each participant is psychologically ready and available to do the work of the group. It provides an opportunity to briefly note any between-session issues that have arisen that may interfere with readiness to participate, as well as issues that may have arisen from the work of the previous group. Cotherapists must stress that this is a brief status report, and not an opportunity to tell in depth; it is a time to set an agenda rather than pursue it. Check-in is initiated by one of the cotherapists. Each member presents a minisummary of his or her current status, comments on brief positive or negative reactions to the previous week's session, or notes something that he or she feels needs to be put on the agenda (e.g., "I want to add something to my response to the question John asked me last week"). Initially, members may misunderstand the function of check-in regarding how they are doing. Events outside of the group may be important in each member's life, but they are relevant only insofar as they will interfere with that member's capacity to do the therapeutic work that day (e.g., "I got news that my brother was killed in a car accident and I don't think I'm up to re-presenting my story today"). One of the early tasks of the leaders regarding check-in is to help members differentiate between the vicissitudes of everyday life and those that have a clear connection to the issues that are to be discussed in the group.

Checkout is designed to assess if each participant has any issues that have been raised by the work of the group which need to be aired so that the intervening week can go by without group issues festering or being left unsaid. Cotherapists should also stress that the checkout can be a time to consolidate insights and summarize the progress made in the session, if no member has a pressing reaction that needs attention. At the end of each session all members must ensure that they leave feeling emotionally stable; a checkout assists in this process. It also allows for the opportunity for members to inform leaders of intense anxiety or suicidal or homicidal ideation if such is occurring. If distress is less serious, as would be anticipated, then the checkout procedure can serve a consolidating function for what transpired in the group during the session.

The process is similar to the check-in; each member briefly takes the floor to articulate something that is important to convey—being unsettled about something that was raised, learning something important, or admiring the manner in which another member proceeded in the session.

Group Member Self-Introductions

Group member self-introductions can be used to structure the first few sessions. Cotherapists should allow members to initiate order of presentation. This allows leaders to determine which members initiate and which prefer going last. In the introductions, cotherapists can begin modeling and shaping group behavior, e.g., allowing silences, asking for missing information, demonstrating open-ended questions to elicit more information, helping to facilitate expression of affect, and uncritically suggesting possible recurring patterns, coping styles, and problems.

Cotherapists must actively limit premature disclosure of traumatic material, which could lead to disorganization, shame, or the telling of "horror stories" (i.e., embellished, devoid of genuine affect, or containing conscious inaccuracies). This may lead to early dropout from group or commit members to a distorted or defensive account of their experiences. The horror stories may involve ancillary topics such as litigation versus restitution, the insurance industry, policies concerning death in the line of duty, or the health care system. Members need to be reminded that there will be time to talk about these aspects in sufficient detail.

The introduction portion of Phase 1 should be structured in such a way that ideally, approximately half of the members present in the first session and the other half present in the second session. These guidelines, however, always need to bow to clinical exigencies. If the leaders conclude that the group is ready to move on to the next segment, that is encouraged. Similarly, if a particular topic needs more exploration, the pace can be retarded so that the individual nuances of the particular group are accommodated.

In their presentations, members should be encouraged to include the following: personal history including current family structure; current occupation; length of time in the area; and previous treatment, if any.

Following introductions, cotherapists may ask: What does each group member see as impediments to treatment from his or her self-knowledge? Based upon knowledge of their individual histories, and the PTSD model presented, members can be asked to identify what behavior patterns and avoidance tactics they are likely to utilize that might interfere with change. There are many examples. One is a newly expressed need for premature termination even though the initial contract was clear and the now-expressed reasons seem understandable. A second is a tendency to isolate and withdraw or become the group scapegoat, thereby making it easier to leave. A third is taking the role of "helper" and attending to everyone else's needs before one's own. A fourth is the creation of diversions to shift attention away from oneself. A fifth is pseudocompliance, or the conscious withholding of key elements of one's story, reactions, feelings, or other matters so as not to rock the boat. Finally, there may be an increase in substance abuse or other self-destructive behaviors.

In the self-introductions, there should be some discussion of pretrauma history. PTSD affects views of self and others depending upon who the person was prior to the trauma. Thus, it should be understood that PTSD is not a function of being "defective" or being a problem teenager. The coping styles that members had when they were younger will yield clues as to how they adapted to, dealt with, and defended against what happened.

A somewhat brief family history with some details about the members' childhood and adolescence will be helpful, however, this is not an invitation to discuss previous abuse or traumatic events in any detail. If relevant, cotherapists may wish to discuss the possibility that these or other childhood problems are a part of the manifestations of PTSD, but at this point, telling stories about any trauma is premature.

Members can be encouraged to describe the role they may have taken in their family: e.g., peacekeeper, black sheep, scapegoat, decoy (i.e., drawing attention to oneself to save a sibling or mother from abuse), or de facto parent. These in turn can then be related to current coping strategies utilized during stress, typical problematic interpersonal patterns or ideas on how the participant may proceed within the group. Also relevant will be any patterns of conflict with father and/or mother and subsequent authority figures since these may well be manifest in the transference.

When appropriate, the cotherapists may offer some general impressions as to members' self-presentation, mostly with the goal of communicating that they have "gotten the message." For example, if a member has clearly signaled that she tends to not like to deal with difficulties, one of the leaders may point out a tendency to minimize problems. Alternatively, other summaries may be more apt: overdramatization; a sense of belonging or of not fitting in; a need to please and be the "good patient"; the need to have a special place in the group; or, taking on the role of unofficial cotherapist. Any parallels to current life problems as they appear in the introductions could be commented upon by the leaders as well.

Presentation of the PTSD Model, Its Symptoms, and the Anticipated Course of Treatment

Much of Phase I is devoted to education about PTSD, because it is our view that a strong starting platform enriches subsequent work. The purpose of devoting much of Phase I to presenting a model of response to traumatic events is twofold. The first is to normalize the symptomatology, facilitate group bonding, and begin the development of group cohesion. In providing a brief overview of the treatment and a description of its likely course, the cotherapists will help alleviate the understandable apprehension accompanying the decision to deal with traumatic memories by joining the group. Such a discussion can increase compliance by helping members anticipate high-risk periods and temporary exacerbation of symptoms, since an increase in symptoms typically follows telling one's story. As well, such a discussion can help members set realistic goals for treatment outcome. A common fantasy, which cotherapists can begin to dispel, is the belief that after treatment a member can once again be who she or he was before the trauma occurred. Explicit discussions can help to demonstrate the cotherapists' understanding of PTSD as well as their capacity to understand the suffering of the group members.

The second purpose is to provide group members with several opportunities to join the process while information about responses to stressful events generally, and PTSD specifically, is presented. Piloting of the format revealed that, without a specific admonition against it, cotherapists would fill most of the time in these first sessions talking or lecturing since there is a lot of material to present. This approach led to group members chiefly being passive recipients, which in turn worked against giving the cotherapists and the group members themselves opportunities to learn about one another. Without such interchanges, establishing the essential sense of cohesion was precluded. In order to help establish the importance of input from the members, providing opportunities for members to describe their experiences illustrates by example that all members have something important to say. In this way, each member participates in the educational process and feels his or her input is valuable. In addition, because members hear firsthand from others about comparable experiences, there is reinforcement of the reality that their experiences are not unique. Finally, members can help cotherapists refine their views, since they have not had the experiences the members have had.

The initial session should begin with a discussion of the puncturing of the bubble of invulnerability. The task for adaptation is the struggle between the knowledge that it really did happen and the wish to pretend that it did not. Cotherapists then connect the signs and symptoms of PTSD (intrusion, avoidance, and hyperarousal) to normal acute stress response reactions in which these mechanisms are natural and affect virtually everyone. The need posttrauma to develop physical, emotional, and cognitive psychological defenses to cope with the impact of potentially life-threatening situations and the understandable desire to avoid dealing with the horror is noted and validated. It may be helpful to point out that many of these functional adaptations are based primarily on avoidance.

Cotherapists must also communicate that in the affective realm psychic numbing is a potent issue. The nearly universal, "I can't believe it happened," which follows the first recognition that a traumatic event has occurred, in months or years becomes, "It happened but wasn't so bad," "It could have been worse, I could have been killed" are all presented as ways to numb the psychological distress associated with confronting the reality of the trauma. Attention to the behavioral realm allows cotherapists to discuss alcohol or substance abuse and other behavioral transformations such as the acting out of fear and grief as rage and anger. The leaders need to explain to members that feelings of power replace feelings of powerlessness when this emotional transformation occurs.

Education also covers the cognitive realm. Group members benefit from learning about magical thinking, all-or-nothing thinking, overgeneralization, or rationalization. Special emphasis should be given to the way social relationships are affected by exposure to trauma and subsequent PTSD or PTSD symptomatology. The level of intimacy and the regulation of closeness and distance to others directly connects to issues of trust—trust of other group members, trust of the leaders, trust of friends and family members, and trust of self are all aspects of this issue.

Also very important is education about possible presentation of the physiological aspects of PTSD—e.g., hyperreactivity, fight-or-flight false alarms, and sleep and concentration disturbances. Participants should know that PTSD frequently involves not merely their thoughts, but also their bodies' responses to threats or perceptions of threats. For many, this will have never been made clear previously.

Part of the special nature of victimization, from terrorist activities or other kinds of intentional human actions includes public and private attitudes about victims. These attitudes frequently involve ideas of shame and/or guilt and promote chronic avoidance. Members believe that how they have fared given what happened is humiliating and shameful and cannot be talked about to anyone; thus, one must proceed as if it hadn't happened. In this domain, the stigma of psychological disruption of any kind can be usefully noted as a general issue needing to be dealt with.

Members of the group may be able to help explore the way in which chronic PTSD can manifest as pseudo-characterological behavior patterns and types that lead to chronic self-defeating life patterns. These include substance abuse, legal, marital, or employment problems, anhedonia, and even a kind of existential meaninglessness. The leaders also need to explain that exposure to traumatic stress can often evoke a cascadelike progression of secondary adversity that involves employment, intimacy, and physical limitations if these are included. Thus the consequences of exposure can be more pervasive and complex than just nightmares and exaggerated startle response.

Make sure group members know that the goal of therapy is not to eliminate either the knowledge of what has happened or the use of some of the coping tools that participants have used, but rather to evaluate their usefulness, assess the costs and benefits of certain choices, and increase the range of coping tools and options available.

At the same time, members need to be told explicitly about the inevitable resistance to processing traumatic events, both within the group and with significant others. This resistance often includes fear of losing control, and coping with anxiety when feelings are either too little or too much. Members may harbor a variety of secret ideas: "If I let myself remember everything, I'll go crazy and never come back"; "I will get so angry I might hurt or kill someone, or maybe myself"; "The pain will be impossible to bear; I'll start crying and never be able to stop"; "When people find out what I really did [and didn't do], who I really am, they will abandon me and I'll be totally alone"; "I'll have to know about something I don't want to know about" [e.g., anger, food, sex, love, betrayal].

The resistance members experience is a key aspect of PTSD's alternating cycle of avoidance and intrusion (i.e., nightmares, intrusive memories, intrusive thoughts, and other forms of reexperiencing). Members should understand that traumatic experiences lead to physiological changes which may produce many "false alarms" in autonomic arousal systems. Nonthreatening events may literally feel like impending life-or-death struggles. These may be felt, or seen by others to be overreactions from which it is difficult to calm down or disengage. These feelings cannot be disregarded, but group members must learn to distinguish between internal and external real and fantasied dangers.

Treatment Goals

Another preparatory idea is the understanding that a realistic goal of the group is not the elimination of all PTSD symptoms. Instead, since PTSD is a waxing and waning disorder whose symptoms may be triggered by reminders and other serious life events, the goal is to minimize frequency and intensity of intrusive symptoms and to learn that it is safe to confront and process traumatic memories instead of avoiding them. One analogy that can be used is that dealing with PTSD is more akin to controlling allergic reactions than removing an appendix.

Another goal of treatment is to increase the repertoire of healthy coping behaviors and reduce the maladaptive ones. In so doing, group members can hope to develop more effective, broader-based support systems. "When you cannot make the trauma any smaller, you have to make the rest of your life a little bigger."

Of central importance, as well, is the goal of modifying how group members process trauma in terms of their view of themselves and their story. This involves integration of all relevant components of traumatic incidents, including affective reactions, and becoming clear about what the nature of affective reactions are. For example, it may not be clear to a participant, especially some years after the event, how terrified he or she actually was during the event. Also important are the special personal meanings ascribed to the incident and the implications about self (and others) that the event and its aftermath have raised. These may be ideas such as, "I am not competent to care for myself"; "I was bad and I deserved some punishment"; "I am flawed because this proves my lack of worth." Modification of these implications usually entails a careful enumeration of the details of events during, preceding, and following the incident.

Successful integration of the components of a traumatic event requires that the patient titrate the experience of vivid imagery while remaining aware and present. Indeed, one task for the cotherapists is to help regulate the speed of storytelling so that group members do not quickly skip over the aspects of the story that are the most traumatic. This requires group members to proceed with "one foot in the here and now" and "one foot back" at the time of the trauma.

Because honesty is an essential component of trauma work, cotherapists must create an atmosphere in which patients do not feel the need to exaggerate or minimize experiences. Every member is respected as having suffered a traumatic event regardless of the fact that the intensity, brutality, danger, or other characteristics are not always exactly the same for everyone. A feeling of loss of control is a fundamental commonality, and usually it is the particular group member, rather than others, who feels that an event is somehow less worthy of need for assistance and help. Despite the hope that honesty is omnipresent, calling attention to minimization, exaggeration, and distortion in the context of ongoing group discussions or presenting the traumatic event when ideas and feelings are psychologically alive makes the importance of being as open as possible much less abstract and empha-sizes why honesty is effective.

Cotherapists must also reiterate that the goal of the group is not necessar-ily friendship among the members, though this may indeed develop. In-stead, the goal is to generate enough respect and caring among members to support and confront one another with honest feedback, and to make clear that there are others who have a kind of understanding about what has hap-pened, and that it is safe to make explicit what has happened in the traumatic event.

Members should anticipate that part of the uncovering and working-through process will likely involve increased intrusive symptomatology. This should be explicitly predicted so that when it occurs it will not be an-other experience that is out of their control and unanticipated. The capacity to anticipate a worsening of intrusive symptoms can be a very powerful brake on the anxiety generated by the symptoms. Cotherapists need to em-phasize that during these periods it will be crucial to refrain from previous avoidance behaviors (e.g., substance abuse, binge eating, compulsive ritu-als), to share content and process with group and utilize support systems outside the group (e.g., individual therapists, friends, AA, church) as well as the other members in the group sessions.

Develop a group culture where there is an expectation that the sessions will begin on time and end on time, thereby providing greater external structure and enabling the members to better self-monitor their disclosure and opening up. This can be used as an example that appropriate amounts of control are helpful and reassuring.

Commitment to the Group

Members should be prepared to make a firm commitment to the group by the end of the third session, if they have not been willing to do so before. If someone is unable to do so, it is preferable to have him or her leave the group at this point, rather than to have him or her leave later, as most who cannot make a commitment by the end of the third session will most likely not remain in the group. Members must make a commitment to see the group through to completion, if they wish to continue attending after a few sessions in Phase I. The guideline of Session 3 may be shortened or length-ened a bit to fit the ongoing group process. Not having to make a commit-ment immediately is recognition by cotherapists that allowing members to assess if the group feels safe is a natural and necessary process. On the other hand, requiring a firm commitment early on is a facet of establishing safety and establishing the willingness to relinquish avoidance. Candid public commitment is an indication of the members' motivation for change and a statement of their awareness that recovery requires sacrifice and dedication. Finally, commitment helps make being in the group real.

Cotherapists should stress that members will not and cannot be forced to stay, but since unanticipated losses are a significant aspect of the genesis of PTSD symptoms, members are asked to take this commitment seriously. Cotherapists need to remind members that the group will at times be very difficult, and though there may be a powerful urge to leave, seeing the group through to completion is typically the wisest course. This unvarnished forecast assists members in weathering the difficult moments in the treatment since they are forewarned. As well, the candidness of the cotherapists shows by example the stance to be taken toward difficult material.

Withdrawal from the Group

In the event one or more of the members is unable or unwilling to make a commitment during the designated session, cotherapists need to urge any individual who has decided not to continue to briefly attend the subsequent meeting. There are several reasons for this request. First, for the individual who is not continuing, it provides an opportunity to consolidate his or her thinking and to become clear about the circumstances that would allow a commitment to be made. Second, the members who remain are afforded an opportunity to show compassion to someone whose struggle is still too overwhelming to examine. Third, a face-to-face farewell offers those leaving as well as those staying the option to deal with loss in a different and more constructive fashion than is likely to have characterized the traumatic event.

After the farewells occur, the cotherapists should make explicit that all who have agreed to see the group through to termination need to devote as much effort as required to finish the group and to adhere to the ground rules. In spite of this, in a delicate but realistic way, the cotherapists should acknowledge the possibility that of those who remain, something may emerge that leads a member to permanently discontinue attending the group. Should this rare event happen, the departing member is also urged to attend one final session to say farewell. The group work can begin after the farewell.

After the group loses a member, whether after committing to attend or before, a gateway for a discussion of loss materializes. As we have described earlier, successfully dealing with loss is the keystone to recapturing the psychological stability and equilibrium that has been severely disrupted by the traumatic event. Cotherapists have two tasks to accomplish in parallel. The first is facilitation of an exploration of the reactions to the loss of members from the group. The discussion may immediately focus on painful aspects of loss, a direction that hopefully will foster a clearer recognition of the realistic meaning of loss for one or more members. More probable, especially in the earliest sessions, the response of the group members will be one of minimization, avoidance, and denial that the departure is a notable event. Therefore, its emotional consequences are nonexistent, warranting no discussion. Such a response in the group process requires comment by the cotherapists, beginning by noting what has occurred. The subsequent direction will be shaped by the response to the comment as well as an assessment of the likelihood that further exploration will be helpful rather than harmful.

The second task for the cotherapists is to carefully note the individual response of each member. Note the stance that each member adopts. As well, member reactions contribute important data to further develop the cotherapists' understanding of the place of loss in the inner working model of each member.

Topic for the Day

In Phase I, except for the first session, each session has one "Topic for the Day," which are vehicles for the delivery of the psychoeducational content. These provide a structure that assists group members in discussing symptoms, revealing personally salient information, and beginning to build trust in the cotherapists and one another. During these discussions members should *not* tell their stories. Instead, the goal is to build group cohesion and develop knowledge and understanding about PTSD to set the stage for disclosure when that occurs. When the group process requires, a second topic may be introduced.

Trust

A variety of types of trust can be discussed. The first is trust in self. The second is trust in the other members of the group. The third is trust in the cotherapists. Make clear that these may not develop at the same time or speed. Care should be taken to ensure that trust be neither premature, that is, headlong before determining that the situation is actually safe, nor unnecessarily withheld, that is, holding back long after determining that it is likely safe to begin to open up.

Social Isolation

Social isolation is a common problem in chronic PTSD. A number of factors contribute to it, and the goal of the discussion is to have members acknowledge if social isolation is a problem, and if so, in what way. Potential factors contributing to social isolation are: concerns for safety; fear of intimacy; avoiding reminders; and shame and guilt. Other factors may contribute and members and leaders are encouraged to generate them.

Cotherapists should point out the negative consequences of social isolation. Choosing to become or remain socially isolated may cause the breakdown of a potentially available support system and render it unavailable. Without a viable support system, social isolation becomes self-reinforcing. Social isolation may reinforce the conception that the world is a dangerous place and since human connections are minimized, contrary evidence is sparse or nonexistent, thus interfering with opportunities to disconfirm assumptions about others and prevent learning new ways of behaving or relating to others. Finally, social isolation may inhibit development of understanding the ways in which views of self can change when one shares one's story, since the story is not shared. The general goal in discussing social isolation is to make explicit why it occurs, what its effects are, what the impediments are to breaking through social isolation, and how these impediments can be overcome.

Fear in General and Fear of Repetition in Particular

The discussion on fear should focus on the impact that fear of repetition has on everyday conduct in terms of limitations on freedom, concerns about safety, ideas about foreigners and terrorists, and the ability of members to go where and when they want to. Impact on issues of agency, competence, and self-care are all recruited into the realm of dealing with fear and fear of repetition. Though it may seem unreasonable, since the Twin Towers have been destroyed for example, fear of repetition is a serious concern for individuals exposed to the 9/11 terrorism. It is precisely because the original trauma was so unpredictable that the idea of another unpredictable horror appears to follow logically when the dominant emotion is fear.

A second topic for discussion is the consequences that fear and the experience of fear evoke. Feelings of guilt, shame, sadness, and anger may be closely related to and triggered by fear. Also important is the self-blame for the seemed irrationality that learned fear spawns. Finally, the overall constriction of life possibilities can be an important aspect of being afraid.

These may well involve themes that will be discussed in more detail in later sessions, but that nonetheless are relevant and connected.

A third feature of fear to discuss may involve concerns over performance. That is, performance may appear to be or actually become diminished in physical, intellectual, or psychological realms. This aspect of the impact of fear is often overlooked.

A final aspect of fear involves the whole issue of revictimization and the dampening of vigilance mechanisms so that the occurrence of fear, when it is an appropriate response, remains extinguished. This can lead to poor ability to judge safety and danger, or more commonly, the standard hypervigilance.

The task of the cotherapists is to ensure that members appreciate the actual impact that the generalization of the fear response can have and has had. Themes of necessary actions to survive, including the fight-or-flight response, need to be reconciled with exaggerated expectations about omniscience for not knowing or anticipating the traumatic event or its sequelae, and omnipotence for not being able to foil terrible outcomes or ensure better consequences for oneself or others. The bubble of invulnerability and its repair or restoration are important underlying themes potentially available for discussion with this topic. The necessity of the fear response and consequent lack of trust can often be a key factor in social isolation and problems with intimacy shared by many survivors today. This cascade of reactions has implications for bonding with other group members in the here and now of the group process.

The group setting provides an opportunity to process transference reactions and work through interpersonal issues by physically recreating an atmosphere of safety with others who are initially strangers, and allows participants an opportunity to have a different experience with proximity to potential danger, e.g., attempting to resolve conflict, following through with their commitment, questioning authority, offering support to others, or demonstrating their moral courage. Here-and-now interpretations of intragroup behavior are a critical element of this aspect of treatment.

Shame and Guilt

As with previous topics discussed in Phase I, the discussion of shame and humiliation should be woven into the life-history narrative. Though presentation of the trauma narrative is still premature, shame and stigmatization about being symptomatic is almost universally a concern. When appropriate, discussion should focus on shame and guilt as distortions of responsibility that stem back to childhood ideas of causation and responsibility. This will require exploration of realistic and unrealistic versions of responsibility in terms of symptomatic presentation and concerns about telling the story to others, and being blameworthy. The link to fear and fear of repetition may be helpful to note.

A very important part of this discussion is an explanation of the reasons for guilt. This can assist the members in understanding how they have developed the ideas that they are somehow to blame. Cotherapists should consider discussion of the following ideas regarding exaggerated responsibility: the wish by group members for restoration of the possibility of control over threats and dangers; the connection between expressed shameful self-images of weakness and members' warded-off self-image themes of strength and destructive potential; the all-too-human tendency to have 20/20 hindsight.

A more difficult topic to broach is the need for complete control often manifested through blaming the victim. By blaming the victims, the members, or others, restore the fantasy that control is possible when in fact it is not. Shame and guilt may be experienced as preferable to rage, or as a consequence of profound self-criticism stemming from distorted ideas that control was possible. Thus, this session also leaves the door open for a discussion of rage.

Addressing Specifics of Group's Common Event

The goal in this session is to have as frank and candid a discussion as possible about the specific ramifications of the common traumatic event. In discussing trauma perpetrated by terrorists, commonalities may be less homogeneous than for traumatic events such as rape or motor vehicle accidents. The ramifications may be harder to specify, since the nature of the event can range from "merely" witnessing, to severe physical injury, being near death, or having a family member die with no hope of recovery of a body to bury and mourn. A variety of different themes can be discussed. The choice is as much up to the group members and their experiences as it is to the leaders. One issue for examination is the incomprehensibility and therefore untrustworthiness of those who could sacrifice innocent people for a cause. Another theme is the fear of helplessness in ordinary daily activities. Also relevant in this domain are issues of betrayal, and the linked and omnipresent issue of trust and the risk of being senselessly victimized.

Another aspect of the discussion may concentrate on more behavioral issues. What is the level of appraisal necessary for trust? What about the problem of reminders of trauma during activities reminiscent of the event— being at work at your desk or stopping for a morning espresso? These topics can raise substantial anxiety, but it would be helpful to have them commented upon explicitly.

A review of the conditions required for calm during charged activities can be offered. This includes the capacity for honesty and vulnerability, along with assertiveness and the need to communicate. A reminder of both the potential dangers and the potential rewards of treatment will be useful.

Cotherapists should stress that all instances of daily activities are not the same as the trauma. This can lead to a discussion of contorted perceptions and over-generalizations as a broad theme. Finally, reinforce that daily activities typically occur without malevolent intent.

Preparation for Transition to Phase II

Toward the end of the session, in which the last topic of the day has been concluded, cotherapists must explain that there will be a transition in the next session from topic-oriented groups to sessions in which members will tell their own stories. In advance of this, members ought to be advised that someone will be going first, and that the structure of the sessions will be check-in, decision about who will start telling his or her story with little or no interruption, and then a somewhat extended checkout, at least for the first run-through for each member. The members should be told that they will have forty-five to fifty minutes to tell their stories. Within their zone of comfort and safety at this point, the presentation should be as detailed as it can possibly be with an emphasis on bodily sensations such as sweating, shortness of breath, lack of pain, as well as indicators of peritraumatic dissociation. Finally, attention to how it smelled, how it sounded, as well as other sense memories are valuable.

Leaders Reduce Degree of Structure Imposed

In Phase II of treatment, cotherapists should allow more group-generated discussion and processing of themes when the time for group discussion occurs during each session. This implies greater flexibility in amount of time available for individual sharing and working through of issues. As cotherapists reduce structure, new stages of group development may emerge, e.g., regression at increased anxiety and ambiguity of task, anger at one or both cotherapists, marginalizing one of the members, increased avoidance and resistance, or, alternatively, increased cohesion.

PHASE II

Main Agenda: Telling the Story and Revising the Meaning

The main agenda in this phase of treatment is detailed recounting, by each member in turn, of the circumstances, context, and events of the traumatic event. Not uncommonly, the story in full has actually never been told; broad outlines may have been presented to a few people in the service of dealing with legal, municipal, or emergency service organizations. Thus, what has been told to others has generally been truncated, disjointed, or in some other way, noncoherent and nonintegrated. The purpose of this phase of the group is to give each member the opportunity to make sense of what happened to him or her and what it meant then and what it means now, and how that meaning can change.

The first presentation of the story should occur uninterrupted. Because the first rendition needs to be paced by the member telling the story, questions and interruptions may seriously derail the story. Also, cotherapists need to keenly observe where the affect wells up, what things seem to be skipped, what things are dwelt on, and so forth. These requirements dictate that the pace of the story be controlled by the group member.

Subsequent versions should be prepped by cotherapists based on the ongoing formulation and style that they have observed. If one member tells the story without the connection of emotion, one of the leaders should introduce the retelling in the group with instruction about what to focus on: "This time, as you tell us what happened, try to include how you felt emotionally from moment to moment." If another member has skipped over a piece of the story, one of the leaders would do well to ask the member to be sure to describe this portion in detail. If a member's story is poorly integrated, the leaders can ask that the story be told starting at the beginning, going through the middle, and ending at the end.

Subsequent presentations usually will be uninterrupted by the cotherapists, though at some critical moments they may briefly intervene in an attempt to deepen the affect, e.g., "you just fought back tears." These observations should be approached very cautiously; usually cotherapists can get back to the warded-off affect when the presentation is complete. Other group members should not interrupt these subsequent presentations.

After the telling of the story, group members and cotherapists help the member to process the meaning of the event, and attempt to help the member revise the meanings that have been attributed, for example: "I will never be safe" "I cannot depend on others to help me" "I disobeyed the instruction to check if the other offices were empty as I had been asked to do" "I can never trust anyone again" "I am responsible for what happened to them." This process will be facilitated by considering the themes and issues discussed as follows

Each session should be devoted to only one member's story. The check-in and group process issues need to be monitored carefully; it is unwise to begin a story with less than an hour for the story, processing, and checkout. If other issues have preempted the time for someone telling his or her story, it should be postponed. Acknowledge that despite everyone's knowing about the agenda and the time frames, somehow the telling did not occur. On the other hand, if the main agenda item in this phase is an issue of group process, and all members are agreeable to discussing the process issue, this takes precedence over telling the story.

Selecting the order of who goes first should be a joint decision of the group and the leaders. If cotherapists think that a member who wants to go first is not appropriate, they should assert their authority in a gentle but firm way.

Common Themes and Issues

Cotherapists should be aware of the following common themes and issues. If not discussed spontaneously by the group, the leaders may wish to suggest them as possible topics to explore during this middle phase. Not all areas will be relevant for all members in any one group, nor will all themes be relevant to the stories that each member tells. Nonetheless, the following themes are likely to be important.

Grief and Mourning

Grief work is likely to permeate all phases of treatment, thus, an important aspect of the storytelling is the mobilization of awareness of what is lost and therefore what must be mourned. Cotherapists must establish a culture wherein expression of grief is tolerated and respected. Actively validate the strength and courage it takes to allow the pain, with its fear of endless tears. Some degree of psychoeducational discussion should occur acknowledging the "training" individuals receive in learning to deny, displace, and suppress negative feelings and negative events in life. At the same time, recognize the healing aspects of grief and mourning. Acknowledge that short-term pain may increase and the memory will not disappear, but that the intensity of the image and dysfunctional avoidance behavior will diminish. Differentiation of authentic grief and sadness from fear of endless depression may facilitate the mourning process. In this domain, as in all others, cotherapists must counter the members' tendencies to short circuit grief work. Members may try to "fix" feelings and quickly shift to a problem-solving mode. The task of the group is to become a safe holding environment for members' feelings, allowing adequate time for tears and appropriate silence. During these moments, if the formulation suggests that parental deprivation or non-responsiveness to affect was prominent in the trauma or before, one of the co-therapists may want to acknowledge the painful affect by saying something like, "I'm sorry that this is so painful for you" or "I wish this wasn't so painful for you." These comments acknowledge the affect without being overwhelmed by it, having to deny it, or having to fix it.

Guilt and Shame

The power of guilt and shame stems from the wish for control and power over the reality of the helplessness of the traumatic event. Identification and differentiation of actions that could have been taken from those that could not, and, if relevant, identification of survivor guilt, are important. Story-telling assists in the reiteration of the context of trauma and the events leading up to it. This allows the current fantasied wishes to be explored and allows special attention to differentiating real from fantasied guilty actions. The higher standards of behavior one participant may hold for himself or herself than he or she does for the other members may be quite important to point out. Can they forgive themselves for acts similar to those for which they forgive others? Acknowledging some actual errors means not having to be totally guilty about all actions taken or not taken.

Traumatic events are chaotic, almost by definition. The task of coping psychologically with noncontingent outcomes, both during the trauma and in the present, e.g., unrealistic causal attributions ("If I hadn't been late, she wouldn't have left without me" "I should have known that the Towers were still a target") is one of the main goals of the retelling, as it makes clear and public the irrational internal expectations. The reality of traumatic exposure is that behaviors and outcomes are often overdetermined or heavily dependent upon a multitude of external factors, e.g., bodily injury was likely or even greater danger was likely without coping responses (e.g., running away), none of which is under the control of the victim. Cotherapists should look for projected guilt that manifests as a kind of suspiciousness or paranoia. This may be seen in a lack of trust in others masking a lack of trust in self or as a fear of retaliation or revictimization if some assertive action or stance is taken.

Excitement or Pleasure

Though empirically extremely rare, there is the possibility of the role of pleasure and excitement at some point in the event, likely in the beginning if the individual was successful in coping (the member was able to rescue or save or reduce injury to someone). If this occurred, it can lead to strong conflicting emotions existing simultaneously, e.g., horror/fascination/excitement all laced with a certain unreality before things got clear. If this meaning or stance appears to have played a role in symptomatology, cotherapists need to address these unpleasant considerations.

Reactivation of Latent Trauma

One prevalent aspect of coming to terms with the traumatic event is that it, or its aftermath, including the reactions or lack of reactions of others, may have been a recapitulation of premorbid experiences, e.g., abandonment, which unleashed displaced or repressed rage. More commonly, the traumatic event is experienced and processed in terms of the preexisting conflicts or working models of self that accentuate vulnerabilities. Thus, a part of the telling of the story is to highlight the views of self that continue to interfere with smooth resolution of the traumatic impact. This may be manifest in derivative kinds of revictimization. Cotherapists can look for unconscious repetitive patterns of abuse, rape, or being taken advantage of. In this instance, situations that should raise alarms do not, thus providing the opportunity for further victimization. These behaviors should be explored for secretly held beliefs that this is "punishment" for being bad or confirmation of defectiveness. The role of the repetition compulsion here may be salient, and may be expressed as possible attempts at symbolic mastery. Alternatively, it may stem from guilt and self-punishment. Avoidance may paradoxically create that which is most feared—that others do not understand or cannot be trusted.

Anniversary Reactions and Other Triggers

Many survivors are aware of this phenomenon. Anniversary reactions can be relatively common, even when there is no conscious recollection of relevant dates. Media attention to terrorism or anniversaries of September 11 may lead to a reexperiencing or heightening of intrusive symptoms. Similarly, civilian disasters and events can trigger increased intrusive, avoidant, and physiological responses, e.g., earthquake or reports of other victimizations may exacerbate PTSD symptoms.

Immediate Aftermath of the Event

The nature of the aftermath is itself often a predictor of subsequent PTSD. In terms of support or resources for the member after the event, this aspect may be quite a loaded part of the story. Was a safe person immediately or subsequently available? The response of helpers, family, and friends may be seen as incompetent, too little, too late, indifferent, or frankly re-traumatizing. These themes may well be played out in the transference reactions to cotherapists or to other group members. Part of clarifying the role of authentic helper is to make explicit that group members can feel like both perpetrator and victim because of the impact the telling of their story can have on those close to them. The absence of adequate support is often experienced as subsequent betrayal and loss of control. If law enforcement should be involved in the aftermath of an event experienced by a group member, but the member is uncertain about whether or not to involve law enforcement, this indecision can often be experienced by the member as yet another mistake made. If law enforcement was intrusive this can be experienced as further victimization. In the cases in which a negative decision regarding law enforcement was made, this decision often helps consolidate a participant's guilt or shame. Alternatively, the medical or law enforcement bureaucracy may have evoked anger at "procedures" after the chaos of the traumatic incident, anger at bureaucracy in the face of suffering, or even simultaneous feelings of being both "less than human" and superior to all those who have not experienced trauma.

The reactions of family and friends may have taken a variety of paths as well. Some members may have experienced misguided support or insensitivity, e.g., "What did it feel like?" or "How come you didn't tell anybody?" The trauma may have led to a change in the family system and concomitant difficulty in establishing a new role or place in the family system. Group members or others may have clearly unrealistic expectations as to the transition period required before things begin to settle down and the disruption and symptoms begin to abate. If physical wounds, disabilities, or disfigurements occurred as a consequence of the trauma, the reaction to these may have been problematic.

Cognitive Restructuring

Cotherapists should help the group members, both the storyteller and the other members, identify dysfunctional behavior patterns and coping styles that offer the illusion of control at the expense of realizing actual risks and potential alternatives. These include "all-or-nothing" or "black-and-white thinking," which simplify choices thereby reducing anxiety but simultaneously eliminate too many potential options. Denial should also be identified: "If I don't acknowledge it, it can't hurt me." Similarly, misattributions are especially ripe for change, i.e., externalization of blame, or, more commonly, unrealistic acceptance of responsibility for events and their outcomes far beyond the control of the individual. Another common distorted idea encompasses overgeneralization: "I was too trusting and paid for it, therefore I am totally incompetent."

Another arena in the restructuring or modification of meanings involves the role of emotion and emotional expression. Problems here include the continuation of the "don't-ask-don't-tell" policy that avoids acknowledgment of the emotions associated with the event or even the existence of the event itself. A similar issue involves emotional closeness and vulnerability expressed in the following assertion: "If you don't allow yourself to care about anything or anyone, then you can't be hurt or disappointed." The added sense of safety that this may bring is actually false, since the assumptions can never be disconfirmed, and can actually be an important but overlooked contributor to depression. Similarly, anger can be used as a tool to push people away and limit intimacy.

Recognition of Cascade Effect of Secondary Adversity

The possibility of a downward spiral after the trauma should be understood and comprehended as a secondary consequence. The actual or perceived sense of continued victimization can lead to employment problems, for example, or intimacy problems, which may then lead to hopelessness, thereby setting the stage for further victimization or further loss of social or financial support. This may form a kind of closed-loop system wherein guilt leads to isolation, which increases loneliness, which in turn encourages self-destructive behaviors such as substance abuse as a way to cope with the loneliness. The recognition of the substance abuse induces more guilt, which sets the loop in action again. Breaking this cycle may entail the acceptance of personal responsibility for change by the group member, regardless of what is "fair" or "right."

Impact of Trauma and Subsequent PTSD on Family and Friends

Another area that can be explored, and which frequently encapsulates the main issues in the struggle around the trauma is the member's relationship to his or her family, including children, whether they are a member's children, younger brothers or sisters, or nieces or nephews. Some description of the post-event relationship histories may illuminate aspects of "second-generation PTSD." The discussion may shed light on the way in which a member's worldview has impacted and perhaps been transmitted to family. These phenomena include an unpredictability of when anger would be expressed and the concomitant loss of control that went with it. The group member may describe a pattern in which she or he has become the family's unexpected and unforeseen out-of-control person, possibly augmented by having had to become the family's only parent. Consequently, the children around him or her may learn that anger and blame is a preferred problem-solving strategy. Similarly, by action, deed, or word, children may receive the message that the world is a dangerous place and mistakes can be extremely serious. This can lead to the member often being overly controlling, which can reduce the children's sense of autonomy and independence. His or her fears, anxiety, and self-doubt can be transmitted to the children, and depression and isolation can interfere with children learning appropriate social skills. Sadly, like the victim himself or herself, the children may inevitably feel responsible for a parent's depression, feel like it is their fault, and feel somehow "defective."

The Emphasis in Phase II

Paradoxically, though Phase II comprises the heart of the psychodynamic group treatment approach, specific instructions about how to proceed through each session is not possible. Tasks for telling the story have been presented earlier. The stance that the cotherapists need to take has been identified. The themes that will likely comprise each member's dynamic formulation have been enumerated. The principles of linking current reactions to past working models of self and others have been catalogued in considerable detail. The typical concerns of any psychodynamic treatment, e.g., defenses, conflict, empathic attunement, transference, irrational thinking, and blocking or warding off of painful emotions, apply to the group approach in all respects. Thus, there is little in the way of precise instructions.

Phase II lasts as long as the total session contract allows, or as long as the group members and cotherapists decide, in those groups without a fixed ending session.

PHASE III

Return to Increased Structure

Phase III's sessions deal explicitly with termination of the group. Direct and symbolic termination issues and themes should be actively addressed and processed. These issues and themes include previous experiences with loss and endings that may parallel the process of disbanding the group. Issues to be considered include unspoken farewells and the costs of not being explicit. Cotherapists should emphasize termination as an opportunity for mastery, creating a farewell fundamentally different from the experience of the loss in the trauma. They should be alert to indications of denial that the group is ending, and in so doing, emphasize that termination does not mean bonds, connections, or relationships must be given up. The alternate view is that a change in thinking about new relationship capacities can be maintained. Contact with other group members after the termination of the group is also an option, but should be neither encouraged or discouraged.

One or several members of the group will likely express worries that the conclusion of the group is premature and that therapeutic work remains unfinished. The cotherapists need to reassure these individuals that it is normal to have apprehensions about losing the support and help the group has provided, but that after a short period of time these concerns abate. (Cotherapists also need to be certain that the concerns do not represent deteriorated functioning, in which case facilitating referral is required.) As well, members will benefit from the cotherapists' explicit understanding of members' disappointment that not all important issues could be addressed in time-limited group treatment. Note that experience teaches that no matter how long the treatment lasts, when the end approaches, there is a wish to continue. This feeling may have emerged regarding the time limit for each session—the sense that if meetings were two hours (instead of one) it would be a better experience.

Exploration of the members' fantasies as to how much longer would have been sufficient to feel that ending was appropriate is an important activity in these sessions. As well, discussion of what additional therapeutic work is needed is appropriate. A likely response may be that additional treatment would make a member a new or different person. Another hope may be that further treatment could accomplish the member's return to being who he or she was before the trauma. This discussion provides an opportunity to reiterate the fundamental precept of the goals of treatment: rather than magically turning back the clock, a positive outcome includes being clearer about the meaning of the event, the permanent changes in one's view of self and others that accompanied clarifying the meaning, and reinforcing the recognition of the strength required to achieve this.

Another task of Phase III is engaging in an honest appraisal of the traumatic event and its impact. In describing their current outlook, members should contrast their initial expectations about the group treatment. As this process unfolds, cotherapists may encourage members to provide one another with feedback, especially if someone is overly self-critical, or unrealistic about his or her ability to control the uncontrollable. In the same vein, cotherapists need to be alert to confront members' distortions, overgeneralizations, or selective memory. These may include all-or-nothing judgments, or focusing on the part of the glass that is still empty rather than noticing that the glass now contains considerably more than it did initially. In this process, the cotherapists' goal is to summarize status rather than explore in-depth.

Cotherapists must manage any countertransference feelings of responsibility for members' recovery or lack of it. To convey this sense runs the risk of communicating to members that they are unable to care for themselves. Exposure to trauma cannot be undone; rather, it can be assimilated and adaptation can occur. Thus, cotherapists must remain alert to avoid interventions that undermine the reality that a lot of good work has occurred.

Group members will likely feel bonded with other members, acknowledging the importance of the group in their lives. Cotherapists may point out that this is based on an implicit need for intimacy, despite the possibility of conscious protests and the inherent risk of loss and vulnerability members may feel. Satisfying this need will be a critical component of their recovery. Similarly, cotherapists should express authentically their own feelings about the impending termination. Usually this will include feeling privileged to have been a part of members' lives, and a statement that they too will miss meeting with the group. These authentic expressions are emotionally powerful for members as well as cotherapists.

If it arises in the process, cotherapists need to interpret and discuss members' anger at feeling abandoned and betrayed by the cotherapists (and parent organization), or feeling that they are getting "kicked out" before they have finished their work. Cotherapists can help anticipate group and individual regression that may be manifest in increased symptomatology, superficial participation covering overemotional withdrawal from the group, or excessive intragroup anger or hostility. Nevertheless, it may well be accurate that some or all members have not completed all the therapeutic work that they could.

As a part of the discussion of what may follow after the end of the group, the cotherapists should evaluate individual support systems and the extent to which they will be used, strategizing with members about how best to utilize their resources. These include further treatment at the site or elsewhere, the use of support groups such as AA, reliance on religious affiliations, friends, family members, doing volunteer work, or maintaining contact with other group members.

It will be important to process what the members feel they did get out of the group, as well as what they did not. This discussion should be as candid as possible. This is a good opportunity for the group to share feelings and offer feedback. Giving their impressions of the changes they have seen in themselves, as well as one another, and clarifying how the meaning of what happened to them has been modified by the group will be helpful.

As part of the termination process, cotherapists and group members may consider some sort of ritual. This could be an exchange of members' insights, a kind of graduation ceremony, or some type of symbolic gesture that marks an important theme which emerged in the course of the group meetings, such as mourning and accepting the permanence of loss. Requests for a farewell dinner, another outing, or any extra-group activity should be discouraged, as it blurs the function of the group treatment and is likely a wish to postpone the real termination. If the members insist, the cotherapists should acknowledge that such decisions are ultimately made by the members, but that they will not participate.

CONCLUSION

The core idea of our psychodynamic approach to treating PTSD symptoms following exposure to traumatic stress is that the primary task for each person who has lived through trauma is to come to terms with what it has meant to him or her. Posttrauma, the typical assumptions of safety about the world are rudely punctured and the psychological bubble of invulnerability is distressingly pierced. The individual suffers tremendous upset of emotions and cognitive equilibrium, since psychologically the world has turned upside down. The malevolence and cold indifference to innocent victims of terrorists may make these reactions even more acutely painful.

The recovery from exposure to traumatic stress involves regaining the psychological equilibrium that was lost as a result of the traumatic exposure. However, unlike the Bobo doll of childhood, whose sand-filled base quickly righted itself with no additional effort or input from the child who knocked it off its pins, traumatic stress for many people results in a loss of psychological equilibrium that does not right itself naturally or easily. In those circumstances, the common responses to exposure are not self-limiting and may result in PTSD or at the least symptomatology that almost meets the formal criteria.

Because most individuals do not develop PTSD after exposure, discovering who develops symptoms has been of keen interest. Recent research regarding predictors of PTSD suggests that psychological processes proximal to the traumatic event such as peritraumatic dissociation, peritraumatic emotionality, or social support are most predictive (Ozer et al., 2003). Given that these factors have been shown to be most predictive, rather than more demographic or historical (e.g., family history of psychiatric disorder, previous history of trauma), it is not surprising that mitigating PTSD requires concentrating on psychological issues—self and other.

The description and guidelines of the group approach we have offered focuses on the same psychological processes presumed to characterize psychological processes in all people. These include conflicts, the effects of early experience, differentiating the meaning of behavior from the behavior itself (e.g., staying silent to forget compared to staying silent to remember), and specific themes observed to characterize loss and mourning. Specification of cotherapists' exact maneuvers at exact moments in exact sequences is not realistic. What is needed for psychodynamic group treatment for trauma is an intimate familiarity with psychodynamic principles of formulation, and experience applying them. This experience provides the tools required to allow members to feel understood. If group members feel understood, the process of recovery and healing will occur naturally.

REFERENCES

Bowlby, J. (1973). *Attachment and loss: Separation: Anxiety and anger* (Vol. 2). New York: Basic Books.

Bowlby, J. (1988). *A secure base: Parent-child attachment and healthy human development*. New York: Basic Books.

Charney, D. S., Deutch, A. Y., Krystal, J. H., Southwick, S. M., & Davis, M. (1993). Psychobiologic mechanisms of posttraumatic stress disorder. *Archives of General Psychiatry, 50,* 295-305.

Foa, E. B., Riggs, D. S., Massie, E. D., & Yarczower, M. (1995). The impact of fear activation and anger on the efficacy of exposure treatment for posttraumatic stress disorder. *Behavior Therapy, 26,* 487-499.

Foy, D. W., Glynn, S. M., Schnurr, P. P., Jankowski, M. K., Wattenberg, M., Weiss, D. S., Marmar, C. R., & Gusman, F. (2001). Group psychotherapy for PTSD. In J. P. Wilson, M. J. Friedman, & J. D. Lindy (Eds.), *Treating psychological trauma and PTSD* (pp. 183-202). New York: Guilford Press.

Goodman, M. & Weiss, D. S. (1998). Double trauma: A group therapy approach for Vietnam veterans suffering from war and childhood trauma. *International Journal of Group Psychotherapy, 48,* 39-54.

Goodman, M. & Weiss, D. S. (2000). Initiating, screening, and maintaining psychotherapy groups for traumatized patients. In R. H. Klein & V. L. Schermer (Eds.),

Group psychotherapy for psychological trauma (pp. 47-63). New York: Guilford Press.

Hiley-Young, B. (1992). Trauma reactivation assessment and treatment: Integrative case examples. *Journal of Traumatic Stress, 5,* 545-555.

Horowitz, M. J. (1986). *Stress response syndromes* (Second Edition). Northvale, NJ: Jason Aronson.

Kessler, R. C., Sonnega, A., Bromet, E., Hughes, M., & Nelson, C. B. (1995). Posttraumatic stress disorder in the National Comorbidity Survey. *Archives of General Psychiatry, 52,* 1048-1060.

Meehl, P. E. (1995). Bootstraps taxometrics. Solving the classification problem in psychopathology. *American Psychologist, 50,* 266-275.

Milbrath, C., Bond, M., Cooper, S., Znoj, H. J., Horowitz, M. J., & Perry, J. C. (1999). Sequential consequences of therapists' interventions. *Journal of Psychotherapy Practice and Research, 8,* 40-54.

Ozer, E. J., Best, S. R., Lipsey, T. L., & Weiss, D. S. (2003). Predictors of posttraumatic stress disorder symptoms in adults: A meta-analysis. *Psychological Bulletin, 129,* 52-73.

Schnurr, P. P., Friedman, M. J., Foy, D. W., Shea, M. T., Hsieh, F. Y., Lavori, P. W., Glynn, S. M., Wattenberg, M., & Bernardy, N. C. (2003). Randomized trial of trauma-focused group therapy for posttraumatic stress disorder: Results from a Department of Veterans Affairs cooperative study. *Archives of General Psychiatry, 60,* 481-489.

Stern, D. N. (1985). *The interpersonal world of the infant: A view from psychoanalysis and developmental psychology.* New York: Basic Books.

Stinson, C. H., Milbrath, C., & Horowitz, M. J. (1995). Dysfluency and topic orientation in bereaved individuals: Bridging individual and group studies. *Journal of Consulting and Clinical Psychology, 63,* 37-45.

Tennant, C. (1988). Parental loss in childhood: Its effect in adult life. *Archives of General Psychiatry, 45,* 1045-1050.

Weiss, D. S. (1993). Psychological processes in traumatic stress. *Journal of Social Behavior and Personality, 8,* 3-28.

Weiss, D. S. (2001). *Coping with trauma.* Available online at: http://psych.ucsf.edu/war_section/war_section.htm.

Weiss, D. S. & Marmar, C. R. (1993). Teaching time-limited dynamic psychotherapy for post-traumatic stress disorder and psychological grief. *Psychotherapy, 30,* 589-591.

Yehuda, R. (2002). Posttraumatic stress disorder. *New England Journal of Medicine, 34,* 108-114.

Zilberg, N. J. & Horowitz, M. J. (1983). Regressive alterations of the self concept. *American Journal of Psychiatry, 140,* 284-289.

Chapter 21

Groups for Mental Health Professionals Working with Survivors

Yael Danieli

In 1930 [1929] Freud wrote,

> No matter how much we may shrink with horror from certain situa-
> tions—of a galley-slave in antiquity, . . . of a victim of the Holy Inqui-
> sition, of a Jew awaiting a pogrom—*it is nevertheless impossible for
> us to feel our way into such people*—to divine the changes which orig-
> inal obtuseness of mind, a gradual stupefying process, the cessation of
> expectations . . . have produced upon their receptivity to sensations of
> pleasure and unpleasure. Moreover, in the case of the most extreme
> possibility of suffering, special mental protective devices are brought
> into operation. *It seems to me unprofitable to pursue this aspect of the
> problem any further.* (p. 89) (Italics author's)

Although this passage was written before the Nazi Holocaust it poignantly foreshadowed psychotherapists' participation in the *conspiracy of
silence* (Danieli, 1982b) that has existed between mental health professionals and Nazi Holocaust survivors and their children. This conspiracy of silence is not confined to psychotherapists but is part of the conspiracy of si-

---

Editor's note: This chapter presents countertransference experiences of therapists
working with the survivors of extreme trauma, in this case the Holocaust. Little has been
written about the experience of therapists working with other survivors of massive
trauma. The research, exercise process, and guidelines presented here thus serve to help
therapists prepare themselves and manage their reactions. The exact nature of the
countertransference will differ depending upon the therapist's own history, the nature of
the trauma for the survivor, and the impact of the trauma (as in a terrorist event) that may
have affected the therapist directly or indirectly. We believe the lessons to be learned
from this chapter are significant. Work with victims of the Holocaust and other traumatic events such as the attacks on 9/11 provide significant material related to countertransference in response to treating the victims of terrorism and other acts of mass
murder.

lence that has characterized the interaction between survivors and society at large since the end of World War II.

Survivors and children of survivors have frequently complained of neglect and avoidance of their Holocaust experiences by mental health professionals. This is corroborated by ample documentation in the literature, primarily clinical, which very often contains the authors' reports of extreme "countertransference reactions." A comprehensive review of the literature on the "countertransference reactions" reported by reparation examiners, psychotherapists, and researchers working with Holocaust survivors and their children can be found in Danieli (1982b). Despite the fact that the need was recognized and the call for help clearly stated by Friedman and others by 1948 (Friedman, 1948), and despite the vast literature on the long-term effects of the Holocaust published in the following decades, until the 1970s (Danieli, 1982a), any attempt to develop a structured program for helping survivors of the Nazi Holocaust and their children to reintegrate into society was abortive. In truth, after liberation, as during the war, the survivors were victims of a pervasive societal reaction comprised of indifference, avoidance, repression, and denial of their experiences. Shunned, abandoned, and betrayed by society, survivors could share the most horrifying and painful period of their lives and their immense losses only with their children, with fellow survivors, or even worse, with no one. The most pervasive consequence of the conspiracy of silence for survivors and their children has been a profound sense of isolation, loneliness, and alienation that exacerbated their mistrust of humanity and made their task of mourning and integration impossible.

Elsewhere I have described some of the negative and obtuse societal attitudes and reactions and some of the survivors' fears that contributed to the long-term conspiracy of silence between Holocaust survivors and society (Danieli, 1981a,b). Also discussed were the harmful long-term effects of this larger-scale silence upon the survivors and their families and upon their subsequent integration into the postwar society, which further impeded the possibility of intrapsychic integration and healing (Danieli, 1985, 1998).

The phrase *conspiracy of silence* has also been used to describe the typical interaction of Holocaust survivors and their children with psychotherapists when Holocaust experiences were mentioned or recounted (for example, see Barocas & Barocas, 1979; Krystal & Niederland, 1968; Tanay, 1968). Originally, Niederland (1964) described the phenomenon as

> the tendency to gloss over [which] appears to be widespread in both doctors and patients—likely enhanced in the latter by denial and guilt, and in the former by anxiety at being brought face to face with the stark horror of the patient's experience. (p. 461)

In 1968, Niederland added, "Insofar as it cannot be true, a kind of tacit agreement is reached between the patient and doctor—an agreement to gloss over, and thereby to ignore the potentially traumatic data" in a "flight from horror" on the part of the psychotherapist (Niederland,1968, pp. 62-63).

Psychotherapists and researchers who have interviewed survivors and their children, and have worked with them after they have been seen by other therapists, have repeatedly observed that Holocaust experiences were almost totally avoided in previous therapy. This professional avoidance is amply documented in the clinical literature that often contains authors' reports of an extreme "countertransference reaction." The term countertransference is used herein as it has been commonly used to describe therapists' own emotional reactions and difficulties experienced when working with this traumatized population. A comprehensive review of the literature on the "countertransference reactions" reported by reparation examiners, psychotherapists, and researchers working with Holocaust survivors and their children can be found in Danieli (1982b). Some material about therapists' difficulties in working with other massive traumata can be found in Wilson and Lindy (1994). It must be recognized that whereas society has a moral obligation to share its members' pain, psychotherapists and researchers have, in addition, a professional contractual obligation. When they fail to listen, explore, and understand, they too inflict the "trauma after the trauma" (Rappaport, 1968) or "the 'second injury' to victims" (Symonds, 1980) by maintaining the conspiracy of silence.

Many survivors suffer amnesia of their lives before the Holocaust, whereas others idealize their pre-World War II life and continue to live psychologically in that time period, being unable to recall their war experiences. This may be true in the long-term for 9/11 and other victim/survivors of terrorism (Danieli, Brom, & Sills, 2005). The therapist is thus confronted with discontinuity and disruption on all levels in the order of living—uprootedness, and loss of families, communities, homes and countries, and values. Recreating a sense of rootedness and continuity and meaningfully integrating the Holocaust into their lives are major struggles for survivors and their children. When psychotherapists focus only on certain periods in the patients' lives to the exclusion of others, they may hinder the recovery process and perpetuate their sense of disruption and discontinuity.

TRAINING PROFESSIONALS TO DEAL WITH TRAUMA

Traditional training generally does not prepare professionals to deal with *massive trauma* and its long-term effects (see also Wallerstein, 1973). One psychotherapist stated, "I think the biggest problem is not having any guidelines to deal with the Holocaust. The fear is of going into uncharted territory where your only guide is your patient, and yet you are in the role of the expert." This has been experienced by therapists working with victim/survivors of terrorism. Following the September 11, 2001, terrorist attacks, therapists heard stories and images that they never had before. Those who suffered double (or more) exposure(s) have their patients' images in addition to their own to cope with. In supervision, a therapist asked, "Whose September 11 is it?" (Danieli, Engdahl, & Schlenger, 2003). Indeed, when we speak of integration for severely victimized people we speak of integrating rupture and the *extraordinary* into one's life—that is, confronting and incorporating aspects of human experiences that are not normally encountered. The task of therapy then is to help survivors and children of survivors achieve integration of an experience that has halted the normal flow of life.

Although information cannot undo unconscious reactions, knowledge about trauma(ta) in their historic context does provide therapists with factual and, for example, gender, ethnic, racial, religious, cultural, and political perspectives that help him or her know what to look for, what may be missing in the survivors' (or offsprings') accounts of their experiences, and what types of questions to ask.

Countertransference reactions interfere in the process of acquiring the knowledge about the trauma. A participant in the Countertransference and Trauma seminar offered by the Group Project for Holocaust Survivors and their Children reported about attempting to read one of the assigned homework books,

> I can read it only a little bit at a time. Otherwise my head tunes off. I can't pay attention. It was too painful for me. He [Des Pres, 1976] writes so vividly, the imagery, that I couldn't read some of it, I just had to skip over sections. I wasn't tuning out. I was like putting the book down and looking for a less intense part of the book to read. It must have been very hard to put something like this into words.

The ensuing seminar discussion, in part, explored the implications of these reactions to her ability to listen fully to her survivor patients.

In another training seminar, a therapist said, "It's sitting at my bed table and I keep not reading it. I have read some [of the] other books. . . . I mean, I keep touching it . . . but I just could not get myself to read it."

After 9/11, one therapist described,

looking on the Internet for descriptions of some of the heroes of the plane crashes, I became so distraught after reading only a few lines that I had to stop reading. The descriptions made them not only real to me but known in detail, like a friend. I was tearfully overwhelmed by images of what they must have gone through. It was too easy to imagine the moments on the plane anticipating their deaths, trying to make peace with what would come, and trying to take the right action. I just couldn't get to know any more of them.

In addition to information about the trauma terrain, familiarity with the growing body of literature on the (long-term) psychological sequelae of the traumata on its survivors and their offspring (see Danieli, 1998) also helps prepare mental health professionals. Nonetheless, they should guard against the simple grouping of individuals as "survivors," who are expected to exhibit the same "survivor syndrome" (Krystal & Niederland, 1968), or the same post-traumatic stress disorder (PTSD) symptom picture (American Psychiatric Association, 1994), and the expectation that children of survivors will manifest a single transmitted "child-of-survivor syndrome" (e.g., Phillips, 1978).

Countertransference reactions are integral to, ubiquitous, and expected in our work. Our work calls on us to confront, with our patients and within ourselves, extraordinary human experiences. This confrontation is profoundly humbling in that at all times these experiences try our view of the world we live in and challenge the limits of our humanity.

EVENT COUNTERTRANSFERENCE AND TRAINING

The Group Project for Holocaust Survivors and their Children has provided short- and long-term "Countertransference and Training" seminars and individual supervision to professionals since 1975. In-house, on-the-job training/supervision has been offered by other agencies. The International Society for Traumatic Stress Studies has begun to ameliorate this lack of training, among other activities, through its *Initial Report of The Presidential Task Force on Curriculum, Education, and Training* (Danieli & Krystal, 1989). This report contained model curricula formulated by leading international specialists in the field organized into subcommittees representing different technical specialties or interests including psychiatry, psychology, social work, nursing, creative arts therapy, clergy, and media; organizations, institutions, and public health; paraprofessionals and other professionals; and undergraduate education. The need to cope with and work through countertransference difficulties was recognized as imperative and necessary to optimize training in this field by all expert groups.

The ensuing literature reflected a growing realization among professionals working with other victims/survivors of the need to describe, understand, and organize different elements and aspects of the conspiracy of silence, however, not always analyzing the phenomena fully as an interaction. Haley (1974), Blank (1985), and Parson (1988) reported on and Lindy (1987) adapted and revised the countertransference categories found by Danieli (1982b) to compare and contrast them with responses from therapists of Vietnam veterans with post-traumatic stress disorder. Comas-Diaz and Padilla (1990), and Fischman (1991) discussed countertransference themes related to torture; Mollica (1988) and Kinzie (1989), to refugees; Chu (1988) and Kluft (1989) to multiple personality disorder (MPD); McCann and Pearlman (1990), who similarly named these phenomena "vicarious traumatization," to adult survivors (see also Pearlman & Saakvitne, 1995), and Herman (1992) to adult survivors of childhood sexual abuse among others, to name but some. Wilson and Lindy (1994) compiled a cross-population volume regarding countertransference, and Dahlenberg (2000) reports of her own work in this context. Others use terms such as *secondary traumatic stress* (Hudnall Stamm, 1995), *burnout* (Maslach, 1982), or *compassion fatigue* (Figley, 1995).

These insights and hypotheses about the ubiquity of countertransference reactions in other victim populations have now moved to the forefront of our concern in the preparation and training of professionals who work with victims and trauma survivors.

Processing Event Countertransference

This section* examines how regarding event countertransference as dimensions of a professional's inner, or intrapsychic, conspiracy of silence about the trauma allows us the possibility to explore and confront these reactions to the trauma events prior to and independent of the therapeutic encounter with the victim/survivor patient, in a variety of training and supervisory settings and by ourselves.[1] Following is an exercise process, which I

---

*Portions of this section appeared previously in Danieli, Y. (Copyright 1993). Countertransference and trauma: Self healing and training issues. In M. B. Williams and J. F. Sommer Jr. (Eds.), *Handbook of Post-Traumatic Therapy*. Westport, CT: Greenwood/Praeger Publishing Co. Reprinted with permission of Greenwood Publishing Group, Inc. Westport, CT; and in Danieli, Y. (Copyright 1994). Countertransference, trauma and training. In J. P. Wilson and J. Lindy (Eds.), *Countertransference in the Treatment of PTSD*. New York: Guilford Press. Reprinted with permission of the Guilford Press.

developed over the past three decades, that has been proven helpful in numerous workshops, training institutes, "debriefing"(for critical review, see Raphael & Ursano, 2002) of "front liners," short- and long-term seminars, and in consultative, short- and long-term supervisory relationships around the world, to work through event countertransference.[2] Although it originally evolved, and is still done optimally, as a part of a group experience, it can also be done alone to assist the clinician working privately. As one veteran traumatologist who uses it regularly stated, "It is like taking an inner shower when I am at an impasse with . . . [a] patient. . . ."

Instructions for Participants

In a group setting, participants are asked to arrange the chairs in a circle. After everyone is seated, without any introductions, the leader gives the following instructions:

The first phase of the process will be private, totally between you and yourself. Please prepare at least two large pieces of paper, a pen or a pencil. Create space for yourself. Please don't talk with one another during this first phase.

1. *Systematic deep relaxation.* The session starts with twenty to thirty minutes of systematic, deep relaxation, including guided imagery, to help participants focus internally.

   At the end of the relaxation period, the leader instructs, "Choose the victimization/trauma experience most meaningful to you. Please let yourself focus in to it with as much detail as possible."

2. *Imaging.* "Draw everything and anything, any image that comes to mind when you focus on the experience you chose. Take your time. We have a lot of time. Take all the time you need."

3. *Word association.* "When you have completed this task, turn the page, and please, write down every word that comes to mind when you focus on this experience."

4. *Added reflection and affective associations.* "When you finish this, draw a line underneath the words. Please look through/reflect on the words you wrote. Is there any affect or feeling word that you may have not included? Please add them now."

   "Roam freely around your mind, and add any other word that comes to mind now."

5. *First memory of the trauma.* "When was the *first time* you ever encountered this experience?"

   "What happened, how did it happen?"

"What did you hear?"

"What was it like for you?"

"Who did you hear it from? Or where did you hear it?"

"Go back and explore that situation in your mind with as much detail as you can: What was it like?"

"How old were you?"

"Where are you in the memory?"

"Are you in the kitchen, in the bedroom, living room, in class, in the movies, in the park? Are you watching TV?"

"Are you alone or with other people? With your parents, family, friends?"

"What are you feeling? Do you remember any particular physical sensations?"

"What are you thinking?"

A psychotherapist in the training reported,

> As a child I remember very vividly, I had been scrounging around and found an old carton of pictures. And discovered that my grandmother had lost a sister, and family. I felt terrible, having brought it to everybody's memory. And I knew nothing about it at the time. And everybody, of course, was crying and very upset when I brought down this box and said "who are all these people?" I was at my grandparents and I asked them. I assumed the pictures were from Europe. And I asked them. And the result was everybody was crying. . . . I still feel guilty and sad. . .

Five sessions later this therapist presented a case and realized that his difficulties with asking patients questions were related to this memory. Working through this memory helped place one source of his difficulty and enabled him to explore patients' issues more freely.

6. *Choices and beliefs.* "Are you making any *choices* about life, about people, about the world, about yourself at the time? Decisions such as

    'Because this happened. . . ,' or, 'This means that life is. . . that people are . . . that the world is. . .' What are you telling yourself, are you coming to any conclusions? This is very important. Stay with that."

7. *Continuity and discontinuity of self.* "Think of yourself today, look at that situation, are you still holding those choices? Do you still believe what you concluded then? Would you say 'this is still me' or, 'this is not me anymore'? What is the difference, what changed and why?"

8. *Sharing with others.* "Have you talked with other people about it? Who did you talk to?" (Both in the past and now.)
   "What was their reaction?"
   "What was your reaction to their reaction?"
9. *Secrets: Not sharing with others.* "Is there anything about this that you haven't told anyone, that you decided is not to be talked about, that it's 'unspeakable'?"
   "Is there any part of it that you feel is totally your secret, that you dealt with all alone and kept to yourself? If there is, please put it into words such as, 'I haven't shared it because . . .' or, 'I am very hesitant to share it because. . .'"
   "Would you please mention the particular people with whom you won't share it, and why?"
10. *Personal knowledge of survivors.* "Moving to another aspect of the interpersonal realm, do you personally know survivors of the trauma you chose to focus on or their family members—as friends, neighbors, or colleagues?"
11. *Self secrets.* "There are secrets we keep from others to protect either ourselves or them, and there are self-secrets. Take your time. This is very important. Imagine the situation of the very first time you ever heard anything about the event. Roam inside your mind, like taking a slow stroll. Is there anything about it that you have never talked to yourself about, a secret you have kept from yourself? An area that you have sort of pushed away or kept at arms length from yourself? Or about which you say to yourself, 'I can't handle that.' If this is too painful, try to breathe through it. Why is it the one thing that was too much for you? What haven't you put into words yet, that is still lurking in that corner of your mind you have not looked into? You can draw it first, and when you are ready, please put it into words."
12. *Personal relationship to the trauma.* "What is your *personal* relationship to the trauma? Please write the answers, because even the way you write makes a difference. Did your place of birth figure in your relationship to the trauma? Does your age figure in your relationship to the trauma? Describe both *when* and *how* you experienced it, and your reactions."
13. *Identity dimensions.* "How do the following dimensions of your identity figure in the choices you made, or influence your relationship to the experience?"
    religious
    spiritual
    ethnic
    family

cultural
political
(socioeconomic) class
racial
gender
health
national
international, identity?

"You can answer these one by one, for both then and now. If there is any dimension that makes sense to you which has not been mentioned, please add it."

14. *Professional relationship to the trauma.* "Let us move to your professional self, what is your professional discipline? How long have you been working in it? What is your professional relationship to the trauma on which you chose to focus? Within your professional practice, have you seen survivors or offspring of survivors of the trauma you chose? How many?"

15. *Therapeutic orientation.* "What therapeutic modality did you employ? Emergency/crisis intervention, short-, long-term, individual, family, and/or group therapy? Was it in an inpatient or outpatient basis? What modality did [or would] you find most useful [for yourself or others], and why?"

16. *Victim and trauma survivor populations.* "Was it the only victim/survivor population you have worked with professionally? Please, list others. It is good to review our work every so often."

17. *Training in trauma work.* "Have you ever been trained to work with victim/survivors of trauma?—In school, on the job? If so, what have you found to be the crucial elements of your training without which you would not feel prepared to do the job?"

One trauma therapist retorted, "Other than my personal experience I really had to go by the seat of my pants and not by what I was taught in school."

UNDERSTANDING THE PROCESS

The process described serves both to begin the group and to map out the issues for the group. In addition to being generated by the experience of trauma itself, it is based conceptually on the author's "Trauma and the Continuity of Self: A Multidimensional, Multidisciplinary, Integrative (TCMI) Framework," which views trauma as affecting the affected individual's identity (Danieli, 1998).

The sequence of the *first phase* of processing event countertransference is from the immediate visual imagery, through free associations to the more verbal-cognitive material. It then moves to articulate how the trauma fits within the therapist's experience, personal and interpersonal development, and the gender, racial, ethnic, religious, cultural, and political realms of her or his life. It begins with one's *private* world of trauma and proceeds through the context of one's interpersonal life to one's professional work. As one psychotherapist described it,

> You reexperience the trauma through this. It takes you from the picture, being very concrete . . . like the way the trauma occurs. You are very shocked and numb, shocked at recognizing your own reactions, their depth and intensity. And then gradually words, and then not stopping there but go into feelings that you don't think of and don't have time to think of. And, like what happens in the retelling, putting things into words, from the impersonal to the inside. But then it pulls you out, to the professional. It lifts you back into reality so the therapist is not stuck in it.

Participants in group settings have frequently remarked on the feeling of intimacy that permeates the room even though the first phase of the process takes place in silence, perhaps reflecting the sense that it is opening ourselves to ourselves that allows for intimacy.

The *second phase* of the process works best in a group setting. This is the sharing and exploring phase. As with victim and trauma survivors, group therapists are able to explore and comprehend the consequences of the traumata they have experienced directly or indirectly in their lives and the conspiracy of silence that frequently follows them, and share their feelings and concerns with one another. The group modality thus serves to counteract their own sense of isolation and alienation about working with trauma. As one psychotherapist described it,

> You are invited for a Saturday night dinner or a picnic by very well-meaning people who want to connect you with other people whom they think you may like, and somebody introduces you as a person working with Holocaust survivors, and then you are expected to make small talk. It's like being in a crazy warp. And you are expected to entertain people with your work. You feel the same as you feel after the death of a close person: Distance. As a result we ourselves devalue small talk because we feel distant, potentially reproaching banter and relief.

When deeply involved in interviewing for the book day and night, I remember feeling like I had a double life. When with friends Saturday night I didn't dare to say anything. Carrying this burden, becoming deeper and deeper involved, do you have the right to disturb other people? R.K. and M.T. [survivor friends] say that the survivor is an irritant. When dealing with traumatized people you begin to think of yourself as possibly an irritant to ordinary folk you interact with, family and colleagues. I recall being referred to in my department as "Holocaust Lady," and described as "obsessed" and "overreacting," as if the material emanates and, like the survivors, you are found to be so irritating that you have to be put in a box, like a freak. It is victimization, not vicarious victimization of the caregiver. Really it is thanks to the International Society for Traumatic Stress Studies that we all began to feel that there is a place where we won't be regarded as irritating, weird people.

A traumatologist who has worked with Vietnam veterans since the early 1970s similarly described being ostracized by his colleagues who,

exempted from the war, . . . projected [their] strong antiwar sentiments toward the veterans themselves, and by extension to me, making nasty remarks such as "those guys shouldn't be allowed in here." Increasingly disturbed and moved by how the Vietnam War affected those who fought it, . . . [and finding it] difficult to convey to others, who were not interested, what I was learning and experiencing, I began to feel alone and isolated . . . [until]. . . I felt more in common with Vietnam veterans than with anyone else . . . offended by insensitive questions such as, "Are they all screwed up?" and being protective of them. Through the grapevine I learned that my colleagues, calling me the "Vietnam Vet" guy, thought I was crazy and had some bizarre reason [for] what I was doing. I thought to myself, "Fuck them, fuck 'em all."

I existed in between different worlds. . . . There were years of loneliness, pain, searching, and self-questioning. However, one thing was clear: I would never again be a traditional academic . . . or clinician. My life had changed forever and there was no turning back. . . . Among the things that made a difference was a growing affiliation with others doing work with trauma survivors [which was] not only reassuring but validating my commitment . . .

The network of colleagues around the country became a kind of family: trusted friends on whom I could call to sort out my feelings and the impact of the work. . . . I now believe that everyone involved in our field have to be profoundly affected by the work because it impacts the soul of helpers in the same way that trauma scars the soul of survivors.

Working privately with incest survivors, a psychotherapist stated,

> Individuals on their own have no place to go with it. I had such a lonely feeling about this work and felt so shameful for having to do it. It was like digging ditches. You don't tell anybody what you do. You don't tell them how dirty your hands and your feelings got. It's a put-down, because I am associated with something so horrible and terrible, and with being so helpless, that I became identified with the survivors. After sharing with colleagues, however, there was a different depth of feeling, attachment, and identity. Before there was no one. After that I felt different.

Elsewhere, when discussing the value of group modalities for the victim/survivor patients, I suggested that groups have been particularly helpful in compensating for countertransference reactions. Whereas a therapist alone may feel unable to contain or provide a "holding environment" (Winnicott, 1965) for his or her patient's feelings, the group as a unit is able to. Although any particularly intense interaction invoked by trauma memories may prove too overwhelming to some people present, others invariably come forth with a variety of helpful "holding" reactions. Thus, the group functions as an ideal absorptive entity for abreaction and catharsis of emotions, especially negative ones, that are otherwise experienced as uncontainable (see Krystal, 1988). Finally, the group modality offers a multiplicity of options for expressing, naming, verbalizing, and modulating feelings. It provides a safe place for exploring fantasies, for imagining, "inviting," and taking on the roles and examining their significance in the identity of the participants. Finally, the group encourages and demonstrates mutual support and caring, which ultimately enhances self-care. These considerations also apply to therapists working in groups.

This training process assumes that the most meaningful way to tap into event countertransference is to let it emerge, in a systematic way, from the unique nature of the therapist's experience. She or he can thus be better able to recognize and become familiar with her or his reactions in order to monitor, learn to understand and contain the reactions, and use the experiences preventively and therapeutically. During the sharing phase, when participants describe the process of selecting and drawing the images, they have already put them into words. For example, one psychotherapist related,

> When you said draw a picture, I had the same reaction as always: that there was nothing that I could put on a piece of paper that could, for me, convey the horror that is what I associate with, like this amorphous, just horribleness. Any of the scenes, whether it's the people on the lines to the gas chambers or the barbed wire, or the, you know, I wouldn't know where to start. Just the horror—that is what hits me. And there is nothing that I could pick out. Except that I, and then as I was sitting here, thinking, the thing that strikes me, of course, is . . . what I later wrote: the faces of deaths. It wasn't so much the death but the always staring into the face of death, and always knowing: we're not now, but are going to be there in the next two minutes, never mind days, weeks, months . . .

Having cried while writing, another therapist explained,

> It gets too close to . . . I am surprised . . . I cannot directly connect with the Holocaust without intense pain. . . . It's copeable with largely by avoidance. The Holocaust comes close to ultimate pain, beyond, associations of beyond endurance. The feelings of strength are also there, but I am not connected with those. I feel most identified with the victimization and the overwhelming powerlessness against the horror.

Space does not permit describing the richness of what can be learned in ongoing, prolonged group supervision processes, nor giving full narrative examples of the crystallization of countertransference reactions through repeated reviews; the interacting tapestries of, among others, event countertransference and person countertransference; the mutual impact of differing adaptational styles to the trauma of therapists and patients (Danieli, 1981a); and examination of mutual (counter)transferences among members played out in the group dynamics. One important instance of the latter is the attempted expulsion of the supervisor—the person leading the exercise process, who thus becomes the symbolic agent of the trauma—by/from the group for "victimizing" them and exposing their vulnerabilities by encouraging them to confront—(re)experience—the trauma.

The exercise may also arouse ambivalence in participants. Claiming an inability to draw, and a preference to "only do the words part" is an obvious example of resistance. One psychotherapist attempting to do the exercise process alone stated,

> Even for people who took a seminar it's very powerful and assumes a degree of training and sophistication. To do it in one clip is very traumatic. It forces you to meet, confront yourself, your feelings and

thoughts with regard to issues you would rather not deal with, that you usually won't do on your own. It's better to do it, part by yourself, and discuss it with another person, and then continue with the next part. You have to stop because even though it's worthwhile, it is so difficult. It's easier and more productive to do it with somebody else because you have to convey a complete thought to another person. When writing it down you may fudge. It's individual. Some people perhaps can be very honest with themselves writing. But since it's such powerful and difficult material you need another person's support. If you fall, somebody will be there to catch you or stabilize you.

Even in a workshop you should be flexible and give people choice—group, pairs, individual—and give them the opportunity to decide what is better for them even if they have to do it over ten times to meet everything.

If you do it individually, do only as much as you can. Patients are entitled to human rather than ideal therapists. It is very powerful. I will do it when I am ready.

The exercise process does not aim to replace ongoing analytical supervisory countertransference work. It does aim to provide a sorely needed focus on and experiential multidimensional framework for the trauma aspects of the patient's and therapist's lives.

The process also helps build awareness of the caregiver's vulnerability to being vicariously victimized by repeated exposure to trauma(ta) and trauma stories and of the toll countertransference reactions take on her or his intrapsychic, interpersonal, and family lives. One supervisee reported that

two people at the agency did survivors' group and stopped after eight to nine sessions because they didn't want to come home every time and cry. It didn't get better with time. It got worse and worse. They couldn't handle it. They had nightmares. They were not in shape to get up and go to work.

The exercise process makes poignantly clear the paramount necessity of carefully nurturing, regulating, and ensuring the development of a self-protective, self-healing, and self-soothing way of being as a professional and as a full human being. The importance of self-care and self-soothing is acknowledged in the exercise by building into the process instructional elements such as "take your time . . . take all the time you need," and caring, respectful attention to every element explored.

The composition of the workshop or seminar group is unpredictable. One can be assured, however, that many of the psychotherapists present have been, directly or indirectly, victims or trauma survivors, whose victimization either inspired and energized their choice of work/career or specialty, or interacted with their patient's trauma(ta) part of their countertransference matrix. Invariably, group members learn about cultures other than their own. They come to finish unfinished business with their patients and with themselves, to explore their wounds, clean the pus, and heal them. They come to seek answers, to find forgiveness and compassion, and ultimately, understanding and camaraderie. They mobilize creative energy and allow themselves to transform as people to be more authentic in their work and more actualized in their personal lives.

SOME PRINCIPLES OF SELF-HEALING

The following principles are designed to help professionals recognize, contain, and heal event countertransferences.

Recognize One's Reactions:

1. Develop awareness of somatic signals of distress—one's chart of warning signs of potential countertransference reactions, e.g., sleeplessness, headaches, perspiration.
2. Try to find words to name accurately and articulate one's inner experiences and feelings. As Bettelheim (1984) commented, "what cannot be talked about can also not be put to rest; and if it is not, the wounds continue to fester from generation to generation" (p. 166).

Contain One's Reactions:

1. Identify one's personal level of comfort in order to build openness, tolerance, and readiness to hear *anything.*
2. Remembering that every emotion has a beginning, a middle, and an end, learn to attenuate one's fear of being overwhelmed by its intensity and try to feel its full life cycle without resorting to defensive countertransference reactions.

Heal and Grow:

1. Accept that nothing will ever be the same.
2. When one feels wounded, one should take time, accurately diagnose, sooth, and heal before being "emotionally fit" again to continue to work.

3. Seek consultation or further therapy for previously unexplored areas triggered by patients' stories.
4. Any one of the affective reactions (i.e., grief, mourning, rage) may interact with old experiences that have not been worked through. Therapists will thus be able to use their professional work purposefully for their own growth.
5. Establish a network of people to create a holding environment (Winnicot, 1965) within which one can share one's trauma-related work.
6. Therapists should provide themselves with avocational avenues for creative and relaxing self-expression in order to regenerate energies.

Being kind to oneself and feeling free to have fun and experience joy is not a frivolity in this field but a necessity without which one cannot fulfill one's professional obligations, one's professional contract.

THE "CONSPIRACY OF SILENCE" BETWEEN PSYCHOTHERAPISTS AND PATIENTS

In this section* I report some of the major findings of a study that systematically examined the nature of the emotional responses and other problems experienced by psychotherapists in working with Nazi Holocaust survivors and their children.[3] Participants in this study included sixty-one psychotherapists, forty women and twenty-one men, with four to forty years of experience. Within this group, twenty-eight were social workers, twenty-three were psychologists, and ten were psychiatrists. Fifty had completed postgraduate training and all but one had undergone psychoanalysis or psychoanalytic psychotherapy. Of the fifty-six Jewish (eight Israeli) participants, ten were themselves Holocaust survivors, and eight were postwar children of survivors. I will review and discuss a comparison between the countertransference reactions of psychotherapists in this sample who were survivors and children of survivors with those of therapists who were not themselves victims or children of victims of the Nazi Holocaust (NVH group).

---

*Some of the material presented in this section appeared previously in Y. Danieli (1984). "Psychotherapists' participation in the conspiracy of silence about the Holocaust." *Psychoanalytic Psychology,* 1(1), pp. 23-42. Copyright 1984 by Lawrence Erlbaum Associates and in Danieli, Y. (1988a)."Confronting the unimaginable: Psychotherapists' reactions to victims of the Nazi Holocaust." In J. P. Wilson, Z. Harel, & B. Kahana (Eds.), *Human Adaptation to Extreme Stress* (pp. 219-238). New York: Plenum. With kind permission of Springer Science and Business.

The reader is encouraged to consider how the following countertransference themes might apply to working with terrorist and other related traumatic events.

COUNTERTRANSFERENCE THEMES

Defense

The various modes of defense against listening to Holocaust experiences and against therapists' inability to contain their intense emotional reactions comprised the most frequent "countertransference phenomena" repeatedly reported by psychotherapists and researchers in working with survivors and their children. Some therapists reacted to feeling overwhelmed by numbing themselves. Others reacted with disbelief and accused their patients of exaggerating. Therapists reported a variety of avoidance reactions: they kept "forgetting," "turning off," "tuning out," and "getting bored with the same story repeated over and over again." Many used distancing. They listened to the stories as though they were "science fiction stories" or "as if it happened 5,000 years ago." Others became very abstract, "professional," and intellectual, frequently lecturing the patient. An extreme "cutting-the-Holocaust-out" behavior on the part of psychotherapists was to refer the children of survivors to therapists in the group project "to take care of the Holocaust part" while continuing to see them "for the rest of their personality problems."

Some psychotherapists defended themselves by *overreliance on available methods, theories, theoretical jargon, and prescribed roles.* They used theoretical rationalizations such as: "Let's talk about the here and now. The past is gone . . . there is no sense in complaining. . . . You are in the United States now." Some stated that "the children were born and raised in America: they behave just like typical American Jews. This is just a variant of narcissism." At other times, they may have focused exclusively on the survivor's pre-Holocaust childhood. The latter is especially true of classical psychoanalysts. For example, Zetzel (1970, p. ___) states: "External events, no matter how overwhelming, precipitate a neurosis only when they touch on specific unconscious conflicts." This avoidance rendered such therapists unable to consider Holocaust traumata as etiologically significant or central to the understanding of their patients' psychodynamics. In many cases, this omission led to a misinterpreted etiology, one that circumscribed the therapists' understanding—and therefore their therapeutic activity—to their familiar psychodynamic orientation.

In supervision, a therapist described a patient, Mr. S., whose presenting problem was compulsive showering and scrubbing, which resulted in severe damage to his skin. The therapist worked under the assumption that Mr. S.'s symptomatology was a manifestation of an anal fixation and kept probing into his childhood. An old intake report stated: "In Auschwitz Mr. S. worked for 10-12 hours a day" without mention of the nature of his work. Following the supervisor's suggestion to explore the nature of the patient's "work detail," the therapist learned that Mr. S. removed corpses from the crematorium. This information served as a breakthrough for both therapist and patient and resulted in a dramatic reduction of the symptoms. Whereas most psychological phenomena are multidetermined, it seems clear that the dramatic result here was related to reviewing the patient's Holocaust experience.

A similar example of theoretical reduction and avoidance was naming the following Holocaust-derived dream imagery reported by a survivor's offspring as "pregenital sadism." The dream contained "pits full of hundreds of corpses . . . mutilated bodies against barbed wire . . . a baby blown to pieces while thrown up into the air . . . a skeleton crying for food."

The distortion caused by insufficient understanding of the meaning and function of the experience of "survivor's guilt" is one of the most poignant instances of how extraordinary human experience exposes the limits of traditional psychological theories of conventional life. The pervasiveness of bystander's guilt among psychotherapists and researchers, described as follows, may account for their overuse, stereotypic attribution, and reductionistic misinterpretation of concepts such as "survivors' guilt," described by Niederland (1961, 1964) and by Krystal and Niederland (1968) as a major feature of the survivors' experiences whose central meanings and functions psychotherapists may miss by responding in the ways described here (Danieli, 1988a,b).

Bystander's Guilt

The most common of the affective reactions therapists reported in their work with survivors and their children is bystander's guilt: "I feel an immense sense of guilt because I led a happy and protected childhood while these people have suffered so much."

Therapists who felt guilty were much more fearful of hurting the patient and used guilt to explain their avoidance of asking questions. Merely asking a question, they feared, would hurt the patient "who has suffered so much already." Some therapists who felt guilty were also afraid that survivors were fragile, that they would "fall apart," overlooking the fact that these were people who had not only survived but also had rebuilt families and lives despite immense losses and traumatic experiences. Therapists tended to do too much for survivors and their children to the point of patronizing them and not respecting their strengths.

Guilt often resulted in the therapist's inability to set reasonable limits; not wanting to hear stories or adopting a masochistic position in relation to the survivor. In some instances, the survivors or their offspring were allowed to call at any time of day or night.

Therapists also felt guilty in reaction to their own rage at these individuals. Some therapists stopped exploring the patient's problems when they saw tears in their eyes despite the fact that tears are a perfectly appropriate reaction. Researchers reported feeling guilty for using survivors as subjects and then trying to put such human suffering into a "cold," objective scientific design. Some therapists feared that demonstrating these individuals' resilience and strengths was equivalent to saying that because people could adapt, "it couldn't have been such a terrible experience, and it is almost synonymous with forgiving the Nazis."

I have proposed that survivor guilt, in part, serves as a defense against the total helplessness and passivity experienced during the Holocaust (Danieli, 1981a). Guilt as a defense against the experience of utter helplessness (Danieli, 1981b, 1985) links both survivors and their offspring to the Holocaust. Children of survivors are helpless in their mission to undo the Holocaust both for their parents and for themselves. This sense of failure often generalizes to "No matter what I do or how far I go, nothing will be good enough" (Danieli, 1985). This applies to many other victimization and trauma survivors. The bystander guilt of therapists also appears as a defense when they experience their helplessness to undo the long-term consequences of the Holocaust for their patients. The pervasiveness of bystander's guilt among psychotherapists may account for their tendency to overuse stereotypic attribution and reductionistic misinterpretation of concepts such as "identification with the aggressor" (Bettelheim, 1943) and "survivor guilt" (Niederland, 1961, 1964). The pervasiveness and the misuse in application of the concept of "survivor guilt" in the treatment of survivors led Carmelly (1975) to divide it into two categories, passive and active. *Passive guilt,* the one actually meant by Niederland (1964) when he coined the term *survivor guilt,* is experienced by those who survived "merely because they happened to be alive at the time of liberation" (Carmelly, 1975, p. 140) as "I was spared the fate of those who were murdered." *Active* guilt stems from having committed immoral acts and/or knowingly having chosen not to help when one could possibly have done so. Asserting that "the greatest majority of concentration camp survivors are 'passive guilt carriers,'" Carmelly (1975) notes that

Therapists have interpreted hostile, aggressive and depressive symptoms [of survivors] as a direct result of unrelieved active guilt feelings . . . [out of their] mistaken belief that any survivor must have committed immoral acts. . . . As a result of the focus on the relief of active guilt feelings (which do not exist in reality), these patients have not been helped to relate constructively to their present life. Instead . . . they developed distorted guilt feelings [and their] already painful life might become more drastically painful. (pp. 143-145)

Therapists working with war veterans who report having committed atrocities may be caught in the opposite attribution of passive guilt when their patients need to resolve their active guilt feelings.

Rage

Rage, with its variety of objects, is the most intense and one of the most difficult affective reactions experienced by therapists in working with survivors and their children. They often reported that they became enraged listening to Holocaust stories and were overwhelmed by the intensity of their own reactions.

Nazi Germany created a reality far worse than any fantasy normally available to the human psyche. However, the Nazis are not present as targets for bystander's rage, and thus the survivors or their offspring may become the symbol of the Holocaust in its totality, available for the displacement of these feelings. Survivors remind therapists of their own anger and destructiveness. Some therapists accused victims of bringing the Holocaust upon themselves. This appears to be a rationalization of their displaced anger.

Other clinicians were seriously distressed by the conflict between feeling angry toward survivors and the meaning they attributed to their anger. "How can I get angry with this person who has already suffered by abuse of the Nazis? That makes me a Nazi." This tendency to identify with the aggressor also contributed to the therapists' fear of further harming their patients and could lead to a cycle of rage and guilt. This pattern seemed to be intensified by compliant and sometimes masochistic behavior of survivors with regard to authorities in general and doctors in particular. As previously noted, guilt often rendered the therapist unable to set limits that then led to conscious or unconscious resentment when patients became more demanding.

During the war, being separated meant total and permanent loss. When separation issues are addressed, especially in family therapy, therapists are often confronted with the family's perception of them as Nazis. When therapists overidentify with the child's rebellious rage against parental clinging, they "victimize" the parents. The latter behavior may be further abetted by the general tendency among mental health professionals to blame parents for their children's problems. Some therapists overidentified with the children in calling survivor parents "Nazis" when they described the parents' interactions with their offspring. When they overidentified with the parents' anxiety and hurt at the child's attempt at separation, they tended to inhibit the child's normal anger by "lecturing" the child to "understand" the parents who "have suffered enough." This dilemma may induce helpless rage in therapists who often reported experiencing murderous feeling toward "these parents" or "these children."

Therapists resorted to counterrage in three major instances: (1) in response to being viewed by survivors or their offspring as Nazis; (2) when survivors did not live up to expectations to rise above hate and prejudice (e.g., "I hate all Germans"); or (3) when they became terrified of the extent of rage they anticipated in survivors.

Therapists' inability to cope effectively with the rage they experienced toward survivors and their children led some to reject them or to shorten their therapy. They often justified their actions with reference to "patient's resistance," which again appears to be a rationalization. Some therapists personally sought further psychotherapy primarily to work through issues surrounding (re)awakened intense rage and related imagery.

Shame and Related Emotions

Two criteria were used to categorize affective reactions related to shame. First, all have the common elements of humiliation and degradation. Second, all assume projective identification of the listener with the protagonist in Holocaust stories. One aspect of shame is derived from therapists' fantasies of what the survivors must have done in order to survive. Shame was also related to the therapist's disgust. Disgust and loathing frequently impelled the therapist to prohibit survivors and their offspring from telling these stories.

Shame was often related to the therapists' acceptance of the myth describing the behavior of the victims during the Holocaust as *going like sheep to the slaughter.* This myth not only implies that they could have fought and that they should have been prepared for the Holocaust but also assumes that Holocaust victims had somewhere to go if they chose to escape. As historical evidence clearly indicates, there was no place to escape to because other countries failed to help or outright aided the Nazis. Therapists who accepted this myth tended to feel contemptuous toward and condemn survivors for having been victims and, as such, weak, vulnerable, and abused. The process usually began with shame and contempt, and when therapists could no longer tolerate their shame, they became enraged. Therapists who indignantly expressed their contempt and rage consequently victimized their patients.

Perhaps the deepest aspect of shame is what I have called the *fourth narcissistic blow*. Freud (1917) speculated about the reasons people rejected and avoided psychoanalysis, stating that Copernicus gave the first (cosmological) blow to humanity's naïve self-love or narcissism, when humankind learned that it was not the center of the universe. Darwin gave the second (biological) blow, when he said that humanity's separation from and superiority to the animal kingdom is questionable. Freud claimed that he gave the third (psychological) blow, by showing that "the ego is not even master in its own house" and that, indeed, we have limits to our consciousness. I believe that Nazi Germany gave humanity the fourth (ethical) blow, by shattering our naïve belief that the world we live in is a just place in which human life is of value to be protected and respected.

A country that was considered the most civilized and cultured in the Western world committed the greatest evils that humans have inflicted on humans and thereby challenged the structure of morality, dignity, and human rights, as well as the values that define civilization. Not only therapists, but all of us, in various degrees of awareness, share this sense of shame. Indeed, this fourth narcissistic blow may have caused many in society to avoid confronting the Holocaust by refusing to listen to survivors and their offspring, (those) who bear witness to the experience and its consequences.

Although all four blows forced confrontation with essential truths about human existence, the ethical blow distinguishes itself by massively exposing the potential boundlessness of human evil and ugliness. Unless humanity is willing to integrate this historical narcissistic blow, the pessimistic prophecies stated by Freud (1930) in "Civilization and Its Discontents" may be fulfilled.

Dread and Horror

Another reaction that occurs frequently among psychotherapists is dread and horror. "I dread being drawn into a vortex of such blackness that I may never find clarity and may never recover my own stability so that I may be helpful to this patient." Therapists felt traumatized as if attacked by their own emotions and fantasies. They also reported horror in reaction to cathartic experiences the survivors tend to relive with much vividness and intensity. Those therapists who attempted to control their own reaction were often drained by these sessions. A few found themselves sharing the nightmares of the survivors they were treating.

One therapist reported experiencing herself tuning out to the point of fainting in reaction to her patient's telling her about her own baby being smashed against a wall in front of her eyes and about other children clinging to their parents' bodies in mass graves. This therapist stated that she was "afraid to share this horror with [her] supervisor."

Dread and horror were also reactions to the sense of total passivity and helplessness conveyed in Holocaust stories, which often led the therapists to prevent the recounting of any Holocaust experiences by using the various evasive and defensive maneuvers described earlier.

Grief and Mourning

Therapists also reported experiencing deep sorrow and grief during and after sessions with survivors and their offspring, especially when losses and suffering were recounted. Some found themselves tearful or actually cried at those times. One therapist reported "becoming progressively crushed to the ground . . . with endless, bottomless sadness" when constructing a family tree in an interview with a child of survivors. Having done his "homework," the patient reported when, where, and how each of the seventy-two family members had perished, leaving only two survivors, his mother and father, whose children were killed before their parents' eyes after being torn from their arms (Danieli, 1992).

Some therapists attempted to avoid listening to pain and suffering by asking such questions as, "How did you survive?" instead of "What happened to you?" Or "What did you go through during the war?" Others spoke of "sinking into despair" and fearing to be "engulfed by anguish."

The anguish they experienced is related to the impossibility of adequately mourning so massive a catastrophe as the Holocaust. "How can one even mourn all of this?" Most, if not all, survivors not only view the destruction of their lives, their families and communities, but also 6 million anonymous, graveless losses and the total loss of meaning as their rightful context for mourning (Danieli, 1989).

Therapists who were unable to contain these powerful, intensely painful—yet appropriate—feelings in themselves and in their patients became intolerant or immobilized. They were, therefore, unable to provide a "holding environment" (Winnicot, 1965) in which survivors and/or their offspring could begin to grieve and mourn personal losses, a necessary healing process for them and their families (Danieli, 1988c).

Murder versus Death

Two related phenomena, albeit more specific, are therapists' use of the words *death* and *dead* as contrasted with (mass) *murder* and *murdered* to describe the fate of the victims and/or the deeds of the perpetrators of the Holocaust. Some of the participants in this study who have worked with survivors of the Nazi Holocaust and with the elderly and/or the terminally ill (some of whom were also survivors) have used these words to differentiate between their reaction to personal "normal death" and to the evils of mass murder and its anonymity in the Holocaust.

Therapists who work with members of survivors' families encounter individuals whom the Holocaust deprived of the normal cycle of the generations and ages. The Holocaust also robbed them, and still does, of natural, individual death (Danieli, 1994a; Eitinger, 1980) and normal mourning. The use of the word *death* to describe the fate of the survivor's relatives, friends, and communities appears to be a defense against acknowledging murder as possibly the most crucial reality of the Holocaust.

Victim/Liberator

Therapists may view survivors as either victims or heroes. When they view survivors as *victims,* they are seen as fragile, helpless martyrs. This image generates bystander's guilt, rage, and shame in the therapist. Ramifications of these countertransference reactions have already been considered in previous sections.

In the context of viewing the survivor as a victim, therapists reported another response that I have labeled *therapist as liberator/savior.* When therapists perceived the survivor as if still living in the camps, passive and helpless, they became "annoyed and impatient" and felt the need to liberate them. This need stemmed from the therapist's intolerance for the patient's survivor guilt, resulting in negative therapeutic reactions. Therapists reported feeling frustrated, angry, and unable to bear the patient's persistent suffering. As stated in previous sections, therapists generalized their view of the survivors to their offspring. When they viewed the child of a survivor as a victim, they tended to respond to the offspring as they did to the parents. Some therapists, however, viewed the offspring as victimized by their parents. These therapists attempted to rescue the children from their survivor parents, compete with their survivor parents, and/or compensate for parental deprivation.

Viewing the Survivor As Hero

When therapists view survivors as heroes, they see them as superhumanly strong, capable, heroic figures to be worshipped and admired. Some therapists were awed by the courage, hope, and sheer determination reflected in Holocaust accounts. A sense of awe led some therapists to glorify the survivors, to conceive of them as special people who, having experienced ultimate evil and destruction, have found the essential truths and meaning of life. Some researchers looked for "superior methods of coping" in them. This, in addition to the historical distortion involved in such a view, also implies derogatory attitudes toward the 6 million dead. The main pitfall in overestimating the strengths of survivors in therapy is the therapist's resulting insensitivity to the pain and suffering and the problems in living, which brought the survivor to therapy.

The idealization of both victims and heroes may humble therapists and lead them to view problems and concerns in their own lives as trivial when compared to the survivors'. Such attitudes may result in envious and competitive feelings toward survivors and in feeling excluded or like an outsider.

Some therapists who were not Holocaust survivors or children of survivors reported feeling envious of the moral stature that has accrued to survivors because of their sufferings. Much like survivors' offspring, they reported feeling inferior to survivors because they believed they would never have survived the situations described by their patients. Some therapists reported envying the fact that survivors' offspring are by definition members of a special group with its own identity, and they condemned the offspring for using their parents' suffering to claim this special status. They stated a preference for working with offspring of only one survivor/parent, assuming that they will share a better cultural rapport: "They are more American."

Most therapists generally preferred working with *heroes* to working with victims. One therapist reported wishing to hear heroic stories and "turning off" when his patients "kept complaining." Most therapists also stated that they would rather lead offspring groups than groups of survivors because "hearing the stories second hand is easier."

Privileged Voyeurism

Privileged voyeurism, in contrast to the "countertransference reactions" described previously, tends to lead therapists and researchers to dwell excessively on the Holocaust. Indeed, some professionals reported feeling privileged to work with survivors. One therapist reported feeling "excitement, glamour, and an extra quality of titillation." Therapists' sadism appears to be a major factor in many such reactions. Another therapist chose to treat survivors as a way to learn and understand his family's history and behavior. These therapists tended to become totally engrossed with the Holocaust and asked numerous questions, many of which may not have been relevant to the particular survivor's war experiences. Because of their zeal, they sometimes totally ignored their patient's present life situation, including their experiences following liberation. Similarly, they tended to neglect the patient's prewar history. A major danger of privileged voyeurism is to neglect the survivor or child of survivors as a whole person.

"Me Too"

A somewhat related reaction among psychotherapists and researchers is what I call the *me-too* reaction, also stated as "We are all survivors." Although this global attitude may stem from a sincere attempt on the therapist's part to empathize with his or her patient, I believe it poses a real danger of blurring distinctions among various kinds of survival experiences, under various conditions and degrees of trauma. Therapists who were not survivors or children of survivors of the Nazi Holocaust have claimed "I am a survivor myself" after having initially felt they "had no right being here. I hadn't shared their experience."

Many therapists who are survivors and/or children of survivors used similarity of experience in the service of empathy and understanding, which they reported to be helpful to their patients. However, it was sometimes used in the service of defense or was otherwise problematic. For example, the me-too reaction that assumed sameness of experience sometimes took the form, on the part of some of these therapists, of foreclosing remarks such as, "I know what you mean, I am a survivor [or, a child of survivors], too."

The defensive me-too response on the part of either group of psychotherapists may interact with the patient's own fears that sharing their traumata would lead to reliving them. As such, this "countertransference reaction" acts to perpetuate the conspiracy of silence, rather than to aid the patient's exploration of his or her own particular experiences. It ignores the uniqueness of both the Holocaust and the particular meaning and consequences these have for the survivor and/or for the survivor's child (see also Danieli, 1981a; Furst, 1978).

Sense of Bond

The bond that the therapist feels because of having gone through the survivor patient experience may help and hinder the therapeutic encounter. Therapists who are survivors and/or children of survivors were uniformly convinced that they were better able to understand and help survivors and their offspring because of their shared complex history and unique experiences, culture, language(s), and customs. For example, "I was there. . . . Nobody [who wasn't there] could really know what hunger was really like. Nobody knows what it's like to emerge out of hell to only find out that every single person you know had perished from the face of the earth." Some acknowledged that "partly, I also wanted to help myself with my own issues and I knew my peers, my 'cousins,' are the right people to do it with."

This sense of kinship and "connectedness" was often related to these therapists' stated need to reestablish their own (extended) families and sense of community. Sharing Carmelly's (1975) belief that professional neutrality and detachment "cannot be helpful in counseling [survivors]" (p. 143), some participants in this study expressed conflict over maintaining professional roles and authority in working with "their people." Elsewhere (Danieli, 1981b) I have pointed out that in addition to self-assertion, "assuming authority was also frightening because it was associated with the possibility of abusing one's power (and acting like a Nazi) or becoming ineffectual and inconsistent (like their parents)" (p. 143). This proved to be an additional component of the conflict for therapists who are children of survivors.

Attention and Attitudes Toward Jewish Identity

Several factors determine whether therapists encourage or even permit their patients to raise and explore unavoidable concerns about the meanings of being Jewish after the Holocaust and the establishment of the State of Israel. The first is whether therapists believe that cultural, political, and religious issues belong in therapy or in psychology in general. The second is their conscious and unconscious attitudes toward these issues in their own lives.

Some participants in this study judged their patients as "ethnocentric" for claiming that the Holocaust was a uniquely Jewish phenomenon. Others were clearly perturbed by the cultural self-hate, inferiority, and shame expressed by their patients. These therapists needed survivors and children of survivors not only to be proud of their heritage and cultural identity but also to (re)establish continuity with, and belongingness to, the whole Jewish history and culture, rather than define their identity and their relationship with the postwar world solely in response to the Holocaust.

Issues of culture and heritage are important aspects of most mass murders, including 9/11 (see Danieli & Nader, Chapter 8, this book).

COMPARISONS BETWEEN SCS AND NVH GROUPS

Earlier in this section I referred to the comparison in the study between the countertransference reactions of psychotherapists in this sample who were survivors and children of survivors (SCS group) and those of therapists who were not themselves victims or children of survivors of the Nazi Holocaust (NVH group). Because space limitations do not permit a full report, I will briefly present the major differences. The full report of these data can be found in Danieli (1982b).

In comparison to psychotherapists who were survivors and children of survivors, those who were not themselves victims or children of survivors of the Nazi Holocaust reported using various modes of defending themselves against listening to Holocaust experiences recounted by their patients and being overwhelmed by their intense emotional reactions to them. In addition, they reported experiencing themselves as outsiders and, to counteract that experience, made statements such as "we are all survivors." They also expressed attitudes, feelings, and myths disparaging to the survivors both as Holocaust victims and as parents while viewing the survivors' offspring as the fragile victims. Furthermore, therapists in the NVH group showed a pattern similar to the one previously described in working with the survivor population in general, with the exception of expressing jealousy at viewing children of survivors as being "special." That is, more than the psychotherapists who were themselves victims of children of victims, they reported ways of defending themselves against Holocaust material and their emotional reactions to it, particularly by distancing and clinging to their professional role. More than their counterparts, they expressed rage and disgust at survivor parents. Experiencing themselves as outsiders, they tended to feel pity and contempt, and to view the survivors as having gone "like sheep to the slaughter."

In comparison to the psychotherapists in the NVH group, psychotherapists in the SCS group expressed a sense of bond, a need, or a "mission" to help "their people," and a belief that they themselves will be helped in the process. The latter may be related to the therapists' feeling more conflict over maintaining their professional authority. They insisted on the need for integrating the Holocaust into the totality of their patients' lives. In addition, both patient and therapist experienced more grief and mourning. Also, more than their counterparts, they often used the words *murder* and *murdered* to describe the deeds of the perpetrators and the fate of the victims of the Holocaust. Psychotherapists in the SCS group also demonstrated essentially the same pattern of responses when relating to children of survivors as they reported while working with the survivor population as a whole.

The differences between the SCS and the NVH groups were tested and found to be independent of both the length of experience of the therapists and the therapists' gender.

Although outside of the direct scope of this study, it is important to mention special countertransference reactions to aging survivors (Danieli, 1994a), to child survivors (Danieli, 1996), and to survivors of cultures different from that of the therapist (Danieli & Nader, Chapter 8, this book). In reality, countertransference reactions are the building blocks of the societal as well as professional conspiracy of silence.

The themes that have been described among psychotherapists and researchers were also observed among other professionals such as lawyers and judges, in their interaction with survivors and their children. As stated previously, I believe that these feelings and attitudes may have contributed, at least in part, the building blocks to the long-term conspiracy of silence between Holocaust survivors and society. Increased awareness of the countertransference reactions revealed in this study and of the different patterns of their frequency in the NVH and SCS groups should assist therapists and investigators to contain and use them preventively and therapeutically.

Many of the countertransference phenomena examined in my study were found to be reactions to patients' Holocaust stories rather than to their behavior. The unusual uniformity of psychotherapists' reactions suggests that they are in response to the Holocaust—the one fact that all the otherwise different patients have in common. Because the Holocaust seems to be the source of these reactions, I suggested that it is appropriate to name them countertransference reactions to the Holocaust (the trauma event/s), rather than to the patients themselves. Therapists' difficulties in treating other victim/survivors populations may similarly have their roots in the nature of their victimization (event countertransference).

In 1981 I noted that these reactions "seem very similar to alexithymia, anhedonia, and their concomitants and components which, according to Krystal, characterize survivors" (Danieli, 1981c, p. 201). In 1989, in the context of training (Danieli & Krystal, 1989) I referred to these phenomena as the "vicarious victimization of the care-giver."

CONCLUDING REMARKS

Countertransference reactions are integral to our work, ubiquitous, and expected. In this chapter I have discussed various forms of countertransference in therapists who have worked with survivors or the children of the survivors of traumata that have contributed to therapists' failure to listen, explore, understand, and help, or to maintaining the *conspiracy of silence* following the trauma. Working through countertransference difficulties is of pivotal importance in order to optimize and make meaningful the necessary, heretofore pervasively absent, training of professionals in the field of traumatic stress. Having previously introduced the concept of *event countertransference* to indicate that the source of these reactions is the nature of the patient's victimization or traumata (stories), the chapter presented an exercise tool for processing event countertransference, emphasizing the value of group modalities as a context for its sharing, experiential exploration, and working through. Acknowledging the profound effect trauma work has on therapists, the chapter then suggested principles of self-healing designed to help professionals recognize, contain, and heal event countertransferences.

Extending this knowledge to the plight and the price paid by peacekeepers, humanitarian aid workers, and the media in the midst of crisis, highlighted in *Sharing the Front Line and the Back Hills* (Danieli, 2002; see also Smith, Agger, Danieli, & Weisaeth, 1996), I emphasized that (their) organizational support and understanding is necessary. The cultures of organizations must be altered to make staff security—both physical and psychological—and support an integral part, rather than a short-lived afterthought, usually following tragedies. Steps must be taken prior to, during, and after missions to ensure, as far as possible, both the safety and well-being of those who put their lives on the line for others. Effective premission training can prepare staff for many of the tasks they will face, and teach them how to minimize their exposure. During the mission, basic security, open communications at all levels, and psychosocial support systems should be in place. Following the mission, attention to their psychological well-being, as well as that of their families, is essential as they adjust to returning to their home environments, often having been profoundly affected by their experiences.

Ideally, adequate security and risk training, as well as psychosocial preparation, should be a precondition particularly for personnel going into high-risk areas. They deserve comprehensive and properly funded security and support systems to enable them to do their inherently dangerous work on behalf of humanity—all of us.

The strong possibility that other extreme events will happen that will require a rapid response from the therapeutic community underscores the need for thorough and continuing training in order to be ready for them. Psychotherapists and organizations alike must accept the possibility that they have both participated in and perpetuated a conspiracy of silence. Only by genuinely exploring, processing, and integrating the conscious and unconscious components of their own responses and transcending them can they be fully prepared to help trauma victim/survivors and their families. Only then will they be prepared to be full partners in the process of healing.

NOTES

1. Although event countertransference and personal countertransference (i.e., reactions to the patient's behaviors) are not mutually exclusive, for training purposes it is useful to differentiate the two.

2. The word *event* was chosen to specify the source of these therapists' reactions, not to imply that it was just one event.

3. For methodological details, see Danieli (1982b).

REFERENCES

American Psychiatric Association. (1994). *Diagnostic and Statistical Manual of Mental Disorders* (Fourth Edition). Washington, DC: Author.

Barocas, H. A. & Barocas, C. B. (1979). Wounds of the fathers: The next generation of Holocaust victims. *International Review of Psycho-Analysis, 6,* 1-10.

Bettelheim, B. (1943). Individual and mass behavior in extreme situations. *Journal of Abnormal and Social Psychology, 38,* 417-452.

Bettelheim, B. (1984). Afterward to C. Vegh, *I Didn't Say Goodbye* (R. Schwartz, Trans.). New York: E. P. Dutton.

Blank, A. S. (1985). Irrational reactions to post-traumatic stress disorder and Vietnam veterans. In S. M. Sonnenberg (Ed.), *The Trauma of War: Stress and Recovery in Vietnam Veterans* (pp. 69-98). Washington, DC: American Psychiatric Association Press.

Carmelly, F. (1975). Guilt feelings in concentration camp survivors. Comments of a "survivor." *Journal of Jewish Communal Services, 2,* 139-144.

Chu, J. A. (1988). Ten traps for therapists in the treatment of trauma survivors. *Dissociation, 1,* 24-32.

Comas-Diaz, L. and Padilla, A. (1990). Countertransference in working with victims of political repression. *American Journal of Orthopsychiatry, 60,* 25-34.

Dahlenberg, C. J. (2000). *Countertransference and the Treatment of Trauma.* Washington, DC: American Psychological Association.

Danieli, Y. (1981a). Differing adaptational styles in families of survivors of the Nazi Holocaust: Some implications for treatment. *Children Today, 10*(5), 6-10, 34-35.

Danieli, Y. (1981b). Families of survivors of the Nazi Holocaust some short- and long-term effects. In C. D. Spielberger, I. G. Sarason, & N. Milgram (Eds.), *Stress and Anxiety* (Vol. 8, pp. 405-421). New York: McGraw-Hill/Hemisphere.

Danieli, Y. (1982a). Group project for Holocaust survivors and their children. Prepared for National Institute of Mental Health, Mental Health Services Branch. Contract #092424762. Washington, DC.

Danieli, Y. (1982b). Therapists' difficulties in treating survivors of the Nazi Holocaust and their children. Dissertation Abstracts International, 42(12-B, Pt 1), 4927. (UMI No. 949-904).

Danieli, Y. (1981c). On the achievement of integration in aging survivors of the Nazi Holocaust. *Journal of Geriatric Psychiatry, 14*(2), 191-210.

Danieli, Y. (1984). Psychotherapists' participation in the conspiracy of silence about the Holocaust. *Psychoanalytic Psychology, 1*(1), 23-42.

Danieli, Y. (1985). The treatment and prevention of long-term effects and intergenerational transmission of victimization: A lesson from Holocaust survivors and their children. In C. R. Figley (Ed.), *Trauma and Its Wake* (pp. 295-313). New York: Brunner/Mazel.

Danieli, Y. (1988a).Confronting the unimaginable: Psychotherapists' reactions to victims of the Nazi Holocaust. In J. P. Wilson, Z. Harel, & B. Kahana (Eds.), *Human Adaptation to Extreme Stress* (pp. 219-238). New York: Plenum.

Danieli, Y. (1988b). Treating survivors and children of survivors of the Nazi Holocaust. In F. M. Ochberg (Ed.), *Post-traumatic Therapy and Victims of Violence,* (pp. 278-294). New York: Brunner/Mazel.

Danieli, Y. (1988c). The use of mutual support approaches in the treatment of victims. In E. Chigier (Ed.), *Grief and Bereavement in Contemporary Society: Vol. 3. Support Systems* (pp. 116-123). London: Freund Publishing House.

Danieli, Y. (1989). Mourning in survivors and children of survivors of the Nazi Holocaust: The role of group and community modalities. In D. R. Dietrich & P. C. Shabad (Eds.), *The Problem of Loss and Mourning: Psychoanalytic Perspectives* (pp. 427-460). Madison: International Universities Press.

Danieli, Y. (1992). The diagnostic and therapeutic use of the multi-generational family tree in working with survivors and children of survivors of the Nazi Holocaust. In J. P. Wilson & B. Raphael (Eds.), *The International Handbook of Traumatic Stress Syndromes.* (Stress and Coping Series, Donald Meichenbaum, Series Editor) (pp. 889-898). New York: Plenum Publishing.

Danieli, Y. (1993). Countertransference and trauma: Self healing and training issues. In M. B. Williams and J. F. Sommer Jr., *Handbook of Post-Traumatic Therapy* (pp. 540-550). Westport, CT: Greenwood/Praeger Publishing, Co.

Danieli, Y. (1994a). As survivors age—Part I. *National Center for Post Traumatic Stress Disorder Clinical Quarterly, 4*(1), 1-7.

Danieli, Y. (1994b). Countertransference, trauma and training (1994). In J. P. Wilson & J. Lindy (Eds.), *Countertransference in the Treatment of PTSD* (pp. 368-388). New York: Guilford Press.

Danieli, Y. (1996). Who takes care of the caretakers? The emotional life of those working with children in situations of violence. In R. J. Appel & B. Simon

(Eds.), *Minefields in Their Hearts: The Mental Health of Children in War and Communal Violence* (pp. 189-205). New Haven: Yale University Press.

Danieli, Y. (Ed.) (1998). Trauma and the continuity of self: A multidimensional, multidisciplinary, integrative (TCMI) framework. In Y. Danieli, *International Handbook of Multigenerational Legacies of Trauma*. New York: Kluwer Academic/Plenum Publishing Corporation.

Danieli, Y. (Ed.) (2002). *Sharing the Front Line and the Back Hills: International Protectors and Providers, Peacekeepers, Humanitarian Aid Workers and the Media in the Midst of Crisis*. Amityville, NY: Baywood Publishing Company, Inc.

Danieli, Y., Brom, D., & Sills, J. B. (Eds.) (2005). *The Trauma of Terrorism: Sharing Knowledge and Shared Care*. Binghamton, NY: The Haworth Press.

Danieli, Y., Engdahl, B., & Schlenger, W. E. (2003). The psychosocial aftermath of terrorism. In F. Moghaddam & A. J. Marsella (Eds.), *Understanding Terrorism: Psychosocial Roots, Consequences, and Interventions* (pp. 223-246). Washington, DC: American Psychological Association.

Danieli, Y. & Krystal, J. H. (1989). *The initial report of the Presidential Task Force on Curriculum, Education and Training of the Society for Traumatic Stress Studies*. Chicago: The Society for Traumatic Stress Studies.

Des Pres, T. (1976). *The Survivor: An Anatomy of Life in the Death Camps*. New York: Oxford University Press.

Eitinger, L. (1980). The concentration camp syndrome and its late sequelae. In J. E. Dimsdale (Ed.), *Survivors, Victims and Perpetrators* (pp. 127-162). New York: Hemisphere Publishing Corporation.

Figley, C. R. (Ed.). (1995). *Compassion Fatigue: Coping with Secondary Traumatic Stress Disorder in Those Who Treat the Traumatized*. New York: Brunner/Mazel.

Fischman, Y. (1991). Interacting with trauma: Clinician's responses to treating psychological aftereffects of political repression. *American Journal of Orthopsychiatry, 61*, 179-185.

Freud, S. (1917). A difficulty in the path of psychoanalysis. In J. Strachey (Ed. and Trans.), *The Standard Edition of the Complete Psychological Works of Sigmund Freud* (Vol. 17). London: Hogarth Press.

Freud, S. (1930[1929]). Civilization and Its Discontents. In J. Strachey (Ed. and Trans.), *The Standard Edition of the Complete Psychological Works of Sigmund Freud* (Vol. 21). London: Hogarth Press.

Friedman, P. (1948). The road back for the DP's: Healing the psychological scars of Nazism (1948). *Commentary, 6*(6), 502-510.

Furst, S. S. (1978). The stimulus barrier and the pathogenicity of trauma. *International Journal of Psycho-Analysis, 59*, 345-352.

Haley, S. A. (1974). When the patient reports atrocities: Specific treatment considerations in the Vietnam veteran. *Archives of General Psychiatry, 30*, 191-196.

Herman, J. L. (1992). *Trauma and Recovery*. New York: Basic Books.

Hudnall Stamm, B. (Ed.) (1995). *Secondary Traumatic Stress: Self-Care Issues for Clinicians, Researchers, & Educators*. Lutherville, MD: Sidran Press.

Kinzie, D. J. (1989). Therapeutic approaches to traumatized Cambodian refugees. *Journal of Traumatic Stress, 2*(1), 75-91.

Kluft, R. P. (1989). The rehabilitation of therapists overwhelmed by their work with MPD patients. *Dissociation, 2*(4), 243-249.

Krystal, H. (1988). *Integration and Self-Healing*. New Jersey: The Analytic Press.

Krystal, H. & Niederland, W. G. (1968). Clinical observations on the survivor syndrome. In H. Krystal (Ed.), *Massive Psychic Trauma* (pp. 327-348). New York: International Universities Press.

Lindy, J. D. (1987). *Vietnam: A Case Book*. New York: Brunner/Mazel Publishers.

Maslach, C. (1982). *Burnout: The Cost of Caring*. Englewood Cliffs, NJ: Prentice-Hall.

McCann, I. L. & Pearlman, L. A. (1990). Vicarious traumatization: A framework for understanding the psychological effects of working with victims. *Journal of Traumatic Stress, 3,* 131-149.

Mollica, R. F. (1988). The trauma story: The psychiatric care of refugee survivors of violence and torture. In F. M. Ochberg (Ed.), *Post-Traumatic Therapy and Victims of Violence* (pp. 295-314). New York: Brunner/Mazel, Publishers.

Niederland, W. G. (1961). The problem of the survivor: Some remarks on the psychiatric evaluation of emotional disorders in survivors of Nazi persecution. *Journal of the Hillside Hospital, 10*(3-4), 233-247.

Niederland, W. G. (1964). Psychiatric disorders among persecution victims: A contribution to the understanding of concentration camp pathology and its aftereffects. *Journal of Nervous and Mental Diseases, 139,* 458-474.

Niederland, W. G. (1968). An interpretation of the psychological stresses and defenses in concentration-camp life and the late aftereffects. In H. Krystal (Ed.), *Massive Psychic Trauma* (pp. 60-70). New York: International Universities Press.

Parson, E. R. (1988). The unconscious history of Vietnam in the group: An innovative multiphasic model for working through authority transferences in guilt-driven veterans. *International Journal of Group Psychotherapy, 38,* 275-301.

Pearlman, L. A. & Saakvitne, K. W. (1995). *Trauma and the Therapist: Countertransference and Vicarious Traumatization and Psychotherapy with Incest Survivors*. New York: W. W. Norton & Company.

Phillips, R. D. (1978). Impact of Nazi Holocaust on children of survivors. *American Journal of Psychotherapy, 32,* 370-378.

Raphael, B. & Ursano, R. J. (2002). Psychological debriefing. In Y. Danieli (Ed.), *Sharing the Front Line and the Back Hills: International Protectors and Providers, Peacekeepers, Humanitarian Aid Workers and the Media in the Midst of Crisis* (pp. 343-352). Amityville, NY: Baywood Publishing Company, Inc.

Rappaport, E. A. (1968). Beyond traumatic neurosis: A psychoanalytic study of late reactions to the concentration camp trauma. *International Journal of Psychoanalysis, 49,* 719-731.

Smith, B., Agger, I., Danieli, Y., & Weisaeth, L. (1996). Emotional responses of international humanitarian aid workers. In Y. Danieli, N. Rodley, & L. Weisaeth (Eds.), *International Responses to Traumatic Stress: Humanitarian, Human Rights, Justice, Peace and Development Contributions, Collaborative Actions and Future Initiatives* (pp. 397-423). Amityville, New York: Published for and on behalf of the United Nations by Baywood Publishing Company, Inc.

Symonds, M. (1980). The "second injury" to victims. *Evaluation and Change, Special Issue,* 36-38.

Tanay, E. (1968). Initiation of psychotherapy with survivors of Nazi persecution. In H. Krystal (Ed.), *Massive Psychic Trauma* (pp. 219-233). New York: International Universities Press.

Wallerstein, R. S. (1973). Psychoanalytic perspectives on the problem of reality. *Journal of the American Psychoanalytic Association, 31*(1), 5-33.

Wilson, J. P. & Lindy, J. D. (Eds.) (1994). *Countertransference in the Treatment of PTSD.* New York: Guilford Publications.

Winnicott, D. W. (1965). *The Maturational Processes and the Facilitating Environment.* London: Hogarth Press.

Zetzel, E. R. (1970). *The Capacity for Emotional Growth.* New York: International Universities Press.

PART IV:
FUTURE DIRECTIONS

Chapter 22

Future Directions

David W. Foy
Daryl A. Schrock

INTRODUCTION

Since the advent of post-traumatic stress disorder (PTSD) into the diagnostic classification system in 1980, hundreds of studies have begun to examine the phenomenology, etiology, and epidemiology of human responses to an exhaustive list of traumatic stressors. By the late 1990s sufficient scientific progress had been made in the study of PTSD and the various treatments applied to it so that a "best practices" approach to cataloging treatments and their demonstrated effectiveness was possible, and the first edited volume on that topic was published (Foa, Keane, & Friedman, 2000). Group therapy for trauma survivors was one of the treatments included in the review of therapies for PTSD, and it was found to be "promising" in its potential for use with various groups of trauma survivors (Foy et al., 2000).

However, the evidence base supporting group therapy for trauma was limited at that time. There were very few randomized controlled trials, and those that were published featured very small numbers of participants. Most studies reported up to then had been conducted with female participants who were survivors of childhood sexual abuse. Fortunately, in the past four years an impressive number of new randomized controlled trials have been reported, and the empirical basis for using group therapy with trauma survivors of all ages has been strengthened considerably.

The purpose of this chapter is to present practical suggestions for future clinical and research efforts in group therapies with terrorist disaster survivors. We begin by presenting a brief history of the development of group therapy for trauma, followed by a review of published studies. Next, we update the current status of our knowledge about group debriefing and group psychotherapy for trauma. Critical issues for using group therapy for terrorist disaster survivors are identified and specific recommendations for addressing these issues are made. Finally, we discuss current limitations in our knowledge of group interventions, and identify areas that require further investigation and development.

DEVELOPMENT OF TRAUMA-RELATED GROUP THERAPY

Group therapy for trauma has a lengthy history, with a published report of its use with combat-related trauma dating back to the World War II era (Dynes, 1945). However, two events in the late 1970s greatly accelerated the development of trauma-related group therapy. First, Congress passed legislation to establish a nationwide network of community-based Vet Centers to serve the readjustment needs of thousands of Vietnam veterans. "Rap groups," led by counselors who themselves were Vietnam veterans, soon became a primary therapeutic modality featured in these centers (Sipprelle, 1992). Second, the advent of PTSD into the psychiatric diagnostic system in 1980 initiated a flurry of research that soon established commonalities in symptom manifestations and pathogenesis across survivors of many kinds of traumatic experiences. Application of group therapy methods spread to a variety of trauma groups, ranging from child survivors of sexual abuse to female adults exposed to domestic violence. To date, the literature on group therapy for trauma contains more that thirty-five empirical studies, nearly equally divided between adult (Foy et al., 2000) and child/adolescent samples (Reeker, Ensing, & Elliot, 1997). By far, the most widely studied are female sexual assault survivors, both for adult and youth groups.

Two somewhat diverse groups, group therapy specialists and trauma specialists, have chronicled the evolution of group therapy for problems associated with recovery from traumatic experiences. One the one hand, there are those mental health professionals who identify themselves primarily as *group specialists* whose expertise stems from training in principles of group psychotherapy and years of experience as practicing group therapists. These individuals are likely to be found among the membership of the professional society, the American Group Psychotherapy Association (AGPA). From this perspective an influential edited book, *Group Psychotherapy for Psychological Trauma* (Klein & Schermer, 2000), was written, providing chapters describing group therapy principles applied to several specific trauma populations.

On the other hand, there are those *trauma specialists* whose training and experience stem from a focus upon the psychological effects of traumatic experiences. For these individuals, the professional society that represents their focus is likely to be the International Society for Traumatic Stress Studies (ISTSS). Representing this perspective, an edited book, *Group Treatments for Post-Traumatic Stress Disorder* (Young & Blake, 1999), has also been published. Extending group therapy principles to include their application with survivors of terrorist disasters goes beyond the scope of these two earlier books. Commendably, the editors of this book have attempted to blend the considerable professional resources from both the "group" and "trauma" perspectives to enrich the field as it relates to developing and applying group methods for survivors of terrorist disasters.

From a "future directions" perspective, it seems clear that positive integration of the stores of knowledge from both sets of experts, the group and trauma specialists, is important if progress is to be made. Although there may be differences between group and trauma specialists in terms of their training backgrounds and specific areas of expertise, they share a very large common factor, i.e., the commitment to using group psychotherapy as a major method of relieving distress among disaster survivors. Ideally, group therapy for trauma survivors will be based upon sound principles derived from both "group" and "trauma" perspectives. Indeed, Chapter 12, "Integrating Small Group Process Principles into Trauma-Focused Group Psychotherapy: What Should a Group Trauma Therapist Know?" by Davies, Burlingame, and Layne, is an admirable first step in that direction. In terms of a methodology by which the two perspectives can inform each other and the broader mental health field, a "best practices" approach is recommended. Using this approach, a "gold standard" is applied to decision making for critical issues in implementing trauma groups, such as choices regarding the particular type of group, assessment strategies, and client-to-group matching criteria.

There are two recent examples whereby "best practices" approaches were applied to evaluating available treatments within the trauma field. In the first instance, Foa, Keane, and Friedman (2000) reported the use of a six-point coding system to evaluate the evidence for specific treatment procedures for PTSD. In this scheme, Level A represents the highest level of support, where evidence was based upon randomized controlled trials. Procedures found effective through nonrandomized group studies are assigned Level B, whereas Level C criteria are met by procedures supported by uncontrolled studies or clinical observations. Levels D and E are used to classify empirically untested procedures in widespread or circumscribed clinical use, and Level F is used for recently developed treatments not yet subjected to clinical or empirical evaluation.

More recently, a second "best practices" scheme was developed by Saunders, Berliner, and Hanson (2003) for use in evaluating existing treatments for physical and sexual abuse in children. This approach also used a similar six-point "gold standard" coding system to classify specific treatments in use in the field. Anchor points ranged from 1 ("well supported and efficacious") to 6 ("experimental or concerning"). Both these best practice approaches identified individual cognitive behavioral therapy, comprised of exposure therapy, cognitive restructuring, and coping skills training, as achieving highest levels of empirical support, both for children and adults (Rothbaum, Meadows, Resick, & Foy, 2000; Saunders et al., 2003).

GROUP PSYCHOLOGICAL DEBRIEFING

In our earlier review we concluded that single-session group debriefing might be useful for education and normalizing reactions to trauma, but there was no evidence for its effectiveness in preventing or reducing trauma-related psychopathology (Foy, Eriksson, & Trice, 2001). In this book, two chapters are devoted to the possible applications of group debriefing or group early intervention with disaster survivors. On the positive side, Mitchell and Everly (Chapter 13) make the case for continuing the use of a broader perspective, critical incident stress management (CISM), of which the controversial critical incident stress debriefing (CISD) is one component, with emergency service personnel. Conversely, Raphael and Wooding (Chapter 14) present a rationale of "first do no harm" to recommend caution in using group psychological interventions in the acute disaster phase, especially with direct victims. Unfortunately, there have been no randomized controlled trials reported on group debriefing to date. However, more unfavorable indirect evidence has emerged in a recent review of eleven randomized trials of debriefing, administered in individual or couples formats (Rose, Bisson, & Wessely, 2003). In eight of the studies reviewed there were either no differences between debriefed and control participants ($n = 6$), or there were negative outcomes in the debriefing groups ($n = 2$).

Contrary to existing evidence, many mental health practitioners currently assume that nearly all those exposed to traumatic events would benefit from some form of debriefing or early intervention. A recent review of research on the range of human reactions following trauma revealed that resilience and recovery may well be different pathways (Bonanno, 2004), and resilient individuals who show little or no distress after trauma are often seen as extraordinary or they are pathologized as cold and unfeeling. Indeed, evidence suggests that there are likely multiple pathways to resilience, including hardiness, self-enhancement, repressive coping, and positive emotion and laughter (Bonanno, 2004). These new data pose a serious challenge to the assumption that there is a near universal need for debriefing, and underscore the need for rigorous screening before debriefing or early intervention to identify resilient individuals who might actually be harmed by the intervention.

Future studies on group debriefing are needed that incorporate randomization of participants to debriefing and control groups, as well as ensuring uniformity in the treatment components of debriefing (Foy et al., 2001; Eriksson, Foy, & Larson, 2003; Litz, 2003; Mitchell & Everly, Chapter 13; Raphael & Wooding, Chapter 14). At this time, the weight of evidence suggests that single-session group debriefing cannot be recommended for routine use.

REVIEW OF GROUP THERAPY FOR TRAUMA STUDIES

Tables 22.1 and 22.2 reflect the growth and maturing of studies of group trauma therapy over the past few years. Our last review of fourteen studies available for group therapy with adult survivors revealed that only two randomized controlled trials (RCTs) had been reported (Foy et al., 2000), and the strength of evidence supporting the use of group therapy with trauma survivors was evaluated as limited, but positive and promising. At that time, group therapy was recommended as "potentially effective, based upon consistent evidence from the studies reviewed" (Foy et al., 2000, p. 170). Fortunately, our current search of the literature reveals that seven more RCTs have been published in a short four-year span from 2000 through 2003, providing a much stronger empirical base of support. In order to inform our recommendations regarding the use of group therapy with survivors of terrorist disasters, we will examine the entire set of existing studies for survivors of all ages (Table 22.1). In addition, we will focus more closely on the methods and findings from the nine RCTs now available (Table 22.2).

Among the thirty-seven studies depicted in Table 22.1, there are twenty-three single group, pre-post designs, five control group designs, and nine RCTs. There is fairly equal representation for studies with adult survivors (seventeen studies) and children/adolescents (twenty studies). Within the set of youth studies, twelve were conducted with preteen children, while adolescents were sampled exclusively in five studies. Three studies featured "mixed" samples that included both latency-aged children and teenagers. In terms of gender representation, most studies featured same-sex samples (females, twenty-five studies; males, five studies), but in seven studies, primarily with preschool children, both genders were included in the group therapies. There have been no studies reporting mixed gender groups with adult survivors.

Studies of group therapy with survivors of childhood sexual abuse (CSA) predominate (twenty-three studies) among the different types of trauma represented in Table 22.1. Female adult survivors of sexual assault represent another frequently studied trauma population (five studies). Two

TABLE 22.1. Overview of Current Group Therapy for Trauma Studies

Study	Trauma population	Developmental stage (age)	Gender (N)	Type of therapy: (No. of sessions)	Design rigor
Alexander et al. (1989)	CSA	Adult	F (65)	Interpersonal, Process (10)	RCT
Ashby et al. (1987)	CSA	Teens (13-17)	F (10)	Integrated: (10)	Pre/Post
Bradley & Follingstad (2003)	CSA/CPA (Incarcerated)	Adult	F (49)	CBT/DBT	RCT
Carver et al. (1989)	CSA	Adult	F (29)	Psychodynamic (15)	Pre/Post
Chemtob et al. (2002)	Disaster	Grades 2-6	F/M (248)	CBT Individual, Group (4)	RCT
Classen et al. (2001)	CSA	Adult	F (171)	Trauma-Focus, Present-centered	RCT
Cryer & Beutler (1980)	SA	Adults	F (9)	Supportive (10)	Pre/Post
Deblinger et al. (2001)	CSA	Children/Mothers	F/M (44 pairs)	CBT, Supportive (11)	RCT
De Luca et al. (1993)	CSA	Latency (10-11)	F (7)	Integrated (10)	Pre/Post
Friedrich et al. (1992)	CSA	(4-16)	M (33)	Group and Individual	Pre/Post
Frueh et al. (1996)	Combat	Adults	M (11)	CBT (10)	Pre/Post
Hack et al. (1994)	CSA	Latency (8-11)	M (7)	Integrated (12)	Pre/Post
Hall et al. (1995)	CSA	Adult	F (71)	Psychodynamic, No treatment (26)	CT
Hall-Marley & Damon (1993)	CSA	(4-7)	F/M (13)	Integrated (13)	Pre/Post

Study	Trauma	Age	Gender (N)	Treatment	Design
Hazzard et al. (1993)	CSA	Adult	F (78)	Psychodynamic (52)	Pre/Post
Hiebert-Murphy et al. (1992)	CSA	(7-9)	F (5)	Integrated (9)	Pre/Post
Hoier et al. (1988)	CSA	(5-15)	F/M (18)	CBT	Pre/Post
Kitchur & Bell (1989)	CSA	(11-12)	F (7)	Integrated (16)	Pre/Post
Lubin et al. (1998)	Multiple	Adults	F (29)	Trauma-Focus, CBT (16)	Pre/Post
MacKay et al. (1987)	CSA	(12-18)	F (5)	Drama therapy (8)	Pre/Post
McGain & McKinzey (1995)	CSA	(9-12)	F (30)	Integrated (25)	Pre/Post
Nelki & Watters (1989)	CSA	(4-8)	F (7)	Integrated (9)	Pre/Post
Ovaert et al. (2003)	CV-Incarcerated	Adolescent	M (45)	Structured Therapy (12)	Pre/Post
Perez (1988)	CSA	(4-9)	F/M (21)	Play therapy (12)	Pre/Post
Resick & Schnicke (1992)	SA	Adult	F (39)	CBT, Wait list (12)	CT
Resick et al. (1988)	SA	Adult	F (50)	CBT, Supportive (6)	CT
Richter et al. (1997)	CSA	Adult	F (115)	Supportive, Wait list (15)	CT
Roth et al. (1988)	SA	Adult	F (13)	Psychodynamic, No Treatment	CT
Rust & Troupe (1991)	CSA	(9-18)	F (25)	Integrated (24)	Pre/Post
Schnurr et al. (2003)	Combat	Adult	M (360)	Trauma-Focus, Present-centered	RCT

TABLE 22.1. *(continued)*

Study	Trauma population	Developmental stage (age)	Gender (N)	Type of therapy: (No. of sessions)	Design rigor
Sinclair et al. (1995)	CSA	Adolescent	F (43)	CBT (20)	Pre/Post
Stalker & Fry (1999)	SA	Adult	F (86)	Group, Individual (10)	RCT
Stauffer & Deblinger (1996)	CSA	(2-6)	F/M (19)	CBT (11)	Pre/Post
Stein et al. (2003)	CV	6th Grade	F/M (126)	CBT (10)	RCT
Tutty et al. (1993)	DV	Adult	F (60)	Supportive (12)	Pre/Post
Verleur et al. (1986)	CSA	(13-17)	F (15)	Integrated (24)	Pre/Post
Zlotnick et al. (1997)	CSA	Adult	F (43)	CBT Affect Management (15)	RCT

Note: Trauma population: CPA = Childhood Physical Abuse, CSA = Childhood Sexual Abuse, CV = Community Violence, SA = Sexual Assault, DV= Domestic Violence; Type of Therapy: CBT = Cognitive-Behavioral Therapy, DBT = Dialectical Behavior Therapy; Design Rigor: CT = Controlled Trial, RCT = Randomized Controlled Trial.

TABLE 22.2. Randomized Controlled Trials of Group Therapy for Trauma

Study	Treatment group (N)	Comparison group	Number of sessions	Population	Major findings
Alexander et al. (1989)	Interpersonal transaction (psychodynamic) versus Process group (65)	Wait-list control	10 weekly sessions	Female CSA survivors	Both treatment groups improved more than controls in depression and distress; only the process group improved in social adjustment; gains were generally maintained at six-month follow-up
Bradley & Follingstad (2003)	9 CBT and 9 DBT sessions (49)	No-contact comparison group	18 sessions	Incarcerated Female CSA and/or CPA	Treatment group demonstrated lower PTSD, mood, and interpersonal symptoms than no-contact comparison group
Chemtob et al. (2002)	CBT individual therapy, CBT group therapy (248)	Wait-list control	4 sessions	Elementary school children (Grades 2-6)-Hurricane Iniki	Both treatment groups improved more than controls; no difference between the active treatments; greater dropout rates for individual therapy

TABLE 22.2. *(continued)*

Study	Treatment group (N)	Comparison group	Number of sessions	Population	Major findings
Classen et al. (2001)	Trauma-focused and Present-centered groups (171)	Wait-list controls	Weekly for six months	Female adult CSA survivors	Both treatment groups improved more than controls; no difference between the active treatments
Deblinger et al. (2001)	Separate CBT groups for children and their mothers (44)	Supportive therapy groups	11 weekly sessions	Children CSA and their mothers	CBT groups > supportive therapy for lower PTSD symptoms (mothers) and knowledge of body safety skills (children)
Schnurr et al. (2003)	CBT Trauma-focus (360)	Present-centered	26 weekly sessions	Male RVN veterans with chronic PTSD	Intent-to-treat analysis yielded no difference between trauma-focus and present-centered; effective-dose analysis favored trauma-focus for reduction in PTSD severity
Stalker & Fry (1999)	Group therapy versus individual therapy (86)	Wait-list control	10 sessions	Female sexual assault survivors	No difference between treatment groups; both better than wait list

Stein et al. (2003)	CBT (126)	Wait-list control	10 sessions	Elementary school (Grade 6) CV	Treatment group lower than wait list on PTSD symptoms, depressive symptoms, and psychosocial dysfunction
Zlotnick et al. (1997)	CBT affect management (33)	Wait-list control	15 weekly	Female CSA survivors	Affect management improved more than controls in PTSD and dissociation

Note: Treatment Group: CBT = Cognitive-Behavioral Therapy, DBT = Dialectical Behavior Therapy; Population: CPA= Childhood Physical Abuse, CSA = Childhood Sexual Abuse, CV= Community Violence, RVN = Republic of Vietnam.

studies each for combat and community violence are shown, while domestic violence and disaster are each represented by one study. The more recent studies have featured diversity in the types of trauma represented, including combat, community violence, and disaster. In addition, most of the studies reported since the year 2000 have used RCT designs and have included male participants with sample sizes greater than 100.

In terms of the types of group therapies represented, trauma focus CBT predominated (thirteen studies). Six studies used supportive/present-centered therapy, and four studies psychodynamic group therapy. Among the child/adolescent studies, "integrated" group therapy, usually including CBT and other child modalities, was used in ten studies. Compared to our earlier review (Foy et al., 2000), CBT methods have been much more frequently used in recent studies.

Turning to the "gold standard" set of studies featuring RCT designs in Table 22.2, it is clear that progress is being made in several important aspects. Two of the studies addressed the key issue of comparing treatment effectiveness on PTSD severity between individual and group forms of therapy. In both studies, one with child survivors of Hurricane Iniki (Chemtob, Nakashima, & Hamada, 2002), and the other with adult survivors of CSA (Stalker & Fry, 1999), results showed that both individual and group therapy were more effective than a wait-list control condition, but not significantly different from each other. In addition, Chemtob and his colleagues (2002) found that there were fewer dropouts from treatment among those children receiving the CBT group therapy.

Three recent RCTs have also addressed the issue of relative effectiveness among the types of trauma group therapies (Classen, Koopman, Nevill-Manning, & Spiegel, 2001; Deblinger, Stauffer, & Steer, 2001; Schnurr et al., 2003). In each of these studies, trauma focus groups were compared directly to present-centered or supportive groups. In one study there was a finding of clear superiority for trauma focus CBT (Deblinger et al., 2001). In the Schnurr et al. (2003) study, findings were more complex such that there was a significant treatment condition by cohort interaction favoring present-centered therapy in the first cohort. Conversely, the trauma focus condition was found to be superior in the second and third cohorts. These findings suggest that there may be a lengthier learning curve for trauma-focus group therapists, especially those without previous experience and skills in the CBT methods incorporated in the groups.

Among the recent RCTs, two studies were conducted in school-based settings with elementary-aged participants following exposure to either community violence or hurricane-related disaster (Chemtob et al., 2002; Stein et al., 2003). Positive findings from these well-designed studies demonstrate the usefulness of mixed gender groups for child trauma survivors in nonclinical community settings. These studies of group therapy, applied to survivors of large-scale community violence or destruction, provide the closest empirical link for using group therapy with terrorist disaster survivors.

In our earlier review (Foy et al., 2000) of the effectiveness of group therapy for trauma, we posed three critical questions that empirical studies needed to address:

1. Does group therapy produce symptom improvement in participants?
2. Are there differences in effectiveness between types of trauma group therapy?
3. How can validated findings from research studies be applied in actual clinical practice?

Recent advances in our empirical base of knowledge from studies of trauma group therapy have been nothing short of dramatic, and the strength of conclusions that can now be drawn about these key issues is much improved. Virtually all of the thirty-seven studies in Table 22.1 found positive changes in trauma-related symptoms, and these findings were consistent across trauma group therapies of all types. Findings from the fourteen studies that included control groups were uniform in showing positive treatment effects specific to trauma group therapy, when compared to wait-list controls. Further findings from those studies with follow-up assessments showed that treatment effects were durable for as long as one year posttreatment (e.g., Schnurr et al., 2003).

We have gained some clarity with respect to the second question, regarding differential effectiveness of trauma group therapies, as well. Several recent studies have directly compared CBT trauma focus with present-centered or supportive group therapy. Although there is support for both types of groups in that findings showed that both produced positive outcomes when compared to wait-list controls, it appears that there may be an advantage to CBT trauma focus, especially when used with children (c.f. Deblinger et al., 2001). Given the positive findings for both types of group therapies, studies that address participant-to-therapy matching issues are now needed.

Encouraging findings were found with respect to the third question regarding practical usefulness of research lab-based methods in "real-world" settings. Three of the recent RCTs have been conducted in field settings such as schools (Chemtob et al., 2002; Stein et al., 2003) and correctional facilities (Bradley & Follingstad, 2003), all with positive results. These findings give empirical support for future efforts to venture out from traditional clinical research settings and samples when providing trauma group therapy.

FROM CURRENT STATUS TO FUTURE GROUPS FOR TERRORIST DISASTER SURVIVORS

How can we translate our empirically derived knowledge of the current status of group therapy for trauma into meaningful directions for applications to future groups designed for terrorist disaster survivors? Our offerings in this section are not intended as a substitute, bypass route, or shortcut from careful reading and consideration of the other chapters in this book that proposed group interventions for terrorist disaster survivors. Rather, we have tried to take a unique perspective based upon the current set of studies on trauma group therapy that includes both child/adolescent and adult applications. In terms of future clinical directions in using group therapy for terrorist disaster survivors, we have chosen to address critical issues from eight relevant domains (Table 22.3).

Assessment/Diagnostic Issues

In reviewing the literature on group therapy for trauma we found many articles that described techniques and presented directions for conducting groups without presenting outcome data or even suggesting an assessment

TABLE 22.3. Critical Issues in Using Group Therapy with Terrorist Disaster Survivors

Domain	Critical issues
Assessment/diagnosis	Minimal assessment requirements; additional research requirements; which diagnoses to consider
Ethical considerations	Informed consent; confidentiality; safety contract; institutional review and approval before dissemination
Life stage	Preteen, adolescent, young adult, senior adult life stage differences
Cultural diversity	Racial prejudice in heterogeneous groups; ensuring respect for diversity
Therapist issues	Skills required for different groups; self-care issues
Group selection factors	Therapists' skills; clients' stage of recovery
Client selection factors	Indications and contraindications; gender; differences for trauma-focus versus present-centered groups
Unique aspects of terrorist disasters	Death, grotesque exposure, and traumatic grief; human perpetration and enduring risk of recurrence

strategy to be used. Obviously, these articles were not included in our tables of studies because they cannot contribute directly to our empirical knowledge. Nevertheless, these reports do suggest that there is widespread acceptance and clinical use of group therapy for survivors of various types of trauma. Conducting assessments at pre- and posttreatment intervals provides a means of evaluating clinical progress toward targeted therapy outcomes among group participants. It also makes it possible to compare results in the current group with published treatment effect sizes obtained in other similar group therapy studies. The need for systematic assessment is particularly critical when group therapy is being used with survivors of a trauma type for which no studies presently exist (e.g., terrorist disasters).

Recommendation #1

As a minimum assessment strategy, therapists conducting groups with terrorist disaster survivors should collect pre- and posttreatment data from each client that includes basic demographic information and a measure of the primary intended outcome of the group. The applicability of findings from the current group to those from other studies is dependent upon similarities in demographics such as age, gender, ethnicity, and educational and income levels. For assessment of intended outcome, the use of a general measure of trauma-related distress, such as the Impact of Event Scale-Revised (IESR; Weiss & Marmar, 1997) is encouraged.

Recommendation #2

For research applications the assessment strategy should include standard diagnostic measures for PTSD and depression, as well as more extensive information on risk and resilience factors for each participant. These factors have been identified, and the importance of considering risk and resilience variables in future trauma-related research has been emphasized (King, Vogt, & King, 2003).

Ethical Considerations

Individuals being considered as possible TD group clients have the right to know the nature of the therapy they are being offered so they can make informed choices about their participation. Information about the evidence supporting the particular kind of group being offered should be provided in appropriate lay language. Maintaining a safe and respectful therapy environment is a key therapist responsibility in clinical work with trauma survivors. Therapists should explain the need and limits for confidentiality regarding information about group members and their experiences that emerge within group sessions. If assessment data from the group will be presented or published beyond the current clinical context, then institutional review and approval of the planned group intervention and the evaluation methods should be obtained in advance.

Recommendation #3

Each of these ethical considerations should be thoroughly discussed with each potential group member in an individual orientation session before the group begins.

Life-Stage Considerations

As the literature review of group therapy for trauma shows (Table 22.1), most groups are conducted with survivors of the same type of trauma who are also members of the same gender and developmental stage. In most published reports for children and adolescents, the age range of participants includes only those within a particular developmental stage. Generally, preteens are not mixed with adolescents, and adult groups typically do not mix young adults with seniors. Early in the course of group therapy, group cohesion is promoted by sharing membership among the cohort of similar survivors. Although seeking recovery from trauma is the essential common element across groups, the content and methods used need to be selected for compatibility with the life stage of the group members.

Recommendation #4

To the extent possible, TD groups should be designed to maintain a cohort effect among participants such that members share survivorship of the same disaster (or type) and the same life stage.

Cultural Diversity

TD group therapists need to be culturally aware and sensitive to these issues, as they exist in the context of the particular terrorist disaster and the ethnic makeup of the group members. Pregroup screening for signs of racial prejudice or unusual racial sensitivity is important, especially for heterogeneous groups. Ground rules should be established to ensure respect for diversity in members' backgrounds, life experiences, and beliefs.

Recommendation #5

Prospective members who cannot agree to adhere to guidelines for respecting diversity should be considered for alternative individual treatment options.

Therapist Issues

What are the *requisite therapist skills* for conducting TD groups? Under ideal circumstances, therapists would have skills in both group therapy and the specialty of trauma assessment and treatment. In addition, therapists for CBT trauma focus groups also need skills in using common CBT techniques (e.g., exposure therapy, cognitive restructuring, SUDS monitoring, thought stopping) used in these groups. When TD group therapists lack requisite skills and experience in conducting trauma groups, training, ongoing consultation, and supervision from a more experienced trauma therapist should be sought.

Are there special *therapist self-care issues* to be considered for TD group therapists? Session management for trauma groups presents the therapist with the additional challenge of managing unique countertransference issues presented by trauma survivors. Many survivors may begin group with limited abilities regarding: (1) trusting group leaders and fellow participants; and (2) taking personal responsibility for recovery. Therapists are subjected to the pull of their own reactions to the dynamics of clients' trauma-related attitudes and interpersonal behavior patterns within the group. Therapists may be particularly vulnerable if they are relatively inexperienced and/or unsure of their skills in managing trauma-related groups. Therapists who are also members of the same community where the disaster occurred (fellow survivors) may need to pay particular attention to their own disaster reactions and need for recovery time before rushing to conduct groups for other survivors. Similarly, therapists with a personal history of prior victimization need to have processed and integrated their traumatic experiences so that their personal recovery needs do not interfere with their ability to manage the group.

Recommendation #6

TD groups should be co-led by two therapists who share facilitation responsibilities and support each other. Individual TD therapists should be trained and experienced in principles of group therapy, trauma dynamics, and relevant CBT techniques. Inexperienced therapists should obtain training, ongoing consultation, and supervision from an experienced trauma group therapist.

Group Selection Considerations

Among the studies of group therapy for adult trauma survivors, three kinds of groups are found, including trauma focus psychodynamic, trauma focus CBT, and present-centered (supportive) types. For children and adolescents, trauma focus CBT and "integrated" types of groups are found. Findings from the thirty-seven studies included in Table 22.1 are consistent in showing a positive effect for treatment of all types. Earlier comparisons of relative treatment effect sizes found for the various types did not clearly and consistently identify a superior group type (e.g., Reeker et al., 1997). Findings from later RCTs that made direct comparisons have tended to favor CBT trauma focus group therapy over present-centered or supportive groups.

In the absence of unequivocal superiority for a particular type, what criteria may be useful for determining which kind of group to offer? One factor that merits consideration concerns the experience and skill level of the available therapists. Present-centered groups require group therapy skills and knowledge of trauma dynamics, but not direct trauma processing or other CBT skills. A second factor involves assessing the stage of recovery (Herman, 1992) for group prospects. Clients in Stage One (immediate recovery phase—establishing safety and self-care; reducing intrusion and hyperarousal) or Stage Three (final recovery—giving up survivor identity; repair of relationships with others) may be more appropriate for present-centered groups, while Stage Two (intermediate recovery—integration of traumatic memories into general life narrative) clients' recovery needs may better fit a trauma focus approach, either CBT or psychodynamic.

Recommendation #7

Offer the type(s) of TD groups that are consistent with available therapists' skills, theoretical orientation (CBT or psychodynamic) and experience, and consider clients' stages of recovery as a factor in the group assignment process.

Client Selection Considerations

What client factors need to be considered for assignment to TD group therapy? Key *indications* for assignment to trauma groups include abilities to establish interpersonal trust and maintain confidentiality with other group members and leaders. Active psychosis or paranoia, active suicidality or homicidality, severe organicity or limited cognitive capacity, and pending litigation or compensation seeking have been identified as *contraindications* for trauma groups (Foy et al., 2000).

Gender is a factor that has been treated differently in the child/adolescent studies, compared to studies of adults. Among youths, groups have often been mixed in gender composition, while all adult studies have been conducted with same-sex members. For adult survivors of terrorist disasters, the rationale for keeping men and women separate in group composition seems less clear than it might have been for the groups in the literature dealing with sexual assault and combat experiences. However, if future TD groups are to be mixed, then therapists need to be attuned to potential gender issues and dynamics as additional factors for management in their groups.

In terms of possible differences in selection factors between trauma focus and present-centered groups, there are key client factors to consider (Wattenberg, Foy, Unger, & Glynn, Chapter 15, this book; Unger, Wattenberg, Foy, & Glynn, Chapter 19, this book) First, the client's acceptance of the rationale for the group and willingness to abide by the requirements for participants in that type of group are important. In addition, in the case of trauma focus types, either CBT or psychodynamic, clients need to understand and accept the importance of trauma processing and be willing to make the necessary self-disclosures about their traumatic experiences in the group.

Recommendation #8

For the near future at least, TD groups should follow the precedents for inclusion and exclusion criteria that have been established for groups with other trauma populations.

Unique Aspects of Terrorist Disasters

Relative to other trauma types, there are unique characteristics of terrorist disasters that need consideration in planning group interventions. Drawing on observations from New York after 9/11 and ongoing terrorism in Israel, two chapters in this book (Freedman & Tuval-Mashiach, Chapter 3, this book; Stovall-McClough & Cloitre, Chapter 6, this book) address the widespread death exposure and intentionality of the perpetrators of terrorist disasters as key features that are also related to other issues.

As noted by Spitz and Spitz (Chapter 2, this book), many TD survivors who escape direct exposure to the death and destruction may become "forgotten people," involved profoundly, although indirectly, through the tremendous "ripple effect" of such a disaster. For these individuals their perception of insignificant involvement may become a personal obstacle that prevents their seeking professional help even when the need is apparent. These issues are similar to those involved in selection of members for combat-related groups, where ensuring equivalence for combat exposure is important (Foy et al., 2002; Schnurr et al., 2003).

Recommendation #9

Compose groups for terrorist disasters so that exposure levels of participants are similar. In particular, participants should share equivalent death exposure. Individuals who were directly affected by having someone close (family or friend) killed in the attack should be assigned to homogeneous groups where other participants were selected based on that factor.

STRENGTH OF EVIDENCE FOR RECOMMENDATIONS

Taken together, the studies in Table 22.1 on the various forms of group therapies for trauma survivors present impressive evidence in support of available interventions. Furthermore, much progress, as evidenced by the RCTs in Table 22.2, has been made toward building a firm base of empirical support for using these group interventions with adults, children, and adolescents. Gains have also been made in the representation of male and mixed gender samples and trauma populations other than sexual abuse or assault.

How confident can we be in the recommendations made for using group therapy with future terrorist disaster survivors? On the side for greater confidence is the weight of evidence provided by uniform positive findings from nearly forty studies, including nine "gold standard" RCTs, with several trauma populations. More cautiously, even though both trauma focus and present-centered types have multiple randomized controlled trials supporting their effectiveness with other populations, they have not been tested specifically with terrorist disaster survivors. Therefore, it remains a formidable task for future clinical outcome research to test the specific effectiveness of these treatments.

FUTURE RESEARCH DIRECTIONS

Despite the many difficulties encountered in conducting treatment outcome research, there have been recent monumental accomplishments in the strengthening of the empirical base for group therapy with trauma survivors. Still, there are many ways in which the current body of literature can be improved. Suggestions for future research include

- conducting studies with terrorist disaster survivors to provide a direct test of utility with this trauma population;
- randomizing cohorts of participants, rather than individuals, to treatment types to allow efficient comparisons between treatments;
- extending the posttreatment follow-up period to six to twelve months;
- comparing different age groups to see if group therapy is more effective for certain ages than others;
- comparing the effectiveness of mixed-gender groups to single-gender groups;
- comparing varying lengths of group treatment, e.g., brief (<10 sessions) to extended (>15 sessions);
- conducting matching studies (participants to treatments) to identify individual factors associated with differential treatment outcomes; and
- identifying essential therapist competencies for standard delivery of treatment and positive treatment outcomes.

REFERENCES

Alexander, P.C., Neimeyer, R.A., Follette, V.M., Moore, M.K., & Harter, S. (1989). A comparison of group treatments of women sexually abused as children. *Journal of Consulting and Clinical Psychology, 57,* 479-483.

Ashby, M., Gilchrist, L., & Miramontez, A. (1987). Group treatment for sexually abused America Indian adolescents. *Social Work with Groups, 10,* 21-32.

Bonanno, G.A. (2004). Loss, trauma, and human resilience: Have we underestimated the human capacity to thrive after extremely aversive events? *American Psychologist, 59,* 20-28.

Bradley, R.G. & Follingstad, D.R. (2003). Group therapy for incarcerated women who experienced interpersonal violence: A pilot study. *Journal of Traumatic Stress, 16*(4), 337-340.

Carver, C.S., Scheier, M.F., & Weintraub, J.K. (1989). Assessing coping strategies: A theoretically based approach. *Journal of Personality and Social Psychology, 56,* 267-283.

Chemtob, C.M., Nakashima, J.P., & Hamada, R.S. (2002). Psychosocial intervention for postdisaster trauma symptoms in elementary school children: A controlled community field study. *Archives of Pediatrics and Adolescent Medicine, 156,* 211-216.

Classen, C.C., Koopman, C., Nevill-Manning, K., & Spiegel, D. (2001). A preliminary report comparing trauma-focused and present-focused group therapy against a wait-listed condition among childhood sexual abuse survivors with PTSD. *Journal of Aggression, Maltreatment and Trauma, 4,* 265-288.

Cryer, L. & Beutler, L.E. (1980). Group therapy: An alternative treatment approach for rape victims. *Journal of Sex and Marital Therapy, 6*(1), 40-46.

Deblinger, E., Stauffer, L.B., & Steer, R.A. (2001). Comparative efficacies of supportive and cognitive behavioral group therapies for young children who have been sexually abused and their nonoffending mothers. *Child Maltreatment, 6,* 332-343.

De Luca, R., Hazen, A., & Cutler, J. (1993). Evaluation of a group counseling program for preadolescent female victims of incest. *Elementary School Guidance and Counseling, 28,* 104-114.

Dynes, J.B. (1945). Rehabilitation of war casualties. *War Medicine, 7,* 32-35.

Eriksson, C.B., Foy, D.W., & Larson, L.C. (2003). When the helpers need help: Early intervention for emergency and relief services personnel. In B.T. Litz, (Ed.), *Early Intervention for Trauma and Traumatic Loss.* New York: Guilford Press.

Foa, E.B., Keane, T.M., & Friedman, M.J. (2000). *Effective Treatments for PTSD: Practice Guidelines from the International Society for Traumatic Stress Studies.* New York: Guilford Press.

Foy, D.W., Eriksson, C.B., & Trice, G.A. (2001). Introduction to group interventions for trauma survivors. *Group Dynamics: Theory, Research and Practice, 5,* 246-251.

Foy, D.W., Glynn, S.M., Schnurr, P.P., Jankowski, M.K., Wattenberg, M.S., Weiss, D.S., Marmar, C.R., & Gusman, F.D. (2000). Group therapy. In E. Foa, T. Keane, & M. Friedman (Eds.), *Effective Treatments for PTSD: Practice Guidelines from the International Society for Traumatic Stress Studies* (pp. 155-175; 336-338). New York: Guilford Press.

Foy, D.W., Ruzek, J.I., Glynn, S.M., Riney, S.A., & Gusman, F.D. (2002). Trauma focus group therapy for combat-related PTSD: An update. *In Session: Psychotherapy in Practice, 3,* 59-73.

Friedrich, W., Luecke, W., Beilke, R., & Place, V. (1992). Psychotherapy outcome of sexually abused boys: An agency study. *Journal of Interpersonal Violence, 7,* 396-409.

Frueh, B.C., Smith, D.W., & Barker, S.E. (1996). Compensation seeking status and psychometric assessment of combat veterans seeking treatment for PTSD. *Journal of Traumatic Stress, 9,* 427-440.

Hack, T., Osachuk, T., & De Luca, R. (1994). Group treatment for sexually abused preadolescent boys. *Families in Society: The Journal of Contemporary Human Services, 4,* 217-228.

Hall, Z., Mullee, M.A., & Thompson, C. (1995). Adult survivors of child sex abuse. *British Medical Journal, 311*(7007), 748a-748.

Hall-Marley, S. & Damon, L. (1993). Impact of structures group therapy on young victims of sexual abuse. *Journal of Child and Adolescent Group Therapy, 3,* 41-48.

Hazzaard, A., Rogers, J.H., & Angert, L. (1993). Factors affecting group therapy outcome for adult sexual abuse survivors. *International Journal of Group Psychotherapy, 43,* 453-468.

Herman, J.L. (1992). *Trauma and Recovery.* New York: Basic Books.

Hiebert-Murphy, D., De Luca, R., & Runtz, M. (1992). Group treatment for sexually abused girls: Evaluating outcome. *Families in Society: The Journal of Contemporary Human Services, 73,* 205-213.

Hoier, T., Inderbitzen-Pisaruk, H., & Shawchuck, C. (1988). *Short-term cognitive behavioral group treatment for victims of sexual abuse.* Unpublished manuscript, West Virginia University, Department of Psychology, Morgantown.

King, D.W., Vogt, D.S., & King, L.A. (2003). Risk and resilience factors in the etiology of chronic posttraumatic stress disorder. In B.T. Litz (Ed.), *Early Intervention for Trauma and Traumatic Loss* (pp. 34-65). New York: Guilford Press.

Kitchur, M. & Bell, R. (1989). Group psychotherapy with preadolescent sexual abuse victims: Literature review and description of an inner-city group. *International Journal of Group Psychotherapy, 39,* 285-310.

Klein, R.K. & Schermer, V.L. (2000). *Group Psychotherapy for Psychological Trauma.* New York: Guilford Press.

Lubin, H., Loris, M., Burt, J., & Johnson, D.R. (1998). Efficacy of psychoeducational group therapy in reducing symptoms of posttraumatic stress disorder among multiple traumatized women. *American Journal of Psychiatry, 155*(9), 1172-1177.

MacKay, B., Gold, M., & Gold, E. (1987). A pilot study in drama therapy with adolescent girls who have been sexually abused. *The Arts in Psychotherapy, 14,* 77-84.

McGain, B. & McKinzey, R. (1995). The efficacy of group treatment in sexually abused girls. *Child Abuse and Neglect, 19,* 1157-1169.

Nelki, J. & Watters, J. (1989). A group for sexually abused young children: Unravelling the web. *Child Abuse and Neglect, 13,* 369-377.

Ovaert, L.B., Cashel, M.L., & Sewell, K.W. (2003). Structured group therapy for posttraumatic stress disorder in incarcerated male juveniles. *American Journal of Orthopsychiatry, 73*(3), 294-301.

Perez, C. (1988). A comparison of group play therapy and individual play therapy for sexually abused children. *Dissertation Abstracts International, 48,* 3079.

Reeker, J., Ensing, D., & Elliott, R. (1997). A meta-analytic investigation of group treatment outcomes for sexually abused children. *Child Abuse & Neglect, 21,* 669-680.

Resick, P.A., Jordan, C.G., Girelli, S.A., Hutter, C.K., & Marhoefer-Dvorak, S. (1988). A comparative study of behavioral group therapy for sexual assault victims. *Behavior Therapy, 19,* 385-401.

Resnick, P.A. & Schnicke, M.K. (1992). Cognitive processing therapy for sexual assault victims. *Journal of Consulting and Clinical Psychology, 60,* 748-756.

Richter, N.L., Snider, E., & Gorey, K.M. (1997). Group work intervention with female survivors of childhood sexual abuse. *Research on Social Work Practice, 7,*
53-69.

Rose, S., Bisson, J., & Wessely, S. (2003). A systematic review of single-session
psychological interventions ("debriefing") following trauma. *Psychotherapy
and Psychosomatics, 72,* 176-184.

Roth, S., Dye, E., & Lebowitz, L. (1988). Group therapy for sexual assault victims.
Psychotherapy, 25, 82-93.

Rothbaum, B.O., Meadows, E.A., Resick, P., & Foy, D.W. (2000). Cognitive-behavioral therapy. In E. Foa, T. Keane, & M. Friedman (Eds.), *Effective Treatments for PTSD: Practice Guidelines from the International Society for
Traumatic Stress Studies* (pp. 60-83; 320-325). New York: Guilford Press.

Rust, J. & Troupe, P. (1991). Relationships of treatment of child sexual abuse with
school achievement and self-concept. *Journal of Early Adolescence, 11,* 420-
429.

Saunders, B.E., Berliner, L., & Hanson, R.F. (Eds.). (2003). *Child Physical and Sexual Abuse: Guidelines for Treatment (Final Report: January 15, 2003).*
Charleston, SC: National Crime Victims Research and Treatment Center.

Schnurr, P.P., Friedman, M.F., Foy, D.W., Shea, T.M., Hsieh, F.Y., Lavori, P.W.,
Glynn, S.M., Wattenberg, M.S., & Bernardy, N.C. (2003). Randomized trial of
trauma-focused group therapy for posttraumatic stress disorder: Results from a
Department of Veterans Affairs cooperative study. *Archives of General Psychiatry, 60,* 481-489.

Sinclair, R., Garnett, L., & Berridge, D. (1995). *Social work and assessment with
adolescents.* London: National Children's Bureau.

Sipprelle, R.C. (1992). A Vet Center experience: Multievent trauma, delayed treatment type. In D. Foy, (Ed.), *Treating PTSD: Cognitive-Behavioral Strategies*
(pp. 13-38). New York: Guilford Press.

Stalker C.A. & Fry, R. (1999). A comparison of short-term group and individual
therapy for sexually abused women. *Canadian Journal of Psychiatry, 44,* 168-
174.

Stauffer, L. & Deblinger, E. (1996). Cognitive behavioral groups for nonoffending
mothers and their young sexually abused children: A preliminary treatment outcome study. *Child Maltreatment, 1,* 65-76.

Stein, B.D., Jaycox, L.H., Kataoka, S.H., Wong, M., Tu, W., Elliott, M.N., & Fink,
A. (2003). A mental health intervention for schoolchildren exposed to violence:
A randomized controlled trial. *Journal of the American Medical Association,
290,* 603-611.

Tutty, L., Bidgood, B., & Rothery, M. (1993). Support groups for battered women:
Research on their efficacy. *Journal of Family Violence, 8*(4), 325-1-19.

Verleur, D., Hughes, R., & de Rios, M. (1986). Enhancement of self-esteem among
female adolescent incest victims: A controlled comparison. *Adolescence, 21,*
843-854.

Weiss, D.S. & Marmar, C.R. (1997). The Impact of Event Scale-Revised. In J. Wilson & T. Keane (Eds.), *Assessing Psychological Trauma and PTSD* (pp. 399-
411). New York: Guilford Press.

Young, B.H. & Blake, D.D. (1999). *Group Treatments for Post-Traumatic Stress Disorder*. Philadelphia: Brunner/Mazel.

Zlotnick, C., Shea, M.T., Rosen, K.H., Simpson, E., Mulrenin, K., Begin, A., & Pearlstein, T. (1997). An affect-management group for women with posttraumatic stress disorder and histories of childhood sexual abuse. *Journal of Traumatic Stress, 10*, 425-436.

Index

Abandonment, 512
Abused children. *See under* Children
Acceptance, 359, 506, 792, 808
Activism, 623
Acute reactions, 132-133, 485-488
Acute stress disorder, 166, 633
Adaptation, 268, 487. *See also* Coping
Adaptive functioning, 246-247. *See also* Functioning
Adaptive reactions, 714-716, 792
Adolescents
    adaptation, to loss, 714-716
    after WTC attack, 141
    aggression, 671, 704
    anger, 704, 717, 723
    assessment, 679-680
    avoidance, 705, 713
    behavioral changes, 94, 704, 715
    in Bosnia, 266, 673
    and CBT, 697-704, 699(f), 700(f), 701(f)
    clinical studies, 884(t), 885(t), 886(t)
    cognitive distortions, 714, 717, 720-722
    coping, 676, 696, 717, 718
    cultural factors, 211
    danger appraisal, 92, 704
    developmental factors
        disturbance manifestations, 678
        group homogeneity, 686
        identity emergence, 719
        relationship reappraisal, 716
        stage challenges, 705
    disaster effects, 85-87
    emotions, 697-700, 699(f), 701(f), 706, 712, 715
    expectations, 672, 678, 704, 720
    and family therapy, 725
    and gender, 687
    and grief, 292, 676, 691, 713-719

Adolescents *(continued)*
    trauma/grief group psychotherapy, 669-726 (*see also* UCLA Trauma/Grief Program for Adolescents)
    and group therapy, 410, 413, 678, 719 (*see also* UCLA Trauma/Grief Program for Adolescents)
    and guilt, 672, 704, 717
    help-seeking, 704
    and helplessness, 672, 718
    hyperarousal case, 670
    and individual therapy, 724-725
    isolation, 95, 671
    in Israeli trial, 39
    and loss, 675, 686, 695, 714-716
    and media, 126
    narratives, 694, 704-712
    and parents, 274, 683, 716
    and peers, 671, 716
    and personal responsibility, 721
    physical sensations, 706
    in Pompeii eruption, 84
    posttraumatic distress reactions, 261
    psychoeducation, 674, 677, 680, 713
    and reenactment, 704, 707
    and relationships, 671, 683, 716-717, 721
    reminders, 675, 695, 702-712, 714
    and reminiscing, 717
    research issues, 105
    and rumors, 92
    and secondary adversity, 677, 719
    separation anxiety, 93
    sexual assault, 410, 884(t), 885(t), 886(t)
    siblings, 721
    and suicide, 292 (*see under* Suicide)
    and WMD attack, 97
Adversity. *See* Secondary adversity

905